M A C R O P H Y S I C S

Second Edition, 1981

Donald E. DeGraaf

University of Michigan - Flint

Crystal Press, *Publishers*

Flint, Michigan

Textbooks in this Series

I *THE PHYSICAL UNIVERSE*, Second Edition, 1979
II *MICROPHYSICS*, Second Edition, 1980
III *MACROPHYSICS*, Second Edition, 1981

Published by: CRYSTAL PRESS
1909 Proctor Street
Flint, Michigan 48504
Phone 313 + 239-8281

ISBN 0-930402-07-3

Library of Congress Catalog Card Number 80-70717

Printed in the United States of America

PREFACE

This textbook can be used in either of two ways. It can be integrated with the companion volume *MICROPHYSICS* in a two-semester course in general college physics following the traditional sequence of mechanics, heat, electricity and magnetism, waves and optics, and quantum physics. Or, it can be used for a one-semester course for students who have completed a prior course using *MICROPHYSICS*. It is assumed that students using this text have ready access to a copy of *MICROPHYSICS*, so they can refer to it when necessary for a lower-level introduction to some topics.

This text follows the design of the third course in the Flint Introductory Physics Sequence (FLIPS), a sequence of four courses whose content is organized in a "helical" pattern as described on the page addressed "To the Student." Collectively, the courses in this sequence are designed to serve the needs of the entire range of students who want or are required to study introductory physics at the college level, from non-science majors to physics majors and engineers. Further information on the concept of helical physics and on the Flint Introductory Physics Sequence may be obtained from the author.

MACROPHYSICS consists of three major portions. Chapters 41 - 51 deal with classical mechanics, Chapters 53-55 discuss topics in wave motion and optics, and Chapters 56 - 61 present electricity and magnetism. The instructor may begin the course with any one of these three portions. If the course begins with electricity, Sections 43.1 and 47.1 on vectors would be a good starting point. SI units are used throughout the text.

In the part on optics, either physical optics (Chapter 55) or geometrical optics (Chapters 53 and 54) may be studied before the other. In Chapter 54, after the study of objects and images in Section 54.1, either lenses or mirrors may be studied next with no loss of continuity.

It may be expected that most students who are well prepared for this text have previously studied much of the material in Chapters 41, 42 and 43. A brief review should be sufficient to refresh the students' ability to work with these topics. It is suggested that the students test themselves on the Minimum Learning Goals in these chapters. Those who can achieve the goals correctly and without hesitation probably have an adequate grasp of the related concepts.

This book has developed from materials used by the author during thirteen years of teaching introductory physics in a helical pattern. The aim is to introduce and elucidate the fundamental *concepts* of physics. The exposition, and the Questions at the end of each section, are intended to emphasize the ideas of physics, and to show that the heart of physics does not consist of "plugging in formulas." The Minimum Learning Goals in each section are intended to assist the student in focussing on the central concepts presented there. They should be regarded as *minimum*, not *maximum* goals for learning. Suggestions for their use are given in the portion on How to Use This Textbook. The symbol ★ is used to denote important equations.

The questions and exercises at the end of each section range from some which are rather simple and straightforward to others which are quite complex and thought-provoking. The questions and exercises assigned to students should be chosen with care, bearing in mind the objectives of the course and the abilities of the students.

It is assumed that students using this text have a thorough knowledge of second-year high-school algebra, and can add vectors using the head-to-tail method. Although the *notation* of calculus is used extensively in the text, the *methods* of calculus are used to a lesser extent. Portions of the text and exercises which *require* the use of analytic methods of calculus are enclosed in special brackets: (( )). For some topics, the useful equations are *expressed* as integrals or derivatives, but the *evaluation* of these expressions can often be done using physical reasoning alone, without using recipes for differentiation or integration. In these cases the material is not enclosed in special brackets.

This text is intended for use simultaneously by students who are, and who are not, concurrently studying calculus. Appendix A contains a brief introduction to the methods of differentiation and integration used in the text. The student who has not completed a course in calculus can nevertheless gain some facility in analytic differentiation and integration through study of this appendix and working the exercises included there. The background of the student and the judgment of the instructor will serve as guides to the selection of exercises and learning goals which should be assigned to the student.

Suggestions and comments are invited from those who use this book.

Many are the teachers and students who have influenced the content and presentation in this text. During the years when the concept of helical physics was being developed and tested, the support and encouragement received from the National Science Foundation and from the members of the Advisory Panel of the Flint Introductory Physics Sequence Curriculum Project played a vital role. Arnold A. Strassenburg was a special source of encouragement and of confidence in the worth of the helical approach. Thoughtful criticism and advice were supplied by physics teachers in the colleges participating in the curriculum project. Donald W. Boys, Mary E. Cox, and Frank E. Rose are colleagues in the UM-Flint Physics and Astronomy Department, who have kindly supplied materials, suggested problems, and offered extensive comments on portions of the manuscript in its early stages.

Most of the Minimum Learning Goals, many of the questions and exercises, and some of the exposition in this text are taken from the *Study Guide to Macrophysics*, published by the FLIPS Project at the University of Michigan-Flint, and are used by permission. The right-hand rule presented in Section 58.4 was found in the textbook *Physics*, prepared by the Physical Science Study Committee and published by D. C. Heath and Co., Boston.

Thanks are expressed to students Florence Kattoula, Bernadine Saxe, and Kat Stanajtis for helpful comments on the first edition. Thanks for supplying photos for some of the illustrations are expressed to the Ealing Corporation, the 3M Company, the LaPine Scientific Company and Leybold-Heraeus GMBH

& Co., and the National Aeronautics and Space Administration.

Four colleagues in the FLIPS Project have had a major influence on my teaching and on the writing of this book. Bill Aldridge persuaded me that Minimum Learning Goals are valid, and valuable as a basis for the planning of teaching strategy and for the evaluation of learning. Arnold Arons has insisted on many occasions that it is not enough to throw before students a string of assertions and a multitude of facts. Following his influence, this book is an attempt to convey to students "how we know" and "why we believe." Richard Sands has served as an example of the use of clear thinking and exact statements in the presentation of physics. Gabriel Weinreich has been a source of insight into the essentials of many physical concepts, as well as an example of style in presentation.

My wife supplied encouragement and patience in abundance as well as the cover design. Son Dave prepared many of the drawings. Their collaboration has been a special source of joy and satisfaction. Finally, a special word of thanks to Frances Abel who showed skill and resourcefulness in typing a difficult manuscript, and to Liz Brooks for timely assistance.

Donald E. DeGraaf

CONTENTS

PART TWO: WAVES AND OPTICS

PART THREE: ELECTRICITY AND MAGNETISM

To The STUDENT

This is a textbook for a course in MACROPHYSICS, designed for all students having some curiosity about the world of nature. It has been written to help you understand mechanical, electrical and optical phenomena which you meet every day. The macroscopic viewpoint is emphasized throughout the course. This text follows the design of the third in a series of four courses which make up the helical-style Flint Introductory Physics Sequence (FLIPS).

Each course in the FLIPS Sequence presents topics from several major areas of physics. The textbook for the first course, *THE PHYSICAL UNIVERSE*, contains an introduction to some of the interesting phenomena of nature. It deals with the basic concepts of motion, forces, and energy, and presents a broad view of the universe from galaxies and quasars to gamma rays and quarks. The textbook for the second course, *MICROPHYSICS*, includes a discussion of thermal physics, electricity, light, atoms, and nuclei. This volume builds on the material presented in the preceding textbooks. Learning in this pattern is something like climbing a helical staircase, as each successive course circles back through the major branches of physics at a more advanced level.

Able students who have recently completed one year of high-school physics with high grades should be prepared to begin their college study of physics using this textbook. Selected chapters in *MICROPHYSICS* offer a lower-level introduction to some topics. A thorough knowledge of second-year high-school algebra and a year of high-school geometry should provide the necessary background in mathematics for this course. Additional quantitative methods presented in the text and in Appendix A include vector products, differentiation and integration of elementary functions, and evaluation of definite integrals and line integrals for simple cases. Portions of the text and exercises which require the use of analytic methods of differentiation and integration are enclosed in special brackets: (( )). For some topics, the useful equations are *expressed* as integrals or derivatives, but the *evaluation* of these expressions can often be done using physical reasoning alone, without using recipes for differentiation or integration. In these cases the material is not enclosed in special brackets. Your instructor will decide whether to include these applications of the methods of calculus in your course.

To take full advantage of this textbook you should read the introductory section, "How to Use This Textbook", carefully. Answers to the even-numbered Questions and Exercises are given at the end of each chapter. Answers to all the Exercises in Appendix A are given at the end of Appendix A. The symbol ★ is used to denote important equations.

Chapters 1 through 13 are found in the textbook *THE PHYSICAL UNIVERSE*. Chapters 21 through 34 are found in *MICROPHYSICS*.

This textbook is offered with the hope that it will make your study of physics rewarding and enjoyable. Your comments and criticisms of this edition are cordially invited.

Donald E. DeGraaf

December, 1980

HOW TO USE THIS TEXTBOOK

1. Studying Physics is Different!

Learning physics may require a pattern of study different from those you have used in other courses. The suggestions given here are offered to help you develop your own effective learning pattern in this course.

Each numbered section of the textbook contains a small unit of material which should be mastered before you go on to the next section of the chapter. Work through each section thoughtfully, including the worked examples and the questions and exercises assigned by your instructor. Some of these questions and exercises have been selected to help you relate the new topic to prior material in the text. The Minimum Learning Goals, which state some things you should be able to do at the completion of your study of the section, will help you to focus on the essential concepts introduced in the section. As you study, keep in mind the suggestions in Section 3, "How to Study Physics."

When you understand the material you have read, you should be able to work out the answers to the learning goals, questions and exercises which appear in that section. Do not try to work *all* the questions and exercises, but only those assigned or suggested by your instructor. (You may, however, choose to answer additional questions and exercises in preparation for a test.) Review Section 4, "Hints for Answering Physics Questions," and Section 5, "Hints for Solving Physics Exercises," until these procedures become a habit with you.

It is worthwhile to note carefully the definition of each new physical quantity encountered, the statement of each physical law studied, and each *important* equation (marked with a ★). These definitions, laws and equations are not an end in themselves. It is the *ideas* and *concepts* of physics which should form the core of your learning in this course.

If you feel that your background in mathematics is weak, there are several resources available to help you. Appendix A contains a brief review of some mathematical methods used in this text, plus exercises in these methods.

2. Introduction to Learning Goals

At the beginning of each section you will find one or more Minimum Learning Goals for that section. These goals state explicity what you are expected to do to demonstrate that you have *learned* some essential aspects of that topic. The goals describe what you should be able to do, any special conditions or restrictions imposed, and most importantly, how *you* will know *when* you have learned at the minimally acceptable level.

When you first read the goals at the start of a section, you probably will not understand them very well. Nevertheless, they offer a focus for your effort in studying the topic. You will find it useful to re-read the goals as you progress in that topic. They will take on greater meaning, and they will continue to guide the direction of your study efforts.

Each goal describes a performance of some kind. The goals ask you to recall from memory and state certain definitions, use those definitions, relate concepts, definitions and laws, use those relationships in solving problems, and demonstrate other abilities related to your learning of physics.

It is important to know that the goals included in this textbook are *minimum* goals for each topic. You will know you have met one of these goals when you can answer a question, work an exercise, or solve a problem in agreement with the answer to a stated example. Each goal has one or more such examples, with correct answers supplied on another page in the text.

To take the greatest advantage of the goals and examples, you should look at the answers only after studying the topic, and when you are reasonably confident that you can meet the goals. Then the examples allow you to test yourself on the minimum goals of the section. If you fail that test the first time, re-study the topic before trying the next example.

An average student should be able to demonstrate that he or she has achieved most of these goals. There is not time to test you over all such goals, so examinations usually test only some sampling. Your success on that sample is taken as a reasonable estimate of how you would do if tested on all of them. In addition to these minimal goals, others, perhaps not stated so explicitly, will be conveyed to you by your instructor in class and in the selection of questions or exercises assigned to you.

The goals included in this book set the lower limit of what is expected of you. Your success in meeting them should motivate you to achieve many others not stated here. It is hoped that your interest will be aroused in your studies of some topics and that you will form goals of your own to achieve. There is no upper limit on what you can learn.

3. How to Study Physics

Here are some suggestions for studying each assignment:

1. Read the assigned material quickly to get an idea of the main topics discussed.

2. Re-read the material carefully, striving for understanding. Every sentence in a physics text is packed with meaning.

3. Learn new definitions of physical quantities, new units, and new physical laws. Do not memorize every equation but only the relatively few key equations, which then become your tools for understanding physical principles and for solving mathematical problems. It may be helpful to collect this basic information in one place in your notebook.

4. Work out the solutions to assigned questions and problems with your classmates in an informal "physics club."

5. Keep your homework up-to-date. The new material presented each day will usually build on the concepts accumulated during prior days.

4. Hints for Answering Physics Questions

1. Read the question. If you don't understand it, you probably need to study the assigned material. See Section 3 above.

2. Look up any new words in an adequate dictionary, if they are not defined in the text.

3. Decide which physical principles or ideas are most closely related to the question.

4. Decide where you need to go to find help in answering the question. Is the answer based on your assigned reading? Do you need to dig into some additional reference material? Does the question call for mental reflection on your own personal experience?

5. Decide the nature of the required answer: Does the question ask for a simple "yes" or "no"? For some scientific or general facts? For an explantion? For a description? For a numerical answer?

6. Be prepared for some questions which do not have an absolute answer, and for others which may eventually have a definite answer but which cannot be fully answered at present because of our inadequate knowledge.

7. Can the question be answered only by doing additional scientific experiments?

8. Don't quit if it's hard. If you get bogged down, turn your conscious attention to something else and let your subconscious work on the question for a while, possibly overnight. Experience students have found that this maneuver often works!

5. Hints for Solving Physics Exercises

You will often be asked to apply physical principles in the solution of numerical problems or exercises. A good problem solution which you have figured out and written out clearly has more value to you than a worked-out example in the textbook. Here are some time-saving suggestions which have helped other students.

1. Read the exercise.

2. Draw a neat, not-too-small sketch or diagram of the physical situation and label the pertinent physical variables.

3. Write down the given information in mathematical form.

4. Decide what area of physics and what physical concept is involved. In this book, you will meet exercises involving particle motion, forces, work and energy, wave motion, optics, electric and magnetic fields, electric circuits, and other topics listed in the table of contents. Some exercises involve more than one of these topics.

5. Review the key laws and equations which apply to this class of exercises.

6. Re-read the exercise. For what do you need to solve?

7. Is the quantity which you seek involved in any of the fundamental laws? If so, write down that law or laws in mathematical form. If not, write down the definition of this quantity in the form of an equation which does involve variables appearing in the fundamental laws or equations, and then write down those laws in equation form.

8. From the above, seek equations which relate the variable you need to find to the known quantities provided. This will permit you to solve for the original unknown variable in terms of known quantities. Quite often an unknown variable will cancel in the final equation, thus eliminating any need to calculate it. In the process of obtaining this final equation, it is usually less work if you refrain from substituting numerical values for the variables until the final equation is obtained.

9. In some exercises, you will need to look up the value of some physical quantity in your textbook or in a reference book, when this information is not given in the problem. Or, you may need to make a reasonable estimate of the value, based on your experience.

10. Check your result for consistency using a different method if possible. Is your answer reasonable? Your critical faculties will improve as you gain experience and these will aid you in avoiding many kinds of mistakes.

CHAPTER 41 DESCRIBING THE MOTION OF A PARTICLE IN ONE DIMENSION

Introduction

The variables used to describe straight-line motion are time, displacement, velocity and acceleration. These variables are defined, and procedures for measuring them are described. Tables of data, graphs, and equations offer alternative methods of describing motion; these methods are contrasted and correlated with each other. Special attention is given to the use of graphs in describing motion, and to equations applicable to the important case of motion with constant acceleration.

When you have completed your study of the material in this chapter, you should be able to answer questions and work exercises like the examples in the minimum learning goals assigned by your instructor.

41.1 Displacement, Average Speed and Velocity

Minimum Learning Goal. (1) Be able to distinguish between the average speed and the average velocity of an object by calculating each of these quantities from a description of the object's motion during a given interval of time. For example:

1 The diagram below shows the path followed by the tip of the second hand of a clock during 30 seconds. If the hand is 10 cm long, what is (a) the average speed of the tip during that time, and (b) the average velocity of the tip during that time? (Answer, page 41-38, Line 8)

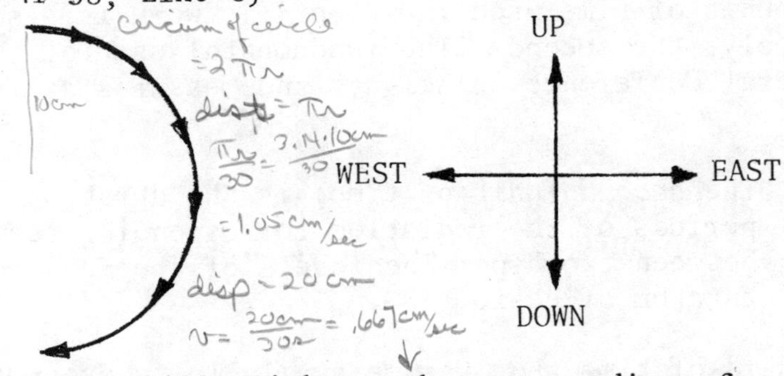

2 A car starts in Lowerville with an odometer reading of 23,498.3 miles, follows a winding road and reaches Upperville, due north of Lowerville, with an odometer reading of 23,745.9 miles. If the trip required 4.00 hours, and the straight-line distance between the two towns is 180 miles, what was (a) the average speed of the car, and (b) its average velocity? (Answer, page 41- 38,Line 3)

The questions we can ask about motion fall into two broad categories. For any object in motion, we may ask: *How* does it move? *Why* does it move that way? Although both of these questions are answered in the scientific study of motion, they are really questions of two different kinds. The question, "How does an object move?", calls for a *description* of the motion in terms of physical variables which give a quantitative account of where the object is at successive points in time, how fast it is going, and so forth. In contrast, the question, "Why does the object move as it does?", asks for an *explanation* of the motion in terms of causative factors acting on the object from the outside, and an internal factor which determines the response of the object to the external agents acting on it.

The branch of science which deals with the description of motion is called *kinematics*. The principles used to explain motion are considered in the branch of physics called *dynamics*. Both kinematics and dynamics are parts of a major area of physics called *mechanics*. In this volume, Chapters 41 through 51 deal with mechanics.

In this chapter, we review elementary mathematical methods used to describe straight-line motions. Chapter 42 presents the physical principles needed to account for these motions. Chapter 43 extends these methods to motion in two and three dimensions, and shows how the methods of calculus can be used to analyze particle motion.

To give an accurate description of the motion of an object, we must be able to say where it is at each of many instants of time. The information needed for a scientific description of motion may be obtained from an experiment. This requires one or more scientific instruments which can be used to measure the object's position at various times. In the International System of Units, the unit used for measuring position (or length) is the *meter*, defined since 1960 as 1,650,763.73 wavelengths of the orange-red light emitted by atoms of the isotope krypton-86. This value was chosen to keep the newly defined meter as close as possible to the value defined earlier in terms of the length between two lines engraved on a platinum-iridium bar kept in a temperature-controlled vault at the International Bureau of Weights and Measures near Paris.

The unit of time used in scientific work is also the unit used in daily life, namely, the second. The fundamental unit of time was defined by the 1967 General Conference on Weights and Measures as follows:

> The *second* (unit of time) is the duration of 9,192,631,770 periods of the radiation corresponding to the transition between two hyperfine levels of the fundamental states of an atom of cesium 133.

The standard of time thus became available to every laboratory having access to a "cesium clock." Laboratories in the United States which do not have one can calibrate their own time standards by using radio signals broadcast from station WWV by the National Bureau of Standards. For laboratory experiments where only moderate accuracy is required, it is usually satisfactory to measure time intervals using an electric clock operated from the electric utility line with its nominal supply frequency of 60 hertz.

A scientific description of motion requires adequate knowledge of the position of the moving object at various times during the motion. While direct measurements on an object usually give its position in some chosen coordinate system, we are often more interested in the *distance* traveled by the object from its starting point, or in the distance from one specified point to another along the path.

The *distance* between two points depends on the path or the route taken between these points. However, if we ask, What is the *shortest distance* between two points, there is only one answer. This shortest distance is called the *displacement*, represented by the symbol $\vec{s}$.

> The *displacement*, $\vec{s}$, between two points is the vector distance (measured in a straight line) from the initial point to the final point.

Displacement is called a *vector* quantity, or *vector* change in position, because a definite direction is associated with it.

One word used for the quantity that tells how fast something is moving is *speed*. The definition of speed, and of other quantities used to describe motion, may be illustrated with reference to the motion of a car moving on a straight, level road. An adequate description of this motion must include a statement of the car's position at many instants of time during the motion. Because the car itself is several meters in length, it is necessary to choose one reference point on the car, such as the center of the front bumper. Then with the aid of a movie camera or other suitable instruments, it is possible to measure the position of the car at any time during its motion.

Actually, position and time are the *only* quantities one can measure directly for the moving car. Any "measurement" of speed requires information obtained from direct measurements of position and time. To utilize this information, we need a precise definition of speed:

> The *speed* of a moving object is the distance traveled divided by the time required for this travel.

If we choose one direction along the road as the $+x$ direction, then the position of the car at any point can be stated by giving the car's value of x, measured from some initial point $x = 0$. We use the symbol $\vec{s}$ to denote the car's displacement from the beginning point of its motion; the scalar quantity s is the magnitude or numerical value of this displacement. For motion in one dimension and only one direction, s is equal to the distance traveled by the car from the beginning of its motion. We use the symbol Δs for the *change in distance* or *change in position*. The symbol for the time variable is t. The time interval, or change in time, while the car moves through a small change in position Δs is written Δt.

We can now write a definition for speed using an equation. We note, however, that in general the car may be speeding up or slowing down while it moves through the interval Δs, so measurements of Δs and Δt do not yield the value of speed at any one instant of time during this motion; rather, they enable us to compute the *average* speed, v_{av}, during this interval.

The *average speed*, v_{av}, during an interval of motion is the distance traveled divided by the time required to travel this distance.

★ In symbols, $v_{av} \equiv \frac{\Delta s}{\Delta t}$. The SI Unit for speed is meter/second.

The *direction of travel* is often just as important as the speed. The direction of the car's motion can be stated as "in the $+x$ direction" or, "toward the north." The physical quantity which indicates *both* the speed and the direction of motion is called *velocity*. The symbol used for velocity is $\vec{v}$, where the small arrow indicates that the velocity includes a direction in space, as well as the magnitude or numerical value of the speed. Because velocity is a vector quantity, it must be defined in terms of a displacement, rather than a distance.

The *average velocity*, $\vec{v}_{av}$, during an interval of motion is the displacement traveled divided by the time required for this displacement.

★ In symbols, $\vec{v}_{av} \equiv \frac{\Delta \vec{s}}{\Delta t}$. The SI Unit for velocity is meter/second.

For a straight-line motion in any one direction, the displacement between two points has the same magnitude as the distance between these two points. For motion in a curved path, however, or motion on a path which consists of a series of straight segments having various directions, the displacement between two points has a smaller magnitude than the distance between these points along the path, because the displacement is defined as the vector distance measured in a straight line between the two points, even though the motion does not take place in a straight line.

Worked Example 41-1

Mark drives from London, Ontario via Detroit to Cleveland, a distance of 510 km, in 5 hours and 54 minutes. The displacement from London to Cleveland is 172 km at an angle of 255°, measured counterclockwise from due east.

(a) What is Mark's average speed during this trip?
(b) What is Mark's average velocity during this trip?

Solution

(a) To compute the average speed, we must divide the total distance traveled by the total time, which is 5.9 hours. Therefore,

$$v_{av} = \frac{510 \text{ km}}{5.9 \text{ hours}} = 86.4 \text{ km/hour}.$$

(b) To compute the average velocity, we must divide the displacement by the time.

$$\vec{v}_{av} = \frac{\Delta \vec{s}}{\Delta t} = \frac{172 \text{ km at } 255°}{5.9 \text{ hours}} = 29.2 \text{ km/hour, at an angle of } 255°.$$

The specification of the displacement of a body at each instant of time is, in principle, a complete description of the motion. From this description, however, we can derive other quantities of motion, namely, velocity and acceleration. One of the reasons for deriving them has to do with the formulation of the laws of physics that govern the motion. These laws can be stated more simply in terms of velocity and acceleration.

Questions and Exercises

Q 41.1-1 What quantities should be measured, and in what units, to adequately describe the motion of an automobile during the period while it travels one block after starting from rest? How many measurements should be made of each quantity?

Q 41.1-2 A stone attached to a small parachute is dropped from the edge of the roof of the physics building on a calm day so that it falls straight down. State the quantities which should be measured, including their units, to adequately describe the motion of this stone. How many measurements should be made of each quantity?

Q 41.1-3 A standard of time has to be something which "repeats at equal intervals." But how can you tell whether it does, if that is your standard of time? For example, suppose you have two clocks. Relative to Clock A, clock B is slowing down. Relative to clock B, A is speeding up. Could you take either one to be your standard of time? If not, how would you determine which one is "good," if indeed either one is? (They could both be bad.)

Q 41.1-4 A ball is released and simultaneously given a push upward along an inclined plane. It rolls up a distance of two feet, then rolls back down again, right past the point where it originally started.

(a) When is the displacement increasing?
(b) When is it decreasing?
(c) What is the value of the displacement at the instant when the ball changes its direction of motion?
(d) What is the displacement at the instant when it is again passing the starting point?

Q 41.1-5 Suppose you are given the magnitude of the average velocity of a car in a certain interval as 43 miles/hour. The length of the interval is 3 hours. Someone says that the car must have traveled 129 miles in this time. Is he using a law of physics, or a law of mathematics? In other words, does this result have to be checked by actual observation? Can you conceive of a situation (whether or not it really occurs) in which an object travels for three hours at an average velocity of 43 miles per hour, but covers a distance that is *not* 129 miles?

E 41.1-1 A race track has a circumference (total length of one lap) of 4.0 kilometers. A racing car completes exactly one lap in 80 seconds. (continued on next page)

(a) What is the distance traveled by the car during this lap?
(b) What is the displacement of the car from the beginning to the end of this lap?

E 41.1-2 A glider is moving in an unusual way on an inclined air track. The position at the extreme upper end of the track is labeled "0 meters," and other positions are *measured starting from the upper end.* At various instants of time the position of the glider is measured and found to be:

time (seconds)	0	1.0	1.1	1.2	1.5	2.0	3.0	4.0	5.0
position (meters)	0.80	0.50	0.481	0.464	0.425	0.40	0.50	0.80	1.30

The glider reaches its highest position at t = 2.0 sec.
(a) What is the total distance traveled by the glider from t = 0 to t = 5 seconds?
(b) What is the displacement of the glider from t = 0 to t = 5 seconds? Is this displacement up the track, down the track, both up and down the track, or neither up nor down?

E 41.1-3 What speed in miles/hour is equal to a speed of one m/s?

E 41.1-4 If you run completely around a 440-yard track in 65 seconds,
(a) What is your displacement? (b) What is your average velocity?
(c) What is your average speed?

E 41.1-5 Refer to the data given with E 41.1-2.
(a) What was the average speed of the glider from t = 0 to t = 2 seconds?
(b) What was the average velocity of the glider from t = 0 to t = 2 seconds?
(c) What was the average speed of the glider from t = 0 to t = 4 seconds?
(d) What was the average velocity of the glider from t = 0 to t = 4 seconds?
(e) What was the average speed of the glider from t = 0 to t = 5 seconds?
(f) What was the average velocity of the glider from t = 0 to t = 5 seconds?

Check your understanding of the material above by testing yourself on the minimum learning goal at the beginning of this section.

41.2 Instantaneous Speed and Velocity

Minimum Learning Goal. Be able to distinguish between instantaneous speed and instantaneous velocity, by calculating each of these quantities from a description of the motion of an object. For example:

1 An auto is traveling along a curved road, as shown on the map at the left below. The map at the right is a "close-up," showing some details of the path in the vicinity of

point *G*. The table of data gives the time, in minutes and seconds, at which the car passes each point.

Point	Time
A	0:00
B	0:41
C	1:12
D	1:41
E	1:48
F	1:52
G	1:54
H	1:56
I	2:00
J	2:06
K	2:12
L	2:43
M	3:12

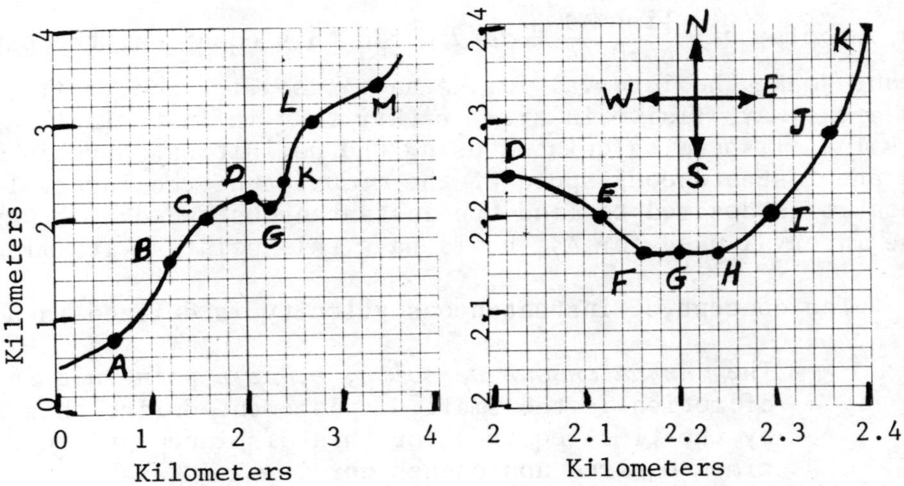

(a) What is the instantaneous speed of the car at point *G*?
(b) What is the instantaneous velocity of the car at point *G*? State both the magnitude and the direction.

(Answer, page 41-38, Line 7)

An automobile driver in a rush-hour traffic jam may drive at an average speed of eight miles per hour, but that figure does not tell us what his speed is at any particular moment. His speed will be zero when his lane of traffic is stalled completely, but it may rise to 40 miles/hour at other times. In this case and in many others, we are interested not only in the average speed of a moving object, but also in its instantaneous speed at various points along its path. How can we obtain the speed at some instant of time, or at some specific point along the path of motion?

To develop a definition and an equation for instantaneous speed, consider again a car moving on a straight level road. If the distance Δs traveled by the car is large, the speed of the car may vary a great deal as it moves through this interval. If Δs is small, however, there is less opportunity for the car's speed to vary markedly in the interval, so the average speed computed using the ratio $\Delta s/\Delta t$ will be closer to the value of speed at any particular point in this interval. To converge on an accurate value for the speed of the car as it is passing some particular point, we need to reduce the value of Δs to a very small interval which includes the point we have chosen. Then we can calculate an approximate value for the instantaneous speed by dividing this small value of Δs by the small time interval Δt required for the car to move through this Δs interval. Actually, this procedure gives us the *average* speed in this small interval, so to obtain the most accurate value of *instantaneous* speed we need to extend this procedure of shrinking Δs and Δt to the limit where Δs and Δt approach zero.

The *instantaneous speed*, v, at a point within a short interval of motion is the small distance traveled in this interval divided by the time required to travel this distance, in the limit as the time required approaches zero.

★ In symbols, $v \equiv \lim_{\Delta t \to 0} \frac{\Delta s}{\Delta t} \equiv ds/dt$. In this equation, the expression $\lim_{\Delta t \to 0}$ means "make the interval for Δs and Δt smaller and smaller." Because Δs and Δt are always finite in size, rather than infinitesimal, as a practical matter a value of speed calculated using the definition above is best interpreted as the instantaneous speed at the *midpoint* of the interval for measurement. When it is desired to find the instantaneous speed at a particular point, the interval used for Δs should have this point as its midpoint, if feasible.

The concept of instantaneous velocity is defined in a similar manner.

The *instantaneous velocity*, $\vec{v}$, at a point within a short interval of motion is the small displacement during this interval divided by the time required for this displacement, in the limit as the time required approaches zero.

★ In symbols, $\vec{v} \equiv \lim_{\Delta t \to 0} \frac{\Delta \vec{s}}{\Delta t} \equiv d\vec{s}/dt$. The instantaneous velocity indicates the direction of the motion, as well as the magnitude of the velocity or speed.

Worked Example 41-2

Shown here is a replica of a spark tape obtained during the motion of a glider down a sloping air track. The spark rate was 10 sparks per second. The spark pits are numbered from zero to 15. Alongside the tape is a distance scale marked in cm. Pit number zero was made at t = 0 seconds. Displacements are measured from the starting point (t = 0).

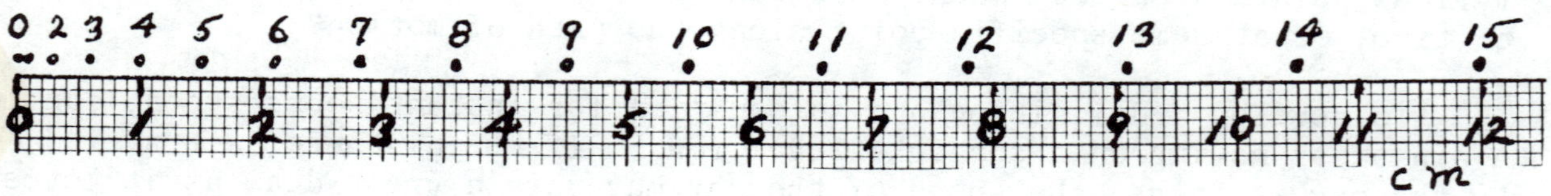

(a) What is the displacement at t = 1.0 second?
(b) What is the average speed during the first 1.5 seconds of this motion?
(c) What is the instantaneous speed at t = 1.0 second?
(d) What is the instantaneous velocity at t = 0.9 second?

Solution

(a) Spark pit number 10 was made at t = 1.0 second. It is located at a displacement of 5.5 cm from the starting point, down the track.

(b) The first 1.5 seconds are encompassed by the distance from spark pit number 0 to pit number 15, a distance of 12.0 cm. Then the average speed is $v_{av} = \frac{\Delta s}{\Delta t} = \frac{12 \text{ cm}}{1.5 \text{ s}} = 8.0$ cm/s.

(c) To calculate v at spark pit number 10, we need to choose an interval having pit 10 in its interior, preferably at its midpoint. A suitable interval is the one from pit 9 to pit 11; this is the smallest such interval which is usable from the data given. For this interval, $\Delta s = 6.6$ cm - 4.5 cm = 2.1 cm, and $\Delta t = 1.1$ sec -0.9 sec = 0.2 seconds. Then $v = \frac{\Delta s}{\Delta t} = \frac{2.1 \text{ cm}}{0.2 \text{ s}} = 10.5$ cm/s.

(d) To calculate $\vec{v}$ at spark pit number 9, we choose the interval from pit 8 to pit 10. For this interval, $\Delta s = 5.5$ cm - 3.6 cm = 1.9 cm, and $\Delta t = 1.0$ s - 0.8 s = 0.2 s. Then $\vec{v} = \frac{1.9 \text{ cm}}{0.2 \text{ s}} = 9.5$ cm/s, down the track.

When calculating an instantaneous velocity, it is not necessary to shrink the interval to an infinitesimal size in order to achieve the limit. In principle, we compute the average velocity in an interval which gets smaller and smaller, but always includes the point at which we want to know the instantaneous velocity. When the interval is small enough, so that making it still smaller does not make any difference in the calculated velocity (to whatever accuracy we have in mind), we have come to the limit.

We have used some care to distinguish between instantaneous velocity and average velocity. In succeeding portions of this textbook, when the work *velocity* is used without a qualifying adjective it should be understood to mean instantaneous velocity.

Questions and Exercises

Q 41.2-1 A stone is tossed straight up. The statement is made, "At the top of its path the stone is instantaneously at rest." What does that mean? Can the velocity of the stone be zero for only one instant? If a measurement of velocity entails observing the stone at two instants of time and comparing the displacements, how can the velocity be zero at one single instant?

Q 41.2-2 Suppose that you are working on a computation of instantaneous velocity of a certain body. You compute the average velocity using intervals of differing size, and centered at various points along the path of the body, and to your surprise you find that you always get the same number. No matter how small or large the interval, you seem to have reached the limit already. What kind of motion does this body have?

Q 41.2-3 How could you measure the instantaneous velocity at various values of time for: (a) an automobile? (b) a flying bird? (c) a walking ant? (d) a swimming dolphin?

E 41.2-1 Refer to the data given with E 41.1-2 on page 41-9.
(a) What was the instantaneous speed of the glider at $t = 1.1$ s?
(b) What was the instantaneous velocity of the glider at $t = 1.1$ seconds?

Check your understanding of the material above by testing yourself on the minimum learning goal at the beginning of this section.

41.3 Acceleration

Minimum Learning Goals. (1) Be able to calculate the average acceleration of an object, given its change in velocity and the time during which the velocity changes. For example:

1 An alpha particle (helium nucleus) moving at 3×10^6 m/s enters a thin sheet of gold foil where it interacts with a nucleus of a gold atom. As a result of this interaction the alpha particle rebounds with (almost) the same speed as it originally had. Although the speed does not change, the velocity changes direction, so that the change in velocity has a magnitude of 6×10^6 m/s. If the change in velocity takes 2×10^{-16} seconds, what is the average acceleration of the alpha particle? (Answer, page 41-38, Line 2)

2 A car is traveling due north at 60 mph (26.8 m/s) when it enters a curve. As the car comes out of the curve 20 seconds later, it is traveling at 60 mph, but due East. What is the average acceleration (magnitude *and* direction) of the car during the 20 seconds it is on the curve? (Answer, page 41-38, Line 1)

(2) Be able to calculate the instantaneous acceleration of an object, from a table of changes in velocity and the times during which these changes occur. For example:

1 A small mass is moving at a constant speed of 0.50 m/s in a clockwise direction around a circle having a radius of 1.0 meter. The table below shows changes in velocity for several time intervals; *each* interval begins when the mass passes point *A*. The direction of $\Delta\vec{v}$ are indicated using the following convention: The reference direction (0°) is toward the *right*, and angles are measured counter-clockwise (ccw) from this direction. What is the instantaneous vector acceleration at point *A*? (Answer, page 41-38, Line 5)

Δt, time interval	Δv, Magnitude of change in velocity	Direction of $\Delta\vec{v}$, measured counterclockwise from the right
0.070 sec	0.0175 m/s	+269°
0.105	0.0262	+268.5°
0.175	0.0436	+267.5°
0.350	0.087	+265°
0.698	0.174	+260°
1.05	0.259	+255°
1.57	0.382	+247.5°
3.14	0.707	+225°

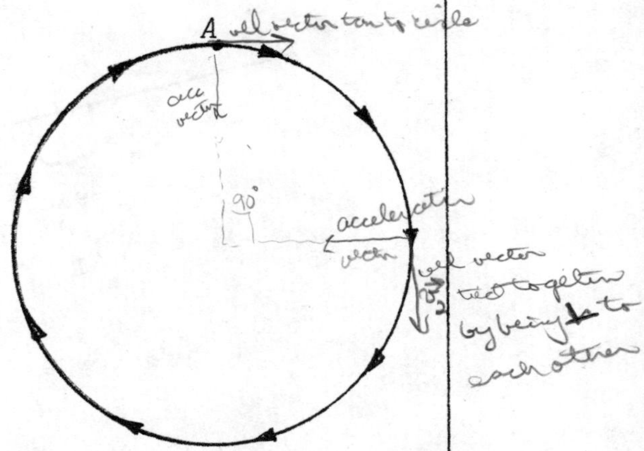

If you watch the speedometer of your car as you start it up and drive down the road, you will notice that the speed of the car increases until the car reaches a more-or-less steady traveling speed. A driver who is simply out for a leisurely ride may be quite content to increase his speed at a very gradual rate. A driver who is in a hurry, however, will "step on the gas" in order to bring his car up to its cruising speed quickly. And the driver of a racing car wants to bring his car up to a high speed at the maximum possible rate. In each of these examples, a car starts from rest and speeds up to 30 miles/hour (or more); the examples differ in the *time-rate* at which the speed changes. This quantity is called acceleration. More exactly,

> The *acceleration* of an object is the time-rate of change of velocity of the object.

In the examples above, the casual driver is content with a small acceleration, while the racing driver wants the maximum acceleration obtainable.

To develop an equation to be used for calculating acceleration, let us return to our consideration of a car moving on a straight level road. We choose two points on the road, numbered *1* and *2*, and we want to compute the acceleration of the car as it moves from point *1* to point *2*. Suppose we have a clock which begins running at the beginning of the car's motion, and which indicates the elapsed time at which the car passes each point on the track; on this clock, t_1 is the time at which the car passes point *1*, and t_2 is the time at which the car passes point *2*. Because we are interested in the rate of change of velocity, we need to find the velocities at point *1* and point *2*. To do this, we need to measure values for Δs_1 and Δt_1 in a small interval having point *1* at its midpoint, and also measure values for Δs_2 and Δt_2 in another small interval surrounding point *2*. From these data, we can calculate $v_1 = \Delta s_1/\Delta t_1$ at point *1*, and $v_2 = \Delta s_2/\Delta t_2$ at point *2*. If we combine the calculated speed v_1 with the direction of the car's motion as it passes point *1*, we obtain the velocity $\vec{v}_1$; in a similar way we can obtain the velocity $\vec{v}_2$ at point *2*.

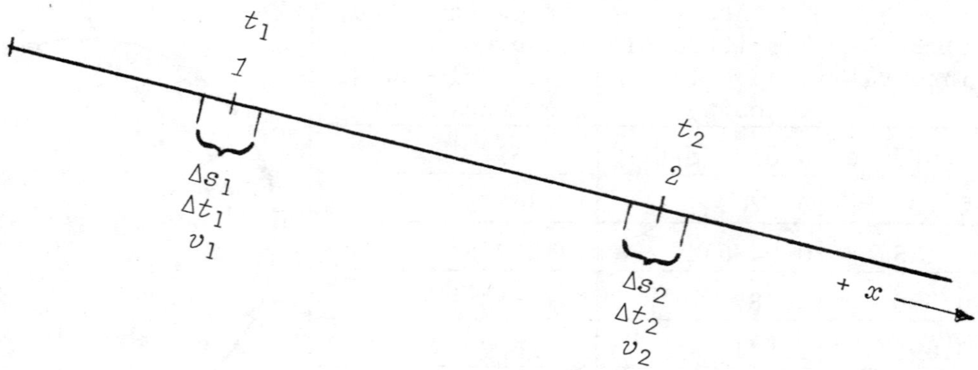

Figure 41-1. Measurements of Δs_1 and Δt_1 for a small interval around point *1* allow one to calculate $\vec{v}_1$ at point *1*. Similar measurements lead to a value for $\vec{v}_2$ at point *2*.

Using these values for velocity, we can calculate the time-rate of change in velocity of the car in moving from point *1* to point *2*. If the distance between these points is not small, the acceleration might actually change in this interval, so the best we can do with the data described above is to compute the *average* acceleration in this interval.

The *average acceleration*, $\vec{a}_{av}$, during an interval of motion is the change in velocity during this interval divided by the time required for this change in velocity.

★ In symbols, $\vec{a}_{av} \equiv \dfrac{\Delta\vec{v}}{\Delta t} = \dfrac{\vec{v}_2 - \vec{v}_1}{t_2 - t_1}$. The SI Unit for acceleration is meter/second per second, or meter/second^2 or m/s^2.

A comment about the unit for acceleration, m/s^2. This should not be thought of as including a separate quantity "seconds-squared." The acceleration of an object is defined as its change in velocity per unit of time, or simply the change in velocity per second. So the unit for acceleration is meter/second, per second, which may be written $\dfrac{\text{m/s}}{\text{s}}$ or m/s^2.

If the car gains speed rather slowly, its acceleration is small. If the car's speed increases rapidly, however, its acceleration is large. The reader is encouraged to observe carefully the moving objects he sees during the day, and to develop skill at recognizing large and small accelerations.

In discussing an example of straight-line motion, it is usually helpful to designate some direction as the *positive* direction. Then all of the vector quantities associated with the motion (such as displacement, velocity, and acceleration) which have this direction are positive quantities, while those having the opposite direction are negative. For our example of a car moving on a straight level road, suppose we choose the direction toward the right along the road as the positive direction. Then for a car which moves toward the right with *increasing* velocity, the acceleration is positive. If the car moves toward the right with *decreasing* velocity, however, its acceleration is negative.

To use the definition of average acceleration, it is necessary to compute a change in velocity, $\Delta\vec{v} = \vec{v}_2 - \vec{v}_1$. This calculation requires that we subtract $\vec{v}_1$ from $\vec{v}_2$. Subtracting $\vec{v}_1$ from $\vec{v}_2$ is the same as *adding* $-\vec{v}_1$ to $\vec{v}_2$. We can therefore regard the quantity $\vec{v}_2 - \vec{v}_1$ as the *vector sum* of $\vec{v}_2 + (-\vec{v}_1)$. The calculation of $\Delta\vec{v}$ therefore becomes a problem in vector addition, and requires a vector diagram in which the vectors $\vec{v}_2$ and $-\vec{v}_1$ can be added to obtain $\Delta\vec{v}$. Worked Example 41-3 shows how $\Delta\vec{v}$ can be calculated when the motion is not on a straight line.

Worked Example 41-3

Twyla is traveling at a constant speed of 7 meter/sec (about 15 miles/hour) on her bicycle, when she turns a corner. She enters the turn heading north, and 2.4 seconds later she completes the turn heading west. What is her average acceleration while making this turn?

Solution

We use $\vec{v}_1$ to denote Twyla's initial velocity of 7 m/s toward the north, and $\vec{v}_2$ to denote her later velocity of 7 m/s toward the west. Her change in velocity is $\Delta\vec{v} = \vec{v}_2 - \vec{v}_1 = \vec{v}_2 + (-\vec{v}_1)$. The figure shows the vector diagram for adding $\vec{v}_2$ and $-\vec{v}_1$, and their sum $\Delta\vec{v}$. Using the scale, we find that $\Delta\vec{v}$ = 9.9 m/s, directed southwest. Using a protractor, we could measure the direction as 225°, following the convention for measuring angles. Twyla's average acceleration is therefore

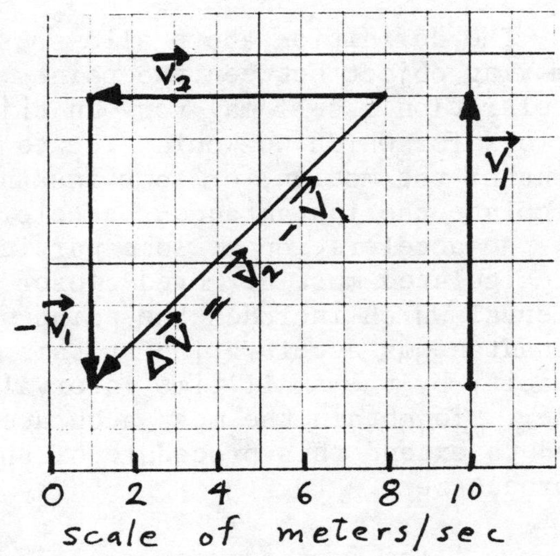

$$\vec{a}_{av} = \frac{\Delta\vec{v}}{\Delta t} = \frac{9.9 \text{ m/s}}{2.4 \text{ s}} = 4.1 \text{ m/s}^2, \text{ toward the southwest.}$$

The definition of average acceleration shows that the direction of this acceleration is the same as the direction of the change in velocity, $\Delta\vec{v}$.

For any example of straightline motion we might study, we can choose some direction of the motion as the forward, or positive, direction. Vector quantities having the opposite direction are then considered as negative. If an object is moving forward and speeding up, its acceleration is positive. If the object is moving forward and slowing down, its acceleration is negative. If the object is moving backward and speeding up, its acceleration is negative, i.e., in the negative direction. When the object is moving backward and slowing down, its acceleration is positive. When the velocity of the object is *constant*, whether it be zero or 100 miles per hour, the acceleration is zero.

Average acceleration is defined in terms of the change in *velocity*, not the change in speed. Because velocity is a vector quantity, having both magnitude and direction, it can change in two rather different ways. For an object falling from rest, the velocity changes in magnitude but not in direction. For a ball thrown straight upward, the velocity changes in magnitude and also in direction as the ball moves first up and then down; throughout this motion, however, the velocity vector is always vertical, in either the positive or negative direction. A velocity can also change in direction by changing its angle with respect to some reference direction, as we saw in Worked Example 41-3. For a car moving around an oval race track, the direction of the velocity changes through an angle of 360° during one lap. An object moving in a circle or an ellipse, such as the earth in its orbit around the sun, is continually changing the direction of its velocity. Whenever the velocity of an object changes in direction, it is necessary to construct a vector diagram in order to find the change in velocity, $\Delta\vec{v}$.

The definition above allows us to calculate the average acceleration for a moving object between two points at which the velocities are known. The acceleration itself may vary in this interval, however. There are some motions for which we would like to know the acceleration at some particular point of the motion, or some instant of time during the motion. How can we calculate the instantaneous acceleration? To converge on an accurate value for the acceleration at some particular point, the points where the velocity is calculated must be moved closer together until they define a very small interval which includes the point we have chosen. Then we can calculate the instantaneous acceleration at this point by dividing the small change in velocity by the small time interval required for motion from one point to the other. To obtain the most accurate value of instantaneous acceleration, we need to extend this procedure of shrinking $\Delta\vec{v}$ and $\Delta\vec{t}$ to the limit where they approach zero.

> The *instantaneous acceleration*, $\vec{a}$, at a point within a short interval of motion is the small change in velocity during this interval divided by the time required for this small change in velocity, in the limit as the time required approaches zero.

★ In symbols, $\vec{a} \equiv \lim_{\Delta t \to 0} \frac{\Delta\vec{v}}{\Delta t} \equiv dv/dt$. In this equation, the expression $\lim_{\Delta t \to 0}$ means "move the points which define the interval for $\Delta\vec{t}$ closer and closer together."

Each measurement of velocity requires the use of a displacement interval $\Delta\vec{s}$, so there is a practical limit to the size of the smallest interval which can be used to measure acceleration. Nevertheless, the definition and equation given above indicate the approach one can use to measure instantaneous acceleration. As we shall see in the next section, a graphical approach to the description of motion is of value in visualizing and calculating changes in velocity and acceleration.

We have made a distinction between average acceleration and instantaneous acceleration. In succeeding portions of this book, when the word *acceleration* is used without a qualifying adjective, it should be understood to refer to instantaneous acceleration. For motions where the acceleration is constant, the acceleration at any instant will be the same as the average acceleration during the entire motion.

We can now consider some further questions about acceleration. Under what conditions does an object have zero acceleration? One possibility, of course, is for the object to remain at rest. While its velocity remains zero, the velocity is not changing so the acceleration is zero. If the object is moving with constant velocity (in a straight line), then also its acceleration is zero.

When a ball is thrown upward, while the ball is moving up its acceleration is down, and later while the ball is moving down its acceleration is also down. What is the acceleration of the ball at the point where it changes the direction of its motion? At this turn-around point, the velocity changes its direction from up to down, so the velocity at the turn-around point is zero. If the velocity at this point is zero, is the acceleration also zero?

To calculate the acceleration at this point, we must choose an interval which begins just before the ball reaches the turn-around point, and ends just after the ball leaves this point. Then the ball is at the turn-around point in the interior of the *time interval*, even though the turn-around point is not within the space interval between the beginning and ending points we are using. The velocity at the beginning of this interval is up, and at the end of the interval it is down. A vector diagram to calculate the *change in velocity* for this interval show that $\Delta\vec{v}$ is down, so the instantanteous acceleration at the turn-around point must also be down. In other words, the acceleration is down at all times during this motion, even though the velocity changes its direction.

If this result seems surprising, consider it from another point of view. Suppose the acceleration were equal to zero at the turn-around point. Then $\Delta\vec{v}$ would also be zero for at least a small interval which includes this point, and since the velocity is itself zero at the turn-around point, it would remain zero for at least a small interval of time. But if the velocity remains zero for an interval, and the ball has no acceleration, then the ball would not begin to move at all - it would simply remain at the hightest point of its motion. Since this result is not what we observe, we must conclude that in fact the acceleration is not zero at the turn-around point.

Questions and Exercises

Q 41.3-1 In discussing straight-line motion, we choose one of the directions along the line as positive. For example, we may choose the direction toward the right as positive. Then a displacement, a velocity, or an acceleration to the right is positive. Quantities to the left are negative. How is it possible to have:

(a) positive velocity but negative acceleration?
(b) negative velocity but positive acceleration?
(c) negative velocity and acceleration?

Q 41.3-2 The word "deceleration" is used for an acceleration of such a sign that it leads to a decrease of speed.
(a) Is deceleration the same as negative acceleration?
(b) Can speed be increasing while the acceleration is negative? How?
(c) Can speed be decreasing while acceleration is positive? How?

Q 41.3-3 As you watch a moving car, how can you tell whether its acceleration is positive, negative, or zero?

Q 41.3-4 Give an example of each of the following types of motion which can actually occur:
(a) velocity is negative while acceleration is positive;
(b) velocity is positive while acceleration is negative;
(c) velocity is zero while acceleration is negative;
(d) acceleration is zero while velocity is positive.

Q 41.3-5 (a) Is it possible to have a body moving with non-zero instantaneous velocity, but zero instantaneous acceleration?
(b) Is it possible to have a body moving with non-zero instantaneous acceleration, but zero instantaneous velocity?
(c) Give examples, if possible, to illustrate your answers.

Q 41.3-6 A ball rolls up an inclined plane until it reaches a certain point, and then rolls down again.
(a) At the highest point, what is the ball's velocity?
(b) At the highest point, is the ball accelerating? Give a reason for your answer.

E 41.3-1 Referring to data given on the spark tape included with Worked Example 41-2 on page 41-8, compute the average acceleration of the glider between spark pit 5 and pit 9.

E 41.3-2 Referring to the spark tape given in Worked Example 41-2 on page 41-8, compute the average acceleration of the glider as it moves from x = 5.5 cm to x = 10.5 cm.

E 41.3-3 A large truck takes 10 seconds to change its speed from 30 mi/hr to 40 mi/hr on a smooth straight road. A small car takes 15 seconds to change its speed from 50 mi/hr to 65 mi/hr in order to pass the truck. Which vehicle has the larger average acceleration?

E 41.3-4 An airplane is traveling due East at 400 mi/hr (179 m/s). It banks around a curve for 40 seconds and emerges going 400 mi/hr heading due West. What is the average acceleration of the airplane during the time it is banking?

E 41.3-5 Refer to the spark record given with Worked Example 41-2 on page 41-8. (continued on next page)

(a) Using the most suitable data, compute the instantaneous acceleration of the glider at some point within the first cm of its motion.

(b) Compare your results with your answers to E 41.3-1 and E 41.3-2. What can you conclude about the acceleration of the glider during this motion?

(c) For this spark tape, at what value of time could the speed of the glider equal zero?

(d) What is the acceleration of the glider immediately after t = 0 seconds?

E 41.3-6 A body in uniformly accelerated motion has an initial velocity of 5 m/s south; 7 seconds later it has a velocity of 9 m/s north. What was the acceleration during this time interval?

Check your understanding of the material above by testing yourself on the minimum learning goal at the beginning of this section.

41.4 Using Graphs to Find Velocity and Acceleration

Minimum Learning Goals. (1) Given a graph of distance versus time for a moving object, be able to read physical information directly from the graph, and be able to draw inferences about the object's motion, including an estimate of the speed of the object at a specified time. For example:

1 The graph below shows the distance traveled by a car along a race track, as a function of time since the start of a race.

(a) How far does the car travel in the first ten seconds?

(b) How far does the car travel between ten and twenty seconds after it starts?

(c) How many seconds are required to travel the first mile (1609 meters)?

(d) At what time did the car reach its maximum speed?

(e) Estimate the maximum speed of the car.

(Answer, page 41-38, Line 12) (Graph at the top of next page.)

(2) Be able to calculate the acceleration of an object at a given time, and be able to infer other information about the object's motion, when given a graph of speed versus time or a table of velocity versus time. For example:

1 The graph below is that of speed versus time for a baseball thrown vertically into the air.

(a) What is the initial upward speed of the ball?

(b) At what time does the ball reach the highest point of its motion?

(c) What is the acceleration at t = 2.0 seconds?

(Answer, page 41-38, Line 6)

Graph for Minimum Learning Goal (1)

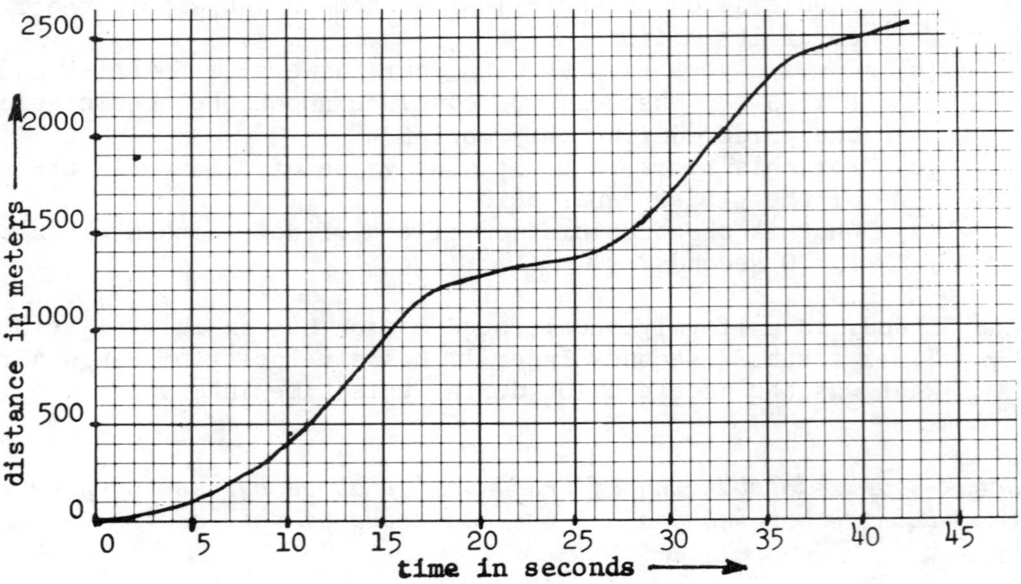

Graph for Minimum Learning Goal (2)

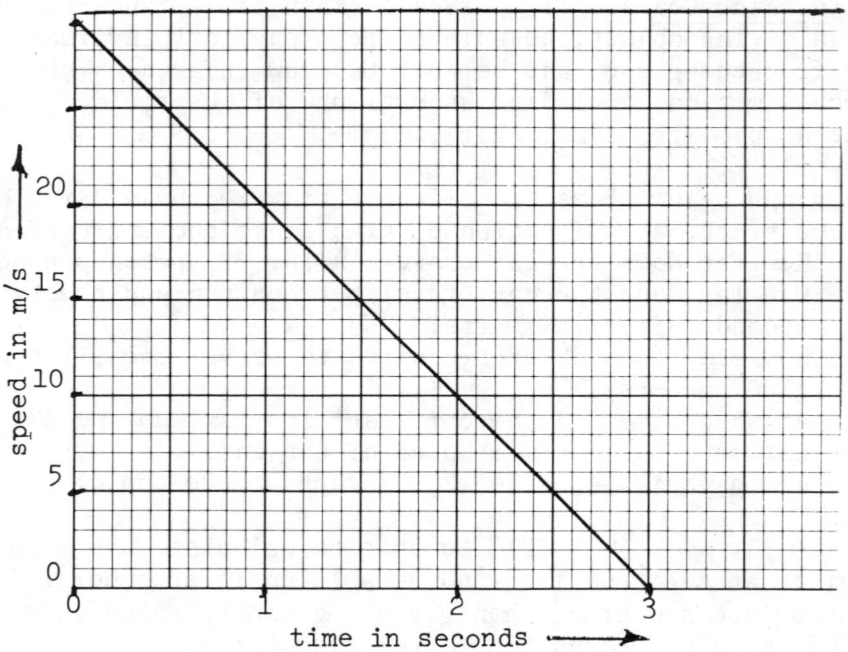

Let us review briefly the steps taken thus far in our efforts to describe motion. We have identified the variables needed to describe motion -- time, displacement, velocity, and acceleration -- and have given precise definitions of the last three of these. Experiments may be done to measure values of an object's position at many values of time, expressed as a table of numerical data.

Reliable data on position as a function of time are basic to any quantitative study of motion, and they do constitute a description of the motion. However, a description in this form is not convenient for visualizing the motion or for answering questions about velocity and acceleration. For these purposes, it is advantageous to describe the motion using graphs. In this Section and the next we discuss the use of graphs to describe motion, for cases when the motion is in a straight line. Section 41.6 presents yet another approach to describing motion, the use of equations.

Table 41-1

Numerical data on the position of a ball moving in the $+x$ direction, as a function of time.

Time, t seconds	Position, x meters
0	0
0.2	0.01
0.4	0.05
0.6	0.11
0.9	0.24
1.2	0.43
1.4	0.59
1.6	0.77
1.9	1.08
2.1	1.32

As a concrete example, Table 41-1 gives some data on the position of a ball moving in a straight line, at various values of time. From these data, one could develop information on the velocity and acceleration at selected points through a series of calculations, but a great deal more can be inferred about this motion by drawing graphs.

To begin, we plot x versus t for the moving ball (Figure 41-2). The graph shows that the position increases with time in a regular pattern. The rate of change of position, or speed of the ball, also increases with time. From the graph, how can we find the velocity at a chosen value of time?

Figure 41-2. A graph of the data in Table 41-1, showing the position of the ball as a function of time. The displacement and the velocity of the ball both increase with time.

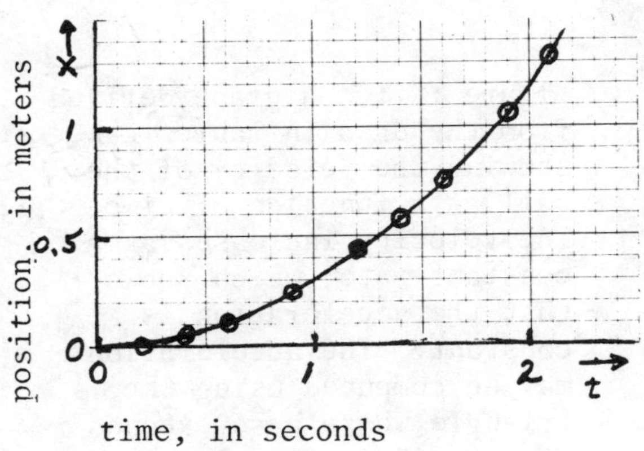

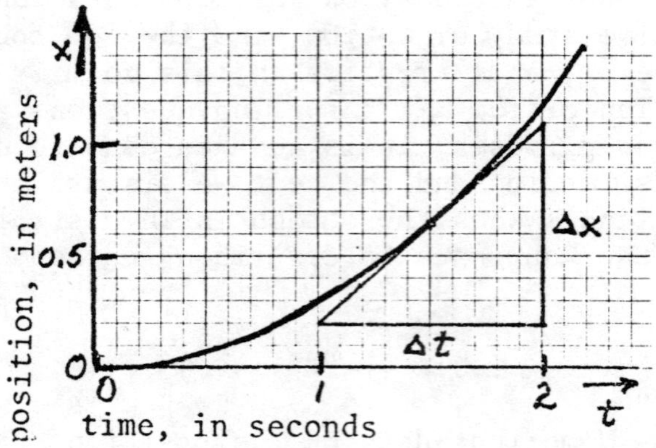

Figure 41-3. The graph of Figure 41-2, for the motion of a ball. To find the slope at t = 1.5 seconds, a tangent line is drawn and a triangle is constructed having sides at Δx and Δt.

The velocity at any point (or any value of time) is the value of $d\vec{x}/dt$ for a very small interval which includes the point. The quantity dx/dt is called the *slope* of the graph. As shown in Figure 41-3, to find the slope at t = 1.5 seconds, a straight line may be drawn tangent to the curve at the point where t = 1.5 seconds. A triangle is constructed on the tangent line such that the sides of the triangle define an interval which includes the point at t = 1.5 seconds. Comparing the sides of this triangle with the scales of the graph, we see that Δx = 1.1 - 0.2 meters = 0.9 meters, and Δt = 2.0 - 1.0 seconds = 1.0 second. The speed at our chosen point is then computed to be

$$\frac{\Delta x}{\Delta t} = \frac{0.9 \text{ meter}}{1.0 \text{ second}} = 0.9 \text{ meter/second},$$

and the velocity at this point is 0.9 m/s in the $+x$ direction.

From Figure 41-2 or Figure 41-3, we can infer that because the slope increases steadily with time, the velocity also increases steadily beginning from a value of zero at t = 0. By a similar procedure, the velocity of the ball can also be computed at any other value of time. If we make these calculations and then plot the results, we obtain the graph shown in Figure 41-4.

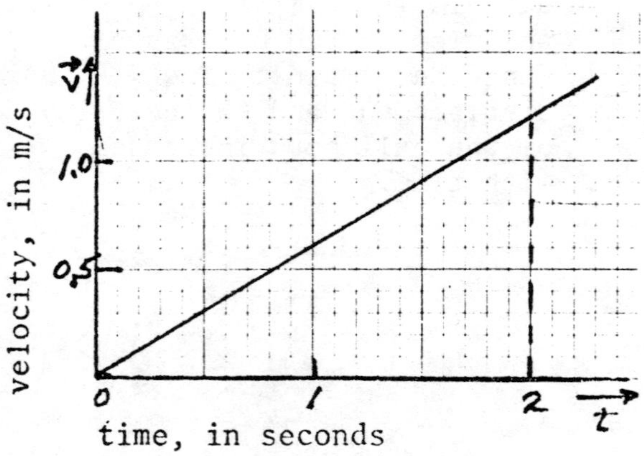

Figure 41-4. A graph derived from the data in Table 41-1, showing the velocity of the ball as a function of time. The velocity increases at a constant rate, which shows that the acceleration is constant. The acceleration may be computed using the triangle whose base extends from t = 0 to t = 2 seconds.

The time-rate of change of *instantaneous* velocity is the acceleration. This rate of change, $d\vec{v}/dt$, is just the slope of a graph of instantaneous velocity vs. time. For the ball whose motion we are discussing, the slope of the $\vec{v}$-vs.- t graph in Figure 41-4 is constant; that is, the velocity increases linearly with time. From this we conclude that for this ball the acceleration is constant. The accleration is computed from the slope, using the triangle indicated in Figure 41-4.

$$\vec{a} = \frac{\Delta\vec{v}}{\Delta t} = \frac{1.2 \text{ meter/second}}{2.0 \text{ second}} \text{ in the } +x \text{ direction}$$

$$= 0.6 \text{ meter/second}^2 \text{ in the } +x \text{ direction.}$$

In this special case the acceleration has a constant value of 0.6 m/s^2. In general however, we find that the acceleration of a moving object also varies with time.

Given a graph of position vs. time, we can use the slope of this graph to obtain data for a graph of velocity vs. time, and then use the slope of the second graph to obtain data for a graph of acceleration vs. time. In other words, we can use graphs to extend our knowledge of motion from position to velocity to acceleration, in that order. In the next section, we consider how to use graphs to go in the reverse order, from acceleration to velocity to position or displacement.

Exercises

E 41.4-1 This is a graph of speed vs. time for a car which stops suddenly. The direction of the car's forward motion is positive. What are the magnitude and direction of the car's acceleration at t = 1.0 second?

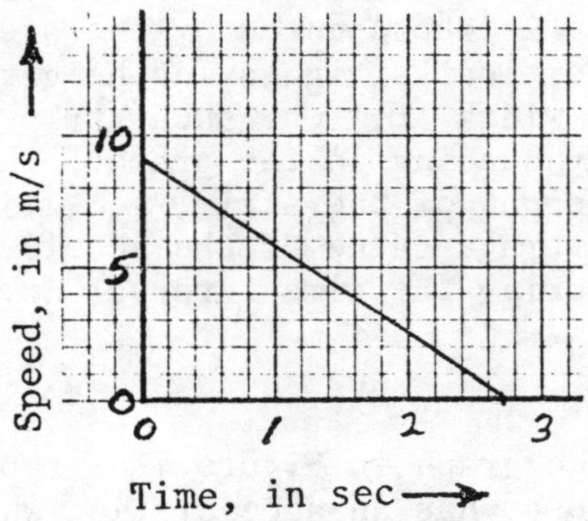

E 41.4-2 (a) Draw a graph of v vs. t for a body whose velocity is +20 m/s at t = 0, decreasing at the constant rate of 4 m/s^2.

(b) Draw the corresponding curve for a vs. t.

E 41.4-3 (a) Draw a qualitative graph of displacement vs. time for a body swinging back and forth like a pendulum. The displacement is zero at the midpoint of the motion.

(b) Draw the corresponding graph of velocity vs. time.

(c) Draw the corresponding graph of acceleration vs. time.

Check your understanding of the material above by testing yourself on the minimum learning goals at the beginning of this section.

41.5 Using Graphs to Find Velocity and Position

Minimum Learning Goal. Given a graph of velocity versus time, be able to estimate the displacement of the moving object during a given interval of time, and be able to infer other information about the object's motion. For example:

1 Refer to the graph in Example 1 of Minimum Learning Goal (2) in Section 41.4.
(a) How high has the ball risen at t = 2.0 seconds?
(b) What was the maximum height reached by the ball?
(Answer, page 41-38, Line 9)

How can we use graphs to extend our description of motion, from acceleration to velocity to position or displacement? We begin by recalling the definition of average acceleration, $\vec{a}_{av} = \Delta\vec{v}/\Delta t$. Multiplying both sides by Δt leads to

$$\Delta\vec{v} = \vec{a}_{av}\ \Delta t.$$

The increment in velocity Δv during a short time interval Δt can be found for any particular motion with the use of a graph of a vs. t for this motion.

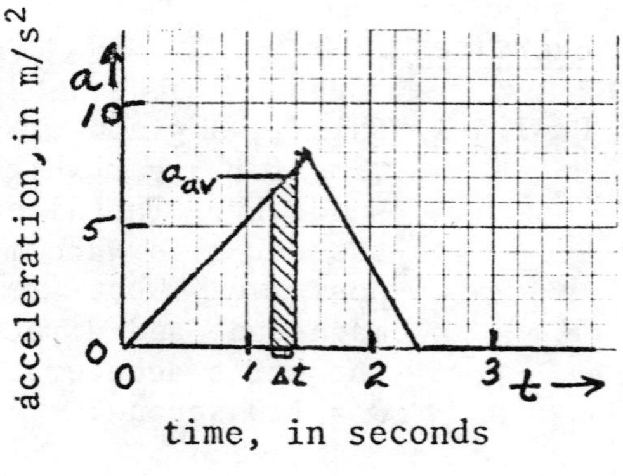

Figure 41-5. On this graph of a vs. t, the product $a_{av}\ \Delta t$ for a time interval Δt is represented by the area of a vertical strip having height a_{av} and a width Δt. One such strip is shown by shading on the graph. The average height of the strip, a_{av}, is the average value of the acceleration during the time interval Δt.

The graph in Figure 41-5 represents the motion of a car which starts from rest, has an acceleration which increases steadily to a peak value at t = 1.5 seconds, and then decreases steadly to zero at t = 2.4 seconds and remains zero thereafter. By t = 2.4 seconds, the driver has accelerated his car to the speed desired and he then continues at this steady speed.

To find the change in velocity during the time interval from t = 1.2 to t = 1.4 seconds, this interval Δt is marked on the time axis, and at the beginning and end of the interval vertical lines are drawn upward to the curve. These lines form the sides of a vertical strip, indicated by shading in Figure 41-5, which has a base Δt and an *average* height a_{av}. The "area" of this strip is the area of a rectangle having base Δt and height a_{av}, given by the formula $a_{av}\ \Delta t$. During Δt, the velocity increases by $\Delta\vec{v} = \vec{a}_{av}\ \Delta t$, that is, by the amount represented by the area of this strip. The base of the strip has a unit of seconds, and the height of the strip has units of meter/second^2, giving a product of seconds × meter/second^2 = meter/second, the correct unit for velocity. During Δt the average acceleration is positive, so the area of the strip is positive and the increment in velocity $\Delta\vec{v}$ is positive during this interval. In a case where the average acceleration during the interval is negative, the area of the strip is regarded as negative and the velocity decreases during that interval.

In this example, the area of the strip gives the increase in velocity during the interval from $t = 1.2$ to $t = 1.4$ seconds. To find the change in velocity over a longer time interval, we can repeat this procedure using successive strips which together fill up the area under the curve during the entire interval under study, and then add the areas of all these strips. In this way, we find that

★ The total area under a curve of acceleration vs. time, from some initial time t_1 to some later time t_2, equals the total change in velocity between t_1 and t_2.

If we know the velocity $\vec{v}_1$ at t_1, then at t_2 the velocity will be

$$\vec{v}_2 = \vec{v}_1 + \Sigma\,(\vec{a}_{av}\,\Delta t) = \vec{v}_1 + (\text{area under curve from } t_1 \text{ to } t_2).$$

Notice that this procedure only enables us to find the *change in velocity* during some time interval. To calculate the velocity $\vec{v}_2$ at t_2, it is necessary to know the initial velocity $\vec{v}_1$ at t_1.

In the example of Figure 41-5, let us assume that the velocity is zero at $t = 0$. By adding the areas of successive strips as described above, we can compute the velocity at any time. The result of such calculations is shown in Figure 41-6, a graph of velocity vs. time which describes the motion whose acceleration is shown in Figure 41-5. Beginning at zero, the velocity increases to 9.6 m/s in the positive direction, then remains constant after $t = 2.4$ seconds.

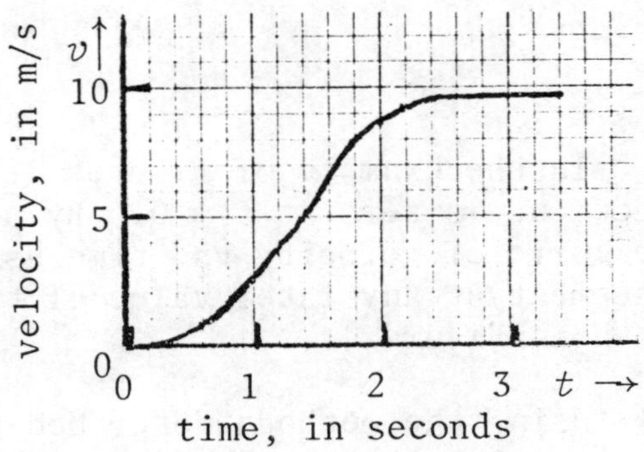

Figure 41-6. A graph of velocity vs. time for the motion described in Figure 41-5 using a graph of acceleration vs. time. The velocity at any time t is found by adding the area under the graph of a vs. t, from $t = 0$ to the desired value of t.

Just as we have used information in a graph of acceleration vs. time to obtain the corresponding graph of velocity vs. time, we can use information in a graph of velocity vs. time to obtain a graph of position vs. time. The procedure is to divide the area under the curve of velocity vs. time into narrow strips having a base Δt and an average height v_{av}, then add the areas of successive strips to obtain the total area between two chosen values of time. For any one of these strips, the increment in distance traveled during Δt is $\Delta s = v_{av}\,\Delta t$. The base of the strip has a unit of seconds, and the height of the strip has units of meter/second, giving a product of second × meters/second = meters, the correct unit for distance or displacement. For an interval where the average velocity is positive, the area of the strip is positive and the increment in distance Δs is positive. For an interval where the average velocity is negative, the area of the strip is negative and the displacement decreases during that interval.

★ The total area under a curve of velocity vs. time, from some initial time t_1 to some later time t_2, equals the total change in displacement between t_1 and t_2.

If we know the displacement $\vec{s}_1$ at t_1, then at t_2 the displacement is

$$\vec{s}_2 = \vec{s}_1 + \Sigma(\vec{v}_{av}\,\Delta t) = \vec{s}_1 + (\text{area under curve from } t_1 \text{ to } t_2\,).$$

Notice that this procedure only enables us to find the *change in displacement* during some interval of time. To calculate the displacement $\vec{s}_2$ at t_2, it is necessary to know the initial displacement $\vec{s}_1$ at t_1.

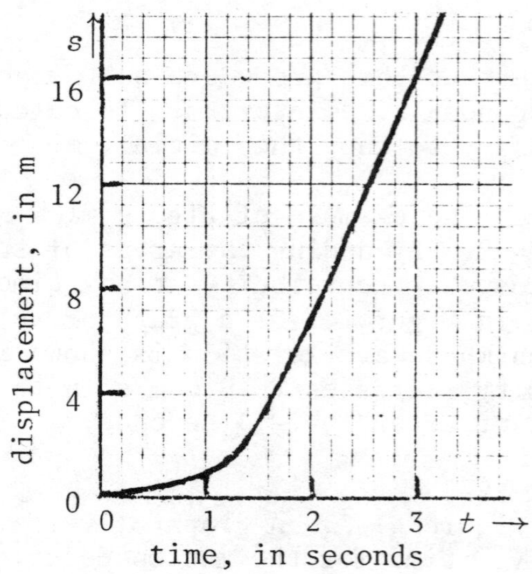

Figure 41-7. A graph of displacement vs. time for the motion described in Figures 41-5 and 41-6. The displacement at any time t is found by adding the area under the curve of v vs. t, from $t = 0$ to the desired value of t.

In the example of Figures 41-5 and 41-6, let us assume that the displacement is zero at $t = 0$. By adding the areas of successive strips under the curve of velocity vs. time as described above, we can compute the displacement at any later value of time. The result of this calculation is shown in Figure 41-7.

Using the methods described above, one can begin with a graph of acceleration vs. time, use the areas of strips under the curve on this graph to obtain a second graph of velocity vs. time, and then use the areas of strips under the second curve to obtain a graph of displacement vs. time.

In Figure 41-8 on Page 41-25 are four sets of graphs, each set describing a different example of motion. For any one set of three graphs, if the initial values of position and velocity are known, one can begin with any one of the three graphs and calculate the other two using the methods described in this section. In Examples 1, 2, and 4 of Figure 41-8, the

Figure 41-8. EXAMPLES OF GRAPHS WHICH DESCRIBE MOTION

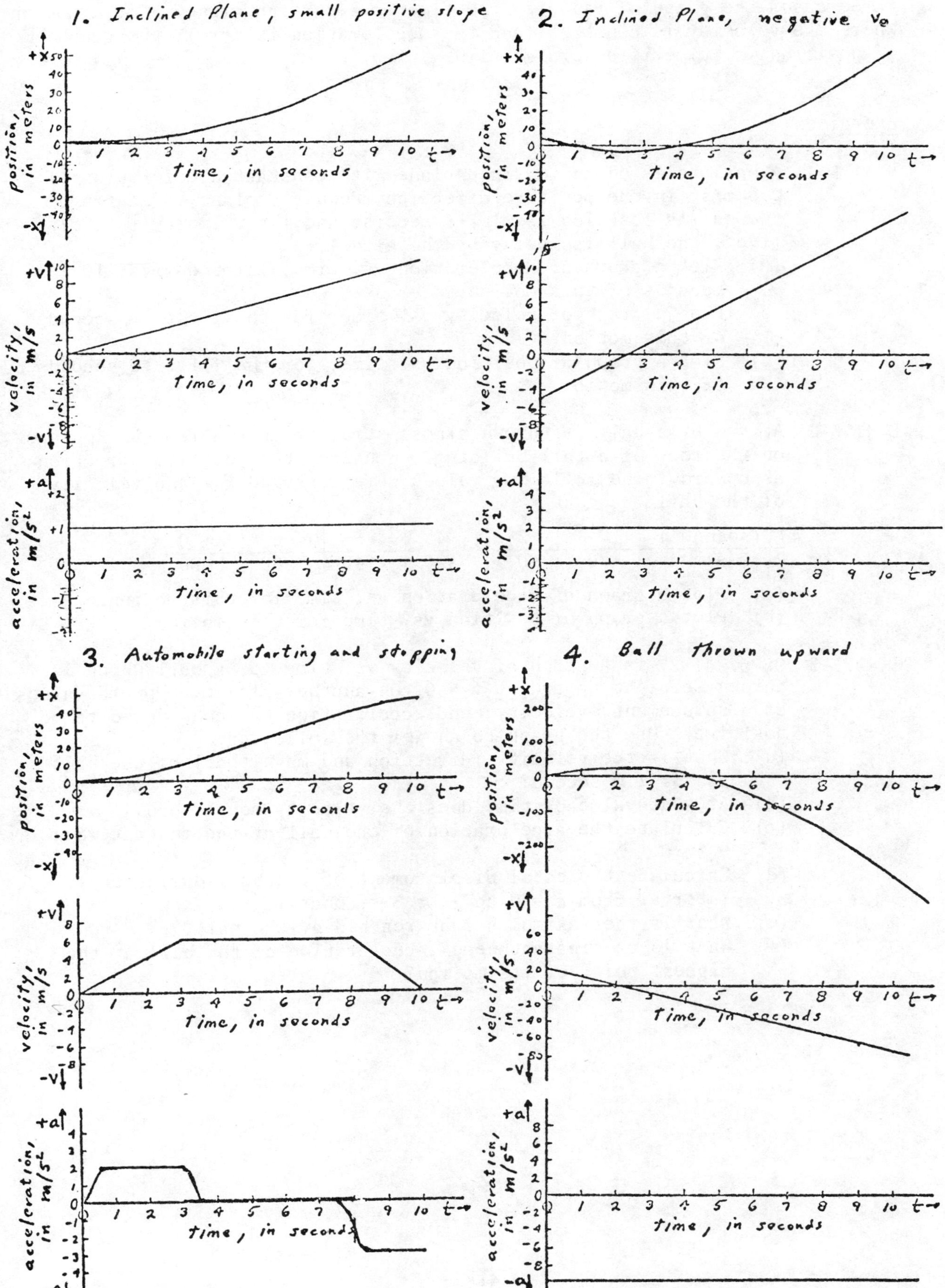

acceleration is a constant throughout the motion and correspondingly in these three cases the graph of velocity vs. time is a straight line. For a motion where the velocity is constant (and the acceleration is zero), the graph of position vs. time is a straight line.

Exercises

E 41.5-1 A ball rolls on an inclined plane with constant acceleration of 2.0 m/s^2 in the positive direction (down the plane). At $t = 0$ the ball's position is $x = +4$ meters, and its velocity is -5m/s (i.e., the ball is moving up the plane).

(a) Plot a graph of acceleration vs. time, for the first 10 seconds of this motion.

(b) Plot a graph of velocity vs. time, for the first 10 seconds of this motion.

(c) Plot a graph of position vs. time, for the first 10 seconds of this motion.

E 41.5-2 At $t = 0$, a ball is thrown almost straight upward from the parapet on the roof of a tall building. Consider the positive direction as upward. The following values were measured for the velocity of the ball:

t (seconds)	0	2	4	6	8	10
v (m/s)	+20	0	-20	-40	-60	-80

(a) Plot a graph of acceleration vs. time for this motion.

(b) Plot a graph of position vs. time for this motion.

E 41.5-3 On p. 41-27 is a graph of velocity vs. time for a ball which is thrown straight upward at $t = 0$, on another planet. Upward values of displacement, velocity, and acceleration are considered to be positive. Use the graph to answer the following.

(a) What is the velocity (direction and magnitude) of the ball at $t = 1.0$ second?

(b) At what value of time does the velocity equal zero?

(c) Calculate the acceleration of the ball at the instant when $v = 0$.

(d) Calculate the total displacement of the ball during the interval from $t = 0$ to $t = 3$ seconds.

(e) What is the maximum height reached by the ball?

(f) What is the instantaneous acceleration of the ball at the highest point of its motion?

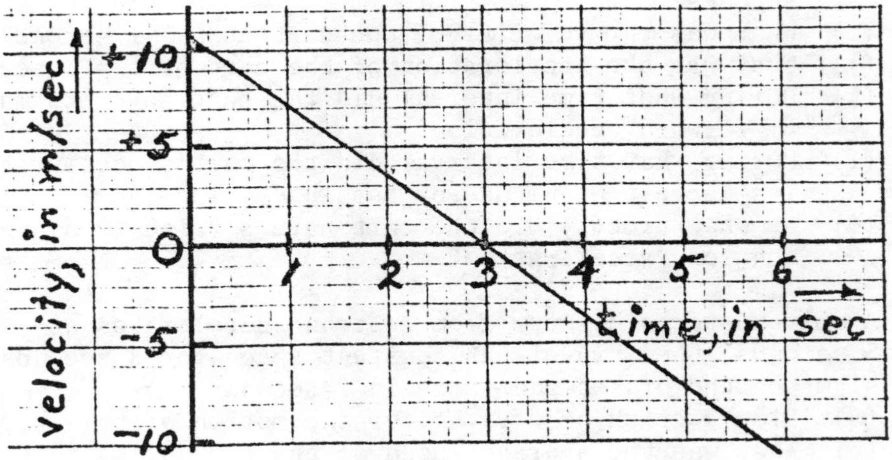

E 41.5-4 Shown here is a graph of velocity vs. time for a moving cart. Positive or forward values of velocity are plotted upward, and negative or backward values are plotted downward. The motion begins at t = 0 seconds. Answer the following questions about this motion, using information obtained from the graph.

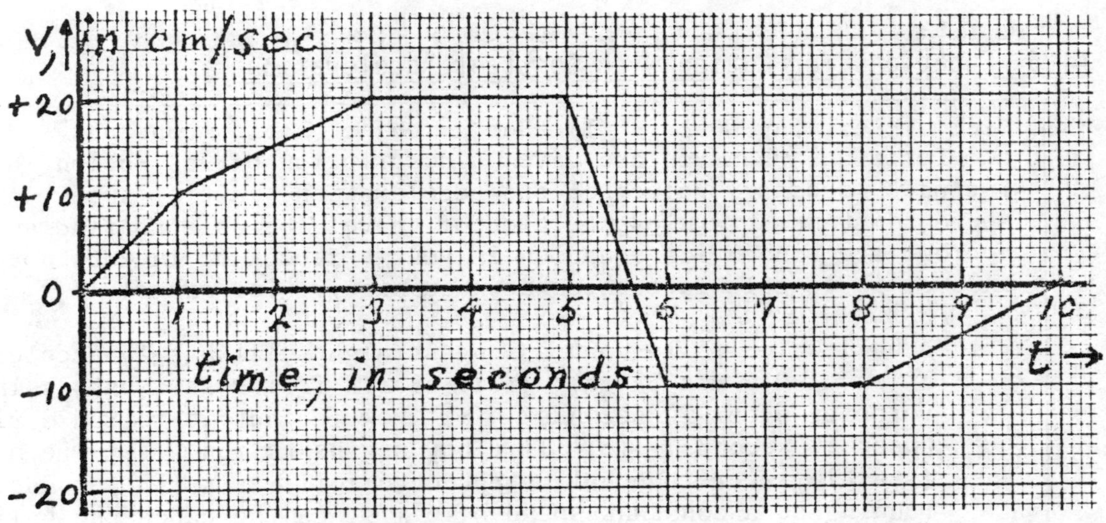

(a) During what period of time was the cart moving in the forward direction?

(b) When did the cart reach its maximum displacement from its starting point?

(c) What was the maximum displacement of the cart from its starting point?

(d) What was the displacement of the cart at t = 10 seconds?

(e) At what time did the total distance traveled by the cart reach its maximum value?

(f) What was the total distance traveled by the cart during the first ten seconds?

(g) What was the average speed of the cart during these ten seconds? (continued on next page)

(h) What was the average velocity of the cart during these ten seconds?
(i) What was the velocity of the cart at t = 1 second?
(j) What was the acceleration of the cart at t = 2 seconds?
(k) During what time interval did the cart have its maximum value of forward acceleration?
(l) During what time interval did the cart's acceleration have its largest magnitude or absolute value?
(m) At what time(s) did the cart have a velocity of zero, but a non-zero acceleration?

E 41.5-5 A driver starts his car from rest and accelerates at 3 m/s^2 for 5 seconds, then travels at constant speed for 8 seconds, then slams on the brakes and stops in 2 seconds.
(a) Plot a graph of v vs. t for the motion of this car.
(b) What was the average speed of the car during this motion?
(c) How far did the car travel?

Check your understanding of the material above by testing yourself on the minimum learning goal at the beginning of this section.

41.6 Equations for Motion with Constant Acceleration

Minimum Learning Goals. (1) For an object having constant acceleration, when you are given a description of its motion which includes necessary values of the acceleration, initial speed, final speed, distance traveled, or time of travel, be able to find an appropriate kinematics equation in your textbook or in your notes, and use it to calculate an unknown quantity. For example:

1 A ball which rolls down a long straight hill has an acceleration of 0.6 m/s^2. (a) If it starts from rest, what is its velocity after 10 seconds? (b) How far has it rolled during these 10 seconds? (c) If a boy begins to run after the ball (starting where the ball started to roll) at t = 5 seconds, running at a constant speed of 6 m/s, can he catch the ball by t = 10 seconds? (Answer, page 41-38, Line 10)

(2) Given a description of the motion of an object moving with the acceleration due to gravity near the earth's surface, including necessary kinematics quantities, be able to find an appropriate equation and use it to calculate an unknown quantity. For example:

1 A little Leaguer throws a baseball straight up, with an initial speed of 20 m/s. (a) How high will the ball rise? (b) How long after he throws the ball will it return to the ground? (Answer, page 41-38, Line 4)

Direct measurements on a moving object can yield a table of numerical data giving the position at many values of time. A graph, however, is more useful than a table of data for describing the motion, because the graph allows us to visualize the motion and to calculate the velocity and the acceleration as a function of time. Now we consider still another method of describing motion, namely, the use of equations.

Equations have the advantage of enabling us to calculate selected quantities of the motion rather quickly, when we have values for enough other variables of the motion. We shall find that the use of equations offers a powerful method for solving many kinds of problems related to motion, and for predicting motions under certain conditions. It must be stressed, however, that understanding the essential physics of motion does not center on the manipulation of equations. The great bulk of this chapter on describing motion has been devoted to presenting the *concepts* associated with motion, with relatively little emphasis on the use of equations or formulas to calculate the variables of motion.

In this section, equations are developed for the special case of *motion in a straight line with constant acceleration*. This is sometimes referred to as *uniformly accelerated motion.* For motion along a straight line from left to right, the displacement increases as the object moves toward the right from its point of beginning. Because all of the quantities of motion (displacement, velocity, and acceleration) are directed along this line, it is not necessary to write these quantities using an arrow ($\vec{}$) to denote them as vector quantities. Rather, we will write the symbols s, v, and a without the vector symbols, but we must be careful to use the correct signs (+ or -) for every quantity we work with. All displacements, velocities, and accelerations directed toward the right are considered *positive*, while those directed toward the left are considered *negative*.

Our equations are simplified, without neglecting any essential elements of physics, if we choose the time as zero at the beginning of any motion we want to study. At $t = 0$, then, the displacement $s = 0$ and the initial velocity is designated by the symbol v_0. At any later value of time t, the displacement is designated by s and the velocity by v. We now use some algebra to obtain a set of equations which can be used to describe any motion having constant acceleration in a straight line.

When the acceleration is constant, the average acceleration during an interval of motion is equal to the instantaneous acceleration at any point. The definition for average acceleration may then be written

$$a_{av} = a = \frac{\Delta v}{\Delta t} = \frac{v - v_0}{t - t_0} .$$

Because $t_0 = 0$, this equation can be multiplied by t and then rearranged in the form

★ $$\boxed{v = v_0 + at.} \qquad \text{Eq. (41-1)}$$

Equation (41-1) can be used to find the value of v at any time t, provided the values of v_0 and a are known. The definition for average velocity may be rewritten, using the notation adopted in this section, as

$$v_{av} = \frac{\Delta s}{\Delta t} = \frac{s - s_0}{t - t_0} = \frac{s}{t}.$$

Multiplying by t to clear of fractions, we obtain

★ $$s = v_{av}\, t. \qquad \text{Eq. (41-2)}$$

From the ordinary meaning of an average quantity, we have

$$v_{av} = \frac{v + v_0}{2}. \qquad \text{Eq. (41-3)}$$

Substituting Eq. (41-1) into Eq. (41-3), we obtain

$$v_{av} = \frac{v_0 + at + v_0}{2} = v_0 + \frac{1}{2}at. \qquad \text{Eq. (41-4)}$$

Substituting Eq. (41-4) into Eq. (41-2) gives

$$s = \left(v_0 + \frac{1}{2}at\right)t, \text{ or } \quad ★\ s = v_0 t + \frac{1}{2}at^2. \qquad \text{Eq. (41-5)}$$

Solving Eq. (41-1) for t, we obtain

$$t = (v - v_0)/a.$$

Substituting this into the first form of Eq. (41-5), we have

$$s = \left[v_0 + \frac{1}{2}a\left(\frac{v - v_0}{a}\right)\right]\left(\frac{v - v_0}{a}\right) = \left(v_0 + \frac{v - v_0}{2}\right)\left(\frac{v - v_0}{a}\right)$$

$$= \left(\frac{v_0}{2} + \frac{v}{2}\right)\left(\frac{v - v_0}{a}\right) = \frac{1}{2a}(v + v_0)(v - v_0) = \frac{1}{2a}(v^2 - v_0^2).$$

Finally, multiplying both sides by $2a$ leads to

$$2as = v^2 - v_0^2, \text{ or } \quad ★\ v^2 = v_0^2 + 2as. \qquad \text{Eq. (41-6)}$$

The derivations above have led us to three important equations, Eqs. (41-1), (41-5), and (41-6), which connect the five variables t, s, v, v_0, and a. In addition to these three, Eqs. (41-2) and (41-3) should be regarded as important aids in solving problems and other calculations on uniformly accelerated motion.

Comparing Eqs. (41-1) and (41-5) with the graphs in Section 41.4, it is seen that each equation corresponds to one of the graphs. The graph of x vs. t in Figure 41-2 is represented by Eq. (41-5), with $v_0 = 0$ and using s in place of x for position. The graph of v vs. t in Figure 41-4 is represented by Eq. (41-1); the curve is a straight line because a is constant, and it begins at the origin because $v_0 = 0$. This correlation of graphs and equations is a result of our choosing a motion having a constant acceleration (Table 41-1) to introduce Section 41.4.

Worked Example 41-4

A car having an initial speed of 31 m/s (70 mi/hr) brakes to a stop in a distance of 65 m (215 feet).

(a) Find the acceleration of this car, assuming it is constant.
(b) Find the time required for the car to come to a stop.

Solution

This solution follows the Hints for Solving Physics Exercises given in Section 5, page 4. This systematic procedure is recommended for solving exercises and problems in physics, especially for students who experience some difficulty or uncertainty in knowing how to go at the solution of these exercises.

We begin by drawing a diagram of the situation described in the Example.

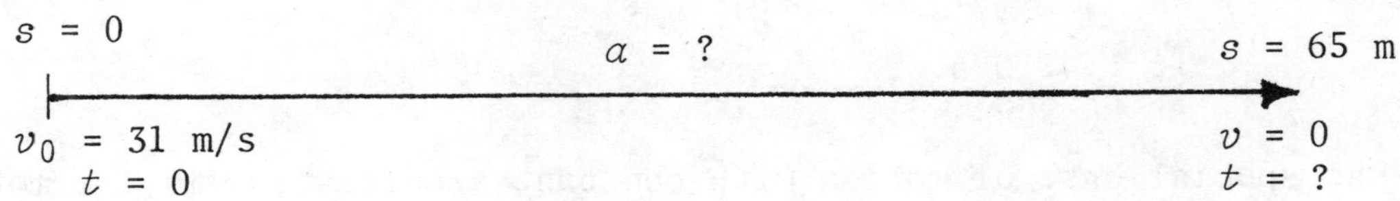

For our purposes in solving this exercise, the motion begins at displacement $s = 0$ when the brakes are applied at time $t = 0$, with the car moving at its initial speed of $v_0 = 31$ m/s. The motion ends when the speed has decreased to zero, after a displacement $s = 65$ meters. Of the five variables of the motion (t, s, v, v_0 and a), we are given values for three. The "unknowns" are the acceleration a, and the total time t required for this braking motion. The equations available to us for solving exercises in uniformly accelerated motion are Eqs. (41-1), (41-2), (41-3), (41-5), and (41-6).

(a) To find the acceleration, we need an equation which includes the variable a but not t, because we do not yet know the time required for the motion. The only equation which satisfies this requirement is Eq. (41-6), so we must use it if we want to solve for a before solving for t. Before substituting numbers into Eq. (41-6), it is desirable to rearrange it into the form $a =$ (something). Dividing both sides of Eq. (41-6) by $2s$ leads to

$$a = \frac{v^2 - v_0^2}{2s} .$$

Substituting numerical values listed on the diagram:

$$a = \frac{0 - (31 \text{ m/s})^2}{2 \times 65 \text{ m}} = \frac{-(31)^2 \text{ (m/s)}^2}{130 \text{ m}}$$

$$= \frac{-961 \text{ m}^2/\text{s}^2}{130 \text{ m}} = -7.39 \text{ m/s}^2 .$$

When working numerical exercises in physics, it is highly desirable to keep track of the physical units at every step; if the units for your answer do not correspond to the correct units for the quantity being solved for, it is quite likely that there is a mistake in the arithmetic as well as in the units.

Our result for the acceleration is negative, which means that the car was slowing down during this motion; that result agrees with the statement of the problem, and indicates (but does not prove) that our solution is correct.

(b) To find the time t, we can use any equation which applies to uniformly accelerated motion and which contains the variable t. Choosing Eq. (41-1) and solving for t leads to

$$t = \frac{v - v_0}{a}$$

$$= \frac{0 - 31 \text{ m/s}}{-7.39 \text{ m/s}} = 4.19 \frac{\frac{\text{m}}{\text{s}}}{\frac{\text{m}}{\text{s}^2}} \times \left(\frac{\text{s}^2}{\text{s}^2}\right) = 4.19 \frac{\text{m} \times \text{s}}{\text{m}}$$

$$t = 4.19 \text{ s.}$$

One special case of motion with constant acceleration is the motion of a freely falling object. By "free fall," we refer to the motion of an object acted on only by its weight. It is found by experimemt that an object in free fall near the earth's surface has a downward acceleration of 9.8 m/s^2. This quantity, sometimes called the *acceleration of gravity,* denoted by the special symbol g.

$$g = 9.8 \text{ m/s}^2.$$

Because the displacement of a falling (or rising) object is often vertical rather than horizontal, the symbol h (suggested by the word *h*eight) is used for this displacement. Using g and h, the three key equations for motion in free fall may be written:

$$h = v_0 t - \frac{1}{2} gt^2 \qquad \text{Eq. (41-7)}$$

$$v = v_0 - gt \qquad \text{Eq. (41-8)}$$

$$v = v_0^2 - 2gh \qquad \text{Eq. (41-9)}$$

In Eqs. (41-7), (41-8), and (41-9), vector quantities directed upward are regarded as positive, while those directed downward are negative. An upward initial velocity would be expressed as a positive value of v_0. The acceleration g is itself a positive number, but because it is downward the terms in these equations which contain g are written with a negative sign.

Questions and Exercises

Q 41.6-1 Rearrange these four methods of describing motion into a logical sequential order: *equation, graph, proportionality, table of numerical data.*

Q 41.6-2 When you slam on the brakes of your car and come to a quick stop (without skidding), does your car have constant acceleration? How could you find out?

Q 41.6-3 Do all objects falling near the earth's surface have constant acceleration? If not, name some which do not.

Q 41.6-4 Which variable of motion changes abruptly when you feel a jerk?

E 41.6-1 A motorcycle starts from rest and reaches a speed of 25 m/s (about 56 mi/hr) in 20 sec.

(a) What was its acceleration (assumed constant) in m/s^2?

(b) How far did the motorcycle travel before reaching the speed of 25 m/s?

E 41.6-2 In some states, the traffic laws require that an automobile must be able to stop within 30 feet (9.2 m) from an initial speed of 20 miles/hour (8.9 m/s). Find the deceleration of a car which can just stop within the required distance.

E 41.6-3 A bicyclist who likes speed found a steep hill 300 m long. Starting at the top, he pedaled all the way down the hill and reached a speed of 20 m/s at the bottom.

(a) What was his average acceleration during his descent?

(b) Assuming constant acceleration, how long did it take for him to go down the hill?

(c) Estimate his speed halfway down the hill.

E 41.6-4 An automobile enters an intersection while moving at 80 ft/sec, when the light changes to amber. The driver hits his brake pedal (which can only stop the car, not give it a speed in the opposite direction) and accelerates at -100 ft/sec^2 uniformly as he crosses the intersection. If the intersection is 19.5 ft. wide, how long is he in it?

E 41.6-5 A workman on the deck of the Mackinac Bridge dropped his wrench. 3.5 s after the wrench slipped out of his hand he saw it splash into the water.

(a) How high above the water surface was he?

(b) With what speed did the wrench hit the water?

E 41.6-6 A ball is thrown upward from the edge of a tall building and released with a speed of 19.6 m/sec. When the ball comes down, it misses the building and falls to the street far below. At the end of 4 seconds, how high is it above (or below) the point where it was released?

E 41.6-7 Consider a stone thrown upward with an initial velocity of 14.7 m/s. Its acceleration is constant, equal to 9.8 m/s^2 downward. It moves for 3 seconds.

(a) What is the total displacement?

(b) What is the average velocity?

(c) What is the final velocity?

(d) What is half the sum of initial and final velocities?

(e) Are the answers to (b) and (d) the same?

E 41.6-8 Ken stands on the roof with a water-filled balloon, which he plans to drop on his budy Jim (2 m tall) as Jim reaches the point X on the ground 17 m directly below Ken. Jim is walking straight toward X at a steady speed of 1.5 m/s. How far from X should Jim be when Ken drops the balloon, if Ken is to hit Jim on the head?

Check your understanding of the material above by testing yourself on the minimum learning goals at the beginning of this section.

41.7 Using Calculus to Describe Motion

It is convenient at this point to change our notation slightly and use x (instead of s) as the symbol for the position of an object in one-dimensional motion. (This change prepares the way for using x, y, and z to describe the position of an object in three-dimensional motion, in Chapter 43.) Then the definitions for instantaneous velocity and acceleration become

★ $v = dx/dt$ and $a = dv/dt$.

We write the symbols x, v, and a without the vector symbol ($\vec{\ }$), so we must take care to use the correct sign (+ or -) for every quantity we work with. All displacements, velocities and accelerations having the $+x$ direction are considered positive, while those having the $-x$ direction are negative.

In Section 41.4 we showed that dx/dt is the slope of a graph of x vs. t, and dv/dt is the slope of a graph of v vs. t. Beginning with data for x at several values of t, one can plot x vs. t and compute the slope of the curve to find v at several values of t, then plot v vs. t and compute the slope to find a at several values of t.

In the language of calculus, dx/dt is called the *derivative of x with respect to t*, and dv/dt is called the *derivative of v with respect to t*. When we have an equation for x as a function of time, $x(t)$, we can use the method of calculus called differentiation to obtain these derivatives and thus obtain equations for $v(t)$ and $a(t)$. The procedures for differentiating elementary functions are straightforward; they are described in Appendix A.2. For the analysis of motion, the methods of calculus are more powerful than those used earlier in this chapter: using them, it is not necessary to plot graphs or make numerical calculations of their slopes at several points. Differentiation can be used to find the velocity $v(t)$ and acceleration $a(t)$ for any functional form of $x(t)$.

In Section 41.5 we showed that the area under a curve of acceleration vs. time, between t_1 and t_2, is equal to the change in velocity during the time interval from t_1 to t_2. The change in position during the interval from t_1 to t_2 is equal to the net area under the curve of velocity vs. time, between t_1 and t_2. Beginning with a graph of a vs. t, one can compute areas to find v at several values of t, then plot v vs. t and compute more areas to find x at several values of t.

In the language of calculus, the area under a curve of a vs. t from t_1 to t_2 is represented by the *definite integral* of adt from t_1 to t_2. This is written

★ $$v_2 - v_1 = \int_{t_1}^{t_2} a\, dt,$$

where the integral sign $\int$ has a meaning similar to Σ, indicating a summation. The area under a curve of v vs. t from t_1 to t_2 is represented by the definite integral of vdt from t_1 to t_2. This is written

★ $$x_2 - x_1 = \int_{t_1}^{t_2} v\, dt.$$

When we have an equation for a as a function of time, $a(t)$, we can use the method of calculus called integration to obtain these integrals and thus obtain equations for $v(t)$ and $x(t)$. The procedures for integrating simple functions are described in Appendix A.4. For the analysis of motion, the method of integration is more powerful than those used earlier in this chapter. Using integration, it is not necessary to plot graphs or make numerical calculations of areas for several intervals. Integration can be used to find velocity $v(t)$ and position $x(t)$ for any functional form of $a(t)$.

<u>Worked Example 41-5</u>

A ball rolls downward along an inclined plane with a constant acceleration of 0.80 m/s^2. At $t = 0$, its velocity is zero and its position is $x = 0.4$ m.

(a) Use calculus to obtain an equation for $v(t)$.

(b) Use calculus to obtain an equation for $x(t)$.

<u>Solution</u>

In this example, we are given $a(t)$ and asked to find $v(t)$ and $x(t)$. Using calculus, this requires evaluation of the integrals in

$$v_2 = v_1 + \int_{t_1}^{t_2} adt \quad \text{and} \quad x_2 = x_1 + \int_{t_1}^{t_2} vdt.$$

(a) The initial time $t_1 = 0$ and the corresponding velocity $v_1 = 0$. The "final" time t_2 may be any value of t, so the final velocity v_2 is the corresponding value of $v(t)$. Because a is constant, the integral of adt is evaluated using rule 1 from Appendix A.4. We obtain

$$v(t) = 0 + \int_0^t a dt = a \int_0^t dt = a \Big[t \Big]_0^t = at$$

$$= (0.80 \text{ m/s}^2)t \ .$$

For any value of t in seconds, this equation gives v in m/s.

(b) The initial position x_1 is given as 0.4 m, at the initial time $t_1 = 0$. As in part (a), the final time t_2 may be any value of t, so the final position x_2 is the corresponding value of $x(t)$. $v(t)$ is a power of t, so the integral of vdt is evaluated using rule 2 from Appendix A.4. We obtain

$$x(t) = 0.4 \text{ m} + \int_0^t vdt = 0.4 \text{ m} + (0.80 \text{ m/s}^2) \int_0^t tdt$$

$$= 0.4 \text{ m} + (0.80 \text{ m/s}^2)\Big[t^2/2 \Big]_0^t$$

$$x(t) = 0.4 \text{ m} + (0.40 \text{ m/s}^2)\ t^2.$$

For any value of t in seconds, this equation gives x in meters.

The equations obtained in the solution to Worked Example 41-5 have the same form as the equations for uniformly accelerated motion derived in Section 41.6 which is to be expected for motion having constant acceleration. The methods of calculus can also be applied to motions where the acceleration is not constant, to solve problems whose solution would be tedious and difficult without calculus.

Worked Example 41-6

A particle moves with a velocity given by

$$v(t) = 3.0 \text{ (m/s)}e^{-bt} + 1.2 \text{ (m/s}^3)\ t^2,$$

in the interval from $t = 0$ to $t = 2$ s. The constant $b = 0.1 \text{ s}^{-1}$. The particle's position at $t = 0$ is -0.6 m.

(a) Find the particle's acceleration at $t = 1.5$ s.

(b) Find the particle's position at $t = 2.0$ s.

Solution

(a) Given $v(t)$, we can use the method of differentiation to find $a(t) = dv/dt$, applying rules 2 and 4 from Appendix A.2.

$$a(t) = \frac{d}{dt}[3.0\ (\text{m/s})\ e^{-bt}] + \frac{d}{dt}[1.2\ (\text{m/s}^3)t^2\ ]$$

$$a(t) = 3.0\ (\text{m/s})(-b)\ e^{-bt} + 1.2\ (\text{m/s}^3)(2t).$$

After obtaining this expression for $a(t)$ valid for $0 \le t \le 2$ s, we substitute $t = 1.5$ s. At $t = 1.5$ s, $bt = (0.1\ \text{s}^{-1})(1.5\ \text{s}) = 0.15$, and

$$a(1.5\ \text{s}) = -3.0\ (\text{m/s})(0.1\ \text{s}^{-1})e^{-0.15} + 2.4\ (\text{m/s}^3)(1.5\ \text{s})$$

$$= -0.3\ \text{m/s}^2\ (0.861) + 3.6\ \text{m/s}^2$$

$$= 3.34\ \text{m/s}^2.$$

(b) Given $v(t)$, we use the method of integration to find $x(t)$, first in terms of algebraic symbols and then substituting numerical values for the constants:

$$x(t) = x_1 + \int_{t_1}^{t_2} v(t)\ dt$$

$$= -0.6\ \text{m} + \int_0^t [3.0\ (\text{m/s})e^{-bt} + 1.2\ (\text{m/s}^3)\ t^2]\ dt$$

$$= -0.6\ \text{m} + 3.0\ (\text{m/s}) \int_0^t e^{-bt}\ dt + 1.2\ (\text{m/s}^3)\int_0^t t^2\ dt$$

$$= -0.6\ \text{m} + 3.0\ (\text{m/s})(-1/b)\left[e^{-bt}\right]_0^t + 1.2\ (\text{m/s}^3)\left[t^3/3\right]_0^t$$

$$x(t) = -0.6\ \text{m} + (3.0\ \text{m/s})(-1/b)[e^{-bt} - e^0] + (1.2\ \text{m/s}^3)[t^3/3 - 0]$$

$$x(2\ \text{s}) = -0.6\ \text{m} + (3.0\ \text{m/s})(-10\ \text{s})[e^{-(0.1\ \text{s}^{-1})(2\ \text{s})} - 1]$$

$$+ (1.2\ \text{m/s}^3)(2\ \text{s})^3/3$$

$$= -0.6\ \text{m} - (30\ \text{m})[e^{-0.2} - 1] + (1.2\ \text{m/s}^3)(8\ \text{s}^3)/3$$

$$= -0.6\ \text{m} - (30\ \text{m})[0.819 - 1] + (9.6\ \text{m})/3$$

$$= -0.6\ \text{m} - (30\ \text{m})(-0.181) + 3.2\ \text{m}$$

$$x(2\ \text{s}) = 8.03\ \text{m}.$$]]

Exercises

E 41.7-1 The motion of a particle along the x axis is described by $x(t) = 3t^2 + 4t^{-2}$, for $t > 0.5$ seconds. Numerical coefficients are given in SI units.

(a) Find $v(t)$ for $t > 0.5$ s.
(b) Find the magnitude and direction of v at $t = 1$ s.
(c) Find the magnitude and direction of v at $t = 2$ s.
(d) Obtain the equation for $a(t)$ for $t > 0.5$ s.
(e) What is the approximate nature of this motion for $t > 5$ s?

E 41.7-2 What functional form is required for $x(t)$ in motion with constant acceleration along the x axis?

E 41.7-3 For each of the following one-dimensional motions, obtain equations for $v(t)$ and $x(t)$. Assume that $x(0) = 0$ and $v(0) = 0$, and that all numerical coefficients are given in SI units.
(a) $a(t) = 4$.
(b) $a(t) = 2\ t$.
(c) $a(t) = 1 + 3\ t^2$.
(d) $a(t) = 0$.

E 41.7-4 A particle moves along the x axis with an acceleration $a(t) = (24\ \mathrm{m/s^3})t$, beginning at $t = 0$.
Its velocity at $t = 0$ is -4 m/s, and its position at $t = 2$ s is $x = 5$ m.
(a) Find $v(t)$.
(b) Find $x(t)$.
(c) At what time does the particle reverse its direction of motion?

E 41.7-5 Consider a particle moving in a vertical line. Beginning with $a_y = -g$ and the initial conditions $y = y_0$ and $v_y = v_{0y}$ at $t = 0$, use the method of integration to compute $v_y(t)$ and $y(t)$.

Answers to Minimum Learning Goals

Line 1 1.895 $\mathrm{m/s^2}$ Southeast

Line 2 $3 \times 10^{22}\ \mathrm{m/s^2}$ in a direction opposite that of the initial velocity.

Line 3 (a) 61.9 mph; (b) 45 mph North

Line 4 (a) 20.4 m; (b) 4.08 seconds

Line 5 0.25 $\mathrm{m/s^2}$ at 269° (which is toward the center)

Line 6 (a) 29.4 m/s; (b) $t = 3$ seconds; (c) -9.8 $\mathrm{m/s^2}$

Line 7 (a) 0.02 km/sec; (b) 0.02 km/s, toward the east

Line 8 (a) 1.05 cm/s; (b) 0.667 cm/s down

Line 9 (a) 39.2 m; (b) 44.1 m

Line 10 (a) 6.0 m/s; (b) 30.0 m; (c) Yes

Line 11

Line 12 (a) 400 meters; (b) 870 meters; (c) 29 seconds; (d) between $t = 12$ and $t = 16$ sec; (e) 110 m/s

ANSWERS to Questions and Exercises in Chapter 41

Section 41.1

Q 41.1-2 Measure the position of the stone (in feet or meters from some reference point, such as the starting point) at several values of time (in seconds) spread throughout the motion. Include enough measurements to reveal any unusual aspects of this motion which may occur.

Q 41.1-4 If displacement is considered positive downward along the plane:
(a) displacement increases whenever the ball rolls downward;
(b) displacement decreases whenever the ball rolls upward.
(c) - 2 feet;
(d) zero.

E41.1-2 (a) 1.30 meters; (b) 0.50 meter down the track.

E41.1-4 (a) zero; (b) zero; (c) 6.77 yard/second

Section 41.2

Q 41.2-2 Motion with constant velocity.

Section 41.3

Q41.3-2 (a) Yes, if the velocity is in the positive direction; otherwise it is not. (b) Yes, if the velocity and acceleration are both in the negative direction. (c) Yes, if the velocity is in the negative direction.

Q41.3-4 There are many examples of each. Some are described in the answers to Q 41.3-1 and Q 41.3-2. Case (d) occurs for motion with constand velocity.

Q41.3-6 (a) Zero. (b) Yes; the ball's velocity is continually changing.

E 41.3-2 10 cm/s^2

E 41.3-4 8.95 m/s^2 toward the west

E 41.3-6 2 m/s^2 toward the north

Section 41.4

E 41.4-2 (a)

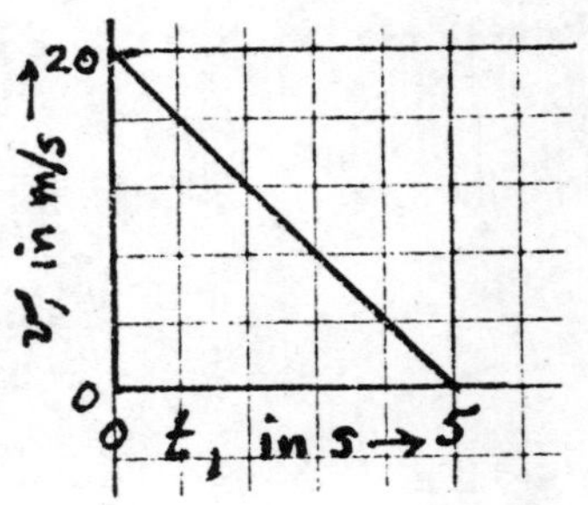

(b)

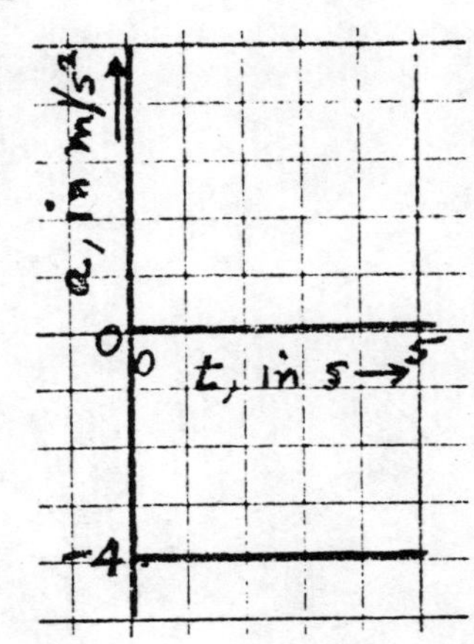

Section 41.5

E 41.5-2 Answers are the graphs in Example 4 of Figure 41-8 on page 41-25.

E 41.5-4 (a) From t = 0 to t = 5.67 s; (b) at t = 5.67 s; (c) +81.7 cm; (d) +50 cm; (e) at t = 10s; (f) 113.4 cm; (g) 11.34 cm/s; (h) 5 cm/s in the positive direction. (i) + 10 cm/s; (j) 5 cm/s^2; (k) from t = 0 to t = 1 s; (l) from t = 5 to t = 6 s; (m) at t = 5.67 s; also just after t = 0 and just before t = 10 s.

Section 41.6

Q 41.6-2 If the roadway is level and has uniform friction, and the car moves in a straight line as it stops, the acceleration is expected to be constant. Verification that a = constant would require measurements of position and time during this motion.

Q 41.6-4 Acceleration

E 41.6-2 4.3 m/s^2

E 41.6-4 0.3 seconds

E 41.6-6 At the same height

E 41.6-8 2.62 meters

Section 41.7

E 41.7-2 $x(t) = c_1t^2 + c_2t + c_3$

E 41.7-4 (a) $v(t)$ = -4 (m/s) + 12 (m/s^3) t^2;

(b) $x(t)$ = -15 m - 4 (m/s)t + 4 (m/s^3)t^3;

(c) $t = +1/\sqrt{3}$ s

CHAPTER 42 EXPLAINING MOTION IN ONE DIMENSION

Introduction

Chapter 41 presented some methods of describing the motion of a particle. This chapter considers the way in which the mass of a particle and forces acting on it determine the particle's motion, in conformity with Newton's three laws of motion. These laws are then applied to specific examples of one-dimensional motion. Methods are presented for computing the resultant of several forces acting on a particle. Some specific kinds of forces, weight and friction, are given special consideration.

When you have completed your study of the material in this chapter, you should be able to answer questions and work exercises like the examples in the minimum learning goals assigned by your instructor.

42.1 Forces and Motion. Newton's First Law of Motion

> Minimum Learning Goals. (1) Be able to state whether the velocity of an object is zero, constant, or changing, when you are given the forces acting on the object. For example:
>
> 1 An automobile is traveling on a long level stretch of highway, the road exerting a forward force of 200 newtons on the car, and the air through which the car is moving exerting on it a backward force of 200 newtons. What is the combined effect of these forces on the velocity of the car? (Answer page 42-43, Line 20)
>
> (2) Be able to state whether the resultant external force on an object is, or is not, zero, when you are given a description of the motion of the object. For example:
>
> 1 A small dust particle falls at a constant speed of 40 cm/s in a closed chamber where there are no air currents. Is there a non-zero resultant force on the dust particle? (Answer, page 42-42, Line 1)

What is required to make an object begin to move? A book lying at rest on a table top can be caused to move by pushing it from the side, or by grasping one corner and pulling it along the table top. If the book is on a sheet of paper, we can pull the paper sideways so the book begins to move when the paper moves; in this case, friction between the paper and the book is causing the book to move along with the paper. So there are at least three ways in which motion can be started: using a push, or a pull, or friction.

What is required to stop a moving object? A child's wagon rolling along the sidewalk can be stopped by pushing on it opposite to its motion, or by pulling on it from behind. If the wagon is rolling on a level sidewalk, sooner or later friction will bring it to a stop even if there is no push or pull. So there are at least three ways in which motion can be stopped: using a push, or a pull, or friction.

When we consider the causes of a change in motion, we see that a push, a pull, and friction produce comparable effects. The word chosen to refer to any one of these, or to any other factor like these which can cause a change in motion, is *force*.

A *force* is that which tends to change the velocity of an object.

Pushes, pulls, and friction are forces we meet continually in everyday life. Other forces in our daily experience include buoyant forces, weight, and upward forces which support objects. A variety of forces are introduced at appropriate points in the study of physics, with equations which apply to them.

What is required for a moving object to continue in motion? If a book is sliding on the table and you exert no forces on it, the book soon comes to a stop, so at first glance it may appear that a force must also be applied to *keep* the book in motion. We note, however, that it is friction between the book and the table which brings the book to a stop. Would the book slow down if there were no friction? We can reduce friction to a negligible value and thus investigate motion in the absence of friction, using a linear air track or air table (Figure 42-1). Then we find that when the force of friction is very small, a glider in motion continues to move for a very long time, even if it receives no additional push. We conclude that if friction were *entirely* absent, a glider in motion would continue to glide indefinitely. To continue with its motion unchanged, an object must be subject to *no forces*, or to no force which is not exactly counterbalanced by an opposite force. (A glider moving near the middle of a level air track is subject to no horizontal forces, but there are two vertical forces, the weight acting downward and the air cushion lifting upward, which exactly counterbalance each other.)

We find by experience that to start an object moving, the push applied must be stronger than the opposing force of friction. More generally, to start an object moving, or to stop one already moving, what is needed is an *unbalanced* force, one that is not offset by a force in the opposite direction. Any unbalanced force, no matter how small, is sufficient to change the motion of an object. When several forces act on the same object, the object responds to the *net* effect or combined effect of all the forces acting together. The special word used for the sum of several forces, or the combination of several forces acting on the same object, is the *resultant*.

The *resultant force* on an object is the vector sum of all the individual forces acting on the object.

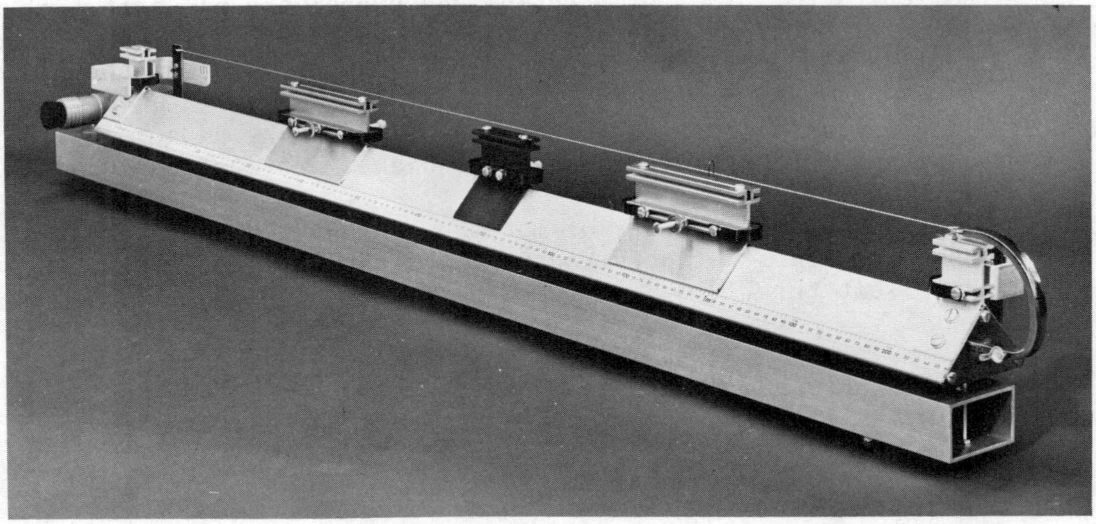

Figure 42-1. A linear air track. A blower (not shown) applies compressed air to the interior of the track at the far end. This air flows outward through a large number of tiny holes in the inclined upper surfaces of the track. A glider, three of which are pictured, can move on the air track with negligible friction by riding on the film of air which flows between the glider and the track. (Photo supplied by courtesy of the Ealing Corporation.)

Section 42.2 discusses methods for finding the resultant of several forces.

The experiments just described can be summarized in the following statement:

> A body at rest remains at rest, and a body in motion continues to move with constant velocity, unless it is acted on by an unbalanced external force.

This generalization is known as *Newton's first law of motion.* It is one of the elementary principles of dynamics, discovered by Sir Isaac Newton about 1666 and now called Newton's laws of motion. The two kinds of motion which the first law mentions, namely no motion (rest) and uniform motion in a straight line, have in common the fact that there is no acceleration. Thus the law sets the stage for dynamics by saying that it is the acceleration of a body, rather than the velocity, which is determined by outside forces. If there are no outside forces, there is no acceleration.

When the sum of the external forces is equal to zero, each of the external forces is "balanced" or offset by one or more of the others, so the resultant force equals zero.

Questions

Q 42.1-1 Name at least six kinds of forces which you meet in everyday life.

Q 42.1-2 The four jet engines of a Boeing 747 propel the plane through the air at 31,000 feet with a constant speed of 550 miles/hour in a straight line. Is there a nonzero resultant force on the plane?

Q 42.1-3 Using Newton's first law of motion, explain why an automobile passenger lurches forward when the brakes are applied suddenly.

Q 42.1-4 Using Newton's first law of motion, explain why an automobile passenger leans abruptly toward the right when the car makes a sharp left turn.

Check your understanding of the material above by testing yourself on the minimum learning goals at the beginning of this section.

42.2 Combining Forces

Minimum Learning Goals. (1) Be able to compute the rectangular components of a force, or velocity, or displacement, when you are given the magnitude of this quantity and its direction with respect to a coordinate system. For example:

1 An electron in a cathode-ray tube is subject to an electric force in the xy-plane, having a magnitude of 4.80×10^{-19} N and making an angle of 80° with the $+x$ axis and 10° with the $+y$ axis. Compute the x and y components of this force. (Answer, page 42-42, Line 5)

(2) Given two or three vector quantities in the same plane, be able to find their resultant using the method of components. For example:

1 A pith ball is acted on by two forces in the same vertical plane: a vertical downward force of 2.40×10^{-2} N and an easterly force of 1.70×10^{-2} N at an angle of 25° above the horizontal. What are the magnitude and direction of the resultant force? (Answer, page 42-43, Line 21)

2 At a point in a horizontal xy plane, the electric field strength is the resultant of the contributions from three point charges in the plane. The individual values of electric field strength are: 42.7 N/C at an angle of 78°, 35.2 N/C at an angle of 145°, and 56.0 N/C at an angle of 287°. Find the resultant field strength at the point. (Answer, p. 42-43, Line 23)

(3) Given the magnitude and direction of two or more forces acting on the same object, be able to determine the magnitude and direction of a single force which, if it

alone were applied to the object, would produce an acceleration equal to the acceleration actually produced by the forces acting together. For example:

1 A particle is acted on by a magnetic force $\vec{F}_m$ = 20 newtons at 0°, an electric force $\vec{F}_e$ = 40 newtons at 45°, and a gravitational force $\vec{F}_g$ = 30 newtons at 270°. What single force would produce the same acceleration as the sum of these three forces? (Answer, p. 42-43, Line 22)

2 An airplane has two forces acting on it, the force of gravity $\vec{F}_g = 1.80 \times 10^5$ lb at 270° and the force of air (a combination of drag, lift, and thrust) $\vec{F}_a = 2.00 \times 10^5$ lb at 70°. What single force would produce the same acceleration if it were applied to the airplane in the absence of these two forces? (Answer, page 42-42, Line 9)

The various physical quantities we meet in the study of physics have great diversity, but they also have some similarities which allow us to group them into two "families." A physical quantity having a numerical value but no particular direction in space belongs to the family called *scalar quantities* or simply *scalars*. Examples of scalars include mass, time, length, volume, pressure, and charge. The other family includes all the quantities which have some numerical value *and* a particular direction in space; these are called *vector quantities* or simply *vectors*. Examples of vectors include displacement, velocity, force, momentum, and electric field strength.

The motion of any object is determined by the combined effect of all the forces acting on it. To analyze motion, we need a method for finding the sum or resultant of several forces acting on the same object. A force is a vector quantity, so in adding a set of individual forces we must use the rules for adding vectors, which are illustrated below. Experiments to show that forces add as vector quantities can be done using a laboratory device known as a force table.

In the English system of units, and in engineering practice and everyday life in the USA, the unit for force is the familiar *pound*. In most scientific work, however, the SI unit for force is used, namely the *newton*. The definition of the newton will be given in Section 42.4, but we begin here to add forces measured in newtons.

Because forces are vector quantities, to add them we use force diagrams. Each individual force is represented by an arrow, having the proper length (to some chosen scale) and direction. The direction of a vector is often specified by giving an angle with reference to an established coordinate system. In physics it is customary to choose the direction of 0° to the right, and then to measure angles counterclockwise from the direction of 0°. Thus the direction of 90° is up, 180° is to the left, and 270° is down.

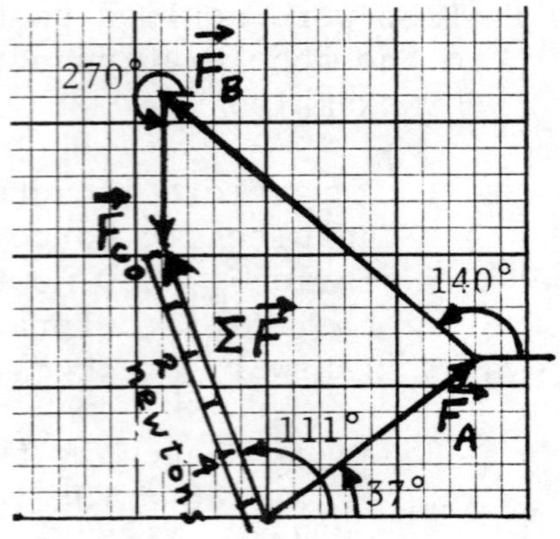

Figure 42-2. A diagram for the addition of three vectors using the head-to-tail method. The vectors are forces $\vec{F}_A$ = 5 newtons at 37°, $\vec{F}_B$ = 8 newtons at 140° and $\vec{F}_C$ = 3 newtons at 270°. The resultant of the three forces is $\Sigma\vec{F} = \vec{F}_A + \vec{F}_B + \vec{F}_C$; it is represented by the arrow drawn with its tail at the tail of the first vector ($\vec{F}_A$), and its head at the head of the last vector ($\vec{F}_C$). The magnitude of $\Sigma\vec{F}$ is measured using the same scale used to draw the arrows for the individual forces: $\Sigma\vec{F}$ is about 5.5 newtons at 111°.

The simplest method of adding vectors is the *head-to-tail method*, illustrated in Figure 42-2. To find the sum of three forces,

$$\Sigma\vec{F} = \vec{F}_A + \vec{F}_B + \vec{F}_C,$$

the head-to-tail method proceeds as follows: Begin by drawing $\vec{F}_A$ to scale and having the proper direction; at the head of $\vec{F}_A$, place the tail of $\vec{F}_B$ and draw $\vec{F}_B$ to scale and pointing in the correct direction; at the head of $\vec{F}_B$ place the tail of $\vec{F}_C$ and draw $\vec{F}_C$ to the correct scale and headed in the proper direction. $\Sigma\vec{F}$, the vector sum of $\vec{F}_A + \vec{F}_B + \vec{F}_C$, is the vector drawn with its tail at the beginning point (the tail of $\vec{F}_A$), and its head at the ending point (the head of $\vec{F}_C$).

Although each force or other vector quantity has its own direction in space, it often happens that we need to find the *component* of a vector along a direction at an angle to the direction of the vector. A general procedure for doing this is described in Section 24.4, which can be applied to any problem where it is required to find the components of a vector along two perpendicular directions, so long as the vector lies in the plane defined by these two directions. This procedure is often referred to as *resolving a vector into components* parallel to two perpendicular axes. To find the components of a vector using the notation of *unit vectors* (Section 43.1), it is necessary to know the angles between the vector and the coordinate axes. In Figure 42-3, the angle between the direction of $\vec{A}$ and the $+x$ axis is labeled θ_x, the angle between $\vec{A}$ and the $+y$ axis is θ_y, and the angle between $\vec{A}$ and the $+z$ axis is labeled θ_z. Using these angles, the cartesian components of $\vec{A}$ are

$$\vec{A}_x = \hat{x}A\cos\theta_x,\ \vec{A}_y = \hat{y}A\cos\theta_y,\ \vec{A}_z = \hat{z}A\cos\theta_z.$$

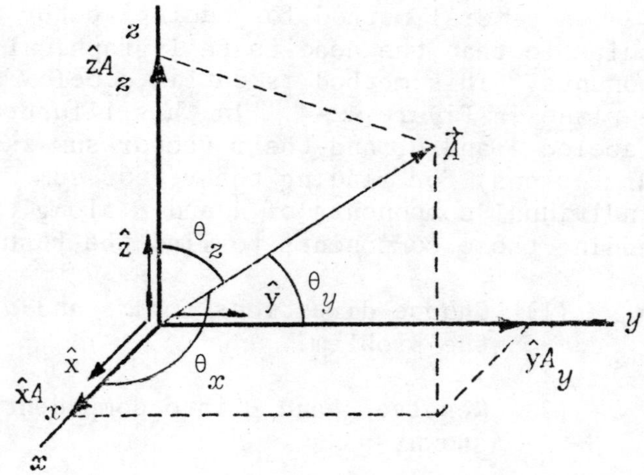

Figure 42-3. The unit vectors $\hat{x}$, $\hat{y}$, and $\hat{z}$ are directed along the $+x$, $+y$, and $+z$ axes, respectively. The three Cartesian components of a vector $\vec{A}$ are labeled $\hat{x}A_x$, $\hat{y}A_y$, and $\hat{z}A_z$. The magnitudes of these components are $A_x = A \cos \theta_x$, $A_y = A \cos \theta_y$, and $A_z = A \cos \theta_z$.

The sum of these components is

★ $$\vec{A} = \vec{A}_x + \vec{A}_y + \vec{A}_z.$$

Applying the Pythagorean theorem leads to

★ $$A^2 = A_x^2 + A_y^2 + A_z^2.$$

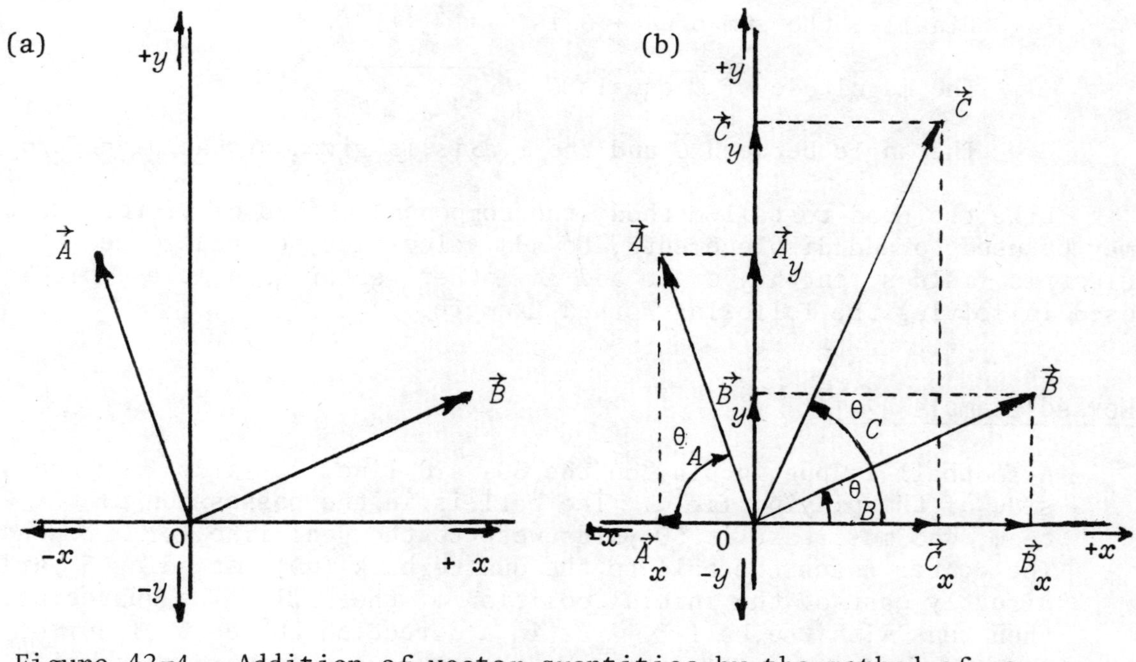

Figure 42-4. Addition of vector quantities by the method of components. (a) The vectors $\vec{A}$ and $\vec{B}$ which are to be added, each drawn with its tail at the origin of the xy coordinate system. (b) The constructions used to find the x and y components of $\vec{A}$ and $\vec{B}$. After the x components are added to obtain $\vec{C}_x$ and the y components are added to obtain $\vec{C}_y$, $\vec{C}_x$ is combined with $\vec{C}_y$ to obtain the resultant $\vec{C}$.

A general method for adding vector quantities, which is often more suitable than the head-to-tail graphical method, makes use of vector components. This method is outlined below and illustrated for two vectors in a plane in Figure 42-4. In this illustration, the vectors to be added are labeled $\vec{A}$ and $\vec{B}$, and their vector sum is labeled $\vec{C}$. The procedure (in two dimensions) for finding the vector sum $\vec{C} = \vec{A} + \vec{B}$ begins with finding the individual components of $\vec{A}$ and $\vec{B}$ along the chosen axes, and continues with adding these components to find the resultant. The method is:

★ (1) Choose directions for x and y axes which are convenient for the problem.

★ (2) Resolve $\vec{A}$ and $\vec{B}$ into components parallel to these axes. As shown,

$$A_x = -A \cos \theta_A; \quad A_y = A \sin \theta_A;$$

$$B_x = B \cos \theta_B; \quad B_y = B \sin \theta_B.$$

It is essential to use the correct sign (+ or -) for each component, which depends on the angle which the vector ($\vec{A}$ or $\vec{B}$) makes with the x and y axes.

★ (3) Add x-components: $A_x + B_x = C_x$.

Add y-components: $A_y + B_y = C_y$.

★ (4) Finally, the sum of $\vec{A} + \vec{B}$ is $\vec{C} = \vec{C}_x + \vec{C}_y$.

The magnitude of $\vec{C}$ equals $C = \sqrt{C_x^2 + C_y^2}$.

The angle between $\vec{C}$ and the x axis is given by $\tan \theta_C = C_y/C_x$.

Like the head-to-tail method, the component method of vector addition may be used to add displacements, to add velocities, to add forces, to add electric field strengths, or to add any other vector quantities. It is used in solving the following Worked Example.

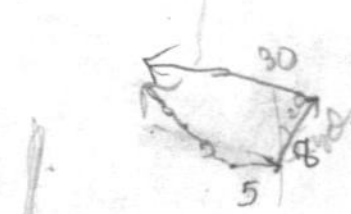

Worked Example 42-1

A football is put in play on the 50-yard line, 20 yards from the south side of the playing field. The ball is in the possession of the Blue team, who must take it 50 yards west to the goal line for a touchdown. The center snaps the ball to the quarterback (qb), standing 5 yards directly east of the initial position of the ball. The quarterback then runs with the ball 8 yards in a direction 20° east of north, while looking for a pass receiver. He then throws a forward pass which is caught at a point 30 yards from where he throws the ball, in a direction 80° west of north. The pass receiver is immediately downed at the point where he caught the ball.

(a) What is the displacement of the ball, from its initial position at the line of scrimmage to the point where the forward pass is caught?

(b) How many yards were gained on the play?

Solution

A diagram of the ball's path is shown below.

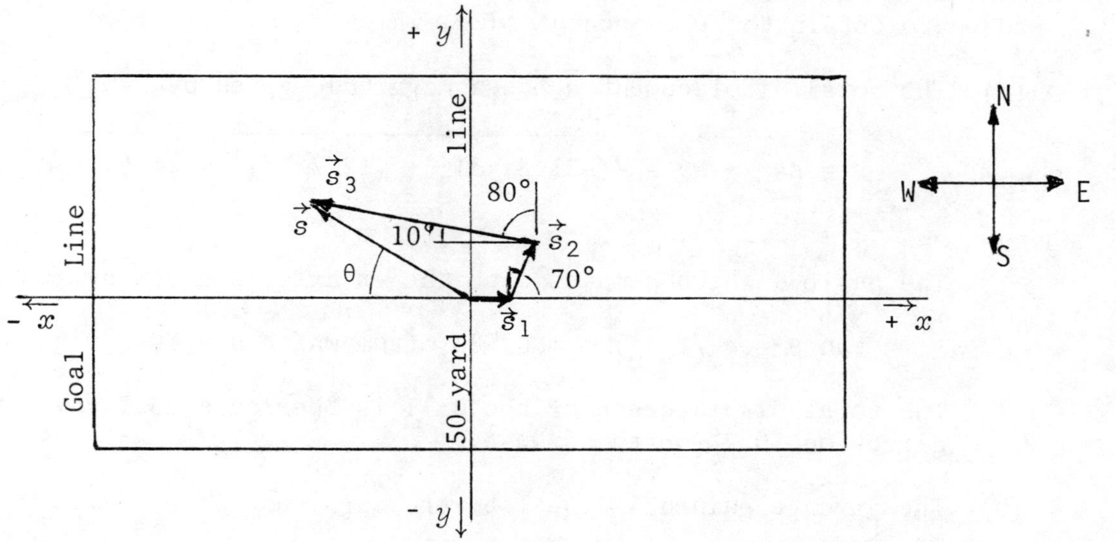

The total displacement $\vec{s}$ of the ball during this play is the sum of the three displacements:

$$\vec{s} = \vec{s}_1 + \vec{s}_2 + \vec{s}_3.$$

It is convenient to choose an x axis in the east-west direction, and a y axis in the north-south direction, with the origin of coordinates at the initial position of the ball. To keep track of the successive displacements of the ball in an orderly manner, we set up a table as follows:

Successive Displacements of the Ball

Action	$\vec{s}_i$	Magnitude of $\vec{s}_i$	Direction of $\vec{s}_i$	x comp.	y comp.
center to qb	$\vec{s}_1$	5 yards	east	+ 5 yd	0
qb runs	$\vec{s}_2$	8 yards	20° E of N	+ 2.7 yd	+ 7.5 yd
forward pass	$\vec{s}_3$	30 yards	80° W of N	- 29.5 yd	+ 5.2 yd
				- 21.8 yd	+12.7 yd

In this table, the subscript i takes the successive values 1, 2, and 3. Data for the third and fourth columns of the table are given in the statement of the Example. The entries in the fifth and sixth columns are computed as follows:

$\vec{s}_i$	x component	y component
$\vec{s}_1$	$5 \times \cos\ 0° = 5$	$5 \times \sin\ 0° = 0$
$\vec{s}_2$	$8 \times \cos 70° = 2.7$	$8 \times \sin 70° = 7.5$
$\vec{s}_3$	$-30 \times \cos 10° = -29.5$	$30 \times \sin 10° = 5.2$

After the x and y components have been tabulated, the individual components are added to obtain the x component of the total displacement $\vec{s}$ of the ball, $s_x = -21.8$ yd. The individual y components are added to obtain the y component of $\vec{s}$, which is $s_y = 12.7$ yd.

(a) The total displacement $\vec{s}$ has a magnitude given by

$$s = \sqrt{s_x^2 + s_y^2} = \sqrt{(-21.8 \text{ yd})^2 + (12.7 \text{ yd})^2} = \sqrt{636.5 \text{ yd}^2}$$

$$= 25.2 \text{ yd.}$$

The angle θ which s makes with the $-x$ axis is given by

$$\tan\theta = s_y/(-s_x) = +0.583, \text{ from which } \theta = 30.2°.$$

The total displacement of the ball is therefore 25.2 yd in the direction 30.2° north of west.

(b) The yardage gained is equal to the magnitude of s_x, 21.8 yd.

Vector components are used in solving exercises and problems which occur in all branches of physics. Some of the Exercises in this section are based on the reader's (assumed) prior knowledge of physics.

Questions and Exercises

Q 42.2-1 Give some examples of situations for which it would be most convenient to choose directions for x and y axes which are not vertical or horizontal.

E 42.2-1 An airplane flies 125 miles in a direction 35° east of north. Find the components of its displacement toward the north and toward the east.

E 42.2-2 An electric force of 4.70×10^{-5} newton is in the xy plane and directed at an angle of 27 degrees to the $+x$ direction and 63 degrees to the $+y$ direction.
(a) What is the x component of this force?
(b) What is the y component of this force?

E 42.2-3 At a certain point in an electric field the electric field strength of 5.79×10^3 N/C lies in the xy plane with a direction of 17 degrees counterclockwise from the $+y$ direction. (continued)

(a) What is the component of field strength along the x axis?
(b) What is the component of field strength along the y axis?

E 42.2-4 A tightly stretched string is located between two uniformly charged parallel plates as shown below. The electric field strength between the plates is 2.5 N/C, directed perpendicular *to the plates*. Choose axes which are respectively parallel and perpendicular *to the string*. In the middle region between the plates, what is the component of the electric field strength: (a) in a direction along the string? (b) in a direction perpendicular to that along which the string lies, and in the plane of the paper?

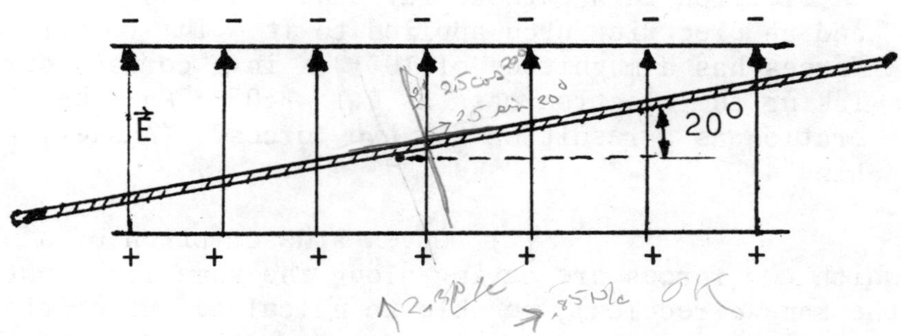

E 42.2-5 A chair is moved 5 feet east and then 12 feet south. What are the magnitude and direction of the resultant displacement?

E 42.2-6 The electric force on a certain charge has an x-component of +2.4 newtons and a y-component of -4.7 newtons. What are the magnitude and direction of the resultant force?

E 42.2-7 A proton is subjected to an electric force having an x-component of -7.3×10^{-12} N and a y-component of $+3.7 \times 10^{-12}$ N. What is the resultant force on the proton?

E 42.2-8 The electric field strength at a point is horizontal, with a component of 2.46×10^3 N/C directed toward the north and a component of 6.42×10^3 N/C directed toward the west. What is the resultant electric field strength at the point?

E 42.2-9 The forces acting on a ball include a gravitational force $\vec{F}_g$ = 1.2 N at 270°, and a drag force $\vec{F}_d$ = 0.5 N at 180°. What single force $\vec{F}$ could replace these two forces and produce the same acceleration of the ball?

E 42.2-10 Find the resultant of these three horizontal forces which act on the same hockey puck: F_1 = 60 lb at θ_1 = 30° cw (clockwise) with respect to North, F_2 = 30 lb at θ_2 = 135° cw with respect to North, and F_3 = 20 lb at θ_3 = 240° cw with respect to North.

Check your understanding of the material above by testing yourself on the minimum learning goals at the beginning of this section.

42.3 Mass. Newton's Second Law of Motion

Minimum Learning Goals. (1) Be able to recall the law relating resultant force, mass, and acceleration, and use it to calculate any one of these quantities, given the other two. For example:

1 Four freshmen are pushing a stalled car having a mass of 2000 kg. If the car accelerates at 0.1 m/s^2, what is the resultant force on the car? (Answer, page 42-42, Line 19)

2 An electron in a cathode-ray tube has both a magnetic force and an electric force applied to it. The vector sum of these forces has a magnitude of 10^{-16} N in a certain direction. Taking the electron mass as 9.1×10^{-31} kg, what is its acceleration as a result of the two forces? (Answer, page 42-42, Line 4)

(2) Given a description of a physical system on which two forces are acting along the same line (but not necessarily in the same direction), be able to calculate one of the forces, when you are given the other force and the acceleration of the system. For example:

1 An old clunker (an aged car which does not run well) having a mass of 1500 kg is being towed by a new car using a chain which pulls the clunker with a forward force of 1100 newtons. Unknown to the driver, the hand brake of the clunker has been set with the result that this motion is opposed by a braking force of unknown amount. If the acceleration of the clunker is 0.2 m/s^2, compute the braking force. (Answer, page 42-42, Line 17)

In Chapter 41 we found that four basic quantities or variables are needed to *describe* motion: time, displacement, velocity, and acceleration. In seeking to *explain* motion, we have found that it is the resultant force which changes the motion of an object. How is the resultant force related to the variables which describe motion?

Even though an object is acted on by zero resultant force, its displacement may change with time and its velocity may be large or small, depending on how it was set into motion. When the net force is zero, the acceleration is the only descriptive variable which remains constant, and that is zero, according to Newton's first law of motion. How can we find the quantitative relationship between acceleration and force?

We need an experiment in which the acceleration of some object is measured in response to a range of applied forces. The use of a linear air track eliminates friction, for practical purposes; a force can be applied to a glider by a string which runs over a pulley and is connected to a falling weight. When the glider's acceleration is measured using several values of applied force, it is found that the acceleration is directly proportional to the force:

$$a \propto F.$$

Figure 42-5. An experiment to measure the acceleration of a glider in response to an applied force. The force is applied to the glider by a string which passes over a pulley and is connected to a falling weight.

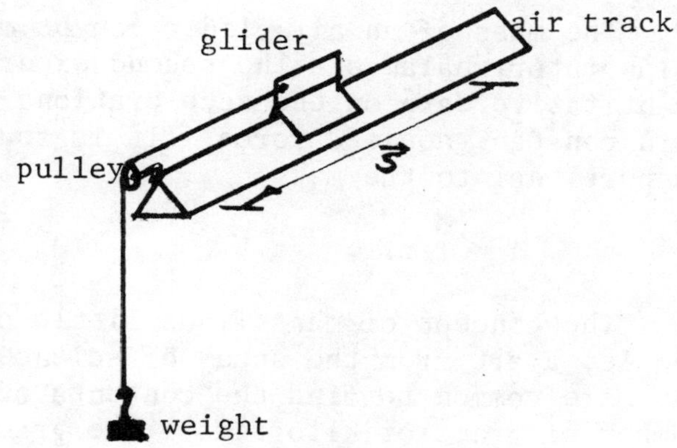

Do all gliders have the same acceleration, for a fixed value of resultant force? This question can be investigated in a similar experiment, this time keeping the falling weight constant but allowing the applied force to act successively on gliders having different sizes. It is found that the resulting acceleration depends on the size of the glider. When the length of the glider is doubled, the resulting acceleration drops to half the prior value. We conclude that there is some physical property of the glider itself which determines its acceleration in response to the resultant force. This property is related to the size of the glider, or more specifically to the amount of material the glider contains. The bigger the glider, the more sluggish is its response to an applied force. This sluggishness, or tendency to resist acceleration, or inertia of a material object is attributed to a specific physical property of the object called *mass*.

> *Mass* is an intrinsic property of an object which is a measure of the quantity of matter in the object, and which determines its acceleration in response to a resultant force.

To state a numerical value for the mass of an object, we need a unit of mass. This unit must be embodied in some *standard* of mass, a physical object which is solid, durable, and preserved in such a way that its mass will remain constant. The SI unit of mass is defined by international agreement as follows:

> The *kilogram* is the mass of a particular cylinder of platinum-iridium alloy, called the International Prototype Kilogram, which is preserved in a vault at Sèvres, France, by the International Bureau of Weights and Measures.

Every scientific laboratory has its own internal standards of mass, labeled pieces of metal which range from one milligram to a kilogram; those of better quality are provided with a certificate which indicates their calibration with reference to the national standard of the country in which they were made, that standard in turn having been calibrated by careful comparison with the International Prototype Kilogram.

The mass of an air glider can be measured in grams or kilograms using a laboratory balance. The second experiment described above can then supply quantitative data on the accelerations of various values of mass, in response to a constant applied force. It is found that the acceleration is inversely proportional to the mass:

$$a \propto 1/m.$$

The concept of mass finds little use in the daily experience of most people, apart from the study of science. Nevertheless, it is becoming more and more common to find the contents of packaged goods described as some number of grams or kilograms. One gram, of course, is equal to 0.001 kilogram. A potential source of confusion is the practice of referring to the contents of a package as the "net weight," and then expressing this quantity in grams or kilograms. In physics weight is a physical quantity quite distinct from mass; it is expressed in a unit of newtons (see the discussion of weight in Section 42-4.) A quantity expressed in kilograms refers to the property which physicists call mass. The relationship between weight and mass is correctly stated as follows:

A mass of one kilogram weighs 2.20 pounds.
A mass of 453.6 grams, or 0.4536 kg, has a weight of one pound.
A mass of 28.35 grams weighs one ounce.

If we combine the proportionalities obtained in the two experiments described above, we have

$$a \propto F/m, \qquad \text{or} \qquad F \propto ma.$$

As far as we know, there are no other factors which affect the acceleration of an object as long as it moves much slower than the speed of light. (The special theory of relativity indicates that when an object moves at a very high speed its mass increases with speed, so under those conditions the equation we are about to write needs to be expressed in a different form.) We therefore change the combined proportionality given above into an equation:

$$F = k\,m\,a.$$

The proportionality constant, k, has a value which depends on the units used for accleration, mass, and force. The SI units for mass and acceleration have already been chosen, but the unit for force has not. We are free, then, to choose the unit of force so the equation becomes as simple as possible; we choose it so that k equals unity. The SI unit of force is called the *newton*, symbol N.

One *newton* is that force which gives to a one-kilogram mass an acceleration of one meter/sec^2.
$1\ \mathrm{N} = 1\ \mathrm{kg{\cdot}m/s^2}$.

The constant k can thus be deleted from the equation.

In using this equation, one must always remember that the acceleration of an object having mass m is determined by the *resultant* force acting on it, and to emphasize this we write

★ $$\Sigma\vec{F} = m\,\vec{a}\,.$$

The student is urged to write this equation with $\Sigma\vec{F}$, as a reminder that *all* forces acting on the object must be taken into account in order to compute the acceleration. This equation is Newton's second law of motion. It lies at the heart of the answer to the question: *Why* does an object move as it does?

Worked Example 42-2

A child's wagon is pushed on a level sidewalk with a horizontal force of 55 newtons. The wagon's motion is opposed by a friction force of 15 N, which acts horizontally. The mass of the wagon and its passenger is 25 kg. At a certain time t_0 the wagon's speed is 0.3 m/s.

(a) What is the resultant force on the wagon?
(b) What is its acceleration?
(c) What will be the wagon's speed two seconds after t_0?

Solution

(a) The resultant force is the vector sum of all the forces on the wagon (and the rider). The earth's force of gravity pulls down on the wagon but the sidewalk pushes up on the wagon's tires with an equal force; we know the upward force must equal the downward force of gravity in this case because the wagon's motion is along the level sidewalk -- it has no acceleration in the vertical direction. These two vertical forces cancel each other and make no net contribution to the resultant force. The horizontal forces are shown in the figure: the forward push of 55 N, and the backward force of friction (acting in part on each wheel) which totals 15 N. These forces are not drawn to scale in the pictorial figure,

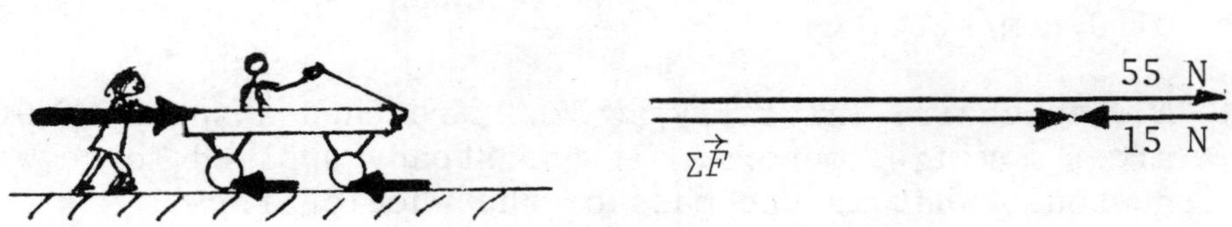

but they are drawn to a scale of 1 cm = 10 N in the force diagram at the right. The resultant of these forces, $\Sigma\vec{F}$, is 40 N to the right, or in the forward direction.

(b) The acceleration can be found using Newton's second law, $\Sigma\vec{F} = m\,\vec{a}$, transposed to the form $\vec{a} = \dfrac{\Sigma\vec{F}}{m}$. Remembering that the resultant force is directed to the right, we substitute numbers and compute $a = \dfrac{40\text{ N}}{25\text{ kg}} = 1.6\ \dfrac{\text{N}}{\text{kg}} = 1.6\text{ m/s}^2$, to the right.

(c) The wagon's speed at t_0 is $v_0 = 0.3$ m/s. Assuming the acceleration is constant, we use Eq. (41-1) from page 41-29 to compute the speed at $t = 2$ s after t_0:

$$v = v_0 + a\,t = 0.3\text{ m/s} + (1.6\text{ m/s}^2) \times 2\text{ s} = 3.5\text{ m/s}.$$

The unit of force commonly used in the U.S.A. is the *pound.* One pound equals 4.45 newtons.

Questions and Exercises

Q 42.3-1 "Two balls, one of lead and one of wood, are dropped. It is observed that they have the same acceleration. Since they are both subject to the same force (gravity), they must have the same mass." Criticize this statement.

Q 42.3-2 What is the intrinsic property of an object which determines the object's response to an unbalanced force? Define the SI Unit of this property.

Q 42.3-3 Criticize the following statement: "A steel ball and a piece of tissue paper are dropped. The steel ball is observed to have the larger acceleration. Since they are both subject to the same force (gravity), the tissue paper must have a larger mass."

Q 42.3-4 Why can it be said that "Newton's first law is a special case of Newton's second law?"

E 42.3-1 Find your own mass, in kilograms.

E 42.3-2 What force is required to give a 0.1-kg apple an acceleration of 0.3 m/s^2?

E 42.3-3 A wheelchair, initially at rest, reached a speed of 0.6 m/s after moving 2 meters. If the steady applied force was 2.0 newtons, what is the mass of the wheelchair?

E 42.3-4 It is desired to accelerate electrons with an acceleration of $2 \times 10^{12}\text{ m/s}^2$. How large is the required electrical force? (The mass of an electron is 9.1×10^{-31} kg.)

E 42.3-5 A net force of 0.4 newtons is applied to a sack lunch whose mass is 0.25 kg, causing it to move the length of a table 2.0 meters long. If the lunch starts from rest,
(a) What is its acceleration?
(b) How many seconds elapse before the lunch falls off the end of the table?

E 42.3-6 Keith and Robin are competing for the use of a tricycle which has a mass of 12 kg. Robin is pulling the tricycle with a force of 30 newtons, but Keith is pulling harder in the opposite direction with the result that the tricycle is accelerating toward him at 1.0 m/s^2. How large is the force of Keith's pull on the trike?

E 42.3-7 The head of an adult human has a mass of about 5 kg. When an auto crashes into an immovable barrier at a speed of 30 miles/hour, the deceleration of the occupants of the car is estimated to be 300 m/s^2 or greater. If a passenger's head is stopped by the windshield with this deceleration,
(a) What force does the windshield exert on his head, in newtons?
(b) What is this force, in pounds?

Check your understanding of the material above by tesing yourself on the minimum learning goals at the beginning of this section.

42.4 Weight

Minimum Learning Goals. (1) Given the acceleration due to gravity at some location near a large and massive object (like the earth), and the mass of some other much smaller object (like a man) placed at that location, be able to calculate the weight of the smaller object. For example:

1 A planet in our solar system has an acceleration of gravity of 24.9 m/s^2 near its surface.
(a) If a man has a mass of 70 kg, what is his weight on that planet?
(b) What is his weight on the earth (taking g = 9.8 m/s^2)?
(Answer, page 42-42, Line 7)

From Newton's second law we know that when an object accelerates, this is the result of an unbalanced force acting on the object. When a baseball is dropped and falls through the air, it accelerates with g = 9.8 m/s^2. What force causes this acceleration? A common answer is "the force of gravity," a brief way of saying "the downward gravitational pull of the earth on the baseball." This gravitational force on the baseball is just one example of the gravitational forces found everywhere in the universe. In physics, we use *weight* to refer to this force of gravity.

The *weight*, $\vec{w}$, of an object is the downward force exerted on the object by the earth's gravitational attraction.

The same word is used when an object is not on or near the earth, but on some other large body such as the moon. We are then talking about the gravitational attraction of the moon. However, the object has different weights on the earth and on the moon, because the gravitational attraction of the two is different. The weight of a body even varies slightly from one place to another on the earth.

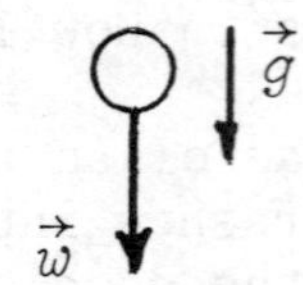

Figure 42-6. A freely falling object has a downward acceleration of $\vec{g}$ because it is acted on by its weight, $\vec{w}$.

We can compute the weight of an object by applying Newton's second law, $\Sigma\vec{F} = m\,\vec{a}$, to a freely falling object such as a baseball. Assuming ideal "free fall," the force of air drag or air resistance can be neglected so the only force on the falling ball is its weight; therefore for the ball $\Sigma\vec{F} = \vec{w}$, and the acceleration $\vec{a} = \vec{g}$. Substituting in Newton's second law, we obtain

★ $$\vec{w} = m\,\vec{g}.$$

If we know the mass of an object in kilograms we can compute its weight in newtons, and vice versa.

In the United States it has been customary to express the weight of an object in pounds. One pound equals 4.45 newtons. As the U.S.A. moves slowly toward metrication (the use of metric units for expressing measurements), we find the "net weight" of groceries and other goods expressed in grams and kilograms. This practice probably derives from the usage customary in chemistry, to measure the amounts of substances in grams, sometimes called "grams weight." The balances commonly used in science laboratories for "weighing" give their indications in grams. In physics, it is important to keep a clear distinction between weight and mass; physicists regard the kilogram as the unit of mass (but not of weight), and the newton as the proper unit for weight.

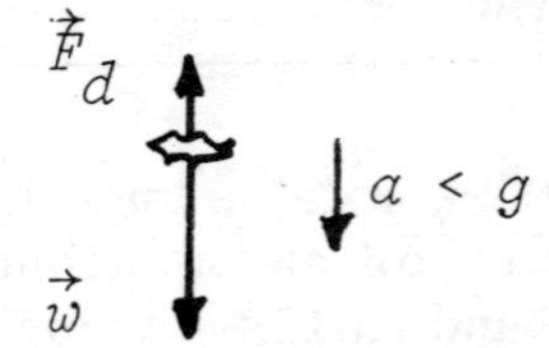

Figure 42-7. A leaf falls in air with an acceleration a smaller than g, because its weight $\vec{w}$ is partially offset by the upward force of air drag, $\vec{F}_d$.

Any object in free fall near the earth's surface is accelerated downward by its weight with an acceleration $g = 9.8\ \text{m/s}^2$. However, when a feather or a leaf or a parachute falls, clearly its acceleration a is much less than $9.8\ \text{m/s}^2$. Since $a < g$, it follows that $m\,a < m\,g$, which in turn means that the resultant force $\Sigma\vec{F}$ on the object is less than $\vec{w}$. The falling object, therefore, must have acting on it some upward force which partially offsets the weight. This force is sometime called air resistance, but a better name for it is *air drag* or simply *drag*. We use the symbol $\vec{F}_d$ for this force. Air drag is one example of a frictional force. Friction is discussed in Section 42.7. When a feather is allowed to fall in a tube from which the air has been evacuated, the air drag disappears and the feather may be seen to fall with the same acceleration as a steel ball.

Worked Example 42-3

A sky-diver is descending vertically with his parachute open. The combined weight of the diver and parachute is 850 newtons, while the upward drag force on them is 800 newtons.

(a) What is the mass of the diver-plus-parachute?

(b) What is their acceleration?

Solution

(a) Mass and weight are related by the equation $w = mg$. To find the mass, we compute

$$m = \frac{w}{g} = \frac{850\ \text{N}}{9.8\ \text{m/s}^2} = 86.7\ \text{kg}.$$

(b) The starting point for calculating the acceleration which results from a set of forces is Newton's second law, $\Sigma\vec{F} = m\vec{a}$. In this Example, $\Sigma\vec{F}$ is the sum of two forces, $\vec{w}$ downward and air drag $\vec{F}_d$ upward. The drawing includes a force diagram from which we can see that the magnitude of $\Sigma\vec{F}$, $|\Sigma\vec{F}|$, is simply the numerical difference between w and F_d. Calculating,

$$|\Sigma\vec{F}| = w - F_d = 850\ \text{N} - 800\ \text{N} = 50\text{N},$$

acting downward. The acceleration is therefore

$$\vec{a} = \frac{\Sigma\vec{F}}{m} = \frac{50\ \text{N downward}}{86.7\ \text{kg}} = 0.577\ \text{m/s}^2,$$

downward.

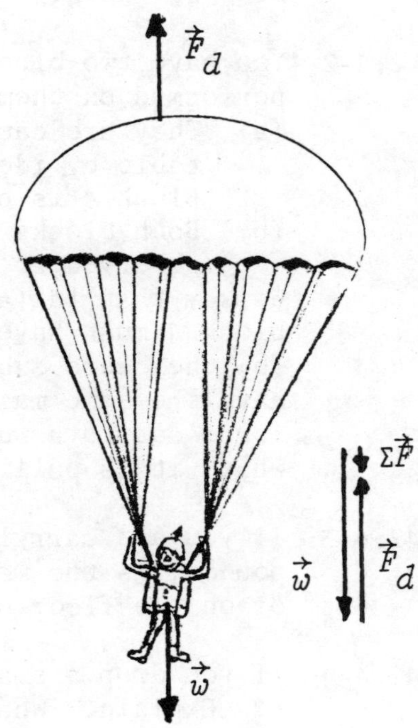

If a baseball is allowed to fall near the moon's surface, its acceleration in free fall is only about 1.6 m/s^2, much less than 9.8 m/s^2. Scientists predicted this result long before astronauts were able to go to the moon and actually do the experiment. The ball's *mass* remains the same wherever it may go in the universe, but its *weight* on the moon is less than on the earth because the moon's gravitational force is smaller than the earth's. If the ball is taken far from the surface of every planet, sun, and star, the gravitational forces on it would be negligible and we could say that the ball was *weightless*. In a spacecraft orbiting the earth in a circular orbit, however, the ball would not actually be weightless because the earth's gravitational attraction would still cause it to fall toward the earth -- or rather, "around the earth" in uniform circular motion (See Chapter 44). In this case, though, the spacecraft is also falling at the same rate, so when the ball's motion is observed from a frame of reference fixed to the spacecraft the ball *appears* not to be falling. Under these conditions, we could say that the ball has *apparent weightlessness*. The astronauts also are apparently weightless, and the TV pictures have shown them floating inside the spacecraft with no need for the floor to hold them up or to support their weight.

Questions and Exercises

Q 42.4-1 Why does the weight of an object always point down? What do you mean by "down?"

Q 42.4-2 You have two blocks, *A* and *B*. The following experiments are performed on them:
(a) They are caused to accelerate horizontally on a frictionless table by identical rubber bands stretched by identical amounts. Block *A* is observed to have a larger acceleration.
(b) Both blocks are dropped so as to fall freely. They are observed to have the same acceleration.
We say that in (a) they were subject to the same force, so that block *A* must have a smaller mass. Why can we not say that in (b) they were subject to the same force, so that both blocks must have the same mass? If weight can depend on mass, why cannot the force due to a rubber band depend on the mass of the object on which it is pulling?

Q 42.4-3 If you are carrying on your shoulders a child who weighs 50 pounds, is the *weight of the child* acting on the child, on you, or on the floor below you?

Q 42.4-4 If you drop a small feather, does it fall with an acceleration of g? Explain. What if the feather is dropped in an evacuated chamber?

E 42.4-2 What is the weight of a bucket which has a mass of 5 kg?

Bill is an 80-kilogram man. The acceleration of gravity on earth is 9.8 m/s^2, while the acceleration of gravity on the moon is 1.6 m/s^2.
(a) What is Bill's mass on Earth?
(b) What is Bill's weight on Earth?
(c) What is Bill's mass on the Moon?
(d) What is Bill's weight on the Moon?

E 42.4-3 A forest-fire-fighter is descending vertically on a parachute with an acceleration of 1.0 m/s^2 downward. The man and parachute have a combined mass of 102 kg, and they are pulled downward by gravity with a force of 1000 newtons. What are the magnitude and direction of the drag force on the parachute?

E 42.4-4 A girl jumps out of a balloon, opens her parachute and falls to the earth. She manages to control her descent so that she falls in a straight vertical line. The girl and parachute have a total mass of 80 kilograms. Air currents and other factors combine to give her a motion described by the following data, beginning at an arbitrary time. The downward direction is considered positive for all vector quantities.

time (seconds)	0	0.1	0.5	0.9	1.0	1.1	1.2	2.0	3.0
position (meters)	0	1.5	7.0	12.0	13.0	14.0	14.9	22.0	30.5

The girl's speed at t = 0 is + 16 m/s. The average resultant force on the girl-plus-parachute during the interval from t = 0 to t = 1 second is 400 newtons. Using these data, compute the following quantities for this motion, and state the units and direction (positive or negative) for each quantity:

(a) The distance traveled by the girl from t = 1 to t = 3 seconds.
(b) The average speed of the girl during the interval from t = 0 to t = 2 seconds.
(c) The instantaneous speed of the girl at t = 1 second, as accurately as you can using the data given.
(d) Her average acceleration during the interval from t = 0 to t = 1 second.
(e) The weight of the girl-plus-parachute during this motion.
(f) The average drag force exerted by the air on the girl-plus-parachute during the interval from t = 0 to t = 1 second.

E 42.4-5 The distance ℓ between the supporting feet at the two ends of an air track is 1.27 m. When a riser block of height $h = \frac{1}{2}$ inch = 1.27 cm is placed under the foot at one end of an initially level track, the track slopes at an angle θ

(a) Find θ and $\sin \theta$.
(b) For a glider of mass m on this air track, find the component of the weight which is parallel to the sloping surface of the track.
(c) Find the acceleration a of this glider, in m/s^2.
(d) Show that for a riser block having any h, $a/g = h/\ell$.

Check your understanding of the material above by testing yourself on the minimum learning goals at the beginning of this section.

42.5 Newton's Third Law of Motion

Minimum Learning Goals. (1) Be able to point out errors in statements sometimes used in discussing Newton's third law. For example:

1 What is wrong with the following statement? When an object has a resultant force applied to it, by Newton's third law the reaction force is equal and opposite to the applied force; therefore, the object should never accelerate. (Answer, page 42-42, Line 14)

2 What is wrong with the following statement? By Newton's third law, the force of a horse pulling on a cart is the same as that of a cart pulling on the horse; therefore, the cart can never accelerate. (Answer page 42-42, Line 2)

(2) Be able to use Newton's third law to find the force on an object when you are given the force exerted *by* that object on something else. For example:

1 A man has a weight of 700 newtons. What is the reaction force to this weight, and to what object is it applied? (Answer, page 42-42, Line 16)

2 A mass is suspended by a spring. A force of 100 N downward is exerted on the mass by its weight, and a force of 100 N exerted upward on it by the spring. What is the force of the mass on the spring? (Answer, page 42-42, Line 11)

Consider a simple pendulum having a heavy bob, suspended from a rod attached to a flat board which in turn is placed on a flat air table (Fig. 42-8). The surface of the air table contains an array of small holes through which air flows upward from the interior of the table when the air supply is turned on. When the air table is level, the board floats on the film of air flowing between the air table and the board, and the board is thus free to move sideways in response to any horizontal force. If the board is held stationary with one hand while the bob is pulled to the right with the other hand, and then both the board and the bob are released simultaneously, it will be observed that the bob swings to the left and at the same time the board moves to the right. A bit later the bob swings back to the right while the board moves to the left, and as the pendulum continues to oscillate the board and the bob always move in opposite directions.

Let us consider the forces causing these motions. For simplicity, we neglect the fact that the pendulum bob is actually moving in a circular arc (relative to its point of support) and consider its motion to be essentially horizontal. Just after the bob is released and while it is accelerating toward the left, it must be acted on by a net force directed toward the left; at the same time, the board is being accelerated toward the right so it must be acted on by a net force toward the right. These forces are transmitted through the string by which the bob is fastened to its point of support.

Figure 42-8. A pendulum bob suspended from a rod mounted on a flat board, which is floating on a cushion of air supplied by an air table. When the bob moves in one direction, the board moves in the opposite direction.

The remarkable fact that we observe is that both forces occur simultaneously; we can't have one without the other. Using the apparatus under the conditions described above, there is no way to place the pendulum in oscillation without having the board which supports it oscillating in the opposite direction at the same time. When the string pulls to the right on the bob, at the same time it pulls to the left on the support attached to the board. The two forces which cause the bob and the board to oscillate always occur as a pair, and the two members of this pair always have opposite directions.

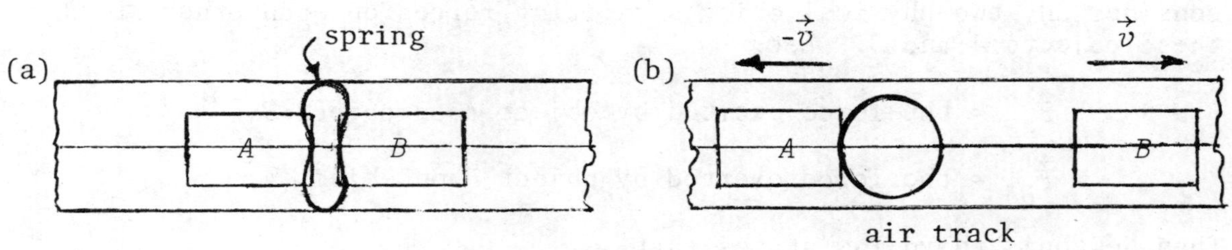

Figure 42-9. A top view of two gliders (*A* and *B*) on an air track. In (a), the air supply is turned off and the gliders are at rest with a large circular spring compressed between them into the shape of a figure eight. In (b), the air supply is turned on and the gliders are moving apart from each other with opposite velocities, while the spring has returned to its normal circular shape.

Does the conclusion drawn from observing the pendulum apply to other physical phenomena as well? Imagine a level air track with its air supply initially turned off, on which are placed two gliders touching each other. One of the gliders is fitted with a large circular spring which can readily be compressed. When the gliders are placed together with the spring between them, the spring is pushed into a figure-eight shape and is ready to spring back to its normal circular shape when the gliders are free to move apart. When the air supply is turned on, we observe that the gliders are accelerated away from each other in response to the forces applied to both of them by the spring. When the gliders in this experiment are both free to move we always observe that *both* gliders are set in motion, and we conclude that when one glider is being accelerated by a force toward the right the other glider is being accelerated by a force toward the left. In this experiment, once again there is no way to arrange for the spring to apply a force only to one glider but not to the other.

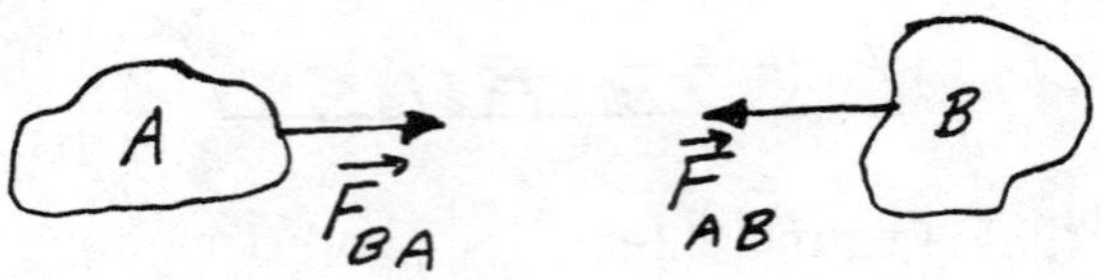

Figure 42-10. Newton's third law states that when any object *A* exerts a force $\vec{F}_{AB}$ on another object *B*, at the same time object *B* exerts a force $\vec{F}_{BA}$ on object *A*. $\vec{F}_{AB}$ and $\vec{F}_{BA}$ are equal in magnitude but opposite in direction.

Additional experiments show that the conclusions about forces drawn from the two examples described above reflect a physical law which applies to every physical phenomenon we examine:

> *Forces always occur in pairs, with the two forces in a pair acting in opposite directions. The two forces in a pair never act on the same object, but always act on two different objects.*

This is Newton's third law of motion. To express it mathematically, consider any two physical objects exerting forces on each other; call these objects *A* and *B*. Let

$\vec{F}_{AB}$ = the force exerted by object *A* on object *B*;

$\vec{F}_{BA}$ = the force exerted by object *B* on object *A*.

Then Newton's third law states that

★ $$\vec{F}_{AB} = -\vec{F}_{BA}.$$

The negative sign in this equation indicates that the two forces in a pair act in opposite directions. The qualitative experiments described above do not themselves prove that the two forces in a pair are equal in magnitude, but this equality can be shown by precise quantitative experiments. The objects *A* and *B* may be any two physical objects, and the forces may be pushes, pulls, friction, electrical forces, or any other kind.

Worked Example 42-4

A car is pulling a trailer on a level road. The hitch of the car exerts a horizontal pulling force of 400 newtons on the trailer.

(a) Consider the pulling force of the car on the trailer as one member of a pair. What is the magnitude of the other force of this pair, and on what object does it act?

(b) If the two forces of this pair are equal and opposite, how can the car move?

Solution

(a) In analyzing pairs of forces using Newton's third law, it is helpful to keep in mind the illustration in Figure 42-10. One of the forces should be described in a statement similar to the following: "The car exerts a force of 400 newtons on the trailer in the forward direction." Then the other force in this pair can readily be identified by simply interchanging the names of the two objects involved (corresponding to *A* and *B* in Figure 42-10), and remembering that the second force must have a direction opposite to the first. Following this guideline, the companion statement which describes the second force becomes: "The trailer exerts a force of 400 newtons on the car in the backward direction."

(b) The two members of an action-reaction pair can never cancel or offset each other, because they always act on two different objects. *Only one* of the two forces described above acts on the car. There *is* another force on the car, however, namely the forward force of the road on the rear wheels (see Sections 42.6 and 42.7 for a further discussion). When this forward force equals the backward forces on the car (exerted by the trailer and by air drag), the car has a constant velocity. When the forward force exceeds the backward forces, the car accelerates in the forward direction.

Newton's third law is sometimes expressed in terms of "action" and "reaction:" *For every action there is an equal and opposite reaction.* This formulation may be useful as long as it is remembered that *action* and *reaction* apply only to physical forces. This physical law cannot properly be invoked as though it applied to social, economic and political phenomena, although it is sometimes misused in that way. Applied to a physical phenomenon, however, *any* force may be referred to as the action force, and the other member of its pair is the reaction to it. The two forces in any action-reaction pair always act simultaneously.

Questions and Exercises

Q 42.5-1 If action and reaction forces are equal in magnitude but opposite in direction, why don't they cancel each other?

Q 42.5-2 A lamp suspended by a lamp cord feels a downward force because of its weight. Which one of the following best describes the reaction to the weight?
(a) the upward force of the cord on the lamp.
(b) the downward force of the lamp on the cord.
(c) the upward force of the lamp on the earth.
(d) the downward force of the earth on the lamp.

Q 42.5-3 What is wrong with the following statement?
"The rear wheel of a bicycle pushes horizontally on the road,

and the road pushes horizontally on the wheel. By Newton's third law, these forces are equal and opposite. Therefore, the bicycle can never accelerate."

Q 42.5-4 A moose pulls a log with force $\vec{F}$; the log, by Newton's third law, pulls the moose with force $-\vec{F}$. The total of these forces is thus zero. Is there any hope that the moose can ever move? Why?

Q 42.5-5 Look at the list of everyday forces in your answer to Q 42.1-1, on page 42-3.
(a) Which of these forces could be the reaction to a push?
(b) Which of these forces could be the reaction to a pull?
(c) Which of these forces could be the reaction to friction?
(d) Which of these forces could be the reaction to weight?

Q 42.5-6 An apple is falling toward the ground. Adam says: "The apple is accelerating downward." Eve says: "No, the ground is accelerating upward."
(a) Is there a meaningful distinction between the two statements?
(b) If not, how do you explain the fact that only one of the two accelerations is measurable, when the forces on the apple and earth are equal in magnitude?

Q 42.5-7 A very heavy truck collides with a very light car.
(a) Assuming the two bodies are made of the same kind of sheet metal, which will be crumpled more?
(b) Who is likely to be hurt more, the person in the truck, or the person in the car? Explain carefully.

Q 42.5-8 What force is indicated by the reading of a bathroom scale? Hint: To see that "weight" is not the correct answer in the most general sense, consider what happens when you grasp at the sides with both hands and squeeze on the scale.

E 42.5-1 An apple having a mass of 160 grams falls to the earth.
(a) What are the magnitude and direction of the force causing this motion?
(b) What are the magnitude and direction of the "reaction" to this force on the apple?
(c) What is the acceleration of the earth in response to this reaction force?

Check your understanding of the material above by testing yourself on the minimum learning goals at the beginning of this section.

42.6 Free-Body Diagrams

Minimum Learning Goals. (1) Given a description of a simple physical situation, be able to draw the free-body diagram of a designated object in this situation. For example:

1 A batter hits a pitched ball on a long fly into the outfield. Draw a free-body diagram of the ball at the instant when it is hit by the bat. (Answer, page 42-42, Line 18)

2 A girl is standing on a bathroom scale in an elevator which is accelerating upward. Draw a free-body diagram of the scale during this motion. (Answer, page 42-43, Line 26)

(2) Given a description of an elevator or an object in an elevator, and the acceleration of the elevator, be able to compute an unknown force acting on the object, constructing and using a free-body diagram if necessary. For example:

1 A 500-kg elevator has two forces applied to it, the force of gravity which has a magnitude w, and the force of the cable pulling on the elevator, called the *cable tension*, T. What is the cable tension when the elevator is:
(a) moving at a constant speed?
(b) accelerating downward at 6 m/s^2?
(c) accelerating upward at 2 m/s^2?
(Answer, page 42-42, Line 6)

2 A 70-kg man riding an elevator has a weight of 686 N. What is the upward force of the elevator on him when the elevator is:
(a) moving up at a constant speed?
(b) accelerating downward at 4 m/s^2?
(c) accelerating upward at 3 m/s^2? (Answer, page 42-42, Line 15)

Newton's third law is a fundamental physical law, and it is useful in analyzing a wide variety of problems in statics and dynamics. When it is necessary to find an unknown force, for example, it is often helpful to find first the reaction force to the unknown. In working out problems of this sort, it is advantageous to draw a *free-body diagram* for one or more of the objects or bodies in the problem.

Whenever one is working out a problem in physics, it is helpful to make a sketch of the situation which shows the objects of interest and their relationship to one another. In contrast to such a sketch showing all the objects in a problem, in a free-body diagram *only one* object is shown together with all of the *forces* acting on this object. Such a diagram thus serves to isolate for study one body or object in the problem; the other objects are replaced by the forces which they exert on the one selected for analysis. The diagram shows the selected body *free* from contact with the other objects which interact with it, and shows *all the forces* acting on the body as arrows acting in the proper direction and having correct labels. When the arrows are drawn with lengths which indicate the approximate relative strengths of the forces, the free-body diagram is an important aid in visualizing the pattern of forces acting on the body. A proper free-body diagram does *not* need to show the velocity or acceleration of the object, but it is sometimes useful to indicate the object's acceleration by drawing a labeled arrow near the free-body diagram.

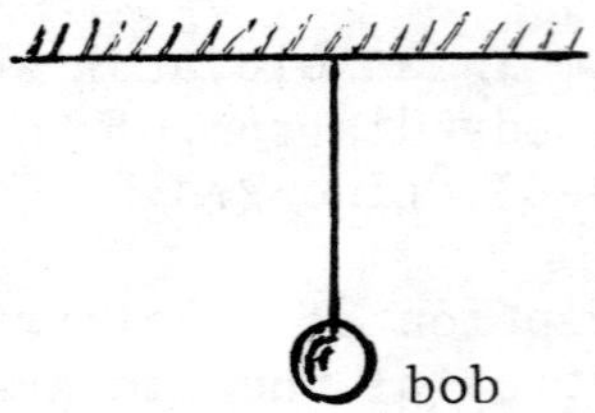

Figure 42-11. A *sketch* of a simple pendulum showing the bob, cord, and ceiling where the cord is attached. No forces are shown in this pictorial drawing.

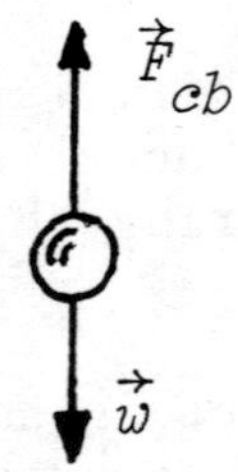

Figure 42-12. A *free-body diagram* of a simple pendulum, in which the bob is the only object shown. The diagram also includes an arrow for each force acting on the bob.

As a simple example, consider a pendulum (or lamp, or mobile) hanging form the ceiling. The sketch of the situation in Figure 42-11 includes the ceiling, the pendulum bob, and the cord which supports the bob. The forces acting on the bob are shown on a free-body diagram of the bob, Figure 42-12. These forces include the weight $\vec{w}$ of the bob, and the upward force of the cord on the bob, $\vec{F}_{cb}$. When the bob is at rest in its equilibrim position, these two forces are equal in magnitude but opposite in direction, so that for the bob $\Sigma\vec{F} = 0$ and the bob remains at rest. These two forces are *not*, however, the members of the same action-reaction pair; they cannot be, because they are acting on the same object. The reaction force to the weight of the bob is the upward force of the bob on the earth. The reaction to the force $\vec{F}_{cb}$ is a downward force $\vec{F}_{bc}$, the force of the bob on the cord. These two reaction forces could be shown in free-body diagrams of the earth and the cord, respectively, but they do not belong in the free-body diagram of the bob.

To introduce a second example, consider the question: What is required to accelerate your car in the forward direction, from a position of rest on a level street? Many people would answer, "You need to turn on the car's engine, put it in gear, release the brake and step on the accelerator pedal." While all of this is correct, it does not answer the question from the standpoint of physics, because the engine and transmission are internal parts of the car and as such can only exert *internal* forces on the car. In order to accelerate, Newton's first law tells us that an *external* force is required. What object can exert the required external force on the car?

We can arrive at the answer with the help of a free-body diagram of the car. As shown in Figure 42-13, the forces of the car include its weight $\vec{w}_c$ and the upward forces $\vec{F}_{rt}$ of the road on the tires which support the car. These, however, are vertical forces which cancel each other, so there must be at least one more force on the car, if it is to accelerate in the forward direction. This forward force can only be exerted by the road on the tires, more specifically on the tires of the driving wheels. This force is labeled $\vec{F}_{rc}$ on the free-body diagram. Because the vertical forces on the car cancel each other, the resultant force on the car is $\vec{F}_{rc}$, and it is this forward force which accelerates the car. It arises from friction between the road and the tires, discussed further in Section 42.7.

Figure 42-13. A free-body diagram of a car accelerating forward on a level road. The forces on the car include its weight $\vec{w}_c$, the upward forces of the road on the tires $\vec{F}_{rt}$, and the forward force of the road on the driving wheels, $\vec{F}_{rc}$. The retarding force of air drag is omitted from the diagram.

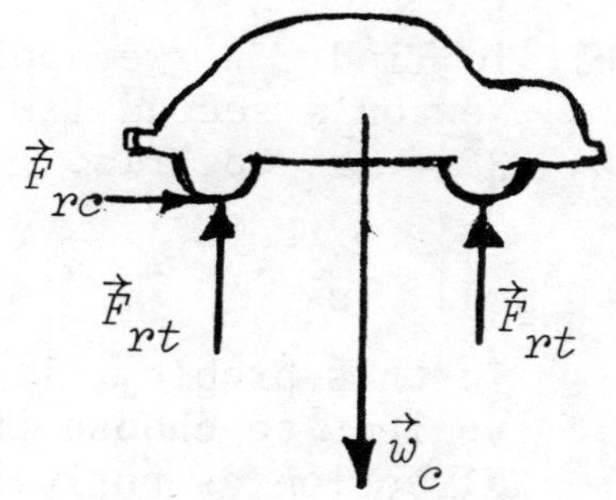

A free-body diagram of the section of the road under the car would include a backward force $\vec{F}_{cr}$ of the car on the road, which is the reaction to $\vec{F}_{rc}$. Does $\vec{F}_{cr}$ really exist? Does the car actually push backward on the road? In our ordinary experience roadways don't move, so we do not think of a force such as $\vec{F}_{cr}$. However, when the road surface *is* free to move, as for example when the cars's tires are resting on loose sand or gravel or snow, then we know from experience that when the car begins (or tries) to move forward, the gravel or snow is pushed backward by the driving wheels of the car. When the road surface is firm and dry, however, it does not move. This is a result of a forward force of the ground on the road, which balances $\vec{F}_{cr}$ and allows the road to remain at rest. In our study of physics we do not need to consider the effects of these forces acting on the roadway, but a civil engineer who is designing highways needs to consider all of these and some more besides.

Worked Example 42-5

Kris is standing on a bathroom scale in an elevator accelerating upward at 1.2 m/s^2. Her mass is 40 kg.
(a) Draw a free body diagram of Kris during this motion.
(b) What is the upward force of the scale on Kris?
(c) What is the downward force of Kris on the scale?
(d) What force does the scale indicate on its dial?

Solution

(a) The free-body diagram must show only Kris, not the elevator or the scale. The forces on Kris include her weight $\vec{w}$ acting downward, and the upward force of the scale on Kris which we will label $\vec{F}_{sk}$. In the diagram, an attempt is made to draw the force arrows in a way that indicates the points at which they act. Kris's weight acts down on a point (sometimes called the "center of gravity") near the center of her body, and the scale pushes upward on her feet. Knowing that Kris is accelerating upward, it is realistic to draw the arrow for $\vec{F}_{sk}$ longer than the arrow for $\vec{w}$.

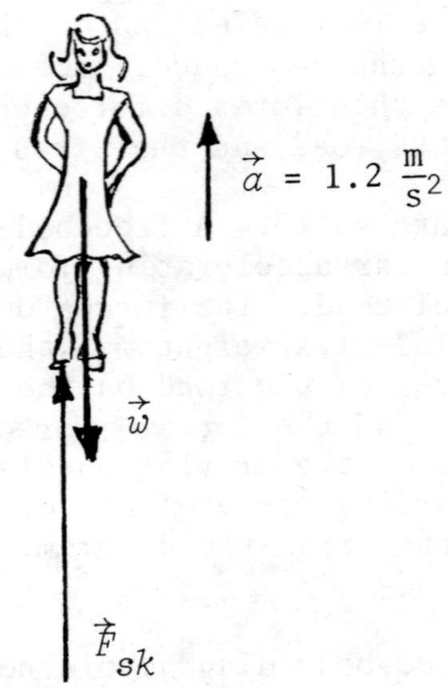

(b) To find $\vec{F}_{sk}$, we apply Newton's second law, $\Sigma\vec{F} = m\vec{a}$, to Kris.

$$\vec{F}_{sk} + \vec{w} = m\vec{a}.$$

In this problem, it is convenient to choose the upward direction as positive. Then the vector quantities (force and acceleration) may be written as scalars, indicating upward-directed quantities as positive and downward vectors as negative.

$$F_{sk} - w = ma$$

The weight of Kris is $mg = 40\text{ kg} \times 9.8\text{ m/s}^2 = 392$ newtons. We solve the equation for F_{sk}, and then substitute:

$$F_{sk} = ma + w = 40\text{ kg} \times 1.2\text{ m/s}^2 + 392\text{ N} = 440 \text{ newtons, upward.}$$

(c) According to Newton's third law, the upward force of the scale on Kris must be equal in magnitude to the downward force of Kris on the scale. Therefore, Kris pushes down on the scale with a force of 440 newtons.

(d) Assuming the scale gives accurate readings, its dial indicates the force pressing down on the scale, 440 newtons.

Some physical systems have a motion which is essentially one-dimensional, even though the system contains two or more objects traveling in different directions. Two systems of this sort are pictured in Figure 42-14. In each case the string connecting m_1 and m_2 passes over a pulley, which may often be assumed to operate without friction. If the string has a fixed length, each mass travels the same distance from its starting position. A single variable, x, may therefore be used to describe the positions of both masses in the system. When Newton's second law is applied to separate free-body diagrams for m_1 and m_2, one obtains simultaneous equations which can be solved to find the acceleration of the system and the tension in the string.

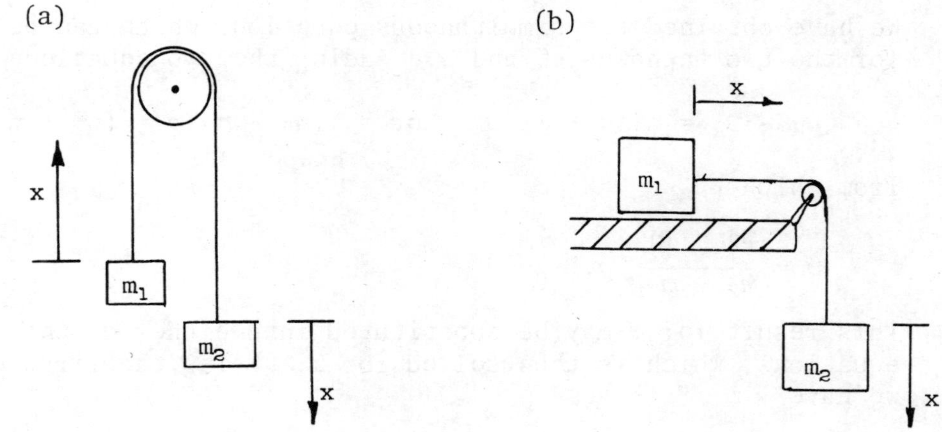

Figure 42-14. (a) Two unequal masses, suspended from the ends of a string looped over a pulley, have the same acceleration. This system, known as Atwood's machine, enables one to measure the acceleration of free fall. (b) Two masses, connected by a string passing over a pulley, have the same acceleration.

Worked Example 42-6

For Atwood's machine (Figure 42-14):
(a) Compute the acceleration of the system.
(b) Compute the tension in the string.

Solution

(a) Assuming m_2 is larger than m_1, m_2 accelerates downward while m_1 accelerates upward at the same rate, a. We draw separate free-body diagrams for m_1 and m_2.

Assuming that the pulley is frictionless and massless, the tension T is constant in magnitude throughout the string. Applying Newton's second law to the free-body diagram for m_1, and choosing the upward direction as positive, we have

$$\Sigma\vec{F} = m_1\,\vec{a} \quad \text{so} \quad T - w_1 = m_1\,a.$$

Applying Newton's second law to the diagram for m_2, and choosing the downward direction as positive, we have

$$\Sigma\vec{F} = m_2\,\vec{a} \quad \text{so} \quad w_2 - T = m_2\,a.$$

We have obtained two simultaneous equations which can be solved for the two unknowns, T and a. Adding the two equations gives

$$w_2 - w_1 = m_1\,a + m_2\,a \quad \text{or} \quad (m_2 - m_1)g = (m_2 + m_1)\,a,$$

from which

$$a = \frac{m_2 - m_1}{m_2 + m_1}\,g.$$

(b) This result for a may be substituted into either of the simultaneous equations, which is then solved for T. Using the first equation, we have

$$T - m_1\,g = m_1\,\frac{m_2 - m_1}{m_2 + m_1}\,g\,,$$

$$T = m_1\,g\left[\frac{m_2 - m_1}{m_2 + m_1} + 1\right] = m_1\,g\left[\frac{m_2 - m_1 + m_2 + m_1}{m_2 + m_1}\right],$$

$$T = 2\,m_1 m_2\,g/(m_2 + m_1).$$

The result obtained for a shows that the acceleration of the system is always less than g, and that it can be made small by choosing m_2 approximately equal to m_1. The equations obtained directly from the free-body diagrams show that the tension is always larger than the weight of the smaller mass, but smaller than the weight of the larger mass.

Questions and Exercises

Q 42.6-1 (a) What forces act on an object at rest on the floor?
(b) What is the reaction to each of these forces?

Q 42.6-2 (a) What forces act on a lamp suspended by a cord?
(b) What is the reaction to each of these forces?

Q 42.6-3 A twenty-newton lamp is hanging from a cord fastened to the ceiling. The cord itself has negligible weight.
(a) What are the magnitude and direction of the force exerted by the cord on the lamp?
(b) What are the magnitude and direction of the force exerted by the cord on the ceiling?

Q 42.6-4 (a) What do you feel in your legs when you are standing in an elevator which accelerates rapidly upward?
(b) What do you feel in an elevator which accelerates rapidly downward?
(c) Explain these experiences, using Newton's laws of motion.

Q 42.6-5 A real pulley has some mass and some friction.

(a) If an Atwood's machine is used to measure g, how do these factors affect the experimental value obtained for g?

(b) How do these factors affect the relative tension which the string applies to the two masses in Atwood's machine?

E 42.6-1 Draw a free-body diagram of a baseball thrown vertically upward, after the ball is thrown but while it is still rising. Neglect air resistance.

E 42.6-2 Suppose you are in an elevator accelerating upward at 0.1 g.

(a) If your mass is 60 kg, with what force do you push down on the floor?

(b) With what force does the floor push up on you?

E 42.6-3 Big Bertha, the fat lady in the circus who weighs 400 pounds, has an appointment with her doctor on the fifth floor of the Medical Arts Building. As she enters the elevator, she steps on a scale (to weigh herself) and remains there until the elevator is nearly up to her floor.

(a) Find Bertha' correct weight in newtons. (You may wish to work the entire problem in SI Units, then convert your answer to pounds.)

(b) As long as Bertha remains in the building, does her actual weight remain constant, or could it change? Explain.

(c) Is Bertha's mass a constant, or does it change? Explain.

(d) What force does a scale reading indicate? Hint: to see that "weight" is not the correct answer, consider what happens when you grasp a scale at the sides with both hands and squeeze down on the scale.

(e) Draw a free-body diagram which shows the forces on Bertha while the elevator accelerates upward at 1.0 m/s^2. What is Bertha's apparent weight (the scale reading) under these conditions?

(f) While the elevator continues upward at constant velocity, what is the scale reading?

(g) The doctor tells Bertha she is overweight, gives her some reducing pills (to reduce her appetite), and she takes one pill immediately. To her amazement, as she begins her descent in the elevator while standing on the scale, she has already "lost" some weight. What is the scale reading while the elevator accelerates downward at 1.0 m/s^2?

(h) Bertha decides to return at once to the doctor and thank him for his marvelous pills, so on the ground floor she remains on the scale in the elevator and asks the operator to "hurry back up to five." What is the maximum upward acceleration of the elevator which will not damage the scale, whose breaking strength is 500 pounds? (Hint: What are the forces acting on Bertha?)

E 42.6-4 Two masses used in an Atwood's machine have values of 200 grams and 220 grams. Compute the acceleration of the system, and the tension in the string.

E 42.6-5 Some students used Atwood's machine with masses of 190 and 200 grams, and measured an acceleration of 2.47 cm/s^2 for this system.
(a) Compute the acceleration of free ball, using these data.
(b) How would you explain the variation of the result from the accepted value for g?

E 42.6-6 In an effort to obtain a controlled acceleration of 0.1 m/s^2, an Atwood's machine is used in which each mass is initially 0.200 kg. Assuming the pulley is massless and frictionless, how much mass must be added to one side to achieve the desired acceleration?

E 42.6-7 A frictionless Atwood's machine is used with masses of 0.110 kg and 0.100 kg. How long does it take for m_2 to descend 1.00 meter, starting from rest?

E 42.6-8 For the system pictured in Figure 42-14(b), assume there is no friction between m_1 and the surface it slides on, and that the pulley is frictionless and massless.
(a) Draw free-body diagrams of the two masses.
(b) Use these diagrams to show that the acceleration of the system is $a = g\, m_2/(m_1 + m_2)$.
(c) Compute the tension in the string when the masses are moving.

Check your understanding of the material above by testing yourself on the minimum learning goals at the beginning of this section.

42.7 Friction and Coefficients of Friction

<u>Minimum Learning Goals</u>: (1) Given a description of a physical situation, be able to state the direction of the force of friction on a designated object, and whether this is an example of static friction, sliding friction, or rolling friction. For example:

1 A packing crate is at rest on the slope of an inclined ramp.
(a) What is the direction of the force of friction on the crate?
(b) Is this an example of sliding friction, rolling friction, or static friction? (Answer, page 42-42, Line 3)

2 A billiard ball is slowing as it rolls without slipping and without colliding on a pool table.
(a) What is the direction of the frictional force on the ball?
(b) Is this an example of sliding friction, rolling friction, or static friction? (Answer, page 42-42, Line 12)

3 An auto is skidding to a stop on an icy surface with its wheels locked.
(a) What is the direction of the force of friction on the tires?

(b) Is this a case of sliding friction, rolling friction, or static friction? (Answer, Page 42-42, Line 8)

4 An auto accelerates in the forward direction on a smooth level road without spinning its wheels on the road.
(a) What is the direction of the frictional force on the tires?
(b) Is this an example of sliding friction, rolling friction, or static friction? (Answer, page 42-42, Line 10)

(2) Given an adequate description of a physical situation, be able to calculate the force of static friction acting on a specified object. For example:

1 A loaded crate having a total mass of 60 kg is at rest on a ramp whose surface is inclined at 20° to the horizontal. What is the frictional force which keeps the crate from sliding down the ramp? (Answer, page 42-42, Line 13)

2 An auto starts from rest on a level road and reaches a speed of 20 m/s in 5 seconds. The auto's mass is 1500 kg. Assuming the acceleration is uniform, what is the total frictional force of the road on the car? (Answer, page 42-43, Line 2 4)

(3) Given an adequate description of a physical situation, be able to calculate the force of sliding friction acting on a specified object. For example:

1 A loaded crate having a total mass of 60 kg is sliding down a ramp whose surface is inclined at 40° to the horizontal, gaining speed at $0.8\ m/s^2$. What is the frictional force which opposes the motion of the crate down the ramp? (Answer, page 42-43, Line 27)

2 An auto having a mass of 1500 kg slides to a stop on a level icy roadway, taking 12 seconds to stop from an initial speed of 15 m/s. What is the frictional force of the ice on the car? (Answer, page 42-43, Line 25)

When a book is given a brief push so it begins to slide along a table top, it slows down and soon comes to rest. If your car runs out of gas on a level road, it rolls to a stop even if you do not apply the brakes. These examples illustrate the general observation that whenever two objects in contact are moving past one another, there is always a force which opposes this motion. The general name given to this kind of force is *friction*. It occurs as sliding friction or rolling friction when two solid objects move past each other, as air drag when an object is moving through the air, and as liquid drag when a solid object moves through or on the surface of a liquid. Under some conditions, friction also occurs as static friction between two objects at rest relative to each other. When two solid objects are separated by a film of air (as in the case of an air track or air table) or a film of liquid (for example, lubricating oil), there is no friction between the solid objects themselves although there is fluid drag between each object and the adjacent fluid film. We consider here some of the basic properties of friction forces, confining our attention to friction between solid objects in contact with each other.

When we consider the *direction* of friction forces, in these examples the deceleration of the sliding book and of the rolling car are parallel to the surfaces on which they are moving. From this we conclude that a friction force acts parallel to the surfaces, at the points where two objects are in contact with one another. In these two examples, friction acts to *oppose* the motion of the book and of the car. We find other examples, however, where the friction *aids* motion or even *causes* motion, as we now illustrate. Referring to Worked Example 42-2 on page 42-15 it is the girl's push that gives the wagon its acceleration (see the figure), but she does not push the boy directly and he accelerates only because of some other force. If he is sitting in the middle of the wagon as shown, the force which accelerates him from rest is the friction force of the wagon on him. If this friction force were not present he would slide backward in the box of the wagon when she pushed the wagon forward. In Section 42.6 we considered a car accelerating from rest on a level road, and found that this motion requires a forward force exerted by the road on the car (Figure 42-13. This forward force $\vec{F}_{rc}$, and the backward force $\vec{F}_{cr}$ which is the reaction to it, are both forces of friction. When the friction is negligible, as on an icy road surface, it is very difficult for the car to accelerate under its own power, and also difficult for the car to decelerate rapidly using its brakes. So, we are quite dependent on friction to set a car into motion, and also to bring it to a stop safely after it is moving.

Combining the properties of friction we have discussed leads to the following definition:

> *Friction* is a contact force, acting parallel to the surface between two objects, which tends to oppose the relative sliding motion of one object past the other.

Under what conditions is there friction between two objects at rest relative to each other? Consider first a box at rest on a level table, with no horizontal force applied to the box. In this case there is no friction force on the box, for if there were it would act in some horizontal direction and would cause the box to begin moving in that direction. The only forces on the box are its weight pulling down, and the opposite force of the table pushing up. This upward force is often called the *normal force* of the table on the box, where "normal" has the meaning of "perpendicular," i.e., perpendicular to the surface of the table.

Next consider a box on a level table, but this time subjected also to a horizontal push (Figure 42-15). As before, the weight is just balanced by the upward normal force. The push or applied force $\vec{F}_a$ tends to set the box in motion to the right, but the friction force $\vec{F}_f$ opposes this motion and thus acts toward the left on the box, as shown in the free-body diagram. If the applied force is small, its magnitude is equalled by that of the friction

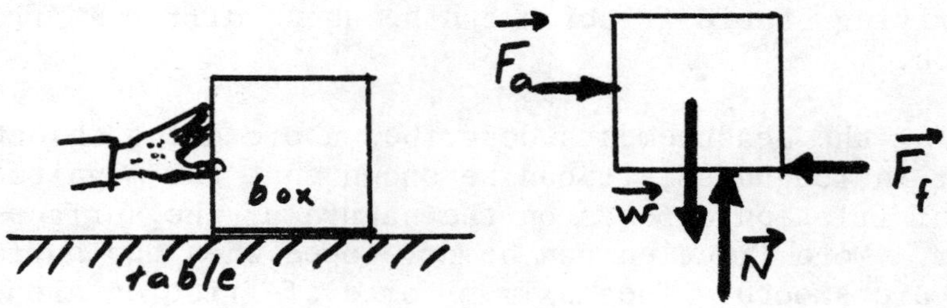

Figure 42-15. A push is applied to a box resting on a table. The free-body diagram above shows the four forces on the box: the push $\vec{F}_a$, the friction force $\vec{F}_f$, the weight $\vec{w}$, and the normal force $\vec{N}$ of the table supporting the box.

force so the box does not move. If the applied force is slowly increased, the friction force increases by the same amount and thus prevents the box from moving, but after the friction force has grown to some maximum value it is not able to increase further. When the applied force exceeds this maximum force of friction, the box begins to move to the right. The behavior just described can be demonstrated by pulling (instead of pushing) the box, using a spring balance to indicates the strength of the pull.

Figure 42-16. When two solid objects are in contact with each other and an applied force F_a acts on one of them as shown in Figure 42-15, the resulting friction force F_f depends on the applied force.

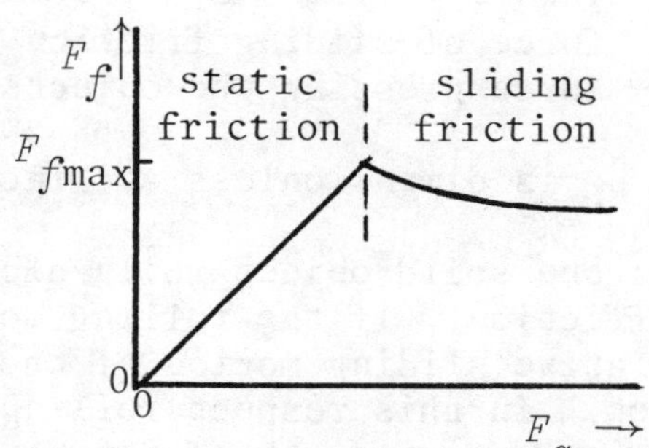

If the friction force F_f pictured in Figure 42-15 is plotted as a function of the corresponding applied force, the result obtained may be something like the graph in Figure 42-16. When F_a is small enough so the box does not move, $F_f = F_a$; this is the regime of static friction. When F_a exceeds the maximum friction force which can occur for the two objects, the box begins to move and it is found that the friction force drops below its maximum value; this is the regime of sliding friction.

If the box in Figure 42-15 is pushed toward the left, the friction force on it acts toward the right, with a magnitude as shown in Figure 42-16. The force of static friction between two objects adjusts itself in magnitude and direction in such a way as to balance any applied forces and thus maintain the state of relative rest between the two objects. For any given situation, however, there is a maximum magnitude which the force of static friction can have; if the sum of the other forces exceeds that value, the objects "break away" from each other and motion ensues. The physical phenomena underlying static friction can be quite different from those for sliding friction.

By repeating the measurements described above using objects whose surfaces differ in roughness, it can be shown that the magnitude of the maximum force of friction depends on the nature of the surfaces in contact with each other. More friction can be developed when the surfaces are rough than when they are smooth. The maximum force of friction also increases with an increase in the normal force pressing the surfaces together; for a box on a level table, this normal force is the weight of the box and its contents. The maximum force of static friction is approximately proportional to the normal force.

★ The *coefficient of static friction*, μ_s, is the ratio of the maximum force of static friction between two objects to the normal force pressing the objects together. $\mu_s = F_{f\ max}/N$.

μ_s is a dimensionless quantity, smaller than unity for most surfaces, which increases with surface roughness.

When a box is in motion along the table top, we have sliding friction or kinetic friction. The force of sliding friction is also approximately proportional to the normal force.

★ The *coefficient of kinetic friction*, μ_k, is the ratio of the force of sliding friction between two objects to the normal force pressing the objects together.

Like μ_s, μ_k is dimensionless and increases with surface roughness.

When one solid object rolls along the surface of another, we have rolling friction. If the rolling motion takes place without slipping, there is no relative sliding motion of the two surfaces at the instantaneous point of contact. In this respect rolling friction is similar to static friction. However, in most cases the friction force that opposes rolling motion between two objects is much smaller than the friction force for sliding motion between the objects.

Worked Example 42-7

Matt is pulling his wagon forward on a level driveway with a force of 20 newtons. The wagon has a mass of 15 kg, and is accelerating in the forward direction at 0.3 m/s^2. Find the direction and magnitude of the friction force on the wagon, and the coefficient of friction.

Solution

The resultant force on the wagon equals the product of the wagon's mass times its acceleration, which is 15 kg × 0.3 m/s^2 = 4.5 newtons. Matt's pulling force is much larger than this, so the wagon's motion must be opposed by a backward friction force, and this friction must be exerted by the driveway on the tires of the wagon. The magnitude of the resultant force is the difference between the pulling force and the friction force. The friction force, therefore, is the difference between the pulling force and the resultant force, which is 20 newtons - 4.5 newtons = 15.5 newtons, in the backward direction.

The coefficient of friction $\mu = \frac{F_f}{N} = \frac{F_f}{mg} = \frac{15.5 \text{ N}}{15 \text{ kg} \times 9.8 \text{ m/s}^2} = 0.11.$

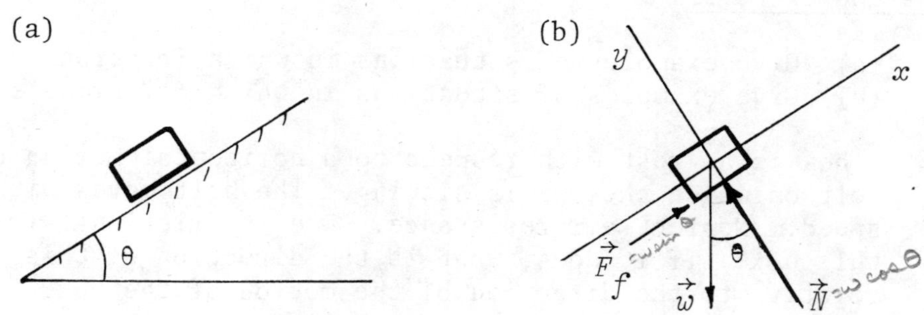

Figure 42-17. (a) A box at rest on an inclined plane. (b) A free-body diagram showing the forces on the box. The component of the weight parallel to the slope balances the friction force, while the component of the weight perpendicular to the slope balances the normal force.

For a box at rest on an inclined plane (Figure 42-17), the most convenient coordinate system is one having axes parallel and perpendicular to the plane. The force of the ramp on the box is at an angle to the surface. The component of this force parallel to the slope is the friction force $F_f = w \sin\theta$, while the component perpendicular to the slope is the normal force $N = w\cos\theta$. The resultant of the friction and the normal force just balances the weight of the box. If the angle θ between the plane and the horizontal is small, $w \sin\theta$ is small and the friction force is sufficient to prevent the box from sliding down the plane. As the angle θ is increased, however, $w\sin\theta$ increases and eventually exceeds the maximum force of static friction $F_{f\max} = \mu_s N$ which can occur for the box and the plane. The maximum angle for which the box remains at rest on the plane is the critical angle θ_c, at which the maximum friction force is

$$w \sin\theta_c = \mu_s N = \mu_s w \cos\theta_c .$$

Solving for μ_s, we obtain

$$\mu_s = \tan \theta_c.$$

With this equation, a plane having an adjustable angle of inclination may be used to measure coefficients of static friction.

Our discussion of friction has dealt with only a few aspects of the nature of friction forces. It must be remembered that friction is a macroscopic force, a way of describing and making sense out of observations on the behavior of macroscopic objects. When we describe the motions and interactions of individual atoms and molecules, there is no friction. Phenomena which involve friction at the macroscopic level are discussed in completely different terms when they are described on the microscopic level.

Questions and Exercises

Q 42.7-1 (a) Give examples of situations in which friction is a hindrance.
(b) Give examples of situations in which friction is a help.

Q 42.7-2 A box is at rest with respect to a horizontal moving conveyor belt on which the box is sitting. The belt moves with constant speed. Neglect air resistance. Does a frictional force act on this box? If it does, what is the direction of this force relative to the direction of the motion of the box?

Q 42.7-3 An automobile is accelerated from rest by the frictional force of the road on the ties. How do you reconcile this with the statement that friction between two objects always tends to decrease their relative motion?

Q 42.7-4 A heavy crate is standing at rest on a level floor. A man comes and leans against its side, but it still does not move. Discuss the frictional force (magnitude and direction) before the man leaned on the crate, and when he is leaning on it.

Q 42.7-5 Can a book resting on a surface be subject to a friction force if no push is applied to the book? Explain.

Q 42.7-6 When a bicycle wheel rolls without slipping on a smooth road surface, does the road exert a frictional force on the tire? If it does, is this static friction, or sliding friction?

Q 42.7-7 Suppose a metal casting is dropped onto a horizontal conveyor belt moving at constant speed which has a slippery, oil-covered surface. (a) Describe the motion of the casting relative to the belt. (b) What kind of friction does the belt exert on the casting just after the casting lands on the belt? (c) What kind of friction does the belt exert on the casting after the casting has reached a steady speed?

Q 42.7-8 Consider a small box resting near the center of a long flat stationary horizontal board. (a) Does the board exert a frictional force on the box under this condition? (b) Now suppose one end of the board is gradually raised so that the angle between the board and the horizontal increases. What kind of friction force initially acts on the box, and in what direction? How does this force depend on the angle of incline of the board? (c) At some angle of tilt, the box begins to slide down the board. Then what kind of friction acts on the box?

Q 42.7-9 A small rocket is launched straight upward and coasts to an altitude of 10 miles, from which it falls back to earth. The rocket is acted on by a drag force which opposes its motion. How does the time required for its upward motion compare with the time required for the rocket to fall back to earth? Explain your reasoning.

E 42.7-1 An automobile is pulling a trailer on a level road at a steady speed of 50 mph. The force of rolling friction on the trailer is 200 pounds. (a) What are the direction and magnitude of the horizontal force which the trailer exerts on the car? (b) What are the direction and magnitude of the net horizontal force which the road exerts on the car?

E 42.7-2 A 500-gram book slides across a horizontal floor with an initial speed of 200 cm/sec; it comes to rest in 4 sec. What is the force of friction between the book and floor?

E 42.7-3 A toy wind-up car having a mass of 0.3 kg is moving in a straight line on a level floor. It starts from rest and gains a velocity of 2.4 m/s in 2.0 seconds. Assume constant acceleration. What is the friction force acting on the car?

E 42.7-4 A box having a mass of 6.0 kg is sliding in the positive direction on a level floor. It comes to rest from an initial speed, v_0, in ten seconds, with a deceleration of 0.3 m/s^2.

(a) What is the magnitude of the friction force which caused the box to decelerate?

(b) What was the initial speed of the box?

(c) How far did the box travel during this sliding motion?

E 42.7-5 A mass is dragged across a horizontal surface by means of a spring balance, a device used to measure force. When dragged at any constant velocity, the balance shows a reading of 0.3 newtons. When pulled harder at a constant acceleration of 1.0 m/sec^2, the balance reads 2.5 newtons. What is the mass of the object?

E 42.7-6 For the situation described in Q 42.7-8, if the box begins to slide at a tilt angle of 52° what is the coefficient of static friction?

E 42.7-7 A wooden box weighing 50 lb rests on the floor. A minimum horizontal push of 30 lb is required to set the box in motion. Compute the coefficient of static friction between the box and the floor.

Check your understanding of the material above by testing yourself on the Minimum Learning Goal at the beginning of this section.

Answers to Minimum Learning Goals

Line 1 No

Line 2 The force of the cart pulling on the horse does *not* affect the acceleration of the cart. We need to know *only* the resultant force *on the cart* and its mass to find its acceleration.

Line 3 (a) parallel to the ramp in the "uphill" direction;
(b) static friction.

Line 4 $1.1 \times 10^{14}\ m/s^2$

Line 5 $F_x = 0.834 \times 10^{-19}$ N, $F_y = 4.73 \times 10^{-19}$ N

Line 6 (a) 4900 N; (b) 1900 N; (c) 5900 N

Line 7 (a) 1743 N; (b) 686 N

Line 8 (a) opposite to the velocity of each tire on the ice;
(b) sliding friction

Line 9 68,900 lbs at 6.6°

Line 10 (a) the forward direction, parallel to the car's velocity;
(b) rolling friction, because the tires do not slip on the road.

Line 11 100 N downward applied to (pulling) the spring by the mass.

Line 12 (a) opposite to the direction of the ball's velocity;
(b) rolling friction, because the ball does not slip on the table.

Line 13 201 newton, upward along the ramp.

Line 14 The reaction force is *not* applied to the object we are considering. Only those forces applied to the object affect its acceleration.

Line 15 (a) 686 N; (b) 406 N; (c) 896 N

Line 16 700 N applied to (pulling) the earth by the man.

Line 17 800 newtons in the backward direction.

Line 18 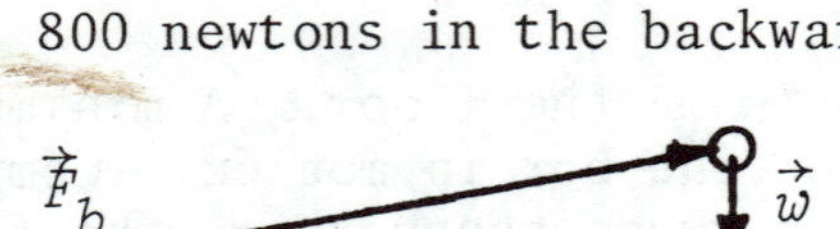

$\vec{F}_b$ = force of bat on the ball

Line 19 200 N

Line 20 The velocity is unaffected by these forces.

Line 21 2.28×10^{-2} N, sloping downward toward the east at an angle of 42.5° to the vertical.

Line 22 48.3 N at 358°

Line 23 9.1 N/C at an angle of 112.6°

Line 24 6,000 newton in the forward direction

Line 25 1.9×10^3 newton, opposite to the car's velocity

Line 26

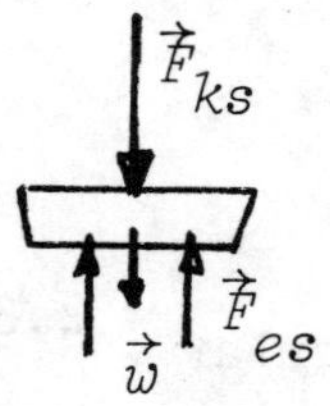

$\vec{w}$ = weight of scale

$\vec{F}_{es}$ = force of elevator on scale

$\vec{F}_{ks}$ = force of girl on scale

Line 27 330 newton, upward along the ramp

ANSWERS to Questions and Exercises in Chapter 42

Section 42.1

Q 42.1-2 No

Q 42.1-4 When the car turns left, the passenger continues with the same forward motion (relative to the ground) until the side of the car pushes on him from the right and causes him to begin moving to the left. His leaning abruptly to the right is his motion relative to the car, which is turning left.

Section 42.2

E 42.2-2 (a) $+4.19 \times 10^{-5}$ N; (b) $+2.13 \times 10^{-5}$ N

E 42.2-4 (a) $E_{parallel} = 0.86$ N/C; (b) $E_{perpendicular} = 2.35$ N/C

E 42.2-6 5.3 N at an angle of 62.9° CW with respect to the $+x$ axis.

E 42.2-8 6.88×10^3 N/C at an angle of 21.0° north of west.

E 42.2-10 39.8 lb at 58.5° cw with respect to north.

Section 42.3

Q 42.3-2 Mass; the *kilogram* is the mass of a particular cylinder of platinumiridium alloy which is preserved in a vault at Sevres, France.

Q 42.3-4 Newton's second law relates the resultant force on an object to the acceleration of the object for *any* value of acceleration. Newton's first law states this relationship for the special case when $a = 0$.

E 42.3-2 0.03 newtons

E 42.3-4 1.82×10^{-18} newtons

E 42.3-6 42 N

Section 42.4

Q 42.4-2 Weight is a gravitational force. Experiments which established the universal law of gravitation show that the weight of an object *is* proportional to its mass. The force of a stretched rubber band results from its internal structure; experiments show that this force depends on the amount by which the rubber band is stretched, and not on the mass of the object attached to the stretched band.

Q 42.4-4 In air a feather falls with $a < g$, because the force of air drag partially offsets the weight.

E 42.4-2 (a) 80 kg; (b) 784 newtons; (c) 80 kg; (d) 128 newtons

E 42.4-4 (a) +17.5 m; (b) +11 m/s; (c) + 10 m/s; (d) $-6\ m/s^2$; (e) 784 N; (f) -1264 N.

Section 42.5

Q 42.5-2 (c)

Q 42.5-4 The two forces act on two *different* objects, and do not cancel each other. The moose will move forward when the force of the ground on his feet exceeds the backward pull of the log.

Q 42.5-6 (a) Both statements are correct, but the acceleration of the earth is *much* smaller than the acceleration of the apple.
(b) The mass of the earth is much greater than the mass of the apple.

Q 42.5-8 The force pushing down on the top of the scale.

Section 42.6

Q 42.6-2 (a) (1) the weight of the lamp; (2) the upward force of the cord on the lamp.
(b) (1) the upward force of the lamp on the earth; (2) the downward force of the lamp on the cord.

Q 42.6-4 (a) Your legs feel weak, like they might collapse. (b) A light sensation, as if you are ready to float. (c) While accelerating upward rapidly, your trunk requires an upward force much greater than its weight, and this force must be supplied by your legs; by Newton's third law, your trunk also pushes down on your legs with a force much larger than normal. While accelerating downward rapidly, your trunk requires an upward force much smaller than its weight, and so your legs feel a load much smaller than normal.

E 42.6-2 (a) 647 N; (b) 647 N

E 42.6-4 $0.467\ m/s^2$; 2.05 N

E 42.6-6 4.12 grams

E 42.6-8 (a)

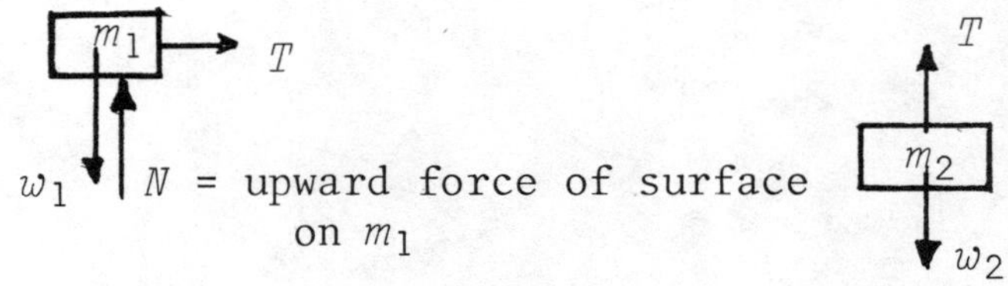

(c) $T = m_1 m_2\, g/(m_1 + m_2)$

Section 42.7

Q 42.7-2 No, because the resultant horizontal force on the box is zero.

Q 42.7-4 Before the lean, there is no frictional force on the crate. While the man leans on the crate he pushes it from the side, but the frictional force of the floor on the crate exactly counterbalances the push, so that the crate does not move.

Q 42.7-6 Yes; it is rolling friction, a form of static friction.

Q 42.7-8 (a) No. (b) Static friction, upward along the board. The force increases as the angle increases. (c) Sliding friction.

E 42.7-2 0.25 newton

E 42.7-4 (a) 1.8 N; (b) + 3 m/s; (c) 15 m

E 42.7-6 1.28

CHAPTER 43 PARTICLE MOTION IN THREE DIMENSIONS

Introduction

Having developed the concepts of kinematics and dynamics for one-dimensional motion in earlier chapters, we now apply these concepts to some examples of particle motion in two and three dimensions. Resolving the vectors of motion into cartesian components enables us to discuss the motion of projectiles. Equations for circular motion are developed and applied to some examples of uniform motion in a circle. The methods of calculus are used to compute the variables of motion in cases where the acceleration is not constant.

when you have completed your study of the material in this chapter, you should be able to answer questions and work exercises like the examples in the minimum learning goals assigned by your instructor.

43.1 Unit Vectors

Many of the topics presented in introductory physics can be considered adequately using only one dimension. To extend the principles of physics in a form applicable to more complex situations, however, it is often necessary to work in two or three dimensions. Vector quantities may have components in three mutually perpendicular spatial directions. In making calculations with vectors, it is often convenient to use unit vectors.

An vector $\vec{A}$ can be written as the product $\vec{A} = A\hat{a}$, where A is the magnitude of $\vec{A}$ and $\hat{a}$ is a *unit vector* in the direction of $\vec{A}$. A unit vector has a magnitude of unity and no physical units, but it does have a direction in space. In this text unit vectors are printed using a circumflex, ^, for example: $\hat{a}$, $\hat{x}$. The symbol $\hat{x}$ is sometimes read "x-hat."

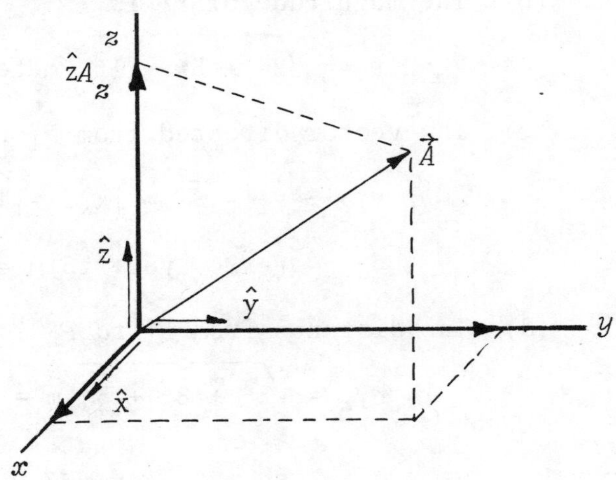

Figure 43-1. The unit vectors $\hat{x}$, $\hat{y}$, and $\hat{z}$ are directed along the $+x$, $+y$, and $+z$ axes, respectively. The three Cartesian components of a vector $\vec{A}$ are labeled $\hat{x}A_x$, $\hat{y}A_y$, and $\hat{z}A_z$.

Because most of our work with vectors in three-dimensional space uses a Cartesian coordinate system, it is convenient to introduce unit vectors $\hat{x}$, $\hat{y}$, and $\hat{z}$, parallel respectively to the x, y, and z coordinate axes (Figure 43-1). (Some texts write these unit vectors as $\hat{i}$, $\hat{j}$ and $\hat{k}$.) These unit vectors may be used to express any vector quantity in terms of its Cartesian components. From Figure 43-1, it is seen that $\vec{A}$ is the vector sum of three mutually perpendicular vectors, $\hat{x}A_x$ and $\hat{y}A_y$ and $\hat{z}A_z$. That is,

★ $$\vec{A} = \hat{x}A_x + \hat{y}A_y + \hat{z}A_z = A_x\hat{x} + A_y\hat{y} + A_z\hat{z}.$$

The position vector of a point having coordinates x, y, and z may be written

★ $$\vec{r} = \hat{x}x + \hat{y}y + \hat{z}z \quad \text{or} \quad \vec{r} = \hat{r}r.$$

The unit vector $\hat{r}$ has the direction of increasing r at any point; the direction of $\hat{r}$ is always radially away from the origin. Using the Pythagorean theorem, the magnitude of $\vec{r}$ is

★ $$r = (x^2 + y^2 + z^2)^{1/2}.$$

Worked Example 43-1

Point P_1 is located at $\vec{r}_1 = (\hat{x}4 + \hat{y}3 - \hat{z}6)$ meters, and point P_2 is located at $\vec{r}_2 = (\hat{x}2 - \hat{y}5 + \hat{z}1)$ meters.

(a) How far is P_1 from the origin?

(b) How far is P_2 from the origin?

(c) Using unit vectors, write the vector from P_1 to P_2.

(d) How far is point P_1 from point P_2?

Solution

(a) Using the Pythagorean theorem, the magnitude of $\vec{r}_1$ is

$$r_1 = \sqrt{4^2 + 3^2 + 6^2} \text{ meters} = \sqrt{61} \text{ meters} = 7.81 \text{ m}.$$

(b) The magnitude of $\vec{r}_2$ is

$$r_2 = \sqrt{2^2 + 5^2 + 1^2} \text{ meters} = \sqrt{30} \text{ m} = 5.48 \text{ m}.$$

(c) The vector directed from P_1 to P_2 may be written

$$\vec{r}_{12} \equiv \vec{r}_2 - \vec{r}_1 = [\hat{x}(2 - 4) + \hat{y}(-5 - 3) + \hat{z}(1 - -6)] \text{ meters}$$

$$= (-\hat{x}2 - \hat{y}8 + \hat{z}5)\text{m} = (-2\hat{x} - 8\hat{y} + 5\hat{z}) \text{ m}.$$

(d) The distance from P_1 to P_2 is the magnitude of $\vec{r}_{12}$, which is

$$r_{12} = \sqrt{2^2 + 8^2 + 5^2} \text{ m} = \sqrt{93} \text{ m} = 9.64 \text{ m}.$$

Questions and Exercises

Q 43.1-1 What are the two unit vectors normal to: (a) the xy-plane? (b) a plane surface parallel to the yz-plane?

Q 43.1-2 What is the direction of the vector $\vec{F} = -\hat{x}3$ newtons?

E 43.1-1 (a) Express the velocity vector $\vec{v}$ in terms of its cartesian components, using unit vectors.
(b) Express acceleration $\vec{a}$ in terms of its Cartesian components, using unit vectors.

E 43.1-2 $\vec{A} = \hat{x}A_x + \hat{y}A_y + \hat{z}A_z$ and $\vec{B} = \hat{x}B_x + \hat{y}B_y + \hat{z}B_z$. Write an analogous equation for $\vec{A} + \vec{B}$.

E 43.1-3 Write an equation for the position vector of the point located at $x = 3\text{m}$, $y = -2\text{m}$, $z = -4\text{m}$, using unit vectors. How far is this point from the origin?

E 43.1-4 Point P_1 is located at $\vec{r}_1 = \hat{x}3\text{m} - \hat{y}2\text{m} + \hat{z}5\text{m}$. Point P_2 is located at $\vec{r}_2 = \hat{x}4\text{m} + \hat{y}1\text{m} - \hat{z}2\text{m}$. Express the vector $\vec{r}_{12}$ from P_1 to P_2 in terms of unit vectors.

43.2 Vector Equations of Motion

In preceding chapters we found that the motion of a particle may be described by giving its position, velocity, and acceleration as functions of time. Unless specifically stated otherwise, in this and the following chapters the term velocity refers to instantaneous velocity, and acceleration refers to instantaneous acceleration. For motion in one dimension, these quantities are defined by the equations.

★ $$\vec{v} \equiv d\vec{s}/dt \quad \text{and} \quad \vec{a} \equiv d\vec{v}/dt,$$

where the vector symbols ($\vec{\ }$) are often omitted when it is understood that the motion is confined to one dimension.

The same definitions apply to motion in three dimensions, but now one cannot simply drop the vector symbols because in general $\vec{s}$, $\vec{v}$ and $\vec{a}$ all have three spatial components, one along each cartesian coordinate axis. When working with three dimensions, it is customary to express the position of a particle relative to the origin of coordinates using the symbol $\vec{r}$, rather than $\vec{s}$. Using unit vectors described in Section 43.1, the position vector of a point having coordinates x, y, and z is written

$$\vec{r} = \hat{x}x + \hat{y}y + \hat{z}z \quad \text{or} \quad \vec{r} = \hat{r}r,$$

where the unit vector $\hat{r}$ at any point has the direction of increasing $\vec{r}$. Using the Pythagorean theorem, the magnitude of $\vec{r}$ is

$$r = (x^2 + y^2 + z^2)^{\frac{1}{2}}.$$

The velocity vector for a particle having velocity components v_x, v_y and v_z is written

★ $$\vec{v} = \hat{x}v_x + \hat{y}v_y + \hat{z}v_z .$$

The magnitude of the velocity is

$$v = (v_x^2 + v_y^2 + v_z^2)^{\frac{1}{2}} .$$

Similarly, the acceleration vector for a particle having acceleration components a_x, a_y and a_z is written

★ $$\vec{a} = \hat{x}a_x + \hat{y}a_y + \hat{z}a_z .$$

The magnitude of the acceleration is

$$a = (a_x^2 + a_y^2 + a_z^2)^{\frac{1}{2}} .$$

In the most general case, each component of each vector listed above varies with time. The unit vectors $\hat{x}$, $\hat{y}$, $\hat{z}$ and $\hat{r}$ are constants which do not depend either on time or on the position of the particle.

Substituting the equation for $\vec{r}$ into the definition of velocity, $\vec{v} \equiv d\vec{r}/dt$, we obtain the following equation for the time-rate of change of position:

$$\vec{v} = \hat{x}\,\frac{dx}{dt} + \hat{y}\,\frac{dy}{dt} + \hat{z}\,\frac{dz}{dt} .$$

Comparing this with our earlier equation for $\vec{v}$, we see that

★ $$v_x = dx/dt, \qquad v_y = dy/dt, \qquad \text{and} \qquad v_z = dz/dt .$$

Substituting our earlier equation for $\vec{v}$ into the definition of acceleration, we obtain the following equation for the time-rate of change of velocity:

$$\vec{a} = \hat{x}\,\frac{dv_x}{dt} + \hat{y}\,\frac{dv_y}{dt} + \hat{z}\,\frac{dv_z}{dt} .$$

Comparing this with our earlier equation for $\vec{a}$, we see that

★ $$a_x = dv_x/dt, \qquad a_y = dv_y/dt, \qquad \text{and} \qquad a_z = dv_z/dt .$$

When a particle moves through space in such a way that its motion is not parallel to any one coordinate axis, we can resolve the vectors which describe this motion into components parallel to the x, y and z axes. In this way, the motion in three dimensions can be analyzed as three one-dimensional motions, one for each coordinate axis. The x motion depends on the initial coordinate x_0 and the initial speed v_{x0}. The y motion begins with y_0 and v_{y0}, while the z motion begins with z_0 and v_{z0}. These motions also depend on the forces acting on the particle, which in the most general case may have components in the x, y, and z directions.

Newton's second law of motion for a particle having a constant mass m is

$$\Sigma\vec{F} = m\,\vec{a},$$

where $\Sigma\vec{F}$ is the vector sum or resultant of all the forces acting on the particle. The resultant force can be resolved into three mutually perpendicular components:

$$\Sigma\vec{F} = \hat{x}\,\Sigma F_x + \hat{y}\,\Sigma F_y + \hat{z}\,\Sigma F_z.$$

If Newton's second law is written with both the force and the acceleration expressed in terms of three vector components, we can obtain a separate equation of motion for each coordinate axis:

★ $$\Sigma F_x = ma_x, \qquad \Sigma F_y = ma_y, \qquad \Sigma F_z = ma_z.$$

In applying these equations to a problem, it is necessary in each case to select the directions of the x, y, and z axes used to analyze the problem. This can often be done so that one or more components of $\Sigma\vec{F}$ are zero. In any case, the three one-dimensional equations of motion are solved to obtain solutions for $x(t)$, $y(t)$ and $z(t)$. Then the results may be combined to obtain resultant values of $\vec{r}$, $\vec{v}$, and $\vec{a}$.

Although the form of Newton's second law of motion is the same as in the one-dimensional case, in fact it says much more, because the concept of vector acceleration in more than one dimension is more complex than in the one-dimensional case. Consider, for example, a particle moving with a certain speed and having a certain force acting on it. If the velocity of the particle is in the direction of the force, Newton's law says that the speed is increasing at a certain rate. If the velocity is perpendicular to the force, however, the law says that the particle's *direction* is changing at a certain rate. Conceptually these two cases are quite different, yet they are quantitatively described by exactly the same law. This illustrates how vectors are useful in expressing physical laws, and how this usefulness is itself an *empirical* fact, not one that could have been discovered by armchair considerations alone.

This discussions shows that the three-dimensional motion of a particle can be thought of as three one-dimensional motions, in which the x-motion is determined by the x-component of the resultant force, and similarly for the y- and z-motions. This does not mean, however, that the three problems can always be solved separately as though they had nothing to do with each other. The reason is that the x-component of force, say, may depend on the y- and z-components of position (or velocity). In the next sections we shall see some examples where this is the case.

Questions and Exercises

Q 43.2-1 A particle has the position $\vec{r} = 3\hat{x} - 4\hat{y} + 6\hat{z}$ meters. What is the vector position of a second particle located exactly on the opposite side of the origin from the first, and at the same distance from the origin?

E 43.2-1 Particle A has the vector position $\vec{r}_A = 4\hat{x} + 6\hat{y} - 5\hat{z}$ meters, and particle B has the vector position $\vec{r}_B = 9\hat{x} - 1\hat{y} - 1\hat{z}$ meters.

(a) How far is *A* from the origin, in meters?
(b) What is the distance of *B* from the origin, in meters?
(c) What is the distance between *A* and *B*, in meters?

E 43.2-2 Show that the position of a particle moving in three-dimensional space with constant acceleration is $\vec{r} = \vec{r}_0 + \vec{v}_0 t + \vec{a}t^2/2$.

E 43.2-3 For each of the following examples of motion make a simple diagram showing the path of the particle, and at several points along the path draw the vectors $\vec{r}$, $\vec{v}$, and $\vec{a}$.
(a) A baseball hit by Willie Stargell into the grandstand.
(b) A simple pendulum.
(c) A ball swinging at the end of a string in a circle in a vertical plane.

43.3 Motion of Projectiles

Minimum Learning Goal: Given a description of the motion of a projectile in two dimensions as a function of time, be able to specify the position versus time separately for either of the two components of the motion. For example:

1 On the moon, the velocity (in m/s) of a particle is given by the expression

$$\vec{v} = (v_0 \cos \theta_0)\ \hat{x} + (v_0 \sin \theta_0 - 1.57t)\ \hat{y},$$

where v_0 is the magnitude of the initial velocity, θ_0 is the angle which the initial velocity vector makes with $\hat{x}$ (the horizontal), and t is the time in seconds. On a moon landing, one of the astronauts kicked up some moon dust with his shoe. A particle of the dust had an initial velocity of 10 m/s at an angle of 60° with the horizontal. Taking the initial position of the dust particle at time zero as the origin:
(a) What was its vertical position as a function of time?
(b) What was its horizontal position as a function of time?
(Answer, page 43-27, Line 1)

A familiar example of two- or three-dimensional motion is that of a *projectile*, a term referring to any object which is thrown, launched, or in some other way projected into the air and allowed to fall back to the earth. Examples of projectiles include footballs, snowballs, boomerangs and ballistic missiles. In the most general case, the path taken by a projectile can be complicated by the effects of wind, spinning motion of the projectile, drag and other factors.

To simplify our analysis, we assume initially that the projectile is not subject to forces resulting from drag, wind, or spinning motion. We further assume that its flight does not go so high that we must consider the variation of g with altitude (Section 44.3), or so far that we must consider the curvature of the earth. In other words, we are assuming that the only force on the projectile during its flight is its constant weight, $w = mg$, and that the projectile's path begins and ends at the surface of a flat earth. In this case, the path remains within the vertical plane containing the initial velocity vector, because there are no sideways forces which deflect the projectile out of this plane. This motion can therefore be analyzed completely using only two dimensions.

Figure 43-2. A projectile launched with an initial velocity $\vec{v}_0$ follows a path described using x and y coordinates. The initial components of velocity are v_{x0} and v_{y0}.

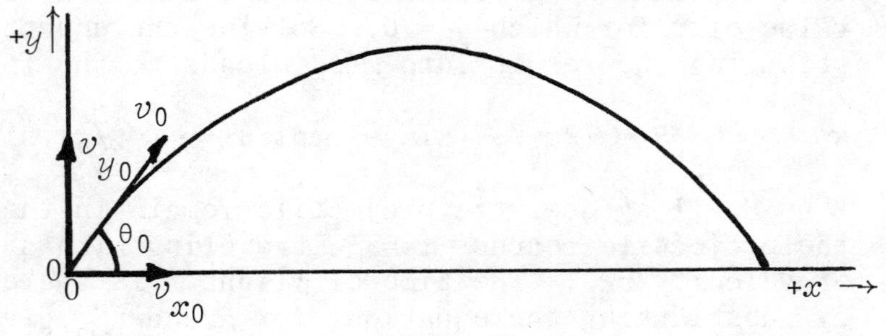

We choose a coordinate system having its origin at the point where the projectile is launched at $t = 0$, with the $+y$ axis vertical upward and the $+x$ axis parallel to the earth's surface in the vertical plane containing the initial velocity vector $\vec{v}_0$, as shown in Figure 43-2. Using θ_0 to designate the angle between $\vec{v}_0$ and the $+x$ direction, the initial components of velocity are

$$v_{x0} = v_0 \cos \theta_0 \qquad \text{and} \qquad v_{y0} = v_0 \sin \theta_0 .$$

Given these initial conditions, some questions of interest are the following: How far from its starting point does the projectile return to earth? How high does it go? What is the duration of its flight?

We begin with the equations of motion developed in Section 43.2:

$$\Sigma F_x = ma_x \qquad \text{and} \qquad \Sigma F_y = ma_y .$$

When the only force on the projectile is its weight, acting in the $-y$ direction,

$$\Sigma F_x = 0 \qquad \text{and} \qquad \Sigma F_y = -mg.$$

It is evident that the x-equation of motion can be solved to find $x(t)$, and the y-equation of motion can be solved separately to find $y(t)$. Because $\Sigma F_x = 0$, we have $a_x = 0$ and as a result the component of velocity in the x direction remains constant at its initial value v_{x0}. The x coordinate of the projectile's position therefore increases linearly with time, beginning with $x = 0$ at $t = 0$, so that

$$x = v_{x0} t.$$

Substituting $\Sigma F_y = -mg$ in the y-equation of motion, we obtain

$$a_y = -g\,.$$

The y motion is therefore the motion of a freely falling object having an initial upward velocity v_{y0}. This motion is described by Equations (41-7, 41-8 and 41-9) on page 41-32. Adapting the notation found there to that used in this section, we have

$$v_y = v_{y_0} - gt \qquad \text{and} \qquad y = v_{y0}t - \frac{1}{2}gt^2.$$

How far from its starting point does the projectile return to earth? This distance is referred to as the range, R, of the projectile; it is the value of x for which $y = 0$. Solving the equation $y(t) = 0$ for t, and substituting the result into $x(t)$, leads to the following equation for the range:

★ $$R = (2v_0^2/g)\sin\theta_0\cos\theta_0 = (v_0^2/g)\sin 2\theta_0.$$

How long does the projectile remain in the air? The flight ends when the projectile reaches $x = R$, traveling with a constant horizontal component of velocity v_{x0}. The time of flight T is therefore given by $T = R/v_{x0}$, and by substituting the equations for R and v_{x0} given above it is found that

★ $$T = 2(v_0/g)\sin\theta_0.$$

What is the maximum height reached by the projectile? The vertical motion is one of free fall, and as we found earlier for such a motion the time spent traveling upward equals the time spent traveling downward. That is, the projectile reaches its maximum height H at the midpoint in time of its flight, which is $t = T/2 = v_{y0}/g$. Substituting this value for t in the equation for $y(t)$ leads to

$$H = v_{y0}^{\ 2}/(2g).$$

By eliminating t from the pair of equations for $x(t)$ and $y(t)$, one obtains the equation for the path of the projectile. This is

$$y = (v_{y0}/v_{x0})\,x - (g/2v_{x0}^{\ 2})\,x^2.$$

This has the mathematical form of a parabola, corresponding to the path shown in Figure 43-2.

If one does not assume the ideal conditions under which the weight is the only force on the projectile, it is necessary to include in the equations of motion a term for each additional force. Including the effects of drag, wind or spin can lead to equations which are much more difficult to solve than the ones solved in this section. Methods of solution for these cases are developed in more advanced textbooks in mechanics. If one considers a projectile launched with a very large initial velocity, such as a ballistic missile, it is necessary to take account of the decrease in g with increasing distance from the center of the earth, as well as the earth's curvature and the earth's rotation about its polar axis. Then one could consider the motion as a problem in satellite motion, rather than projectile motion. In satellite motion (Section 44.4), the typical orbit is an ellipse having the center of the earth at one focus. The parabolic path of a short-range projectile can be thought of as the very tip of a long ellipse which would loop through the earth, around its center, and back up again -- if the earth itself were not in the projectile's way.

Questions and Exercises

Q 43.3-1 Give an example of short-range projectile motion which takes place in three dimensions.

Q 43.3-2 Considering the effect of air drag, how would the actual range of a projectile compare with that computed in this section?

Q 43.3-3 Considering the effect of air drag, how would the duration of the rising motion of a projectile compare with the duration of the descending motion?

Q 43.3-4 Some examples of objects moving through the air include: a golf ball driven from a tee; a baseball hit in a line drive; a ping-pong ball served in a ping-pong game; a shuttlecock hit in a badminton game; a maple seed falling from a tree; a kite rising as the string is given out; a Frisbee sailed to another player.
(a) For which of these motions does the force of air resistance have the least effect, in comparison with the force of gravity?
(b) For which of these motions does the force of air resistance have the greatest effect, in comparison with the force of gravity?

Q 43.3-5 Suggest an appropriate coordinate system for analyzing the motion of an inter-continental ballistic missile.

E 43.3-1 Using the initial conditions and equations of motion given in this section, complete the derivations of the following equations for projectile motion:
(a) $x = (v_0 \cos \theta_0)\, t$;
(b) $y = (v_0 \sin \theta_0)\, t - gt^2/2$;
(c) $R = (v_0^2/g) \sin 2\theta_0$;
(d) $T = 2\, v_0 \sin \theta_0/g$;
(e) $y = (v_{y0}/v_{x0})\, x - (g/2v_{x0}^{\,2})\, x^2$;
(f) $H = v_{y0}^{\,2}/(2g)$.

E 43.3-2 For a given initial speed, show that the range of a projectile is a maximum when $\theta_0 = 45°$, in the absence of air drag.

E 43.3-3 A slugger hits a baseball 140 m (460 ft) at an initial angle of 20° to the ground.
(a) What was the initial speed of the ball?
(b) How long was the ball in the air?
(c) What was the greatest height of the ball above the ground during its flight?

E 43.3-4 A bundle of propaganda leaflets is dropped from a plane while it is traveling horizontally at 10^3 km/hr. Neglect air drag.

(a) If the plane is 400 m above the ground and the bundle remains intact, what is the horizontal distance traversed by the bundle before it hits the ground?

(b) What is the velocity of the bundle when it hits the ground?

E 43.3-5 For the ideal projectile motion analyzed in this section, for what value(s) of t is the speed a maximum?

E 43.3-6 A football punted directly into a headwind has a vector position during the time it is in the air given approximately by

$$\vec{r} = \hat{x}(20t - 2t^2) + \hat{y}(20t - 4.9t^2),$$

where the numerical coefficients have (unexpressed) units such that when t is expressed in seconds, r is given in meters. The x axis is on the ground, and the $+y$ direction is upward.

(a) Compute the vector position of the football at $t = 0, 1, 2,$ 3, and 4 seconds, and plot these points on a graph of y vs x. On your graph, draw and label these position vectors, from the origin(kicking position)to the position of the ball at each of these times.

(b) At what value of t does the ball return to the ground ($y = 0$)?

(c) What physical force accounts for the term $-\hat{x}\,2t^2$ in the description of this motion?

(d) What physical force accounts for the term $-\hat{y}\,4.9t^2$ in the description of this motion?

(e) What additional information is needed to compute the resultant force on the football during this motion?

E 43.3-7 A baseball hit on a long fly to the outfield has a vector position during the interval $0 \leq t \leq 3$ seconds given approximately by

$$\vec{r} = \hat{x}\,8t + \hat{y}\,(30t - 1.0t^2) + \hat{z}\,(15t - 5t^2),$$

where the numerical coefficients have units such that when t is expressed in seconds, r is given in meters. The x and y axes are on the ground, and the $+z$ direction is upward.

(a) What is the position of the ball at $t = 0$?

(b) How far is the ball from its starting point at $t = 3$ seconds?

(c) Make a quantitative graph of the height of the ball above the ground, as a function of time.

(d) What is the maximum height of the ball during this motion? At what value of t does this occur?

(e) Which components (if any) of the ball's position change sign during this motion?

(f) What physical force accounts for the term $\hat{z}(-5t^2)$ in the description of this motion?

(g) What physical force could account for the term $\hat{y}(-1.0t^2)$ in the description of this motion?

(h) Is the motion of the ball after $t = 1.5$ seconds a mirror image of the motion before $t = 1.5$ seconds? Explain.

Check your understanding of the material above by testing yourself on the minimum learning goal at the beginning of this section.

43.4 Centripetal Force

Minimum Learning Goal. (1) Be able to identify the physical origin of a centripetal force, from a description of the motion of some object traveling in a curved path. For example:

1 What is the physical origin of the centripetal force for (a) a car going around a curve on the highway, and (b) the moon traveling in orbit around the earth? (Answer, page 43-27, Line 4)

2 What is the physical origin of the centripetal force for (a) a ball whirling on the end of a string, and (b) an electron in orbit around an atomic nucleus? (Answer, page 43-27, Line 6)

When an object is moving in a straight line, *no force* is required to maintain that motion at constant speed. In fact, what is specifically required is that the resultant force on the object is zero, a result summarized in Newton's first law of motion. A ball attached to a string, therefore, moves in a straight line if the net force on it is zero. If one end of the string is held in the hand, however, the ball can be caused to move in a circle. Because it is moving in a circle, we infer that some force must be acting on it which causes the ball to deviate from a straight line and follow a curved path. This force must be that of the string pulling inward on the ball, because if this inward force is "turned off" by letting go of the string, the ball immediately stops moving in a circle and flies off in a straight line tangent to the circle.

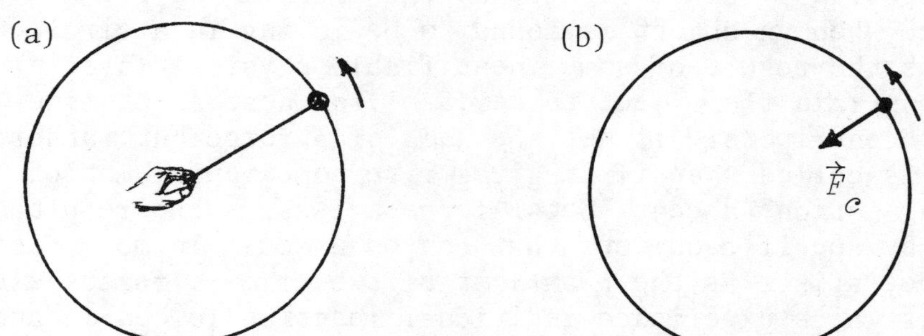

Figure 43-3. (a) A ball moving in a circular path at the end of a string. (b) To move in a circular path, the ball must be acted on by a centripetal force, $\vec{F}_c$.

As long as the string pulls inward on the ball, the ball continues in its circular path. The string pulls the ball toward the center of the circle, so the force exerted by the string on the ball is called a *centripetal force* -- it is directed toward the center of the circle. (This word comes from the Latin *peto, petare,* to seek; it is a force which "seeks" the center.)

> A centripetal force on an object is required to keep the object moving in a circle.

We use the symbol $\vec{F}_c$ for centripetal force.

In this discussion we are viewing the moving ball from a stationary point of observation, or frame of reference. Sometimes people speak as if there were an outward force on the ball moving in a circle, which they call a "centrifugal force." If such a centrifugal force were acting in addition to the centripetal force, and the two just balanced each other, then the resultant force would be zero and the ball would move in a straight line rather than a circle. So from our stationary frame of reference, it is not correct to say that a centrifugal force acts on the ball. (If one considers the motion of the ball as viewed from a frame of reference which revolves with the ball itself, however, it is useful to introduce the idea of a "fictitious" centrifugal force.)

According to Newton's third law of motion, forces occur in pairs, so the inward pull of the string on the ball is accompanied by the outward pull of the ball on the string. As is always the case with Newton's third law, the two forces which belong to a pair act on two *different* objects; in this example, one force acts on the ball while the other force acts on the string. This outward pull *on the string* may be described as a *centrifugal force.* (This word comes from the Latin *fugo, fugare,* to flee; it is a force which "flees" the center.)

Other examples of motion in a curved path are given in the following Questions and Exercises. In every case, a centripetal force is required to keep an object moving in a circular path. A centripetal force is not something which mysteriously and magically appears when an object moves in a circle. Rather, when an object is found to be moving in a circle this motion is always the result of some identifiable physical force or set of forces which constrain the object to deviate from a straight line. One might say that "centripetal" is not the *name* of a force but rather its *occupation;* it is called that if it gives rise to circular motion. In every case of circular motion, a centripetal force arises as the resultant of forces exerted by specific objects that can be named. In most cases, the centripetal force arises as the *resultant* of two or more forces exerted by specific objects which experience individual reaction forces in accord with Newton's third law of motion. In the case of an airplane looping the loop, for example, the centripetal force is the resultant of the weight and the force exerted by the surrounding air.

Except for some simple cases, therefore, it is not appropriate to talk about *the* reaction force to the centripetal force, because the latter is a *resultant* force, so that its reaction is in general spread over a number of other objects. There is no reason to expect any of the individual reactions to be parallel to a radius of the circle; only their resultant is.

When a ball is whirled in a circle at the end of a string held in the hand, actually the string does not pull the ball exactly toward the center of the circle. The reason for this is that the force of air drag tends to slow the ball, so to maintain the ball's motion with a constant speed the string must exert a small forward force on the ball. To provide this forward force it is necessary to move the hand in a small circle, so that the string always "leads" the motion of the ball by a small amount. In the absence of air drag, this forward force would not be required. In the rest of this chapter we neglect the drag force and the forward force which counterbalances it, and continue the discussion as if the string pulled the ball directly toward the center.

Questions

Q 43.4-1 If a ball on a string is whirling in a circle and the string breaks, what happens to the ball? What law does this illustrate?

Q 43.4-2 For each of the following, state which object is *exerting* the centripetal force and the name of this force, that is, what kind of force it is:
(a) the moon traveling in orbit around the earth.
(b) a bobsled going around a steeply banked curve on an icy bobsled run.
(c) a jogger rounding a curve on a level running track.

Q 43.4-3 If the centripetal force continually pulls a ball in circular motion toward the center, why doesn't the ball go to the center?

Check your understanding of the material above by testing yourself on the minimum learning goal at the beginning of this section.

43.5 Centripetal Acceleration

Minimum Learning Goal. Given a description of the motion of some object along a curved path, including velocities at two different times, be able to calculate the average vector acceleration. For example:

1 As a car heading due North enters a curve, it is moving at 55 mph (miles per hour). It slows down to 40 mph on the curve but speeds up again to 55 mph, coming out of the curve in a direction of 60° West of North. (a) What was the change in velocity of the car? (b) If the car was on the curve for 5 seconds, what was its average vector acceleration? (Answer, page 43-27, Line 2)

According to Newton's second law of motion, for a ball moving in a circle the centripetal force $\vec{F}_c$ causes acceleration $\vec{a}_c = \vec{F}_c/m$, where m is the mass of the ball. An acceleration must have the same direction as the force which causes it, so this acceleration is directed toward the center of the circle and is therefor called a *centripetal acceleration*, $\vec{a}_c$.

How does the centripetal acceleration affect the velocity of the ball? For a ball moving in a circular path, the velocity vector at any point on the circle is tangent to the circle at that point. As the ball travels around the circle, the direction of the velocity vector is continually changing. This change in velocity is the result of the acceleration of the ball, as we see from the definition of instantaneous acceleration:

$$\vec{a} = \lim_{\Delta t \to 0} \frac{\Delta \vec{v}}{\Delta t} .$$

Even when the speed is constant, the change in velocity is related to the centripetal acceleration by this equation.

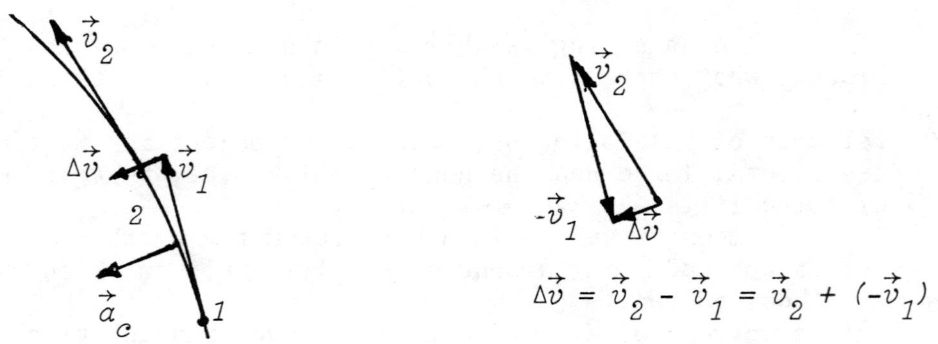

Figure 43-4. As an object moves from point *1* to point *2* on a curved path, its velocity vector changes direction. The change in velocity, $\Delta\vec{v}$, between point *1* and point *2* is found using a vector diagram.

Consider the change in velocity of the ball as it moves from some point *1* to a nearby point *2* on the circle (Figure 43-4). At point *1* the velocity is $\vec{v}_1$, and at point *2* it is $\vec{v}_2$. As the ball moves from point *1* to point *2*, it is continuously accelerated toward the center of the circle. The direction of the $\vec{a}_c$ vector changes slightly along the arc from point *1* to point *2*; its direction at the midpoint is shown in Figure 43-4. This acceleration is related to the change in velocity $\Delta\vec{v}$ during this interval by $\vec{a}_c = \Delta\vec{v}/\Delta t$, so $\Delta\vec{v}$ has the same direction as $\vec{a}_c$. When the initial velocity $\vec{v}_1$ is combined with $\Delta\vec{v}$, their vector sum is just the velocity $\vec{v}_2$ at the end of the interval; the equation is

$$\vec{v}_2 = \vec{v}_1 + \Delta\vec{v}, \qquad \text{or} \qquad \Delta\vec{v} = \vec{v}_2 - \vec{v}_1 .$$

There is a useful equation which relates the variables a_c, v, and r for an object moving in a circle. To derive this equation, we construct two triangles (see Figure 43-5). The distance triangle is formed by drawing radii from the center to two points, *1* and *2*, on the circle, and then drawing the straight line (called the chord) joining points *1* and *2*; this chord is denoted by the symbol $\overline{12}$. The arc of the circle between points *1* and *2* is the distance traveled by the moving object from point *1* to point *2*, given by $v\Delta t$. The velocity triangle is essentially the one used in Figure 43-4; the lengths of its sides are v_1, v_2, and Δv. Both v_1 and v_2 are equal to the constant speed, v. The radius to point *1* is perpendicular to the velocity $\vec{v}_1$, and the radius to point *2* is perpendicular to the velocity $\vec{v}_2$, so these two triangles are similar triangles. In the limit as $\Delta t \to 0$, points *1* and *2* come closer together, and the length of the chord $\overline{12}$ becomes $v\Delta t$. Then, making use of the property of similar triangles that the ratios of corresponding sides are equal, we have the result

$$\frac{v\Delta t}{r} = \frac{\Delta v}{v} .$$

Multiply both sides of this equation by $v/\Delta t$ leads to

$$\frac{\Delta v}{\Delta t} = \frac{v^2}{r} ,$$

and in the limit as $\Delta t \to 0$ this may be written in the form we will use:

★ $$a_c = \frac{v^2}{r} .$$

When an object moves on a circular path with a speed which varies from point to point, its acceleration also varies. At every point, however, the instantaneous speed and acceleration are related by $a_c = v^2/r$.

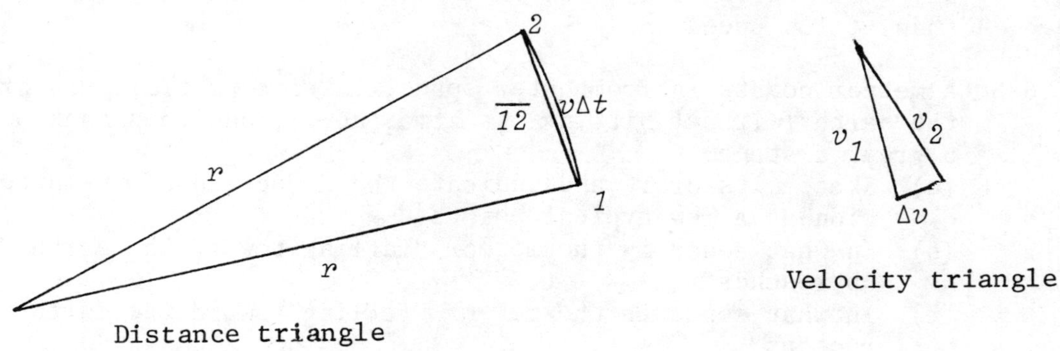

Figure 43-5. Similar triangles for an object which moves from point *1* to point *2* in uniform circular motion.

Worked Example 43-2

The speed of the earth in its (nearly circular) orbital motion around the sun is 3.0×10^4 m/s, and the radius of the orbit is 1.49×10^{11} meters. What is the centripetal acceleration of the earth in its orbital motion?

Solution

We have the earth's speed and the radius of the circle in which it is traveling, so we can compute the centripetal accelerating using $a_c = v^2/r$. Substituting numerical values, we have

$$a_c = \frac{(3.0 \times 10^4\ \text{m/s})^2}{1.49 \times 10^{11}\ \text{m}} = \frac{9.0 \times 10^8\ \text{m}^2/\text{s}^2}{1.49 \times 10^{11}\ \text{m}}$$

$$= 6.0 \times 10^{-3}\ \text{m/s}^2,\ \text{directed toward the sun.}$$

Questions and Exercises

Q 43.5-1 If the acceleration of a moving object is perpendicular to the velocity, is the speed changing? Is the direction of motion changing?

Q 43.5-2 The batter in a baseball game hits a high fly ball into the outfield.
(a) If the air drag and wind are ignored, at what point of its path is its acceleration perpendicular to its velocity?
(b) Does its behavior at that moment comply with your answer to Q 43.5-1?

Q 43.5-3 (a) List several examples of motion in a circular path.
(b) List several examples of motion in a curved path which is not circular motion.

Q 43.5-4 Why is it harder for a car to make a given turn at high speed than at low speed?

Q 43.5-5 A meteor coasts in from outer space, takes a partial turn around the earth (without hitting its atmosphere), and goes back out to a great distance.
(a) Sketch its orbit and indicate the direction of its acceleration at a few typical points.
(b) In what sense is the meteor "falling toward the earth" as it approaches?
(c) In what sense is the meteor "falling toward the earth" as it recedes?

E 43.5-1 The line at the right is an aerial (top) view of the path followed by a motorbike traveling *at constant speed*. Place the letter *A* next to the path at two points where the magnitude of the acceleration is greatest. At each of these points draw an arrow which shows the direction of the acceleration at that point.

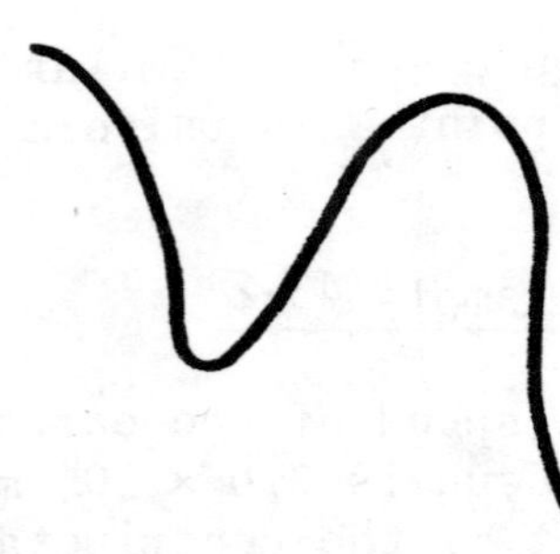

E 43.5-2 A 140-gram ball is whirling in a *horizontal* circle at the end of a strong cord at a speed of 6 m/s. The radius of the circle is 1.2 meter. Compute the force exerted on the ball by the cord, assuming the cord remains horizontal during this motion.

E 43.5-3 A centrifuge whirls an astronaut trainee in circular motion with a speed of 30 m/s at a radius of 10 meters. What is his centripetal acceleration?

Check your understanding of the material above by testing yourself on the minimum learning goal at the beginning of this section.

43.6 Uniform Circular Motion

Minimum Learning Goals. (1) Given sufficient data in a description of some object in uniform circular motion, be able to calculate the centripetal acceleration, speed, frequency, period, or radius, as required. For example:

1 Some of America's astronauts were trained to withstand large accelerations by being whirled in a centrifuge at the Naval Air Development Center at Johnsville, Pennsylvania. The astronaut is whirled in uniform circular motion at a radius of 50 feet (15.3 m).

(a) What is his tangential speed for an acceleration of 16 *g*'s? (one $g = 9.8\ m/s^2$)
(b) What frequency of revolution is required to give an acceleration of 16 *g*'s?
(c) What frequency of revolution is required to give an acceleration of 40 *g*'s? (Answer page 43-27, Line 5)

2 According to the Bohr planetary model of the hydrogen atom, the single electron of that atom revolves about the proton nucleus at a radius of 5.3×10^{-11} m, with a speed of 2.2×10^6 m/s. What are the period and frequency of this motion? (Answer, page 43-27, Line 7)

(2) Given the mass, speed and radius (or equivalent information) of an object moving in uniform circular motion, be able to calculate the magnitude of the force needed to produce this motion. For example:

1 A man weighing 160 pounds (mass = 72.6 kg) is riding in the front seat of an automobile next to the driver when the car goes around a curve having a radius of curvature of 1283 feet (391 m). If the car is traveling at 60 mph (26.8 m/s), with what force must the car push the man toward the center of the curvature, if he is to follow the same circular path as the car? (Answer, page 43-27, Line 3)

Circular motion is a special type of motion in a plane. Circular motion for which the speed is constant is often called *uniform circular motion.* Note, however, that when an object is moving in a curved path which is *not* a circle, its motion is not uniform circular motion even though its speed may be constant.

In addition to the variables discussed thus far, there are two others which are useful for describing uniform circular motion. These are defined as follows:

> The *period,* τ, of uniform circular motion is the time in seconds required for one revolution.
>
> The *frequency,* ν, of uniform circular motion is the number of revolutions in one second.

The symbol for period is the Greek letter *tau,* τ; the symbol for frequency is the Greek letter *nu,* ν. (Caution is required in writing the symbol ν so it is distinct from v and v.) Because each of these quantities is just the reciprocal of the other, we immediately have the relation between frequency and period:

★ $$\tau = \frac{1}{\nu} .$$

The speed of an object in uniform circular motion is the distance traveled in one revolution divided by the time required for one revolution. In symbols,

★ $$v = \frac{2 \pi r}{\tau} = 2 \pi r \nu .$$

Two of the equations obtained earlier also apply to all cases of uniform circular motion:

★ $$\vec{F}_c = m \vec{a}_c \quad \text{and} \quad a_c = \frac{v^2}{r} .$$

While a centripetal force is always directed toward the center of the circle, not every centrally directed force gives rise to uniform circular motion. To produce uniform circular motion, a force must be constant in magnitude, perpendicular to the velocity at all times, and always directed toward the same central point. This requirement is satisfied in various physical phenomena by electric, magnetic, gravitational, frictional, and other forces.

Worked Example 43-3

Dave has an old-fashioned sling which he uses to whirl a smooth stone in a horizontal circle above his head. The stone and sling have an effective mass of 0.2 kg, the circle has a radius of 0.8 meter, and the motion has a frequency of 120 rpm (revolutions per minute). Neglect the weight of the stone and the sling, and the fact that the sling will slope downward from Dave's hand to the stone; assume that the sling is horizontal.

(a) Find the period of this motion.
(b) Compute the speed of the stone in its circular motion;
(c) Compute the centripetal acceleration of the stone;
(d) Compute the centripetal force required to maintain this motion.

Solution

(a) The frequency $\nu = \dfrac{120 \text{ revolutions}}{1 \text{ minute}} \times \dfrac{1 \text{ minute}}{60 \text{ seconds}} = 2 \text{ sec}^{-1}$.

Therefore the period $\tau = 1/\nu = 1/(2 \text{ sec}^{-1}) = 0.5$ second.

(b) The speed $v = 2 \pi r \nu = 2 \pi \times 0.8 \text{ m} \times 2 \text{ sec}^{-1} = 10.1$ meter/second.

(c) The acceleration $a_c = v^2/r = (10.1 \text{ m/s})^2/(0.8 \text{ m}) = 1.28 \times 10^2 \text{ m/s}^2$.

(d) The centripetal force $F_c = m\, a_c = 0.2 \text{ kg} \times 1.28 \times 10^2 \text{ m/s}^2$

$= 25.6$ newton.

Uniform circular motion is an example of two-dimensional motion for which it is not useful to separate the motion into x, y, and z components. For example, if a car is going around a banked circular track, the x component of the force exerted by the track depends on all three components of the position of the car. It is better to consider the resultant velocity, acceleration, and force at each point along the path. The resultant force must be a force directed toward the center of the circle, even though the individual forces making up the resultant may point in various other directions. By constructing a free-body diagram and resolving the forces into suitable components, we can analyze many instances of circular motion.

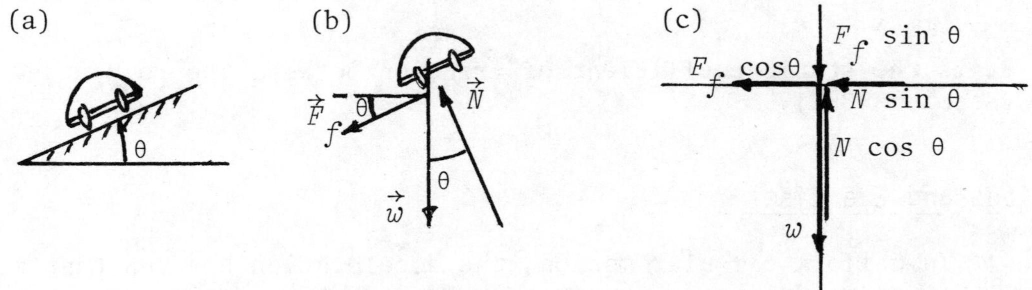

Figure 43-6. (a) Front view of a car on a banked road. (b) The car is acted on by several forces: its weight $\vec{w}$, the normal force $\vec{N}$ of the road on the tires, and the friction force $\vec{F}_f$ of the road on the tires. The friction force may be directed either upward or downward along the slope of the road surface. If the car is traveling in a curve, the required centripetal force is the resultant of the individual forces on the car. (c) If the resultant centripetal force is horizontal, it is convenient to resolve the individual forces into horizontal and vertical components.

As an example, consider an auto rounding a horizontal curve where the highway is banked at an angle θ (Figure 43-6). In addition to its weight, $\vec{w} = m\vec{g}$, the forces on the car include the normal force $\vec{N}$ and the frictional force $\vec{F}_f$ exerted by the road on the tires. In general, a normal force and a frictional force act on each tire, but the magnitudes of these forces may vary from one wheel to another. Figure 43-6b shows the resultant normal and friction forces on the car for the case when the car is traveling at a fairly high speed v relative to the angle of banking, so that the horizontal component of the normal force must be supplemented by the horizontal component of the friction to provide a centripetal force large enough to keep the car on the road. When the car's speed is small compared with the design speed of the curve, the friction force on the car acts upward along the road, to partially offset the horizontal component of the normal force and keep the car from moving downward along the slope.

When the curved path followed by the highway is horizontal and the car remains in its lane of traffic, the centripetal force on the car must be horizontal. It is then convenient to resolve the forces on the car into horizontal and vertical components, as shown in the free-body diagram of Figure 43-6c. The car has no vertical acceleration, so the resultant vertical force on it equals zero:

$$N \cos \theta - F_f \sin \theta - mg = 0.$$

The resultant horizontal force serves as the centripetal force:

$$N \sin \theta + F_f \cos \theta = F_c = mv^2/r,$$

where r is the radius of the curve. The normal force and maximum friction force are related by

$$F_{f\text{max}} = \mu_s N,$$

where μ_s is the static coefficient of friction between the road and the tires (Section 42.7).

Questions and Exercises

Q 43.6-1 In uniform circular motion, the acceleration has constant magnitude and is always perpendicular to the velocity. Is the converse true? In other words, if we know that a particle's acceleration has a constant magnitude and is always perpendicular to its velocity, can we conclude that it is in uniform circular motion?

Q 43.6-2 (a) In any specific example of uniform circular motion, which "variables" or parameters of the motion remain constant?
(b) Which vary during the motion?

E 43.6-1 The precision clock in a physics laboratory has a second hand 10.7 cm long, which makes one revolution/second.
(a) Find the tangential speed of the tip.
(b) Find the centripetal acceleration of the tip.

E 43.6-2 (a) Compute the centripetal acceleration of a body on the earth at the equator.
(b) Compare your result with g, and explain any discrepancy.

E 43.6-3 (a) What centripetal force is required to keep a mass of 0.2 kg rotating in a horizontal circle of radius 50 cm at 120 rpm?
(b) What is the period of this motion?

E 43.6-4 The speed of the moon in its orbit is 1.02×10^3 meter/second. The radius of the moon's orbit about the earth is 3.84×10^8 meter. The moon's mass is about 7.35×10^{22} kg.
(a) Compute the centripetal acceleration of the moon in its orbital motion around the earth.
(b) How much time is required for one revolution of the moon in this orbit?
(c) Compute the centripetal force which keeps the moon in its orbital motion around the earth.

E 43.6-5 Assume that a hydrogen atom consists of a proton at the center, with an electron traveling around it in a circular orbit having a radius of 5.3×10^{-11} meter.
(a) Compute the electric force of the proton on the electron.
(b) Compute the orbital speed of the electron, using an equation for the centripetal force.
(c) Compute the time required for the electron to complete one orbit around the proton.

E 43.6-6 A conical pendulum is a simple pendulum having length ℓ, whose bob moves in a horizontal circle. If the pendulum makes an angle θ with the vertical, what is the period of its motion?

E 43.6-7 A small metal sphere having a mass of 0.1 kg is suspended at the end of a string of length 1.0 m. The sphere is whirled in uniform circular motion in a horizontal plane, so that the string describes a cone. If the sphere has a speed of 6 m/s, what force must the string exert on the sphere?

E 43.6-8 A 2000-kg automobile is traveling along a horizontal (unbanked) highway which curves with a radius of 400 m. Assuming the car has a speed of 40 m/s while on the curve, what force must the road exert on the car for it not to slide sideways, either to the right or left?

E 43.6-9 If a driver wants to round a horizontal curve (not banked at all) having a radius of 1/4 mile at a speed of 70 mi/hr, what is the required value of the coefficient of static friction between the tires and the road?

E 43.6-10 For a horizontal highway curve banked at an optimum angle θ_0, no lateral friction force is required for a car to go around the curve at a speed v.
(a) Show that $\tan \theta_0 = v^2/rg$, where r is the radius of the curve.
(b) Compute the required θ_0 for $v = 31$ m/s (70 mi/hr) and $r = (1/8)$ mile; (continued on next page)

(c) Compute the required θ_0 for v = 70 mi/hr and r = (1/2) mile.

(d) On a curve constructed with the optimum θ_0 for 70 mi/hr, what is the direction of the required friction force on a car going 80 mi/hr, so the car will remain in its traffic lane?

(e) On a curve designed for 70 mi/hr, what is the direction of the required friction force on a car going 60 mi/hr, so the car will remain in its lane of traffic?

(f) If the curve is ice-covered and slippery, what happens to a car rounding the curve at a speed much less than the design speed?

(g) If the curve is ice-covered and slippery, what happens to a car rounding the curve at a speed much greater than the design speed?

E 43.6-11 For a car weighing 9000 N (about 2000 lb) and moving at 70 mi/hr on a horizontal curve having a radius of 0.5 km banked at an angle of 10°, what lateral friction force is required to keep the car in its lane of traffic?

E 43.6-12 A highway curve having a radius of curvature r = 400 m is banked at an angle θ = 8°. If the static coefficient of friction is 0.5, what is the maximum speed with which a car can go around the curve without skidding?

Check your understanding of the material above by testing yourself on the minimum learning goals at the beginning of this section.

43.7 Applications of Calculus

The methods of calculus were applied to the description of one-dimensional motion in Section 41.7. For motion along the x axis, if the position $x(t)$ is known the velocity and acceleration may be obtained by differentiation:

★ $$v(t) = dx/dt \qquad \text{and} \qquad a(t) = dv/dt.$$

If the acceleration $a(t)$ is known, the velocity and position may be obtained by integration:

★ $$v(t) = v_0 + \int_{t_0}^{t} a(t)dt \qquad \text{and} \qquad x(t) = x_0 + \int_{t_0}^{t} v(t)dt.$$

The complete calculation requires a knowledge of the position x_0 and velocity v_0 at some initial time t_0. Procedures for differentiating and integrating elementary functions are given in Appendices A.2 and A.4.

These procedures may be extended to motion in three dimensions by regarding the motion as the vector sum of three one-dimensional motions, one along each cartesian coordinate axis. As we found in Section 43.2, using unit vectors the position of a particle is given by

$$\vec{r}(t) = \hat{x}\, x(t) + \hat{y}\, y(t) + \hat{z}\, z(t).$$

In general, the x, y, and z components of the position vector vary with time, but the unit vectors $\hat{x}$, $\hat{y}$ and $\hat{z}$ are constants (as long as we use only stationary cartesian coordinates). In differentiation and integration, therefore, $\hat{x}$, $\hat{y}$ and $\hat{z}$ are handled like numerical constants. The velocity of a particle is defined as

$$\vec{v}(t) \equiv \frac{d\vec{r}}{dt} = \hat{x}\frac{dx}{dt} + \hat{y}\frac{dy}{dt} + \hat{z}\frac{dz}{dt} = \hat{x}\, v_x + \hat{y}\, v_y + \hat{z}\, v_z.$$

The acceleration of a particle is defined as

$$\vec{a}(t) \equiv \frac{d\vec{v}}{dt} = \hat{x}\frac{dv_x}{dt} + \hat{y}\frac{dv_y}{dt} + \hat{z}\frac{dv_z}{dt} = \hat{x}\, a_x + \hat{y}\, a_y + \hat{z}\, a_z.$$

If the position $\vec{r}(t)$ is known, the velocity and acceleration may be obtained by differentiating each component separately:

$$v_x(t) = dx/dt \qquad a_x(t) = dv_x/dt\ ;$$

$$v_y(t) = dy/dt \qquad a_y(t) = dv_y/dt\ ;$$

$$v_z(t) = dz/dt \qquad a_z(t) = dv_z/dt\ .$$

If the acceleration $a(t)$ is known, the velocity and position may be obtained by integrating each component separately:

★ $$v_x(t) = v_{x0} + \int_{t_0}^{t} a_x(t)dt \qquad x(t) = x_0 + \int_{t_0}^{t} v_x(t)dt \; ;$$

★ $$v_y(t) = v_{y0} + \int_{t_0}^{t} a_y(t)dt \qquad y(t) = y_0 + \int_{t_0}^{t} v_y(t)dt \; ;$$

★ $$v_z(t) = v_{z0} + \int_{t_0}^{t} a_z(t)dt \qquad z(t) = z_0 + \int_{t_0}^{t} v_z(t)dt \; .$$

The complete calculation requires a knowledge of the position

$$\vec{r}_0 = \hat{x}\, x_0 + \hat{y}\, y_0 + \hat{z}\, z_0$$

and the velocity

$$\vec{v}_0 = \hat{x}\, v_{x0} + \hat{y}\, v_{y0} + \hat{z}\, v_{z0}$$

at some initial time t_0, or equivalent information.

The acceleration of a particle is related to the resultant force on it by Newton's second law of motion, $\Sigma\vec{F} = m\vec{a}$. In component form:

$$\Sigma F_x = m\, a_x, \qquad \Sigma F_y = m\, a_y, \qquad \Sigma F_z = m\, a_z.$$

Worked Example 43-4

A football is punted directly down-field with an initial velocity (at $t = 0$) of 14.0 m/s at an angle of 40° above the horizontal. Just as it is kicked a gust of wind arises to oppose the motion with a horizontal time-dependent force $-Wt$. To describe this motion, it is convenient to choose a coordinate system so the ball's motion is in the xz plane, with the $+x$ axis horizontal in the direction of the ball's motion and the $+z$ axis vertical upward, with the origin at the point where the ball was when kicked. Then the ball's velocity is given by

$$\vec{v}(t) = \hat{x}\,[v_{x0} - (0.45 \text{ m/s}^3)t^2] + \hat{z}\,(v_{z0} - gt).$$

(a) Compute the x and z components of the initial velocity.

(b) Obtain the equation for $\vec{a}(t)$.

(c) Obtain the equation for $\vec{r}(t)$.

Solution

The path of the football is shown at the right. The head wind causes the ball to travel inside the wind-free parabolic path (Section 43.3) and fall short of the wind-free range.

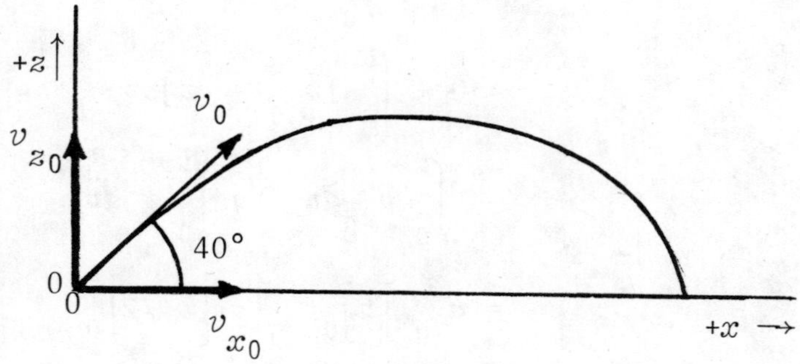

(a) $v_{x_0} = v_0 \cos 40° = (14.0 \text{ m/s})\ 0.766 = 10.7 \text{ m/s}.$

$v_{z_0} = v_0 \sin 40° = (14.0 \text{ m/s})\ 0.643 = 9.0 \text{ m/s}.$

(b) The two components of $a(t)$ may be obtained by differentiating $v(t)$. In this example, there is no sideways motion so the y components of $\vec{v}$, $\vec{a}$, and $\vec{r}$ are equal to zero.)

$$a_x(t) = \frac{dv_x}{dt} = \frac{d}{dt}\left[v_{x_0} - (0.45 \text{ m/s}^3)t^2\right]$$

$$= -(0.45 \text{ m/s}^3)2t = -(0.90 \text{ m/s}^3)t\ .$$

$$a_z(t) = \frac{dv_z}{dt} = \frac{d}{dt}\left[v_{z_0} - gt\right] = -g\ .$$

$$\vec{a}(t) = -\hat{x}\ (0.90 \text{ m/s}^3)\ t - \hat{z}\ 9.8 \text{ m/s}^2.$$

(c) The two components of $\vec{r}(t)$ may be obtained by integrating the components of $\vec{v}(t)$.

$$x(t) = x_0 + \int_0^t v_x(t)dt$$

$$= 0 + \int_0^t \left[v_{x_0} - (0.45 \text{ m/s}^3)t^2\right] dt$$

$$= \int_0^t v_{x_0}\ dt - (0.45 \text{ m/s}^3) \int_0^t t^2\ dt$$

$$= v_{x_0}\left[t\right]_0^t - (0.45 \text{ m/s}^3)\left[\frac{t^3}{3}\right]_0^t$$

$$= v_{x_0}\ t - (0.45 \text{ m/s}^3)\left(\tfrac{1}{3}\right) t^3$$

$$x(t) = (10.7 \text{ m/s})\ t - (0.15 \text{ m/s}^3)\ t^3.$$

$$z(t) = z_0 + \int_0^t v_z(t)dt$$

$$= 0 + \int_0^t [v_{z_0} - gt]dt$$

$$= \int_0^t v_{z_0} dt - g \int_0^t t\ dt$$

$$= v_{z_0} \Big[t\Big]_0^t - g \Big[t^2/2\Big]_0^t$$

$$= v_{z_0} t - (g/2)\ t^2$$

$$z(t) = (9.0\ \text{m/s})\ t - (4.9\ \text{m/s}^2)\ t^2.$$

Finally,

$$\vec{r}(t) = \hat{x}\ [(10.7\ \text{m/s})\ t - (0.15\ \text{m/s}^3)\ t^3]$$

$$+ \hat{z}\ [(9.0\ \text{m/s})\ t - (4.9\ \text{m/s}^2)\ t^2]\ .$$

]]

Exercises

E 43.7-1 Consider a particle having initial position $\vec{r}_0$ and initial velocity $\vec{v}_0$, and moving with constant acceleration $\vec{a}$. Integrate each component of $\vec{a}(t)$, and combine the results to show that $\vec{r} = \vec{r}_0 + \vec{v}_0 t + \vec{a}t^2/2$.

E 43.7-2 An oscilloscope has a cathode ray tube (CRT) equipped with parallel plates for deflecting the electron beam. During the time that an electron is between the plates, its velocity (in m/s) is given by the expression

$$\vec{v} = 10^7\ \hat{x} + (8 \times 10^9\ t)\ \hat{y},$$

where $\hat{x}$ is a horizontal unit vector and $\hat{y}$ is a vertical unit vector. Take the initial position of the electron as the origin, just as it enters the space between the plates at $t = 0$.
(a) What is its vertical position as a function of time?
(b) What is its horizontal position as a function of time?

E 43.7-3 Consider the motion of the football described in E 43.3-6.
(a) Compute the vector velocity of this football as a function of time.
(b) Compute the vector velocities of the football at $t = 0$, 1, 2, 3, and 4 seconds, and draw and label these vectors on your graph obtained for E 43.3-6
(c) Compute the vector acceleration of the football as a function of time. (continued on next page)

(d) Compute the vector acceleration of the football at $t = 0, 1, 2, 3,$ and 4 seconds, and draw and label these vectors on your graph obtained for E 43.3-6.

(e) Which two vectors associated with the motion of this football do not change in magnitude or direction during this motion?

E 43.7-4 Consider the motion of the baseball described in E 43.3-7.
(a) Compute the velocity of this baseball as a function of time.
(b) Which component of this velocity has the largest value at $t = 0$?
(c) Which components of the velocity (if any) change sign during this motion?
(d) At what point of its motion does the ball have its maximum speed? What is the ball's velocity at this point?
(e) Compute the acceleration of the ball as a function of time.
(f) What is the maximum acceleration of the ball?

E 43.7-5 In Section 43.3 an equation for the path of a projectile was obtained in the form $y(x)$. The slope of this path at any point is dy/dx.
(a) Use differentiation to obtain a general equation for the slope, as a function of x.
(b) What is the slope at the highest point on the path?
(c) Use the result of part (b) to compute the value of x at the highest point on the path.
(d) Use the result of part (c) to compute the highest elevation reached by the projectile.

Check your understanding of the material above by testing yourself on the minimum learning goals at the beginning of this section.

Answers to Minimum Learning Goals

Line 1 (a) $y = 8.66t - 0.785t^2$ meters, when t is in seconds.
(b) $x = 5t$ meters, when t is in seconds.

Line 2 55 mph in a direction 30° south of west; 11.0 mph/s at 30° south of west.

Line 3 133.4 N

Line 4 (a) Frictional force on the tires, and a component of the car's weight when the curve is banked;
(b) The gravitational force of the earth on the moon.

Line 5 (a) 49 m/s; (b) 0.51 s^{-1}; (c) 0.81 s^{-1}

Line 6 (a) The string pulling on the ball;
(b) The coulomb force on the electron due to the nucleus.

Line 7 Frequency = 6.6×10^{15} s^{-1}; period = 1.5×10^{-16} s.

ANSWERS to Questions and Exercises in Chapter 43

Section 43.1

Q 43.1-2 The negative x direction

E 43.1-2 $\vec{A} + \vec{B} = \hat{x}(A_x + B_x) + \hat{y}(A_y + B_y) + \hat{z}(A_z + B_z)$

E 43.1-4 $\vec{r}_{12} = \hat{x}1\text{m} + \hat{y}3\text{m} - \hat{z}7\text{m}$.

Section 43.3

Q 43.3-2 The actual range is smaller than the range computed using the formula for an ideal projectile.

Q 43.3-4 (a) the motions of a golf ball and a baseball;
(b) the motions of a maple seed and a kite.

E 43.3-4 (a) 2.51 km; (b) 291.6 m/s at an angle of 17.7° below the horizontal.

E 43.3-6 (a) $\vec{r}(0) = 0$; $\vec{r}(1\text{ s}) = (\hat{x}\ 18 + \hat{y}\ 15.1)\text{m}$; $\vec{r}(2\text{ s}) = (\hat{x}\ 32 + \hat{y}\ 20.4)\text{m}$; $\vec{r}(3\text{ s}) = (\hat{x}\ 42 + \hat{y}\ 15.9)\text{m}$; $\vec{r}(4\text{ s}) = \hat{x}(48 + \hat{y}\ 1.6)\text{m}$.
(b) 4.08 s; (c) air drag from headwind; (d) weight of the football;
(e) mass of the football.

Section 43.4

Q 43.4-2 (a) The earth exerts a gravitational force on the moon.
(b) The ice bank exerts a normal force (a push normal to the surface) on the bobsled;
(c) The tract exerts a friction force on the jogger's feet.

Section 43.5

Q 43.5-2 (a) At the highest point. (b) Yes

Q 43.5-4 The centripetal force required to keep the car on the road increases with the square of the speed.

E 43.5-2 4.2 newtons

Section 43.6

Q 43.6-2 (a) The radius, speed, period, frequency, and magnitude of the acceleration remain constant.
(b) The directions of velocity and acceleration.

E 43.6-2 (a) 0.0337 m/s^2
(b) $a_c \ll g = 9.8$ m/s^2; the earth's gravitation is much more than enough to supply the a_c required to keep objects on the earth's surface from flying off into space.

E 43.6-4 (a) 2.7×10^{-3} m/s^2
(b) 2.37×10^6 s = 27.4 days
(c) 1.99×10^{20} N

E 43.6-6 $\tau = 2\pi \sqrt{\ell \cos \theta / g}$

E 43.6-8 8.0×10^3 N

E 43.6-10 (b) 26.0°; (c) 6.9°;
(d) downward along the banked highway surface, perpendicular to the car's velocity;
(e) upward along the banked highway surface, perpendicular to the car's velocity;
(f) the car slides inward toward the center of the curve;
(g) the car skids outward away from the center of the curve.

E 43.6-12 52.0 m/s = 116 mi/hr.

Section 43.7

E 43.7-2 (a) $y = 4 \times 10^9\ t^2$ meters, when t is in seconds;
(b) $x = 10^7\ t$ meters, when t is in seconds.

E 43.7-4 (a) $\vec{v}(t) = \hat{x}\ 8 + \hat{y}\ (30 - 2.0\ t) + \hat{z}\ (15 - 10\ t)$ m/s, when t is in seconds;
(b) v_y ; (c) v_z
(d) At the origin, when the ball is first hit;
$\vec{v}(0) = 8\ \hat{x} + 30\ \hat{y} + 15\ \hat{z}$ m/s; $v(0) = 34.5$ m/s
(e) $\vec{a}(t) = -2.0\ \hat{y} - 10\ \hat{z}$ m/s^2;
(f) $\vec{a}$ is constant; $a = 10.2$ m/s^2.

CHAPTER 44 GRAVITATION

Introduction

We use the term *resultant force* to refer to the sum of several forces acting on a particular object. A *fundamental force* is a force which is not itself the resultant of other forces. This chapter begins with an overview of some aspects of the four fundamental forces of nature. One of these, the universal force of gravitation, is then discussed in some detail. This force accounts for the motions of all astronomical bodies, including the orbital motions of planets around the sun and satellites around a planet.

When you have completed your study of the material in this chapter, you should be able to answer questions and work exercises like the examples in the minimum learning goals assigned by your instructor.

44.1 Fundamental Forces

Minimum Learning Goal. Given a statement about some characteristics of fundamental forces, be able to state which of the four fundamental forces have this characteristic. For example:

1 Which of the fundamental physical forces are not resultants of other forces? (Answer, p. 44-17, Line 1)

2 Which of the fundamental forces of nature are observable only on the microscopic scale? (Answer, p. 44-17, Line 6)

3 Which of the fundamental forces is dominant in the motions of bodies in the solar system? (Answer, p. 44-17, Line 9)

As used in physics, the word *force* refers to one or another of the physical forces which have been found in the study of nature. The physical forces considered in Chapter 42 were all forces which we meet repeatedly in our daily experience. We can refer to them as "everyday forces." They include pushes, pulls, friction, weight, upward forces which support the weight of an object, and others. These forces manifest themselves in the physical behavior of *macroscopic* objects, that is, objects large enough to be seen with the unaided eye. The macroscopic domain of physics includes the study of physical phenomena which can be observed directly by our senses, or which can be described in terms of their effect on macroscopic objects. Roughly speaking, macroscopic physics is concerned with phenomena which take place on a scale of size which is 0.1 mm (10^{-4} meter) or larger.

In contrast with the macroscopic domain is the realm of *microscopic* physics, the study of phenomena and objects which require the use of a microscope of some kind for their observation. Phenomena which require us to take into account the behavior of microscopic particles such as individual molecules, atoms, and nuclei lie in the domain of microscopic physics. Nearly all of the material in this book is concerned with phenomena of macroscopic physics. The textbook *Microphysics* is primarily concerned with microscopic physics.

We use the term *resultant force* to refer to the vector sum of several forces acting on a particular object. If we now consider the individual macroscopic forces which act on an object, we find in many cases that each of them is itself a resultant of other forces. If you throw a ball, for example, during the throwing motion the ball is in contact with many parts of your hand so the total force of your hand on the ball is the resultant of many individual macroscopic forces. When we consider other macroscopic forces between two objects, we find in every case that such a force is the resultant of many individual forces. If we pursue this approach and ask about the individual macroscopic forces, we find that every macroscopic force turns out to be the resultant of *microscopic* forces between the atoms of the macroscopic objects which are interacting with each other. The every-day forces we observe are not themselves fundamental forces, but are resultants of individual fundamental forces between atomic particles. The complexity of forces on the macroscopic level is a result of the fact that so many atomic or fundamental forces are acting at the same time. On the atomic level, the nature of the forces is much simpler.

We use the term *fundamental force* to refer to a force which is not itself the resultant of other forces. This means that fundamental forces are the ones which act on individual particles of matter (electrons, protons, neutrons and other elementary particles). Fundamental forces cannot be "explained" by attributing their origin to some other kind of interaction. Because the fundamental particles of matter are so small, and often come within a short distance of one another, all of the fundamental forces in nature are microscopic forces. They must be taken into account in order to understand physical phenomena on the microscopic scale.

Whether a fundamental force also manifests itself on the macroscopic scale depends on the *range* of the force, that is, on the distance to which the force "reaches out" from a particle that can exert the force. Experiments in this and succeeding chapters show that of the four forces we call fundamental, two have a "long range" and manifest themselves clearly on the macroscopic scale of distances. The other two are never observed on the macroscopic scale, but reveal themselves only at very short range (less than 10^{-14} m), in fact, only within the immediate vicinity of an atomic nucleus!

Physical forces never occur in the abstract. Rather, a physical force is always exerted by some physical object on another physical object. The two objects exert forces on each other, or interact with each other, according to some definite law or laws of force. The word *interaction* is often used in physics to refer to the mutual actions of two particles on one another through a specific kind of fundamental force.

Four fundamental forces or interactions are known today. The two which are long-range forces are those of gravitation (discussed in this chapter) and electromagnetism (introduced in Chapters 56 and 58). The remaining two are called the *strong interaction* (sometimes called the nuclear interaction or nuclear force), and the *weak interaction*. The weak interaction manifests itself in the radioactive decay of nuclei and other particles.

When we examine the behavior of atomic particles we find that electromagnetic interactions are very much stronger than gravitational ones. As a result, the structure of atoms is governed entirely by electromagnetic interactions; gravity has a negligible effect. On the other hand, when we talk of interactions of large bodies like the ones encountered in astronomy, gravitational forces predominate. This is so because gravitational effects are cumulative when many atoms are present, whereas electromagnetic forces tend to cancel out because of the presence of both positive and negative charges in each body.

On the macroscopic scale, the forces we meet include pushes and pulls, as well as other kinds. When we consider forces on the microscopic scale, instead of referring to a push or a pull we describe a particular force as "attractive" or "repulsive." In order to find out whether any particular fundamental force is attractive, or repulsive, or both, experiments are required. It turns out that on the microscopic scale both attractive and repulsive forces occur, but there is no microscopic analog to the macroscopic force of friction.

In presenting this survey of physical forces, we have made some assertions about fundamental forces without giving the basis for these assertions. The scientific basis for our understanding of fundamental forces is developed more fully in subsequent discussions of each kind of fundamental force. Our purpose here is simply to introduce the concept of a fundamental force, and to give an overview of some aspects of these forces.

Continuing this broad-brush presentation, we may ask what holds the world together. From the standpoint of physics, the answer is that the physical world is held together by fundamental physical forces. In particular, the galaxies and our solar system are held together by gravitational forces. Buildings, people, and indeed all macroscopic objects on earth are held together by electromagnetic forces. Molecules and atoms, too, are held together by electromagnetic forces. Atomic nuclei are held together by the strong interaction or nuclear force.

There may be other fundamental interactions, in addition to the four we have mentioned, which haven't been discovered yet. It is also possible that there are fewer than four if it turns out that some of them can be combined into a single unified theory. The search for a successful unified theory of fundamental interactions is an important field of theoretical physics today.

Questions

Q 44.1-1 Here is a list of "forces": dramatic force, electromagnetic force, emotional force, ethnic force, frictional force,

gravitational force.
(a) Which of these are fundamental physical forces?
(b) Which of these are resultants of fundamental physical forces?
(c) Which of these are not physical forces at all?

Q 44.1-2 (a) Name the four forces presently thought by physicists to be fundamental.
(b) Give an example of a physical interaction or phenomenon where each predominates.

Q 44.1-3 Which of the fundamental forces can you experience directly in everyday life? Give an example of each.

Q 44.1-4 Since the nuclear force between protons in the nucleus is about 100 times as strong as the electrical force between protons in the nucleus, why don't we observe effects of the nuclear force directly in our daily experience?

Check your understanding of the material above by testing yourself on the minimum learning goal at the beginning of this section.

44.2 Universal Gravitation

Minimum Learning Goal. (1) Given the names of two objects in our solar system, be able to find necessary data in your text or another reference, and calculate the gravitational force between the two objects. For example:

1 Compute the gravitational force exerted on the earth by the moon. (Answer, page 44-17, Line 2)

2 Compute the gravitational force of the earth on the sun. (Answer, page 44-17, Line 4)

A freely falling object near the earth's surface accelerates downward with $g = 9.8\ m/s^2$, when the object is dropped or is thrown horizontally with a high speed. This acceleration is the result of the *weight* of the object, a downward force often called the force of gravity. Everywhere on the earth there is a tendency for objects to fall down, so it is natural to conclude that the downward force on an object is a pull or attractive force exerted on it by the earth.

Figure 44-1. When a bullet, a long-range projectile or a satellite is launched from the earth's surface, it falls toward the earth with the acceleration of gravity, g.

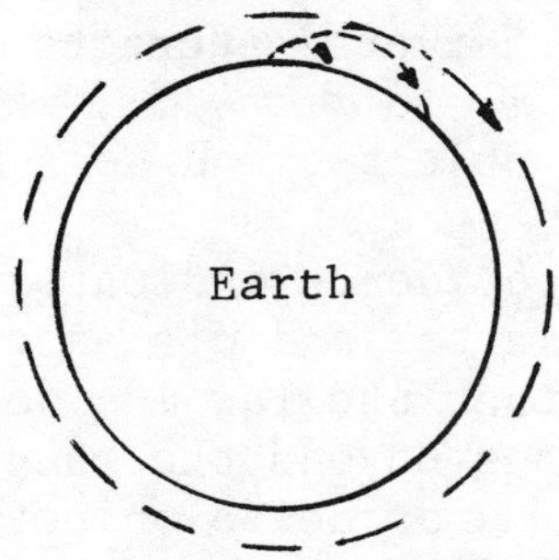

An object thrown or lunched at a *very* high speed, such as a rocket launched from Cape Kennedy, also falls toward the earth with the acceleration of gravity. Even a satellite in an orbit around the earth is pulled toward the earth by the gravitational force, which causes a downward acceleration so the satellite falls toward the earth, or perhaps we should say it "falls around the earth."

These examples show that no matter where we place or throw an object, the object is always pulled toward the earth. Even the Moon, nearly a quarter of a million miles from us, is in an orbit around the earth, so it too is being pulled toward the earth. We conclude that there is some physical property of the earth which produces a force on every other physical object and causes that object -- if it is free to move -- to accelerate toward the earth. The more distant planets in the solar system are so far from the earth that the pull of the earth's gravity on them is too small to be observed, but these planets *are* accelerating toward the sun. Therefore, we conclude that it is the sun's gravity which pulls on the planets and keeps them in the solar system. On a still larger scale, our solar system is in motion around the center of the Galaxy, so it is reasonable to suppose that this acceleration, too, is the result of a gravitational force.

Generalizing these observations, we find that wherever we look in the universe there is evidence for gravitational forces. Further study confirms the view that there is a Universal Law of Gravitation operating throughout the universe. This law was discovered by Isaac Newton about 1666. He found that the gravitational force on an object is directly proportional to the *mass* of the object. When two objects exert gravitational forces on each other, these forces are therefore proportional to the product of the two masses, $m_1 m_2$. From his studies of the motion of the moon around the earth, Newton concluded that the gravitational force decreases as the distance, d, between the two bodies increases, and is inversely proportional to the square of the distance. Combining these two results,

$$F \propto m_1 m_2 \quad \text{and} \quad F \propto \frac{1}{d^2},$$

Newton obtained the

★ *Universal Law of Gravitation:* Every material object in the universe attracts every other material object with a force $F = G\, m_1\, m_2/d^2$, where m_1 and m_2 are the masses of the two objects, and d is the distance between their centers.

Available evidence indicates that this force law is valid over the entire range of masses and distances which we can observe. We cannot be certain, however, that the law may not need to be modified to account for future observations beyond the range of our experience to date, either for exceedingly massive or dense objects, or for fundamental particles separated by very small distances.

The constant of proportionality in the universal law of gravitation is the universal gravitational constant,

$$G = 6.67 \times 10^{-11}\ \mathrm{N{\cdot}m^2/kg^2}.$$

Because it is so small, it is very difficult to measure the gravitational constant with high precision. The measurement was first made by Henry Cavendish in the 1790's. Using great care and patience, the "Cavendish experiment" can be repeated in an undergraduate physics laboratory.

In essence, the universal law of gravitation deals with the force between two *particles*, that is, between bodies so small that their spatial extent can be considered negligible. There is no difficulty in defining what we mean by the distance between two particles, nor in determining the exact direction along which the force acts, namely, along the line between the particles. When we deal with bodies of extended size, however, the total gravitational force between them must be viewed as the vector sum of the various attractions between all possible pairs of particles that constitute the bodies. This is usually a much more complicated problem.

One special case has a simple solution (though not a simple proof that the solution is correct). This is the case of two massive bodies each having *spherical symmetry*. Spherical symmetry means that, measuring from the center of each object, the mass distribution is independent of the direction along which we go. An example of such a body is a solid sphere having uniform density, or one whose density varies only with the distance from the center but not with "latitude" or "longitude" within the sphere. In the case of spherically symmetric bodies which are not overlapping, it can be shown (using methods of integral calculus) that the equation for the gravitational force law is the same as for particles, except that the distance d in this equation must be taken as *the distance between the centers of the two bodies*. The resultant gravitational force on an object having spherical symmetry acts on the center of the object.

In practice, these limitations on the validity of the simple equation for the gravitational force do not prevent us from applying the equation to cases of interest where the gravitational force is significant. In considering the gravitational force of the earth or the sun on some other astronomical body, the distance *between* the two bodies is so much greater than the diameter of either body that when one body is viewed from the other, it "looks like" a small object in the distance, so that any irregularities or inhomogeneities in the distribution of mass within either body become insignificant.

Compared with the other three fundamental forces, the gravitational force has some features which give it special interest. For one thing, it is the only fundamental force which acts on *all* the particles and objects in nature. In fact, recent work has shown that it even influences non-material particles such as photons (Section 30.9). Gravitational forces are only attractive, whereas electromagnetic forces can be attractive or repulsive. It is a long-range force, decreasing to zero only for two objects whose distance of separation approaches infinity, so it operates on both the macroscopic and microscopic scales of distance. Because the gravitational constant is so small, however, the force has a noticeable effect only when at least one of the two bodies has a very large mass, comparable with the mass of a moon or larger. The gravitational force between any two objects on the earth is negligible in comparison with the weight of either object. The earth's mass is so great, however, that the gravitational force of the earth on any object, including you, is so large that it is easily the most conspicuous of all the fundamental forces we encounter in daily life.

The gravitational force law may be used to find the mass of a planet having a natural or artificial satellite in orbit around it. What is required is a measurement of the radius of the orbit, and a measurement of the time required for the satellite to complete one orbit. These data allow one to compute the speed of the satellite in its orbital motion, and also the mass of the planet. If a planet has no moons or artificial satellites, its mass can still be estimated by observing the effect of the planet in perturbing, or changing slightly, the motion of another planet in its orbit around the sun, as a result of the gravitational force between the two planets. In fact, the planet Neptune was discovered as a result of perturbations observed in the motion of Uranus. The motion of a satellite in response to the gravitational force on it is discussed in Section 44.4.

For astronomical bodies such as stars, planets, and satellites, the gravitational force is by far the largest force and in fact it completely dominates their motions. Appendix G contains a table of numerical astronomical data useful in calculations of graviatational forces.

Worked Example 44-1

(a) Compute the gravitational force of the earth on the moon, to two significant figures. Numerical data are given in Appendix G.
(b) Is the force computed in part (a) larger than, equal to, or smaller than, the gravitational force of the moon on the earth?

Solution

(a) Using the subscripts E for Earth and M for Moon, the force is

$$F = G\frac{m_E m_M}{d_{EM}^{\ 2}}$$

$$= 6.7 \times 10^{-11}\ \frac{\mathrm{N{\cdot}m^2}}{\mathrm{kg^2}} \times \frac{6.0 \times 10^{24}\ \mathrm{kg} \times 7.4 \times 10^{22}\ \mathrm{kg}}{(3.8 \times 10^{8}\ \mathrm{m})^2}$$

$= 2.1 \times 10^{20}$ newtons, directed toward the center of the earth.

(b) The two forces are equal. This can be seen by writing the equation for the force of the moon on the earth and calculating, or simply by applying Newton's third law of motion (Section 42.5).

Questions and Exercises

Q 44.2-1 (a) What is the greatest distance from Earth that an object can have, yet still be acted on by the earth's gravitational force?
(b) In qualitative terms, how far into space is the gravitational force of the earth the major portion of the total gravitational force on an object?

Q 44.2-2 The sun's mass is about 330,000 times the earth's mass. Is the gravitational force of the sun on the earth less than, the same as, or more than the gravitational force of the earth on the sun? Explain.

Q 44.2-3 It is sometimes said that the gravitational force on an object is the result of other material objects "in the vicinity." What exactly is the "vicinity"?

Q 44.2-4 Discuss the possibility of constructing a shield which would screen out the earth's gravitational field. Suggest some uses for such a shield.

E 44.2-1 Using the universal law of gravitation, show that for any two objects the gravitational forces $\vec{F}_{21}$ and $\vec{F}_{12}$ which the objects exert on each other satisfy Newton's third law.

E 44.2-2 (a) Compute the gravitational force of the sun on the moon.
(b) Compare your answer with part (a) of Worked Exampled 44.1. Which exerts the dominant force on the moon: the earth, or the sun?
(c) Would you expect the moon to follow an orbit around the earth, or around the sun, or both?
(d) It is an interesting fact that the path of the moon in space is always concave toward the sun.

E 44.2-3 Compute the gravitational force of the sun on Mars.

E 44.2-4 Estimate reasonable values for the physical quantities involved, and compute the gravitational force between two opposing linemen when both are crouched at the line of scrimmage before the snap of the football.

E 44.2-5 Lead has a density of 11.3×10^3 kg/m^3. If we could have two spheres of lead in contact with each other, what diameter should they have to attract each other with a force of one pound (4.45 newtons)?

E 44.2-6 Compare the gravitational forces due to the earth and due to the sun at the place where you are. Do you think that your apparent weight on a scale is a little less at noon, when the sun is pulling up on you, than at midnight, when the sun is pulling down on you? (For a complete answer to this question, one must consider the construction of the scale and also the effects of rotating reference frames.)

Check your understanding of the material above by testing yourself on the minimum learning goals at the beginning of this section.

44.3 Gravitational Field of a Planet

Minimum Learning Goals. (1) Given the name of a planet or moon in our solar system, be able to find necessary data and calculate the acceleration of gravity near the surface of that planet or moon due to its own mass. For example:

1 Calculate the acceleration of gravity near the surface of Jupiter. (Answer, page 44-17 , Line 7)

2 Calculate the acceleration of gravity near the surface of the moon (of our earth). (Answer, page 44-17 , Line 5)

(2) Be able to describe by means of a graph the variation with altitude of the acceleration of gravity due to a planet. For example:

1 A planet has an acceleration of gravity of 22.7 m/s^2 at its surface. Correctly label the axes and draw a graph showing the acceleration of gravity (on the vertical axis) as it varies with the distance in units of planet radii from the surface of the planet (on the horizontal axis). Plot the graph for distances out to five times the radius of the planet. (Answer, page 44-17 , Line 8)

The universal law of gravitation, $F_g = G\, mM/d^2$, may be used to compute the gravitational force of a planet having mass M) on some other object (mass m) located at any distance d from the center of the planet. This force decreases as the distance of m from the planet increases. We have also used $w = mg$ to compute the gravitational force of the earth on a mass m, when m is located near the earth's surface where the acceleration of free fall is $g = 9.8\ m/s^2$. The use of the equation $w = mg$ may be extended to any altitude above the planet's surface, if we find the variation of the acceleration of free fall g with altitude.

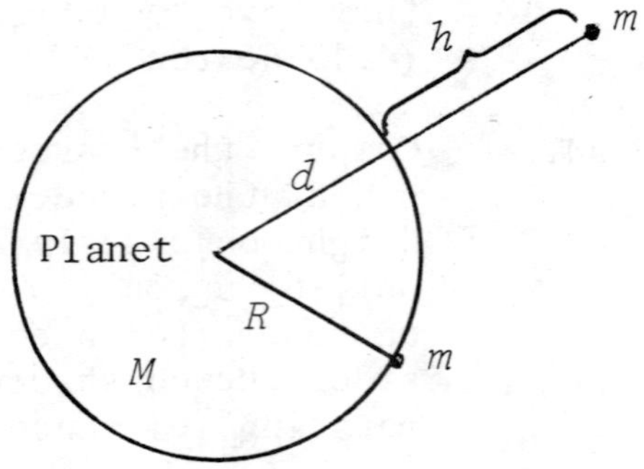

Figure 44-2. An object having mass m located near a planet's surface, and another mass m located at an altitude h above the surface. The planet has a mass M and a radius R.

We first derive an equation for g as a function of d (the distance from the center of the planet, *not* the diameter). Figure 44-2 shows two objects having mass m located at different distances from the planet's center. The gravitational force of the planet on either m may be calculated using the universal law, or simply using $F_g = mg$:

$$F_g = m\,g, \qquad \text{and also} \qquad F_g = \frac{G\,Mm}{d^2}.$$

Equating these two expressions for F_g, and then solving for g, we obtain

★ $$g = \frac{Gm}{d^2}, \qquad \text{for} \quad d > R.$$

This result applies at the planet's surface where d is equal to the radius R of the planet, and also at any altitude h above the surface, where $d = R + h$. The equation $g = Gm/d^2$ can be used to compute the value of g at any d, or the value of d required for any g, for any planet, satellite, or star whose mass is known.

If we introduce the symbol g_0 for the value of g *at the surface* of the planet, where d equals R, then by substitution we obtain

$$g_0 = \frac{GM}{R^2}.$$

This result shows that the value of g at the planet's surface depends simply on the universal constant G, and also on the mass and radius of the planet.

Some values of g at various altitudes are tabulated in Figure 44-3, and are also plotted there in a graph of g vs. altitude. As an object rises some distance above the surface of the planet (or other large body) which is attracting it, the acceleration of the object in free fall becomes weaker because of the inverse-square law.

Figure 44-3. The acceleration of free fall (g) at various altitudes above the surface of a planet. The planet's radius is R, and the altitude is h.

h	d	g
0	R	g_0
R	$2R$	$g_0/4$
$2R$	$3R$	$g_0/9$
$3R$	$4R$	$g_0/16$

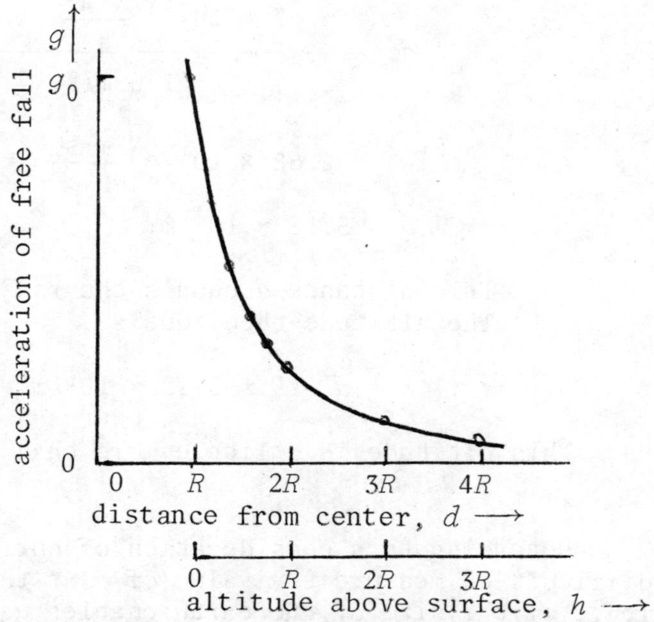

Worked Example 44-2

(a) Compute the acceleration of free fall at the surface of Mars.

(b) At what altitude above the surface of Mars does the acceleration of free fall have a value of 1.6 m/s^2?

Solution

(a) To find the value of g at the *surface* of the planet, we need to know the mass and radius of Mars. In Appendix G, we find that for Mars the mass equals 6.42×10^{23} kg, and the radius equals 3.39×10^6 meters. Substituting in $g_0 = GM/R^2$, we calculate the acceleration of free fall at Mars' surface:

$$g_0 = \frac{GM}{R^2} = \frac{6.67 \times 10^{-11}\ \frac{\text{N}\cdot\text{m}^2}{\text{kg}^2} \times 6.42 \times 10^{23}\ \text{kg}}{(3.39 \times 10^6\ \text{m})^2}$$

$$= 3.73\ \frac{\text{N}}{\text{kg}} = 3.73\ \text{m/s}^2.$$

The units of N/kg are converted to m/s^2 by substituting $1\ N = 1\ kg{\cdot}m/s^2$, obtained from Appendix D inside the front cover.

(b) We are given a value for g, and asked to find the corresponding altitude. Before we can find the altitude h above the surface, we need to find the total distance d from the center of Mars. To do this, we solve the equation $g = GM/d^2$ for $d^2 = G\,M/g$. Substituting numerical values, we have

$$d^2 = \frac{6.67 \times 10^{-11}\ \frac{\text{N}\cdot\text{m}^2}{\text{kg}^2} \times 6.42 \times 10^{23}\ \text{kg}}{1.6\ \text{m/s}^2}$$

$$= 2.68 \times 10^{13}\ \frac{\text{N}\cdot\text{m}\cdot\text{s}^2}{\text{kg}} = 26.8 \times 10^{12}\ \text{m}^2, \quad \text{from which}$$

$$d = 5.18 \times 10^6\ \text{m}.$$

This distance d equals the radius of Mars plus the altitude. The altitude then equals

$$h = d - R = 5.18 \times 10^6\ \text{m} - 3.39 \times 10^6\ \text{m} = 1.79 \times 10^6\ \text{meters}.$$

This altitude is a little more than half the radius of the planet.

Returning to a consideration of our planet, the earth, it is not difficult to measure the value of g at the earth's surface. Then a knowledge of the size of the earth enables us to calculate the product GM_E. After the value of G was found in the Cavendish experiment, it was possible to compute the mass of the earth.

There is some variation in the value of g from place to place on the surface of the earth. This is partly due to the fact that the earth does not have perfect spherical symmetry. In addition to surface irregularities such as mountains, there are geological irregularities in density; moreover, the overall shape of our planet is not spherical but somewhat "squashed at the poles.

There is another cause for variation in g which is apparent rather than real, but must be taken into account if the rate of acceleration of falling bodies is measured relative to the earth which is itself rotating. As we found in Section 43.5 , because of the earth's rotation the surface of the earth is actually accelerating inward. If an object is dropped toward a point on the earth's surface, that point toward which it is falling is itself accelerating and this will influence any relative acceleration measurement and affect the measured value of g. Thus to an observer fixed on the earth it appears that the acceleration of gravity (for example, at the equator) is less than it would be if the earth were not rotating. This "centrifugal" diminution of g depends on latitude and disappears completely at the poles.

Questions and Exercises

Q 44.3-1 Name four physical causes for the slight variations in g which occur from place to place on the earth's surface.

E 44.3-1 What is the acceleration of a freely falling object at an altitude of 6.4×10^6 meter (about 4000 miles) above the earth's surface?

E 44.3-2 What is the gravitational acceleration of an artificial earth satellite when it is 9×10^6 meters from the center of the earth?

E 44.3-3 (a) Compute the acceleration g_m of a freely-falling object near the surface of the moon.
(b) Compute g_v near the surface of Venus.
(c) Compute g_j near the surface of Jupiter.

E 44.3-4 An astronaut with his space suit and portable life-support system weighs 350 pounds (1550 N) on earth. What would be his weight:
(a) on the Moon?
(b) on Venus?
(c) on Mars?
(d) on Jupiter?
Hint: Use the results of E 44.3-3 and Worked Example 44-2.

E 44.3-5 Using information from Appendix G, compute the altitude above the surface of Jupiter at which the acceleration of gravity is 6.6 m/s^2 (one-fourth the value at the planet's surface).

E 44.3-6 Sketch the value of g vs. altitude h above the earth's surface, for values of h from 0 to 20 kilomiles. How high above the earth's surface can you consider the value of g to be the same as that at the earth surface, with an error not greater than 1%?

Check your understanding of the materials above by testing yourself on the minimum learning goals at the beginning of this section.

44.4 Satellite Orbits

Minimum Learning Goal. Given a quantitative description of an actual circular orbit of an earth satellite, be able to state whether a proposed orbit which differs from the first only in one or two parameters is also possible as an actual orbit, and if it is not possible, state the reason in terms of a physical principle. For example:

1 A satellite having a mass of 200 kg is in a circular orbit around the Earth at an orbital radius of 4400 miles, with a speed of 17,000 miles/hour and a period of 97.5 minutes.
(a) Can this satellite be speeded up so that the period is only 90 minutes while remaining in a circular orbit at the same radius? If it cannot, state the reason in terms of a physical principle.
(b) Can this satellite be separated into two satellites, one having a mass of 50 kg and the other having a mass of 150 kg, with both of these new satellites remaining in the same orbit as the original satellite? If it cannot, state the reason in terms of a physical principle.
(Answer, page 44-17 , Line 3)

Since the Soviet Union launched Sputnik I in 1957, many hundreds of artifical earth satellites have been placed in orbits around the earth. The motions of these satellites, as well as the motions of the moon around the earth and of the planets around the sun, are examples of the motion of a particle under the action of the force of universal gravitation and no other. Although the great majority of these orbits are in fact elliptical, some of them are close to being circular, and we can understand many aspects of satellite motion by considering an earth satellite in a circular orbit. To do this, we need to combine the physics of circular motion discussed earlier in this chapter with the universal law of gravitation.

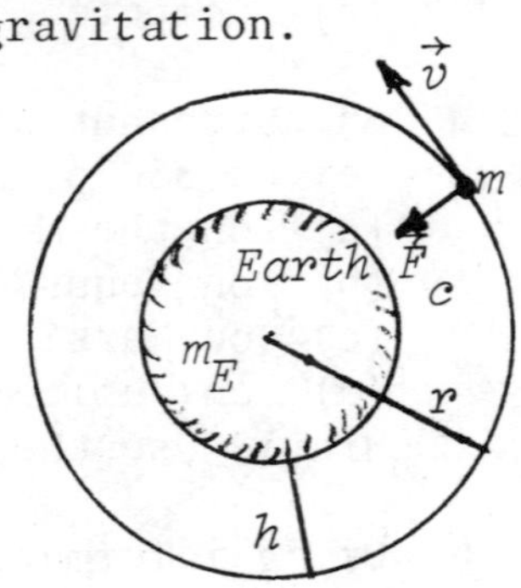

Figure 44-4. The circular orbit of an earth satellite, having a radius r measured from the center of the earth, and an altitude h measured from the earth's surface.

The altitude of the satellite above the earth's surface is h, and the radius of its orbit measured from the earth's center is r (Figure 44-4). The mass of the satellite is m, and the earth's mass is m_E.

The centripetal force required for uniform circular motion is

$$F_c = m\,a_c = \frac{m\,v^2}{r} .$$

The force which pulls the satellite toward the center of its orbit is the gravitational force of the earth on the satellite,

$$F_{grav} = \frac{G\,m_E\,m}{d^2} .$$

The distance d between the centers of the earth and the satellite is equal to r, the radius of the orbit. To maintain the satellite in a circular orbit, the gravitational force of the earth must exactly equal the required centripetal force:

★ $$\underbrace{\frac{G\,m_E\,m}{r^2}}_{\text{force exerted}} = \underbrace{\frac{m\,v^2}{r}}_{\text{force required}} , \qquad \text{which leads to} \qquad v^2 = \frac{G\,m_E}{r} .$$

The last equation above shows that for a given orbital radius the satellite can have only one value of speed, which is independent of the mass of the satellite. For a satellite in a circular orbit having radius r, the gravitational force of the earth is exactly equal to the centripetal force required at this radius. An attempt to speed up the satellite while maintaining the *same* radius would fail, because the force exerted by the earth would remain constant while the centripetal force required would be larger for the higher speed. On the other hand, an attempt to increase the radius while maintaining the same speed would also fail because the available gravitational force would diminish according to the square of the radius while the required force would decrease only as the first power of the radius. The period of a satellite's motion around the earth is given by $\tau = 2\pi r/v$, and since there is only one speed for a given radius, it follows that there can be only one value of period for a given radius. Furthermore, the period does not depend on the mass of the satellite. If a very massive satellite and a very light one were placed in circular orbits at the same altitude, they would travel with the same speed and take the same amount of time to encircle the earth.

When a satellite is launched and placed in orbit around the earth, after the engine of the final stage of the launching rocket assembly burns out the satellite has some velocity relative to the earth. Regardless of the direction of this velocity vector, there is a plane which contains the velocity vector and the earth's center. This plane also contains the vector force of the earth's gravitational pull on the satellite, and therefore it contains the centripetal acceleration vector of the satellite. As a result, the satellite's motion continues in this plane which contains the earth's center. Because the earth's equatorial diameter is larger than the polar diameter, the plane of the satellite's orbit does not remain constant in space but rather rotates slowly around the earth's polar axis. Nevertheless, it is correct to visualize the satellite's orbit as lying in a plane through the center of the earth.

Our discussion of satellite orbits has assumed that the force of atmospheric drag on the satellite is negligible. This assumption is valid for satellites having an altitude of more than 500 km above earth's surface, and it is approximately correct for orbits located somewhat lower. When a satellite's orbit brings it within approximately 100 miles (160 km) of the earth's surface, however, the atmospheric drag becomes quite significant with the results that the satellite is heated because of its friction with the air, and its orbit drops still lower. Interestingly, although the drag is directly opposed to the motion of the satellite, an indirect result of the drag is that the satellite actually gains speed and travels faster and faster. The heating also increases, so that a satellite or space capsule not protected by a heat shield will "burn up" as its path approaches an altitude of about 87 miles above the earth.

Questions and Exercises

Q 44.4-1 For a circular satellite orbit at a specified altitude above the earth, how many values of speed are possible?

Q 44.4-2 How do the period and speed of a satellite orbit depend on the mass of the satellite?

Q 44.4-3 What propulsive force is required to maintain a satellite in its orbit?

Q 44.4-4 Why does the orbit of an earth satellite always lie in a plane through the earth's center?

Q 44.4-5 If an astronaut "walking in space" accidentally detaches himself from the rope which holds him to the space craft, so that he floats freely in space, what motion will he have? Will he be in orbit? If he is, will he move faster, or slower, than the more massive space capsule?

Q 44.4-6 Is an orbiting satellite in a state of free fall? Explain.

Q 44.4-7 If the gravitational force of the moon on the earth is always directed toward the moon, is the earth in circular motion about the moon? Explain.

E 44.4-1 An astronaut wants to put himself into a circular orbit around his space capsule. The mass of the capsule is 10,000 kg, and he wants the radius of the orbit to be 10 m. How long will it take for him to complete one revolution about the capsule?

E 44.4-2 Show that for an earth satellite in a circular orbit, $a_c = GM_E/r^2$.

E 44.4-3 Show that for an earth satellite in a circular orbit having radius r meters, the period is $\tau = 2\pi r\sqrt{r}/\sqrt{GM_E} = 3.15 \times 10^{-7}\, r\sqrt{r}$.

E 44.4-4 (a) Compute the speed of the Apollo command module in a circular orbit 69 miles above the moon's surface.
(b) What is the period of this space ship in its lunar orbit?

E 44.4-5 (a) Calculate the period of a satellite circling the earth 22,300 miles above the earth's surface. (Hint: Use the result of E 44.4-3.)
(b) Calculate the orbital speed of this satellite.
(c) (This is a THINK problem and does not require a numerical calculation if you have solved part (a) correctly.) Find the altitude required for a satellite which would appear to remain stationary overhead.
(d) Could this satellite be placed in an orbit such that it would appear to remain stationary over your college?

E 44.4-6 (a) Compute the gravitational force of the sun on the earth.
(b) Compute the centripetal acceleration of the earth in its orbit, using the result of part (a).
(c) Compute the approximate speed of the earth in its orbit, using the result of part (b).
(d) Compute the period of the earth's revolution about the sun in seconds and in days, using the result of part (c). Compare with your calendar.

E 44.4-7 Two stars having equal mass m exert gravitational forces on each other with the result that each star moves with uniform circular motion in a circular orbit of radius r about a center point midway between the two stars, Each star has a speed v. (a) If $m = 3 \times 10^{30}$ kg and $r = 3 \times 10^{9}$ meter, what is the gravitational force which one star exerts on the other? (b) Compute the speed of each star.

Check your understanding of the material above by testing yourself on the minimum learning goals at the beginning of this section.

Answers to Minimum Learning Goals

Line 1 All four fundamental forces.

Line 2 2.0×10^{20} N

Line 3 (a) No; for a circular orbit at a given radius, only one speed is possible, namely, the speed at which the required centripetal force just equals the available gravitational force.
(b) Yes

Line 4 3.6×10^{22} N

Line 5 1.67 m/s^2

Line 6 Strong interaction and weak interaction.

Line 7 26.5 m/s^2

Line 8

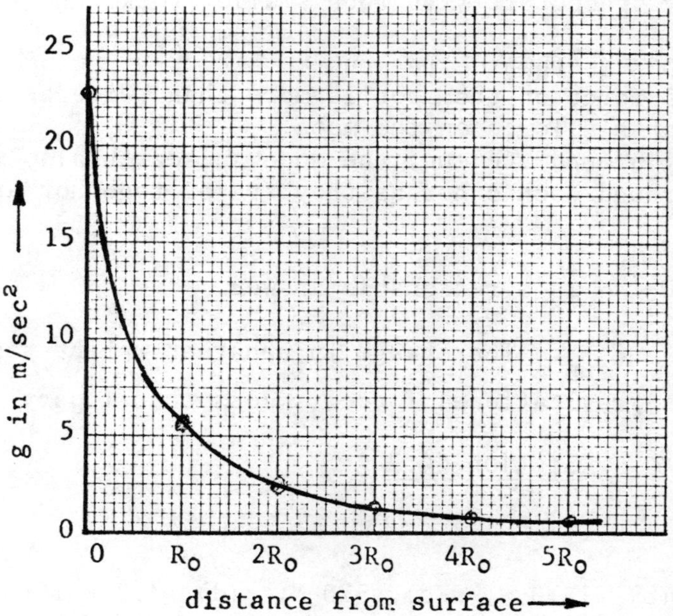

Line 9 Gravitation

ANSWERS to Questions and Exercises in Chapter 44

Section 44.1

Q 44.1-2 (1) Gravitational; interaction between Earth and Sun.
(2) Electromagnetic; interaction between atomic nucleus and atomic electrons.
(3) Strong or nuclear; interaction between protons and neutrons within a nucleus.
(4) Weak; the radioactive process in which a nucleus ejects a beta particle.

Q 44.1-4 Because the nuclear force is limited to a short range of less than 10^{-14} m.

Section 44.2

Q 44.2-2 The same. Each of these forces depends on the same product of (Earth's mass) × (Sun's mass).

Q 44.2-4 Such a shield would require a repulsive gravitational force, which (as far as is known) does not occur in nature. If such a shield could be constructed, it would permit the launching of interplanetary spacecraft with much smaller rocket engines.

E 44.2-2 (a) 4.35×10^{20} N, toward the sun.
(b) The sun's force on the moon is more than twice as large as the earth's force on the moon.
(c) Primarily around the sun; the moon and earth also move in orbits around each other as they travel together around the sun.

E 44.2-4 A few times 10^{-7} N is a reasonable answer.

E 44.2-6 $F_{earth} = 1655 \times F_{sun}$. The change in apparent weight resulting from the sun's gravitational force is less than 1 part in 1655.

Section 44.3

E 44.3-2 4.9 m/s^2

E 44.3-4 (a) 264 N; (b) 1410 N; (c) 570 N; (d) 4.2×10^3 N.

E 44.3-6 If g_0 equals the value of g at Earth's surface, then $g = g_0$ up to an altitude of 20 miles if we can tolerate an error of 1%.

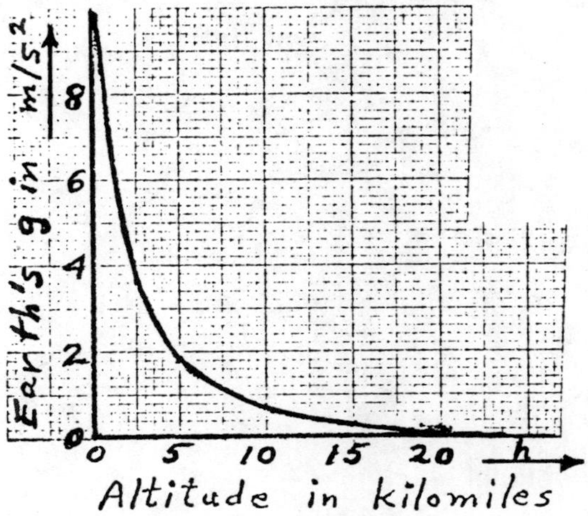

Section 44.4

Q 44.4-2 They are independent of the mass.

Q 44.4-4 See discussion in the next-to-last paragraph of this section.

Q 44.4-6 Yes; it is moving under the action of only one force, the earth's gravitation.

E 44.4-4 (a) 7.13×10^3 s = 119 minutes; (b) 1.63×10^3 m/s

E 44.4-6 (a) 3.5×10^{22} N; (b) 5.8×10^{-3} m/s^2; (c) 3.0×10^4 m/s; (d) 3.15×10^7 s = 365 days.

CHAPTER 45 FRAMES OF REFERENCE

Introduction

When a casual observer sees something happen, or a scientist makes measurements on a physical phenomenon, he or she does so from a particular physical frame of reference. Each physical frame of reference includes (at least implicitly) devices for measuring the spatial locations and times of physical events. A class of reference frames having special importance are the inertial frames, those in which Newton's first law of motion is valid. For many practical purposes, the earth serves as an inertial frame. Consideration of the motion of objects as viewed from two different inertial frames leads to the concepts and equations of Newtonian relativity. When motion is viewed from an accelerating frame of reference, "fictitious forces" appear. Some fictitious forces considered in this chapter include inertial forces, centrifugal forces, and Coriolis forces.

When you have completed your study of the material in this chapter, you should be able to answer questions and work exercises like the examples in the minimum learning goals assigned by your instructor.

45.1 Events and Frames of Reference

Minimum Learning Goals. (1) When you are given a description of something which is purported to be an *event*, be able to state whether this is a single event, and if not, state why not. For example:

1 A pianist strikes the note G-sharp as she begins to play Beethoven's *Moonlight Sonata* in Carnegie Hall. Is this a single physical event? If not, state why not. (Answer, p. 45-31 Line 13)

2 The 274 members of the graduating class from Einstein College parade across the platform to receive their diplomas from the dean. Is this a single physical event? If not, state why not. (Answer, p. 45-31, Line 6)

3 On June 21 it is noon when at a longitude of 90° west, the sun crosses the local meridian. Is this a single physical event? If not, state why not. (Answer, p. 45-31, Line 7)

4 A cartesian coordinate system in Room 2053 of Randall Laboratory at the University of Michigan has an origin (the point where $x = y = z = 0$). Is this a single physical event? If not, state why not. (Answer, p. 45-31, Line 17)

(2) When you are given a description of equipment which is to be used to make measurements on an event, be able to state whether this equipment constitutes an adequate frame of reference for these measurements, and if not, state why not. For example:

1 A physics laboratory contains a cartesian coordinate system with meter sticks positioned along the axes beginning from the origin, and a set of synchronized clocks positioned at several locations in the laboratory. Jim intends to use this equipment to measure the time and position at which a falling ball strikes the taple top. He uses a small iron ball and an electric magnet. The ball is originally supported by the magnet and he is able to release the ball by disconnecting the power source of the magnet. Does this equipment constitute an adequate frame of reference for this purpose? If not, state why not. (Answer, p. 45-31, Line 10)

2 Fred wants to photograph a lightning bolt. He mounts a movie camera so that he can take moving pictures of the sky during an electrical storm, with a precision electric clock positioned in the field of view of the camera. During a storm he succeeds in getting one picture which shows a lightning flash at 10:18:47.5 pm. Does this equipment constitute an adequate frame of reference for the measurement of a lightning event? If not, state why not. (Answer, p. 45-31, Line 15)

3 A team of physics students is required to plan and perform an experiment in which they make measurements on an event in everyday life. They decide to measure the precise time and position at which a touchdown is scored in the next college football game. To do this, they plan to use the gridiron markings as their spatial coordinate system, and to take moving pictures of the ball carrier as he crosses the goal line, using a camera mounted in a helicopter which flies above the playing field. The helicopter also carries a precision clock positioned in the field of view of the camera. While the fullback carries the ball on a long run to a touchdown, the helicopter flies above him with the camera running. Does this combination of the gridiron markings and the helicopter constitute a frame of reference? If not, state why not. (Answer, p. 45-32, Line 19)

Suppose you are studying the motion of a freely falling ball, and you decide to take photos of the ball using a stroboscopic light source which flashes at regular intervals of 0.05 second. The ball is dropped from a stand fixed to one wall of the laboratory. The camera is fastened to a carrier which can glide on a long straight horizontal track parallel to this wall (Figure 45-1). The room is darkened, the strobe light is turned on so that the ball is illuminated while it is falling, and the camera shutter is opened for one second just as the ball is released from rest and allowed to fall vertically. How will the motion of the ball appear in the photo?

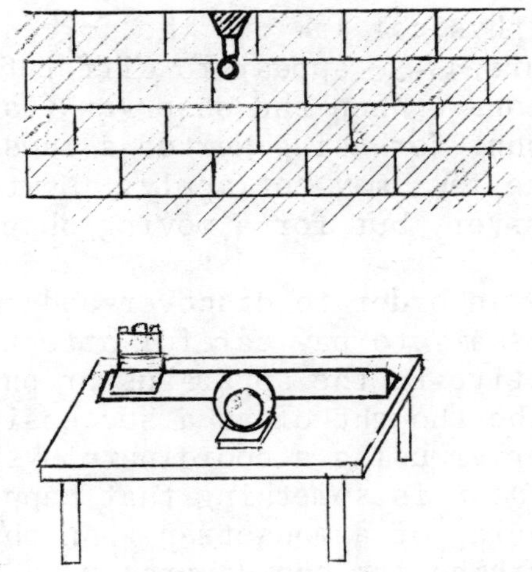

Figure 45-1. Equipment for measuring the positions of a falling ball at successive instants of time. The ball can be dropped from its supporting stand attached to the laboratory wall. The camera can move parallel to the wall on an air glider. A stroboscope on the table provides flashes of illumination at regular intervals of time.

If the camera is stationary while the ball is falling, the path of the ball as shown in the photo appears as a straight vertical line (Figure 45-2a). If the camera and the cart are moving at constant speed, however, the path appears to be curved (Figure 45-2b); the mathematical form of the curve is a parabola. If the camera and cart are accelerating while the ball is falling the curvature of the path differs from that observed when the camera's velocity is constant; if the camera has a constant acceleration, the path shown in the photo is a straight line (Figure 45-2c) provided the camera and the ball start from rest at the same time.

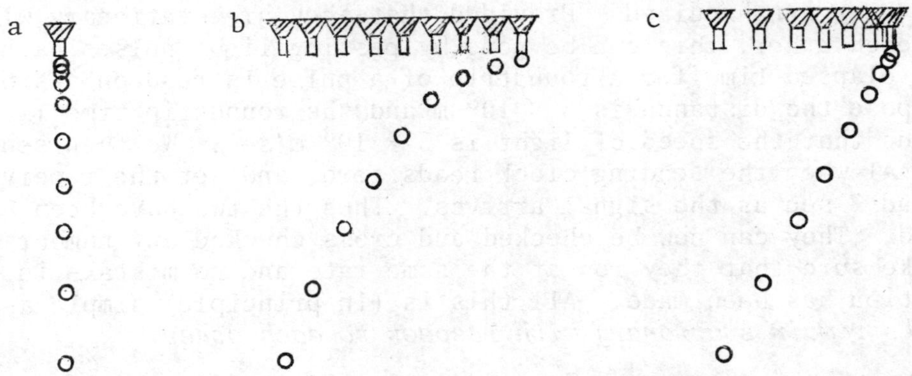

Figure 45-2. Paths of a falling ball obtained using the equipment shown in Figure 45-1. On each path, positions of the ball are shown at intervals of 0.05 second. (a) The path in a stationary frame of reference. (b) The path in a frame of reference moving to the right with constant velocity. (c) The path in a frame moving to the right with constant acceleration; frame and ball begin moving at the same time.

It is evident from this example that the motion observed for the ball depends not only on the acceleration of free fall, but also on the motion of the camera, that is, of the observer. In Chapter 41 the motion of a falling ball was chosen as one of the simplest motions we could study, yet even for this simple motion the description (using a table of data obtained from a photo, or a graph, or equations) depends on the motion of the observer. For the three motions observed in Figure 45-2a, 2b, and 3c, the descriptions differ and therefore the explanations will also differ. That is, the physical law and the forces needed to account for the motion of Figure 45-2c appear to differ from the laws required for the other two motions. When the observer (camera) is stationary the motion is easy to account for using Newton's laws of motion, but when the observer is moving these laws may not apply. Newton's laws are valid for a stationary observer, but for a moving observer their validity is in question.

In order to discover and to apply dependable physical laws, it is necessary to pay careful attention to the state of motion of the observer relative to the apparatus or phenomenon being observed. The phenomenon may be thought of as a succession of events, which are studied by an observer using a coordinate system located in a particular laboratory. An *event* is something that happens at one place in one instant, like the snapping of a mousetrap. An observer locates an event in space and time by giving its coordinates x, y, z and the time t when it occurred. An event must be thought of as something that actually happens, by which a point in *spacetime* can be identified. The actual values of x, y, z, t, on the other hand, will depend on the coordinate system or, as we say in this case, the *frame of reference* in which they are measured.

The concept of a frame of reference must also be carefully defined. Consider a situation where a pulse of light is propagating and we wish to measure its speed. To do this, we must know its positions at two different times. This requires some type of light-detecting equipment to be stationed at two places along its path, each piece of equipment having a clock attached to it so that the time when the pulse passes can be recorded.

Obviously, the whole arrangement is useless unless we know that the clocks have been synchronized. Provided that they are stationary with respect to each other, this can be done by passing light pulses back and forth. The elapsed time for a roundtrip of a pulse is read on a single clock. Suppose the distance is 6×10^8 m and the roundtrip time is 4 sec; this tells us that the speed of light is 3×10^8 m/sec. We then send out another signal when the sending clock reads zero, and set the receiving clock to read 2 sec as the signal arrives. Thus the two have been synchronized. They can now be checked and cross checked any number of times to make sure that they run at the same rate and no mistake in synchronziation has been made. All this is (in principle) simple as long *as the clocks remain stationary with respect to each other.*

> *A frame of reference* is an arrangement of physical equipment, including a coordinate system and synchronized clocks, which is suitable for measuring the locations in space and time of events of interest.

We assume that a frame of reference has whatever equipment is required at whatever place for the measurements under consideration, but that these various pieces of apparatus are *at rest relative to each other.*

Suppose that in the experiment shown in Figure 45-1 a meter stick is fastened to the wall in a vertical position with its upper end near the stand from which the ball is dropped. Then this meter stick and the other apparatus used constitute an adequate frame of reference for this experiment *as long as the camera remains stationary.* If a meter stick were fastened to the wall, it would appear in the photo of Figure 45-2a alongside the successive positions of the falling ball, and it could be used to measure the positions of the ball at the times of successive flashes of the strobe light (which serves as the clock in this experiment). The meter stick, strobe light and camera thus constitute a frame of reference for measurements based on the photo of Figure 45-2a, but *not* for measurements based on Figures 45-2b and 45-2c because in those cases the meter stick is moving relative to the clock. If the camera and clock were supplemented by meter sticks which were at rest relative to the camera (e.g., fastened to the moving cart), then this equipment would serve as a frame of reference for all three photos in Figure 45-2.

Relativity is the branch of physics that is concerned with relationships among quantitites measured in different frames of reference, but referring to the same physical events. This chapter deals with relative motions in which the moving objects have speeds which are small compared with the speed of light. (That is why we can allow the strobe light to be moving with respect to the camera in the frames of reference described above.) When an object or a frame of reference has a speed which is not small compared with the speed of light, it is necessary to use a more complicated theory of relativity to account for physical observations. This is the *special theory of relativity* formulated by Albert Einstein in 1905.

Questions

Q 45.1-1 Lightning strikes the top of the Capitol dome in Washington, D. C. Is this a single physical event? If not, state why not.

Q 45.1-2 During an electrical storm, the electric power to the entire town of Ether goes off at 5:18:32 p.m. Is this a single physical event? If not, state why not.

Q 45.1-3 An electric current of 0.35 ampere flows through a lamp located 1.2 meter east and 1.7 meter north of the southwest corner of the ceiling of Laboratory Room 142 in the Fitzgerald Science Building at Shamrock College. Is this a single physical event? If not, state why not.

Q 45.1-4 The traffic signal facing south and located above the center of the intersection of Space Street and Time Avenue in the town of Gamow turns green at regular intervals of one minute. Is this a single physical event? If not, state why not.

Q 45.1-5 How can a physicist measure the coordinates x, y, and z and the time t of an event, such as the emission of an alpha particle by a radioactive source?

Q 45.1-6 Give an example of a device which could be used to measure the position and time of arrival of a light pulse at a specific point in space.

Q 45.1-7 In doing a free-fall experiment, a student wants to measure the space and time coordinates of the falling ball as it impacts on the floor. He carefully sets up a three-dimensional cartesian coordinate system having one axis vertical. Does this equipment constitute an adequate frame of reference for his purpose? If not, state why not.

Q 45.1-8 Using an astronomical telescope which has a time calibrated photo-electric recording attachment, an astronomer records an apparent eruption of smoke from a volcano on the moon. Does this equipment constitute an adequate frame of reference for the measurement of this eruption? If not, state why not.

Q 45.1-9 In order for an observer to obtain adequate measurement of events which interest him, why is it essential for his clocks and his spatial coordinate system to be located in the same frame of reference?

Q 45.1-10 A laboratory contains two parallel tracks one meter apart, and on each track is a cart which rolls along the track. Clock A is mounted on Cart 1, while on Cart 2 is a magnet holding a ball which can be dropped from a height h, an xyz coordinate system, and Clock B. The clocks are connected to electronic devices so that Clock B records the time at which the ball is released and Clock A records the time when the ball hits the cart.

For which of the situations described below, is this equipment an adequate frame of reference for measuring the time and position at which the ball strikes the cart? If the equipment does not constitute an adequate frame of reference, state the reason.

(a) The two carts pass position P simultaneously and move forward with the same constant velocity. The clocks are synchronized.

(b) The two carts pass position P simultaneously and move forward with the same constant velocity. The clocks are not synchronized.

(c) The clocks are synchronized. The two carts pass position P simultaneously. Cart 1 moves with constant velocity, while cart 2 has an acceleration of 0.1 m/s^2.

Q 45.1-11 How can an event be in two different frames of references at the same time?

Check your understanding of the material above by testing yourself on the minimum learning goals at the beginning of this section.

45.2 Inertial Frames of Reference

Minimum Learning Goals. (1) Given a description of a frame of reference, be able to state whether the frame is inertial, and state the reasons for your decision. For example:

1 A train having a mass of 200,000 kg moves at a constant velocity of 7.5 m/s. The train contains an observation car equipped with synchronized clocks, a three-dimensional

coordinate system, and other apparatus. Is this car an inertial frame? Explain your answer. (Answer, p. 45-31, Line 3)

2 A balloon moves upward with constant mass, increasing volume, and decreasing velocity. The gondola of the balloon contains synchronized clocks, a coordinate system, a motion-picture camera, and other apparatus. Is this an inertial frame of reference? Explain your answer. (Answer, p. 45-31, Line 8)

(2) When given a description of an experiment as seen by an observer, be able to state whether or not the observer is in an inertial frame of reference. For example:

1 A glider having a mass of 1.0 kg resting on a horizontal air track is pushed with a horizontal force of 0.5 N, and is observed to move along the track with an acceleration of 1.0 m/s^2, under the influence of only this force. Is the observer in an inertial frame? (Answer, p. 45-31, Line 1)

2 A baseball thrown straight upward is observed by Kathy to move with an initial velocity of 19.6 m/s^2, rising straight upward and reaching the highest point of its motion 2.0 sec after it was thrown. Is Kathy in an inertial frame? (Answer, p. 45-31, Line 11)

(3) Given a description of several frames of reference, one of which is an inertial frame, be able to state which of the others are also inertial frames of reference. For example:

1 Assuming that the earth is an inertial frame of reference, state which of the following are also inertial frames, assuming that each contains synchronized clocks, a coordinate system, and other equipment needed to measure the spacetime positions of events:
(a) A train traveling at a constant speed of 60 mph on a curved track.
(b) A satellite in a circular orbit around the earth.
(c) A car accelerating at 2.0 m/s^2 on a flat, level road.
(d) A spacecraft headed for the great galaxy in Andromeda at a constant velocity of one-half the speed of light.

(Answer, p. 45-31, Line 5)

Suppose we have a horizontal linear air track with the air supply turned on, and a glider at rest on the track. The upward force of the air on the glider just balances the weight of the glider acting downward, and so the glider remains at rest. Now if the glider is pushed at one end so that it moves along the track, after the push we would observe that the glider moves with constant velocity (until it strikes the bumper at the end of the track). We can account for this motion at constant velocity by noticing that the upward forces on the glider again cancel each other, and because there is no horizontal force on the glider the net force on the glider is zero. This is an application of Newton's first law of motion (Section 42.1).

Figure 45-2a showed the motion of a ball falling vertically downward with an acceleration of 9.8 m/s^2. In this case the ball is subjected to the downward force of its weight, *mg*, which gives the ball (mass *m*) an acceleration of *g*. We account for this acceleration by applying Newton's second law of motion (Section 42.3).

Newton's second law states that the acceleration of an object is equal to the resultant force on the object divided by the object's mass. Newton's first law of motion is a special case of the second law, and states that the acceleration of an object is zero when the resultant force on the object is zero. Newton's first law is sometimes called the *law of inertia*. For the motions of the glider and the falling ball described above, Newton's first and second law hold. Implicit in this statement about Newton's laws is the fact that these motions were observed from a stationary frame of reference, that is, stationary with respect to the air track and to the laboratory room. Because Newton's laws of motion are fundamental in any attempt to account for observed motions, frames of reference in which they hold are given a special name; they are called *inertial frames of reference*. For motion at speeds approaching the speed of light, Newton's second law takes a different form, but Newton's first law does not, so it is customary to define an inertial frame of reference simply as one in which Newton's first law of motion is valid. When an object is observed to accelerate in an inertial frame, this acceleration can be accounted for using the proper form of Newton's second law.

In order to find out whether a particular frame of reference is an inertial frame, the fundamental approach is to observe the state of motion of some object in that frame of reference, and the forces acting on the object. If the acceleration is zero when the resultant force is zero, Newton's first law is valid and the frame is an inertial one. Alternatively, if the resultant force is not zero and the resulting acceleration can be accounted for using Newton's second law, the frame is inertial. We have already pointed out that the motion of the falling ball shown in Figure 45-2a conforms to Newton's second law, so that the stationary camera and coordinate system used to obtain Figure 45-2a are an inertial frame. Figure 45-2b was obtained using a camera moving with constant velocity in the laboratory; relative to the camera, the horizontal velocity of the ball is constant while its downward acceleration is again 9.8 m/s^2. Figure 45-2b, therefore, shows a ball acted on by the downward force of its weight and having an acceleration of *g* downward. Once again the motion is exactly accounted for by Newton's second law, and so we conclude that the camera moving at constant velocity (together with an attached coordinate system) is also an inertial frame of reference.

Figure 45-2c was obtained using a camera moving with increasing speed. It shows the motion of a ball acted on only by its weight, so that we would expect the acceleration to be 9.8 m/s^2 directed straight downward. The increasing spacing of the successive positions of the ball shows, however, that the ball's acceleration has components in *both* the horizontal and vertical directions. This motion does not conform to Newton's second law, and so we conclude that the accelerating camera (with attached coordinate system) does *not* constitute an inertial frame.

Suppose we consider only the stand fixed to the wall (Figure 45-1) from which the ball was dropped in obtaining Figures 45-2abc. Because the stand is fixed in the laboratory, the resultant force on it is zero and its acceleration will be zero in every inertial frame. Figures 45-2a and 2b show that its acceleration is zero when the camera is stationary and when it is moving with constant velocity, which confirms our conclusion that the frames of reference in which the camera is stationary and in which it moves with constant velocity are inertial frames. In Figure 45-2c, however, the stand is accelerating to the left, so we again conclude that the accelerating frame of reference used to observe Figure 45-2c is not an inertial frame. From the point of view of an observer who is stationary in the laboratory, the stand has no motion but the camera is accelerating to the right. From the point of view (or frame of reference) of an observer moving with the camera, however, the camera is stationary but the stand and indeed the entire laboratory are accelerating to the left. Thus the question of whether an object is accelerating or not depends on the observer's frame of reference. Newton's first law cannot possibly hold in all frames of reference, because the same object may be unaccelerated as seen from one frame but accelerated as seen from another.

The same thing can happen if two observers use different clocks, one of which is speeding up or slowing down with respect to the other. If one of these observers sees a particle in uniform motion in a straight line, the other will see it in non-uniform motion, because the particle will cover unequal distances in what this observer considers equal time intervals.

We have seen that Newton's laws of motion are valid in a stationary frame of reference and also in a frame moving with constant velocity relative to the stationary one. From this result we may infer that all reference frames moving at constant velocity relative to one another are equivalent: if one of these frames is an inertial frame, all of them are inertial frames. However, the statement "All frames of reference moving with constant velocity relative to each other are equivalent" is true only if the coordinate axes in these frames maintain fixed directions in space. If one of the frames is rotating relative to an inertial frame, the rotating frame is not inertial. Rotating frames of reference are considered in Section 45.6.

We have assumed in this section that the earth is an inertial frame, because measurements on objects moving in a laboratory building on the earth can be accounted for using Newton's laws of motion. We know, however, that the earth rotates about its own axis once each 24 hours. The earth as a frame of reference is discussed further in the next section.

Questions

Q 45.2-1 Margarita claims to be in an inertial frame of reference. What experiment could she do which would convince you that she is indeed in an inertial frame?

Q 45.2-2 Which one of the following frames of reference is most nearly an inertial frame?
- (a) A spacecraft traveling from Earth to Moon with its engines off.
- (b) A car traveling at 55 miles per hour from Detroit to Cleveland.
- (c) The physics laboratory in your school
- (d) The elevator car in a downtown building.

Q 45.2-3 Newton's first law says that a particle with no forces acting on it will move without acceleration when measured in an inertial reference frame. An inertial reference frame is one in which Newton's first law holds. This sounds like reasoning in a circle! If we pick our reference frame so as to make the law work, it is hardly surprising that in this frame the law works. Can you explain why, in spite of that, Newton's law is an important law of physics, subject to experimental verification, and not just a semantic trick?

Q 45.2-4 How can you determine experimentally whether your physics lab is an inertial frame? Do you have any theoretical reasons for believing that your physics lab is or is not an inertial frame? What are they?

Q 45.2-5 Observer *A* sees a certain particle subject to no forces, moving uniformly in a straight line. Observer *B* is sitting on a swinging pendulum. He also has a kooky clock which slows down and speeds up in synchronism with the swinging pendulum in such a way that, when he studies the motion of the same particle *A* is studying, he, too, comes to the conclusion that the motion is uniform in a straight line. Would you be willing to call *B*'s frame of reference inertial? Explain.

Q 45.2-6 Suppose you are inside a windowless and perfectly quiet jet plane moving at uniform velocity. To amuse yourself you begin bouncing a super ball up and down.
(a) Can you tell from watching the ball whether you are moving?
(b) Suppose that you cut a hole in the floor and let the ball fall out. Neglecting air resistance, would you see the ball fall straight down? What path would the ball follow according to an observer on the ground?
(c) Now suppose the pilot begins to accelerate the plane in the forward direction. You again release the ball from rest and allow it to fall to the floor. What path does it follow? Does this allow you to say that the plane is accelerating?

Check your understanding of the material above by testing yourself on the minimum learning goals at the beginning of this section.

45.3 The Earth as a Frame of Reference

<u>Minimum Learning Goal</u>. Considering the rotation of the earth, be able to distinguish between an inertial frame of reference and a non-inertial frame of reference by selecting from a descriptive list those which most closely approximate inertial frames. For example:

1 Which two of the following five frames of reference are most nearly inertial frames? (Answer, p. 45-32, Line 20)
(a) The surface of the earth at a latitude of 40° North.

(b) An airplane flying smoothly due West along the Equator at a constant altitude measured from sea level and at a ground speed of 1038 mph (the speed of rotation of the earth).
(c) A railroad car traveling East along level ground at 100 mph at 40° North latitude.
(d) A railroad car traveling West along level ground at 100 mph at 40° North latitude.
(e) A level platform at the North pole rotating relative to the earth once each day in a direction from East to West.

At sea level on the earth's equator, the acceleration of gravity which results from the gravitational pull of the earth is $g = 9.78\ m/s^2$, directed toward the center of the earth. Accordingly, we might expect that the measured acceleration of free fall at the equator would also be $9.78\ m/s^2$. However, a precise measurement of the acceleration of a freely falling body at the equator yields a value of $9.75\ m/s^2$, $0.03\ m/s^2$ smaller than expected. This discrepancy of $0.03\ m/s^2$ shows that the observed motion of free fall at the equator cannot be accounted for simply by using Newton's laws and the force of gravity, and we conclude that the earth is not exactly an inertial frame of reference. The measured acceleration of free fall differs by only about three parts in a thousand from the expected value, so that the departure of the earth from being an inertial frame is not noticed when measurements are made having a lower degree of precision. That is why the earth may be regarded as an inertial frame for most practical purposes, as we did in Section 45.2. When measurements of motion having a high precision are desired, however, the non-inertial nature of the earth as a frame of reference must be taken into account.

The departure of the earth from being an inertial frame is a result of the earth's rotation about its own axis. This can be shown by calculating the centripetal acceleration of an object on the earth's surface near the equator (Section 43.5) which is $0.033\ m/s^2$. One can then account for the fact that the measured acceleration of free fall at the equator is less than the expected value by considering that while the ball is falling, the earth's surface is *also* falling away from it with an acceleration of 0.033 m/s^2. Such a statement, of course, presupposes a frame of reference fixed in space from which this (centripetal) acceleration of the earth's surface can be observed. An alternative explanation may be given using a frame of reference fixed to the earth's surface; then one may say that in addition to the weight mg of the ball pulling it toward the earth, there is an outward centrifugal force equal to 0.033 times the mass of the ball which opposes the force of gravity. A centrifugal force invoked in this way is not a real physical force, however, but is rather a "fictitious force" which is invented in order to continue using Newton's laws in a non-inertial frame. The role of fictitious forces in non-inertial frames is discussed in Sections 45.5 and 45.6.

The statement that the earth is rotating about its own axis presupposes some "fixed" frame of reference with respect to which this rotation can be measured. Such a frame is provided by those stars whose distance from the earth is so great that their apparent motion relative to each other is not observable from Earth. These are sometimes referred to as "fixed stars," although we should not suppose that they are really fixed or stationary in space. When the stars (and the sun) are observed from Earth to move across the sky from East to West in circles around the Earth's polar axis, this apparent motion results from the Earth's rotation on its axis from West to East within the frame of reference provided by the stars.

Experimental evidence for the Earth's rotation about its axis is also provided by the motion of a Foucault pendulum. This is a massive spherical bob supported by a long wire, preferably ten meters or more in length. A pendulum of this sort which is properly supported and is protected from disturbing air currents tends to maintain its plane of oscillation in a plane *fixed in space* (that is, relative to the fixed stars) even though the earth rotates. As a result, the plane of oscillation is observed to rotate about a vertical axis through the pendulum's point of support with a period of several hours, in the sense *opposite* to that of the earth's own rotation. The exact value of the period depends on the latitude of the pendulum's location, and it differs from the Earth's period of 24 hours because of the Coriolis force (Section 45.6). Carefully constructed Foucault pendulums may be found in several of the major science museusm and in a few university physics departments.

If the Earth did not rotate on its axis, it would still depart from being an inertial frame because of the Earth's orbital motion around the sun. This motion has a centripetal acceleration of 0.006 m/s^2 (see Worked Example 47-1).

Check your understanding of the material above by testing yourself on the minimum learning goal at the beginning of this section.

45.4 Newtonian Relativity. Addition of Velocities

Minimum Learning Goal. (1) Be able to use the Newtonian velocity addition law to find the velocity of an object in one inertial frame of reference, when you are given the velocity of the object in another inertial frame and the velocity of the second frame relative to the first. For example:

1 A passenger on a moving train is walking in the direction of motion of the train with a speed of 2.1 m/s relative to the train. The velocity of the train on the track is 12.2 m/s westward.
- (a) What is the velocity of the passenger relative to the ground?
- (b) What would be his velocity relative to the ground if he were walking in the opposite direction on this train?

(Answer, p. 45-31, Line 2)

2 A boat, Belinda, is anchored and at rest in a lake while a second boat, Marcy, is crossing the lake with a velocity of

1.6 m/s in the westward direction. Albert walks on the deck of the Marcy with a velocity of 0.8 m/s toward the north relative to the Marcy. What is the velocity of Albert relative to the Belinda? (Answer, p. 45-31, Line 14)

3 When its anchor is raised, the Belinda drifts southward with a velocity of 0.5 m/s relative to the ground below, while the Marcy continues with a velocity of 1.6 m/s westward relative to the ground. What is the velocity of the Belinda relative to the Marcy? (Answer, p. 45-31, Line 9)

If you swim or row a boat across a river so that your motion relative to the water is perpendicular to the direction of water flow, then your motion relative to the river banks is *not* perpendicular to the direction of water flow because the water carries you some distance downstream while you are crossing the river. When an airplane is flying in one direction through (or relative to) the air while the wind is blowing the aircraft in a different direction, the velocity of the aircraft relative to the ground is different from its velocity relative to the air; we say that the moving air carries the plane "downwind." While a ball player is riding his bicycle in a straight line with constant speed, if he drops a baseball it appears to him that the ball falls straight down, whereas another person watching from the side would see that the ball's motion has a forward component as well as a downward component relative to the ground. In each of these examples, the velocity of a moving object relative to the gound is different from its velocity relative to the frame of reference of the moving medium (the water, or the air or the bicycle).

To calculate a velocity for motion of the type described above, one must use an equation or law for the addition of velocities. In order to derive such an equation, we note that in each example some object is moving simultaneously with respect to two frames of reference. The boat has one velocity in the frame of reference which moves with the water, and another velocity in the frame attached to the ground. The plane has one velocity relative to the air, and a different velocity relative to the ground. The baseball moves with respect to the bicycle (one frame of reference), but it has a different motion with respect to the ground (another frame).

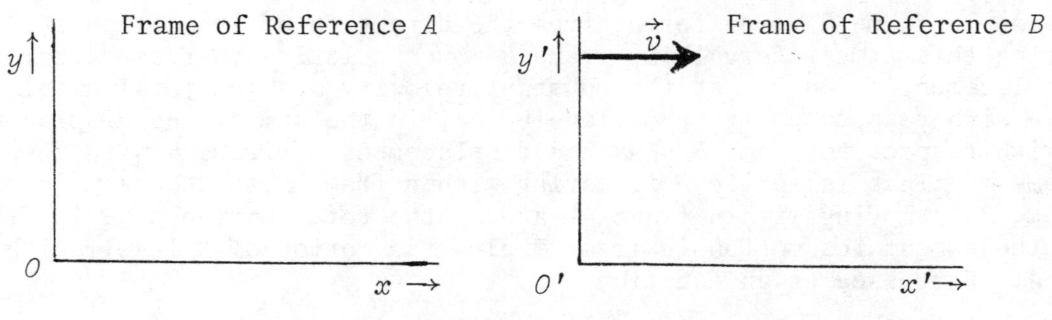

Figure 45-3. Two frames of reference. Frame *B* moves with velocity $\vec{v}$ in the $+x$ direction, relative to Frame *A*. The x' axis is parallel to the x axis, and the y axis is parallel to the y' axis.

To develop a general formula for velocity addition, we will label the first frame of reference *A*, and the second one *B*. As shown in Figure 45-3, for each frame we will consider two coordinate axes. (The third axis, perpendicular to these, can be included in a more advanced discussion but it is not necessary for our purposes.) In reference frame *A* the axes are labeled x and y; in frame *B* they are labeled x' and y'. Reference frame *B* has a velocity $\vec{v}$ in some direction with respect to frame *A*. We choose the directions of the x and x' axes to be parallel to $\vec{v}$ (and thus parallel to each other), and we also choose the directions of the y and y' axes to be parallel to each other. In Figure 45-3 it appears that frame *A* is stationary while frame *B* is moving, but because all inertial frames have equal standing, an observer who finds himself in frame *B* has as much right to consider that he is in the stationary frame while it is frame *A* that is moving (in the direction opposite to $\vec{v}$).

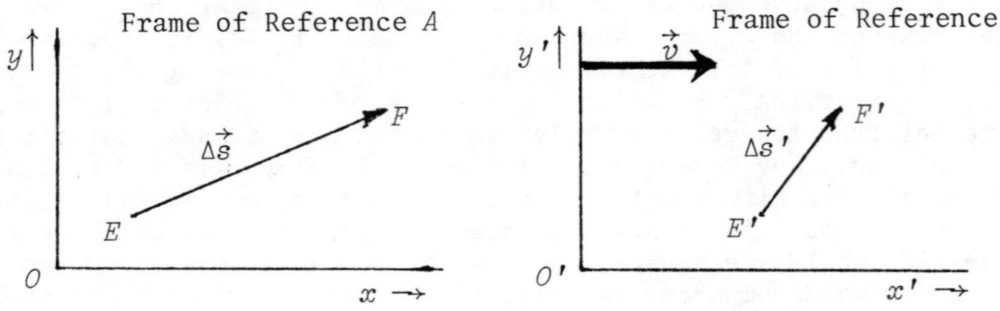

Figure 45-4. An object which moves from point *E* to point *F* in Frame *A*, moves during the same time interval from point *E′* to point *F′* in Frame *B*. Frame *B* moves through a displacement $\vec{d} = v\Delta\vec{t}$ during this same time interval.

Now consider a particle *P* which is moving with respect to frame *A* and also moving with respect to frame *B*. *P* may represent any moving object, such as a boat, an aircraft, a baseball or an atom. During some specific time interval Δt, *P* moves with constant velocity from point *E* to point *F* in frame *A*, and at the same time it is moving with a different constant velocity from point *E'* to point *F'* in frame *B*. The displacement from *E* to *F* is labeled $\Delta\vec{s}$, while the displacement from *E'* to *F'* is $\Delta\vec{s}'$. Considering *A* as the stationary frame, we note that while *P* is moving from *E* to *F* (and from *E'* to *F'*), frame *B* is also moving, with the result that the direction of *P*'s motion in frame *B* is different from the direction of its motion in *A*. During this time interval Δt, frame *B* moves relative to frame *A* through a displacement $\vec{d} = \vec{v}\,\Delta t$, at the constant velocity $\vec{v}$. The total displacement of *P* with respect to frame *A*, labeled $\Delta\vec{s}$, is the sum of the displacement of *P* with respect to frame *B* *plus* the displacement of frame *B* with respect to frame *A*. That is, while *P* is moving within frame *B*, at the same time frame *B* is moving within frame *A*, and so the total motion of *P* in frame *A* is the sum of its motion in frame *B* plus the motion of *B* itself with respect to *A*. Expressed as an equation

$$\Delta\vec{s} = \Delta\vec{s}\,' + \vec{d}.$$

The diagram for the addition of these displacement vectors is given in

Figure 45-5a. The equation above enables us to convert or transform the displacement of P in one frame of reference to its corresponding displacement in the other frame; it is one form of an equation called the *Newtonian transformation* (for Isaac Newton) *or Galilean transformation* (for Galileo Galilei).

Figure 45-5. (a) The displacement vectors $\Delta\vec{s}\,'$ and $\vec{d}$ are added to obtain the displacement $\Delta\vec{s}$. (b) The velocity vectors $\vec{v}_{PB}$ and $\vec{v}$ are added to obtain the velocity $\vec{v}_{PA}$.

In order to obtain the equation for transforming the velocity of P from one frame to another, we divide each term in the equation above by the time interval Δt. We obtain

$$\vec{v}_{PA} = \vec{v}_{PB} + \vec{v}_{BA}, \qquad \text{velocity addition law}$$

where

$\vec{v}_{PA} = \Delta\vec{s}/\Delta t$ is the velocity of P relative to frame A, $\vec{v}_{PB} = \Delta\vec{s}\,'/\Delta t$ is the velocity of P relative to frame B, and $\vec{v}_{BA} = \vec{d}/\Delta t$ is the velocity $\vec{v}$ of frame B relative to frame A. The equation above for velocities is called the *velocity addition law*, and the corresponding vector diagram is given in Figure 45-5b.

From this equation it is easy to see that if $\vec{v}_{PA}$ and $\vec{v}_{BA}$ are both constant in time, then $\vec{v}_{PB}$ will also be constant in time. Therefore, if the relative velocity $\vec{v}$ of the two frames is constant, then a particle moving in unaccelerated motion in one frame will also be moving with unaccelerated motion relative to the other frame.

If we apply this fact to a particle that is not interacting with anything else, that is, it has no forces acting on it, then if Newton's first law is valid in frame A it will also be valid in frame B. In other words, if A is an inertial frame, then B is also an inertial frame. This shows how, if we have found one inertial frame, we can immediately construct an infinity of others -- namely, all the frames that are moving with uniform velocity relative to the first.

Worked Example 45-1

Airplane A is flying in a northwesterly direction relative to the ground, at an angle of 122° measured ccw from east with a velocity of 320 mi/hr. Plane B is flying at a lower altitude in a southwesterly direction relative to the ground, at an angle of 238° measured ccw from east with a velocity of 190 mi/hr. A passenger in plane A looks out the window and watches plane B fly below him. What is the apparent velocity of plane B as seen by this passenger?

Solution

This Example has two moving objects, airplanes A and B. There is also the ground, which may be used as one frame of reference. Our chief interest lies in the apparent motion of plane B as seen from plane A, so we will regard plane A as a frame of reference and in fact plant A is stationary with respect to the observer (passenger) seated on the plane. In the frame of reference of the passenger on A, the second frame of reference (the ground, G) is moving at constant velocity $\vec{v}_{GA}$ relative to A; this velocity is exactly the opposite of the velocity $\vec{v}_{AG}$ of plane A relative to the ground. That is, $\vec{v}_{GA} = -\vec{v}_{AG}$ = 320 mi/hr in the direction 122° + 180° = 302° ccw from east.

We are given that the velocity of plane B relative to the ground is $\vec{v}_{BG}$ = 190 mi/hr in the direction 238° measured ccw from east. The quantity we want to find is the velocity of plane B relative to plane A, which we label $\vec{v}_{BA}$. The three velocities we use in this Example are thus $\vec{v}_{BA}$, $\vec{v}_{BG}$, and $\vec{v}_{GA}$. The equation which connects them may be obtained in either of two ways.

First, it may be obtained from the velocity addition law derived in this section by transposing subscripts in a systematic way. In the law, P represents the object which is moving in *both* reference frames; in our example this is plane B. In the law, A is the "stationary" reference frame, that is, the frame of reference of the observer of direct interest, which in this Example is the passenger on plane A. Finally, in the law B represents a frame of reference which is moving with respect to frame A; in this example this frame is the ground, G. We can therefore apply the general velocity addition law by replacing subscript P by B, subscript A by A, and subscript B by G. The velocity equation for this Example then becomes

$$\vec{v}_{BA} = \vec{v}_{BG} + \vec{v}_{GA}.$$

A second way to obtain the correct velocity addition law in a form suitable for this Example is to notice that in the general form of the law derived in this section, the velocity term on the left side has two subscripts; the first of these subscripts refers to the moving object of interest to the primary observer, while the second subscript refers to the reference frame of this observer. In our Example, the primary observer is the passenger on plane *A*, while the moving object of interest to him is plane *B*. The term on the left side of the velocity addition law is therefore $\vec{v}_{BA}$, the velocity of plane *B* relative to plane *A*. Returning to the general form of the law, we notice that the *first* subscript on the right side is the same as the *first* subscript on the left side, and also the *last* subscript on the right side is the same as the *last* subscript on the left side. The two "*inner*" subscripts on the right side are the same, and they are different from both the subscripts on the left side; they refer to a frame of reference which is moving with respect to the frame of the primary observer. This pattern of the first, last, and inner subscripts is always found in a correct statement of the velocity addition law, and it allows one to quickly write a correct equation provided the subscripts on the left side are written correctly and the second or "moving" frame of reference is identified. Applying this procedure to this Example yields the last equation above.

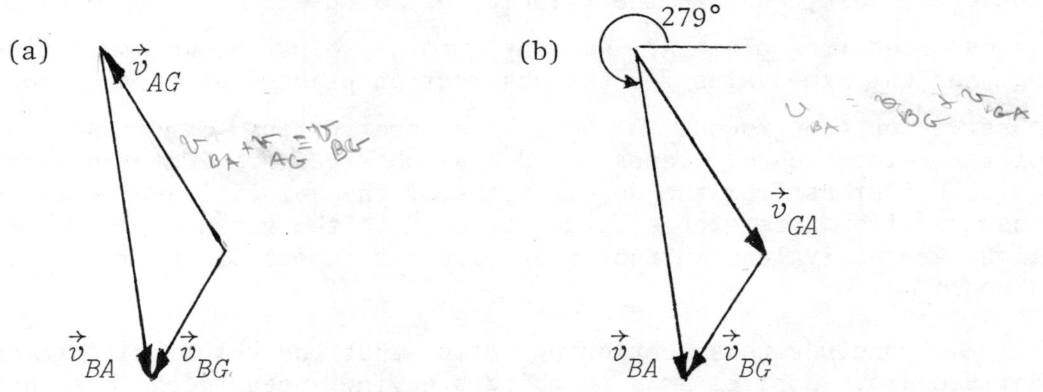

Velocities relative to the ground, *G* Velocities relative to airplane *A*

In order to visualize the velocities in this Example, we draw two vector diagrams.
Diagram (a) shows the velocities of the two planes relative to the ground, which is the frame of reference in which the data are supplied in this Example. $\vec{v}_{AG}$ is the velocity of plane *A* relative to the ground, while $\vec{v}_{BG}$ is the velocity of plane *B* relative to the ground. Diagram (a) also shows the vector velocity $\vec{v}_{BA}$, the velocity of *B* relative to *A*. While initially it may not be obvious that one can draw $\vec{v}_{BA}$ in this position, if we apply the rule for vector addition to this diagram we obtain

$$\vec{v}_{BG} = \vec{v}_{BA} + \vec{v}_{AG} ,$$

an equation which we know is physically correct because the order of the subscripts conforms to the standard pattern described above.

A more traditional approach to drawing a velocity addition diagram for this Example is used in diagram (b). This diagram begins with the given vector $\vec{v}_{BG}$ just as in diagram (a), and also the velocity

$\vec{v}_{GA}$ which is just the negative or reverse of the given velocity $\vec{v}_{AG}$. Using the equation we derived above for the unknown $\vec{v}_{BA}$, we see that $\vec{v}_{BA}$ must be the velocity vector drawn from the tail of $\vec{v}_{GA}$ to the head of $\vec{v}_{BG}$. In order to find the magnitude and direction of $\vec{v}_{BA}$, diagram (b) must be drawn carefully to scale and using a protractor. Then the magnitude and direction of $\vec{v}_{BA}$ may be measured directly from the drawing. The diagrams above have been drawn to a scale of 10 mi/hr per mm. Measuring $\vec{v}_{BA}$, we find that $\vec{v}_{BA}$ is about 437 mi/hr in the direction of 279° ccw from east. This result shows that the *apparent* speed of plane B, in the frame of reference of the passenger in A, is much larger than the speed of either plane A or plane B.

We could consider a more complicated Example in which the observer in plane A sees *two* other planes out of his window, plane B and plane C. We could then ask, what is the velocity $\vec{v}_{BC}$ of plane B relative to plane C, as observed from plane A? Working out the solution, we would find that $\vec{v}_{BC}$ has the same value for the observer on plane A as it does for an observer on the ground. Although the descriptions (magnitude and direction) of the velocities of planes B and C as observed from reference frame A are quite different from the descriptions of the velocities of these planes as observed from a reference frame attached to the ground, these two descriptions are equivalent in that they give the same *relative* motion of planes B and C.

We conclude this section by noting that the velocity addition law derived above applies *only* to objects having speeds much less than the speed of light. To deal correctly with the velocities of objects which travel at extremely high speeds (say, greater than 1% of the speed of light), it is necessary to use the special theory of relativity. The relativistic velocity addition law is much more complex than the one derived in this section.

Questions and Exercises

Note: When it is necessary to construct vector diagrams for the exercises below, a scale drawing will be satisfactory.

Q 45.4-1 An observer sees a particle moving with constant vector velocity. Does that mean that the observer's frame of reference is inertial?

Q 45.4-2 A particle with no forces acting on it has constant velocity relative to an inertial frame A. Will it have the same velocity relative to a different inertial frame B?

Q 45.4-3 A particle has constant acceleration relative to an inertial frame A. Will it have the same acceleration relative to a different inertial frame B?

Q 45.4-4 Suppose you are in an elevator which is controlled from outside and which has no lights or dials which indicate the floor reached by the elevator. Try to describe a method by which you could determine with certainty whether the elevator is at rest or is moving with constant speed.

Q 45.4-5 Consider a stationary *observer* (we'll call him Sam), a scientist or a student sitting on a laboratory stool, who is carefully watching some physical phenomenon. When Sam observes a physical event, he describes it using a coordinate system or *frame of reference* which is *stationary* with respect to his laboratory. We will also consider a moving observer, Mike, who in our imagination is stationed in a *moving* laboratory and who describes the same physical event with reference to his own coordinate system, moving with constant velocity relative to Sam's system. Is Mike entitled to say that his frame of reference is stationary, and that Sam's is moving? Explain.

E 45.4-1 Refer to the terminology used in Q 45.4-5. Sam describes an event, such as the instantaneous position of a moving particle or the impact of a stroke of lightning, by saying that the event occurred at the position x and at the instant t. Mike describes the same event by specifying the coordinate x' and the time t' at which the event occurred in his system. If Sam and Mike synchronize their clocks, and choose the origins of their coordinate systems so that O coincides with O' at the instant when $t = 0 = t'$, show that $x' = x - vt$, where v is the speed of Mike's frame relative to Sam's (in the $+x$ direction).

E 45.4-2 Refer to E 45.4-1. Consider a particle having mass m which is moving with constant velocity parallel to the x- and x'-axes. Sam measures the particle's velocity u in his frame, and Mike measures the particle's velocity u' in his frame.
(a) What is the relation between the values of u and u'?
(b) What is the relation between the values of the particle's acceleration as determined by Sam and by Mike?

E 45.4-3 Refer to E 45.4-1. (a) If Mike's system were *accelerating* in the $+x$ direction relative to Sam, how would the accelerations of the particle as computed by the two observers compare?
(b) Could Sam and Mike agree on the correctness of Newton's first law?

E 45.4-4 Refer to E 45.4-2. Derive the classical law of velocity addition:
$\vec{u} = \vec{u}' + \vec{v}$, where $\vec{u}$ = velocity of particle relative to Sam,
$\vec{u}'$ = velocity of particle relative to Mike,
and $\vec{v}$ = velocity of Mike relative to Sam.

E 45.4-5 A man leans out of a window high above the ground and throws a ball vertically upward. The ball rises until it is about 30 feet above the man (call him A), and then falls.
(a) Give a short description of the motion of the ball as seen by A, in the form of a qualitatively correct graph of position versus time. Think about the physical meaning of the slope of this graph, and draw the graph with reasonable slopes.

(b) At the instant that *A* throws the ball, an elevator (with a window) running up the outside of the building is passing *A* with the same upward speed that *A* gives the ball. The elevator continues upward with constant speed, carrying an observer, *B*, who watches the ball. Describe the motion of the ball as seen by *B*, in the form of a qualitatively correct graph of position versus time. (*B* forgets that he is moving and thinks that all the motion he observes belongs to the ball.)

(c) Another elevator runs beside the first carrying an observer, *C*, steadily upward with smaller constant speed. *C* is also just passing the window when *A* throws the ball. Describe *C*'s observations of the ball's motion, in the form of a qualitatively correct graph of position versus time.

(d) Note that *A*, *B*, and *C* all agree that the ball has the usual *acceleration*. For which variable(s) of motion do they disagree about the motion of the ball?

(e) Recall the definition of an inertial frame of reference. Which, if any, of these observers is in the most-nearly inertial frame of reference? If they are equally inertial, explain why.

E 45.4-6 Diana wants to run her motorboat from Point *A* on one side of the Nicki River to Point *B* directly across the river. The river has straight parallel sides, a width of 100 meters, and a speed of 6 km/hr. The speed of the small motorboat is 8 km/hour.

(a) If Diana heads her boat straight across the river, what will be the actual velocity of the boat relative to the river bank?

(b) If Diana heads her boat straight across the river, how far from point *B* will the boat be when she reaches the opposite bank?

(c) In what direction should she head her boat, relative to the line from *A* to *B*, in order to reach point *B* in the shortest distance?

E 45.4-7 A Boeing 747 has an air speed of 550 mi/hr while flying from Tucson to Chicago. The most direct route from Tucson to Chicago requires the plane to travel in the direction 29° ccw from east. Over the whole of his route, the pilot (Lee) encounters a north wind (blowing directly toward the south) which has a velocity of 100 mi/hr relative to the ground.

(a) If Lee does not correct his heading (from 29° ccw from east) to allow for the wind, what will be his actual velocity relative to the ground?

(b) If Lee is to fly to Chicago over the shortest distance in spite of the wind, what heading should he give the plane? That is, what should be the direction of the plane's velocity relative to the air?

E 45.4-8 Sue is riding her bike on a straight level road with a speed of 10 m/s when the strap holding her transistor radio suddenly breaks and the radio drops to the ground, falling straight down to the ground in Sue's frame of reference which continues forward with a constant speed of 10 m/s.

(a) If the radio falls for 0.4 second before it hits the ground, what is its velocity in Sue's frame of reference when it hits?
(b) What is the radio's velocity (magnitude and direction) in a frame of reference attached to the ground, when it hits the ground?

E 45.4-9 A car is traveling along a straight road 30 feet behind a truck, both vehicles going at a speed of 40 feet/sec. The driver of the car decides to pass the truck and applies a constant acceleration of 6.0 ft/sec^2, swinging into line again ahead of the truck and leaving a safe distance of 30 feet between his car and the truck. Assume the car's length is 16 feet, and the truck's length is 50 feet.
(a) How far did the car travel relative to the truck in this time?
(b) How long did the entire maneuver take, neglecting any effect of side-to-side motion?
(c) What frame of reference did you use in part (a)?
(d) How far did the *truck* travel along the road in this time?
(e) How far did the *car* travel along the road in this time?

Check your understanding of the material above by testing yourself on the minimum learning goal at the beginning of this section.

45.5 Accelerating Frames of Reference. Inertial Forces

Minimum Learning Goal. Given a description of a linearly accelerating frame of reference, and a description of a motion observed in this frame of reference, be able to state and calculate the forces required to account for this motion in the accelerating frame. For example:

1 Figure 45-2c shows the path of a ball which is dropped, as the ball's motion is observed in a frame of reference accelerating at 5.8 m/s^2 to the right. The ball (mass = 0.080 kg) appears to have an acceleration of 5.8 m/s^2 to the left while it is falling. The moving frame of reference (camera and glider) have a mass of 2.5 kg.
(a) Assuming that Newton's laws are valid in this frame of reference, what is the direction of the force required to account for this horizontal acceleration of the ball?
(b) What agent or object causes this force?
(c) Calculate the magnitude of this force. (Answer, p. 45-31, Line 16)

2 An elevator, ascending with an acceleration of 1.2 m/s^2 relative to the ground, is a frame of reference. A passenger on the elevator drops a ball and measures the acceleration of its falling motion to be 11.0 m/s^2. The ball has a mass of 0.14 kg, the passenger's mass is 70 kg, and the elevator's mass is 1450 kg.

(a) Assuming that Newton's laws are valid in this frame of reference, what is the direction of the force required (in addition to the weight of ball) to account for the ball's motion?
(b) What agent or object causes this force?
(c) Compute the magnitude of this force. (Answer, p. 45-31, Line 4)

The earlier sections of this chapter have concentrated on inertial frames of reference, those in which Newton's laws of motion are valid. For most practical purposes, the earth and every frame of reference moving with constant velocity relative to the earth may be regarded as an inertial frame. This section deals with frames of reference which have a linear or straight-line acceleration with respect to an inertial frame. Section 45.6 deals with frames of reference which are rotating with respect to an inertial frame. These accelerating frames of reference are not inertial frames, so we cannot expect Newton's laws of motion to apply in them, as long as we limit ourselves to real physical forces such as those presented in Chapter 42.

Consider a person seated in a car waiting at a red traffic light. When the light turns green and the driver "steps on the gas," the car accelerates forward and people seated in the car feel that they are being pushed back into their seats. Once the car is in motion, if the driver then suddenly applies the brakes the passengers in the car will lurch forward. If someone standing outside the car drops a ball just as the car begins to move and while the car is accelerating in the forward direction, the path of the ball as seen from the car is similar to that in Figure 45-2c (reproduced below in Figure 45-6a), showing that the ball accelerates backward relative to the car. If a ball is dropped outside the car while the car is decelerating (i.e., accelerating backward), the falling ball appears to accelerate in the forward direction. How can we account for these motions observed in accelerated frames of reference?

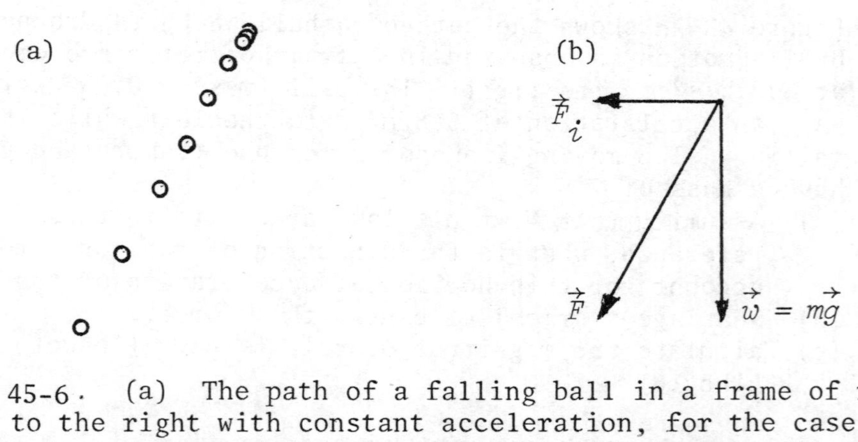

Figure 45-6. (a) The path of a falling ball in a frame of reference moving to the right with constant acceleration, for the case when the frame begins to move at the same time the ball is dropped. Positions of the ball are shown at intervals of 0.05 second. (b) The forces on the ball include its weight $\vec{w}$ and an inertial force $\vec{F}_i$. These combine to give the resultant force $\vec{F}$, which has the same direction as the ball's acceleration in the accelerating frame.

In Chapter 42 it was emphasized that the motion (or lack of motion) of every object is the result of the forces acting on the object. In keeping with this fundamental principle of physics, we would like to identify the forces responsible for the motions described above. To focus on a specific example, we consider the motion of a falling ball pictured in Figure 45-6a. The increasing intervals between successive positions of the ball show that the ball begins from rest and accelerates downward and to the left. From this, we expect that the resultant force $\vec{F}$ on the ball must be downward and to the left, as shown in Figure 45-6b. From our earlier study of forces we know that the one physical force acting on the ball during this motion is the weight of the ball, mg, which acts downward. The weight accounts for the downward component of the acceleration, but not for the leftward component. What is needed is an additional force, directed to the left. However, there is no real physical force directed to the left which acts on the ball. There is nothing in contact with the ball to push or pull it, nor can we attribute this force to some invisible physical agent such as a hidden magnet. We are in a quandary: Newton's law tell us there must be a force acting to the left in order to account for the acceleration to the left, but our knowledge of forces tells us that in fact there is no actual physical force acting on the ball and directed to the left.

In order to preserve the validity of Newton's laws in an accelerating frame of reference, we need to "invent" a force directed to the left. This is shown in Figure 45-6b, labeled $\vec{F}_i$. This is not a real force, but rather a *fictitious force* which we make up, simply pretending that it exists in order to "save" Newton's laws in an accelerated frame. In cases where the frame has a linear or straight-line acceleration, the fictitious force is called an *inertial force;* that is why it is labeled $\vec{F}_i$. An inertial force always has the same direction as the apparent acceleration of the object moving within the accelerated frame (or the direction this object would have if it were free to move), which is the direction *opposite* to the actual acceleration of the moving frame as observed from an inertial frame. The magnitude of an inertial force on a mass m is just the product ma required to account for its apparent acceleration a. As shown in Figure 45-6b, the resultant of the forces $\vec{w}$ and $\vec{F}_i$ is the force $\vec{F}$, having the same direction as the total acceleration of the ball in the inertial frame.

Using a similar approach, we can identify inertial forces for other motions in accelerating frames of reference. In the non-inertial reference frame of a car accelerating forward, the inertial force on the passengers acts in the backward direction, which they might describe as "pushing them back" in their seats. To an observer watching these passengers from an inertial frame, the fictitious inertial force is non-existent but the seats push forward on the passengers while they are being accelerated forward. In the reference frame of a car accelerating backward (or decelerating) while the driver applies the brakes, the inertial force on the passengers acts in the forward direction, giving them the experience of lurching forward in their seats when the car slows suddenly. To an observer watching these passengers from an inertial frame, again the so-called inertial force is non-existent but in accord with Newton's first law the passengers continue their forward motion when the car begins to go slower because the passengers are not decelerating as rapidly as the car.

Worked Example 45-2

The driver of a car applies the brakes suddenly, giving the car a deceleration of 6 m/s^2. A passenger (mass = 80 kg) in the car does not have his safety belt fastened and so when the car decelerates he continues forward (in the earth's frame of reference) with the same velocity which the car had when the brakes were first applied.

(a) Assuming that Newton's laws are valid in the frame of reference attached to the car, what is the direction of the force required to account for the continuing forward motion of the passenger as he slides off the seat?

(b) What agent or object causes this force?

(c) Calculate the magnitude of this force.

Solution

(a) Relative to the car's frame of reference, when the brakes are applied the passenger continues forward with an acceleration of 6 m/s^2. The force responsible for this acceleration must therefore be in the forward direction.

(b) There is no physical agent causing this force, because the force is not a real physical force. Rather, it is a fictitious inertial force.

(c) Using Newton's second law, the inertial force must supply the *ma* of the passenger: F_i = 80 kg × 6 m/s^2 = 480 N.

We may note that if the passenger does not brace himself with hands and feet, his forward motion will be halted very quickly when his body slams into the dashboard and windshield, with a force much larger than 480 N.

One may wonder why we bother to consider the physics of motions in non-inertial frames of reference, with the unreal fictitious forces which such an analysis requires. We could simply pass over non-inertial frames and do all of our analyses using inertial frames and real physical forces. However, it often happens that a motion is observed from an accelerated frame, so that becomes the natural frame to use in a physical analysis of the motion. Because of the power of Newton's laws, it is advantageous to be able to apply them in an accelerated frame, and we can do this by the simple expedient of including an appropriate fictitious force.

Questions and Exercises

Q 45.5-1 Suppose a pencil is resting on the approximately level surface above the instrument panel in the front of your car. Describe the motion of the pencil for each of the following motions of the car:

(a) The car starts from rest and accelerates forward at a high rate.
(b) The car is in motion and turns sharply to the left.
(c) In the car's frame of reference, what kind of force accounts for the motions of the pencil described above?

Q 45.5-2 Suppose you are in an elevator that is rising with an acceleration of 2.0 m/s^2 upward.
(a) What do you feel in your body as a result of this acceleration?
(b) Is the inertial force on you in the same direction as your weight, or in the opposite direction?

Q 45.5-3 Suppose you are in an elevator that is rising but has an acceleration of 2.0 m/s^2 downward.
(a) What do you feel in your body as a result of this acceleration?
(b) Is the inertial force on you in the same direction as your weight, or in the opposite direction?

Q 45.5-4 Why are fictitious forces not used to account for motions in an inertial frame?

Q 45.5-5 Real forces can cause an object to accelerate in a frame of reference. Fictitious forces can also cause an object to accelerate in a frame of reference. Then what is the fundamental distinction between real and fictitious forces?

E 45.5-1 Kelly, whose mass is 20 kg, is seated in a child's car-seat fastened to the front seat of a car. She holds above her head a toy having a mass of 0.8 kg. When the car accelerates forward at 5.3 m/s^2, the toy slips from Kelly's hand and accelerates backward relative to the car at 5.3 m/s^2.
(a) Assuming that Newton's laws are valid in the frame of reference attached to the car, what is the direction of the force required to account for this acceleration of the toy?
(b) What agent or object causes this force?
(c) Calculate the magnitude of this force.

E 45.5-2 A pendulum consists of a ball (mass = 0.20 kg) attached by a string one meter in length to the ceiling in a jet aircraft. The plane has a forward acceleration of 3.2 m/s^2 during its run for takeoff.
(a) Assuming Newton's laws are valid in the frame of reference attached to the plane, what is the direction of the inertial force on the pendulum bob during takeoff?
(b) What agent or object causes this force?
(c) Compute the magnitude of this inertial force.
(d) Compute the magnitude and direction of the resultant force on the pendulum, in the plane's frame of reference.
(e) What is the angle which the string makes with the vertical, just before takeoff (when the plane's velocity is still horizontal)?

E 45.5-3 For the path of the ball pictured in Figure 45-2c, show that the slope of the path is g/a, where g is the acceleration of free fall and a is the acceleration of the frame of reference.

Check your understanding of the material above by testing yourself on the minimum learning goal at the beginning of this section.

45.6 Rotating Frames of Reference. Centrifugal Forces

Minimum Learning Goal. Given a description of a rotating frame of reference, and a description of an object at rest in this frame, be able to describe and calculate the forces required to account for this state of rest in the rotating frame. For example:

1 A child having a mass of 30 kg is riding on a rotating merry-go-round on the playground. She is seated two meters from the center of the merry-go-round and is moving with a speed of 2 m/s relative to the ground.
(a) Assuming that Newton's laws are valid in the frame of reference rotating with the merry-go-round, what are the directions of the horizontal forces required to account for the child's remaining at rest on the seat?
(b) What agent or object exerts each of these forces?
(c) Compute the magnitude of each force. (Answer, p. 45-31, Line 12)

2 A ball having a mass of 140 gram is attached to the end of a cord and is whirling in a horizontal circle having a radius of 1.2 m . The speed of the ball is 4 m/s relative to the ground.
(a) Assuming that Newton's laws are valid in the frame of reference rotating with the ball, what are the directions of the horizontal forces required to account for the fact that the ball remains at rest in this frame?
(b) What agent or object exerts each of these forces?
(c) Compute the magnitude of each force. (Answer, p. 45-32, Line 18)

When a ball attached to a string is set into motion by whirling the ball in a horizontal circle, the ball continues in its circular path only as long as the string continues to pull it toward the center. When the string breaks or is released by the person holding it, the ball flies off on a path which takes it farther and farther from the center, seemingly as if it were being pulled outward by some invisible force. The strength of this apparent outward force is shown in a more dramatic way when a pail partly filled with water is swung in a vertical circle: if the motion is fast enough, the water remains in the pail even when the pail is upside down on the upper part of its circular path. This apparent outward force can be experienced personally on a ride in an amusement park called the "Rotor," in which people are whirled in a horizontal circle with their backs against a vertical wall. The "outward force" presses each rider against the wall so strongly that he does not slide to the ground even when the floor is removed, and he finds it difficult to push an arm or leg away from the wall toward the center of the circle.

In each of the examples just described, some object (a ball, the water, or a person) is moving in a circle, remaining in a fixed position within the rotating frame of reference. How can we account for the fact that these rotating objects remain fixed in their rotating frames? According to Newton's first law, an object remains at rest when the *resultant* force on it is zero. We found in Section 43.4 that an *inward*-directed centripetal force is required to keep an object moving in a circle. This centripetal force is supplied by the string, the pail, or the rotor wall in the examples given above. In order to apply Newton's first law to these examples, in each case there must also be a force directed *outward* which just balances the centripetal force, in order to produce a resultant force of zero on the object in the rotating frame of reference.

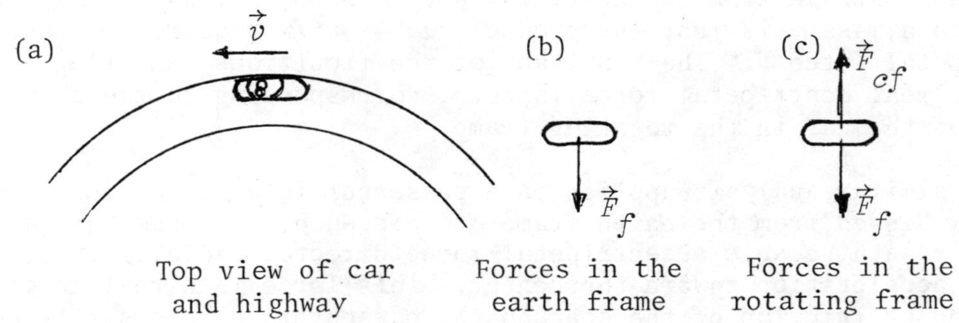

Figure 45-7. (a) A car moving around a curve on a level (unbanked) highway is an example of a rotating frame of reference. The center of curvature of the highway is the center of rotation of this frame. (b) In the stationary earth frame of reference, the only horizontal force on the car is the force of friction $\vec{F}_f$, which is directed toward the center of rotation and serves as the centripetal force. (c) In the frame of reference rotating with the car, the centripetal force is balanced by a fictitious centripetal force $\vec{F}_{cf}$ directed outward from the center.

Figure 45-7 will help us to consider in greater detail a specific example of motion in a rotating frame of reference. An automobile moving around a curve, whether at the corner of the block or on the highway, may be regarded as part of a rotating frame of reference while it remains on the curved path. In our study of circular motion viewed from a fixed or stationary frame of reference (Chapter 43), we found that the changing direction of the car's velocity is a result of a centripetal accleration $a_c = v^2/r$ in the earth frame, and this acceleration in turn results from a force $F_c = ma_c$ directed toward the center of the circle. For a car, this centripetal force is supplied by friction of the road acting on the tires. In a stationary frame of reference, this centripetal force is the only horizontal force on the car (assuming the speed is constant, and road surface is not banked toward the center). In the rotating frame of reference attached to the car, however, the car is at rest and so we expect that the resultant force on the car must be zero in order for Newton's first law to apply in the rotating frame.

The centripetal force on the car is a real physical force (of friction) directed inward; what is needed is an opposite force of the same magnitude, directed outward. However, there is no real physical force directed outward which acts on the car. In order to apply Newton's laws in a rotating frame of reference, we need to "invent" a force directed radially outward. This is shown in Figure 45-7c, labeled $\vec{F}_{cf}$. This is not a real force, but rather a fictitious force which we make up, simply pretending that it exists in order to "save" Newton's laws in a rotating frame. This fictitious force, directed outward on an object at rest in a rotating frame, is called a *centrifugal force*. (This word comes from the Latin *fugo, fugare,* to flee; it is a force which "flees" the center.) A centrifugal force is always directed radially outward, opposite to the centripetal force and *opposite* to the actual acceleration of the moving frame as observed from an inertial frame. The magnitude of a centrifugal force on a mass m is just the product $ma_c = mv^2/r$ required to cancel the centripetal force . The resultant of the fictitious centrifugal force and the real centripetal force is zero, corresponding to the zero acceleration of the car in the rotating frame.

A similar analysis applies to a passenger in a car moving around a curve. Viewed from the earth frame of reference, the passenger moves in a circular path because a centripetal force directed radially inward gives him an acceleration toward the center. This force is a real force, supplied by friction of the seat on the passenger, by his safety belt, and by the door of the car. In the rotating frame of reference attached to the car, the passenger is at rest and so in order to apply Newton's first law we invent a fictitious force directed outward, which just cancels the centripetal force. If the speed of the car is high enough in relation to the radius of the curve, the passenger will be aware of a force acting on him and he may say that he "feels" the centrifugal force. What he actually experiences, however, is a movement of his body *relative to the car* toward the outside of the curve, which causes him to lean toward the outside until the door of the car pushes on him with a force directed toward the center of the circle, in order to keep him inside the car. What the passenger feels is the centripetal force of the door on him, which is a real physical force. He may think, however, that he is being "thrown" toward the outside of the curve because when one is in a rotating frame of reference, one tends to view his own motion from the frame in which this motion appears the simplest, i.e., the rotating frame.

Using a similar approach, we can identify centrifugal forces for other objects at rest in a rotating frame of reference. The ball attached to a string is pulled inward toward the center by the real force of the string, and is "pulled outward" by the fictitious centrifugal force. When the string breaks, the reason the ball flies off on a tangent is *not* because of the centrifugal force, but rather because the centripetal force has vanished and so the ball moves (with *no* horizontal acceleration) in a straight line tangent to the circle at the point where the break occurred. The water in the pail is pulled inward by the real centripetal force of the pail on the water, and is "pulled outward" by the fictitious centrifugal force. The rider on the Rotor is pushed inward by the real force exerted by the wall of the Rotor, and "pushed outward" by the fictitious centrifugal force.

Worked Example 45-3

Beth, who has a mass of 56.0 kg, is riding flat against the wall of a Rotor in an amusement park. She is 4.0 meters from the center of rotation and moving with a speed of 12.0 m/s.

(a) Assuming that Newton's laws are valid in the frame of reference attached to the Rotor, what are the directions of the horizontal forces required for Beth to remain at rest against the wall?

(b) What agent or object exerts each of these forces?

(c) Compute the magnitude of each force.

Solution

(a) There is a centripetal force directed toward the center of rotation, and a centrifugal force of equal magnitude directed outward away from the center.

(b) The centripetal force is a real physical push exerted by the wall of the Rotor against Beth. The centrifugal force, however, is not exerted by anything; it is a fictitious force.

(c) Each force has a magnitude $F_c = ma_c = mv^2/r$

$= 56.0\ \text{kg} \times (12.0\ \text{m/s})^2/(4.0\ \text{m}) = 2.0 \times 10^3\ \text{N}.$

For a body viewed from a rotating frame of reference, the centrifugal force is the only fictitious force required to satisfy Newton's laws of motion as long as the body has no radial motion in the rotating frame. When the distance of the body from the center of rotation is changing, however, a second fictitious force is also required, called the *Coriolis force*. The need for this force arises because the radial component of the body's motion takes the body from a position where the tangential speed of the rotating system has one value, to a larger or smaller radius where the tangential speed of the system has a larger or smaller value, respectively. Because the equation required to compute this force is rather complicated, a quantitative discussion of it is usually deferred to a specialized textbook in mechanics. In the earth's atmosphere, when air flows from regions of high atmospheric pressure toward a central region of low pressure, the Coriolis force deflects the air flow from a simple radial pattern and causes the circulation of winds around the low pressure center. When the low pressure is especially low, these circulating winds attain high speeds and the system becomes a cyclonic storm such as a hurricane, or in extreme cases a tornado. The Coriolis force causes these winds to rotate clockwise in the earth's southern hemisphere, and counterclockwise in the northern hemisphere.

Exercises

E 45.6-1 An astronaut trainee is in a frame of reference rotating about a vertical axis, attached to a large centrifuge having a rotational period of 2.0 seconds. The astronaut's mass is 84 kg and he is located 15.3 m from the center of rotation.

(a) Assuming Newton's laws apply to his motion in this rotating frame, what are the directions of the horizontal forces on the astronaut required to account for his remaining at rest in this frame?

(b) What agent or object exerts each of these forces?

(c) Compute the magnitude of each force.

E 45.6-2 A frame of reference attached to the moon has a rotational period of 2.37×10^6 seconds about its center of rotation (located near the center of the earth). The moon has a mass of 7.35×10^{22} kg and is located 3.8×10^8 meter from the center of rotation.

(a) Assuming Newton's laws of motion are applicable, what are the directions of the forces required to account for the moon's remaining at rest in this rotating frame?

(b) What agent or object exerts each of these forces?

(c) Compute the magnitude of each force.

E 45.6-3 A pendulum consists of a ball (mass = 0.20 kg) attached by a string 0.30 m in length to the ceiling of an automobile. The car is part of a rotating frame of reference as it moves at a steady speed of 18.0 m/s around a horizontal curve on the highway having a radius of curvature of 100 m.

(a) What are the directions of the horizontal forces required to account for the fact that the pendulum bob remains in a fixed position within the car during this motion, assuming Newton's laws must apply?

(b) What agents or objects exert each of these forces?

(c) Compute the magnitude of the centrifugal force.

E 45.6-4 A pail of water containing 2.4 kg of water is being swung in a vertical circle at a constant speed just sufficient to keep the water from spilling out. The water is 1.2 m from the center of the circle.

(a) Draw a free-body diagram of the forces on the water when the pail is in its lowest position, as seen from the frame of reference rotating with the pail.

(b) Draw a free-body diagram of the forces on the water when the pail is in its highest position, as seen from the frame of reference rotating with the pail.

(c) What speed is required for the water at the top of its motion, in order to prevent the water from spilling from the pail?

Check your understanding of the material above by testing yourself on the minimum learning goal at the beginning of this section.

Answers to Minimum Learning Goals

Line 1 No. If the frame were inertial, the glider's acceleration would be only 0.5 m/s^2 for the forces given.

Line 2 (a) 14.3 m/s westward.
(b) 10.1 m/s westward.

Line 3 Yes, because it is moving with constant velocity.

Line 4 (a) Downward; (b) none; this is a fictitious inertial force;
(c) 0.168 N.

Line 5 Only (d) is an inertial frame. Each of the other frames is accelerating relative to the earth.

Line 6 No, because this action takes place over a period of time and therefore no precise point in time can be identified for it.

Line 7 No, because this crossing occurs at all points having a longitude of 90° west, and therefore no precise point in space can be identified with it.

Line 8 No, because the balloon is accelerating (its velocity is decreasing)

Line 9 1.68 m/s in the direction 17.4° south of east.

Line 10 Yes.

Line 11 Yes. The ball's acceleration is 9.8 m/s^2 downward, which is to be expected when the only force acting is the weight, in an inertial frame.

Line 12 (a) The centripetal force is directed inward toward the center, and the centrifugal force is directed outward away from the center.
(b) The centripetal force is exerted by the seat and the bar which the child grasps with her hands. The centrifugal force is not exerted by anything; it is a fictitious force.
(c) 60 N.

Line 13 Yes.

Line 14 1.79 m/s in the direction 26.6° north of west.

Line 15 No, because although Fred is able to pinpoint the time of the flash accurately, his apparatus does not enable him to pinpoint its spatial location.

Line 16 (a) To the left; (b) None; this is a fictitious inertial force;
(c) 0.464 N.

Line 17 No, because nothing is happening which can be associated with a definite time.

Line 18 (a) The centripetal force is directed inward toward the center, and the centrifugal force is directed outward away from the center.

(b) The centripetal force is exerted by the cord. The centrifugal force is not exerted by anything; it is a fictitious force.

(c) 1.87 N.

Line 19 No, because the clock is not at rest relative to the spatial coordinate system (the gridiron) during this measurement.

Line 20 (b) and (e)

ANSWERS to Questions and Exercises in Chapter 45

Section 45.1

Q 45.1-2 No, because it is not confined to a single spatial location.

Q 45.1-4 No, it is a succession of events occurring at various times.

Q 45.1-6 A photoelectric cell having a small opening for the admission of light, electrically connected to a clock or to a chart recorder having a calibrated time scale.

Q 45.1-8 No, because the "clock" on earth is moving relative to the spatial coordinate system on the moon. However, assuming the astronomer knows accurately the distance from earth to moon during his experiment, he can compute the time at which the eruption occurred on the moon and in this sense his equipment is an adequate frame of reference.

Q 45.1-10 (a) An adequate frame, because the clocks and other equipment are at rest relative to each other.
(b) Not an adequate frame, because the clocks are not synchronized.
(c) Not adequate, because the clocks are moving relative to each other.

Section 45.2

Q 45.2-2 (c)

Q 45.2-4 One could observe an object whose resultant force is zero, and if its acceleration is also zero the observations were made from an inertial frame. A physics laboratory fixed to the earth is not precisely an inertial frame, because of the earth's rotation. Unless measurements are made with high precision, a laboratory building on earth serves as a satisfactory inertial frame.

Q 45.2-6 (a) No. (b) Yes, the ball would appear to fall straight down. An observer on the ground would see the ball falling in the path of a parabola, with constant forward velocity but increasing downward velocity.
(c) The ball follows a slanting path, similar to that in Figure 45-2c. Because the ball does not fall straight down inside the plane, you know the plane is accelerating.

Section 45.4

Q 45.4-2 No, unless frame *A* is at rest in frame *B* (with some distance between the origins of the two frames.)

Q 45.4-4 If the elevator has some motion from side to side as it ascends or descends, this motion would show that the elevator is traveling. A sensitive air-pressure gauge on the elevator would indicate a changing elevation if the elevator is rising or falling.

E 45.4-2 (a) $u = u' + v$. (b) The acceleration is zero for both observers.

E 45.4-6 (a) 10 km/hr, at an angle of 53.1° to the riverbank (in the downstream direction).
(b) 75 meters.
(c) upstream at an angle of 48.6° to the line from A to B.

E 45.4-8 (a) 3.92 m/s² downward.
(b) 10.7 m/s in a direction at an angle of 68.6° from the vertical.

Section 45.5

Q 45.5-2 (a) Your body feels heavier, pushing down with greater force on your legs, so that you might feel your knees tend to buckle. In addition, you may have a "sinking" feeling in your stomach.
(b) In the same direction.

Q 45.5-4 All motions observed from an inertial frame of reference can be explained or accounted for as the result of the action of real physical forces caused by physical objects, so that fictitious forces are not needed.

E 45.5-2 (a) backward. (b) None, because it is a fictitious force.
(c) 0.64 N. (d) 2.06 N at an angle of 18.1° to the vertical; the pendulum string "leans backward" in the aircraft.
(e) 18.1°; the string is parallel to the force of part (d).

Section 45.6

E 45.6-2 (a) The centripetal force is directed toward the center of rotation, and the centrifugal force is directed away from the center.
(b) The earth exerts the (gravitational) centripetal force. The centrifugal force is fictitious, not exerted by anything.
(c) 2.0×10^{20} N.

E 45.6-4 (a)

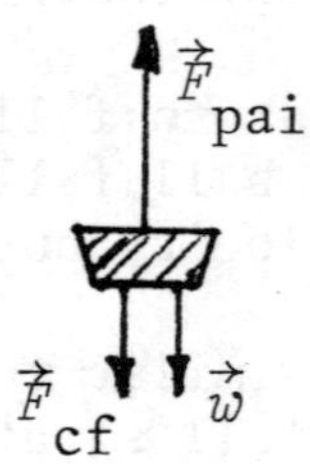

(b)

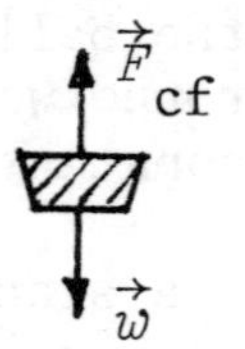

(c) 3.4 m/s

CHAPTER 46 EQUILIBRIUM

Introduction

For the design, analysis and study of engineering structures and their structural members it is often convenient to use the model of a rigid body. For many purposes we can consider that all the mass of the body is located at a point called its center of mass. The motion of a body depends on the resultants of the external forces and torques acting on the body. When both the external forces and the external torques add to zero, the body is in a state of equilibrium. The concept of torque is defined for the simple case where all forces and their position vectors lie in a single plane, and for the more general three-dimensional case where torques are computed using the vector cross product. The two conditions for static equilibrium are applied to structural systems containing ropes, knots, pulleys, struts, beams, and ladders.

When you have completed your study of the material in this chapter, you should be able to answer questions and work exercises like the examples in the minimum learning goals assigned by your instructor.

46.1 Rigid Bodies

Minimum Learning Goal. (1) Given descriptions of several physical systems, be able to identify the system which is most nearly a rigid body. For example:

1 Which of the following systems is most nearly a rigid body? (Answer, page 46-36, Line 13)
(a) a salt crystal at room temperature
(b) the earth's atmosphere
(c) a vat of lead in the molten state.

2 Which of the following systems is most nearly a rigid body? (Answer, page 46-36, Line 14)
(a) a meter stick at room temperature.
(b) 5 cm^3 of water at 1000 °C and 1 atmosphere pressure.
(c) a bucket of molten iron.

Every mechanical system can be thought of as consisting of particles whose individual dimensions are negligibly small. In this sense, any system of finite extent contains a very large number of particles. The complete description of the motion of such a system would be very complicated, because it would be necessary to specify the motion of every constituent particle.

There are some systems, however, for which the problem of describing the motion is simplified. One of these is the case of a rigid body.

> A *rigid body* is a system of particles held together by forces such that the relative positions of the particles remain fixed.

Because all the interparticle distances remain fixed, the size, shape and distribution of mass remain constant for a rigid body, regardless of the motion of the body as a whole. This concept is an idealization, because no perfectly rigid body exists. There are, however, many instances for which this model provides an excellent approximation to the actual situation.

The description of the motion of a rigid body is immensely simpler than that of a general system containing many particles. Once the motions of three non-collinear particles in a rigid body are known, the motion of every other particle is determined by the relative spacings of the various particles in the body. To see how the positions of just three particles determine the positions of all the other particles in the body, consider an ordinary chair having four feet. If we do not know or do not specify the position of even one particle in the chair, then we do not know where any of the particles are. Now let us specify that the rear of the bottom of the left rear foot of the chair must be located at a point *A* on the floor. This specifies the location of one particle of the chair but not of any other particle, because the chair can pivot around point *A* in any way. Any particle not at point *A* can occupy an infinite number of positions in space.

Next, in addition, let us specify that the rear of the right rear foot of the chair must be located at a point *B* on the floor. With the particles fixed at points *A* and *B*, we have also fixed the location of every particle in the chair which lies on the straight line connecting *A* and *B*. The positions of all other particles in the chair are still not specified, however, because the chair is still free to rotate about the line connecting *A* and *B*. The front of the right front foot of the chair, for example, can have any position on a circle centered on the line *AB*.

Finally, we can specify that the front of the right front foot of the chair must be located at a point *C* on the floor. With three particles fixed at the non-collinear points *A*, *B* and *C*, we have now specified the location of every particle in the chair, assuming the chair is a rigid body. Any attempt to move the chair would displace at least one of the particles from point *A*, *B* or *C*. We could choose any other set of three non-collinear particles in the chair, and use the same line of reasoning to obtain a similar result.

So, we find that fixing the positions of any three non-collinear particles in a rigid body determines the position of every particle in the body. If the whole body is in motion, it is only necessary to keep track of the motions of three of its particles in order to know (in principle) the motion of every particle in the body.

Questions

Q 46.1-1 For each of the following physical systems describe a situation in which the system could be regarded as a rigid body, and another situation in which it could not. (continued on next page)

(a) a hula hoop: (c) a balloon
(b) a tennis ball (d) a baseball bat
(e) the human body (f) a glacier

Q 46.1-2 For each of the following systems describe a situation in which the system could be regarded as a rigid body, and another situation in which it could not.
(a) the earth (d) a horseshoe
(b) a dime (e) the Empire State Building
(c) an ice cube

Check your understanding of the material above by testing yourself on the minimum learning goals at the beginning of this section.

46.2 Center of Mass

Minimum Learning Goal. (1) Be able to calculate the position of the center of mass of a system of particles, given their masses and relative positions. For example:

1 A water molecule has an O-H distance of 0.096 nm, and an H-O-H angle of 105°. Find the position of the center of mass of a water molecule, relative to the oxygen atom. ($m_H = 1.67 \times 10^{-27}$ kg; $m_O = 2.67 \times 10^{-26}$ kg) (Answer, p. 46-36, Line 9)

2 Relative to the earth's center, where is the center of mass of the system consisting of the earth and the moon? (See data in Appendix G.) (Answer, page 46-36, Line 7)

We often discuss the motion of a macroscopic object, such as an automobile or the earth, as if the object were a particle having all of its mass concentrated at one point. It is also useful in many cases to compute the forces on a stationary object assuming all its mass is concentrated at one point. Which point in the object is most appropriate? For many purposes the mass of the object may be regarded as concentrated at the *center of mass*. The definition of the center of mass of a system, given below, can be applied to any system or object, whether it is a solid, liquid, gas, or combination of these.

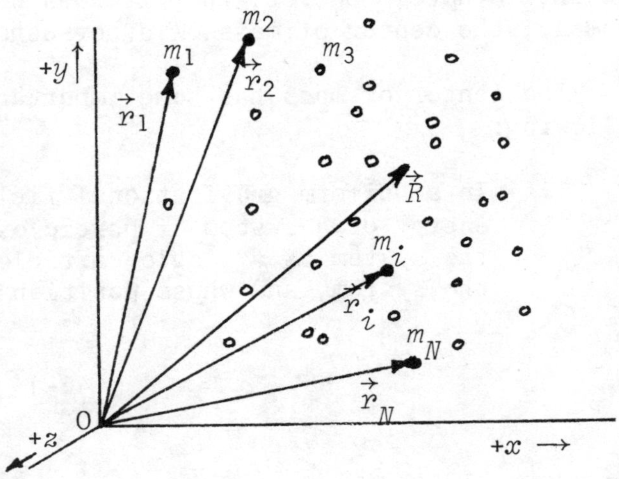

Figure 46-1. A system of particles, numbered from 1 to N. The position of the ith particle is $\vec{r}_i$. The position of the center of mass of the system is designated by $\vec{R}$.

Consider a system as an assembly of individual particles having masses m_1, m_2, m_3, etc., as shown in Figure 46-1. The total mass of the system is

★ $$M = \Sigma m_i = m_1 + m_2 + m_3 + \ldots ,$$

there the sum includes all the particles in the object. Relative to some origin of coordinates, the position of m_1 is $\vec{r}_1$, the position of m_2 is $\vec{r}_2$, and so on. The center of mass of the system is defined as the point located by the vector $\vec{R}$ given by

★ $$\vec{R} \equiv \frac{m_1\vec{r}_1 + m_2\vec{r}_2 + m_3\vec{r}_3 + \ldots}{M} = \frac{\Sigma m_i\vec{r}_i}{M} ,$$

or $$M\vec{R} = \Sigma m_i\vec{r}_i ,$$

where the sum includes all the particles of the system.

The positions of the particles may be expressed in terms of their x, y, and z coordinates by writing

$$\vec{r}_1 = \hat{x}x_1 + \hat{y}y_1 + \hat{z}z_1 ,$$

$$\vec{r}_i = \hat{x}x_i + \hat{y}y_i + \hat{z}z_i ,$$

etc. The position of the center of mass is given by

$$\vec{R} \equiv \hat{x}x_{\text{cm}} + \hat{y}y_{\text{cm}} + \hat{z}z_{\text{cm}} .$$

Combining the two equations for $\vec{R}$, it is not hard to show that the individual coordinates of the center of mass are then given by

★ $$x_{\text{cm}} = \frac{\Sigma m_i x_i}{M} , \quad y_{\text{cm}} = \frac{\Sigma m_i y_i}{M} , \text{ and } z_{\text{cm}} = \frac{\Sigma m_i z_i}{M} . \qquad \text{Eq. (46-1)}$$

The position of the center of mass thus has the form of a weighted average of the positions of the various particles of the system, in which the importance of each particle is in proportion to its mass. Although the position vectors of the particles and the center of mass depend on the location of the origin of coordinates, the position of the center of mass relative to the system is independent of the location chosen for the origin.

For a rigid body, the center of mass is fixed relative to the body. One should note, however, that it does not have to be *inside* the body. For example, the center of mass of a horseshoe is not in the iron.

The center of mass has some important properties, among which are the following:

1. In a uniform gravitational field, the gravitational potential energy of a system of particles is correctly obtained by regarding the system as a single particle whose mass is the total mass of the system, and whose position is the position of the center of

mass of the system. Gravitational potential energy is discussed is Section 48.1.

2. The forces acting on a rigid body can be correctly analyzed by assuming that the total weight of the body acts at the center of mass. To prove this, it is necessary to use the concept of torque (Section 46.6).

3. For a system in motion, the center of mass moves as if it were a single particle having the total mass of the system, subjected to a force equal to the total external force applied to the system.

$$\Sigma\vec{F}_{ext} = M\vec{a}_{cm} \ .$$

The distinction between internal and external forces is considered in the next section.

For an object in a gravitational field, the resultant of the gravitational forces on the particles of the object acts at a point called the *center of gravity*. The total gravitational force on the body is the same as if all the mass of the body were concentrated at the center of gravity. When the gravitational field is uniform, so that $\vec{g}$ is the same for all the particles in the body, the center of gravity coincides with the center of mass. For an object near the earth's surface, where $\vec{g}$ is constant, the total weight of the body acts at the center of mass. (For a body in a non-uniform gravitational field the center of gravity is not necessarily at the center of mass.) If an object is suspended from a string so that the string supports the entire weight of the object, the center of gravity and center of mass of the object lie in the vertical line of the string.

There are some situations for which it is not correct to regard the entire weight of an object as being applied at the center of mass. One must consider the weight of the object and the effect the weight has on the object. Imagine, for example, a very long stick supported by two supports near its center. If the whole weight were really applied at the center of mass, the center of the stick would be pulled down between the supports, and the stick would be deformed into something of a "V" shape. In fact, however, the actual distortion is just the opposite. The end sections of the long stick droop because of their own weight, and because the stick is not a rigid body. This shows how one can get a wrong answer by pretending the whole weight is applied at the center of mass.

The center of mass of a solid object can be located experimentally by suspending the object from a string attached to two or more points of the object in turn. As stated above, for each point of suspension the center of mass then lies in the plane of the string. The point of intersection of two of these lines fixes the location of the center of mass.

For a system in which the relative locations of all the constituent particles are known, the center of mass can be calculated using the equations given above. For a system composed of two or more parts whose centers of mass are known, the center of mass of the system can be computed by regarding its component parts as single particles located at their respective centers of mass. Examples are given in the Worked Examples below.

The center of mass is used in the study of the equilibrium of solid objects, discussed later in this chapter, and in the study of the motions of rigid bodies, discussed in more advanced books.

Worked Example 46-1

Some molecules of heavy water have the chemical structure HOD, where D represents a deuterium atom. The O-H and O-D distances are both 0.096 nm, while the H-O-D angle is 104.4°. Find the center of mass of an HOD molecule.

Solution

For our purposes we may regard the mass of each atom as concentrated in a small particle at the nucleus. The three nuclei determine a plane, which we take as the *xy* plane. We choose the origin of coordinates at the oxygen nucleus, with the $-y$ axis along the line which bisects the angle H-O-D, as shown in the diagram. The angle $\theta = 104.4°/2 = 52.2°$. The x and y coordinates of the atoms are computed as follows:

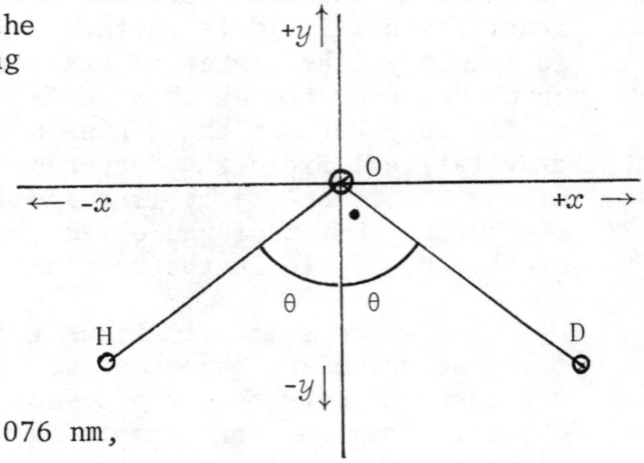

$$x_O = 0, \quad y_O = 0 .$$

$$x_H = -0.096 \text{ nm} \times \sin 52.2° = -0.076 \text{ nm},$$

$$y_H = -0.096 \text{ nm} \times \cos 52.2° = -0.059 \text{ nm},$$

$$x_D = +0.096 \text{ nm} \times \sin 52.2° = +0.076 \text{ nm},$$

$$y_D = -0.096 \text{ nm} \times \cos 52.2° = -0.059 \text{ nm}.$$

The masses of the atoms are

$$m_O = 16.0 \text{ amu} \times 1.66 \times 10^{-27} \text{ kg/amu} = 2.66 \times 10^{-26} \text{ kg},$$

$$m_H = 1.01 \text{ amu} \times 1.66 \times 10^{-27} \text{ kg/amu} = 1.68 \times 10^{-27} \text{ kg},$$

$$m_D = 2.01 \text{ amu} \times 1.66 \times 10^{-27} \text{ kg/amu} = 3.32 \times 10^{-27} \text{ kg}.$$

The total mass is

$$M = m_O + m_H + m_D = 3.16 \times 10^{-26} \text{ kg}.$$

The coordinates of the center of mass are:

$$x_{cm} = \frac{\Sigma m_i x_i}{M} = \frac{m_O x_O + m_H x_H + m_D x_D}{M}$$

$$= \frac{0 + 1.68 \times 10^{-27}\ \text{kg} \times (-0.076\ \text{nm}) + 3.32 \times 10^{-27}\ \text{kg}\ (0.076\ \text{nm})}{3.16 \times 10^{-26}\ \text{kg}}$$

$$= 0.0039\ \text{nm}.$$

$$y_{cm} = \frac{\Sigma m_i y_i}{M} = \frac{m_O y_O + m_H y_H + m_D y_D}{M}$$

$$= \frac{0 + 1.68 \times 10^{-27}\ \text{kg} \times (-0.059\ \text{nm}) + 3.32 \times 10^{-27}\ \text{kg}\ (-0.059\ \text{nm})}{3.16 \times 10^{-26}\ \text{kg}}$$

$$= -0.0093\ \text{nm}.$$

In the diagram above, the location of the center of mass is indicated by the dot (•) at the lower right of the oxygen atom.

Worked Example 46-2

A carpenter's framing square is made of metal having a uniform thickness. It has an L shape, with dimensions in inches as shown in this diagram. Assuming the square lies in the xy plane as shown, what are the coordinates of its center of mass?

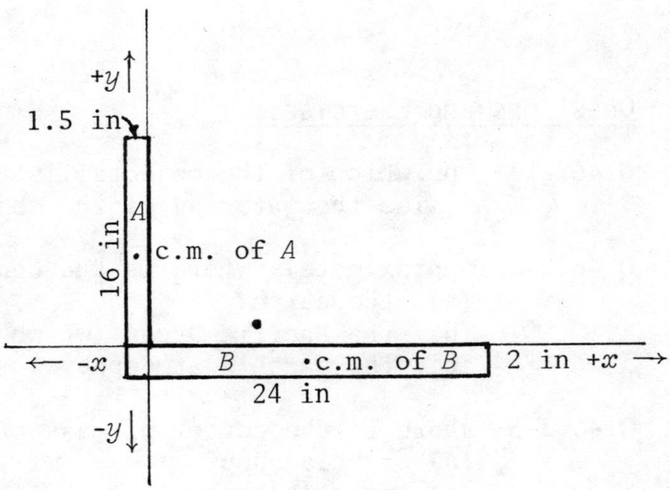

Solution

We consider the tool as made of two rectangles, labeled A and B. From the symmetric shape of each rectangle, it can be seen that its center of mass lies at the center of the rectangle. A has sides of 14 in and 1.5 in, so from the diagram we see that its center of mass is at $x_A = -0.75$ in and $y_A = +7.0$ in. B has sides of 24 in and 2 in, so its center of mass is at $x_B = +10.5$ in and $y_B = -1.0$ in.

The tool has uniform thickness, so the mass of each rectangle is proportional to its area in the xy plane. Using σ for the mass per unit area, the mass of A is $m_A = \sigma \times 1.5\ \text{in} \times 14.0\ \text{in} = 21.0\ \sigma\cdot\text{in}^2$ units of mass, and the mass of B is $m_B = \sigma \times 24\ \text{in} \times 2\ \text{in} = 48.0\ \sigma\cdot\text{in}^2$ units of mass.

The total mass of the system is then

$$M = m_A + m_B = 21.0\ \sigma\cdot\text{in}^2 + 48.0\ \sigma\cdot\text{in}^2 = 69.0\ \sigma\cdot\text{in}^2.$$

The coordinates of the center of mass of the system are:

$$x_{\text{cm}} = \frac{\Sigma m_i x_i}{M} = \frac{m_A x_A + m_B x_B}{M}$$

$$= \frac{21.0\ \sigma\cdot\text{in}^2 \times (-0.75\ \text{in}) + 48.0\ \sigma\cdot\text{in}^2 \times 10.5\ \text{in}}{69.0\ \sigma\cdot\text{in}^2}$$

$$= \frac{4.88 \times 10^2\ \sigma\cdot\text{in}^3}{69.0\ \sigma\cdot\text{in}^2} = 7.08\ \text{in}.$$

$$y_{\text{cm}} = \frac{\Sigma m_i y_i}{M} = \frac{m_A y_A + m_B y_B}{M}$$

$$= \frac{21.0\ \sigma\cdot\text{in}^2 \times 7.0\ \text{in} + 48.0\ \sigma\cdot\text{in}^2 \times (-1.0\ \text{in})}{69.0\ \sigma\cdot\text{in}^2}$$

$$= \frac{99.0\ \sigma\cdot\text{in}^3}{69.0\ \sigma\cdot\text{in}^2} = 1.43\ \text{in}.$$

In the diagram, the location of the center of mass is indicated by the dot (•).

Questions and Exercises

Q 46.2-1 For which of the objects listed in Q 46.1-1 is the center of mass inside the material of the object?

Q 46.2-2 Approximately where is the center of mass of:
(a) the earth?
(b) the Pacific Ocean, which has a maximum depth of about 7 miles?
(c) the earth's atmosphere?

Q 46.2-3 Where is the center of mass of:
(a) a hula hoop?
(b) a hollow spherical shell having uniform thickness?
(c) a meter stick?

Q 46.2-4 (a) When you are standing upright, approximately where is the center of mass of your body?
(b) How could you position your body so your center of mass lies outside your body?

Q 46.2-5 The position vectors $\vec{r}_1$, $\vec{r}_2$, etc., of the particles in a system depend on the origin chosen for the coordinate system. Why then can we say that the position of the center of mass of an object is independent of the origin of the coordinate system?

Q 46.2-6 Consider a system consisting of two gliders at rest on a horizontal air track, tied together with a string, and having a spring compressed between them. (continued on next page)

(a) What is the motion of the center of mass?
(b) What is the resultant force acting on the system?
(c) If the string is burned through, the spring will push the gliders apart. Then what is the motion of the center of mass?
(d) What is then the resultant force acting on the system?

E 46.2-1 Where is the center of mass of the system consisting of the earth and the sun, relative to the center of the sun?

E 46.2-2 In the HCN molecule the three atoms lie along a straight line. The H atom is 0.1064 nm from the C atom, and the N atom is 0.1156 nm from the C atom. How far is the center of mass of HCN from the N atom? Atomic masses: H, 1.008 amu; C, 12 amu; N, 14.003 amu.

E 46.2-3 In the ammonia molecule, the three H atoms lie at the corners of an equilateral triangle 0.1628 nm on a side, while the N atom is equidistant from the three hydrogens and is 0.0381 nm from the plane containing the hydrogens. How far from the N atom is the center of mass of NH_3, and in what direction? Atomic masses: H, 1.008 amu; N, 14.003 amu.

E 46.2-4 Fill in the missing steps in the derivation of Eq. (46-1).

E 46.2-5 A diving board is clamped to hold one end (*A*) down, and is supported by a bar at *B*, as shown in this diagram. Assuming all the mass of the board is concentrated at its center, draw a free-body diagram of the board.

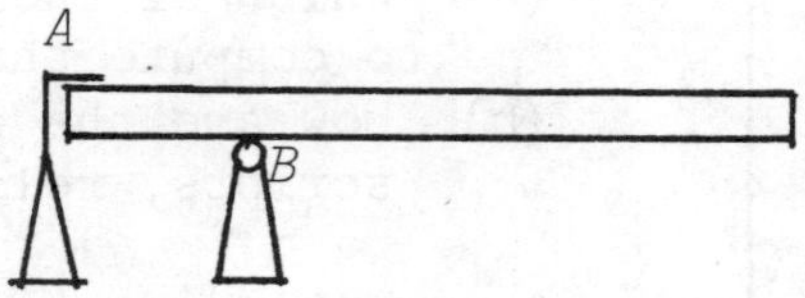

Check your understanding of the material above by testing yourself on the minimum learning goals at the beginning of this section.

46.3 External and Internal Forces

Minimum Learning Goals. (1) Given a description of interacting objects, be able to designate a physical system appropriate for studying the motion of one or more of the objects in response to internal and external forces on the system. For example:

1 (a) Considering the earth, sun, and moon as a set of objects isolated from the rest of the universe, which of the three would you include in a physical system to study the motion of the moon with respect to the earth?
(b) What are the external forces on your system?
(c) What are some of the internal forces, and their resultant?
(Answer, page 46-36, Line 15)

2 Consider a piece of salt consisting of 2×10^{23} sodium ions and 2×10^{23} chlorine ions in a crystalline structure, resting on a table top on earth.
(a) Select a physical system to study the dynamics of the salt crystal relative to the table top.
(b) What are the external forces on your system?
(c) What are the primary internal forces, and what is their resultant? (Answer, page 46-36, Line 11)

(2) Given a description of the interactions among two or more systems, be able to select the forces which determine the acceleration of one of the systems, and be able to relate each of these forces to its reaction force. For example:

1 A box is placed on an incline. When the box is released, there is a force $\vec{F}_1$ exerted by the incline on the bottom of the box, directed parallel to and up the incline. There is a force $\vec{N}_1$ exerted by the incline on the box normal to the incline surface, and there is a force $\vec{w}_1$ exerted downward by the earth on the box. In addition, there is a force $\vec{F}_2$ exerted on the incline by the bottom of the box directed parallel to and down the incline. There is a force $\vec{N}_2$ exerted by the box on the incline normal to the incline surface, and there is a force $\vec{w}_2$ exerted on the earth by the box.
(a) Which of these forces must be included in Newton's second law to compute the acceleration of the box?
(b) How are the forces with the same symbol, but different subscripts, related to each other? (Answer, page 46-36, Line 8)

2 A man standing in an elevator is acted on by two forces: a downward force of gravity $\vec{w}_1$ and an upward force $\vec{F}_1$ exerted by the elevator floor. In addition, a force $\vec{w}_2$ is exerted upward on the earth by the man, and a force $\vec{F}_2$ is exerted downward on the floor by the man.
(a) Which of these forces must be taken into account to compute the acceleration of the man?
(b) How are the forces with the same symbol, but different subscripts, related to each other? (Answer, page 46-35, Line 1)

When we discuss a system of particles, we can distinguish between internal forces and external forces. *Internal forces* include all forces exerted by one particle of the system on another particle of the system. *External forces* include all those exerted on a particle of the system by something else outside the system. To apply this distinction, we need to clearly identify the physical system we are dealing with. We are often free to choose the objects we regard as the system. From all the physical objects in the universe, we select one or a few that we are interested in and designate these as the system. For example, if we regard the moon as the system, the gravitational force of the earth on the moon is an external force. However, if we take the earth-moon combination as the system, this gravitational force is an internal force.

As we saw in Section 42.5, forces between particles are governed by Newton's third law, which says that the force exerted by particle *A* on particle *B* is the exact negative of the force exerted by particle *B* on particle *A*. It follows from Newton's third law that if we add all the internal forces on all the particles in a system, we must get zero. That's because every interaction appears twice--once as the force by *A* on *B*, and once as the force by *B* on *A*. Each of these forces is the negative of the other, so each pair adds up to zero.

The resultant internal force on a system is always zero, so this resultant cannot cause an acceleration of the system as a whole. That is, the center of mass of the system has no acceleration in response to internal forces. This center of mass accelerates only when the resultant of the *external forces* differs from zero. Each external force on the system also belongs to an action-reaction pair, but in this case the reaction force is exerted by the system on something outside the system, so this reaction force does not affect the acceleration of the system.

Questions

Q 46.3-1 Some of the forces associated with an automobile traveling around a banked curve on a highway are: (1) the resultant force exerted by driver's hands on the steering wheel; (2) the resultant force by the driver's seat on the driver; (3) the force by the driver on his seat belt; (4) the horizontal force by the highway on the tires; (5) the vertical force by the highway on the tires; (6) the force by the tires on the highway in a direction normal to the road surface; (7) the force by the rear axle and housing on the right rear wheel of the auto; (8) the weight of the car; (9) the force by the wind on the car including air resistance to the car's motion; (10) the gravitational force by the auto on the police car following it around the curve.

(a) For the system of consisting of this one auto (including its contents), which of the forces listed above are *internal* forces?

(b) For the system consisting of this one auto, which of the forces listed above are *external* forces on the auto?

(c) For the system consisting of the driver, which of the forces listed above are *internal* forces? Which are *external* forces?

(d) For the system consisting of the right rear wheel of the car, which of the forces listed above are *internal* forces? Which are *external* forces?

(e) For the system consisting of the car and the portion of the highway with which it is in contact, which of the forces listed above are *internal* forces? Which are *external* forces on the system?

Q 46.3-2 Some of the forces associated with a water molecule include: (1) the electric force between the proton in a hydrogen nucleus and a proton in the oxygen nucleus; (2) the gravitational force between the two hydrogen nuclei; (3) the nuclear force between two neutrons in the oxygen nucleus; (4) the electric force between an electron in the inner electron shell of the oxygen atom and

an electron close to one of the hydrogen atoms; (5) the nuclear force between a neutron in the oxygen nucleus and the proton in a hydrogen nucleus; (6) the electric force between a proton in the oxygen nucleus and an electron in the inner electron shell of the oxygen atom; (7) the gravitational force of the earth on the oxygen nucleus; (8) the Coulomb force of this molecule on a neighboring water molecule; (9) the electric force between two protons in the oxygen nucleus; (10) the magnetic force on an electron in the inner electron shell of the oxygen atom, as a result of the earth's magnetic field.
(a) For the system consisting of this one water molecule, which of the forces listed above are *internal* forces? (b) For the system consisting of this one water molecule, which of the forces listed above are *external* forces? (c) For the system consisting of the oxygen atom in this water molecule, which of the forces listed above are *internal* forces? (d) For the system consisting of the oxygen atom in this water molecule, which of the forces listed above are *external* forces? (e) For the system consisting of the oxygen nucleus in this water molecule, which of the forces listed above are *internal* forces? Which are *external* forces?

Q 46.3-3 Newton's third law says that forces occur in pairs. For each force listed in the answer to part (b) of Q 46.3-2, what is the other force of the action-reaction pair, and on what object does it act?

Q 46.3-4 For each external force listed in the answer to part (c) of Q 46.3-1, what is the other force of the action-reaction pair, and on what object does it act?

Q 46.3-5 What force is indicated by the dial on a bathroom scale?

Check your understanding of the material above by testing yourself on the minimum learning goals at the beginning of this section.

46.4 First Condition for Static Equilibrium

Minimum Learning Goals. (1) Given a description of the motion of a system of particles relative to an inertial frame, be able to state whether or not the system is in equilibrium, justifying your decision on the basis of the defining criterion for equilibrium. For example:

1 A man is riding a train traveling at a uniform velocity of 30 m/s relative to a "fixed" observer. If the man is sitting quite still, so that every particle of his body has an average velocity of 30 m/s parallel to the train track, is he in equilibrium? Justify your answer. (Answer, page 46-35, Line 6)

(2) Given a description of a system in equilibrium, be able to draw a free-body diagram of the system, identify the forces acting on the system, and calculate one of these forces when you are given enough information. For example:

1 A man who weighs 700 N is riding in a 2000-kg elevator moving upward at a steady speed of 4 m/s. Draw a free-body diagram showing the forces acting on the man, then calculate these forces. (Answer, page 46-36, Line 10)

In this chapter we are concerned primarily with rigid bodies in *equilibrium*. A body is said to be in *static equilibrium* in an inertial frame of reference in which the body remains at rest. A sled moving with constant velocity is not at rest relative to the snow, but is at rest in a frame of reference attached to the sled, so we can say the sled is in equilibrium. A body in equilibrium but moving with constant velocity is sometimes said to be in *dynamic equilibrium* in an inertial frame not moving with the body. A ball thrown straight upward comes to rest momentarily at the top of its path, but such a ball is not in equilibrium because it does not *remain* at rest for more than an instant. The ball *is* in equilibrium while it is lying at rest on the floor. Most cases of static equilibrium which are of interest to scientists and engineers are concerned with bodies or structures at rest relative to the earth's surface.

A rigid body at rest in an inertial frame has neither translational motion along a line or rotational motion about an axis. This means that for a rigid body in equilibrium the acceleration of every particle of the body is zero, and the net force on every particle is zero. Therefore, the sum of the forces acting on all the particles of the body is also zero. We have already seen that the resultant internal force on a system is always zero. It follows, then, that for a rigid body in equilibrium *the vector sum of the external forces on the body is zero.* In equation form:

★ $$\Sigma\vec{F}_{ext} = 0$$ First condition for equilibrium.

While the first condition for equilibrium is sufficient to ensure that the body has no translational motion, it does not guarantee that the body has no rotation. If the forces on the body are not all acting on the same point, their combined effect may produce a torque (Section 46.6) which imparts rotational motion to the body. Rotational motion is eliminated by the second condition for equilibrium (Section 46.7). If the line of action of every force on a body passes through the same point, the forces are said to be *concurrent.*

In every problem dealing with the equilibrium of a rigid body, it is essential to draw one or more correct free-body diagrams before one applies the equations for equilibrium.

In analyzing examples of static equilibrium, it is often necessary to find the forces exerted on a body by structural members such as cables, struts, and beams. A cable, rope, or string in such a system usually exerts a force much greater than its own weight, so at the elementary level it is customary to assume that this weight is negligible. A rope or cable cannot push an object attached to it; it can only exert a pull parallel to its length. That is, a rope can exert a force of tension, but not one of compression or shear. In elementary problems such as those considered in this chapter, it is also assumed that a rope or cable is *inextensible*, that is, its elongation in response to the tension in it is negligible.

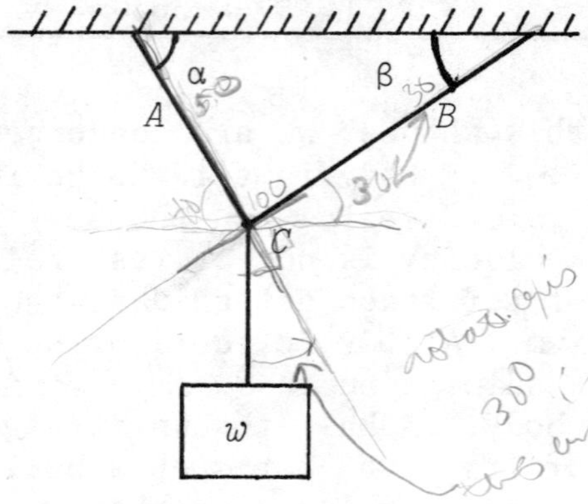

Figure 46-2. A weight w supported by a vertical rope, which in turn is supported by two other ropes attached at a common point, C.

Sometimes a structural system contains three or more ropes fastened together at a common point, which we may call a *knot*. Figure 46-2 shows three ropes connected in a knot at C. If the tension in one of the three ropes is known, the tensions in the other two can always be found by constructing a suitable free-body diagram and applying the first condition for equilibrium,provided all the angles between ropes are known. When the angles are fixed, the lengths of the individual ropes do not affect the forces in them.

Worked Example 46-3

A weight of 400 N is suspended from a rope attached to two other ropes by a knot as shown in Figure 46-2. The angle $\alpha = 50°$ and $\beta = 30°$. Find the forces in ropes A and B.

Solution

The knot at C is in static equilibrium, so the resultant force on C equals zero. The three forces pulling on the knot, one exerted by each rope, are shown in this free-body diagram of the knot. These forces must lie in a common plane for the knot to be in equilibrium. In making this diagram, initially one does not know how long to draw the arrows for $\vec{F}_A$ and $\vec{F}_B$, but this does not matter because the magnitudes of these forces will be obtained in the solution.

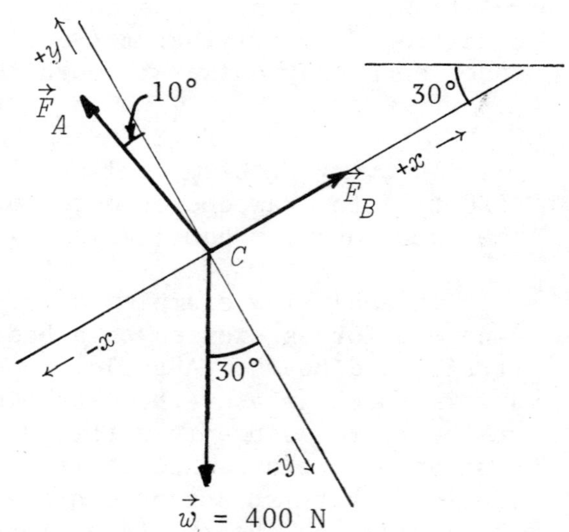

After the forces are drawn, one needs to choose directions for the x and y axes, so the forces can be resolved into components. The solution is simplified a bit if one axis is chosen parallel to one of the unknown forces. We choose axes as shown, with the origin at C and the $+x$ axis parallel to F_B.

For the knot $\Sigma\vec{F} = 0$, so the resultant forces along the x and y axes are also zero:

$$\Sigma\vec{F}_x = 0 \qquad \text{and} \qquad \Sigma\vec{F}_y = 0.$$

To apply these equations, we must find the components of the three forces along the axes. The x axis is 30° from the horizontal, so the weight acts 30° from the y axis. The drawing at the right shows the x and y components of the three forces. From this diagram, we see that

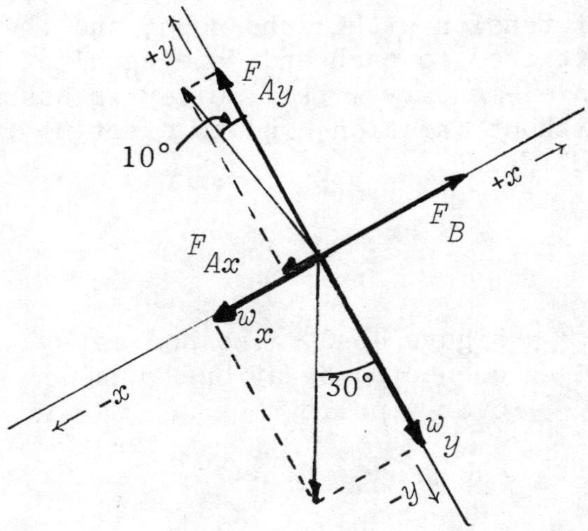

$$w_y = -w \cos 30° = -400 \text{ N} \times 0.866 = -346.4 \text{ N},$$

$$w_x = -w \sin 30° = -400 \text{ N} \times 0.500 = -200 \text{ N},$$

$$F_{Ay} = F_A \cos 10° = 0.985\, F_A\,,$$

$$F_{Ax} = -F_A \sin 10° = -0.174\, F_A,$$

$$F_{Bx} = F_B.$$

Substituting in $\Sigma\vec{F}_y = 0$, we have

$$F_{Ay} + w_y = 0, \qquad \text{or} \qquad 0.985\, F_A - 346.4 \text{ N} = 0,$$

from which

$$F_A = 351.7 \text{ N}.$$

Substituting in $\Sigma\vec{F}_x = 0$, we then have

$$F_{Ax} + F_B + w_x = 0,$$

$$-0.174 \times 351.7 \text{ N} + F_B - 200 \text{ N} = 0,$$

from which

$$F_B = 261.2 \text{ N}.$$

This method for finding the forces in ropes connected in a knot may be extended to any number of ropes connected by knots in a net. Provided only three ropes are joined in each knot and provided all the angles are known, one needs to know initially only one force in the system. Then the forces in all ropes of the net may be computed, by solving one knot at a time.

A rope sometimes runs over a pulley, which allows the rope to change its direction while maintaining a constant tension in the rope. The force of tension within the rope, and the force exerted by the rope on the object attached to each end, are parallel to the length of the rope at each point. In elementary work a pulley is assumed to have no mass, to turn on its shaft without friction, and to exert no friction on the rope running through the pulley.

Figure 46-3. Two bodies connected by a rope running over a pulley.

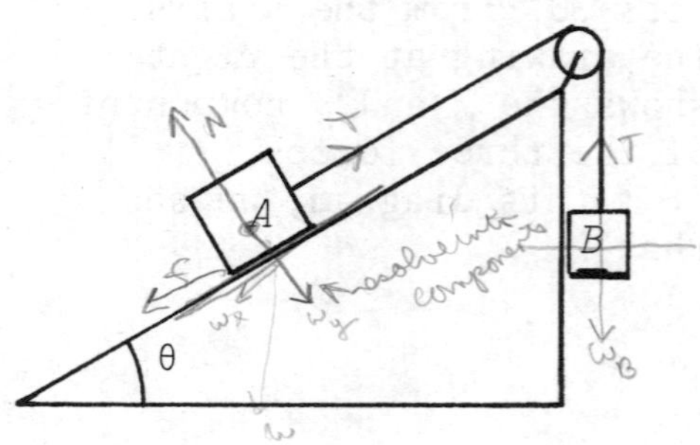

Consider the system shown in Figure 46-3, where one of the bodies is on an inclined plane. In the general case, the system cannot be in static equilibrium unless there is friction between body *A* and the plane. The total force of the plane on *A* can be resolved into two perpendicular components: (1) the friction force $\vec{F}_f$ parallel to the plane, and (2) the normal force $\vec{N}$ perpendicular to the plane. The friction force on *A* may act either upward or downward along the plane; when body *A* is in equilibrium, the direction of $\vec{F}_f$ is determined by the other forces on *A*. The maximum friction force which can occur is given by $F_{f_{max}} = \mu_s N$, where μ_s is the coefficient of static friction for body *A* on the surface of the plane (Section 42.7).

Worked Example 46-4

A box *A*, at rest on an inclined plane, is connected by a rope to another box *B* as shown in Figure 46-3. *A* has a mass of 40 kg, *B* has a mass of 30 kg, and θ = 25°.

(a) What friction force is required to keep *A* in equilibrium?

(b) What is the smallest value of μ_s which can provide this force?

Solution

Assuming the pulley is frictionless, the rope exerts a force F_{BA} on

A equal to the weight of B, so

$$F_{BA} = w_B = m_B g = 30\ \text{kg} \times 9.8\ \text{m/s}^2 = 294\ \text{N},$$

upward along the plane. A free-body diagram of A includes the forces shown at the right. We do not know in advance whether F_f is upward or downward along the plane. We guess it acts downward, but if the solution yields an F_f acting upward we will know our guess was wrong. We choose an x axis parallel to the plane and a y axis normal to it. Then the forces on A can be resolved into components as shown in the drawing at the right.

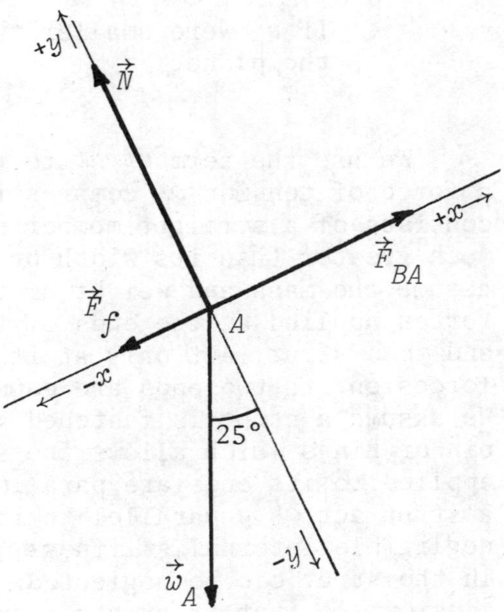

(a) The weight of A has components:

$$w_{Ax} = -w_A \sin 25° = -m_A g \sin 25°$$
$$= -40\ \text{kg} \times 9.8\ \text{m/s}^2 \times \sin 25°$$
$$= -166\ \text{N}.$$

$$w_{Ay} = -w_A \cos 25° = -m_A g \cos 25°$$
$$= -40\ \text{kg} \times 9.8\ \text{m/s}^2 \times \cos 25°$$
$$= -355\ \text{N}.$$

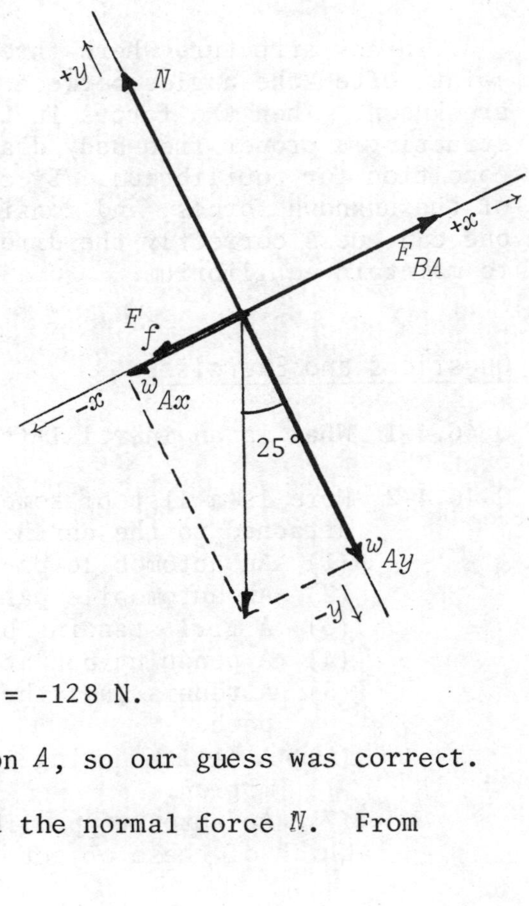

Body A is in equilibrium, so

$\Sigma \vec{F}_x = 0$ and $\Sigma \vec{F}_y = 0.$

Substituting in $\Sigma \vec{F}_x = 0$,

we then have

$$F_{BA} + F_f + w_{Ax} = 0, \text{ or}$$

$$F_f = -F_{BA} - w_{Ax} = -294\ \text{N} + 166\ \text{N} = -128\ \text{N}.$$

F_f does act in the $-x$ direction on A, so our guess was correct.

(b) To compute μ_s, we must first find the normal force N. From $\Sigma \vec{F}_y = 0$, we have

$$N + w_{Ay} = 0.$$

Then $N = -w_{Ay} = -(-355\ \mathrm{N}) = +355\ \mathrm{N}$.

The required $\mu_s = \frac{|F_f|}{N} = \frac{128\ \mathrm{N}}{355\ \mathrm{N}} = 0.361$.

If μ_s were smaller than 0.361, body *A* would slide upward along the plane.

We use the term *strut* to refer to a structural member which can exert a force of tension *or* compression parallel to its length. A strut often consists of a straight member such as a solid rod or pole having a length much greater than its width or thickness. For our elementary analyses, we assume the mass and weight of the strut to be negligible in comparison with forces applied at the ends of the strut. Then the forces applied by a strut, and to a strut, act only at its two ends. For a strut in equilibrium, the forces on its two ends are equal in magnitude but opposite in direction. We assume a strut is fastened at each end using a connection such as a smooth pin or hinge which allows the strut to rotate to a position where the forces applied to its ends are parallel to its length, so the forces exerted by a strut act *only* parallel to its length. That is, we assume each strut has negligible internal stiffness, so that shear (transverse or crosswise) forces in the strut can be neglected. For any particular strut, whether it is in tension (so that it exerts a pull at each end) or compression (so that it exerts a push at each end) is determined by the other forces acting at the two ends of the strut.

In any structure where three members (struts and/or ropes) meet at one joint, often the angles between the members and the force in one of them are known. Then the forces in the other two can always be found by constructing a proper free-body diagram of the joint and applying the first condition for equilibrium. By choosing one coordinate axis parallel to one of the unknown forces, and considering the direction of the known force, one can guess correctly the direction required for the second unknown force to maintain equilibrium.

Questions and Exercises

Q 46.4-1 What is an inertial frame of reference?

Q 46.4-2 Here is a list of some objects at rest in a coordinate system attached to the earth:

(1) An automobile parked on a level street.
(2) An automobile parked on a steep hill.
(3) A girl standing balanced and motionless on a pair of stilts.
(4) A pendulum bob at one extreme of its swing.
(5) A tennis ball thrown vertically upward, at the top of its path.
(6) A ball bouncing in a vertical line, at the bottom of its motion.
(7) An auto stopped for a red light.

Which of these objects are in equilibrium?

Q 46.4-3 For each object listed in Q 46.4-2 which is *not* in equilibrium, what is the unbalanced force acting on the system?

Q 46.4-4 Suppose we add all the external forces on the particles of a system, and find that the sum is zero. What conclusions can we draw about the motion of the system?

Q 46.4-5 If every particle of a system has zero resultant force acting on it, does it necessarily follow that: (a) the center of mass of the system has zero acceleration? (b) the center of mass of the system is at rest? (c) the system is in equilibrium? (d) the system is a rigid body?

E 46.4-1 Draw a free-body diagram for box *A*, and another for box *B*. Neglect air pressure and gravitational forces other than weight.

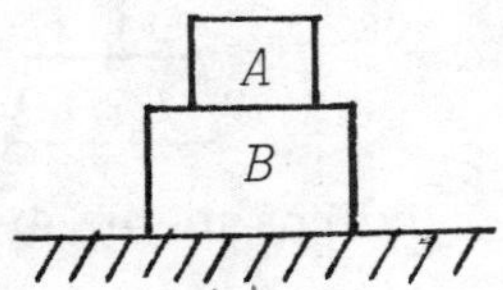

E 46.4-2 Draw a free-body diagram for box *C*, and another for box *D*. Consider that both boxes move under the applied force $\vec{F}_a$, and include friction.

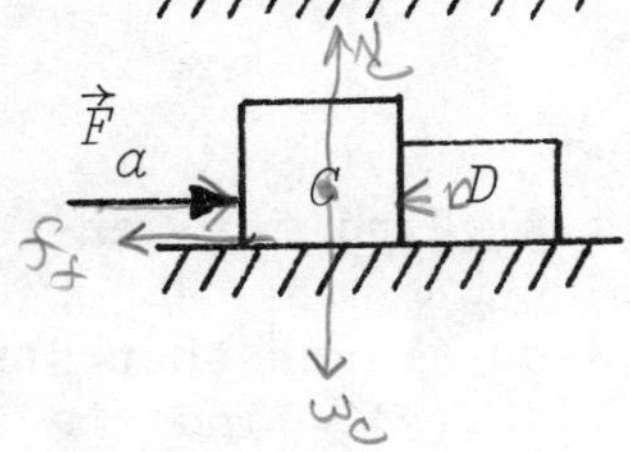

E 46.4-3 Draw a free-body diagram of an auto having rear-wheel drive which is:

(a) accelerating forward on a straight level road;

(b) applying its brakes while traveling down hill.

E 46.4-4 A clothesline sags so that each segment of it dips 5° below the horizontal when a concentrated weight of 50 N is hung from the center of the line. What is the tension in the clothesline?

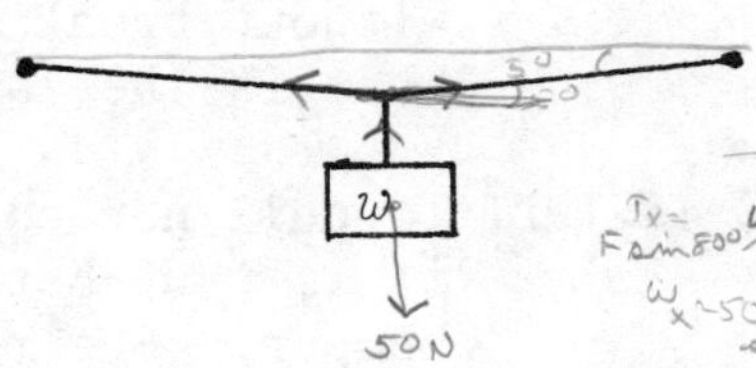

The diagrams below apply to the next three exercises.

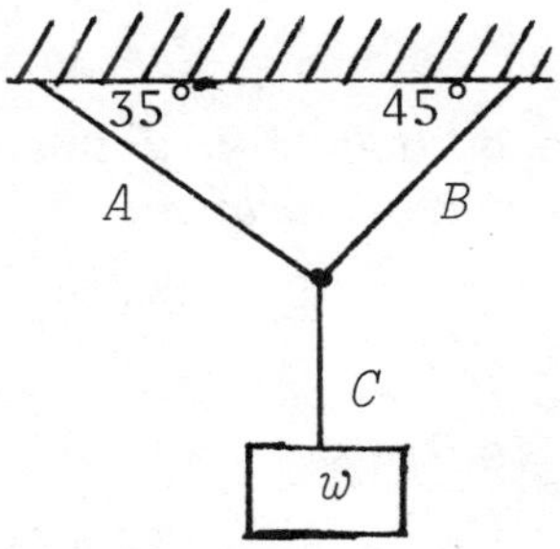

Exercise 46.4-5

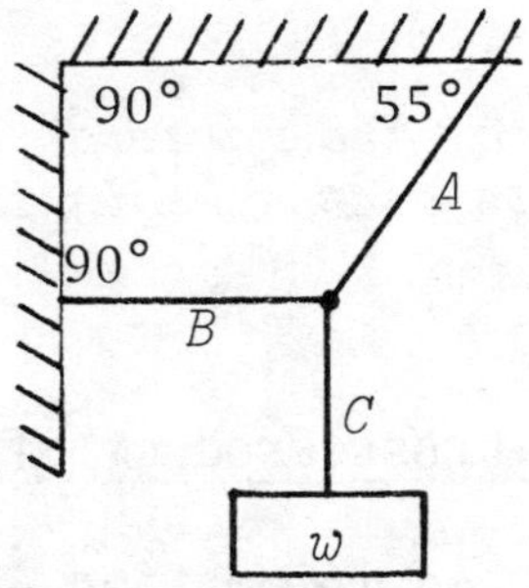

Exercise 46.4-6

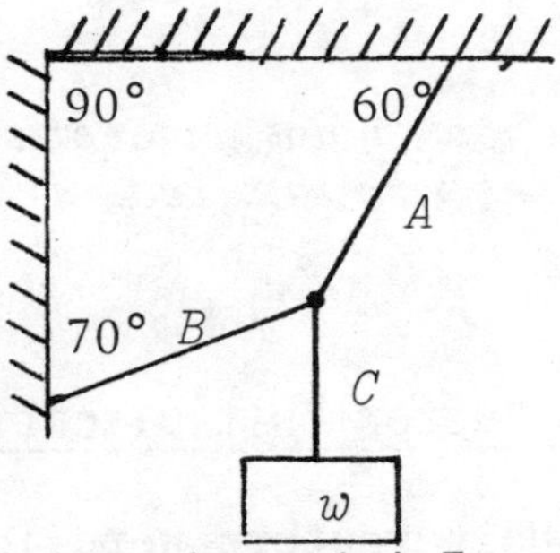

Exercise 46.4-7

E 46.4-5 Find the tension in each rope when w = 2400 N.

E 46.4-6 Find the tension in each cable when w = 15,000 N.

E 46.4-7 Find the tension in each string when w = 60 N.

The diagrams below apply to the next three exercises.

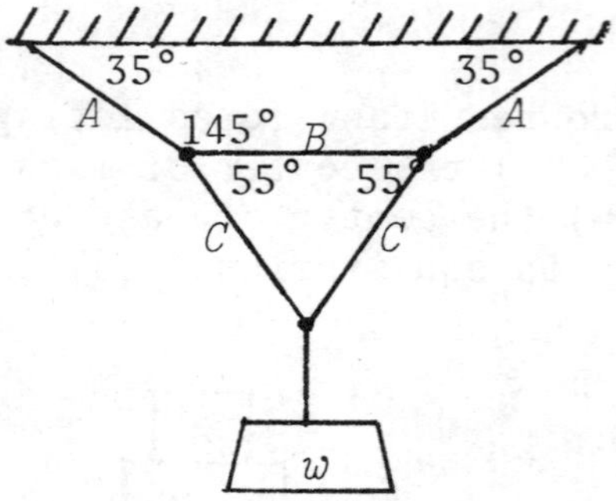

Exercise 46.4-8

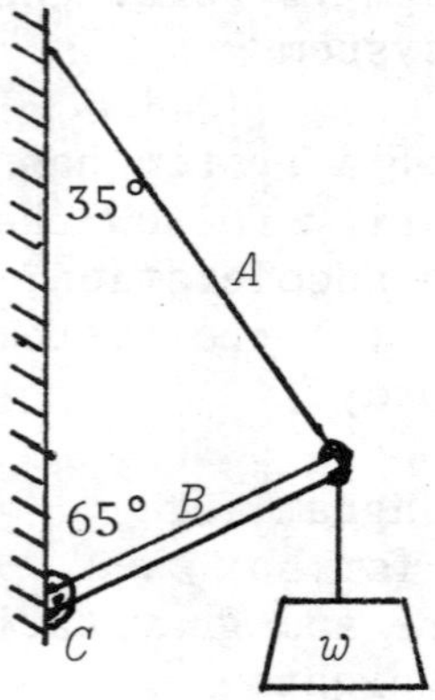

Exercise 46.4-9

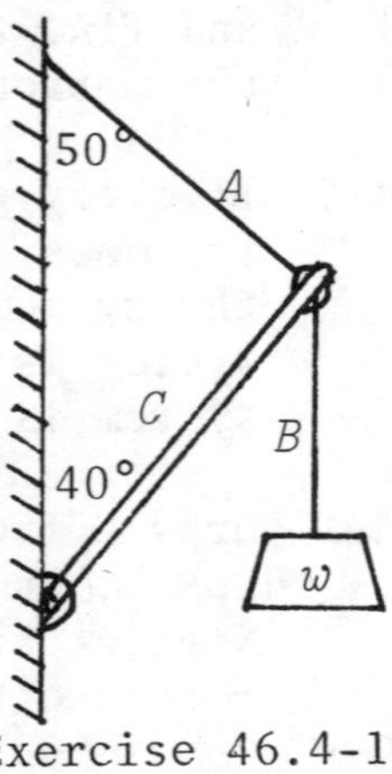

Exercise 46.4-10

E 46.4-8 Find the tension in each cord when w = 320 N.

E 46.4-9 Assume that strut B has no mass, and w = 2000 N.
(a) Find the tension in cable A.
(b) Find the force in strut B.
(c) What are the magnitude and direction of the force exerted on the strut by the smooth pin at C?

E 46.4-10 For the system shown, rope A can sustain any force up to 1800 N without breaking. Strut C can sustain any force up to 2600 N without breaking. Cable B can sustain any force up to 8000 N. What is the largest value of w which the system can support?

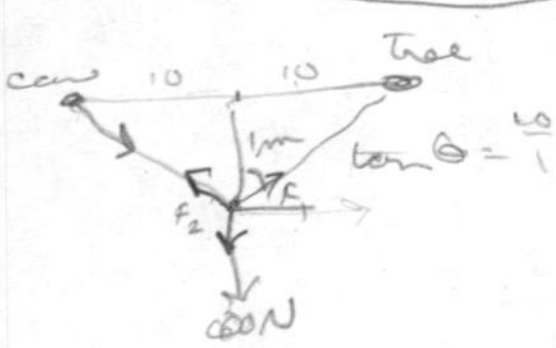

E 46.4-11 To pull her car out of a ditch, a physics major ties a rope so it is taut between her car bumper and a tree 20 m from the bumper. She then pushes on the rope at its midpoint with a force of 600 N perpendicular to the rope, displacing the midpoint by one meter. What force does the rope exert on the car?

Check your understanding of the material above by testing yourself on the minimum learning goals at the beginning of this section.

46.5 Vector Multiplication. Cross Product of Two Vectors

Unit vectors were introduced in Section 43.1, and the procedure for multiplying two vectors when their product is a scalar quantity is given in Section 47.1. In cases where the product of two vector quantities is another vector, one must use the rule for multiplication which yields the *vector product* or *cross product* of the two vectors. Examples of a cross product include the magnetic force on a charged particle moving in a magnetic field (Section 58.4), and the torque exerted by a pair of forces acting on the object (Section 46.6).

The rule for finding the vector product of two vectors is formulated using a cross (×) as the symbol for multiplication.

★ The *cross product* or *vector product* of the vectors $\vec{A}$ and $\vec{B}$ is defined as a vector perpendicular to the plane of $\vec{A}$ and $\vec{B}$, having magnitude $AB \sin\theta$, and having the direction of advance of a right-handed screw rotated from $\vec{A}$ to $\vec{B}$ through an angle θ of less than 180°.

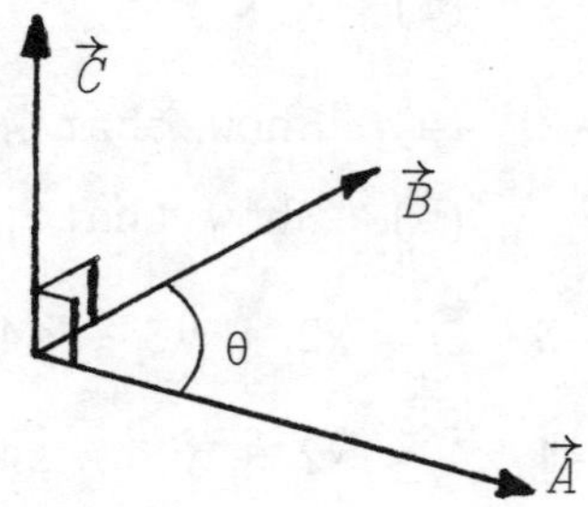

To remember the direction of advance of a right-handed screw, one may think of an ordinary lamp bulb which advances this way when turned into its socket. If the vector product of $\vec{A}$ and $\vec{B}$ is designated by $\vec{C} = \vec{A} \times \vec{B}$, then the magnitude of this product is $C = AB \sin\theta$. The right-hand rule included in the definition of the cross product requires that the *first* vector in the product be rotated into the *second*. Reversing the order will reverse the direction of the vector product:

$$\vec{A} \times \vec{B} = -\vec{B} \times \vec{A}.$$

From the definition, it is evident that the vector product of any vector with itself is equal to zero, because the angle θ is zero. In a physical equation, a term in which one vector is multiplied by another is written as a cross product when the use of this rule gives the correct value for the physical quantity represented by the product.

In terms of components, the cross product of $\vec{A}$ and $\vec{B}$ may be computed as follows:

$$\begin{aligned}\vec{A} \times \vec{B} &= (\hat{x} A_x + \hat{y} A_y + \hat{z} A_z) \times (\hat{x} B_x + \hat{y} B_y + \hat{z} B_z) \\ &= \hat{x} \times \hat{x} A_x B_x + \hat{x} \times \hat{y} A_x B_y + \hat{x} \times \hat{z} A_x B_z \\ &+ \hat{y} \times \hat{x} A_y B_x + \hat{y} \times \hat{y} A_y B_y + \hat{y} \times \hat{z} A_y B_z \\ &+ \hat{z} \times \hat{x} A_z B_x + \hat{z} \times \hat{y} A_z B_y + \hat{z} \times \hat{z} A_z B_z.\end{aligned}$$

Using the rules for the cross product of two unit vectors (Exercise 46.5-1), we see that each cross product of a unit vector with itself is equal to zero because $\sin\theta = 0$. Each cross product of two different unit vectors is equal to the third unit vector or its negative. Collecting the results, we obtain

$$\begin{aligned}\vec{A} \times \vec{B} &= \hat{y} \times \hat{z} (A_y B_z - A_z B_y) + \hat{z} \times \hat{x} (A_z B_x - A_x B_z) \\ &+ \hat{x} \times \hat{y} (A_x B_y - A_y B_x) \\ &= \hat{x} (A_y B_z - A_z B_y) + \hat{y} (A_z B_x - A_x B_z) + \hat{z} (A_x B_y - A_y B_x).\end{aligned}$$

Exercises

E 46.5-1 Compute the following cross products:

(a) $\hat{x} \times \hat{x} =$; (b) $\hat{y} \times \hat{y} =$; (c) $\hat{z} \times \hat{z} =$;

(d) $\hat{x} \times \hat{y} =$; (e) $\hat{y} \times \hat{z} =$; (f) $\hat{z} \times \hat{x} =$.

E 46.5-2 (a) Show that $\vec{A} \times \vec{B} = 0$ if $\vec{A}$ is parallel to $\vec{B}$.

(b) Show that $|\vec{A} \times \vec{B}| = AB$ if $\vec{A}$ is perpendicular to $\vec{B}$.

E 46.5-3 $\vec{A} = \hat{x}2 + \hat{y}3 + \hat{z}4$. $\vec{B} = \hat{x}4 - \hat{y}2 - \hat{z}5$. Compute $\vec{A} \times \vec{B}$.

E 46.5-4 $\vec{A} = \hat{x}2 + \hat{y}3 + \hat{z}4$. $\vec{B} = \hat{x}3 + \hat{y}2 - \hat{z}3$. Compute $\vec{A} \times \vec{B}$.

E 46.5-5 Which (if either) of the following two expressions for kinetic energy is correct: $KE = \frac{1}{2} m\vec{v} \cdot \vec{v}$, or $KE = \frac{1}{2} m\vec{v} \times \vec{v}$? Explain.

E 46.5-6 Which (if either) of the following two expressions for the work done against friction is correct: $W = \vec{F}_f \cdot \vec{s}$, or $W = \vec{F}_f \times \vec{s}$? Explain.

E 46.5-7 $\vec{C} = \hat{x}3 + \hat{y}4 + \hat{z}2$. $\vec{D} = -\hat{x}2 + \hat{y}5 + \hat{z}4$. Find a vector perpendicular to the plane defined by $\vec{C}$ and $\vec{D}$.

E 46.5-8 Given two non-zero vectors $\vec{A}$ and $\vec{B}$.

(a) Is it possible for *both* $\vec{A} \times \vec{B}$ and $\vec{A} \cdot \vec{B}$ to equal zero? Explain.

(b) Is it possible for *neither* $\vec{A} \times \vec{B}$ nor $\vec{A} \cdot \vec{B}$ to equal zero? Explain.

46.6 Torque

Minimum Learning Goal. (1) Given a force applied at a point and the vector position of this point with respect to a specified origin, be able to calculate the resulting torque. For example:

1 Two children are sitting on a level seesaw as shown in Figure 46-4. Child *A* has a weight of 280 N, and a center of mass located 1.8 m from an origin chosen at the center of the plank. Compute the torque about the origin which results from the downward force of *A* on the plank. (Answer, page 46-35, Line 5)

2 A mechanic is tightening the head bolts on an automobile engine without the benefit of a torque wrench. As he proceeds in the correct sequence, he attempts to calculate torques by attaching a spring scale to the end of the wrench, measuring the distance from that end to the center of the bolt, and pulling on the scale. Just as he completes the tightening of a bolt, the radius vector from the center of the bolt to the end of the wrench is given (in feet) by $\vec{r} = 1.2\ \hat{x}$, and the force he exerts is given (in pounds) by $\vec{F} = -20\ \hat{x} - 66\ \hat{y}$. What torque does he apply to this bolt at that instant? (Answer, page 46-35, Line 3)

Sometimes a force is applied to a body in such a way that it tends to change the state of rotation of the body. For example, if a seesaw is balanced and at rest on the bar supporting its center, and then a child pushes down on one end of the seesaw, this end begins to move down while the opposite end begins to move upward. If the child pushes down on the center of the seesaw just above the support bar, the seesaw remains at rest and no rotation occurs. It is not the force by itself which causes the rotation, but the combination of the force and its distance from the bar about which the seesaw can rotate. The physical quanity which tends to cause rotation, or change the state of rotation of an object, is called *torque*.

When we consider a torque associated with some force, we always need to take account of the distance between the point where the force is acting and the point about which the torque is applied. In each instance where we are concerned with torques, we choose some point as the *origin* of a coordinate system, and compute torques about this origin. Then each torque acts with reference to the origin or an axis of rotation passing through the origin. The torque associated with a force $\vec{F}$ depends on the magnitude and direction of the force, and on the vector distance $\vec{r}$ from the origin to the point where the force is applied.

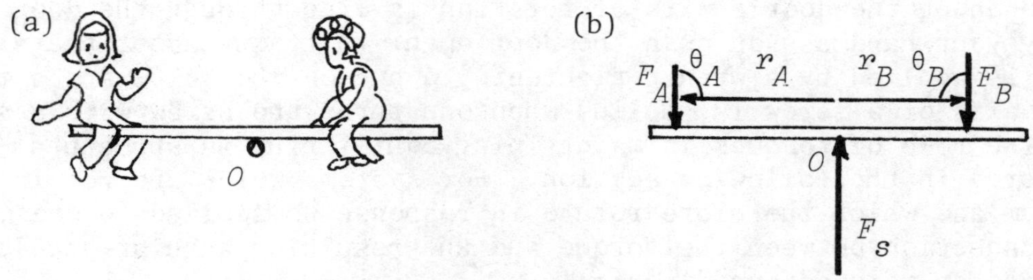

Figure 46-4. (a) Each child on a seesaw exerts a force on the plank equal to the child's weight. To compute the resulting torques, a point near the center of the plank on its axis of rotation is chosen as the origin. (b) A free-body diagram of the seesaw. The torque applied by a force depends on the magnitude of the force, its vector distance from the origin, and the angle between the force and the position vector. The torque about O resulting from $\vec{F}_A$ is positive (ccw), but the torque about O resulting from $\vec{F}_B$ is negative (cw). The torque about O resulting from the support force $\vec{F}_s$ is zero.

We first consider the case where both $\vec{r}$ and $\vec{F}$ lie in a plane which we may choose as the plane of the drawing (usually the xy plane), as shown in Figure 46-4. In this case, each torque tends to cause angular motion about the origin which is either clockwise (cw) or counterclockwise (ccw) in the plane of the drawing. We designate ccw torques as positive and cw torques as negative, using the following symbols when this is convenient:

↶ + and ↷ − .

The angular direction of a torque acting on a body may be found by considering the direction of the rotation which would result from the associated force if the body were free to rotate. To fully specify a torque, one must state both its angular direction and its magnitude. The symbol we use for torque is τ (Greek, tau).

★ A force $\vec{F}$ acting at a vector position $\vec{r}$ from the origin exerts a torque about the origin having a magnitude $\tau = r\,F\,\sin\theta$, where θ is the angle between the directions of $\vec{r}$ and $\vec{F}$.

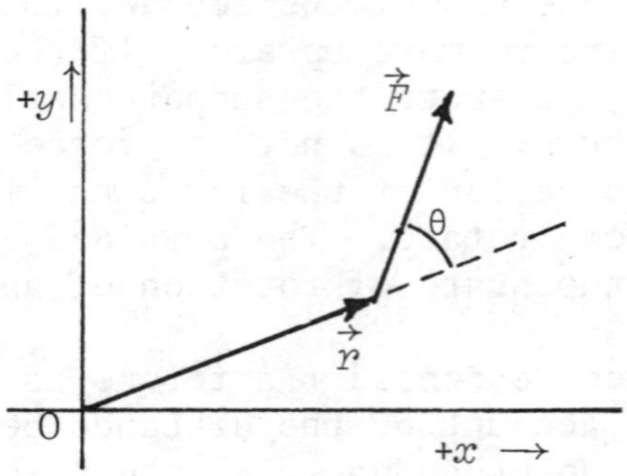

The SI unit of torque is the meter-newton. This is *not* equal to the joule, for reasons given below.

Examples of torques from our daily experience include the torque about the axis of a doorknob applied by your hand as it turns the doorknob, and the torque about the door's axis of rotation (a line through the door hinges) applied by your hand as you push the door open. A torque about the axis of a bolt is applied by a wrench tightening a nut on the bolt, and a torque about the axis of a screw is applied when one turns the screw with a screw-driver. The role of torques in maintaining equilibrium in structural systems is considered in the following section. For systems which are not in equilibrium and which therefore rotate in response to applied torques, the precise connection between the torque and the resulting angular acceleration is developed in more advanced texts.

Worked Example 46-5

A pencil having a mass of 6.0 grams and a total length of 18.0 cm has its point on a table top as shown here. What is the torque about the tip of the pencil point resulting from the pencil's weight, when the pencil leans at an angle $\phi = 40°$?

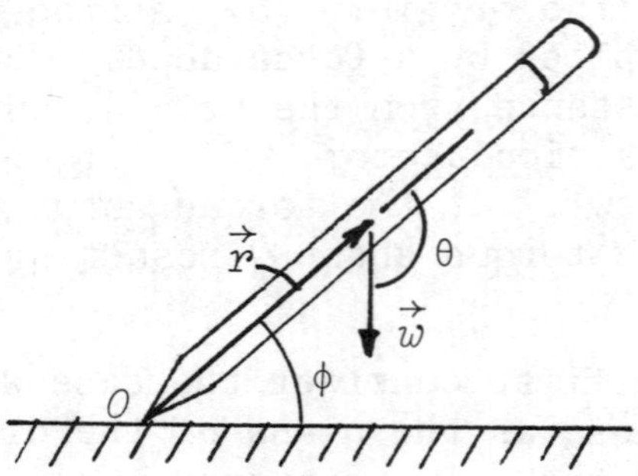

Solution

We choose the tip of the pencil point as the origin. The weight acts downward at the center of mass of the pencil, which we take to be at the midpoint of the pencil, 9.0 cm from the origin. Relative to the origin, this force applies a torque to the pencil which is clockwise, or negative. The angle between the directions of $\vec{r}$ and $\vec{w}$ is $\theta = 90° + \phi$.

The magnitude of the torque is

$$\tau = r\,F \sin\theta = r\,mg \sin\theta$$

$$= 9.0 \times 10^{-2}\ \text{m} \times 6.0 \times 10^{-3}\ \text{kg} \times 9.8\ \text{m/s}^2 \times \sin 130°$$

$$= 4.1 \times 10^{-3}\ \text{m·N, clockwise.}$$

If there is no opposing torque on the pencil, the torque resulting from its weight tends to make the pencil rotate clockwise until it hits the table top.

In general, the position vector $\vec{r}$ and the force $\vec{F}$ which combine in a torque have directions in three-dimensional space such that it is not feasible to include both of them in the plane of the drawing. It follows that the resulting torque cannot be described as simply cw or ccw in the plane of the drawing. For this general three-dimensional case the torque is defined using the cross product of the vectors $\vec{r}$ and $\vec{F}$.

★ A force $\vec{F}$ acting at a vector position $\vec{r}$ from the origin exerts a *torque* about the origin given by $\tau \equiv \vec{r} \times \vec{F}$.

The drawing given above with our earlier definition of torque also illustrates this more general definition, if we remember that the vectors $\vec{r}$ and $\vec{F}$ need not be confined to the *xy* plane. This definition shows that torque is a vector quantity, directed perpendicular to the plane containing $\vec{r}$ and $\vec{F}$. The torque is about the origin, so it may be drawn as an arrow through the origin having the direction given by the cross product. If one thinks of this torque vector as collinear with an axis of rotation through the origin, then the torque tends to rotate the system on which it acts about that axis. The sense of rotation is related to the direction of the torque vector by a right-thumb rule: If the thumb of the right hand is extended while the fingers are curled (as in hitch-hiking), when the thumb is pointed in the direction of the torque the fingers indicate the direction in which the system tends to rotate around the axis.

The definition of the magnitude of torque given earlier in this section is consistent with the definition of torque as a cross product. If $\vec{r}$ and $\vec{F}$ are in the *xy* plane, the torque is either clockwise in the *xy* plane ($\vec{\tau}$ vector directed along the -*z* axis) or counterclockwise in the *xy* plane ($\vec{\tau}$ vector directed along the +*z* axis). Note that a force directed parallel to a position vector $\vec{r}$ produces zero torque about the origin.

Both torque and work are defined as the product of a force times a distance or displacement. However, work is a scalar quantity (Section 47.2) while torque is a vector quantity. The unit of joule = newton·meter is used only for work and energy, not for torque. The SI unit of torque should be written as meter·newton to avoid confusion with the joule.

Worked Example 46-6

A force $\vec{F} = (\hat{x}\ 4.2 + \hat{y}\ 6.9)$ newtons is applied at the point

$\vec{r} = (\hat{x}\ 0.36 - \hat{z}\ 0.48)$ meters relative to an origin. What is the resulting torque?

Solution

Using the definition $\vec{\tau} = \vec{r} \times \vec{F}$, we compute

$$\vec{\tau} = (\hat{x}\ 0.36 - \hat{z}\ 0.48)\ \text{m} \times (\hat{x}\ 4.2 + \hat{y}\ 6.9)\ \text{N}$$

$$= -(\hat{z}\ 0.48\ \text{m}) \times (\hat{y}\ 6.9\ \text{N}) - (\hat{z}\ 0.48\ \text{m}) \times (\hat{x}\ 4.2\ \text{N})$$

$$+(\hat{x}\ 0.36\ \text{m}) \times (\hat{y}\ 6.9\ \text{N})$$

$$= -\hat{z} \times \hat{y}\ 3.3\ \text{m}\cdot\text{N} - \hat{z} \times \hat{x}\ 2.0\ \text{m}\cdot\text{N} + \hat{x} \times \hat{y}\ 2.5\ \text{m}\cdot\text{N}$$

$$= (\hat{x}\ 3.3 - \hat{y}\ 2.0 + \hat{z}\ 2.5)\ \text{m}\cdot\text{N}.$$

Note that the $\hat{z} \times \hat{z}$ term equals zero, because $\hat{z}$ is parallel to $\hat{z}$.

Questions and Exercises

Q 46.6-1 What is the SI unit of torque?

Q 46.6-2 How could one apply a force to the seesaw described in Figure 46-4 without exerting a torque on it about the origin?

Q 46.6-3 Consider a high-altitude satellite in uniform circular motion about the earth. Does the earth exert a torque about the center of the earth on this satellite?

Q 46.6-4 Can an object at rest be set into rotational motion without exerting a torque on it?

Q 46.6-5 A meter stick floats on the surface of a pool in such a way that it can rotate about a vertical axis through the center of the stick (at the 50-cm mark). A single horizontal force of 5 newtons applied to the stick does not cause any rotation of the stick, but when a second force of 5 newtons is applied the stick begins to rotate. Explain.

Q 46.6-6 For the pencil discussed in Worked Example 46-5, does the torque about the pencil tip increase, or decrease, while the pencil is falling to the table top?

E 46.6-1 With reference to Figure 46-4, what is the torque about the origin which results from a force $F_A = 320$ N applied 2.0 m from the origin?

E 46.6-2 In Figure 46-4, $F_B = 410$ N acting downward at a point 1.7 m from the origin. Compute the resulting torque about the origin.

E 46.6-3 In Figure 46-4, the plank is 4.2 m in length and supported at its center by a force of 750 N. What is the torque of this support force about the left end of the plank?

E 46.6-4 A ladder having a mass of 29 kg and a length of 7.3 meters is leaning against a building at an angle of 18° from the vertical. Assume the ladder's center of mass is at the midpoint of its length.
(a) What is the torque about the base of the ladder, resulting from the ladder's weight?
(b) What is the torque about the upper end of the ladder, exerted by the ladder's weight?
(c) What is the torque about the ladder's center of mass, exerted by the ladder's weight?

E 46.6-5 A diving board 6.0 m in length and having a uniform cross-section is supported by a bar 1.2 m from the left end. The board has a mass of 70 kg. What is the torque about the support bar resulting from the weight of the board?

E 46.6-7 A force $\vec{F} = -\hat{x}\,2 + \hat{y}\,6$ newtons acts on a body at the point $\vec{r} = \hat{x}\,5 + \hat{y}\,3$ meters.
(a) Compute the resulting torque.
(b) Will the body tend to rotate clockwise, or counterclockwise, as viewed from a point on the $+z$ axis looking toward the origin? Assume a right-handed coordinate system.

E 46.6-8 A force $\vec{F} = \hat{x}\,3.2 - \hat{y}\,5.9 - \hat{z}\,2.8$ N acts on a body at the point $\vec{r} = \hat{x}\,1.7 - \hat{y}\,4.2 + \hat{z}\,6.2$ m. What is the resulting torque about the origin?

E 46.6-9 A force $\vec{F} = \hat{x}\,7.1 + \hat{y}\,3.6 - \hat{z}\,4.7$ N acts on a rigid body at the point $\vec{r} = \hat{x}\,4.8 - \hat{y}\,1.2 + \hat{z}\,3.5$ m. What is the resulting torque about the origin?

E 46.6-10 A baseball player swings a bat and hits a home run. During the time the ball is in contact with the bat, the radius vector from a point (chosen as the origin of the coordinate system) on the bat between the batter's hands to the point where the bat touches the ball is given approximately by

$$\vec{r} = 0.50\,\hat{x} + (20t - 0.10)\,\hat{y} + (8t - 0.04)\,\hat{z} \text{ meters}$$

for $0 \leqq t \leqq 0.01$ seconds.
During that same interval of time the force exerted on the ball by the bat is given approximately by

$$\vec{F} = e^{-3.2 \times 10^5 (t - 0.005)^2} (743\,\hat{y} + 298\,\hat{z}) \text{ newtons}$$

for $0 \leqq t \leqq 0.01$ seconds.
What is the torque on the ball at the time $t = 0.007$ seconds?

E 46.6-6 A force $\vec{F} = -\hat{y}\,5.9 \times 10^{-2}$ N acts on a pencil at the point $\vec{r} = \hat{x}\,6.9\text{ cm} + \hat{y}\,5.8\text{ cm}$ relative to the pencil tip. What is the resulting torque about the tip?

Check your understanding of the material above by testing yourself on the minimum learning goal at the beginning of this section.

46.7 Second Condition for Static Equilibrium

Minimum Learning Goal. (1) Given a description of a physical system in equilibrium, be able to use two conditions for equilibrium to calculate an unknown force or torque. For example:

1 A metal bar and a piece of pipe are used to lift one side of a refrigerator, as shown below. The total force exerted by the bar on the refrigerator is vertical upward and is represented by $\vec{F}_1$. Assume that point *B* is the center of mass. Neglect the weight of the bar. Calculate: (a) the force exerted by the floor on the refrigerator at point *A*; (b) the torque about point *A* resulting from the refrigerator's weight; and (c) the force *F* needed to hold the refrigerator as shown. (Answer, page 46-36, Line 12)

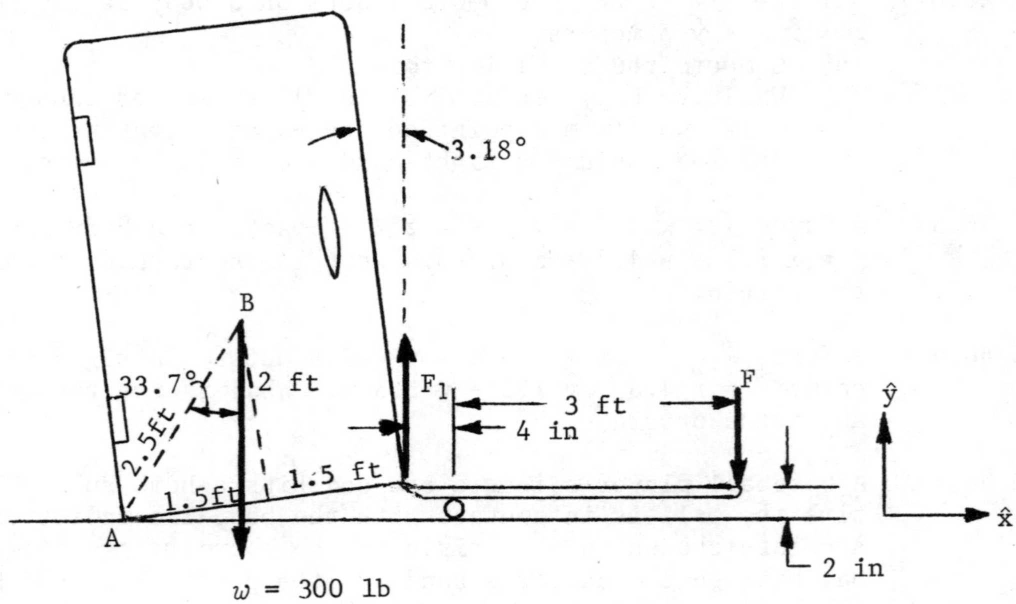

2 A 60-pound box of canned soup is being unloaded from a truck when one of the workers stops briefly, holding the box on edge on an inclined ramp as shown below. Calculate: (a) the force *F* (parallel to the ramp) which he exerts on the upper edge of the box; (b) the torque about point *A* exerted by the force of gravity on the box; and (c) the force exerted by the ramp on the box at point *A*. Assume that the center of mass of the box is at point *B*, the center of the box. (Answer, page 46-35, Line 4)

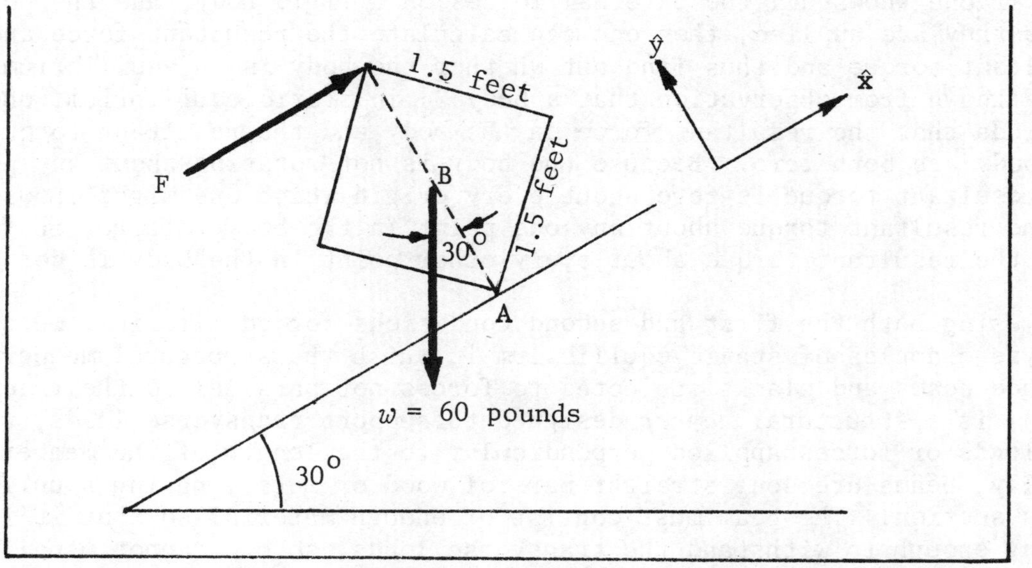

In Section 46.4 we considered examples of bodies in static equilibrium under conditions where the first condition for static equilibrium was enough to ensure equilibrium and to enable us to compute unknown forces on the body. Even when the sum of the forces on a body is zero, however, if not all of these forces are concurrent (that is, directed along lines passing through a common point) they can give rise to a resultant torque on the body. In this case, the resultant torque causes a change in the rotational motion of the body, so the body is not in equilibrium.

Could the *internal* forces between the particles of a body produce a net torque on the body as a whole? Newton's third law (Section 42.5) expresses the fact that the force $\vec{F}_{AB}$ of particle A on particle B is always equal in magnitude but opposite in direction to the force $\vec{F}_{BA}$ of particle B on particle A. Now if these two forces are directed along the line joining the particles, then any torque resulting from $\vec{F}_{AB}$ is exactly cancelled by the corresponding torque resulting from $\vec{F}_{BA}$. Fundamental gravitational and electrical forces between particles are always directed along the line joining the particles, and so are all the macroscopic forces which are resultants of these fundamental forces. Under these conditions, the internal torques sum to zero for any system of particles.

A body is in static equilibrium in an inertial frame of reference in which the body remains at rest. For a rigid body in equilibrium the angular acceleration of every particle in the body is zero, so the net torque on each individual particle is zero. Therefore, the sum of the torques acting on all the particles of the body is also zero. This sum consists of the resultant internal torque plus the resultant external torque on the body. We found above that the resultant internal torque on a system is always zero. We conclude, then, that for a rigid body in equilibrium *the vector sum of the external torques on the body is zero*. In equation form:

★ $\Sigma \vec{\tau}_{ext} = 0$ Second condition for equilibrium.

If one knows all the external forces on a rigid body, and the points where they are applied, then one can calculate the resultant force and the resultant torque and thus find out whether the body is in equilibrium. If it is known from observation that a body *is* in static equilibrium, one can conclude that the resultant force on the body and the resultant torque on the body are both zero. Because the body is not rotating about any point, the resultant torque is zero about every origin which one might choose. So, if the resultant torque about any one point in the body is zero, it follows that the resultant torque about every other point in the body is zero.

Using both the first and second conditions for equilibrium, we can analyze examples of static equilibrium in which the structural members include beams and planks subjected to forces not parallel to their length. A *beam* is a structural member designed to support transverse loads, that is, loads or forces applied perpendicular to the length of the member. Usually, beams are long straight bars of wood or metal, having a uniform cross section. The beam must consist of enough material so that it is strong enough to withstand the transverse loads, so one cannot usually assume that the beam's own weight is negligible in comparison with other forces on the beam. For a rigid beam having a uniform cross section, we can assume that the weight acts at the center of mass. In a uniform gravitational field, the total gravitational torque on a body is correctly obtained by regarding the total weight of the body as being applied at the center of mass.

Worked Example 46-7

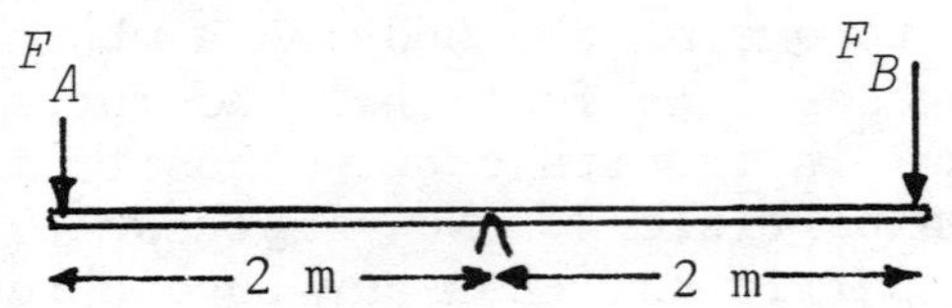

A horizontal plank 4.0 m in length is supported by a transverse bar at its center which serves as a fulcrum. Loads F_A = 300 N and F_B = 500 N are placed on the end of the plank, as shown. Where should one place an additional load F_C = 250 N, to achieve equilibrium? The plank has a uniform cross section and a weight of 400 N.

Solution

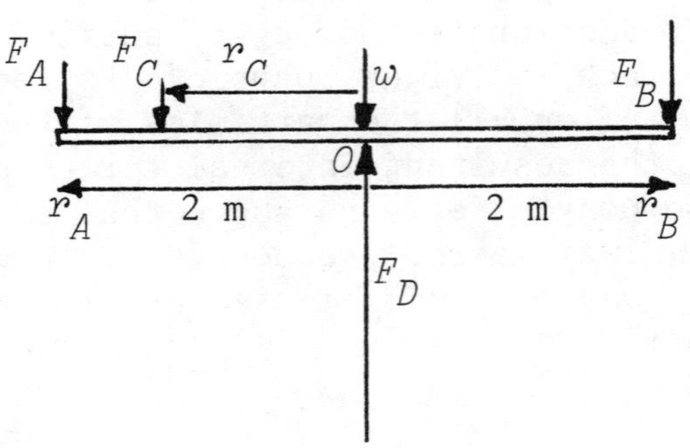

The equation $\Sigma \vec{F} = 0$ does not enable us to solve this problem. We must compute torques about some convenient point. To help us choose this point, we estimate the approximate position of $\vec{F}_C$, and draw the free-body diagram shown here. w = 400 N is the weight of the plank, and F_D is the force of the bar supporting the plank.

Which origin shall we choose, about which to compute torques? The most convenient origin is the fulcrum, because then the position vector $\vec{r}$ equals zero for both w and F_D. In equilibrium, the resultant torque about the origin equals zero. Each force applies a torque so we write

$$\vec{\tau}_A + \vec{\tau}_C + \vec{\tau}_w + \vec{\tau}_D + \vec{\tau}_B = 0.$$

For each torque we substitute the corresponding $\vec{r} \times \vec{F}$:

$$\underbrace{\vec{r}_A \times \vec{F}_A}_{+} + \underbrace{\vec{r}_C \times \vec{F}_C}_{+} + \underset{0}{\vec{r}_w} \times \vec{w} + \underset{0}{\vec{r}_D} \times \vec{F}_D + \underbrace{\vec{r}_B \times \vec{F}_B}_{-} = 0.$$

The sense of each torque (ccw or cw) is shown by a curved arrow below it. Both r_w and r_D equal zero, because the forces $\vec{w}$ and $\vec{F}_D$ act at the origin. Regarding ccw torques as positive and cw torques as negative, and noting that in each cross product $\vec{r}$ is perpendicular to $\vec{F}$, we can substitute numerical values to obtain

$$+2 \text{ m} \times 300 \text{ N} + r_C \times 250 \text{ N} - 2 \text{ m} \times 500 \text{ N} = 0.$$

Solving for r_C, we obtain $r_C = 1.6$ m.

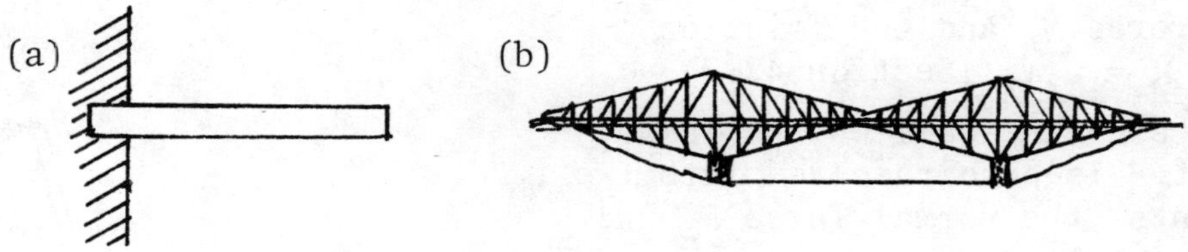

Figure 46-5. (a) A cantilever beam. (b) A cantilever bridge.

A *cantilever* is a beam or structural member supported at only one end (Figure 46-5). The support must supply the torque needed to offset the opposing torque resulting from the weight of the cantilever, in order to maintain equilibrium. A diving board is an example of a cantilever. Some large bridge spans are built as double cantilevers in which the two arms of the bridge meet in the center of the span. Airplane wings are cantilevers supported by the fuselage while the plane is on the ground, but the wings support the fuselage while the plane is in flight.

When a ladder is in use as it leans against a vertical wall, the ladder remains in equilibrium only when both the resultant torque and the resultant force on it are zero. To find an unknown force on a ladder one can apply both conditions for equilibrium.

Worked Example 46-8

A ladder makes an angle of 75° with a horizontal floor and leans against a vertical wall. The ladder has a weight Mg = 200 N and a length $L = 6.0$ m, and supports a man having a weight $mg = 700$ N at a distance $\ell = 4.0$ m from the floor (measured along the ladder). The coefficient of static friction between the ladder and the wall is $\mu_B = 0.30$.

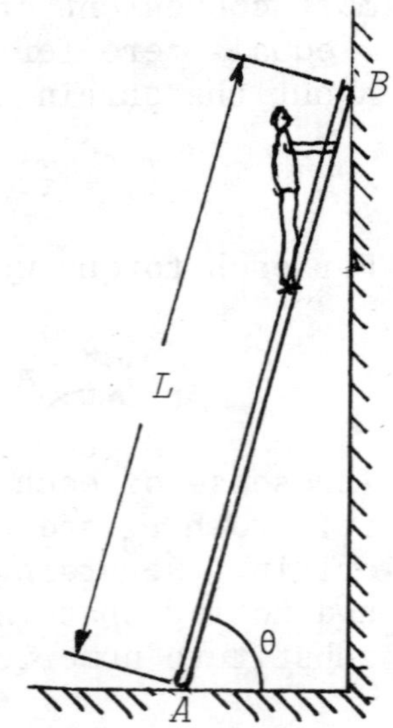

(a) What coefficient of static friction μ_A between the ladder and the floor is required to prevent the ladder from slipping?

(b) What will happen to the ladder if μ_A equals the minimum value found in part (a), and the man then climbs higher on the ladder?

Solution

The first step is to draw a free-body diagram of the ladder, shown at the right. The forces are not drawn to the same scale. The force of the floor on the ladder at point A is resolved into two components, the normal force N_A and the friction force $F_{fA} = \mu_A N_A$ (Section 42.7). The force of the wall on the ladder at point B is also resolved into two components, the normal force N_B and the friction force $F_{fB} = \mu_B N_B$. The weight of the ladder acts at its center of mass, assumed to be the midpoint $L/2 = 3.0$ m from A. Both friction forces act in directions which tend to prevent the ladder from falling.

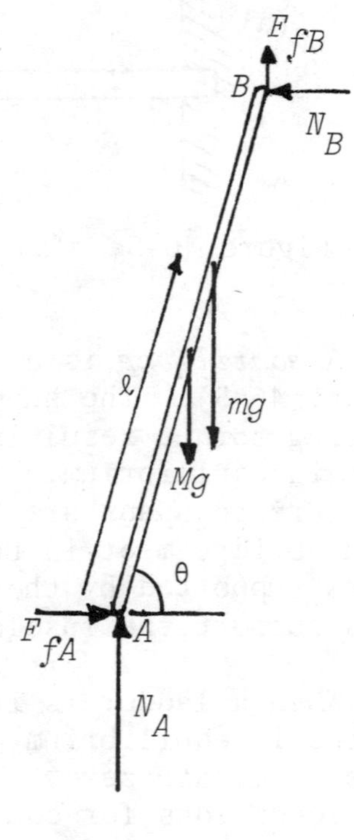

(a) To find μ_A, we first need to obtain F_{fA} and N_A. By applying the condition $\Sigma \vec{F}_x = 0$ to the ladder, we obtain

$$F_{fA} = N_B. \qquad \text{Eq. (1)}$$

By applying the condition $\Sigma \vec{F}_y = 0$ to the ladder, we obtain

$$NA + F_{fB} = Mg + mg. \qquad \text{Eq. (2)}$$

In these equations, only the forces Mg and mg are known. The four unknown quantities are N_A, $F_{fA} = \mu_A N_A$, N_B, and $F_{fB} = \mu_B N_B$.

We can find N_B by applying the condition $\Sigma \vec{\tau} = 0$ to the torques about an origin at A. Each torque has a magnitude written in the form $rF \sin \alpha$, where α is the angle between the directions of $\vec{r}$ and $\vec{F}$. As usual, ccw torques are considered positive while cw torques are negative. Each distance vector $\vec{r}$ is measured from A to the point where the force acts. Then

$$\Sigma \tau_A = -Mg\,(L/2) \sin(\theta + 90°) - mg\ell \sin(\theta + 90°) + N_B L \sin\theta + \mu_B N_B L \sin(90° - \theta) = 0. \quad \text{Eq. (3)}$$

We note that $\sin(\theta + 90°) = \cos\theta$ and $\sin(90° - \theta) = \cos\theta$. Eq. (3) has only one unknown, N_B. Solving for N_B and substitituting, we obtain

$$N_B = \frac{Mg(L/2)\cos\theta + mg\ell\cos\theta}{L(\sin\theta + \mu_B\cos\theta)} \quad \text{Eq. (4)}$$

$$= \frac{200\text{ N} \times 3\text{ m} \times \cos 75° + 700\text{ N} \times 4\text{ m} \times \cos 75°}{6\text{ m}\,(\sin 75° + 0.3\cos 75°)}$$

$$N_B = 1.41 \times 10^2\text{ N}.$$

Then

$$F_{fB} = \mu_B N_B = 0.3 \times 1.41 \times 10^2\text{ N} = 42.3\text{ N}.$$

Also, from Eq. (1),

$$F_{fA} = N_B = 1.41 \times 10^2\text{ N}.$$

From Eq. (2),

$$N_A = Mg + mg - F_{fB}$$

$$= 200\text{ N} + 700\text{ N} - 42.3\text{ N} = 857.7\text{ N}.$$

Then the required coefficient of friction is

$$\mu_A = \frac{F_{fA}}{N_A} = \frac{1.41 \times 10^2\text{ N}}{857.7\text{ N}} = 0.164.$$

If μ_A is smaller than 0.164, the ladder will slip and fall.

(b) With the man standing on the ladder at $\ell = 4.0$ m and $\mu_A = 0.164$, the ladder is in equilibrium. If the man climbs higher, ℓ increases. Then it is seen from Eq. (4) that N_B also increases, because ℓ occurs in the numerator of this fraction. When N_B increases, $F_{fB} = \mu_B N_B$ also increases. Then Eq. (2) shows that N_A decreases, because F_{fB} supports a greater share of the load $Mg + mg$. When N_B increases, to maintain $\Sigma \vec{F}_x = 0$ it is also necessary that F_{fA} increase. But in fact when N_A decreases the friction force $F_{fA} = \mu_A N_A$ also decreases, so this friction force will be too small to maintain equilibrium and the ladder will fall.

Questions and Exercises

Q 46.7-1 Is it possible to have a body acted on by a resultant torque, even though the resultant force on the body is zero? Explain, using an example.

Q 46.7-2 Is it possible for a person to stand on his two feet and lean forward so far that a vertical line through his center of mass passes in front of his toes, and still keep his balance? Explain.

Q 46.7-3 If every particle of a body has zero resultant torque acting on it, does it necessarily follow:
(a) that the body *as a whole* has no rotational motion? Explain.
(b) that the body as a whole is in equilibrium? Explain.
(c) that the body is a rigid body? Explain.

Q 46.7-4 Where is the center of mass of a diving board? Why doesn't the board fall into the water?

Q 46.7-5 (a) When an empty tin can is placed in a lake, why does it float at an angle?
(b) Under what conditions does it float in an upright position?

Q 46.7-6 When turning a corner, a car has a tendency to roll outward, but a motorboat has a tendency to roll inward. Explain.

E 46.7-1 A seesaw is supported by a transverse bar placed below its center of mass. Dave (weight 270 N) is on the seesaw 2 m east of the support bar. Dan (weight 400 N) is on the seesaw 2 m west of the support.
(a) Compute the resultant torque on the seesaw, about the support bar.
(b) If the boys start from rest with the seesaw horizontal, which boy will first rise above his initial position?
(c) How does the weight of the seesaw affect your analysis?
(d) Where should Gwen (weight 450 N) sit on the seesaw in order to establish stable equilibrium with the seesaw horizontal?

E 46.7-2 A bamboo fishpole having uniform cross-section is 5 m long and weighs 12 N. Izaak holds the pole near one end.
(a) What torque must be apply to hold the pole horizontal?
(b) What forces must he apply if his hands are effectively 15 cm apart?

E 46.7-3 A meter stick has a mass of 170 grams and is partially supported by a fulcrum at the 40-cm mark, as shown here. It is desired to maintain the meter stick in equilibrium using only one additional applied force.

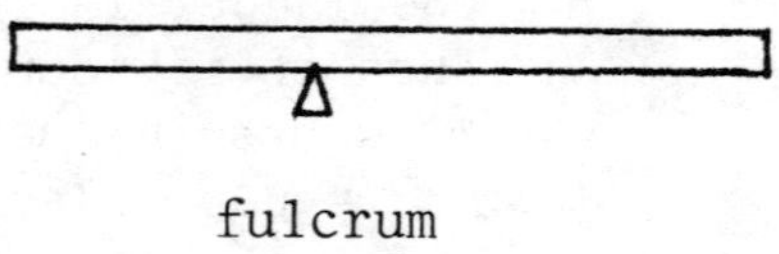

(a) If the applied force acts upward, what is the minimum value of applied force which will suffice, and where should it be applied?
(b) What are the magnitude and the direction of the torque about the fulcrum which results from the force applied in part (a)?
(continued)

(c) If the applied force acts downward, what is the minimum value of applied force which will suffice, and where should it be applied?

(d) What are the magnitude and direction of the torque about the fulcrum which results from the force applied in part (c)?

(e) If the meter stick is moved so that the fulcrum is at a different location, where should the fulcrum be so the applied force required to maintain equilibrium will be approximately zero?

E 46.7-4 A uniform plank 7.0 m in length has a weight of 520 N. The plank rests on two supports 4.0 m apart, each located 2.0 m from the center of the plank. A girl weighing 800 N begins at the center of the plank and walks toward one end. How far beyond the support can she walk before the plank begins to tip?

E 46.7-5 A ladder rests on a horizontal floor and leans against a vertical wall, making an angle of 72° with the floor. The ladder has a mass of 40 kg and a length of 6.6 m, and supports a mass of 80 kg at a distance of 4.9 m from the floor (measured along the ladder). The coefficient of static friction between the ladder and the wall is 0.42, while the coefficient of static friction between the ladder and the floor is 0.60.

(a) Compute the normal and friction forces of the wall on the ladder.

(b) Compute the normal and friction forces of the floor on the ladder.

E 46.7-6 Work E 46.7-5, assuming the ladder makes an angle of 60° with the floor.

Check your understanding of the material above by testing yourself on the minimum learning goal at the beginning of this section.

Answers to Minimum Learning Goals

Line 1 (a) $\vec{w}_1$ and $\vec{F}_1$; (b) $\vec{w}_1 = -\vec{w}_2$, and $\vec{F}_1 = -\vec{F}_2$.

Line 2

Line 3 $-\hat{z}$ 79.2 pound-ft (79.2 pound-ft clockwise as seen from above.)

Line 4 (a) 15 pounds; (b) 31.8 pound-ft , ccw;
(c) $(15\ \hat{x} + 52\ \hat{y})$ pounds

Line 5 504 m•N, ccw

Line 6 Yes; because his center of mass has an acceleration of zero relative to an inertial frame of reference.

Line 7 4.6×10^6 m from the center of the earth (1.8×10^6 m *inside* the earth), on the line between centers of earth and moon.

Line 8 (a) $\vec{F}_1$, $\vec{N}_1$, and $\vec{w}_1$; (b) $\vec{F}_2 = -\vec{F}_1$, $\vec{N}_2 = -\vec{N}_1$, and $\vec{w}_2 = -\vec{w}_1$.

Line 9 The center of mass is in the plane of the three atoms, 0.0065 nm from the O-atom on the line which bisects the angle between the two O-H bonds.

Line 10

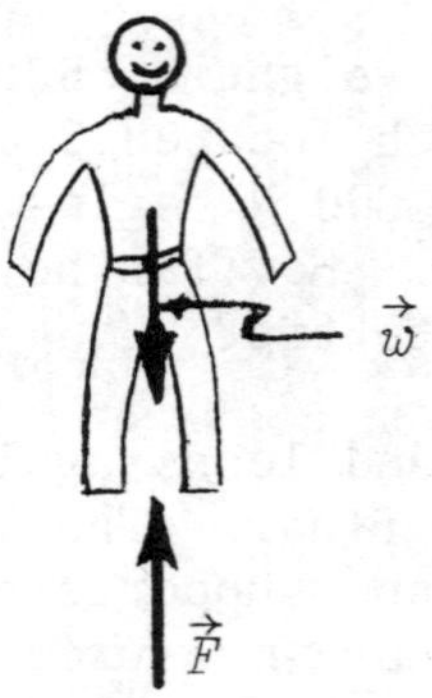

$\vec{w}$ is weight of man; $\vec{F}$ is force exerted by floor on man.
w = 700 N, F = 700 N.

Line 11 (a) System is the piece of salt. (b) External forces are weight of the salt, and normal and frictional forces exerted on the salt by the table. (c) Internal forces are the electric forces holding the ions together in the crystal structure. Their resultant is zero.

Line 12. (a) 162 lb upward; (b) 416 lb-ft, clockwise; (c) 15.3 lb downward.

Line 13 (a)

Line 14 (a)

Line 15 (a) The moon only is the system. (Note, however, that for precise calculations one must consider the combined motion of both Earth and Moon about their common center of mass.)
(b) External forces are force of Earth on Moon and force of Sun on Moon.
(c) Internal forces are those holding the moon together. Their resultant is zero.

ANSWERS to Questions and Exercises in Chapter 46

Section 46.1

Q 46.1-2
(a) Rigid when viewed simply as an object rotating on its axis; non-rigid when considering an earthquake.
(b) Rigid under most conditions; non-rigid when struck by a hammer.
(c) Rigid while at rest in a freezer; non-rigid when it is being crushed or melted.
(d) Rigid while at rest; non-rigid while being shaped in a forge.
(e) Rigid on a quiet day; non-rigid while swaying in gusty winds.

Section 46.2

Q 46.2-2 (a) and (c) at the center of the earth; (b) about two thousand miles below the center of the Pacific Ocean surface.

Q 46.2-4
(a) Near the center of the pelvic or intestinal region.
(b) One way is to bend at the waist and lean forward as far as possible.

Q 46.2-6 (a) It is at rest. (b) Zero. (c) It is at rest. (d) Zero.

E 46.2-2 0.0596 nm, toward C atom

Section 46.3

Q 46.3-2
(a) (1), (2), (3), (4), (5), (6), (9);
(b) (7), (10);
(c) (3), (6), (9);
(d) (1), (4), (5), (7), (10);
(e) internal forces are (3), (9); external forces are (1), (5), (6), (7).

Q 46.3-4 The reaction to (2) is the resultant force of the driver on his seat.

Section 46.4

Q 46.4-2 (1), (2), (3), (7)

Q 46.4-4 The system as a whole is in equilibrium so its center of mass moves with constant velocity.

E 46.4-2

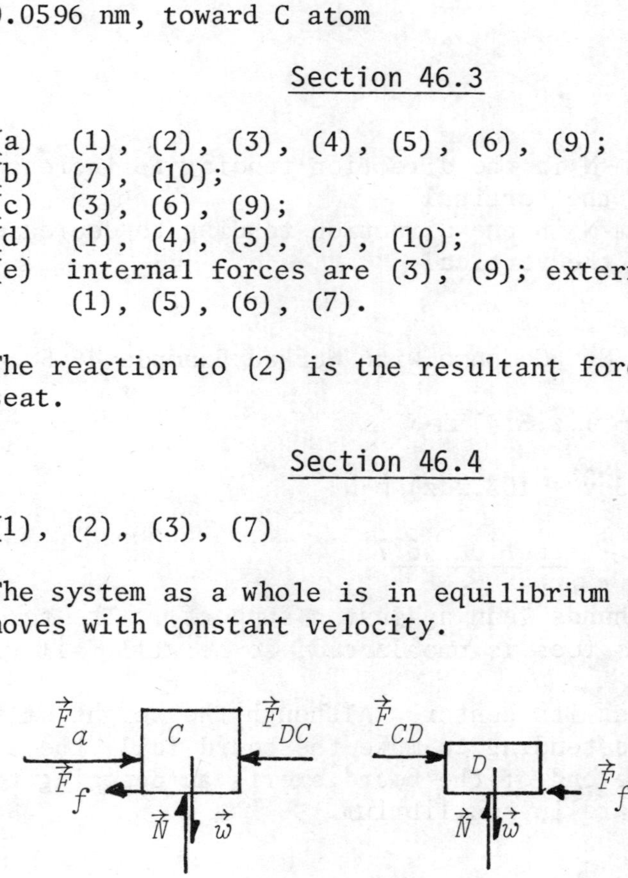

E 46.4-4 2.87×10^2 N

E 46.4-6 $F_A = 1.83 \times 10^4$ N; $F_B = 1.05 \times 10^4$ N

E 46.4-8 $F_A = 278.9$ N; $F_B = 116.5$ N; $F_C = 195.3$ N.

E 46.4-10 2.8×10^3 N

Section 46.5

E 46.5-4 $-17\ \hat{x} + 18\ \hat{y} - 5\ \hat{z}$.

E 46.5-6 The dot product is correct, because work is a scalar.

E 46.5-8 (a) No, because $\vec{A}$ and $\vec{B}$ cannot be perpendicular to each other and parallel to each other at the same time.
(b) Yes, if $\vec{A}$ and $\vec{B}$ are neither parallel nor perpendicular to each other.

Section 46.6

Q 46.6-2 Apply a force directed along a line through the origin.

Q 46.6-4 No

Q 46.6-6 It increases

E 46.6-2 697 m•N cw

E 46.6-4 (a) 3.2×10^2 m•N in the direction tending to increase the ladder's angle with the vertical.
(b) 3.2×10^2 m•N in the direction tending to decrease the ladder's angle with the vertical
(c) zero

E 46.6-6 $-\hat{z}\ 4.1 \times 10^{-3}$ m•N. Compare with Worked Example 46-5.

E 46.6-8 $(\hat{x}\ 48.3 + \hat{y}\ 24.6 + \hat{z}\ 3.4)$ m•N

E 46.6-10 $(0.009\ \hat{x} - 41.43\ \hat{y} + 103.2\ \hat{z})$ m•N

Section 46.7

Q 46.7-2 No (unless his hands lean against a support). The torque of his weight about his toes is unbalanced, so he will fall forward.

Q 46.7-4 In the board near its center. Although the weight of the board applies a torque tending to make the board fall, the supporting structure at the end of the board exerts an opposing torque to maintain the board in equilibrium.

Q 46.7-6 Relative to the car, the "centrifugal force" acts on the center of mass of the car, which is some distance above the road. This force exerts a torque about the bottom of the tires which causes the car to tip outward until there is an equal opposing torque resulting from the car's weight acting downward at the center of mass. For a motorboat, the center of mass lies below the center of buoyancy (the center of mass of the displaced water). When the boat turns, relative to the boat there is a "centrifugal force" on its center of mass which exerts a torque which causes the hull to tip outward until this torque is balanced by an opposite torque applied by the buoyant force.

E 46.7-2 (a) 30 m•N; (b) 200 N upward with one hand, and 200 N downward with the other.

E 46.7-4 1.30 m

E 46.7-6 (a) Normal force = 361.1 N; friction force = 151.7 N;
(b) Normal force = 1024.3 N; friction force = 361.1 N.

CHAPTER 47 WORK AND ENERGY

Introduction

Long before concern for energy resources and energy conservation moved to the center of public affairs, energy was a central concept in the study of the physical world. The law of conservation of energy is perhaps the most fundamental principle of physics. As far as we know, this law applies to every natural process that occurs anywhere in the universe.

Energy is an unusual physical concept, however, in that it is difficult to arrive at a fully satisfactory, yet simple, definition of what we mean by *energy* itself. One reason for this is that it is often difficult to visualize the *total energy* associated with some physical body. It is easier to focus on the *individual forms of energy* which go to make up this total. Another reason is that the energy of a system at any particular time is not nearly as interesting as the *transformations of energy* which occur when the system gains or loses energy, or when the energy of the system changes from one form to another.

The full value of the law of conservation of energy and its companion, the work-energy principle, are not evident until one can make quantitative calculations using these principles. Applying them successfully requires an ability to visualize or identify the energy changes which take place during a variety of physical processes. This ability, in turn, requires an understanding of individual forms of energy and the physical factors on which each depends.

This chapter begins by describing the method of vector multiplication to be used when the product of two vectors is a scalar, such as work. The concept of physical work is introduced and applied first to simple situations, then to more difficult problems requiring the use of calculus. The time-rate of doing work is called power. The concept of kinetic energy is introduced in connection with the work-kinetic energy theorem, which relates the work done by the resultant force on a particle to the change in kinetic energy of the particle.

When you have completed your study of the material in this chapter, you should be able to answer questions and work exercises like the examples in the Minimum Learning Goals assigned by your instructor.

47.1 Vector Multiplication. Dot Product of Two Vectors

The equations used in introductory physics contain symbols for two kinds of physical quantities: scalar quantities and vector quantities. A scalar quantity has no direction in space, so when a scalar is multiplied by one or more other scalars the product also has no direction. Therefore, the rule for obtaining the product of two or more scalars is the same as the familiar rule for multiplying ordinary numbers. A vector quantity does have a direction in space, so when it is multiplied by one or more scalars the product has the same direction (or exactly the opposite direction if the multiplier is negative).

When one vector quantity is multiplied by another vector, however, one must use special care to find out whether the product of the two vectors also has a direction in space, and if it does, to indicate what this direction is. In cases where the product of two vectors is a scalar quantity, we use the rule for multiplication which yields the *scalar product* or *dot product* of the two vectors, discussed in this section. In cases where the product of two vectors is another vector, one must use the rule for multiplication which yields the *vector product* or *cross product* of the two vectors, discussed in Section 46.5.

The rule for finding the scalar product of two vectors is formulated using a dot (·) as the symbol for multiplication.

★ The *dot product* or *scalar product* of the vectors $\vec{A}$ and $\vec{B}$ is defined as

$\vec{A} \cdot \vec{B} \equiv AB \cos\theta = A(B\cos\theta) = B(A\cos\theta)$,
where θ is the angle between $\vec{A}$ and $\vec{B}$.

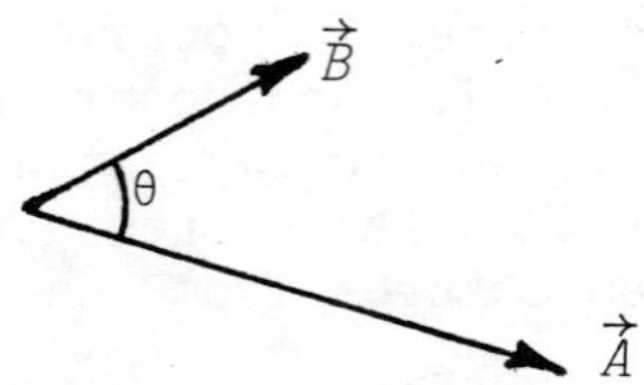

From the definition, it is evident that the dot product of any vector $\vec{A}$ with itself is equal to the square of the magnitude of the vector, A^2. When one vector is multiplied by another in a physical equation, in cases where the product is a scalar quantity one must use the dot product of the two vectors.

One example of a physical quantity defined in terms of a dot product is work.

> The *work* dW done by a force $\vec{F}$ acting through a small displacement $d\vec{s}$ is $dW \equiv \vec{F}\cdot d\vec{s}$.

For a finite displacement $\vec{s}$ the work is $W = \vec{F}\cdot\vec{s}$ if $\vec{F}$ is constant during the displacement. Notice that when work is defined as a dot product, the one definition applies regardless of the angle between the force and the displacement.

Worked Example 47-1

Two students carry a mattress having a mass of 25.0 kg up a stairway having an effective slope of 40° from the horizontal. Assuming the mattress is moved with constant velocity for 4.3 meters along this slope, how much work is done in moving it?

Solution

While the mattress travels at constant velocity, the resultant force on it is zero. Therefore, the lifting force must be equal to the weight of the mattress but acting vertically upward; this force equals

$$F = mg = 25.0\ \text{kg} \times 9.8\ \text{m/s}^2 = 2.45 \times 10^2\ \text{N}.$$

The angle between this force and the displacement of 4.3 m is 50°, derived from the diagram shown here. The work is therefore

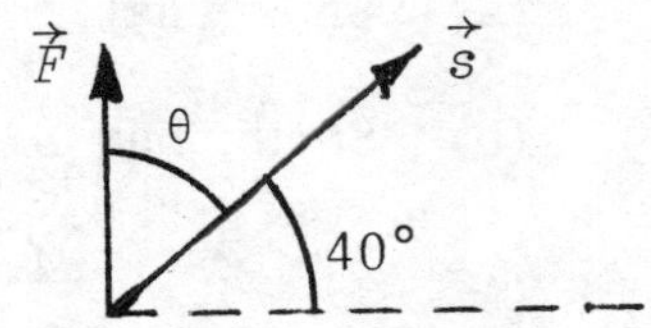

$$W = \vec{F}\cdot\vec{s} = F\cdot s\cos\theta =$$

$$= 2.45 \times 10^2\ \text{N} \times 4.3\ \text{m} \times \cos 50°$$

$$= 6.77 \times 10^2\ \text{N}.$$

A second example of a dot product occurs in the equation for the potential difference dV between two points separated by a small displacement $d\vec{s}$ in an electric field whose strength is $\vec{E}$ (Section 25.5). This potential difference is given by the dot product

$$dV = -\vec{E}\cdot ds.$$

There is a similarity between this equation for potential difference and the equation above for work, because the electric potential is just the potential energy per unit test charge while the electric field strength is just the force per unit test charge.

From the definition of a dot product, it is seen that the two vectors in the product may be multiplied in either order:

$$\vec{A}\cdot\vec{B} = \vec{B}\cdot\vec{A}.$$

In terms of components,

$$\vec{A}\cdot\vec{B} = (\hat{x}A_x + \hat{y}A_y + \hat{z}A_z)\cdot(\hat{x}B_x + \hat{y}B_y + \hat{z}B_z)$$

$$= \hat{x}\cdot\hat{x}A_xB_x + \hat{x}\cdot\hat{y}A_xB_y + \hat{x}\cdot\hat{z}A_xB_z$$

$$+ \hat{y}\cdot\hat{x}A_yB_x + \hat{y}\cdot\hat{y}A_yB_y + \hat{y}\cdot\hat{z}A_yB_z$$

$$+ \hat{z}\cdot\hat{x}A_zB_x + \hat{z}\cdot\hat{y}A_zB_y + \hat{z}\cdot\hat{z}A_zB_z$$

$$= A_xB_x + A_yB_y + A_zB_z.$$

The final result above follows from the rules for the dot products of two unit vectors (Exercise 47.1-1), which may be readily obtained using the definition of the dot product.

Exercises

E 47.1-1 Evaluate the following dot products:

(a) $\hat{x} \cdot \hat{x} =$ (b) $\hat{y} \cdot \hat{y} =$ (c) $\hat{z} \cdot \hat{z} =$ (d) $\hat{x} \cdot \hat{y} =$

(e) $\hat{x} \cdot \hat{z} =$ (f) $\hat{y} \cdot \hat{z} =$

E 47.1-2 From the definition of the dot product:

(a) Show that $\vec{A} \cdot \vec{B} = AB$ if $\vec{A}$ is parallel to $\vec{B}$.

(b) Show that $\vec{A} \cdot \vec{B} = 0$ if $\vec{A}$ is perpendicular to $\vec{B}$.

E 47.1-3 $\vec{A} = \hat{x}2 + \hat{y}3 + \hat{z}4$. $\vec{B} = \hat{x}4 - \hat{y}2 - \hat{z}5$.

(a) Compute $\vec{A} \cdot \vec{B}$.

(b) Find the angle between $\vec{A}$ and $\vec{B}$.

E 47.1-4 $\vec{A} = \hat{x}2 + \hat{y}3 + \hat{z}4$. $\vec{B} = \hat{x}3 + \hat{y}2 - \hat{z}3$.

(a) Compute $\vec{A} \cdot \vec{B}$.
(b) Find the angle between $\vec{A}$ and $\vec{B}$.

E 47.1-5 Jim pushes on a lawnmower handle with a force of 40 N at an angle of 30° to the horizontal. How much work does he do while pushing the mower for 20 m on a horizontal lawn?

E 47.1-6 Which of the following two expressions for kinetic energy is correct: $KE = \frac{1}{2} m\vec{v}\cdot\vec{v}$, or $KE = \frac{1}{2} m\vec{v} \times \vec{v}$? Explain.

E 47.1-7 Which of the following two expressions for the work done against friction is correct: $W = \vec{F}_f \cdot \vec{s}$, or $W = \vec{F}_f \times \vec{s}$? Explain.

E 47.1-8 Given two non-zero vectors $\vec{A}$ and $\vec{B}$.

(a) Is it possible for both $\vec{A} \times \vec{B}$ and $\vec{A} \cdot \vec{B}$ to equal zero? Explain.

(b) Is it possible for neither $\vec{A} \times \vec{B}$ nor $\vec{A} \cdot \vec{B}$ to equal zero? Explain.

47.2 Work

Minimum Learning Goal. (1) Given a description of the path over which a particle moves and the force on the particle at each point along the path, be able to calculate the work done on the particle by the force. For example:

1 For each of the following situations, find the work done on the object to which the given force is applied:
(a) A 2000-kg automobile travels 200 m on level ground. How much work is done on the car by the force of gravity during this motion?
(b) A man is carried up 6 stories by a series of escalators, a vertical distance of 20 m. If an escalator exerts an average upward force of 800 N on the man, how much work is done by the escalators on him as he is raised the 6 stories?
(c) In part (b), how much work is done on the man by the force of gravity? (Answer, p. 47-22, Line 6)

2 In each of the following situations, calculate the work done on the object to which the given force is applied:
(a) A centripetal force of 2 N is used to keep a 0.1-kg ball in horizontal uniform circular motion at the end of a string 50 cm long. How much work does this force do on the ball during each revolution of the ball?
(b) A man lifts a radio weighing 20 N through a height of 2 m at a constant speed. How much work does he do on the radio?
(c) In part (b), how much work is done on the radio by the force of gravity? (Answer, p. 47-22, Line 2)

Much of our interest in energy is centered in understanding *changes* of energy from one form to another, or the *transfer* of energy from one object to another. As we shall see in more detail in the following sections, energy is transferred by means of *work*. For this reason, we begin our study of the broad concept of energy by examining the concept of physical work.

When a force acts on a moving object, the force does work on the object. Physical work is defined in terms of the force and the displacement through which it acts:

★ The *work dW* done by a force $\vec{F}$ acting on an object which moves through a small displacement $d\vec{s}$ is $dW \equiv \vec{F} \cdot d\vec{s}$.

★ When a force $\vec{F}$ remains constant while it acts on an object moving through a macroscopic displacement $\vec{s}$, the *work* done is $W = \vec{F} \cdot \vec{s}$.

These definitions of work have some important implications. Note that simply exerting a force, even a very large force, does not accomplish any work from the physical point of view. If you extend your arm while holding a brick in your hand, it won't be long before your arm becomes tired and you need to lower your hand or let go of the brick. From the standpoint of physics, however, your arm and hand have not done any work. Your muscles were doing something, and they became fatigued, but while you were holding the brick stationary you were not transferring any energy, so no physical work was done on the brick. Even if you carry the brick for a mile, holding it in your hand while you walk with constant velocity on a horizontal path, you would not do any physical work on the brick. In this case the force you exert on the brick is upward (to counterbalance the downward weight of the brick) while the displacement of the brick is horizontal and thus perpendicular to the force.

Physical work is done on an object only when a force is exerted on the object, and there is a component of this force in the direction in which the object is moving. If an object does not move, no work can be done on the object no matter how large the forces. To contribute to the work done on an object, the force applied must have a component parallel to the object's displacement. When a particle is in uniform circular motion, for example, the centripetal force on the particle is always in a direction perpendicular to the path of the particle, so the centripetal forces does not do any work on the particle nor transfer any energy to it.

Although force and displacement are both vector quantities, work is a *scalar* quantity, so it is defined as the scalar product or dot product (Section 47.1) of the force and displacement. The SI unit of work is called the *joule*, symbol J, after James Prescott Joule.

> One *joule* is the work done by a force of one newton acting parallel to a displacement of one meter. That is, 1 joule = 1 newton·meter or 1 J = 1 N·m.

Joule was an English Scientist who did most of his scientific experiments between 1835 and 1861. He was a clever experimenter who continually devised ways to improve the accuracy of his measurements. He was the first to accurately demonstrate several examples of conservation of energy.

Worked Example 47-2

A packing crate is pushed in a straight line upward along an inclined ramp through a distance of 7.5 meters. The ramp is inclined at an angle of 10 degrees to the horizontal, and the pushing force of 480 N is applied at an angle of 20 degrees to the ramp. How much work is done by this force during this motion?

Solution

The angle between the force and the displacement is 20°, so the work is

$$W = \vec{F} \cdot \vec{s} = F\, s \cos 20°$$

$$= 480\text{ N} \times 7.5\text{ m} \times 0.940 = 3.38 \times 10^3\text{ J.}$$

It is instructive to plot a graph of force vs. displacement for the motion described in Worked Example 47.2. The force that contributes to the work is not the total force, but only the component of force parallel to the displacement, which we write as $\vec{F}_{\parallel}$. Figure 47-1 is a graph showing a constant force of 480 × 0.940 = 451 newtons acting through a displacement of 7.5 meters.

Figure 47-1. A graph of force vs. displacement, showing a constant force of 451 newtons acting through a displacement of 7.5 meters. The area under the curve represents the work done by this force, a total of 3.38×10^3 joules.

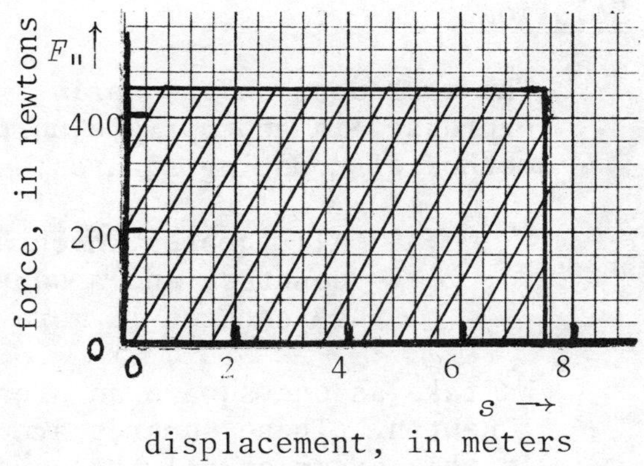

From the geometrical point of view, the area under the curve in Figure 47-1 is the area of a rectangle having a height of 451 newtons and width of 7.5 meters, which equals 3.38×10^3 joules. Note that the units of the area under a curve are *not* meter², but rather the product of the unit on the vertical axis times the unit on the horizontal axis.

Generalizing from this example, when we have a graph of force vs. the displacement through which the force acts, *the area under a curve of force vs. displacement equals the work done by the force during this displacement, provided the force acts parallel to the displacement.* The method of computing work by finding the area under a curve can be applied even when the curve is not a straight line, by counting squares as in Worked Example 47-3 or by integration as in Worked Example 47-4 in Section 47.3.

Worked Example 47-3

A spring mounted at the end of an air track is compressed by applying a force which gradually increases, until the spring is fully compressed when the movable end is displaced 11.8 cm. The force required to compress the spring increases with displacement as shown in the graph below. How much work is required to compress the spring fully?

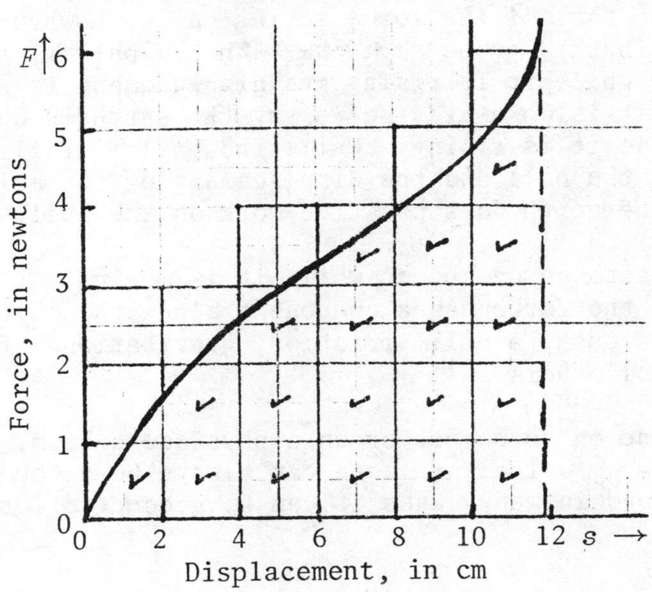

Solution

The work done on the spring is equal to the area under the curve, expressed in appropriate units. The area may be found by the method of counting squares:

> All squares which are completely under the curve are counted, and a square which is partially under the curve is counted if more than half of its area is under the curve.

We take as one square an area having a width of 2 cm and a height of 1 newton. These squares are indicated on the graph by dark lines. In this example, all the squares to be counted are checked (✓); there are 19 such squares. The area of each individual square has the value of $(1\ N) \times (2\ cm) = 2\ N{\cdot}cm = 2 \times 10^{-2}\ N{\cdot}m$. The total area of the 19 squares is then

$$19 \times 2 \times 10^{-2}\ N{\cdot}m = 38 \times 10^{-2}\ J = 0.38\ J.$$

That is, a total of 0.38 joules of work is required to fully compress the spring.

It must be recognized that the result obtained is only approximate. It is really quite accurate, however, as may be verified by repeating the solution, this time taking as one square a smaller area having a width of 1 cm and a height of 0.5 newton. The reader who does this should obtain an area of about 78 small squares, which represents a work of 0.39 J.

When we calculate the work done in some physical situation, we need to have clearly in mind the object or agent which is doing the work, that is, which is exerting the force we are considering. We must also have in mind some other object on which work is being done, that is, an object moving while the force is acting on it. When the object which "receives" the work is moving in the same direction as the force exerted on it, the work done on this object is considered to be *positive*. When an object moves in a direction opposite to that of the force acting on it, however, the work done on that object is said to be *negative*. When a ball is thrown straight upward, for example, while it is rising its displacement is upward but the force of gravity on it is downward, so we say the earth is doing negative work on the ball while it is rising. While the ball is falling, however, the earth's force on the ball and the displacement of the ball are in the same direction, so the earth does positive work on the ball.

Any force which acts on an object while it is moving may do work on the object, provided the force has a component along the object's path. Work can be done by a push, a pull, friction, gravitational forces, electrical forces, and others.

When work is done on an object or on a physical system, it always shows up as some form of *energy*. The work done can always be accounted for, at least in principle, and in many cases it can be accounted for quantitatively to a high degree of accuracy. The fact that nature "keeps track" of all transactions involving work and energy with exquisite exactness is reflected in the Work-Energy Principle and the Law of Conservation of Energy (Section 49.2). We will study these principles quantitatively later. In this section, we simply consider the question, "Where does the work go?," on a qualitative basis.

When we examine a variety of examples in which work is done, we find the remarkable fact that if the same amount of work is done in each case, there is something about the outcome which is also the same. There is a physical property of the system which has the same amount of change in each case. The general term used to refer to this property is *energy*. In every situation where work is done, the work transfers a definite quantity of energy to the object receiving the work. The differences between one example and another lie only in the *form of energy* acquired as a result of the work. This general result is summarized in the

> *Work-Energy Principle:* The work done on a physical system equals the increase in energy of the system.

A corollary of this principle is that energy cannot be transferred to an object or a system without doing work on it. Work transfers energy.

In presenting the conclusion given above, we have actually reached ahead to state some results which can only be asserted with confidence on the basis of experiments. The discussion here does not prove anything, but is intended to provide a conceptual framework for interpreting quantitative experiments. We next examine various kinds of work-energy transformation in detail, dealing with only one form of energy at a time. In Section 49.2 we will return to the comprehensive Work-Energy Principle in equation form.

Questions and Exercises

Q 47.2-1 As a pendulum swings back and forth, is work done by the string on the bob? What is the reason for your answer?

Q 47.2-2 A car is towing a trailer at constant speed.
(a) Is the car doing work on the trailer?
(b) If the trailer were sliding on a frictionless surface, would any work be done on it by the car?

Q 47.2-3 A car is towing a trailer with increasing speed while the trailer slides on a frictionless surface. Is work done by the car on the trailer?

Q 47.2-4 When a satellite moves around the earth in an elliptical orbit, the point where it is closest to earth is called the *perigee* and the point where is is farthest from the earth is called the *apogee*.
(a) Is work done on the satellite by the gravitational force as it moves from apogee to perigee? (continued on next page)

(b) Is work done on the satellite by the gravitational force as it moves from perigee to apogee?
(c) Is a net amount of work done in the course of a complete orbit?

Q 47.2-5 While a magnetic force is acting on a moving charged particle, does this force do work on the particle? What is the basis for your answer?

Q 47.2-6 A tugboat is towing a barge at constant speed.
(a) Is the force exerted on the barge by the tow cable doing positive work, negative work or no work at all?
(b) Is the drag force of the water on the barge doing positive work, negative work, or no work at all?
(c) What is the net amount of work on the barge while it travels one kilometer?

E 47.2-1 How much work is required to raise a mass of 10 kilograms through a height of 4 meters near the earth's surface?

E 47.2-2 How much work is done by the force of gravity on a 1500-kg auto as it goes around a horizontal curve having a length of 300 meters?

R 47.2-3 A golf club exerts an average force of 1200 newtons on a golf ball during the interval of 0.001 second while the club is in contact with the ball. During this interval the ball moves 0.03 meter. How much work does the club do on the ball?

E 47.2-4 A bucket of mortar weights 160 newtons. It is raised at a constant speed of 0.5 m/sec through a vertical height of 4 meters to a scaffold where bricklayers are working. How much work is done on the bucket of mortar?

E 47.2-5 A steady horizontal force of 50 N pushes a box along the floor through a distance of 2 m.
(a) Calculate the work done.
(b) If there is no friction between box and floor, and the box has a mass of 20 kg, what is the acceleration of the box?
(c) If there is a friction force such that the box suffers no acceleration, what must that friction force be?

E 47.2-6 Jim pushes on a lawnmower handle with a force of 40 newtons at an angle of 30° to the horizontal. How much work does he do while pushing the mower horizontally for 20 m?

Check your understanding of the material above by testing yourself on the minimum learning goal at the beginning of this section.

((47.3 Work by a Variable Force

Minimum Learning Goals. (1) Given a description of a path along which a particle moves and a force on the particle which varies along the path, be able to calculate the work done on the particle by the force. For example:

1 The force in newtons acting on a particle moving along a path from (-1,-3,1) to (2,6,1) is given by

$$\vec{F} = 3xy\hat{x} - 2z\hat{y} + 5x^2y\hat{z},$$

where x, y and z are in meters. Suppose the path is described by

$$y = 3x \qquad \text{and} \qquad z = 1,$$

so that $d\vec{r} = \hat{x}dx + \hat{y}dy = \hat{x}dx + \hat{y}3dx$. Compute the work done on the particle by this force. (Answer, p. 47-22, Line 10)

2 The force on a particle moving along a path from $\vec{r}_0$ to $\vec{r}_1$ is given by

$$\vec{F} = -\frac{GMm}{r^2}\hat{r}.$$

Calculate the work done on the particle by this force for an arbitrary path between the two points. (Answer, p. 47-22, Line 3)

When a force varies in magnitude but remains parallel to the displacement of the object on which it acts, the work done by the force can be expressed in terms of an integral. The integral sign $\int$ is used to indicate a summation. The total work W_{12} done by the force while the object moves from point *1* to point *2* is the sum of all the small increments of work during this motion:

$$W_{12} = \int_1^2 dW.$$

For motion in one dimension it is convenient to use x as the position variable, so the work dW during the small displacement dx is given by $dW = Fdx$.

★ When a force $F(x)$ varies with position while acting on an object parallel to the direction of its motion, the work done on the object while it moves from x_1 to x_2 is $W_{12} = \int_{x_1}^{x_2} F(x)dx.$

To compute W_{12} it is necessary to evaluate this definite integral. Methods for doing this are given in appendix A.4, and illustrated in Worked Example 47-4.

Worked Example 47-4

The force F required to compress the spring described in Worked Example 47-3 varies with the displacement s of the movable end of the spring, approximately according to the equation

$$F = ks + b(s - a) + c(s - a)^3,$$

where $k = 0.54$ N/cm, $a = 5.9$ cm, $b = -0.21$ N/cm, and $c = 0.0061$ N/cm³. Use integration to compute the work required to compress the spring by 11.8 cm.

Solution

The work required is represented by the area under the curve of force vs. displacement of this equation, from $s = 0$ to $s = 11.8$ cm. This curve is sketched in the illustration for Worked Example 47-3. The work is expressed by the following integral:

$$W = \int_0^{s_1} F(s)ds, \quad \text{where } s_1 = 11.8 \text{ cm.}$$

This integral is evaluated using the procedure and table of integrals given in Appendix A.4:

$$W = \int_0^{s_1} [ks + b(s - a) + c(s - a)^3]\, ds$$

$$= \int_0^{s_1} ks\, ds + \int_0^{s_1} b(s - a)ds + \int_0^{s_1} c(s - a)^3\, ds$$

$$= \int_0^{s_1} ks\, dx + \int_0^{s_1} bs\, ds - \int_0^{s_1} ba\, ds$$

$$+ c\int_0^{s_1} (s^3 - 3s^2a + 3sa^2 - a^3)\, ds$$

$$= k[s^2/2]_0^{s_1} + b[s^2/2]_0^{s_1} - ba[s]_0^{s_1} + c[s^4/4]_0^{s_1}$$

$$- 3ca[s^3/3]_0^{s_1} + 3ca^2[s^2/2]_0^{s_1} - ca^3[s]_0^{s_1}$$

$$W = ks_1^2/2 + bs_1^2/2 - bas_1 + cs_1^4/4$$

$$-3ca\, s_1^3/3 + 3ca^2s_1^2/2 - ca^3s_1.$$

Substituting numerical values and rounding each product to two significant figures, we obtain

$$W = 38\ \text{N}\cdot\text{cm} - 15\ \text{N}\cdot\text{cm} + 15\ \text{N}\cdot\text{cm} + 30\ \text{N}\cdot\text{cm}$$
$$- 59\ \text{N}\cdot\text{cm} + 44\ \text{N}\cdot\text{cm} - 15\ \text{N}\cdot\text{cm}$$
$$= 38\ \text{N}\cdot\text{cm} = 0.38\ \text{J},$$

in good agreement with the result of Worked Example 47-3.

When a force varies in magnitude and also in its direction relative to the displacement of the object on which it acts, the work done by the force can be expressed as a *line integral* (Appendix A.5),

★ $$W = \int_1^2 \vec{F} \cdot d\vec{r},$$

where the integral is to be evaluated along the path taken by the object from the initial point *1* to the final point *2*. In many cases the value of the integral, that is, the work done, depends on the path from point *1* to point *2*, but there are some important forces for which the work done does *not* depend on the path (Section 48.5). Examples of forces for which the work is independent of the path include the gravitational force and the electrostatic force. A procedure for evaluating line integrals is outlined in Appendix A.5.

Exercises

E 47.3-1 Write (but do not evaluate) the integrals which represent:
(a) The work W_{34} done by a variable force $\hat{x}F(x)$ as it acts on a car moving in a straight line from x_3 to x_4;
(b) The work W_{ab} done by a force $\hat{y}F(y)$ as it acts on a ball moving along the y axis from y_a to y_b.

E 47.3-2 Write (but do not evaluate) the line integrals which represent:
(a) the work W_{34} done by a force $\vec{F}(x,y,z)$ as it acts on a bird moving from $\vec{r}_3$ to $\vec{r}_4$ along a path C;
(b) the work W_{of} done by a force $\vec{F}(x,y,z)$ as it acts on a ball moving from $\vec{\ell}_o$ to $\vec{\ell}_f$ along a path C_1.

E 47.3-3 Compute the work done by the force $\vec{F}(x) = 3x^2\hat{x}$ newtons on a particle as the particle moves from $x_3 = 1$ meter to $x_4 = 3$ meters.

E 47.3-4 A massless linear spring having spring constant k is securely fastened at one end. The equilibrium position of the other end is $x = 0$. When a force $\vec{F}_a$ in the $+x$ direction is applied to the free end, the spring is compressed.
(a) Write the equation for $\vec{F}_a$ as a function of x.
(b) By integration, compute the work W_{02} required to compress the spring from $x_0 = 0$ to x_2.
(c) By integration, compute the work W_{23} required to compress the spring from x_2 to $x_3 = 2x_2$.
(d) Plot a graph of $\vec{F}_a(x)$ vs. x, and show on this graph the areas which correspond to W_{02} and W_{23}.
(e) If the free end of the spring is pulled from $x_0 = 0$ to $x_4 = -x_2$, compute the work W_{04} which would be required. Compare this work with W_{02}. Show this work on your graph.

E 47.3-5 Evaluate the line integrals:

(a) $\int_{r_1}^{r_2} \vec{F} \cdot d\vec{r}$, where $\vec{F} = \hat{x}F_0$, $\vec{r}_1 = \hat{y}\ 3$ meters and $\vec{r}_2 = [\hat{x}5 + \hat{y}3]$ meters, and you may choose the path of integration.

(b) $W = \int_{r_1}^{\infty} (q_1 q/r^3)\vec{r} \cdot d\vec{r}$, where you may choose the path of integration.

E 47.3-6 The magnitude of the gravitational force of the Earth on a mass m at a distance r from the center of the Earth is $GM_e m/r^2$, where M_e is the Earth's mass. Suppose that an applied force $F_a = GM_e m/r^2$ acts so that m moves in a radial line at constant speed from $r = \infty$ to a final position $r = r_1$. Since the Earth attracts m, $\vec{F}_a$ must be directed away from the Earth, opposite to the displacement of m. Set up the appropriate integral, and show that the work done by $\vec{F}_a$ during this motion is $W = -GM_e m/r_1$.

Check your understanding of the material above by testing yourself on the minimum learning goal at the beginning of this section.

47.4 Power

Minimum Learning Goal. (1) Given the work done on an object by a force, or information from which you can calculate that work, be able to find the average rate at which energy is transferred to that object while that work is being done on it. For example:

1 A slave is used to lift stones and to shovel dirt in the construction of a dam. In a ten-hour day, he exerts an average force of 50 pounds (222 N) in 4000 tasks of lifting and moving these substances an average of 0.50 m (in the direction of the average force).
  (a) At what average rate does the slave transfer energy to the materials on which he does work?
  (b) If the price of electric power is 3¢ per 1,000 watts for one hour, how much is the slave worth for his ten-hour day? (Answer, p. 47-22, Line 7)

2 A horse pulling a plow does 1.44×10^7 J of work on the earth it pushes aside in an eight-hour day. A tractor pulling 4 plows can do 9×10^6 J of work on the earth it pushes aside in 1 hour. Find the rate at which energy is transferred to the earth by the plows on the tractor, and compare it with the rate at which energy is transferred to the earth by the horse-drawn plow. (Answer, p. 47-22, Line 11)

Every activity in which we engage requires the transfer of energy by means of work. There was a time when all the work people wanted to accomplish was done by human muscles. Then people learned to harness natural forces such as those of wind and water to work for them, and also learned how to train animals to do work. A man can do as much work as a draft horse or a windmill, if the man lives long enough. But in technologically advanced nations men are not employed primarily to produce mechanical work, because they cannot work fast enough nor compete economically with machines. We don't use many draft animals, either; we use machines when we want to produce a large amount of physical work in a short time.

The time-rate of doing work is called *power*.

★ The *power* P delivered when the work ΔW is done during a time Δt is $P \equiv \Delta W/\Delta t$.

The SI unit for power is the *watt*, symbol W.

One *watt* = 1 joule/second. 1 W = 1 J/s.

If a total amount of work W is completed in the time t, the average power supplied by this work is $P = W/t$.

Because work transfers energy, we can also say that power is the time-rate of transferring energy or of supplying energy from one object or system to another. From the point of view of the user of energy, power is the time-rate of consuming energy. Heat (Chapter 22) is a form of work, so it is also correct to say that the time-rate of supplying heat is power.

Among the most familiar devices used to accomplish work are electric motors and internal-combustion engines used in automobiles and other vehicles. Motors and engines are rated in terms of their ability to deliver power, in units of watts, or kilowatts, or horsepower. The *horsepower* is an engineering unit which is equal to 746 watts. The horsepower rating of an automobile engine as stated by the manufacturer represents the actual power delivered to the driving wheels with the engine connected directly to a dynamometer; the rating is given at the speed for which it is a peak. Because the efficiency of a modern gasoline engine is only about 30%, the power delivered to the wheels is much less than the power developed in the cylinders of the engine. Unfortunately, this low value for efficiency is dictated by a fundamental law of nature called the second law of thermodynamics (Section 22.13). It can be used to show that the maximum attainable efficiency is limited by the maximum temperature achieved in the cylinders of the engine. This law cannot be circumvented by ingenious engineering design. All internal combustion engines, or heat engines, are limited in this way by the second law of thermodynamics. An electric motor is not a heat engine, so its efficiency is much higher, usually above 95%. This does not mean that electric motors provide a more efficient overall use of energy however, because the electric power which drives them is generated (in most cases) in a power plant using a stream turbine which is also limited by the second law of thermodynamics.

Electric lamps and household appliances are rated in terms of the power they *consume* when they are operating, usually stated in watts.

Questions and Exercises

Q 47.4-1 (a) Give an example in which a lot of work is done with very little power.
(b) Give an example in which very little work is done although the power is large.

Q 47.4-2 What is the power rating of the highest-power electric appliance in your residence?

Q 47.4-3 The monthly bill from your local electric power company contains a charge for the number of *kilowatt-hours* of electricity you have used.
(a) Are you being billed for a quantity of energy, or a quantity of power?
(b) Explain how the unit kilowatt-hour corresponds to your answer to part (a).

E 47.4-1 If a weight of 10 N is raised through a height of 4.0 m in 4 seconds, what power is required?

E 47.4-2 A student who weighs 800 N is late for class. She runs up two flights of stairs, a total vertical distance of 8 m, in 12.8 seconds. What is her power output?

E 47.4-3 Measure your own weight and the height of two or three flights of stairs, then determine your own horsepower rating by running up the stairs while someone times your ascent.

Check your understanding of the material above by testing yourself on the minimum learning goal at the beginning of this section.

47.5 Kinetic Energy. The Work-Kinetic Energy Theorem

Minimum Learning Goals. (1) Given the speed of a moving body and its mass, be able to calculate its kinetic energy in joules. For example:

1 What is the kinetic energy of a satellite having a mass of 100 kg orbiting at a speed of 17,000 mph (7.6×10^3 m/s)? (Answer, p. 47-22, Line 1)

2 A 1500-kg automobile is traveling at 60 mph (26.8 m/s). What is its kinetic energy? (Answer, page 47-22, Line 5)

(2) Be able to compute the change in kinetic energy of an object as a result of the net work done on the object by all applied forces, given information from which the work can be computed. For example:

1 A 20-gram bullet is fired vertically up from the earth's surface with an initial velocity of 600 m/s.
   (a) Calculate the work done on the bullet by the force of gravity during the first two seconds of its motion.
   (b) Ignoring air drag, what is the change in kinetic energy of the bullet during those two seconds?
   (Answer, p. 47-22, Line 9)

2 An electron ($m = 9.11 \times 10^{-31}$ kg) is accelerated from rest to some speed by a uniform electric field. If the acceleration is 2×10^{12} m/s², what is the kinetic energy of the electron when it has moved 1.0 m? (Answer, p. 47-22, Line 4)

3 A baseball is thrown vertically upward with an initial speed v_0. While it is rising two forces act on the ball: its weight $-mg\hat{z}$ and a drag force given by $\vec{F}_d = -b\hat{z}$, where b is a constant. If the ball has an initial height h_0 and a later height h, what is the change in kinetic energy of the ball? (Answer, p. 47-22, Line 8)

4 A mass m suspended by a spring is set into oscillation along a vertical line. The forces acting on the mass include its weight $-mg\hat{z}$ and the force of the spring on the mass, $\vec{F}_s = -k(z - z_0)\hat{z}$, where k is the spring constant and z_0 is the position of the mass when the spring is relaxed. What is the change in kinetic energy of the mass as it moves from the height z_0 to the position $z = z_0 - h$? (Answer, p. 47-22, Line 12)

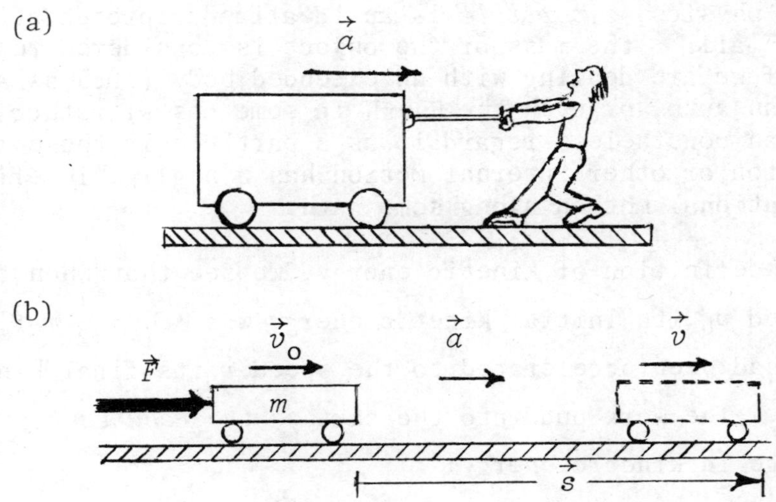

Figure 47-2. (a) A man pulls a cart, causing it to accelerate. The net work done on the cart equals the increase in the cart's kinetic energy. (b) The net force $\vec{F}$ does work on the cart, accelerating it from an initial velocity $\vec{v}_0$ to a final velocity $\vec{v}$.

Figure 47-2(a) shows a man pulling a cart on a horizontal surface with a constant horizontal force which causes the cart to accelerate. The pulling force is opposed by friction, so we use $\vec{F}$ to denote the *resultant force* on the cart, shown in Figure 47-2(b). During a displacement $\vec{s}$ the work done on the cart or put into the cart by the resultant force is

$$W_{\text{in}} = \vec{F} \cdot \vec{s} = Fs.$$

The acceleration of the cart (which has a mass m) in response to this force is $a = F/m$, so $F = ma$. At the beginning of this work the speed is v_0, and at the end it is v. Because this acceleration is constant, Eq. (41-6) on page 41-30 applies; we transpose this equation to obtain

$$2as = v^2 - v_0^2.$$

Substituting these expressions for F and for $2as$ successively, we obtain

$$\begin{aligned} W_{\text{in}} &= ma \cdot s = \frac{m}{2} \cdot 2as \\ &= \frac{m}{2}(v^2 - v_0^2) \\ &= \frac{m}{2}v^2 - \frac{m}{2}v_0^2. \end{aligned}$$

We have thus derived the result that the work on the cart by the resultant force under these conditions goes into a change in the quantity $\frac{1}{2}mv^2$. This quantity is called *kinetic energy.*

★ The *kinetic energy* of a particle having mass m and moving with speed v is $KE \equiv \frac{1}{2}mv^2$.

In macroscopic physics, a *particle* is an idealized representation of an object in which all the mass of the object is considered to be located at a point. If we are dealing with an extended body (such as a box, a cart, a ball, an auto, or even the Earth in some cases) rather than a particle, we can nonetheless regard it as a particle in those instances when its rotation or other internal motion has a negligible effect on the overall translational motion along some path.

Using the definition of kinetic energy, we see that when the cart had an initial speed v_0 its initial kinetic energy was $KE_i = \frac{1}{2}mv_0^2$, and later when the cart had been accelerated to the speed v its final kinetic energy was $KE = \frac{1}{2}mv^2$. The work put into the cart *by the resultant force* went into this change in kinetic energy:

★ $$W_{\text{in}} = KE - KE_i \quad \text{or} \quad W_{\text{in}} = \Delta KE.$$

((To show that this result obtains for the general case where a particle moves on a curved path while acted on by a resultant force which is not necessarily parallel to the path, we consider the situation sketched in

Figure 47-3. A particle having mass m moves under the influence of a resultant force $\Sigma\vec{F}$ along a curved path from point *1* to point *2*. The work on the particle by the resultant force is equal to the change in kinetic energy of the particle.

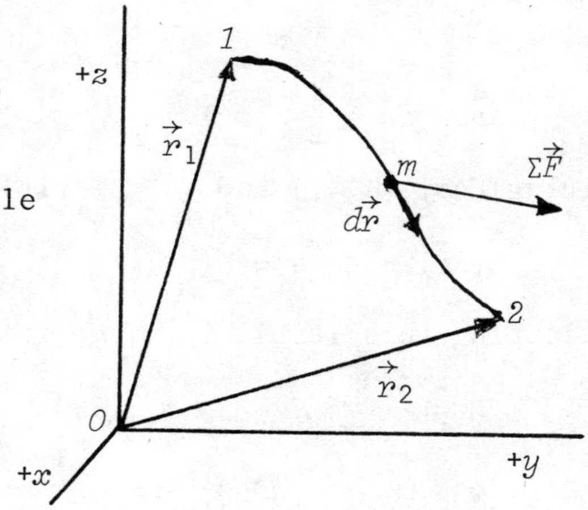

Figure 47-3. The work done by the resultant force ΣF on the particle while it moves from point *1* to point *2* is given by

$$W = \int_1^2 (\Sigma\vec{F}) \cdot d\vec{r},$$

where $d\vec{r}$ is a small displacement along the path. The equation for W is a line inegral which may be evaluated using the procedures given in Appendix A.5 if one has an equation for $\Sigma\vec{F}$ at all points along the path. For our present purpose of deriving the Work-Kinetic Energy Theorem, however, we do not need an explicit function for $\Sigma\vec{F}$ along the path. Rather, we note that at every point the particle's acceleration is given by

$$\Sigma\vec{F} = m\vec{a} = m\ d\vec{v}/dt.$$

From the definition of velocity $\vec{v} = d\vec{r}/dt$, we obtain $d\vec{r} = \vec{v}dt$. Using these results, the integrand in the integral above may be written

$$(\Sigma\vec{F}) \cdot d\vec{r} = m\frac{d\vec{v}}{dt} \cdot \vec{v}\ dt = m(d\vec{v}) \cdot \vec{v} = m\ \vec{v} \cdot d\vec{v}. \qquad \text{Eq. (47-1)}$$

The last step in Eq. (47-1) follows from the fact that in a dot product the two vectors may be written in either order (Section 47.1).

Next consider the quantity

$$v^2 = \vec{v} \cdot \vec{v}.$$

If we compute the time-derivative of this equation using the applicable rules from Appendix A.2, we have

$$\frac{d(v^2)}{dt} = 2v\frac{dv}{dt} = \vec{v} \cdot \frac{d\vec{v}}{dt} + \frac{d\vec{v}}{dt} \cdot \vec{v} = 2\vec{v} \cdot \frac{d\vec{v}}{dt}. \qquad \text{Eq. (47-2)}$$

In taking the derivative of $\vec{v} \cdot \vec{v}$, we have used the rule which requires that the first factor in the product is held constant while the second is differentiated, and then the second factor is held constant while the first is differentiated. Multiplying the first and last terms in Eq. (47-2) by dt leads to

$$d(v^2) = 2\vec{v} \cdot d\vec{v}$$

or

$$\vec{v} \cdot d\vec{v} = \frac{1}{2}\, d(v^2). \qquad \text{Eq. (47-3)}$$

Combining Eqs. (47-1) and (47-3) yields

$$(\Sigma\vec{F}) \cdot d\vec{r} = \frac{m}{2}\, d(v^2).$$

Substituting this result into the integral for work then yields

$$W = \int_1^2 \frac{m}{2}\, d(v^2) = \frac{m}{2}\, [v^2]_1^2 = \frac{m}{2}\, (v_2^2 - v_1^2)$$

$$W = \frac{m}{2}\, v_2^2 - \frac{m}{2}\, v_1^2.$$

]]

This result has been obtained for the most general case, where the resultant force may have any variation along the path of the particle. It is called the Work-Kinetic Energy Theorem.

★ *Work-Kinetic Energy Theorem*: The total work done by the resultant force acting on a particle along any section of its path is equal to the change in kinetic energy of the particle along that section of the path.

Depending on the relative directions of the particle's path and the resultant force, the work done may be positive, negative, or zero. When net positive work is done on an object its speed increases; when the net work is negative, its speed decreases. When several forces act simultaneously, the total work may be computed using the resultant force, or by computing separately the work done by each individual force and then adding the results.

Kinetic energy is associated with every individual particle that is moving. When we consider a mechanical *system*, that is, a collection of particles, the kinetic energy of the system is just the sum of the kinetic energies of the particles making up the system.

For many physical systems which contain moving particles, not only the kinetic energy but also the values of other forms of energy (internal energy, gravitational potential energy, etc.) change during the motion. However, these changes in energy compensate each other so that the work done by the *resultant* force always equals the change in kinetic energy. Other forms of energy and the forces associated with them are discussed in the following sections.

Questions and Exercises

Q 47.5-1 A soapbox derby car is accelerating down a hill. The forces acting on it are (1) its weight, and (2) the contact force of the road, which consists of (2a) normal force perpendicular to the road surface, and (2b) a friction force parallel to the road surface.

(a) For each of the three forces, state whether it does positive work, negative work, or no work on the soapbox racer.
(b) Is a net amount of work being done on the car?

Q 47.5-2 Under what conditions is the equation $W_{in} = \Delta KE$ an adequate statement of the work-kinetic energy theorem?

Q 47.5-3 A particle moves along a path from point *A* to point *B*, and it has the same kinetic energy at *B* as at *A*.
(a) Which of the following statements *must* be true?
(b) Which of the following statements *cannot* be true?
(c) Which of the following statements could be true, but are not necessarily true, for this motion?
The statements are:
(1) No force acted on the particle between *A* and *B*.
(2) Some non-zero resultant force acted on the particle during at least part of this motion.
(3) The resultant force acting on the particle between *A* and *B* was constant in magnitude.
(4) The resultant force acting on the particle between *A* and *B* was constant in direction.
(5) The net work done on the particle during this motion was zero.
(6) No work was done on the particle during any part of this motion.

Q 47.5-4 If an enclosed vessel of air is at rest on the table, the kinetic energy of the vessel is zero. However, each molecule of air inside the vessel has some speed, and the sum of all the individual kinetic energies of the molecules in the vessel is far from zero. How than can we say that the kinetic energy of the vessel of air equals the sum of the individual kinetic energies of the particles which comprise it?

E 47.5-1 Compute the translational kinetic energy of a golf ball (mass 0.030 kilogram) as it leaves the tee with a speed of 50 meter/sec.

E 47.5-2 A football having a mass of 0.4 kg is kicked and receives an initial speed of 20 meter/sec. Compute the kinetic energy of the football.

E 47.5-3 A certain radioactive isotope emits beta particles (high-speed-electrons) having a speed of 10^9 cm/sec. Compute the kinetic energy of one of these beta particles.

E 47.5-4 What is the kinetic energy of a snowball which has a mass of 0.6 kg and travels at a speed of 20 m/sec?

E 47.5-5 A racing car having a mass of 1800 kg is driven with a constant net forward force of 1300 newtons through a distance of 70 meters on a straight horizontal track. If the car starts from rest, what is its kinetic energy at the end of this distance?

E 47.5-6 An object having a mass of 4 kg moves with an initial velocity of 1 m/s.
(a) What is the kinetic energy of this object?
Thirty joules of work are performed on this object by a force exerted parallel to its path. (Continued on next page)

(b) What is the final velocity of the object?
(c) What is the increase in velocity of the object?
(d) If the same object is moving with an initial velocity of 7 m/s and 30 J of work are done again in the same way, what is the final velocity of the object?
(e) What is the new increase in velocity of the object? Why is Δv smaller than in part (c)?

E 47.5-7 A 0.05-kg ball is at rest before a resultant force of 5×10^{-3} N is applied. In the absence of friction, what is the speed of the ball after it has gone 5 m?

((E 47.5-8 A force $\vec{F}(x) = 3x^2\hat{x}$ newtons acts on a particle as it moves from $x_3 = 1$ meter to $x_4 = 3$ meters. (See E 47.3-9). The particle has a mass of 0.1 kg and starts from rest at x_3.
(a) What is the kinetic energy of the particle at x_4?
(b) What is the speed of the particle at x_4?))

Check your understanding of the material above by testing yourself on the minimum learning goal at the beginning of this section.

Answers to Minimum Learning Goals

Line 1 2.89×10^9 J

Line 2 (a) 0 J; (b) 40 J; (c) -40 J.

Line 3 $GMm\left(\frac{1}{r_1} - \frac{1}{r_0}\right)$.

Line 4 1.82×10^{-18} J

Line 5 5.39×10^5 J

Line 6 (a) 0 J; (b) 1.6×10^4 J; (c) -1.6×10^4 J

Line 7 (a) 12.4 W; (b) 0.37¢

Line 8 $-(mg + b)(h - h_0)$

Line 9 2.31×10^2 J

Line 10 9 J

Line 11 Tractor: 2,500 W (3.35 HP); Horse: 500 W (0.67 HP).

Line 12 $mgh - kh^2/2$

ANSWERS to Questions and Exercises in Chapter 47

Section 47.1

E 47.1-4 (a) 0; (b) 90°

E 47.1-6 $KE = \frac{1}{2}\, mv \cdot v$, because kinetic energy is a scalar.

E 47.1-8 (a) No, because $\vec{A}$ and $\vec{B}$ cannot be perpendicular to each other and parallel to each other at the same time.
(b) Yes, if $\vec{A}$ and $\vec{B}$ are neither parallel nor perpendicular to each other.

Section 47.2

Q 47.2-2 (a) Yes; (b) no.

Q 47.2-4 (a) Yes (positive work; (b) Yes (negative work);
(c) No, if the satellite is at the same altitude at the beginning and at the end of the orbit.

Q 47.2-6 (a) Positive work; (b) negative work; (c) zero.

E 47.2-2 Zero.

E 47.2-4 640 J

E 47.2-6 6.93×10^2 J

Section 47.3

E 47.3-2 (a) $W_{34} = \int_{C\,\vec{r}_3}^{\vec{r}_4} \vec{F}(x,y,z) \cdot d\vec{r}$; (b) $W_{0f} = \int_{C_1\,\vec{\ell}_0}^{\vec{\ell}_f} \vec{F}(x,y,z) \cdot d\vec{\ell}$

E 47.3-4 (a) $\vec{F}_a = \hat{x}kx$; (b) $W_{02} = kx_2^2/2$; (c) $3kx_2^2/2$

(d) (e) $W_{04} = kx_2^2/2 = W_{02}$.

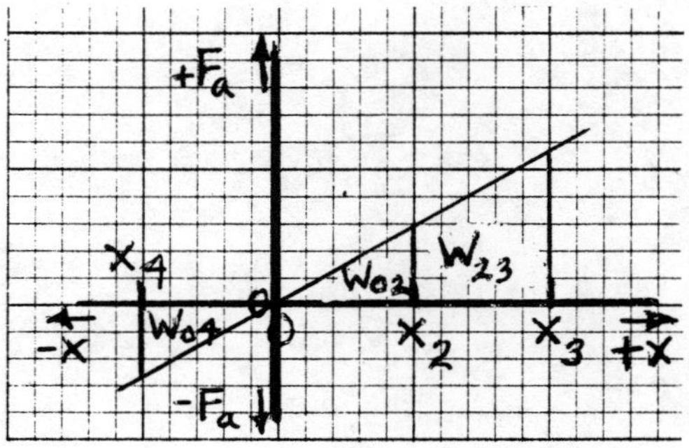

Section 47.4

Q 47.4-2 Some typical ratings of high-power electric appliances are: portable 115-volt electric heater, up to 1650 W; electric toaster, 1200 to 1600 W; electric range, up to 12,000 W total input; gas or electric clothes dryer, 4400 to 6000 W; central air-conditioning system for a house, up to 8000 W.

E 47.4-2 500 W

Section 47.5

Q 47.5-2 When W_{in} refers to the net work done on a system by the resultant external force, and ΔKE refers to the resulting change in kinetic energy of the system.

Q 47.5-4 The kinetic energy of the vessel of air, taken as a whole, is a macroscopic quantity which equals zero. The individual kinetic energies of the particles are microscopic quantities, not equal to zero, but from the macroscopic viewpoint these energies belong to the internal energy of the system rather than to the kinetic energy of the system as a whole.

E 47.5-2 80 joules

E 47.5-4 120 joules

E 47.5-6 (a) 2 J; (b) 4m/s; (c) 3m/s; (d) 8 m/s;
(e) 1 m/s; kinetic energy increases as the square of the speed, so with a higher initial speed a given increase in kinetic energy produces a smaller increase in speed.

E 47.5-8 (a) 26 J; (b) 22.8 m/s.

CHAPTER 48 POTENTIAL ENERGY

Introduction

For any physical system the various forms of energy fall into two major categories, kinetic energy and potential energy. In contrast with kinetic energy which is energy of motion, potential energy is stored in a system even if the system has no motion. Specific forms of potential energy discussed in this chapter include gravitational potential energy and elastic potential energy. All forms of potential energy share several important properties which are developed throughout the chapter.

When you have completed your study of the material in this chapter, you should be able to answer questions and work exercises like the examples in the Minimum Learning Goals assigned by your instructor.

48.1 Gravitational Potential Energy Near the Earth

Minimum Learning Goal. Be able to calculate the potential energy, or change of potential energy, of a system consisting of the Earth and an object near the Earth's surface, given some zero reference level, the mass of the object, and its elevation or change in elevation. For example:

1 A box having a mass of 12 kg is lifted from the floor to a table top 70 cm from the floor. Taking the floor as the zero level of potential energy, what is the potential energy of the box-earth system when the box is:
(a) on the floor?
(b) on the table?
(c) What is the change in potential energy from the floor to the table top?
Taking the ceiling (4.0 meters above the floor) as the zero level of potential energy, what is the potential energy of the system when the box is:
(d) on the floor?
(e) on the table top?
(f) What is the change in potential energy from the floor to the table top? (Answer, page 48-25, Line 2)

2 A 2000-kg automobile is driven to the top of a mountain having an elevation of 14,000 feet (4,270 m) above sea level. At the base of the mountain, the elevation is 6,000 feet (1830 m). Taking sea-level as the zero of potential energy, what is

(a) the potential energy at the base?
(b) the potential energy at the top?
(c) What is the change in potential energy from the base to the top of the mountain? (Answer, page 48-25, Line 5)

When a ball is picked up from the ground and raised to a higher elevation, a lifting force is required. This lifting force acts through a displacement, so work is done on the ball while it is being raised. Where does this work go? Assuming the ball is at rest both before and after it is raised, the work does *not* go into a change in kinetic energy of the ball, because this kinetic energy is the same after the work is done as it was before. The work produces a change in the position of the ball. Because the lifting force is required to offset the downward gravitational force of the earth on the ball, it is more significant to say that the work increases the separation of the ball and the earth. In the framework of the Work-Energy Principle we could say that the work put into the ball goes into "energy of relative position" associated with a change in the relative positions of the ball and the earth. The accepted term for this kind of energy is *gravitational potential energy*.

Although the work put into the ball does not have the form of kinetic energy while the ball is in its raised position, once the ball has been raised the energy which was given to it *will be* converted into kinetic energy if the person who lifted it simply lets go of the ball. As the ball falls back toward the ground, the work put into it earlier is transformed into *actual* kinetic energy of the ball. While the ball was in a raised position, the work which was put into it did not reside in the form of actual kinetic energy, but it has the possibility or *potential* to become kinetic energy (when the ball was dropped), so we say that while the ball was in a raised position it had *potential* kinetic energy, or (to use the conventional term) simply *potential energy*.

It is appropriate, and helpful, to think of potential energy as *stored energy*. The energy transferred to the ball while work was done on it remains stored in the ball, as long as it is in the raised position. In order to get this energy into the form of kinetic energy, all that is needed is to let go of or release the ball. It is a characteristic of potential energy that it can always be converted to kinetic energy if the objects which are holding the stored energy are simply "released." The specifics of what it means to "release" depend on the particular physical situation in which potential energy is present.

The reason why the ball acquires kinetic energy when it is let go is that the earth pulls it down with the force of gravity and accelerates the ball. Without this force, the potential energy would have no tendency to become kinetic energy, and in fact there would not even be any "potential kinetic energy" or potential energy. Every form of potential energy is inseparably associated with some kind of force. (The converse is not true, however; not every kind of force is associated with a form of potential energy.)

The force which converts the potential energy to kinetic energy is exerted on the ball by some other object, in this case by the earth. Without the earth, there would be no potential energy when the ball is raised, so we need to associate the potential energy with the *combination* of the ball *and* the earth. The ball and the earth interact with each other and form a physical *system*. The potential energy does not simply belong to the ball in its raised position, but rather *it belongs to the system* of the ball-and-the-earth in their changed *relative* positions, or their changed *configuration*.

Figure 48-1. An applied force F_a does work on a ball, lifting it through a height Δh.

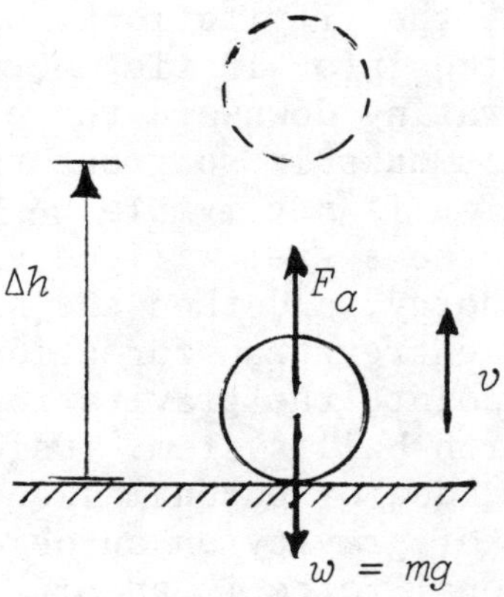

Let us compute the work required to lift the ball through a change in height Δh. The required lifting force F_a is directed upward and has a magnitude equal to the weight of the ball, mg (Figure 48-1). If the lifting force exceeds the weight the ball will accelerate upward, but if the lifting force is smaller than the weight the ball accelerates downward. The resulting changes in kinetic energy do not enter our calculation of the work required to change the elevation of the ball. This work is

$$W_{in} = \vec{F}_a \cdot \vec{s} = F_a s = mg\Delta h.$$

Using calculus to compute the work as $\int F_a dy$ leads to the same result. All of this work goes into a change in gravitational potential energy, *GPE*.

★ The increase in *gravitational potential energy* $\Delta GPE = mg\Delta h$ for an object-earth system, when an object having mass m is raised through a vertical distance Δh near the earth's surface (where g may be regarded as a constant).

If the kinetic energy of the object changes while it is being raised, or if there is friction or air drag which opposes the object's motion, additional contributions to the work are required. In any case, the work required to supply the increase in gravitational potential energy is always given by the product $mg\Delta h$, as long as the object remains near the earth's surface where g may be regarded as constant.

The gravitational potential energy of a system depends on the height of the object under consideration. What was the value of *GPE* before the object was raised? We may call it zero, if we choose to do so. Any absolute value which we choose to assign to *GPE* at some elevation is arbitrary. We may regard the lowest point of the motion as the arbitrary zero level, or the earth's surface as the zero level, or choose any other zero for the *GPE* which is convenient for the problem at hand. The reason we are always free to choose a convenient zero level for *GPE* in any physical situation, is that it always turns out that it is only the *change in GPE*, ΔGPE, which is significant.

If a ball which has been raised is then allowed to fall back toward the earth, the gravitational potential energy of the ball-earth system is converted into kinetic energy of the ball. (Actually, while the ball is accelerating downward the earth is also accelerating upward, but because the earth's mass is so great the speed acquired by the earth during this acceleration is negligible, and therefore the earth's share of the kinetic energy of the system will be vanishingly small.) We can thus compute the kinetic energy, and then the speed acquired by a falling ball, directly from this energy transformation without using the concept of work. From this viewpoint, the gravitational force on the ball is an *internal force* of the earth-ball system. Using an alternative point of view, one can avoid the idea of potential energy but instead compute the work done by the force of gravity on an object while it is falling. In this view, the gravitational force is an *external force* on the ball. Either point of view is applicable to motions of this kind. One may either regard the falling object as the system and compute the work done by the force of gravity on the system, or regard the object-and-earth-together as the system and compute the decrease of gravitational potential energy of the system as the object falls. In either case, one computes the product $mg\Delta h$ and this equals the increase in kinetic energy of the falling object.

Worked Example 48-1

Annette likes to play "jump the rope." Her mass is 45 kilograms, and each jump raises her body an effective distance of 12 cm.

(a) What is the change in gravitational potential energy when Annette jumps from the ground to her raised position?

(b) How much work is required to raise her body, for each jump?

(c) What supplies the force which accomplishes this work?

(d) Into what kind of energy does the gravitational potential energy change while she is falling from her highest position back to the ground?

Solution

(a) Annette remains near the earth while she jumps, so her change in gravitational potential energy may be computed as follows:

$$\Delta GPE = mg\Delta h = 45 \text{ kg} \times 9.8 \text{ m/s}^2 \times 0.12 \text{ m}$$

$$= 52.9 \text{ joules.}$$

(b) The work required is just equal to the change in *GPE*, which is 52.9 joules.
(c) Her leg muscles supply the required force.
(d) While she is falling, the *GPE* decreases and her downward speed increases, which means that the *GPE* is transformed into kinetic energy.

As we saw in Section 44.3, the acceleration of free fall, g, decreases as the elevation above the earth's surface increases, so the equation $\Delta GPE = mg\Delta h$ is limited to the region where g may be regarded as having a constant value of 9.8 m/s^2. For practical purposes, this region may be considered to extend upward from the earth's surface for about 32 km or 20 miles, which is the elevation at which g has decreased by 1% from its value at the earth's surface. In the next section we take up the problem of computing ΔGPE at greater distances from the earth's surface.

Questions and Exercises

Q 48.1-1 If a weight is raised to the top of a building, would you say that the potential energy *of the weight* has increased? Or is it the potential energy *of the earth*? Where does this potential energy actually reside?

Q 48.1-2 (a) Can a baseball have a negative value of kinetic energy?
(b) Can a baseball have a negative value of gravitational potential energy? Explain.

E 48.1-1 A girl is poised at the top of a slide, 3 meters above ground level. Her weight is 250 newtons. Compute the potential energy relative to the ground of the girl-earth system.

E 48.1-2 A parachutist having a mass of 70 kilograms descends to the ground through a distance of 400 meters. What is the change in gravitational potential energy?

E 48.1-3 Assume that your mass is 80 kg. How much does the potential energy increase when you ride an elevator from the ground floor of a building to another floor 11 meters higher?

E 48.1-4 (a) At what point in its motion is the kinetic energy of a pendulum bob a maximum? What is the gravitational potential energy of the bob-earth system at this point?
(b) At what point is the gravitational potential energy a maximum? What is the kinetic energy of the bob at this point?
(c) A 2-kg pendulum bob is pulled away from its center position until it has risen a height of 5.1 cm; it is then allowed to swing freely. Find its speed as it passes through the center position.
(d) How does this speed depend on the length of the pendulum?

E 48.1-5 (a) How much work is required to raise a baseball (m = 0.14 kg) to a height of 15 m (about 50 feet) above the ground?
(b) Does this work depend on the path followed by the ball?
(c) Suppose John wants to throw a baseball to Mike who is on the roof of his dormitory, 15 m above the ground. Does it require a lower initial speed if John throws the ball straight up, or he throws the ball at an angle from a spot a few meters from the base of the dormitory?
(d) If John throws the ball approximately straight up, what initial kinetic energy must the ball have? What initial speed v_0?
(e) If John throws the ball from a point 15 m from the base of the dorm, what minimum initial kinetic energy must the ball have? What initial speed?

Check your understanding of the above material by testing yourself on the minimum learning goals at the beginning of this section.

48.2 Gravitational Potential Energy in Space

Minimum Learning Goal. Be able to determine the gravitational potential energy when an object is at some distance above the surface of a planet having a known mass, given the distance from the center of the planet to the object. Also, be able to calculate the change in potential energy as the object moves from one position to another. For example:

1 A 100-kg satellite is in an orbit around the earth having a radius which varies from 2.2×10^7 m at its highest point to 1.27×10^7 m at its lowest point. Using data from Appendix G, compute the potential energy when the satellite is at
(a) its hightest point;
(b) its lowest point.
(c) What is the change in potential energy as the satellite moves from the lowest to the highest point in the orbit?
(Answer, page 48-26, Line 8)

The expression for gravitational potential energy developed in Section 48.1 is applicable only when the weight of an object is independent of its position. If the object is not confined to a region near the earth's surface then both the magnitude and direction of the gravitational force on it vary with position, and this must be taken into account in calculating the potential energy.

One of the primary applications of the concept of gravitational potential energy at high altitudes is to the motions of satellites and other space vehicles. To obtain an equation applicable to such problems, we consider an object having mass m (which may be a satellite or a spacecraft) moving in the gravitational field of a planet having mass M (Figure 48-2). Let us suppose that the object moves from an initial distance r_i to a final distance r_f, both measured from the center of the planet. Assuming there is no friction which opposes this motion, and the kinetic energy of the object at the end of the motion is the same as at the beginning, all of the work required to move the object from its initial to its final position goes into a change in the gravitational potential energy of the system. To calculate this work, one must use the general expression for the gravitational force F_g of the planet on the object at any distance r from the center of the planet, which is given in Section 44.3. This force attracts the object to the planet, so to indicate this attraction we write

$$\vec{F}_g = -\hat{r}\,\frac{GMm}{r^2}\,.$$

Figure 48-2. An object having mass m moves from an initial distance r_i from the center of a planet to a final distance r_f. An increment of length along the path is denoted by $d\vec{\ell}$, while the unit vector $\hat{r}$ is directed radially away from the center of the planet.

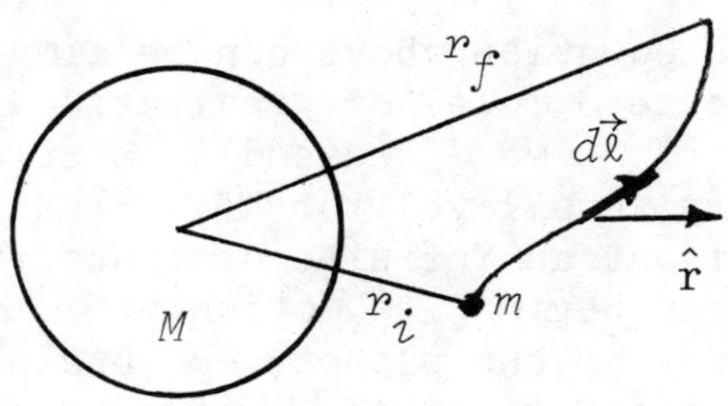

The total work W may be evaluated by constructing a graph of F_g vs. r and finding the area under the graph, or by an elaborate algebraic calcultion, but it may be found simply by evaluating a line integral (Appendix A.5). To do this, we compute

$$W = -\int_{r_i}^{r_f} \vec{F}_g \cdot d\vec{\ell},$$

where the negative sign is needed because the force required to move the object is exactly the opposite of the gravitational force pulling the object toward the planet. Therefore,

$$W = -\int_{r_i}^{r_f} -\frac{Mm}{r^2}\,\hat{r}\cdot d\vec{\ell}\,.$$

For each segement $d\vec{\ell}$ along the path, the dot product $\hat{r}\cdot d\vec{\ell}$ is equal to the length of the component of $d\vec{\ell}$ in the direction of $\hat{r}$, which is just dr, the change in distance from the planet along this segment. We therefore write

$$W = GMm \int_{r_i}^{r_f} \frac{1}{r^2}\,dr$$

$$= GMm\left[\frac{-1}{r}\right]_{r_i}^{r_f} = GMm\left[-\frac{1}{r_f} + \frac{1}{r_i}\right].$$

This work equals the change in *GPE* of the system, so for an object which moves (or is moved) from an initial distance r_i to a final distance r_f from the planet's center, the change in gravitational potential energy is

★ $$\Delta GPE = GMm\left(\frac{1}{r_i} - \frac{1}{r_f}\right).$$

When an object moves to a greater distance from the planet, so that r_f is greater than r_i, the ΔGPE calculated using this equation is positive. This is what we would expect, because to move an object farther from the planet work must be put into the object to overcome the force of gravity.

The equation above can be simplified to some extent by adopting a suitable zero level of gravitational potential energy for an object out in space. As we discussed in Section 48.1, we are free to choose an arbitrary zero level for *GPE*. The zero position chosen is one where the object is at an infinite distance from the planet. If we suppose that the object begins its motion at $r_i = \infty$ and moves to a final distance r from the center of the planet, we obtain an expression for the gravitational potential energy of the system at this distance.

★ The *mutual gravitational potential energy* of two objects, separated by a center-to-center distance r and having masses m and M, is $GPE = -GMm/r$.

Note that the negative sign in this equation is simply a consequence of our choice of the zero level of *GPE*. As is also true for objects which change position near the earth's surface, it is only the *change in GPE* which is significant for analyzing the physics of objects moving in space.

Worked Example 48-2

How much energy must be supplied to a spacecraft having a mass of 2000 kg, to lift it from the earth's surface at Cape Kennedy to an altitude where it is 3.5×10^8 meters from the center of the earth? (At this distance, it could reach the moon with no additional increase in gravitational potential energy.)

Solution

The energy required is just the amount by which the potential energy increases as the spacecraft moves from the earth's surface to the designated altitude. (We neglect here the additional *GPE* required to raise the booster and later stages of the rocket propulsion system.) Using the earth's mass M from Appendix G, this increase in potential energy is

$$\Delta GPE = GMm\left(\frac{1}{r_i} - \frac{1}{r_f}\right)$$

$$= 6.67 \times 10^{-11}\ \text{N}\cdot\text{m}^2/\text{kg}^2 \times 5.98 \times 10^{24}\ \text{kg} \times 2 \times 10^3\ \text{kg}$$

$$\times \left(\frac{1}{6.38 \times 10^6\ \text{m}} - \frac{1}{3.5 \times 10^8\ \text{m}}\right)$$

$$= 7.98 \times 10^{17}\ \text{N}\cdot\text{m}^2\ (1.57 \times 10^{-7}\ \text{m}^{-1} - 2.86 \times 10^{-9}\ \text{m}^{-1})$$

$$= 1.23 \times 10^{11}\ \text{N}\cdot\text{m} = 1.23 \times 10^{11}\ \text{joule}.$$

Questions and Exercises

Q 48.2-1 In Section 48.1 we said that it does not matter from which point h is measured. When you calculate gravitational potential energy in space, does it matter from which point r is measured?

Q 48.2-2 What is the significance of the fact that gravitational potential energy in space is always negative? Since in principle a particle near the earth is a special case of a particle in space, why is it that the formula of Section 48.1 does *not* necessarily give a negative potential energy?

Q 48.2-3 (a) What is the work, W, required to move an object having mass m through in a radial line from $r = \infty$ to a final position $r = r_1$ in the earth's gravitational field?
(b) What happens to the work described in part (a)?

Q 48.2-4 How can the earth and the moon have a mutual gravitational potential energy, *GPE*, equal to $-GM_e m/r$, if the moon did not arrive at its present position by traveling from $r = \infty$?

E 48.2-1 Observations show that the average orbit radius r of an earth satellite (mass m) decreases each day, and the *rate* of decrease becomes greater as the satellite orbit moves to lower altitudes.
(a) As r decreases, does the kinetic energy of the satellite increase, or decrease? (Hint: see Section 44.4)
(b) As r decreases, does the gravitational potential energy of the satellite increase, or decrease?
(c) The total mechanical energy of the satellite is *TME* = *GPE* + *KE*. Derive the equation *TME* = $-GM_e m/2r$, where M_e is the mass of the earth.
(d) As r decreases, does *TME* increase, or decrease? Answer by reference to equations(s), or by considering a physical mechanism which changes the energy of the satellite.
(e) What force causes this change in energy?
(f) As the satellite spirals into smaller values of r and its *GPE* changes, half of the change in *GPE* goes into internal energy (see Section 49. 3). Where does the other half go?

(continued on next page)

(g) The force of atmospheric drag tends to cause the satellite to slow down. What is the force which causes the satellite to speed up as its altitude decreases?

(h) Why do earth satellites usually "burn up" when their orbital radius decreases to the point where the orbital period is about 87 minutes?

E 48.2-2 What is the average gravitational potential energy of the Earth-Moon system?

Check your understanding of the above material by testing yourself on the minimum learning goals at the beginning of this section.

48.3 Escape Velocity

Minimum Learning Goal. Be able to calculate the escape velocity of an object from a planet or moon within our solar system, or the escape velocity from the sun, given the initial location of the object. For example:

1 Compute the escape velocity from the moon for an object on the surface of the moon. (Answer, page 48-26 ,Line 11)

2. Compute the escape velocity from the sun for an object on the surface of the earth. (Answer, page 48-25, Line 3)

We can now apply the energy concepts we have developed to an interesting problem in space physics: What is the *minimum* speed of a spaceship at the earth's surface which will enable it to "escape" from the earth's gravitational field? A spaceship will escape if it can leave the earth's surface using rocket engines and acquire sufficient speed to coast to a great distance r (approaching infinity) from the earth, even if its speed approaches zero at this great distance. The required minimum velocity near the earth's surface is called the *escape velocity*, v_e, from the earth.

During the first few minutes after the spacecraft leaves its launching pad at the earth's surface, its engines are supplying an upward force or *thrust* which accelerates the spacecraft upward, increasing both the kinetic energy of the spacecraft and the gravitational potential energy of the spacecraft-earth system. When all stages of the launching rocket engines have exhausted their fuel supply, the spacecraft can acquire no more energy. The point at which the fuel is exhausted and the engines cease to function is called *burnout*. From this point on, a spaceship which is on a voyage of "escape" will continue to coast upward, gaining gravitational potential energy at the expense of its kinetic energy. From another point of view, the earth's graviational force does negative work on the rising spacecraft, thus decreasing its kinetic energy; the absolute value of this negative work is just the increase in gravitational potential energy of the system, as we saw in Section 48.1.

For the spacecraft to escape entirely, its original kinetic energy must be greater than, or at least equal to, the potential energy which the earth-spacecraft system acquires in going from a small separation to an infinite separation. As we saw in Section 48.2, this gain in potential energy is

$$\Delta GPE = GM_e m\left(\frac{1}{r_e} - \frac{1}{\infty}\right) = \frac{GM_e m}{r_e},$$

where M_e is the mass of the earth and r_e is the earth's radius. For chemical-fuel engines, burnout occurs within a few hundred miles of the earth's surface, so it is a good approximation to use r_e for the distance of the spacecraft from the center of the earth at burnout. Setting the kinetic energy at burnout equal to the required increase in gravitational potential energy, we have

$$\frac{1}{2} mv_e^2 = \frac{GM_e m}{r_e}.$$

Cancelling m and solving for v_e, we obtain

$$v_e = \sqrt{2GM_e/r_e}.$$

Notice that the escape velocity does not depend on the mass m of the spacecraft; it is the same for a hundred-ton spacecraft as it is for a molecule of hydrogen. Calculation shows that the numerical value of v_e from the earth's surface is 1.12×10^4 meter/second or 7.0 miles/second. The equation above for v_e is similar to the equation for the speed of a satellite in a circular orbit around the earth, $v = \sqrt{GM_e/r}$ (Section 44.4). The difference is the factor $\sqrt{2}$, which shows that even if a satellite in a circular orbit could steer itself so that it is heading away from the earth, it does not have enough speed or kinetic energy to escape.

A spacecraft which has enough speed to escape from the gravitational field of the earth does *not* necessarily have enough speed to escape from the gravitational field of the sun. For a spacecraft which begins its journey at a distance from the sun equal to the radius of the earth's orbit around the sun, r_{es}, the speed required to escape from the sun is given by an equation similar to the last one above, but containing the sun's mass M_s instead of the earth's mass:

$$v_e = \sqrt{2GM_s/r_{es}}.$$

If a spacecraft which is intended to escape from the sun's gravitation begins its journey at some other initial distance from the sun, that distance must be used in place of r_{es} in order to compute the required escape velocity.

Suppose it is desired to launch a spacecraft from the earth which will escape from *both* the earth and the sun; what is the required velocity at burnout? It is neither the value of v_e from the earth's surface nor the value of v_e from the sun beginning at a distance equal to the earth's orbit radius, nor is it the sum of these escape velocities. To find the required escape velocity for this case, it is necessary to retrace the argument from the beginning of this section, using the fact that the initial kinetic energy of the spacecraft (at burnout) must equal the *sum of the increases in gravitational potential energy* of the spacecraft-earth-sun system as the spacecraft moves to an infinite separation from both the earth and the sun. The result shows that for a spacecraft which begins its journey at the earth's surface, the required escape velocity is

★ $$v_e = \sqrt{2G\left(\frac{M_e}{r_e} + \frac{M_s}{r_{es}}\right)}.$$

The Pioneer and Voyager spacecraft launched from the U. S. A. are now headed for regions beyond our solar system.

The equations developed in this section can be adapted to compute the escape velocity from any planet or satellite, provided appropriate values of mass and distance are used in the equations. Physical data on bodies in the solar system are found in Appendix G.

Worked Example 48-3

Suppose people in a space colony on the moon's surface desire to launch a spacecraft which will escape from the solar system. What escape velocity is required?

Solution

The spacecraft must escape from the combined gravitational fields of the Moon, Earth and Sun, so the equation for the excape velocity must include for each of these bodies a term having the initial distance of the spacecraft from the center of the body. The equation is

$$v_e = \sqrt{2G\left(\frac{M_m}{r_m} + \frac{M_e}{r_{em}} + \frac{M_s}{r_{sm}}\right)} ,$$

where r_{em} is the distance from the Earth to the Moon and r_{sm} is the distance from the Sun to the Moon (approximately the same as the distance from Earth to Moon). Calculating, we obtain

$$v_e^2 = 2 \times 6.67 \times 10^{-11} \frac{\text{N}\cdot\text{m}^2}{\text{kg}^2}$$

$$\times \left(\frac{7.35 \times 10^{22}\ \text{kg}}{1.74 \times 10^{6}\ \text{m}} + \frac{5.98 \times 10^{24}\ \text{kg}}{3.84 \times 10^{8}\ \text{m}} + \frac{1.99 \times 10^{30}\ \text{kg}}{1.50 \times 10^{11}\ \text{m}}\right)$$

$$= 1.33 \times 10^{-10} \ \frac{N{\cdot}m^2}{kg^2}(4.22 \times 10^{16} + 1.56 \times 10^{16} + 1.33 \times 10^{19}) \ \frac{kg}{m} \ ,$$

from which

$$v_e = 4.20 \times 10^4 \text{ m/s.}$$

It is evident that for this spacecraft the kinetic energy required to escape from the Sun is far greater than the kinetic energy needed to excape from either the Earth or the Moon.

Questions and Exercises

Q 48.3-1 Does every spaceship or rocket which acquires the escape velocity v_e actually escape from the earth? Explain.

Q 48.3-2 What is required for a molecule to escape from the earth?

E 48.3-1 By equating the total (kinetic plus potential) energy of a spaceship when its rocket engines burn out a few miles from the earth to the minimum total energy of the spaceship at $r = \infty$, show that the escape velocity $v_e = \sqrt{2GM_e/r_e}$.

E 48.3-2 Work out the complete derivation of the last equation in this section, for the speed required to escape from both Earth and Sun for a spacecraft launched from Earth's surface.

E 48.3-3 Using the results of Section 44.3, show that the escape velocity from Earth's surface is given by $v_e = \sqrt{2gr_e}$.

E 48.3-4 (a) Compare the speed which a satellite needs in order to skim the surface of the earth in a circular orbit, with the speed it needs to escape the earth entirely.
(b) If a satellite is launched with a velocity tangent to the earth's surface and having a magnitude somewhere between the two values computed in part (a), what kind of orbit would it have?

E 48.3-5 Some astronauts are planning to launch a spaceship from Cape Kennedy which will leave our solar system and travel to Proxima Centauri, the nearest star.
(a) As the spacecraft moves outward from the solar system, what object exerts the dominant gravitational pull back toward the solar system?
(b) How far from this object is the spacecraft located, when it begins its journey at Cape Kennedy?
(c) What minimum speed (in m/s) must this spaceship have?

E 48.3-6 Compute the minimum speed required to escape from the moon and the earth, for a spacecraft starting from the moon's surface.

Check your understanding of the material above by testing yourself on the minimum learning goals at the beginning of this section.

48.4 Elastic Potential Energy

Minimum Learning Goal. Be able to compute the potential energy stored in a linear spring, given the amount by which the spring is stretched or compressed and information from which you can find the spring constant. For example:

1 A (linear) coil spring from an automobile requires a force of 5000 N to compress it 10 cm. What is the elastic potential energy of the spring when it is compressed 5 cm? (Answer page 48-25, Line 6)

2 The spring of a bathroom scale is compressed 0.5 inches (1.27 cm) when a 230-pound (1023 N) man stands on the scale. How much elastic potential energy is stored in this spring when a girl standing on the scale compresses the spring 0.228 inches (0.58 cm)? (Answer 48-26, Line 9)

Figure 48-3. When a spring is pulled or pushed into a deformed configuration, the work done goes into elastic potential energy (as long as the spring does not become permanently deformed).

In Figure 48-3 a child is pulling on a playground spring, bending it into a distorted configuration. Assuming she does not pull hard enough to cause a permanent distortion, when she lets go the toy will spring back to its original upright position, swing past that position and oscillate for a time before coming to rest in its normal position. When the girl pulls on the spring, her pulling force acts through some displacement and so she does work on the spring. Where does this work go?

Before the girl touched the spring, it was in its normal upright "relaxed" configuration. When work is put into the spring, this work leads to a change in the spring's configuration. We think of this work as being stored in the deformed spring as long as it remains in its bent configuration. Actually, the work--or rather the energy supplied by the work--is not stored in the spring alone but in the system consisting of the spring and the ground or base to which it is secured. Because the energy supplied to the spring is stored in the system while the spring is at rest in its deformed state, this energy is a form of potential energy. It could be called "spring potential energy," but the term often used is *elastic potential energy*. The force associated with this potential energy is the internal force of the spring which gives the spring its "springiness" or *elasticity*.

The work put into the system is stored in the form of elastic potential energy as long as the spring remains in its distorted configuration. If the girl pulls on the spring with a small force, its distortion is small and the stored energy is small; if she uses a large force, the deformation is large and the stored energy is large. In this section we develop an equation for the amount of energy stored in a spring.

The playground spring pictured in Figure 48-3 is just one type of spring; there are many other kinds. Coil springs are found in automobile suspensions, in mattresses and innersprings for beds, in upholstered furniture, in ballpoint pens, and many other devices. The rear suspensions of many automobiles use leaf springs. An air track in the physics laboratory may have springs in the form of a thin steel band bent into a circle.

A rubber band is also a kind of spring, and so is a block of foam rubber. Any kind of spring can be used to store potential energy; the physical property that qualifies something as a spring is its internal elasticity. From the multitude of different styles of springs, we choose a coil spring which can either be stretched or compressed as a representative type of spring for further discussion. The results we obtain are applicable to many kinds of elastic bodies.

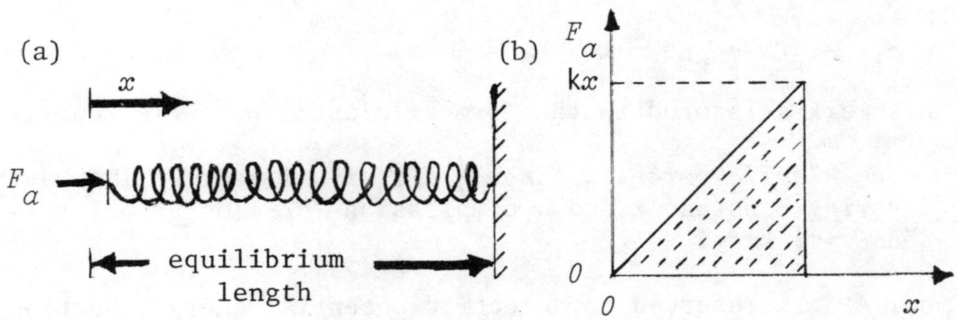

Figure 48-4. When an applied force F_a compresses a linear spring, the required force is directly proportional to the displacement x of the free end of the spring.

Consider a steel coil spring lying on a smooth (frictionless) table, as shown in Figure 48-4(a). One end is fixed to a wall; the other end can be moved so as to stretch or compress the spring. There is a natural length which the spring assumes if the end is free. If the end is displaced from this natural position, it exerts a force on whatever is attached to it.

If the free end is pushed with an applied force $\vec{F}_a$ so as to compress the spring, the spring pushes with an equal force, according to Newton's third law (Section 42.5). The amount of force required to compress or stretch the spring by 1.0 cm, say, depends on the stiffness of the spring. A spring which is very stiff and therefore not very flexible requires a large applied force to compress it by 1.0 cm, while a spring which is easily distorted and therefore highly flexible requires only a small force for the same compression. The physical property of any spring which is a measure of its internal stiffness is called its *spring constant*, k.

If the applied force, F_a, required to compress the spring by an amount x from its natural or equilibrium length is measured for several values of x, the results may be plotted in a graph as shown in Figure 48-4 (b). In general, the plot of these data will not be a straight line, but for some springs the graph is a straight line to a good approximation. A spring for which x is directly proportional to F_a is called a *linear* spring.

★ For a linear spring, the *spring constant* k is defined as the ratio of the applied force to the resulting displacement of the free end of the spring. $k \equiv F_a/x$.

The spring constant is therefore just the slope of the graph of F_a vs. x, and it has units of newton/meter.

The total work put into a linear spring in compressing it from its natural length by a displacement x is represented by the area under the curve of F_a vs. x, as discussed in Section 47.2 and shown by the dashed area in Figure 48-4 (b). This work thus corresponds to the area of a triangle, so

$$W_{in} = \frac{1}{2}\, x \cdot F_a = \frac{1}{2}\, x \cdot kx = \frac{1}{2}\, kx^2.$$

All of this work is stored in the form of elastic potential energy.

★ The *elastic potential energy LPE* of a linear spring having a spring constant k and a compression (or elongation) x is $LPE = \frac{1}{2}\, kx^2$.

(The symbol *EPE* is reserved for electric potential energy, Section 24.6). Implicit in this equation for elastic potential energy is the arbitrary choice that the potential energy stored in the spring is zero when the spring is in its natural, relaxed configuration. Strictly speaking, a spring by itself cannot store energy; a support of some kind is required to secure the stationary end of the spring. The stored energy resides in the system consisting of the spring and its support.

Summarizing this discussion, the work required to compress or stretch a spring is equal to the elastic potential energy stored in the spring-support system. If there is also a net change in the kinetic energy or gravitational potential energy of the system as a result of the applied force, additional work is required to supply these forms of energy.

Worked Example 48-4

A linear coil spring stretches 10 cm from its relaxed length when a force of 1.5 newton is applied to it. How much elastic potential energy is stored in the spring when a total force of 3.0 newton is applied to it?

Solution

To compute the stored energy, it is convenient first to find the spring constant and the displacement of the free end of the spring when 3 newtons of force are applied. (One could find the *LPE* without these preliminary steps, but we will work out the solution in a straightforward manner even though it is a bit longer that way.) Using the data supplied, the spring constant is

$$k = \frac{F_a}{x} = \frac{1.5\ \text{N}}{0.1\ \text{m}} = 15\ \text{N/m}.$$

The displacement of the end of the spring under a force of 3.0 N is

$$x = \frac{F_a}{k} = \frac{3.0\ \text{N}}{15\ \text{N/m}} = 0.2\ \text{meter}.$$

Finally, the stored energy for $x = 0.2$ meter is computed as

$$LPE = \frac{1}{2}\,kx^2 = 0.5 \times 15\ \text{N/m} \times (0.2\ \text{m})^2$$

$$= 0.3\ \text{joules}.$$

If a ball is placed against the end of a compressed spring and the spring is then released, the ball will be projected and acquire kinetic energy, while the potential energy of the spring decreases. This is just one of the ways in which potential energy stored in a spring can be converted into kinetic energy.

A more advanced discussion of the energy associated with an elastic body such as a coil spring would give more attention to the elastic properties of the material from which the body is made. This topic is generally referred to as the study of *elasticity.*

Questions and Exercises

Q 48.4-1 Compare a stiff spring and a "loose" or "floppy" spring. Which has the higher spring constant k?

Q 48.4-2 Is it possible to transform elastic potential energy directly into gravitational potential energy with no intervening form of energy? Explain.

E 48.4-1 A spring whose force constant is 40 N/m is compressed 10 cm, beginning from its equilibrium configuration.

(a) What is the maximum force required?

(b) Compute the total work required.

(c) If the compression were only 5 cm instead of 10 cm, how much work would be required?

E 48.4-2 A linear coil spring stores 15 joules of potential energy when it is stretched 10 cm. How much potential energy is stored when the spring is stretched a total of 20 cm?

E 48.4-3 A "large" bumper spring, having a spring constant of 3 N/m, is compressed with a maximum force of 0.36 N by a glider on an air track.

(a) Compute the energy stored in the compressed spring.

(b) The glider (mass 0.2 kg) in contact with the compressed spring is released from rest on a level track. How much kinetic energy is acquired by the glider?

(c) What is the speed acquired by a 0.2-kg glider released as in part (b)?

(d) How do your answers depend on the mass of the glider?

E 48.4-4 A toy bazooka has a spring (k = 500 N/m) which is compressed 12 cm.

(a) Find the potential energy stored in the spring.

(b) If all this stored energy is transferred to kinetic energy in the launching of a bazooka "shell" having m = 0.2 kg, what initial velocity will the shell acquire?

(c) What is the maximum height to which the shell can be launched by this bazooka?

((E 48.4-5 Use the method of integration to compute the total work required to compress a linear spring from $x = 0$ to $x = x$, by evaluating the integral

$$W_{\text{in}} = \int_0^x \mathrm{k}x\,dx.$$

))

E 48.4-6 A block sitting on a frictionless table is attached to a wall through a spring. If the block is moved so as to compress the spring, and then released, it will oscillate back and forth around the equilibrium point of the spring. Examine qualitatively the variation in kinetic energy as a function of time in the course of this motion and, knowing that total energy has to remain constant, describe how the potential energy has to vary with spring length. Does the expression $LPE = \mathrm{k}x^2/2$ this way?

Check your understanding of the material above by testing yourself on the minimum learning goals at the beginning of this section.

48.5 The Concept of Potential Energy

Minimum Learning Goals. (1) Given an adequate description of a physical system, be able to state whether a potential energy can be defined for the system, and if it cannot, state why not. If it can, be able to state what kind of force is associated with this form of potential energy. For example:

1 The planet Mars travels in an elliptical orbit with the sun at a focus. As the planet moves between any two points on this orbit which are the same distance from the sun, the net work done by gravitation on the planet is zero. Can a potential energy be defined for the Mars-Sun system? If not, why not? If it can, what kind of force is associated with this potential energy? (Answer, page 48-26, Line 12)

2 Linus drags his blanket behind him from his bedroom to the kitchen.
(a) Is a form of potential energy associated with the system consisting of Linus, the blanket, and the floor? If not, why not?
(b) If there is, what force is associated with this potential energy? (Answer, page 48-25, Line 1)

3 A mass suspended from the end of a spring is displaced from equilibrium and released so that it oscillates in a vertical line. As the mass moves away from and then returns to any given location the net work done by gravitation is zero and the net work done by the spring is zero. Can a potential energy be defined for the mass-spring-earth system? If not, why not? If it can, what forces are associated with this potential energy? (Answer, page 48-25, Line 4)

4 For a mass-spring system similar to that described above in example 3, during the motion from a position z_1 through one complete oscillation and back to z_1 again, it is found that the net work done by the spring is negative. Can a potential energy be defined for the mass-spring-earth system? If not, why not? If it can, what forces are associated with this potential energy? (Answer, page 48-25, Line 7)

(2) Given some statements about energy, be able to select those which correctly use or describe the concept of conservation of energy. For example:

1 Which of the following statements about energy are correct?
(a) The potential energy of an isolated mechanical system remains constant at all times.
(b) For an isolated physical system, the sum of the kinetic and potential energies remains constant. (continued)

(c) In an isolated physical system, the sum of the kinetic energy and the potential energy of each individual particle remains constant.
(d) The law of conservation of energy is a logical and necessary consequence of the work-kinetic energy theorem.
(Answer, page 48-26, Line 10)

Earlier in this chapter we discussed two kinds of potential energy: gravitational and elastic. These two kinds have some characteristics in common with all other kinds of potential energy, and which we summarize here.

An elementary definition of potential energy is the following:

> *Potential energy* is energy stored in a system because of its configuration.

The systems we have considered are a ball and the earth (not just the ball alone, Section 48.1), and a spring and its support (not simply the spring alone, Section 48.4). In each case, the amount of potential energy stored in the system is some definite numerical function of its configuration, that is, of the relative positions of the particles or objects making up the system. The system acquires its stored energy as a result of work done on the system by *external forces*, that is, forces acting on the system from outside rather than forces between parts of the system (which are called internal forces). The change in energy of the system as a result of work put into the system is expressed in the Work-Energy Principle (p. 47-9, discussed more fully in Section 49.2).

When a system contains potential energy greater than the minimum value it could have, there is always a tendency for the system to change its configuration in such a way that its potential energy decreases. To prevent this change and preserve the potential energy in the system, it is necessary to arrange or constrain the system in some way. A ball raised from the ground must be held or supported in its elevated position; if the ball is released, the gravitational potential energy of the system is transformed to kinetic energy as the ball falls. A deformed spring must be held in its distorted position; if a person holding it lets go, the spring returns to its relaxed position and the elastic potential energy of the system is converted to kinetic energy of the oscillating spring. Whenever a system containing potential energy is "let go" or released in an appropriate way, some or all of its potential energy is transformed into kinetic energy.

This transition from potential to kinetic energy results from *internal forces* within the system, which always have a tendency to move the parts of the system to a new configuration having less potential energy. Each kind of potential energy is associated with a particular kind of force, which is an internal force for the system having the potential energy. Gravitational potential energy is always associated with gravitational forces between the parts of the system, and elastic potential energy is always associated with some elastic force within the system. The system includes all the objects which are interacting with each other through the force associated with that form of potential energy.

In addition to the two kinds of potential energy discussed above, other kinds of potential energy are associated with other physical systems. If two bar magnets, or toys or other devices which contain bar magnets, are pushed close to each other in a certain manner and then released, the magnets repel each other and the objects acquire some kinetic energy. The work done in pushing the magnets close to each other is stored in the system of the magnets while they are held in a configuration near each other. Such a system has *magnetic potential energy* associated with the magnetic forces between the poles of the magnets (see Section 58.2). This is not the same magnetic force which acts on a moving charged particle; the latter force does *not* have a potential energy associated with it.

Another form of potential energy is associated with the electric force between charged particles or charged objects. This *electric potential energy* is discussed in Chapter 24.

By way of contrast, consider what happens when a box is pushed along the floor against the opposing force of friction. In this case, work is put into the box-floor system while the box is being pushed, but when the box is then "let go" it does *not* begin to move back toward its original position. The reason is that in this case the work put into the system does not go into macroscopic potential energy; instead, it goes into microscopic forms of energy associated with the molecules of the box and the floor, which are collectively called the internal energy of the system. This kind of energy is discussed in Section 49.3.

Someone who is unacquainted with money would be surprised to find out that coins and bills are "the same thing," since casual inspection shows no resemblance between them at all. Yet we know that a person can go into a bank and change one into the other at will, even though the total amount of money after the exchange is the same as before.

The relation of potential energy to kinetic energy is analogous to the coins and bills. If we look only at the definitions of the two energies no resemblance is apparent, and we may wonder why the word "energy" is used for both concepts. However, observations show that the two forms of energy can change into each other, but only in such a way that the sum of the two energies remains constant. This is the *law of conservation of energy*, having consequences which we consider at several places in this chapter. At this point, however, we emphasize that the law of conservation of energy is *not a theorem* like the work-kinetic energy theorem. The latter can be *derived* from something we have already accepted--namely, Newton's laws; but conservation of energy cannot be derived from Newton's laws. We can imagine a universe in which Newton's laws hold, but conservation of energy does not, without running into any logical contradictions. *Experimentally*, however, we find that in our universe energy is always conserved.

Let us now consider an isolated physical system. We want to formulate a definition of its potential energy in relation to the internal forces of the system, in such a way that the total energy of the system is conserved.

For a physical system acted on by no external forces, the law of conservation of energy says that any change ΔKE in the kinetic energy is matched by a corresponding change ΔPE in the potential energy, such that the sum of these changes equals zero:

$$0 = \Delta KE + \Delta PE \qquad \text{(for an isolated system).}$$

This means that the total of the kinetic plus potential energies remains constant for an isolated system. As time goes on, the velocities and positions of the various parts of the system change, so that the kinetic energy and potential energy do not individually remain constant; their sum, however, does remain constant.

For a system which receives work through the action of external forces, the law of conservation of energy (or the work-energy principle) says that the net work put into the system equals the net change in energy of the system:

$$W_{\text{(by ext F)}} = \Delta KE + \Delta PE.$$

From the work-kinetic energy theorem (p. 47-20), we know that the increase in kinetic energy of the system is equal to the work done on the system by the resultant of all forces, internal plus external:

$$W_{\text{(by all F)}} = W_{\text{(by ext F)}} + W_{\text{(by int F)}} = \Delta KE.$$

Subtracting this last equation from the preceding one, we obtain

$$-W_{\text{(by int F)}} = \Delta PE, \quad \text{or} \quad W_{\text{(by int F)}} = -\Delta PE.$$

Therefore, the potential energy of the system must be defined in such a way that its *decrease* is equal to the work done on the particles of the system by the internal forces.

★ The *potential energy* of a system is a function of the configuration of the system having the property that the work done by the forces within the system is equal to the decrease of this function.

For such a potential energy to exist, the forces associated with this potential energy must obey certain restrictions. To see what the restrictions are, imagine that a system is put through a sequence of changes which at the end brings it back to its initial configuration. Because the potential energy is a function of configuration, the potential energy at the end is the same as at the beginning. The work done by the internal forces of the system must be equal to the decrease in potential energy, so it follows that in such a process the total work done by the forces of the system must be zero. In other words, the forces associated with a potential energy are always such that when they act in a system while its configuration is changing from any state *A* to *B* and then back to *A*, the initial potential energy at *A* is the same as the final potential energy at *A*. An internal force associated with the potential energy of a system is called a *conservative force*.

As an example, consider the force of gravity. If an object is removed from its place on a shelf, carried around the room, taken upstairs, dropped through an elevator shaft, and finally taken back to its original place, the total work on the object by the force of gravity will be exactly zero. We can see this by noting that the force of gravity is constant and vertical; therefore, the work done in any small displacement is just proportional to the vertical component of the displacement. But if the particle ends up exactly where it started, the sum of the vertical components of all its displacements must be zero.

It is easy to imagine forces for which this would not be true. In fact, frictional forces seemingly violate the reqirement of zero work on a closed path. If we drag a box around a floor in such a way that it ends up in its original place, the total work done by friction will *not* be zero. This does not really violate the law of energy conservation, because the final configuration of the system is (contrary to appearances) not exactly the same as the initial one: the molecules of the floor and of the box are in more violent motion than they were originally. No energy was lost, but some macroscopic energy was transformed to microscopic energy. When considering problems in *macroscopic* physics, as we do in this textbook, we try to limit our analysis to a few *macroscopic variables*, and ignore the motion of the system on a molecular and atomic scale. This represents a helpful simplification of the situation, of course, but as part of the price for this simplification we often find that *macroscopic* energy is not conserved. The energy "lost" or dissipated from macroscopic forms is transformed to microscopic energy, often called internal energy (Section 49.3).

When a system changes from a state *A* to a different state *B*, the change in potential energy is equal to the negative of the work done by the internal forces. If the system than returns to state *A* by any other sequence of changes, or by any other *path*, the potential energy returns to its original value.

This means that the change in potential energy in going from state *A* to state *B* does *not* depend on the particular sequence of steps followed; we say that *the change in potential energy between any two states is path-independent*. Its follows that the work done by the internal forces of the system during any transition from state *A* to state *B* is also path-independent. If the potential energy of the system is assigned a particular numerical value in some reference state, then the potential energy of the system in any other state or configuration has a definite value which does not depend on how the system got into that configuration. Examples of reference states are the relaxed configuration of a spring, and the state of two bodies which attract each other through the gravitational force when these bodies are infinitely far apart.

Questions

Q 48.5-1 An isolated physical system consists of two particles which move relative to one another along a straight line. It is found that for each distance of separation the particles have definite values of speed, regardless of whether they are moving toward each other or away from each other, and that the speeds decrease as the particles move toward each other. Does the potential energy of this system increase, or decrease, as the particles move toward each other?

Q 48.5-2 In addition to coins and bills, we often use money in a third form: checks. A person can go into a bank and exchange a check for cash, or he can bring bills into a bank and come out with a check for the same amount. How could we determine whether nature also has a third form of energy, different from both potential and kinetic energies, but interchangeable with either of them?

Q 48.5-3 If you drop a tennis ball on a flat surface, after each succeeding bounce the ball has a reduced velocity (and hence a reduced kinetic energy) and it rises to a diminished height. How then can we say that the sum of the kinetic energy and the potential energy remains constant?

Q 48.5-4 To measure the energies of particles in a system, we must use measuring instruments which are not part of the system. How then can we ever obtain experimental data which enable us to make any statement about an "isolated" physical system?

Q 48.5-5 If the law of conservation of energy cannot be derived, how *can* it be reliably established?

Q 48.5-6 How can work be done on the particles of a system if the system is *isolated* from the rest of the physical world?

Q 48.5-7 How can potential energy be defined for a system which is not isolated?

Q 48.5-8 Consider an isolated physical system having a cyclic or periodic motion such that at the end of each cycle the system returns to the same configuration.
(a) Can we be certain that this system has a potential energy?
(b) Can we be certain that there is a force associated with this potential energy?

Q 48.5-9 What form of macroscopic potential energy (if any) is associated with each of the following forces? (a) restoring force of an ideal spring; (b) friction; (c) gravitational force; (d) electric force; (e) magnetic force on a moving charged particle; (f) nuclear force; (g) air drag force.

Q 48.5-10 As an example of an isolated system, consider the moon plus a frictionless pendulum placed on the moon by an astronaut and left in motion when he departs.
(a) At what points of its motion is the kinetic energy of this pendulum zero?
(b) What force acts to increase the kinetic energy of the pendulum?
(c) What force does work on the bob while it is descending from the high point of its path?
(d) What happens to the potential energy of the system while the kinetic energy is increasing?
(e) What is the connection between the force associated with the change in kinetic energy of this system, and the force associated with the change in potential energy of this system?

Q 48.5-11 We used the fact that the force of gravity near the earth is constant in magnitude and direction to show that there is a potential energy associated with this force.
(a) Does every internal force which is constant in magnitude and direction, regardless of the motions of particles within the system, have a potential energy associated with it?
(b) Is a potential energy associated only with those forces which are constant in magnitude and direction?

Q 48.5-12 Of the forces listed in Q 48.5-9, (a) Which appear in macroscopic physical systems? (b) Which appear in microscopic physical systems?

Check your understanding of the material above by testing yourself on the minimum learning goal at the beginning of this section.

Answers to Minimum Learning Goals

Line 1 (a) No, because when he lets go of the blanket it does not gain kinetic energy and return to its original position.

Line 2 (a) 0 J; (b) +82 J; (c) +82 J; (d) -470 J; (e) -388 J; (f) + 82 J.

Line 3 4.35×10^4 m/s

Line 4 Yes; the gravitational force between the earth and the mass-and-spring, and the elastic force of the spring on the mass.

Line 5 + 3.59×10^7 J at the base; +8.37×10^7 J at the top; +4.78×10^7 J

Line 6 62.5 J

Line 7 No, because the total work done by the spring on the mass, as the mass moves through one complete oscillation, is not equal to zero.

Line 8 (a) -1.42×10^9 J at highest;
(b) -2.09×10^9 J at lowest; (c) $+0.67 \times 10^9$ J

Line 9 1.35 J

Line 10 (b)

Line 11 2.37×10^3 m/s

Line 12 Yes; gravitational force.

ANSWERS to Questions and Exercises in Chapter 48

Section 48.1

Q 48.1-2 (a) No; (b) yes, if the zero level of *GPE* is chosen to be higher than the position of the ball.

E 48.1-2 -2.74×10^5 J

E 48.1-4 (a) At the lowest point of its swing; *GPE* = 0, a convenient arbitrary choice.
(b) At either end of the swing, where *KE* = 0.
(c) 1.0 m/s; (d) it does not depend on the length.

Section 48.2

Q 48.2-2 The negative sign is a result of choosing the position of zero *GPE* at an infinite separation of two bodies. The formula in Section 48.1 is based on a different choice of zero *GPE*.

Q 48.2-4 The mutual *GPE* is independent of the paths traveled by two objects in arriving at a distance r from one another.

Q 48.2-2 -7.63×10^{28} J

Section 48.3

Q 48.3-2 A speed at least equal to the escape velocity, and a path which avoids loss of kinetic energy through collisions with other molecules.

E 48.3-4 (a) 7.92×10^3 m/s for a circular orbit; 1.12×10^4 m/s to escape.
(b) An ellipse, with the earth at one focus.

E 48.3-6 2.78×10^3 m/s

Section 48.4

Q 48.4-2 No; kinetic energy is an essential intermediate form of energy.

E 48.4-2 60 J

E 48.4-4 (a) 3.6 J; (b) 6.0 m/s; (c) 1.84 m.

E 48.4-6 As the block oscillates, the kinetic energy varies from zero (at the extremes of maximum stretching and maximum compression) to a maximum (at the neutral point) to zero. Therefore, the *LPE* also varies from zero (at the neutral point) to a maximum (at the extremes). This variation corresponds to the formula *LPE* = $\frac{1}{2}$ k x^2, because the stored energy is the same for $+x$ as for $-x$, and is zero when $x = 0$.

Section 48.5

Q 48.5-2 If experiments show that in a closed system there is a decrease in the total amount of the known kinds of kinetic plus potential energy, we could conclude that some energy was transformed into a third form of energy. Investigation of this third form would be required to find out whether it was a "new" kind of potential energy or some totally new form of energy.

Q 48.5-4 When measurements are made on a system, the energy transfer between the measuring instruments and the system must be taken into account in order to draw conclusions about the energy changes within the system. Assuming the energy principles deduced from such measurements remain valid for closed systems, these principles can be applied to compute energy transfers in an isolated system.

Q 48.5-6 Work cannot be done on an isolated system by *external* forces, but it can be done by *internal* forces at the expense of the potential energy of the system.

Q 48.5-8 (a) yes; (b) yes.

Q 48.5-10 (a) at the end points;
(b) the gravitational force of the moon on the pendulum;
(c) the gravitational force of the moon on the pendulum;
(d) it decreases;
(e) they are the same force.

Q 48.5-12 (a) Forces (a), (b), (c), (d), (e), and (g).
(b) Forces (c), (d), (e), and (f).

CHAPTER 49 CONSERVATION OF ENERGY

Introduction

The fundamental law governing all transformations of energy is the law of conservation of energy. When work transfers energy to a system, this work can be accounted for quantitatively using the work-energy principle. The work may be done by conservative forces, or non-conservative forces.

Each form of potential energy is associated with a conservative force. If the force function is known the potential energy function can be found, and vice versa. The energy of a particle moving under a conservative force can be visualized using a potential well. Non-conservative forces include friction and fluid drag. Work done against friction is dissipated into microscopic forms of energy comprised by the internal energy of the system.

When you have completed your study of the material in this chapter, you should be able to answer questions and work exercises like the examples in the Minimum Learning Goals assigned by your instructor.

49.1 Conservative Forces and Potential Energy

Minimum Learning Goals. (1) Given a description of the forces acting between particles or systems of particles, be able to designate those which are conservative forces, and those which are non-conservative forces. For example:

1 A system consists of two particles which act on each other with a force such that the work done by this force equals the decrease in the potential energy of the system. Is this a conservative force? (Answer, page 49-18, Line 5)

2 A system consists of two particles which act on each other with a force which varies inversely with the square of the distance between the particles and is directed along the line joining them. Is this a conservative force? (Answer, page 49-19, Line 16)

3 A system consists of objects which exert forces on each other such that regardless of how these objects are set into motion, all of them come to rest after a certain time. Are these conservative forces? (Answer, page 49-19, Line 11)

(((2) Given the potential energy of a one-dimensional system as a function of the coordinate which specifies the configuration of the system, be able to compute an expression for the force associated with this potential energy, and be able to evaluate this force for a given system configuration. For example:

1 A toy car sitting on a table top has rubber bands attached to the front and rear, which are stretched along a straight line so that a potential energy of this system can be defined within certain bounds. The potential energy is given by the expression

$$PE = 110\ (\mathrm{J/m^2})\ x^2 - 22\ (\mathrm{J/m})\ x + 1.1\ \mathrm{J}, \text{ for } |x| \leq 0.20 \text{ meter.}$$

(a) What is the force on this car as a function of x?
(b) For what value of x is the force zero? (Answer, page 49-18, Line 1)

2 The potential energy of a certain diatomic molecule is given by the expression

$$PE = -9.5 \times 10^{-29}\ (\mathrm{J{\cdot}m})\ y^{-1} + 1.1 \times 10^{-88}\ (\mathrm{J{\cdot}m^7})\ y^{-7},$$

where y is the interatomic separation in meters.
(a) What is the force of one atom on the other, as a function of y?
(b) For what value of y is the force zero? (Answer, page 49-18, Line 3)

(3) Given an expression for the internal forces of a physical system as a function of the configuration of the system, be able to compute an expression for the potential energy of the system relative to a specified reference configuration. For example:

1 The force $\vec{F}$ on a particle carrying an electric charge q, exerted by another particle bearing a charge Q, is given by the expression

$$\vec{F} = \hat{r}Qq/(4\pi\varepsilon_0 r^2),$$

where $\vec{r}$ is the position vector of q relative to Q, and $\hat{r}$ is a unit vector in the direction of $\vec{r}$. What is the potential energy function for this system, taking the reference position of zero potential energy at $r = \infty$? (Answer page 49-19, Line 12)))

The term *conservative force* used in this section refers to what might better be called a macroscopically conservative force, that is, a force whose associated potential energy can be specified in terms of macroscopic coordinates alone. Gravitational and elastic forces are examples of conservative forces.

There are many situations in which one particle of a system moves relative to the rest of the system. In such a case the configuration of the system is specified by the position of the moving particle, so the potential energy is a function of the coordinates that specify this position. If the potential energy function is known, one can use calculus to compute the conservative internal force associated with it, and conversely if the force function is known one can use calculus to compute the potential energy. In this section we show how this is done for functions of one spatial coordinate.

((Suppose the particle moves from position x to a slightly different position $x + dx$, that is, it has a small displacement in the x direction. The work done on the particle by forces internal to the system is $F_x dx$, where F_x is the force exerted on the particle by the rest of the system. This work equals the decrease in potential energy, so we write

$$dPE = -F_x \, dx$$

or

★ $$F_x = -\frac{dPE}{dx} .$$

The force is the negative of the derivative of thepotential energy with respect to x. Using this equation, one can compute the force at every point if the potential energy is known as a function of x.

In the inverse problem the force is known as a function of position, and the potential energy is to be found. We do that in two steps. First, we arbitrarily choose some position x_0 at which the potential energy is assigned to be zero; this is called the *reference point*. Then to find the potential energy at some point x, we apply the definition of potential energy on p. 49-38. The *decrease* in potential energy as the particle moves from x_0 to x is the work done by the internal forces during this motion, so the *increase* in potential energy is the *negative* of this work:

★ $$PE(x) - PE(x_0) = -\int_{x_0}^{x} F_x dx .$$

Interchanging the limits on the integral is equivalent to computing the work as the particle moves in the opposite direction, from x to x_0; this is the negative of the work done as the particle moves from x_0 to x, so the potential energy at x can also be computed using

$$PE(x) = \int_{x}^{x_0} F_x \, dx .$$))

The reference point may be chosen arbitrarily because the absolute value of potential energy is not a quantity which can be observed experimentally. Only *changes* in potential energy can be observed. The change in potential energy as the system moves between two configurations does not depend on the choice of reference point. The force is computed from the potential energy by taking its derivative, and since the derivative of an additive constant is always zero the force function does not depend on the choice of reference point.

There are many physical situations in which a particle is able to move, but only within certain limits. An example is the bob of a simple pendulum,which moves back and forth between the end points of its motion but does not go beyond the end points. Other examples are the motion of a marble rolling inside a punch bowl, and a mass oscillating at the end of a coil spring. In each case, the kinetic energy of the particle is highest when it passes through the central point of its motion, but the potential energy of the system is a maximum when the particle is momentarily at rest at the end points. The energy relationships of such a system can be visualized using a graph of energy vs. position of the particle.

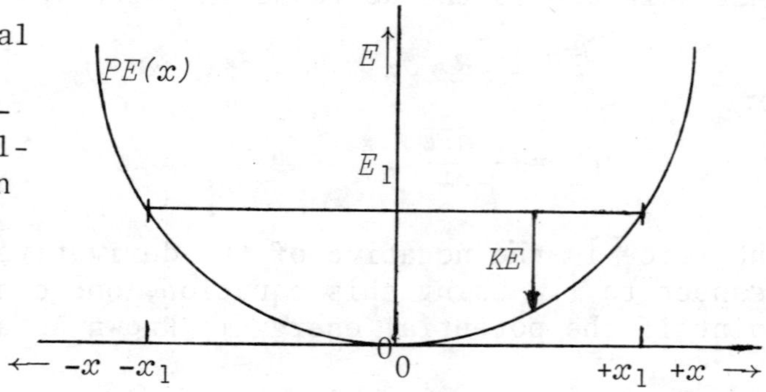

Figure 49-1. The gravitational potential energy of a simple pendulum varies with displacement of the bob from its equilibrium position $x = 0$ as shown on this graph of $PE(x)$ vs. x. When the total energy is E_1, the motion of the bob is confined between the turning points $-x_1$ and $+x_1$.

Figure 49-1 shows how the gravitational potential energy of a simple pendulum varies with position, taking the equilibrium position ($x = 0$) as the position of zero potential energy. The graph of $PE(x)$ vs. x is an arc of a circle, corresponding to the circular path of the pendulum bob. At the end points of its motion where the bob is momentarily at rest, all the energy of the system is potential energy. If the total energy of the system is E_1, the end points are $-x_1$ and $+x_1$, where the potential energy function $PE(x)$ has the value E_1. When the bob is between the end points, the difference between the total energy and the potential energy is the kinetic energy of the system. The kinetic energy is a maximum as the bob passes through its equilibrium psoition, and diminishes to zero as the bob approaches either end point. The moving particle (bob) cannot move beyond either end point; its motion is confined to the region where the kinetic energy is greater than or equal to zero, and it is customary to say the particle is confined in a *potential well*. Any physical system for which the motion is confined within a definite range of positions may be described using a potential well, depicting energy relationships similar to those listed above.

When the pendulum bob in our example has a positive displacement x, the internal force of the system tends to pull the bob back toward the equilibrium position, toward the left or in the negative direction. This negative force is closely related to the positive slope of the curve of $PE(x)$ vs. x; the connection is the equation derived earlier,

$$F_x = - dPE/dx.$$

When the bob's displacement is negative, the internal force is toward the right, in the positive direction; this positive force is associated with the negative slope of the potential energy curve in the region where $x < 0$. The increasing (positive or negative) slope of the curve as the displacement increases shows that the internal restoring force increases as the distance from equilibrium increases. Because the restoring force (the internal force associated with the potential energy) always pulls the bob back toward the equilibrium position, the bob oscillates about the position of minimum potential energy; this is another feature common to all cases of motion in a potential well. If there are no dissipative forces in the system, this oscillation continues indefinitely.

Questions and Exercises (Note: All these Exercises require calculus.)

Q 49.1-1 For which of the following motions are we justified in regarding all physical objects but one as fixed in space, for the purpose of computing the potential energy of the system?
- (a) The motion of a baseball hit to the outfield.
- (b) The motion of a spaceship as it travels from Earth to the Moon.
- (c) Two gliders equipped with strong magnets which repel each other, moving on an inclined air track.
- (d) The motion of the molecules in an enclosed container of gas.
- (e) The motion of a simple pendulum mounted on the moon.

Q 49.1-2 For the motions listed below, (1) state the origin and list the coordinates of a coordinate system suitable for writing the potential energy function; and (2) write the potential energy as a function of these coordinates.
- (a) The motion of a baseball hit to the outfield.
- (b) The motion of a simple pendulum mounted on the moon.

Q 49.1-3 State in words the origin of the minus sign in the equation $dPE = -F_x dx$.

Q 49.1-4 (a) What is the chosen reference position at which gravitational potential energy in space is usually set equal to zero?
(b) What is the reason for this choice?

CC E 49.1-1 For a particle moving only under the action of a conservative force, show that the work done by this force on the particle when it moves from point *1* to point *2* by any path *A* is the same as the work done by the force on the particle when it moves from point *1* to point *2* by any other path *B*.

E 49.1-2 For a particle moving only under the action of a conservative force, show that the net work done on the particle by the force during a round trip from some point *1* to any point *2* and back to point *1* by any combination of paths, is zero.

E 49.1-3 The restoring force exerted by a linear spring is $F_s(x) = -kx$.
(a) By integration, show that the potential energy of a distorted linear spring is $LPE = \frac{1}{2}kx^2$.
(b) By differentiation, check that the spring force equals $-dLPE(x)/dx$.

E 49.1-4 A force parallel to the x axis, described by $\vec{F}(x) = 1.0\ (\text{N/m}^3)\ \hat{x}x^3$, acts on a particle as it moves along the x axis.
(a) Compute the work done by this force *on* the particle as it moves from $x = -3$ meters to $x = 0$.
(b) Compute the work done by this force *on* the particle as it moves from $x = 0$ to $x = -3$ meters.
(c) What is the total work done by this force *on* the particle as it moves from $x = -3$ meters to $x = +3$ meters?
(d) Is this a conservative force?
(e) What is the change in potential energy of this system when the particle moves from $x = -3$ meters to $x = +3$ meters?

E 49.1-5 The charged spherical dome of an electrostatic generator repels a charged ping-ping ball with a radial force given by $F = q_1q_2/(4\pi\varepsilon_0 r^2)$, where q_1 and q_2 are the values of the two charges and r is the distance between the center of the dome and the center of the ball. Choose the center of the dome as the origin of coordinates, and an infinite separation as the reference point.
(a) Compute the work done by the force when the ball moves from a position r_1 to the reference position.
(b) How does the answer to part (a) depend on the path taken by the ball during this motion?
(c) Compute the electric potential energy of the system when the ball is at position r_1.
(d) If the reference position is chosen as r_2, what is the electric potential energy of the system when the distance is r_1?
(e) Is the difference between the answers to parts (c) and (d) a variable (dependent on r), or a constant?
(f) Rewrite your answer to part (d), replacing the symbol r_1 by r to indicate that this is a variable quantity. Then compute the force function associated with this potential energy.
(g) How does the constant term in the potential energy function of part (f) affect the expression for the associated force? How does this force compare with that given at the beginning of this exercise?

E 49.1-6 (a) Compute the potential energy $PE(x)$ associated with the force $F(x) = -bx^2$ for $x > 0$ and $F(x) = +bx^2$ for $x < 0$, where b is a positive constant. Assume $PE(0) = 0$.
(b) Sketch the potential well for this potential energy function.
(c) For which values of x is $F(x)$ in the $-x$ direction?
(d) For which values of x is $F(x)$ in the $+x$ direction?
(e) Suppose a particle having mass m moving in this well has zero speed at $x = 3$ m. What is the kinetic energy for this particle at $x = 0$?
(f) For the particle described in part (e), what is the kinetic energy at $x = -3$ m?
(g) Indicate on your sketch of the well the value of the total macroscopic energy of the particle described in part (e).
(h) Is the $F(x)$ a conservative force?
(i) For the particle of part (e), at what values of x does the kinetic energy equal the potential energy of the system?))

Check your understanding of the material above by testing yourself on the minimum learning goals at the beginning of this section.

49.2 The Work-Energy Principle. Conservation of Energy

Minimum Learning Goal. (1) Given a description of a simple situation in which (positive or negative) work is done on a mechanical system, be able to state for each form of energy associated with the system whether it increases, decreases, or remains constant. For example:

1 A carton of physics textbooks is raised by a freight elevator from the first floor to the third floor. For the system consisting of the textbooks, as it moves from the first floor level to the second floor level:
(a) Which forms of energy increase?
(b) Which forms of energy decrease?
(c) Which forms of energy remain constant? (Answer, p. 49-19, Line 15)

2 Hiawatha paddles his canoe across a lake at constant speed, doing work every time he pushes his paddle through the water. For the system consisting of the canoe and the lake:
(a) Which forms of energy increase?
(b) Which forms of energy decrease?
(c) Which forms of energy remain constant? (Answer, p. 49-18, Line 2)

3. At the construction site for a new highway bridge, a crane lifts a steel beam from a truck and raises it several feet, at the same time moving it to the side with a speed of 1.0 m/s. For the system consisting of the beam alone:

(a) Which forms of energy increase?
(b) Which forms of energy decrease?
(c) Which forms of energy remain constant?
(Answer, p. 49-18, Line 6)

(2) Given values of the energy changes in a system, be able to specify whether or not an external force is acting on the system, and when such a force is acting, be able to compute the work done on the system by that force. For example:

1 A boy and his sled at the top of a snow-covered hill have a gravitational potential energy of 5000 joules relative to the bottom of the hill. The boy and sled are released and move from rest, sliding down the hill. On the level ground some distance from the bottom of the hill, the boy-sled combination has a kinetic energy of 3000 joules. (a) Was a force other than gravity doing work on the boy-sled system? (b) What was its direction, if it existed? (c) How much work was done by this force? (Answer, p. 49-18, Line 4)

2 A baseball is thrown vertically upward with an initial kinetic energy of 6.0 joules. When it reaches its highest position, the gravitational potential energy is 5.8 joules relative to its initial position. (a) Was a force other than gravity doing work on the baseball? (b) If so, how much work was done by this force and what was its probable origin? (Answer p. 49-19, Line 10)

(3) Given adequate information on the external forces acting on a system during some change in the system, be able to calculate the change in total energy of the system. For example:

1 Eleven air pucks are in motion on a level frictionless air table at time t_1. Assume that no energy is exchanged when a puck strikes a side rail of the table. Each puck contains a magnet, so that it repels each other puck with a force which depends on the distance between magnets. A student uses a billiard cue to accelerate one of the pucks by striking it with an average force of 60 N over a distance of one centimeter. A few seconds later at time t_2, after the struck puck had 17 collisions with other pucks, and these in turn collided with others, each puck has an energy value different from its energy at t_1. What is the change in the total energy of the system of pucks from t_1 to t_2? (Answer, p. 49- 19 Line 13)

(4) Given a description of the forces acting in or on a physical system, be able to use energy concepts to account for or predict the motion of particles in the system. For example:

1 A simple pendulum has a length of 50 cm and a bob whose mass is 0.08 kg. The pendulum is released from rest at an initial

angle of 40° to the vertical. During its motion the only forces on the bob are the tension in the string and the weight of the bob. What is the speed of the bob at the points where the string makes an angle of 20° with the vertical? (Answer, p. 49-19, Line 14)

(5) Given a description of a simple mechanical system which has changes of energy within the system, be able to state which forms of energy are increasing, which are decreasing, and which are remaining constant. For example:

1 Two scouts are canoeing downstream, not paddling but riding with the current as the river slowly drops in elevation. Then they come to a backwater behind a dam, where the canoe loses speed and finally comes to a stop. For the system consisting of the canoe, the stream, and the earth, what forms of energy are increasing, which are decreasing, and which remain constant during the motion described above. Answer, p. 49-19, Line 8)

As we have seen, the work done on a physical system by the conservative *internal* forces of the system is equal to the decrease in the potential energy of the system. Work by *external* forces has a different effect on the system, so in making energy calculations it is necessary to distinguish clearly between internal and external forces. This requires that we clearly identify the *physical system* we are dealing with. Of all the physical objects in the universe, we need to select one or a few that we are interested in, and designate these as "the system" (Figure 49-2). The rest of the universe lies outside the system, for the purposes of analyzing the problem immediately at hand. An object outside the system can put work into the system by exerting on some part of the system a force which acts through a displacement. Conversely, an object *within* the system can do work on something *outside* the system; in accounting for the work, this would be equivalent to *negative work* done *on* the system.

An equation which enables us to account for the changes in energy which result from work put into the system begins with the term W_{in}. This represents all the work put into the system by forces acting *from outside* the system, which we call *external* forces; it does not include any work which may be done by forces acting wholly within the system. Any work done by *internal* forces is reflected as a change in some form of potential energy within the system.

Figure 49-2. In analyzing a physical problem, it is necessary to select the objects of immediate interest and designate one or more of these as the *system*. The rest of the universe lies outside the boundary of the system. The dashed line encloses the system in this figure.

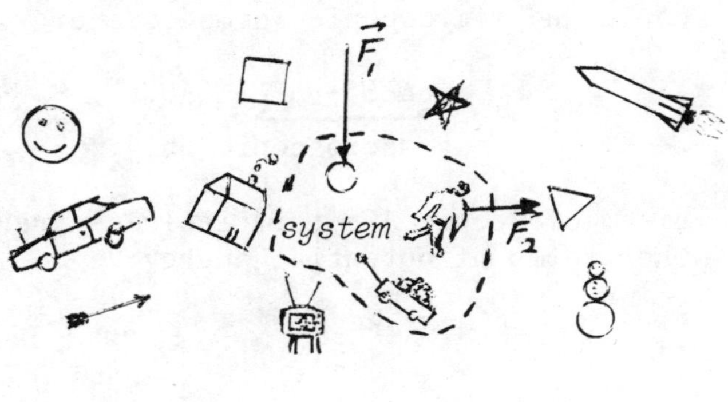

It is found by experiment that the work put into a system always shows up as a change in one or more forms of energy within the system. This law is known as the work-energy principle.

> *Work-Energy Principle:* The work done on a physical system by the external forces alone equals the increase in the total energy of the system.

Every kind of energy may be regarded as a form of either potential energy or kinetic energy, so an equation for the work-energy principle is

$$W_{\text{in}} = \Delta KE + \Delta PE. \qquad \text{Work-Energy Principle}$$

In this equation, *KE* refers to *both* macroscopic and microscopic forms of kinetic energy, and *PE* refers to *both* macroscopic and microscopic forms of potential energy.

When we are dealing with a macroscopic physical system, it is usually not difficult to measure or take into account each distinct macroscopic form of energy. Elastic potential energy is associated with the deformation of a spring or other elastic object, gravitational potential energy is computed using the masses and positions of objects, etc. The various microscopic forms of energy, however, cannot be measured or controlled separately. When the work put into a system goes into microscopic forms of energy, the experimenter cannot decide how much he will put into translational kinetic energy of the molecules, or rotational kinetic energy, or vibrational kinetic energy, or intermolecular potential energy. The partition of the work supplied into these various forms of energy is controlled by the quantum properties of the atoms and molecules which make up the system. For this reason, it is convenient for our present purposes to lump together all the microscopic forms of energy, both kinetic and potential, in the system. The term *internal energy* is used to refer to the sum of the kinetic energy and potential energy of all the individual microscopic particles of the system. This internal energy is not associated with the bulk motion or position of the system as a whole, but with the disorganized and random motions of the particles which make up the system. The interrelationships of work, internal energy and heat are studied in the branch of physics called *thermodynamics*, some aspects of which are presented in Chapters 22 and 23. In thermodynamics the symbol for internal energy is U, and we use this symbol here to include all the microscopic forms of energy in a system.

We can rewrite the work-energy principle to show separately the macroscopic and microscopic forms of energy:

★ $$W_{\text{in}} = \underbrace{\Delta KE + \Delta PE}_{\text{macroscopic only}} + \Delta U. \qquad \text{Work-Energy Principle}$$

This macroscopic ΔPE term includes gravitational, elastic, electric and other forms of potential energy:

★ macroscopic $\Delta PE = \Delta GPE + \Delta LPE + \Delta EPE + \ldots$

The sum of macroscopic kinetic energy and macroscopic potential energy is often referred to as *total mechanical energy*. This terminology goes back historically to the former idea that there were many kinds of energy: kinetic, potential, thermal, chemical, and so on. Of these, kinetic and potential energies were thought of as "mechanical," while the others were not. Actually however, on an atomic scale particles have only two kinds of energy: kinetic and potential. The so-called thermal energies or chemical energies are still made up of kinetic and potential energies on an atomic and molecular scale. The correct distinction is, accordingly, not between mechanical and non-mechanical energies, but between macroscopic and microscopic forms of energy.

In the preceding, we have considered how work put into a system from the outside leads to changes in the various forms of energy within the system. What if a system is effectively isolated from energy transfer by external forces, so that *no* work is put into the system? In that case, the work term in the work-energy principle equals zero. The compact form of this principle then becomes

$$0 = \Delta KE + \Delta PE, \qquad \text{if no work is put into a system.}$$

In using this equation, one must remember that *both macroscopic and microscopic* terms are included. If there is no change in the total energy of a system, then the equation above expresses the

★ *Law of Conservation of Energy:* The total energy of an isolated physical system is constant.

Even though no energy is transferred into the system from the outside, there may be transformations of energy from one form to another *within* the system. Examples of approximately isolated systems which have internal changes of energy are all around us. As the bob of a simple pendulum swings back and forth, the energy of the bob repeatedly changes from kinetic energy to potential energy and back to kinetic energy; in the absence of air drag the motion of the pendulum would continue indefinitely, but when drag is present we must enlarge the system under consideration because the energy of the bob is gradually transferred to internal energy of the bob and the surrounding air. When a ball is dropped so that it hits the floor and rebounds upward, the gravitational potential energy of the ball at its highest point is transformed to kinetic energy and then to elastic potential energy while the ball is in contact with the floor, then back to kinetic energy and then to gravitational potential energy again, after which the cycle repeats itself; because of air drag and internal friction within the flexing ball, the energy of the ball will gradually be transformed to internal energy of the ball and the surrounding air and floor. When a ball rolls down a slope the initial gravitational potential energy is transformed to kinetic energy, but this in turn will gradually become internal energy as the ball rolls to a stop on the level surface beyond the bottom of the slope.

In using an equation for the law of conservation of energy, it is often convenient to include separate terms for gravitational potential energy, elastic potential energy, and internal energy. Other forms of energy which are sometimes stated separately include chemical energy such as that possessed by fuel or explosives, and nuclear energy which resides in atomic nuclei. Both chemical energy and nuclear energy are associated with kinetic and potential energies on a microscopic level. On the macroscopic level, it is sometimes useful to define and keep account separately of electrical energy, magnetic energy, rotational energy, and other special forms which play an identifiable role in certain physical problems. In the further study of physics, these forms of energy are defined and used when it is advantageous to do so.

From the standpoint of science, the universe as a whole is an isolated physical system. It then follows from the law of conservation of energy that the total amount of energy in the universe remains constant. The law of conservation of energy may well be regarded as the most fundamental law of nature that science has discovered. As far as we know, it applies to every physical process that takes place anywhere in the universe.

Questions and Exercises

Q 49.2-1 Some of the forces associated with an automobile traveling around a banked curve on a highway are: (1) the resultant force exerted by driver's hands on the steering wheel; (2) the resultant force by the driver's seat on the driver; (3) the force by the driver on his seat belt; (4) the horizontal force by the highway on the tires; (5) the vertical force by the highway on the tires; (6) the force by the tires on the highway in a direction normal to the road surface; (7) the force by the rear axle and housing on the right rear wheel of the auto; (8) the weight of the car; (9) the force by the wind on the car including air resistance to the car's motion; (10) the gravitational force by the auto on the police car following it around the curve.

(a) For the system of consisting of this one auto (including its contents), which of the forces listed above are *internal* forces?

(b) For the system consisting of this one auto, which of the forces listed above are *external* forces on the auto?

(c) For the system consisting of the driver, which of the forces listed above are *internal* forces? Which are *external* forces?

(d) For the system consisting of the right rear wheel of the car, which of the forces listed above are *internal* forces? Which are *external* forces?

(e) For the system consisting of the car and the portion of the highway with which it is in contact, which of the forces listed above are *internal* forces? Which are *external* forces on the system?

Q 49.2-2 In Q 49.2-1 there is a list of forces associated with an automobile. For the system consisting of the auto and its contents, which of these forces could effect a change in the total energy of the system?

Q 49.2-3 Consider a simple pendulum moving at the end of a string whose support is rigidly fastened to the earth.

(a) For the system consisting only of the pendulum bob, what force(s) account(s) for the changes in kinetic energy of the swinging pendulum? Are these external, or internal, forces for this system? Which, if any, of these forces is associated with a potential energy of the system?

(b) For the system consisting of the pendulum, support, and the earth, what force(s) account(s) for the changes in kinetic energy of the swinging pendulum? Are these external, or internal, forces for this system? Which, if any, of these forces is associated with a potential energy of the system?

Q 49.2-4 List all the successive forms of energy which are involved when:

(a) You shoot a projectile upward from a spring-operated toy gun, and the shell lands in a pond.

(b) You buy some gasoline and drive your car to the top of Mt. Washington.

(c) Your car runs out of gas at 60 mph on a level road.

(d) An auto rolls downhill with its engine turned off.

Q 49.2-5 What forms of energy does your body acquire or use during a day's activity?

Q 49.2-6 Check all of the correct completions.
If an automobile runs out of gas while traveling at 70 miles per hour on the highway:

(a) the kinetic energy of the car will all be transformed into potential energy if the car is traveling uphill.

(b) None of the kinetic energy of the car will be transformed into thermal energy unless the driver applies the brakes.

(c) the speed of the car will *always* immediately begin to decrease.

(d) energy will be conserved regardless of what the driver does.

Q 49.2-7 Check all of the correct completions.
When an earth satellite is on the launching pad, its kinetic energy is zero. The kinetic energy of the satellite when it is in orbit comes from:

(a) the earth's gravitational force acting through the altitude h.

(b) gravitational potential energy given up by the booster rocket.

(c) chemical energy stored in the rocket fuel.

(d) the energy of many photons absorbed from sunlight by solar batteries.

E 49.2-1 A car having a weight of 9000 N (about 1 ton) is pushed along a level road by two sophomores exerting a steady combined horizontal force of 450 N, against a friction force of 450 N.

(a) If the speed of the car is 1.0 m/s at $t = 0$, what is the speed after one minute?

(b) How much work is done by the boys during this minute?

(c) Is the total mechanical energy of the car conserved during this minute?

E 49.2-2 Tom wanted to ride his motorbike straight up a steep hill having a slope of 40°, so he accelerated up to a speed of 27 m/s (60 mi/hr) on the level just as he reached the bottom of the hill. He then ran out of gas, but coasted up the hill as far as he could. The combined weight of Tom and his bike is 2000 N, and the force of friction on his bike as he rolled up the hill was 60 N.

(a) What was the resultant force on Tom-plus-bike as they went up the hill?

(b) How far up the hill was Tom able to go after he ran out of gas?

Check your understanding of the material above by testing yourself on the minimum learning goals at the beginning of this section.

49.3 Work Done by Friction

Minimum Learning Goal. (1) Be able to calculate the amount of energy dissipated, given a frictional force and a distance through which that force acts. For example:

1 A gymnast pulls himself 30 feet (9.15 m) up to the top of a rope; then he slides down, moving at a constant speed by taking advantage of the friction between his hands and the rope. If he has a weight of 748 N, how much energy is dissipated between his hands and the rope? (Answer, p. 49-18, Line 7)

2 A motorcycle rider falls off his bike when it is moving. He slides 120 feet (36.6 m), with the frictional force between his body and the pavement being 200 N. How much energy is dissipated between his body and the pavement? (Answer, p. 49-19, Line 9)

When a box is pushed along a level floor at constant speed, the pushing force is required to offset the opposing force of friction and maintain the constant speed. During this motion the pushing force puts work into the system consisting of the box and the floor, but neither the kinetic energy of the box nor the gravitational potential energy of the system changes. What kind of energy change results from the work put into the system?

The work leads to a change in position of the box, so at first glance it might seem that this work goes into a change in macroscopic potential energy as it does when a ball is lifted (Section 48.1) or a spring is compressed (Section 48.4). There is a fundamental difference in this case, however, because when the ball or spring is released it returns to its original position, whereas when the pusher lets go of the box it remains in the place where it was pushed. Although the work does not go into a macroscopic form of energy, it is found that because of friction the box and floor have become warmer where they rubbed past each other. This indicates that the work done in opposing friction has gone into internal energy of the molecules in the box and in the floor. From the macroscopic point of view it is sometimes said that the work has been lost or *dissipated,* because it is no longer available as useful mechanical energy, but from the microscopic point of view the work has gone into increased kinetic and potential energy associated with the random motions of these molecules.

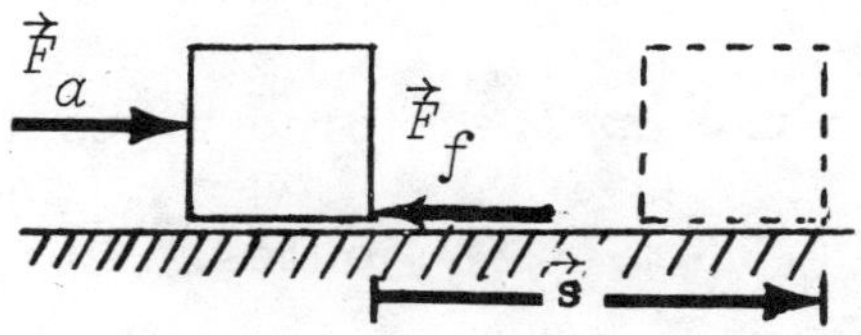

Figure 49-3. A box is pushed on a horizontal floor by an applied force which is just balanced by an opposing friction force. The applied force does work on the box-floor system.

Figure 49-3 shows a box being pushed through a displacement $\vec{s}$ on a horizontal floor. The constant applied force $\vec{F}_a$ is opposed by a friction force $\vec{F}_f$ which is equal in magnitude, so the box moves with constant velocity. The work done by $\vec{F}_a$ is $W = F_a \cdot s$, but since F_a equals F_f, this work is also given by $F_f \cdot s$.

★ The *increase in internal energy* ΔU which results when a friction force F_f acts through a displacement s is $\Delta U = F_f \cdot s$.

When the friction force is not constant, the work which goes into internal energy must be computed using the integral $\int F_f(s)\, ds$ over the path where the friction is present.

To gain a clearer picture of internal energy, we need to remember that both the box and the floor consist of an immense number of molecules. Viewed at the molecular level, the friction forces between the floor and the box are the resultants of a very large number of electric forces between the molecules in the floor and the molecules in the box. As the box moves along the floor these forces alter the motions of the molecules, giving them added kinetic energy and potential energy. The molecules are moving relative to one another "every which way" in a random manner; this *disorganized motion* of the molecules is in addition to the *organized motion* of the box as a whole (including its molecules) as it moves along the floor. This leads us to a qualitative definition of internal energy:

> The *internal energy* of a system is the sum of the kinetic and potential energies carried by the microscopic particles constituting the system in their random motions within the system.

In the branch of physics called *statistical mechanics* the thermal properties of matter are related to the microscopic behavior of the molecules of the system. Chapter 23 is an introduction to statistical mechanics at an elementary level.

Worked Example 49-1

A bus-boy in the student union cafeteria pushes a cart of dirty dishes for 16 meters with a horizontal push of 52 newtons. The opposing friction force is 47 newtons.

(a) How much work does he do during this motion?
(b) How much internal energy is developed in the cart and the floor?

Solution

(a) The work done is the *applied* force times the displacement:

$$W_{in} = 52 \text{ N} \times 16 \text{ m} = 832 \text{ joules.}$$

(b) The internal energy developed is *not* simply equal to the work input, because the cart is not moving with constant velocity but rather is accelerating; we know this because the applied force is greater than the friction so the net force causes the cart to accelerate. The internal energy produced equals the *friction* force times the displacement:

$$\Delta U = 47 \text{ N} \times 16 \text{ m} = 752 \text{ joules.}$$

In this example, the work which does not go into internal energy goes into increasing kinetic energy of the cart.

A *nonconservative force* is defined as one capable of doing work but with which no potential energy can be associated; an example is friction. In reality, there are no nonconservative forces; that's what we mean when we say that conservation of energy is a law of nature. There are *seemingly* nonconservative forces, however; they appear in macroscopic physics because we choose to use an incomplete specification of the configuration of the system, by ignoring the microscopic molecular motion. When we leave out some of the coordinates that describe the system and expect the potential energy to be a function only of the remaining ones, it is not surprising that we do not succeed in our expectations!

Sometimes we encounter a situation where a system moves under the action of forces all of which are macroscopically conservative. In that case, *macroscopic energy* -- that is, kinetic energy specified in terms of macroscopic velocities, plus potential energy which is a function of macroscopic coordinates -- is conserved. This fact can be a great help in solving problems, and therefore the ability to recognize such a situation is very important.

Total macroscopic energy may be a useful concept even in some situations where it is not conserved. Its changes must then be attributed to work done by nonconservative forces. For example, a pendulum set swinging will slowly lose total macroscopic energy because of work done by friction.

Questions and Exercises

Q 49.3-1 If someone claimed that, as a swinging pendulum slowly loses its motion, the energy has not disappeared but has "gone to heaven," most people would consider this a scientifically empty statement. Why is the statement that the energy has not disappeared but has "gone into internal energy" any different?

Q 49.3-2 If someone claimed to have found a special paint with which he can coat the top of a level table such that a book sliding on it speeds up instead of slowing down, would you believe him? Why, or why not?

Q 49.3-3 Give examples of energy transformation processes in which friction is not present.

Q 49.3-4 Which of the forces listed in Q 48.5-9 are conservative forces? Which are non-conservative forces?

Q 49.3-5 What is meant by a "dissipative force?"

Q 49.3-6 When macroscopic kinetic energy is dissipated by friction, e.g., when the amplitude of oscillation of a pendulum gradually decreases, in what microscopic form(s) does the "lost" energy appear?

Q 49.3-7 (a) Which are the most common non-conservative forces which you have encountered in experimental work in physics? (b) What could be done to eliminate them from laboratory experiments, so that you could study the behavior of a macroscopically conservative system?

E 49.3-1 The tires of a car are subjected to a constant total frictional force of 400 newtons while the car rolls to a stop in a straight line through a distance of 90 meters. How much energy is dissipated during this motion?

E 49.3-2 A horizontal applied force of 3 newtons pushes a box of No-Nods in a straight line on a horizontal table through a distance of 0.8 meter in 4 seconds. The motion of the box is opposed by a friction force of 1.2 newton.
(a) How much work is done by the applied force?
(b) How much internal energy is developed through friction?

E 49.3-3 A small electronic calculator is pushed for 5 seconds in a straight line across a horizontal table by a horizontal applied force of 0.8 newton. The motion is opposed by a friction force of 0.2 newton. The calculator moves 0.5 meter during this five-second interval. Compute the internal energy produced by friction.

E 49.3-4 A box having a mass of 100 gm slides with an initial horizontal velocity of 20 cm/sec onto a horizontal surface where the friction force on the box is 4.0×10^{-3} N. How far does the box slide before coming to rest?

E 49.3-5 Ben starts from rest at the top of a slide and slides to the bottom, going off the end with a horizontal velocity $\vec{v}_f$. Ben's weight is 196 N. His starting position is 3 m higher than the bottom of the slide.

(a) Compute the *decrease* in Ben's potential energy during this descent on the slide.

A frictional force having a constant magnitude of 40 N opposes Ben's motion along his 5.70-meter path on the slide.

(b) How much energy was dissipated?
(c) If Ben is not pushed at the start, what is his kinetic energy at the end of the slide?
(d) What is his final speed?
(e) If he receives an initial push (parallel to his motion) of 500 N for 0.26 m, what is his final speed?
(f) His initial *forward* velocity was zero; his final *forward* horizontal velocity is $\vec{v}_f$. What is the direction of his total change in velocity, $\Delta\vec{v}$, during this motion?
(g) What is the direction of his acceleration, a, during *this* Δv?
(h) What is the direction of the force required to produce *this* acceleration?
(i) What object exerts *this* force on Ben, if he receives no push? (It is *not* the Earth!)
(j) In the spirit of Newton's third law, what is the reaction to this force? What effect does it have on the slide?

Check your understanding of the material above by testing yourself on the minimum learning goal at the beginning of this section.

Answers to Minimum Learning Goals

Line 1 (a) $F(x) = -220$ (N/m) x + 22 N; (b) $F(0.1 \text{ m}) = 0$.

Line 2 (a) The wave energy of the water (a combination of kinetic and gravitational potential energy) increases, but this is transformed to internal energy as the waves die away.
(b) None; (c) kinetic energy of the canoe, and other forms of energy.

Line 3 (a) $F(y) = -9.5 \times 10^{-29}$ $(\text{N}\cdot\text{m}^2)$ y^{-2} + 7.7×10^{-88} $(\text{N}\cdot\text{m}^8)$ y^{-8};
(b) $y = 1.42 \times 10^{-10}$ m.

Line 4 (a) yes; (b) opposite to the sled's motion; (c) -2000 J

Line 5 Yes

Line 6 (a) Gravitational potential energy and kinetic energy; (b) none;
(c) all other forms of energy.

Line 7 6.84×10^3 J

Line 8 Kinetic energy and gravitational potential energy decrease, internal energy increases, and other forms of energy remain constant.

Line 9 7.32×10^3 J

Line 10 (a) yes; (b) -0.2 J; air drag

Line 11 No

Line 12 $EPE(r) = Qq/(4\pi\varepsilon_0 r)$

Line 13 +0.6 joule

Line 14 1.30 m/s

Line 15 (a) Gravitational potential energy and kinetic energy increase; (b) none; (c) all other forms.

Line 16 Yes

ANSWERS to Questions and Exercises in Chapter 49

Section 49.1

Q 49.1-2 (a) Origin could be at home plate or anywhere on the playing field; use z for the vertical coordinate; $GPE = mgz$.
(b) Origin could be at the equilibrium position of the pendulum bob; use z for the vertical coordinate. $GPE = mgz$, where g is the acceleration of free fall at the moon's surface.

Q 49.1-4 (a) A position or configuration in which the interacting masses are infinitely far apart.
(b) This choice leads to the simplest expression for the GPE.

E 49.1-4 (a) -81/4 J; (b) +81/4 J; (c) zero; (d) yes; (e) zero.

E 49.1-6 (a) $PE(x) = bx^3/3$ for $x \geqq 0$, and $PE(x) = -bx^3/3$ for $x \leqq 0$.
(b)

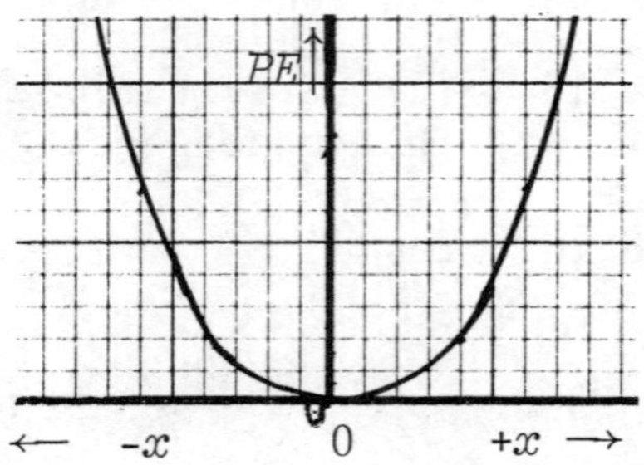

(c) for $x > 0$;
(d) for $x < 0$; (e) 9 b joules;
(f) zero; (h) yes; (8) ±2.381 m.

Section 49.2

Q 49.2-2 (4), (5), (8), (9).

Q 49.2-4 (a) LPE of the gun spring → KE → GPE → KE → U
(b) Chemical energy → U in the engine → KE → U (of car and road) + GPE.
(c) KE → U
(d) GPE → KE + U of car and road

Q 49.2-6 (d)

E 49.2-2 (a) 1.35×10^3 N, downward along the slope of the hill;
(b) 55.1 m along the slope.

Section 49.3

Q 49.3-2 For this claim to be true, the paint would need to continuously supply energy to any book sliding on the table. This is not believable.

Q 49.3-4 Conservative forces are (a), (c), (d), (f); the others are nonconservative.

Q 49.3-6 Dissipated energy goes into internal energy of atoms and molecules.

Q 49.3-2 (a) 2.4 J; (b) 0.96 J.

Q 49.3-4 50 cm

CHAPTER 50 IMPULSE AND MOMENTUM

Introduction

Our study of motion began in Chapter 41, where we met the physical variables and some mathematical methods which are useful in describing motion. The explanation of motion in terms of the forces acting on moving objects was discussed in Chapter 42. In Chapters 47-49 we studied the concepts of work and energy, and saw how the approach to motion from the standpoint of energy provided new insight and enabled us to solve some problems in motion which could not be solved using the methods we studied earlier. Now we are ready to examine motion and the interaction of moving bodies from still another point of view. In this chapter we will meet new concepts and a fundamental law of nature which provide another vantage point for the study of motion.

When you have completed your study of the material in this chapter, you should be able to answer questions and work exercises like the examples in the minimum learning goals assigned by your instructor.

50.1 Momentum

Minimum Learning Goals. (1) Be able to recall from memory and state the defining equation for the momentum of a material particle, telling what quantity is represented by each symbol and the units in which it is expressed. (Answer, p. 50-20, line 12)

(2) By recalling and using the definition, calculate the magnitude of the momentum of an object (regarded as a particle) from a description of the object and of its motion. For example:

1 Compute the linear momentum of the earth in its orbital motion (using data in Appendix G). (Answer, p. 50-20 line 6)

2 Compute the momentum of a proton moving at one-tenth the speed of light (using data in Appendix F). (Answer, p. 50-20 line 2)

(3) Be able to use the definition of momentum to determine the magnitude and direction of the momentum of an object, given a description of the object and its motion. For example:

1 A particle having a mass of 1.2 grams follows the path shown below; its speed is shown at different points on the path. What is the magnitude and direction (measured in degrees referenced to east) of its momentum at the point marked D? (Answer, p. 50-20 line 10)

point	Speed
A	0
B	5 m/s
C	20 m/s
D	12 m/s
E	10 m/s
F	6 m/s

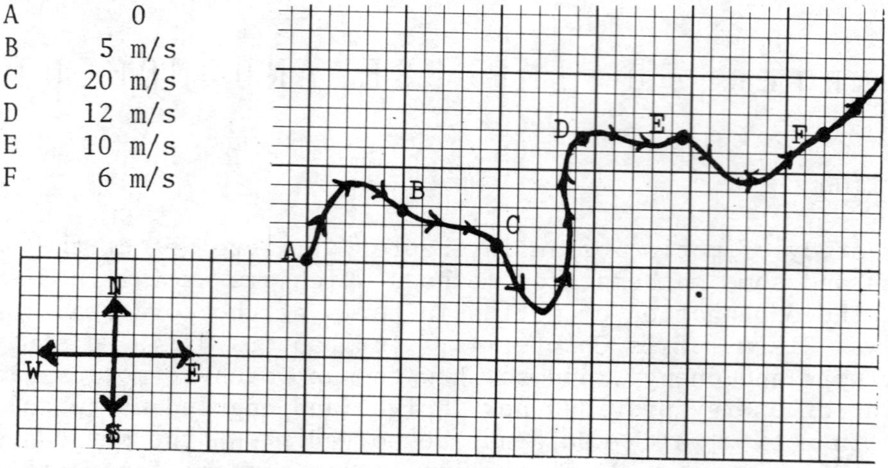

Most of us have observed collisions in which one physical object bumps into another one. A billiard game, for example, offers a fine opportunity to observe collisions between objects (the billiard balls) having equal mass. In a football game there are other collisions as the players are blocked and tackled. And on the roads there are thousands of collisions between vehicles, and sometimes a vehicle and a pedestrian.

Collisions between vehicles allow us to develop some intuition about the physical factors which are important in determining what happens to the objects which collide with each other. Let us consider two different collisions between a heavy truck and a light-weight sub-compact automobile, assuming that the vehicles have bumpers which are at the same height above the roadway so that the bumpers meet each other when the vehicles collide. Firstly, suppose that the truck is stopped at a traffic light and the car is traveling at 30 miles/hour when it hits the truck in the rear. In a collision like this, based on our experience we would expect the truck to move forward very little, if at all, as a result of the impact, but we would expect that the car might bounce backward after it receives some damage in its front end. Secondly, suppose that it is the car which is stopped for the traffic light when the heavy truck (traveling at 30 miles/hour) fails to stop and strikes the car in the rear. In this case, the impact on the car would send it ahead for some distance while the truck would continue forward after the impact with only slightly diminished speed.

In comparing these two collisions, what factors account for the differences in motion of the car after the collisions, and for the differences in motion of the truck after the collisions? The major differences lie in the speeds and the masses (or weights) of the two vehicles. It should be clear from consideration of collisions such as these that both the mass and the velocity of colliding objects play a part in determining the outcome of the collision.

When quantitative measurements are made on colliding objects, some of which will be described in the next section, it is found that the product of the mass times the velocity of an object is a key physical property in collisions. The name given to this property is *momentum*, with the symbol $\vec{p}$.

★ The (linear) *momentum* $\vec{p}$ of a material object having mass m and velocity $\vec{v}$ is $\vec{p} \equiv m\vec{v}$.

The SI unit for momentum is kg·m/s; it is not given a special name. The plural of momentum is *momenta*. This quantity is sometimes called *linear momentum* to distinguish it from a different quantity called angular momentum, which is used in the study of rotational motion. The outcome of a collision clearly depends on the *directions* of motion of the colliding objects, which requires that momentum be defined as a vector quantity.

Worked Example 50-1

Craig is riding northwest on his motorcycle which has a mass of 420 kilograms. His speed is 27 meter/second (about 60 mph). What is the momentum of the motorcycle?

Solution

Using the definition of momentum we compute

$$\vec{p} = m\vec{v} = 420 \text{ kg} \times 27 \text{ m/s} = 1.13 \times 10^4 \text{ kg·m/s, toward the northwest.}$$

It may be mentioned here that there is also a linear momentum associated with non-material particles called photons (Chapter 30); this momentum is defined in a different way from the momentum of a material particle, but it obeys the same laws.

Exercises

E 50.1-1 A golf ball (m = 0.025 kg) has a speed of 60 m/s immediately after being hit. Calculate the magnitude of the momentum of the ball at this instant.

E 50.1-2 A 0.2-kg glider is moving initially to the right on an air track with a speed of 0.5 m/s. It strikes the end and rebounds with a final velocity equal in magnitude but opposite in direction to the initial velocity. (a) Calculate the magnitude and direction of the *initial* momentum of the glider.
(b) Calculate the magnitude and direction of the *final* momentum of the glider.
(c) Calculate the total change in the momentum of the glider.

E 50.1-3 Compute the momentum of a baseball (mass = 140 grams) moving due north with a speed of 30 meters/second.

E 50.1-4 Compute the momentum relative to Earth of an astronaut (mass = 75 kg) in an Apollo spacecraft moving toward the moon with a speed of 3 km/sec relative to Earth.

Check your understanding of the material above by testing yourself on the minimum learning goals at the beginning of this section.

50.2 Conservation of Momentum

Minimum Learning Goals. (1) Be able to state in words the law of conservation of momentum. (Answer, p. 50-20, line 8)

(2) Be able to use the law of conservation of momentum to determine the speed of one or the other of two bodies immediately after a head-on collision between the two, given the masses of both, their initial speeds, and sufficient information about the after-collision speed of one or both of them. For example:

1 The driver of a 7000-kg truck approaching an intersection sees a 1000-kg car running a red light. The drivers hit the brakes, but the truck skids on at a speed of 16 m/s, hitting the car (momentarily stopped) broadside. The two vehicles are entangled and they continue together. What is the speed of the wrecks immediately after the collision? (Answer, p. 50-20, line 5)

2 A 2.0-kg puck moving at 5 m/s on an air table approaches a 0.5-kg puck at rest in what will be a head-on collision. The pucks have a gummy substance on their edges so that they stick together upon colliding. What is the speed of the pucks immediately after the collision? (Answer, p. 50-20, line 9)

What physical principles apply to collisions? What natural law governs exchanges of momentum during collisions? In order to answer these questions, we need some data obtained from experiments on collisions. As we found in Chapter 42,the presence of friction can have a major effect on the motions of objects and it can easily obscure the fundamental principles which govern motion. In order to minimize the effects of friction, we consider some simple collision experiments which can be carried out using linear air tracks in the laboratory.

Experiment A

Figure 50-1. Two gliders which are initially at rest on an air track spring apart from each other. The gliders acquire equal momenta, but in opposite directions.

In Experiment A (Figure 50-1) two gliders are both at rest before the interaction, held in a position where they are in contact with each other and with a small spring compressed between the gliders. While this is not a "collision" experiment in the usual sense of the word, it is an experiment where two objects (the gliders) interact with one another and impart momentum to each other. While the spring is compressed the two gliders exert forces on each other, and when the gliders are released simultaneously they will move in opposite directions. Careful measurements show that after the gliders move apart their speeds are inversely proportional to their masses. In symbols, the results of the experiment show that

$$\frac{m_A}{m_B} = \frac{v_B}{v_A}, \quad \text{or} \quad m_A v_A = m_B v_B.$$

These equations describe the outcome of the experiment, regardless of the values of the masses of the two gliders.

Experiment B

Figure 50-2. A collision in which glider *C* strikes glider *D*, initially at rest. The combined momentum of gliders *C* and *D* after the collision equals the initial momentum of glider *C* before the collision.

In Experiment B (Figure 50-2) one glider *(D)* is at rest before the collision while the other glider *(C)* is moving with speed v_o toward the one at rest. During the collision the two gliders lock together, and afterward they move as a unit with speed v_f. Measurements of the masses and speeds lead to the following equation for this collision:

$$m_C v_o = (m_C + m_D) v_f \ .$$

Once again, the equation holds for any values of m_C, m_D, and v_o. As the equation shows, the value of v_f is determined by the other three variables and by the law for momentum which lies behind this equation. We now want to deduce this law from the results of these two collisions.

Experiments A and B are quite different, but the equations obtained from these experiments are similar in that the terms in these equations are momentum terms. In order to transpose these two equations into a form which reflects the law of momentum, we will do four things to each equation. To exhibit the similarity of the two equations we will derive, we develop the equation for Experiment A on the left side of the page, and the equation for Experiment B on the right side of the page.

(1) Our first step is to distinguish momentum terms *before* the collision from momentum terms *after* the collision:

For Experiment A	For Experiment B
$m_A v_A(\text{after}) = m_B v_B(\text{after})$	$m_C v_o(\text{before}) = (m_C + m_D) v_f(\text{after})$.

(2) The next step is to express the speeds as *velocities*, which means not only that we write a vector symbol ($\vec{\ }$) above each v, but also that we take note of the *direction* of each velocity. We will consider velocities directed toward the right as positive, and velocities directed toward the left as negative. Doing this for each equation, we obtain:

$$-m_A \vec{v}_A(\text{after}) = m_B \vec{v}_B(\text{after}) \qquad m_C \vec{v}_B(\text{before}) = m_C \vec{v}_f(\text{after}) + m_o \vec{v}_f(\text{after}).$$

(3) Next, we write each momentum term using the symbol $\vec{p}$ instead of $m\vec{v}$:

$$-\vec{p}_A(\text{after}) = \vec{p}_B(\text{after}) \qquad \vec{p}_C(\text{before}) = \vec{p}_C(\text{after}) + \vec{p}_D(\text{after}).$$

(4) Finally, we rewrite each equation with all the "before" terms on the left side, and all the "after" terms on the right side:

$$0 = \vec{p}_A(\text{after}) + \vec{p}_B(\text{after}) \qquad \vec{p}_C(\text{before}) = \vec{p}_C(\text{after}) + \vec{p}_D(\text{after}).$$

Looking at the equation for Experiment A, we recall that both gliders were at rest before the collision and therefore the total momentum of the gliders before the collision was zero; the left side of this equation, therefore, represents the total momentum of the system before the collision. The same is true of the equation for Experiment B, because glider *D* was at rest before the collision. We thus arrive at the following result, which applies equally to both of the collisions we have studied:

★ $$\Sigma \vec{p} \text{ (before)} = \Sigma \vec{p} \text{ (after)}$$

This equation states the law which governs exchange of momentum, one of the physical principles which controls the outcome of collisions. It is called the

> *Law of Conservation of Momentum:* If the external force on a physical system is zero, then the total momentum of the system remains constant with time.

Certainly, we cannot claim to have established the general validity of this law on the basis of only two experiments. The basis on which this principle is now accepted as a general law of nature is many thousands of experiments involving all kinds of interactions between physical objects.

Worked Example 50-2

A small glider having a mass of 200 grams and a speed of 32 cm/s strikes a stationary glider having a mass of 600 grams on a linear air track. The small glider has a spring clip which clamps onto the large glider during the collision, so that after the collision the two gliders move as a unit. What is the speed of the combination after the collision?

Solution

The description of the collision matches the figure in Experiment B. It is convenient in this example to work out the calculations in units of grams and cm/s, rather than changing to SI units. The total momentum of the system before the collision is:

$$p = m_C v_o = 200 \text{ g} \times 32 \text{ cm/s} = 6.4 \times 10^3 \text{ g·cm/s}.$$

After the collision this momentum is shared by the two gliders, which have a combined mass m of 800 grams. The final momentum of the system equals mv. Solving this equation for v, we obtain

$$v = \frac{p}{m} = \frac{6.4 \times 10^3 \text{ g·cm/s}}{800 \text{ g}} = 8.0 \text{ cm/s}.$$

The direction of the final velocity will be the same as the direction of the initial momentum of the small glider.

Not only is momentum a useful concept in studying collisions, but momentum turns out to be a very special kind of physical quantity, one which is *conserved* in collisions, and for that matter, also in every other kind of physical interaction. Momentum thus joins charge and energy in the elite category of physical quantities which obey universal conservation laws. As we shall see in Section 50.5, the conservation laws of energy and momentum together are very powerful in analyzing and predicting the results of collisions.

Questions and Exercises

Q 50.2-1 (a) State the law of conservation of momentum, in words and in equation form.
(b) For what physical situations is this law valid?

Q 50.2-2 If a man sits in a stationary car and pushes on the dashboard, can he make the car move? How is this different from sitting on a bicycle and pushing on the pedals?

Q 50.2-3 In terms of momentum, explain why a man standing still on the middle of a *frictionless* frozen lake would have a hard time getting off the lake without help from the shore. There is, however, a way for him to get to shore without help. What is it?

Q 50.2-4 (a) Could a sailboat be made to move by mounting a large (battery-operated) electric fan on deck and aiming it at the sail?
(b) Could the boat be made to move by aiming the fan over the stern? Analyze the momentum exchange in each case.

Q 50.2-5 Explain how the linear momentum of the earth-moon system can remain (nearly) constant over a period of one month. Why doesn't the momentum of the earth-moon system remain *exactly* constant?

Q 50.2-6 What motion does an originally stationary but untethered rowboat have when you walk from one end of the boat to the other? Explain.

Q 50.2-7 List several examples of situations in which the momentum of an object is not constant. Are these violations of the Law of Conservation of Momentum? Explain.

E 50.2-1 A 0.20-kilogram glider moving at 0.4 meter/sec on a linear air track collides with a 0.40-kilogram glider initially at rest. If the two gliders stick together, what is their final speed after the collision?

E 50.2-2 An attorney asked a physicist to help him prepare a court case. The attorney's client was driving a sedan at night in a 55-mph speed zone and hit a station wagon (which was making a U-turn on the highway and which, at the instant of impact, had no motion parallel to the highway) broadside. From the skid marks of the tangled wreckage of the two cars after the collision it was estimated that the speed of the cars just after impact was 35 miles per hour. Each car had a mass (including occupants) of 2000 kg. What was the speed of the sedan just before the crash?

E 50.2-3 When an Apollo spacecraft is launched toward the moon by a Saturn V rocket, the 750-ton (6.8×10^5 kg) vehicle will acquire an upward speed of 6000 mph (2700 m/s) at burnout of the first stage of the rocket engine. What is the change in velocity of the earth when the vehicle is launched?

E 50.2-4 One way (in fact, the most fundamental way) of measuring the mass of a particle is by having it interact with another particle whose mass is known. We measure the change in velocity of each particle; and since the total momentum of the two must remain constant, the ratio of masses must (except for sign) be the inverse of the ratio of velocity changes. Note that this measurement can be done without any knowledge of the actual law of force that governs the interaction. One might well feel, however, that the whole procedure is meaningless, in that we are now defining the mass as that quantity which makes the total momentum remain constant. If that is the case, the law that momentum remains constant turns into a mere definition without physical content.

Carefully discuss why this is not so. In other words, even though we define mass by such a measurement, show in the following way that the law that total momentum is conserved is capable of predicting things we could not know without it.

(a) Consider a series of collisions between a set of three objects having masses m_1, m_2 *and* m_3. The mass m_1 may be arbitrarily defined as the unit of mass. The mass m_2 is then found by measurements on a collision between m_1 and m_2. Let $\Delta\vec{v}_1$ be the change in velocity of m_1 in this collision, and $\Delta\vec{v}_2$ be the change in velocity of m_2 in this collision. Using the law of conservation of momentum, show that

$$m_2 = -m_1 \frac{\Delta\vec{v}_1}{\Delta\vec{v}_2} .$$

In this equation, which quantities are measured in the experiment? Which quantities are calculated after the experiment?

(b) Now consider a collision between m_1 and m_3 and use the law of conservation of momentum to show that:

$$m_3 = -m_1 \frac{\Delta\vec{v}_1}{\Delta\vec{v}_3} .$$

In this equation, which quantities are measured in the experiment? Which quantities are calculated after the experiment?

(c) Finally, consider a collision between m_2 and m_3. Again use the law of conservation of momentum to derive a similar equation for this collision. Using the results of parts (a) and (b), which quantities in this equation are known *before* the experiment? What quantity can be predicted *before* the experiment? Could this quantity be predicted without using the law of conservation of momentum?

Check your understanding of the material above by testing yourself on the minimum learning goals at the beginning of this section.

50.3 Impulse

Minimum Learning Goals. (1) Be able to state the defining equation for impulse, telling what physical quantity is represented by each symbol used and the units in which each is expressed. (Answer, p. 50-20, line 1)

(2) Be able to calculate the impulse transmitted to some object when you are given the average force exerted and the time for which the force acts. For example:

1 A model-rocket engine has an average thrust of 7 N which persists for 0.17 s. What is the total impulse delivered by this rocket engine? (Answer, p. 50-20, line 13)

2 A jet engine exerts an average thrust of 100,000 pounds (445,000 N) for 10 seconds. What impulse is delivered to the plane by this engine? (Answer, p. 50-20, line 4)

(3) Be able to compute the impulse applied to an object by obtaining data from a graph of force versus time; then be able to use that impulse to find the change in speed of the object, given its mass. For example:

1 A model-rocket engine has a thrust represented by the graph of force versus time given below.
(a) Find the impulse. (b) Given that the mass of the rocket is 3.1 kg (ignoring the mass of the fuel ejected), find the resulting change of speed of the rocket. (Answer, p. 50-20, line 11)

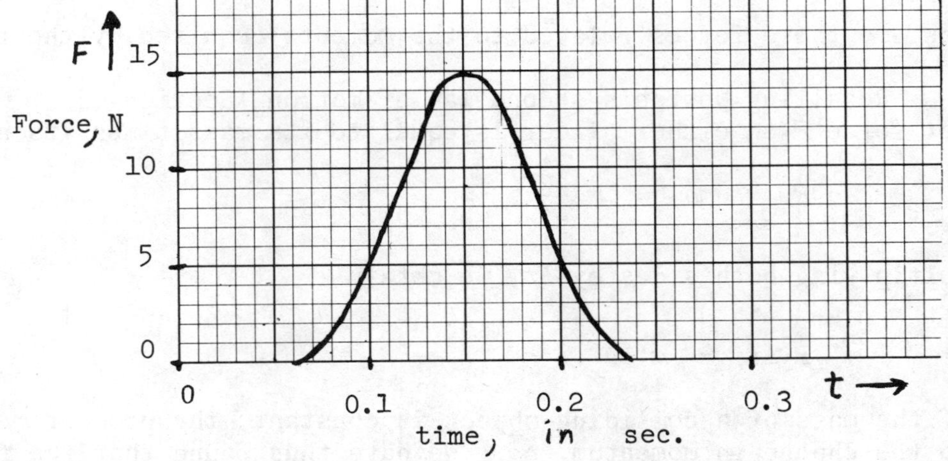

2 A jet engine has a thrust represented by the graph of force versus time given below. (a) Find the impulse. (b) Given the mass of the airplane as 8.9×10^4 kg (ignore the mass of the fuel ejected), what is the change of speed of the airplane? (Answer, p. 50-20, line 3)

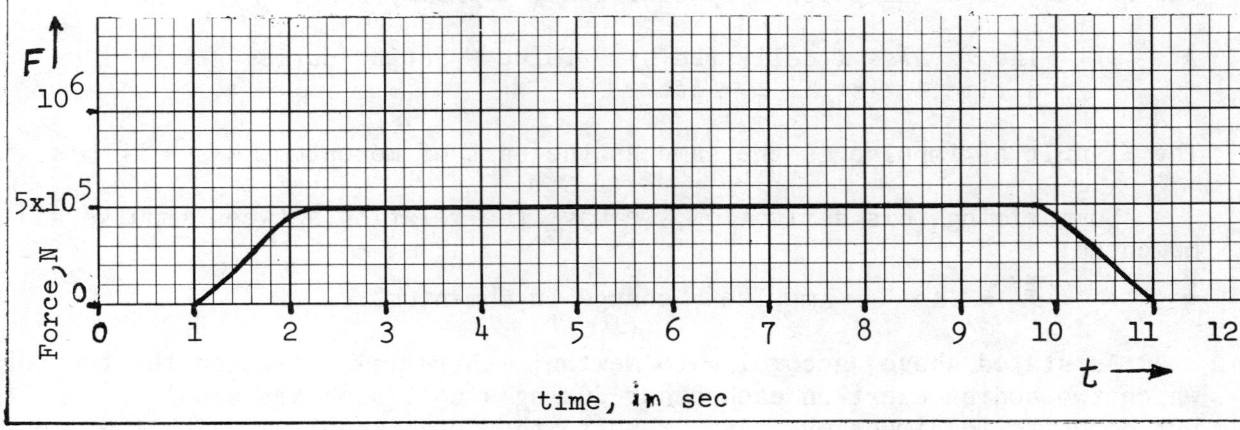

In Section 50.2 we have been studying collisions "from the outside", so to speak. Now let us look at a collision "from the inside". What happens at the moment of impact? What causes the colliding objects to stop moving toward each other and to begin moving apart from each other?

We know from our study of dynamics that whenever an object changes its state of motion, a force is required. In Experiment A of Section 50.2, when gliders *A* and *B* began to move it was because glider *A* was exerting a force $\vec{F}_{AB}$ on glider *B*, and glider *B* was exerting a force $\vec{F}_{BA}$ on glider *A*.

Newton's third law of motion (Section 42.5) ensures that these two forces will be equal but opposite at every instant of time during the interaction:

$$\vec{F}_{AB} = -\vec{F}_{BA} .$$

How are these forces related to the momenta acquired by the two gliders?

Recalling Newton's second law of motion (Section 42.3), we see that the net force $\vec{F}$ on either glider is equal to its mass times its acceleration:

$$\vec{F} = m\vec{a} = m\frac{\Delta\vec{v}}{\Delta t} .$$

Multiplying both sides by Δt, we obtain

$$\vec{F}\cdot\Delta t = m\cdot\Delta\vec{v} .$$

If the mass of a colliding object is constant, the product $m\Delta\vec{v}$ is just equal to the change in momentum, $\Delta\vec{p}$. We have thus found that the force acting on one of the colliding objects, multiplied by the time interval during which it acts, equals the change in momentum of the object. In equation form, this relation is

$$\vec{F}\cdot\Delta t = \Delta\vec{p} .$$

The quantity $\vec{F}\cdot\Delta t$ is given a special name, *impulse*, symbol $\vec{I}$.

★ The *impulse* $\vec{I}$ delivered by a force $\vec{F}$ acting during a time Δt is $\vec{I} \equiv \vec{F}\Delta t$.

The SI unit of impulse is the same as the unit of momentum, N·s = kg·m/s.

Summarizing this derivation, we have the relation between impulse and momentum:

★ $\vec{I} = \Delta\vec{p}$ impulse = change in momentum.

As stated above, according to Newton's third law of motion the two forces which two bodies exert on each other during a collision are equal in magnitude but opposite in direction. As a result, the total impulses which the two bodies deliver to each other are also equal in magnitude but opposite in direction, and therefore the changes in momentum of the two bodies are also equal in magnitude but opposite in direction. In other words, the change in momentum of one body is exactly offset by the change in momentum of the other, so that the total momentum of the two bodies is the same after the collision as it was before. In this way, Newton's third law of motion leads immediately to the law of conservation of momentum; in fact, it can be said that these two laws are simply alternative ways of formulating the same fundamental law of nature.

The concept of impulse is applicable to all interactions between bodies, but it is especially useful when a very large force acts for a very short time. Such forces occur, for example, when a nail is hit by a hammer, or a golf ball is hit by a club; they are sometimes called impulsive forces. In these cases it may be impossible to specify the detailed time dependence of the actual force, but the total impulse delivered by the force still has meaning and has a definite numerical value.

When the variation of force with time is known, this can be plotted on a graph and then the impulse can be found from the area under the curve. This method applies whether the impulse is delivered by a steady force, or by an impulsive force which rises quickly to a large value and then quickly falls to zero.

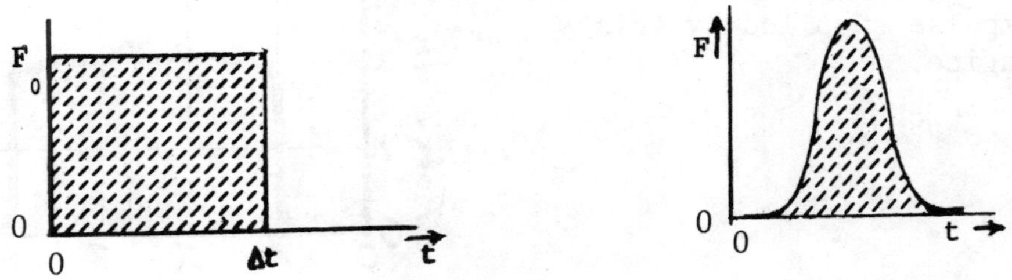

Figure 50-3. The impulse delivered by a force is represented by the area under the curve on a graph of force vs. time.

Practical applications of impulse arise in sports, where a player often desires to hit a ball and give it the maximum possible speed. From the standpoint of physics, in order to impart a large speed or momentum to the ball it must receive a large impulse, which requires that the force applied to the ball must be sustained for as long a time as possible. This same idea is referred to by the coach as "follow-through"; he tells the player to swing her racquet (or bat, or golf club) beyond the point where it hits the ball all the way through to the end of its arc.

Questions and Exercises

Q 50.3-1 In Chapter 47 we defined *power* as the time rate at which work is done, that is, the time-rate of flow of energy. What quantity is equal to the time-rate of flow of momentum, in terms of concepts that we already know?

Q 50.3-2 If it took some time for a force to "travel" from one particle to another, Newton's third law would not hold instantaneously. Instead, the force on one particle *at some instant* would be equal and opposite to the force on the other particle *a little earlier* (or later). Would the law of conservation of momentum hold under those circumstances? Discuss carefully.

Q 50.3-3 To elaborate on the previous question, imagine two baseball players some distance apart on frictionless ice. One throws a ball, and the other catches it. One might be tempted to say that the thrower feels the reaction force *before* the catcher feels the force. Does the law of conservation of momentum hold?

Q 50.3-4 List several examples of strong forces of brief duration (often called impulsive forces) which are not mentioned in the text.

E 50.3-1 A long-burning model-rocket engine has the thrust vs. time curve shown here. From this graph, compute the total impulse supplied by this engine.

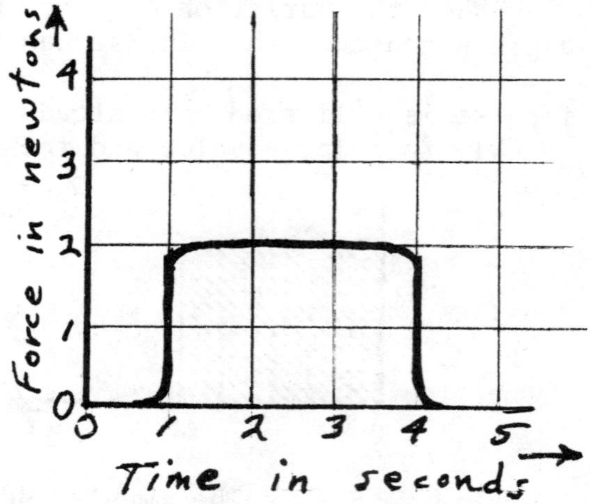

E 50.3-2 A golfer hits his 0.025-kilogram golf ball with a three-iron, giving it an impulse of 1.25 newton-sec. What speed does the ball acquire?

E 50.3-3 (a) A 2-kilogram object is moving with an initial speed of 10 meter/sec. Compute the net impulse necessary to raise the speed of the object to 30 meter/sec.
(b) If this impulse is delivered to the object over a time interval of 0.2 sec, what average force is required?

E 50.3-4 A golf-ball (m = 0.025 kg) has a speed of 60 m/s immediately after Palmer hits a drive. If the golf club strikes the ball with an average force of 200 N, how long is the club head in contact with the ball?

E 50.3-5 Imagine that you are a passenger seated next to the driver in a car which has a head-on collision at 70 mph. Suppose that your seat belt isn't fastened, and your head is stopped by the dashboard.
(a) Estimate the time required for you to come to rest under these conditions.
(b) Estimate any other physical quantities which you will need, and compute the *average* force on your head during the impact.
(c) Estimate the *maximum* force on your head during this impulsive event.

E 50.3-6 A model-rocket engine has an average thrust of 7 N which persists for 0.17 s. If the engine is attached while it burns to an air-supported glider (initially at rest) having a mass equal to 0.2 kg, what speed will the glider acquire at burnout?

E 50.3-7 A 1000-kg car moving at an initial speed of 40 m/sec crashes into a massive wall and stops. From the instant the car hits the wall to the time that it stops the center of the car moves forward 0.1 m. Calculate the average force which decelerates the car during this collision.

Check your understanding of the material above by testing yourself on the minimum learning goals at the beginning of this section.

50.4 Rocket Propulsion

Minimum Learning Goal. Be able to use the law of conservation of momentum to calculate the change in speed of a vehicle when you are given the initial mass of the vehicle, the masses of a number of ejected particles, and their speed relative to the vehicle. For example:

1 A glider at rest at one end of a level air track has a special gun on it which shoots ball bearings in a backward direction. Initially the glider, gun, and ball bearings have a total mass of 400 g. If the gun fires three ball bearings successively at 10 m/s relative to the glider, and if each bearing has a mass of 50 g, what is the speed of the glider and attached gun immediately after the third ball bearing has been fired? Assume that the frictional force between the air track and the glider is negligible. (Answer, p. 50-20, line 14)

2 A man sits on a sled which is at rest on a frozen lake. He fires rifle bullets repeatedly in a backward direction. Suppose the man, sled, rifle, and bullets have a total initial mass of 100 kg, and that he fires the rifle once each second for a total of ten shots. If each bullet has a mass of 40 g, and the muzzle velocity of the rifle is 1200 m/s, what is the approximate speed of the sled and man after the tenth shot is fired? Assume that the frictional force between the sled runners and the ice can be neglected. (Answer, p. 50-20, line 15)

The law of conservation of momentum tells us that it is impossible to impart a forward momentum to a vehicle without at the same time imparting a backward momentum to something else. If the vehicle is in frictional contact with solid ground, the easiest thing is to impart the backward momentum to the ground. In the case of an airplane, the ground is not available, so the backward momentum has to be imparted to the surrounding air -- either by means of a propeller or a jet engine. The jet engine is not, from this point of view, radically different from a propeller; the main distinction is that, in the jet engine, the process by which the chemical energy of the fuel is converted to kinetic energy of the air is much simpler.

If a vehicle is to travel in outer space, where it is not in contact with anything, the material to which the backward momentum is going to be imparted has to be carried on board and ejected in a backward direction. This arrangement is called a rocket engine. It carries its own fuel and oxidizer. When the fuel is burned, the combustion products are expelled to the rear from the exhaust nozzles of the engine at a high speed relative to the rocket. The force which expels the exhaust is exerted by the rocket on the exhaust in the backward direction; the reaction force is exerted by the exhaust on the rocket in the forward direction. This forward reaction force is called *thrust*.

Exercise

E 50.4-1 A piece of a rocket is ejected backwards in such a way that the *relative velocity* between two pieces is u. (The ejected "piece" is some oxidized fuel.) The rocket was initially at rest in interstellar space.

(a) If the ejected piece is 50% of the whole rocket, what will be the velocity of the rocket after ejection, in terms of u?

(b) Suppose that the ejected piece is 99% of the mass of the rocket; what will be the rocket's final speed in terms of u?

(c) Suppose the ejected mass is 99.9999% of the total mass?

(d) From the preceding calculations you should be coming to the conclusion that no matter how large the ejected mass is, the rocket cannot attain a speed larger than u if the mass is ejected *all in one piece*. Actually, if it is ejected slowly and continuously as the rocket itself is gaining speed, much larger final velocities can be attained. For example, if 99% of the initial mass is continuously and uniformly ejected over a period of time, the rocket will attain a final speed of $4.6u$. Give a qualitative argument for why that should be so.

Check your understanding of the material above by testing yourself on the minimum learning goal at the beginning of this section.

50.5 Elastic and Inelastic Collisions

Minimum Learning Goal. Be able to distinguish between an elastic collision and an inelastic collision, when you are given information about the energy transfers in collisions. For example:

1 Which of the following collisions are elastic?

(a) A bullet (KE = 30 J) is fired into a bag of sand suspended by a long rope. After the collision the bag has a kinetic energy of 20 J.

(b) An ion ($KE = 4.8 \times 10^{-19}$ J) collides with a nucleus at rest, after which the nucleus has a kinetic energy of 3.1×10^{-19} J and the ion has a kinetic energy of 1.7×10^{-19} J.

(c) A ball dropped from a height of 1.3 m bounces from the floor to a height of 1.1 m.

(d) A molecule strikes a wall with an energy of 5.0×10^{-21} J and rebounds with the same energy.

(Answer, p. 50-20, line 7)

We have stated that the conservation laws for energy and momentum are universal laws which apply to all physical phenomena, including collisions. Because collisions involve objects which are moving and which therefore have kinetic energy, we may inquire whether there might be another conservation law which prevails in collisions, namely, a conservation law for kinetic energy.

Let us consider again Experiment B in Section 50.2, for the case when $m_C = 2\ m_D$. Then, using the equation $m_C v_o = (m_C + m_D)v_f$, we would have $v_o = (3/2)v_f$.

The total kinetic energy of the gliders *before* the collision was $(1/2)m_C v_o^2$, which is equal to $(9/4)m_D v_f^2$. *After* the collision, the total kinetic energy is $(1/2)(m_C + m_D)v_f^2$, which may be expressed as $(6/4)m_D v_f^2$. We thus find that in this collision some of the macroscopic kinetic energy disappears during the collision; the combined kinetic energy of the gliders after the collision is less than the combined kinetic energy of the gliders before the collision.

We find that there are other collisions, however, in which no kinetic energy is lost during the collision. One example is a collision between gliders on an air track in which the gliders are equipped with elastic bumpers so that the gliders do not stick together after the collision. Another example from the laboratory is a collision between air pucks which float on a film of air on an air table.

It is convenient to give names to the two categories of collisions. A collision in which the total kinetic energy of the colliding bodies remains unchanged is called an *elastic collision.* In an elastic collision, not only the total momentum before and after, but also the total kinetic energy before and after, must be the same. At the other extreme is a *totally inelastic collision,* defined as one in which the two bodies stick together, so that they move off with a common velocity after the collision. It can be shown that a totally inelastic collision represents the largest loss of kinetic energy that is consistent with conservation of momentum.

The general name given to any collision in which the total kinetic energy changes is *inelastic collision.* In an inelastic collision the kinetic energy never increases, but always decreases. The "lost" kinetic energy does not disappear, of course; it is transformed into some other form of energy. The particular form of energy depends on the nature of the collision. In a collision between automobiles, some kinetic energy is transformed into internal energy when the vehicles are deformed. In an inelastic collision between atoms or molecules, kinetic energy can be transformed to energy of electronic excitation (see Chapter 32).

Note that in an elastic collision it is not necessary that the kinetic energy remain constant *during* the collision, while the colliding bodies are in contact with each other. In an elastic collision in which two bodies approach each other head-on with equal but opposite momenta, both bodies will actually come to rest momentarily before they rebound, so that the total kinetic energy is zero at this instant. This kinetic energy has been stored, of course, in the form of elastic potential energy (Section 48.4) of the colliding objects. Only objects which are able to store elastic potential energy in some way can have elastic collisions.

In studying collisions, it is helpful to draw a diagram showing the masses and velocities of the colliding objects *before* the collision, and another diagram showing the masses and velocities of the objects *after* the collision. For an inelastic collision, it is then possible to write one equation which expresses the conservation of momentum. For an elastic collision, two equations can be written: one for conservation of momentum and the other for conservation of kinetic energy. These equations can then be solved simultaneously to compute unknown factors in the collision.

Questions and Exercises

Q 50.5-1 (a) What quantities are conserved in an elastic collision?
(b) What quantities are conserved in an inelastic collision?

Q 50.5-2 Consider a collision between two gliders on a linear air track. The quantities needed to describe the gliders' motions before and after the collision are the initial speeds u_1 and u_2, and the final speeds v_1 and v_2. The masses are m_1 and m_2.
(a) If the collision is elastic, how many independent equations can be written relating these variables? How many numerical values of these variables must be known initially, in order to permit a complete solution for the values of the unknowns?
(b) Repeat part (a) for a totally inelastic collision.
(c) Repeat part (a) for a collision which is neither elastic nor totally inelastic.

Q 50.5-3 Give three realistic examples of totally inelastic collisions.

Q 50.5-4 What information would you need in order to predict the motion of an automobile immediately after it is hit in the rear by a truck?

E 50.5-1 A 0.2-kg glider moving at 0.5 m/s on an air track strikes an initially stationary 0.4-kg glider in such a way that the two gliders lock together and move as a unit after the collision.
(a) What is the glider speed after the collision?
(b) What fraction of the initial kinetic energy is lost during the collision?

E 50.5-2 A 10^4-kg truck moving at 10 mph (4.5 m/s) strikes a stationary 2×10^3 kg car in the rear. Find the speed of the car immediately after the (assumed elastic) collision.

E 50.5-3 Consider a head-on elastic collision between two billiard balls *A* and *B*, having equal masses. Show that in this case the initial velocity of *A* equals the final velocity of *B*, and the initial velocity of *B* equals the final velocity of *A*.

E 50.5-4 Two cars, an Alpha and a Beta, collide head-on and lock together. Before the collision, the Alpha has a momentum of 2×10^4 kg.m/s toward the East, while the Beta has a momentum of 3.2×10^4 kg·m/s toward the West. The mass of the Beta is two times the mass of the Alpha.
(a) What is the total momentum (magnitude and direction) of the two cars after the collision?
(b) What is the momentum of the Alpha after the collision?
(c) What impulse was delivered to the Alpha during this collision?

E 50.5-5 Consider a head-on *elastic* collision between masses m_1 and m_2, in which m_2 is initially at rest. Let u_1 equal the initial speed of m_1, v_1 equal the final speed of m_1, and v_2 equal the final speed of m_2. The values of u_1 and v_2 are always positive; v_1 is positive if its direction is the same as that of u_2, but v_1 is negative if its direction is opposite to u_1.

(a) Show that $u_1 + v_1 = v_2$, independent of the values of m_1 and m_2.

(b) Show that $v_1 = u_1(m_1 - m_2)/(m_1 + m_2)$

(c) Give a realistic example of such a collision in which $m_1 >> m_2$. Simplify the result of part (b) for this example and compare with your experience.

(d) Give an example of such a collision in which $m_1 << m_2$. Simplify the result of part (b) for this example and compare with your experience.

E 50.5-6 Two cars, a Pion and a Muon, collide head-on and lock together. Before the collision, the Pion has a momentum of 4×10^4 kg·m/s toward the North, while the Muon has a velocity of 16 m/s toward the South. Immediately after the collision, the total momentum of the two cars equals 8×10^3 kg·m/s toward the North. The mass of the Muon equals the mass of the Pion = 2000 kg. For each part below, state the magnitude and direction of your answer.

(a) What is the total (combined) momentum of the two cars before the collision?
(b) What is the momentum of the Pion after the collision?
(c) What impulse was delivered to the Pion during this collision?
(d) What was the velocity of the Pion just prior to the collision?
(e) What was the kinetic energy of the Muon, just prior to the collision?
(f) What was the change in momentum of the Muon during the collision?
(g) If the total kinetic energy of the system changes during this collision, state where the difference in kinetic energy comes from or where it goes.

Check your understanding of the material above by testing yourself on the minimum learning goals at the beginning of this section.

Answers to Minimum Learning Goals

Line 1 $\vec{I} = \vec{F}\,\Delta t$, where $\vec{I}$ is the vector impulse in newton-seconds (N-s), and $\vec{F}$ is the average vector force in newtons exerted over the time interval Δt in seconds.

Line 2 $p = 5 \times 10^{-20}$ kg·m/s

Line 3 $I = 4.4 \times 10^{6}$ N•s; $\Delta v = 49$ m/s

Line 4 $I = 4.45 \times 10^{6}$ N•s

Line 5 $v = 14.0$ m/s

Line 6 $p = 1.8 \times 10^{29}$ kg·m/s in a direction tangential to the orbit (assumed circular)

Line 7 (b) and (d)

Line 8 The total momentum of an isolated system remains constant.

Line 9 $v = 4.0$ m/s

Line 10 $\vec{p} = 0.0144$ kg·m/s in a direction of about 45° north of east.

Line 11 (a) $I = 1.2$ N•s; (b) $\Delta v = 0.39$ m/s

Line 12 $\vec{p} = m\vec{v}$, where $\vec{p}$ is the vector linear momentum expressed in kg•m/s, m is the mass in kg and $\vec{v}$ is the vector velocity in m/s.

Line 13 $I = 1.19$ N•s

Line 14 $v = 5.1$ m/s

Line 15 $v = 4.8$ m/s

ANSWERS to Questions and Exercises in Chapter 50

Section 50.1

E 50.1-2 (a) 0.1 kg•m/s to the right
(b) 0.1 kg•m/s to the left
(c) 0.2 kg•m/s to the left

E 50.1-4 2.25×10^5 kg•m/s

Section 50.2

Q 50.2-2 No. The pedals of a bicycle are connected by gears and a chain to the rear wheel; when you push on the pedals the rear wheel pushes against the road, and since forces come in pairs, the road pushes forward on the bicycle with an equal and opposite force, thus causing the bicycle to move.

Q 50.2-4 (a) No; the fan (attached to the boat) pushes forward on the sail, but the sail pushes backward on the fan, so these two internal forces both act on the boat and cancel each other.
(b) Yes; in this case the fan pushes backward on the air, and so the air pushes forward on the fan (with boat attached).

Q 50.2-6 If you walk forward in the boat, the boat will move backward (and vice versa) so that the total momentum relative to the ground of the boat-plus-person remains equal to zero.

E 50.2-2 70 miles/hour

E 50.2-4 (a) $\Delta\vec{v}_1$ and $\Delta\vec{v}_2$ are measured; m_2 is calculated.
(b) $\Delta\vec{v}_1$ and $\Delta\vec{v}_3$ are measured; m_3 is calculated.
(c) m_2 and m_3 are known before the third experiment. The ratio $\Delta\vec{v}_1/\Delta\vec{v}_2$ can be predicted, but only by using the law of conservation of momentum.

Section 50.3

Q 50.3-2 Yes. The momentum supplied by one particle would be carried by the medium, or the field, which occupies the space between the two particles.

Q 50.3-4 Forces between cars, or billiard balls, during a collision; force of a bat on a baseball; force of a racquet on a tennis ball; force of exploding powder on a bullet when a gun is fired; etc.

E 50.3-2 50 m/s

E 50.3-4 7.5×10^{-3} s.

E 50.3-6 5.95 m/s

Section 50.5

Q 50.5-2 (a) Two equations, one for kinetic energy and one for momentum. Four numerical values must be known; the two equations can then be solved for the other two variables.

(b) One equation (for momentum) can be written, but there are only five variables because $v_1 = v_2$; four numerical values must be known initially.

(c) One equation (for momentum), but there are six variables and so five of them must be known initially.

Q 50.5-4 The masses of car and truck, the speeds of car and truck just before the collision, and whether the collision is elastic or totally inelastic. If the collision is neither of these, the motion of the car after collision cannot be predicted quantitatively unless the velocity of the truck after the collision is also known.

E 50.5-2 7.5 m/s

E 50.5-4 (a) 1.2×10^4 kg·m/s, toward the west

(b) 0.4×10^4 kg·m/s, toward the west

(c) 2.4×10^4 newton·second, toward the west

E 50.5-6 (a) 8×10^3 kg·m/s, north;

(b) 4×10^3 kg·m/s, north;

(c) 3.6×10^4 kg·m/s, south;

(d) 20 m/s, north

(e) 2.56×10^5 J

(f) 3.6×10^4 kg·m/s, north

(g) Some kinetic energy goes into internalenergy during deformation of the vehicles.

CHAPTER 51 MECHANICAL OSCILLATIONS

Introduction

A mechanical system which has a position of stable equilibrium can oscillate back and forth about that position when energy is supplied to the system. The restoring force which returns the displaced system to its equilibrium position is associated with a potential energy well, so the energy of the system continually alternates between kinetic and potential

energy as the system vibrates about equilibrium. The special case of a linear restoring force is considered in detail, using as an example the oscillations of a mass attached to a linear spring. Such a system is said to have simple harmonic motion; the position, velocity and acceleration vary sinusoidally with time. The amplitude of vibration depends on the total energy supplied to the system, while the frequency is determined by the mass and the spring constant. When an oscillating system is subject to a dissipative force, the mechanical energy of the system is transformed to internal energy at a rate described by the damping coefficient of the system. When energy is supplied to the system by a periodic applied force, the amplitude rises until the rate at which energy is supplied equals the rate of energy dissipation. The maximum amplitude occurs when the forcing frequency equals the natural frequency of the system, a condition called resonance.

51.1 The Physical Basis of Oscillatory Motion

Minimum Learning Goals. (1) Given an example of a system in motion,
(a) be able to state whether this is an oscillating system, and
(b) if it is be able to identify the position of stable equilibrium or state how to find this position. For example:

1 For each example of mechanical motion given below, state whether it is an oscillatory motion, and identify the position of stable equilibrium or state how to find it.
- (a) A ball suspended from a string is swinging as a simple pendulum.
- (b) A ball on a string is being swung in circular motion in a vertical plane.
- (c) A parked car is set into up-and-down motion by a heavy man who jumps on the front bumper and then immediately jumps off again.
- (d) An ordinary lead pencil is carefully balanced with its point resting on the table top, but it then moves from its vertical position.

(Answer. page 51-24, Line 18)

(2) Given the force on a system as a function of position, be able to use the conditions necessary for stable equilibrium to specify whether the system will oscillate, and if it will, state the equilibrium position. For example:

1 A toy car on a table top has rubber bands attached to the front and rear, as shown below. Each rubber band is stretched with one end held fixed. The net force on the car by the two rubber bands is given approximately by

$$\vec{F} = [22\text{ N} - 220\text{ (N/m) } x]\ \hat{x}.$$

If the car is moved to $x = 0.038$ m and released, will it oscillate? If it will, what is the equilibrium point about which it oscillates? (Answer, page 51-24, Line 8)

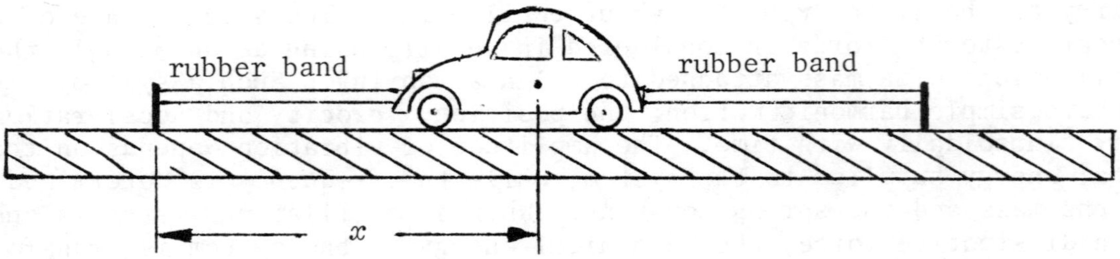

2 An automobile has coil springs, but no shock absorbers. The force in newtons on the car frame due to the front springs is given approximately by

$$\vec{F} = [4{,}000\text{ N} - 10{,}000\text{ (N/m) } z]\ \hat{z}\ ,$$

where z is the distance up from the pavement to the lower edge of the front bumper. If this bumper is pushed to $z = 0.30$ m and released, will the front of the car oscillate? If it will, what is the equilibrium point about which the bumper oscillates? (Answer, page 51-24, Line 2)

In earlier chapters we have studied the motion of a particle under constant acceleration, both on a straight line and in a plane, as well as motion in a circle and some more general motions. In this chapter we consider in some detail another common type of motion: oscillatory or vibrational motion, in which a particle or system moves back and forth about some central position in a repeating pattern. Examples from mechanics include the motion of a pendulum and the vibration of a mass attached to a spring. Analogous oscillations studied in other branches of physics include oscillations in electric circuits (Chapter 61) and acoustic oscillations.

Oscillating mechanical systems differ from one another in their frequency of oscillation, their amplitude (maximum displacement from the equilibrium position), and the rate at which the oscillations decrease in amplitude when the system is left to itself. Every macroscopic mechanical oscillation can be observed to decrease in amplitude as time goes on, unless added energy is supplied to the system. As the amplitude diminishes, in time the system settles into a rest position or *equilibrium position*. For a system where more than one coordinate is required to describe the positions of parts of the system when it is in oscillation, we might better speak of the *equilibrium configuration* of the system. For systems for which the oscillation can be described using only *one* position variable, however, we refer simply to the equilibrium position; it is systems of this sort that are analyzed in detail in this chapter.

When a system capable of oscillatory motion is disturbed or moved away from its equilibrium position, the internal restoring force in the system always tends to *return* the system to equilibrium. An equilibrium position of this kind is called a configuration of *stable equilibrium*. It is generally true for a system having a position of stable equilibrium, that motion of the system near this position is oscillatory motion. For oscillatory motion of a particle along a straight line, we choose the x axis along the line of motion and designate the equilibrium position as x_0. This means that when the particle is at x_0 the internal force of the system on it is zero, whereas at points near x_0 the force is always in a direction opposite to the displacement. Restoring force functions of this kind are illustrated in Figure 51-1. Each of the curves in this figure represents a restoring force for an oscillatory system.

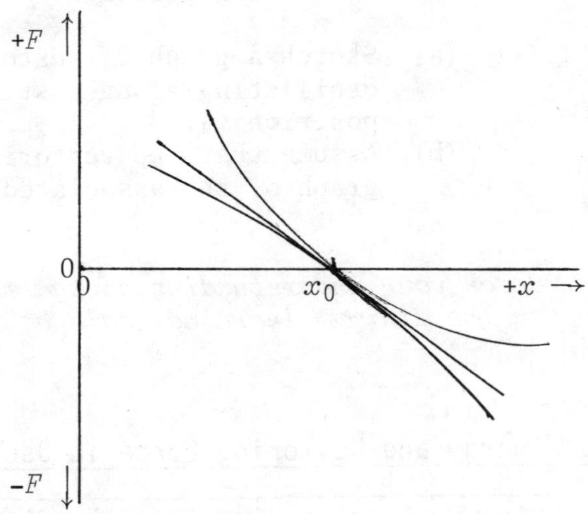

Figure 51-1. The restoring force for an oscillatory system is in the positive direction where the displacement from equilibrium is negative, but in the negative direction where this displacement is positive. At the point of stable equilibrium, x_0, the force equals zero.

If the particle is released at some point near x_0, it will be accelerated toward the equilibrium point, pass through it with some acquired speed, and then be decelerated until it comes momentarily to rest on the other side of x_0. It will then begin to acquire speed back toward x_0, and the process will repeat. This is the typical pattern of oscillatory motion.

Questions and Exercise

Q 51.1-1 List some examples of oscillatory or vibrational motion in your daily experience. What is the restoring force in each of your examples?

Q 51.1-2 Give an example of a macroscopic oscillating system in which the displacements from equilibrium are angular, rather than on a straight line.

Q 51.1-3 Give an example of a physical system which is not usually in oscillation, but which can be set oscillating by something you do to the system.

Q 51.1-4 How can you find the position of stable equilibrium for a macroscopic oscillatory system which you can observe in your daily experience?

Q 51.1-5 What causes the oscillation of a physical system to "die down" to a negligible amplitude after a while?

Q 51.1-6 If an oscillating system is in stable equilibrium at x_0, why doesn't it stop moving the instant it reaches x_0?

Q 51.1-7 Give an example of a macroscopic system in *unstable* equilibrium, such that if it is displaced slightly from equilibrium the force on the system always tends to move the system still farther from the equilibrium position.

E 51.1-1 (a) Sketch a graph of restoring force vs. position for a system oscillating along a straight line about the equilibrium position x_0.
(b) Assume that the restoring force is linear, and sketch a graph of the associated potential energy vs. position.

Check your understanding of the material above by testing yourself on the minimum learning goals at the beginning of this section.

51.2 Energy and Restoring Force in Oscillatory Motion

Minimum Learning Goals. (1) Given the mass of an object in oscillatory motion and an expression for its speed as a function of position, be able to calculate its kinetic, potential, and total energies at any point. For Example:

1 A pendulum has a length of 24.9 cm and consists of a string having negligible mass and a bob having a mass of 100 g. When the pendulum is supported from the ceiling, and the bob is displaced from equilibrium and released, the speed of the bob is found to be given by

$$v = \sqrt{(19.6\ \text{m/s}^2)(0.10\ \text{m} - h)},$$

where v is the speed in m/s and h is the height (in meters) of the bob above its lowest point. When the bob is 0.05 m above its lowest point, what is (a) the bob's kinetic energy, (b) its potential energy relative to the value at its lowest point, and (c) its total energy? (Answer, page 51-24, Line 5)

2 A 100-g mass is suspended by a spring having negligible mass. When the mass is set into oscillatory motion, its speed is given approximately by the expression

$$v = \sqrt{300\ \frac{\text{N}}{\text{m}}\ (0.01\ \text{m}^2 - y^2)},$$

where v is the speed in m/s and y is the height (in meters) of the mass above its equilibrium position. Neglect gravitational potential energy. When the mass is 0.05 meters below its equilibrium position, what is (a) its kinetic energy, (b) its potential energy relative to the equilibrium position, and (c) its total energy? (Answer, p. 51-24, Line 16)

(2) Given an expression for the force on a particle moving in a potential well as a linear function of its position, be able to calculate the force constant. For example:

1 The toy car in example 1 of Minimum Learning Goal (2) of Section 51.1 has a force acting on it given by

$$\vec{F} = [22\ \text{N} - 220\ (\text{N/m})\ x]\ \hat{x}.$$

What is the force constant? (Answer, page 51-24, Line 10)

2 The automobile in example 2 of Minimum Learning Goal (2) of Section 51.1 has a force on the car frame due to the front springs given by

$$\vec{F} = [4{,}000\ \text{N} - 10{,}000\ (\text{N/m})\ z]\ \hat{z}.$$

What is the force constant? (Answer, page 51-24, Line 15)

For an oscillatory system in which the dissipative forces are negligible, the restoring force function $F(x)$ is associated with a potential energy function $PE(x)$ according to the equations obtained in Section 49.1:

$$F(x) = -\frac{dPE}{dx} \quad \text{and} \quad PE(x) = -\int_{x_0}^{x} F(x)dx.$$

Near the position of stable equilibrium x_0, the potential energy of the system has the characteristic form of a potential well: the potential energy is a minimum at x_0, and rises to higher values on either side. Figure 51-2(a) shows the potential well for the case when the restoring force is linear, that is, directly proportional to the displacement from equilibrium.

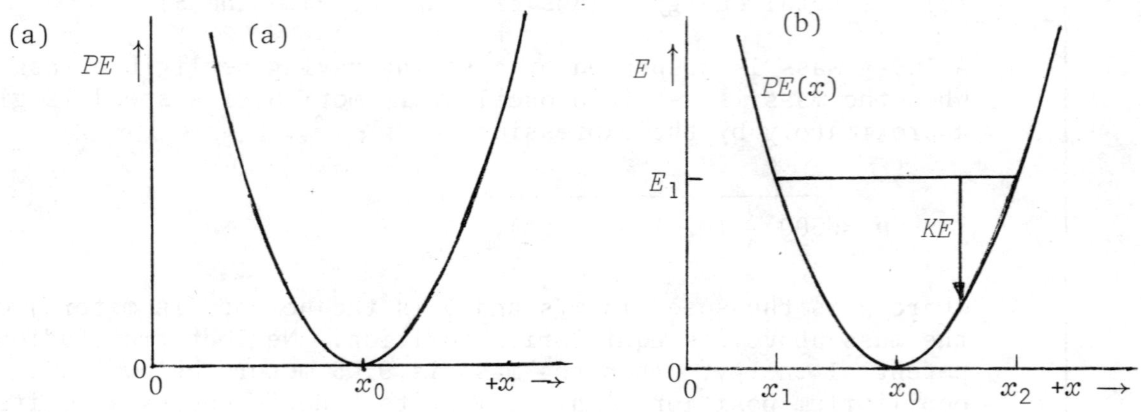

Figure 51-2. (a) The potential well associated with an oscillating system having a linear restoring force $F(x) = -k(x - x_0)$. The well is a parabola described by $PE(x) = k(x - x_0)^2/2$. (b) When the system has a total energy E_1, its oscillations are bounded by the points x_1 and x_2 where $PE(x) = E_1$. At any point x within the region of oscillation, the difference between E_1 and $PE(x)$ is the kinetic energy of the system.

It is instructive to examine oscillatory motion in terms of energy. If the system is released from rest at some point other than the equilibrium position, it initially has zero kinetic energy but a potential energy greater than the minimum. In Figure 51-2(b), x_1 and x_2 are such beginning points (or end points) of the motion for a total energy E_1. As the system begins its motion toward the equilibrium point, its potential energy decreases while its kinetic energy increases. As it passes through the equilibrium point, the system has its maximum kinetic energy (equal to the total energy of the system) and therefore also its maximum speed. On the other side of the equilibrium point the system begins to "climb uphill" again, trading its kinetic energy for potential, until it comes to rest momentarily at a point where the potential energy is again equal to its initial value. The system then moves in the opposite direction until the initial point is reached, after which the whole motion is repeated.

Figure 51-2 shows the potential well for the special case when the restoring force is linear for *all* values of displacement from equilibrium. Many other potential wells have a shape that is similar for small values of displacement, but quite different for larger displacements; examples are shown in some exercises at the end of this section. For these other wells the restoring force is non-linear, with the deviation from linearity increasing as the displacement increases. Some non-linear restoring force functions are shown in Figure 51-1, where it is seen that even though a force function may curve quite far from a straight line at large displacements, it lies very close to a straight line for a small interval surrounding the origin. This means that for oscillations whose amplitude is not too large, it is a good approximation to assume that the restoring force is proportional to the displacement of the system from equilibrium. (This approximation does not apply for a potential well which has a flat or horizontal bottom of some extent where the restoring force is zero everywhere in the level region at the bottom.) Expressed mathematically, any smooth curve (such as those in Figure 51-1) can be approximated by a straight line if we are interested in only a small section of the curve. A straight line through the equilibrium point has precisely the property of giving a restoring force proportional to the displacement from that point.

A restoring force which varies linearly with the displacement from an equilibrium position x_0 is expressed by

★ $$F(x) = -\,\mathrm{k}(x - x_0),$$

where the proportionality constant k is called the *force constant*, having SI units of newton/meter. When any (linear or non-linear) restoring force is plotted as a function of the displacement from x_0, the slope of the curve where it passes through x_0 is equal to the negative of the force constant:

Figure 51-3. A mass m supported by a frictionless horizontal surface and connected to a linear spring. The mass can oscillate about its equilibrium point $x = 0$.

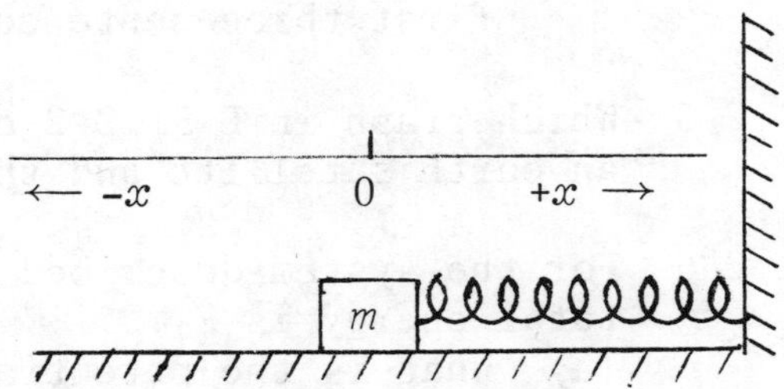

$$\mathrm{k} = -\,(dF/dx)_{x_0},$$

where the subscript x_0 means that after the slope $dF(x)/dx$ is computed as a function of x, it is to be evaluated at the specific point $x = x_0$. For a restoring force which is linear, the equation may be put into the form

$$F(x) = -\,\mathrm{k}x + b,$$

after which the force constant is easily evaluated.

An example of a system having a linear restoring force is a mass m which can slide without friction on a table and which is connected to a fixed point by a massless linear spring, as shown in Figure 51-3. If x

is the distance of (some point on) m from the equilibrium point $x = 0$, and k is the force constant of the spring, the restoring force is $F(x) = -kx$. As shown on page 48-16, the potential energy of this system is

★ $$PE(x) = \frac{1}{2}kx^2.$$

The potential well for this system is like the parabola in Figure 51-2, but shifted so its equilibrium point is at the origin. The motion of a mass-spring system is discussed in the next section.

Questions and Exercises

Q 51.2-1 Which of the potential energy curves in E 51.2-2 represent a potential well? For each of these, identify the point(s) of stable equilibrium.

Q 51.2-2 For which of the potential energy curves in E 51.2-2 can oscillatory motion occur?

Q 51.2-3 Which graph in E 51.2-2 could represent the system of the electron and the proton in a hydrogen atom?

Q 51.2-4
(a) Which graph(s) in E 51.2-2 could represent the system described in the first two sentences of E 51.2-1?
(b) Which graph(s) could represent the system described in the first three sentences of E 51.2-1?

Q 51.2-5 Which graph in E 51.2-2 could represent the system comprising an earth satellite and the earth?

Q 51.2-6 For the system described by graph (a) of E 51.2-2, suppose the total energy is E_1.
(a) What is the potential energy of the system at x_1? at x_2? at x_3?
(b) What is the kinetic energy of the system at x_1? at x_2? at x_3?
(c) What is the range of x values within which this system oscillates?
(d) At what point(s) will the speed of oscillation be greatest?
(e) At what point(s) will the speed of oscillation be zero?

Q 51.2-7 For the system described by graph (d) of E 51.2-2:
(a) What is the range of total energy values for which oscillatory motion can occur?
(b) If the total energy is E_3, what is the potential energy at x_2? at x_3? at x_4?
(c) If the total energy is E_3, what is the kinetic energy at x_2? at x_3? at x_4?
(d) If the total energy is E_3, at what point(s) will the speed of oscillation be greatest?

Q 51.2-8 For the system described by graph (b) of E 51.2-2:

(a) What are the points of stable equilibrium?

(b) For each point of stable equilibrium, what is the range of values of total energy for which the system can oscillate about this point?

(c) When the total energy is E_4, what is the range of x values within which the system oscillates?

(d) When the total energy is E_4 and x is increasing, what is the range of x values for which the *kinetic energy* is increasing?

Q 51.2-9 Shown here are some graphs of force vs position. A positive value of F represents a force in the $+x$ direction.

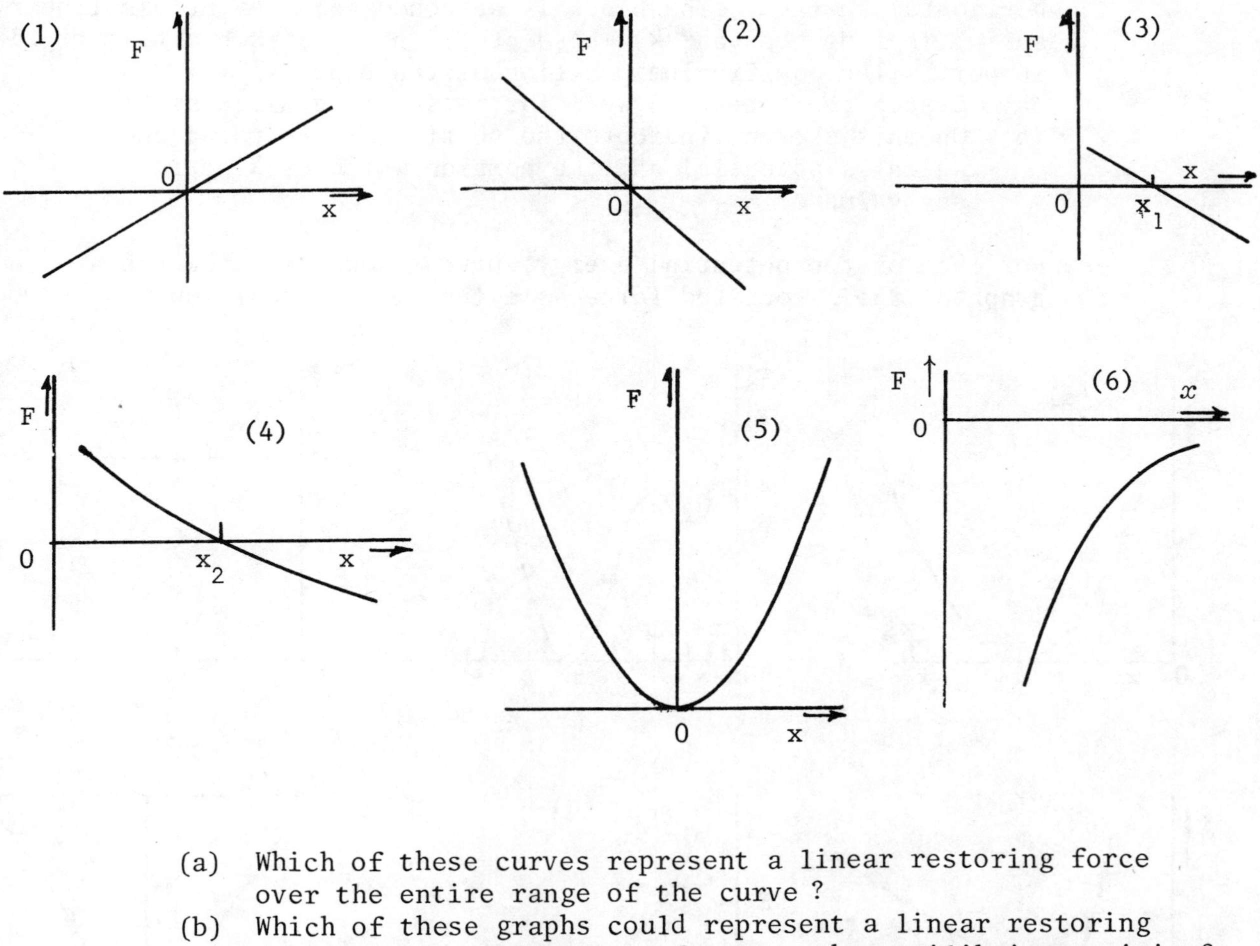

(a) Which of these curves represent a linear restoring force over the entire range of the curve ?

(b) Which of these graphs could represent a linear restoring force over a small range of x near the equilibrium position?

(c) For each graph which displays an equilibrium position, what is that position?

Q 51.2-10 For a restoring force assocated with oscillatory motion, what is the value of this force at the equilibrium position?

Q 51.2-11 (a) If a physical system is subject to a force which is a function only of the position (or configuration) of the system, is that force necessarily a restoring force? Explain. (Cont.)

(b) Which of the graphs in Q 51.2-9 do not represent a restoring force?

(c) Which graph(s) in Q51.2-9 represent a system which has a point of unstable equilibrium?

Q 51.2-12 Which of the graphs in Q 51.2-9 corresponds to the mass-spring system described in this section?

Q 51.2-13 Ignore the subscripts on the x values labeled on the graphs when answering this question. Which of the graphs in Q 51.2-9 corresponds to: (a) graph (a) in E 51.2-2? (b) graph (d) in E 51.2-2? (c) graph (h) in E 51.2-2?

E 51.2-1 Consider a block having mass m which slides on a frictionless horizontal surface. The block is attached to a weightless linear spring (spring constant k) whose other end is attached to a rigid support. The equilibrium position of the block is $x = 0$.

(a) Sketch the potential well for this spring-block system.

(b) On this sketch, indicate the portion of the total energy which is potential and the portion which is kinetic, for any value of x.

E 51.2-2 For each of the potential energy curves shown here, sketch a graph of the associated force as a function of position.

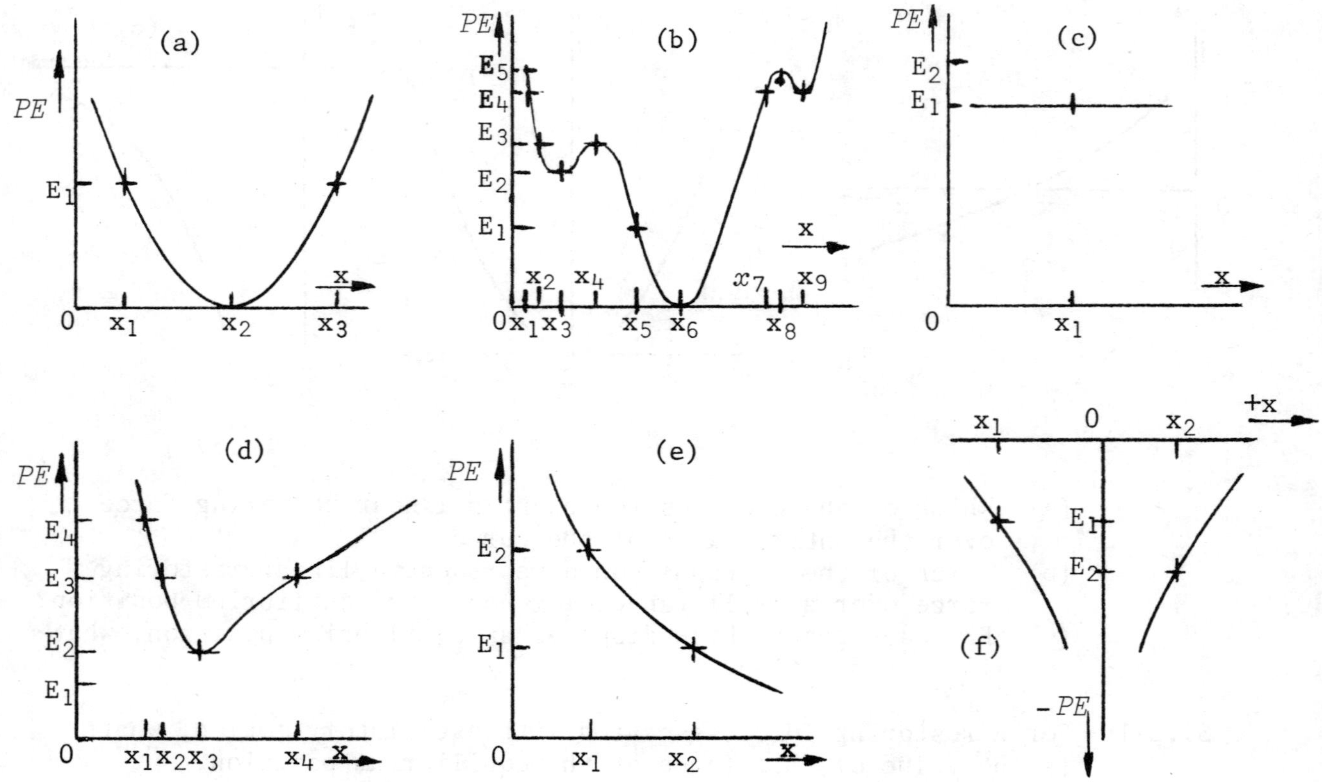

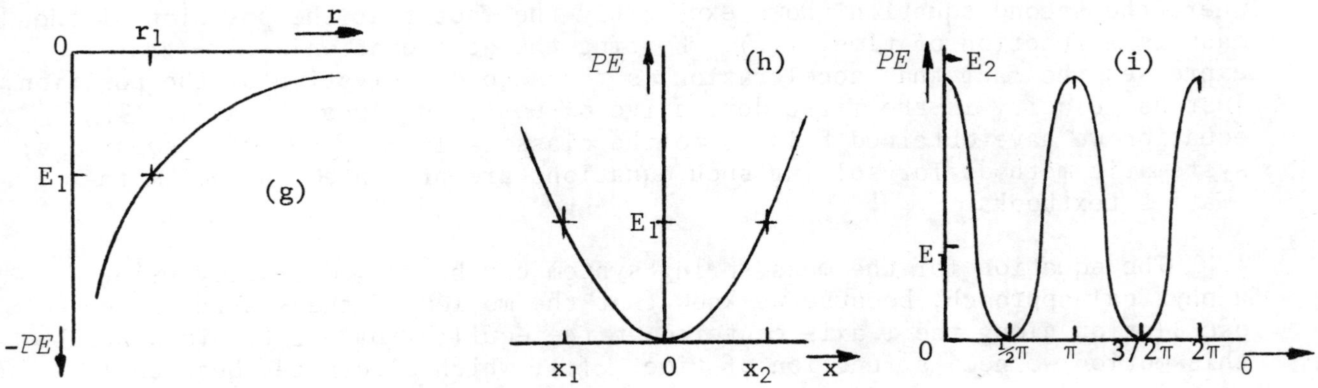

E 51.2-3 For each graph in Q51.2-9 which represents a linear restoring force over (at least) a small range of x, write an equation for this force as a function of x in the linear region.

Check your understanding of the material above by testing yourself on the minimum learning goals at the beginning of this section.

51.3 Simple Harmonic Motion

> Minimum Learning Goal. Given the values of force constant and mass for a system having a linear restoring force, be able to calculate the frequency or period of oscillation of simple harmonic motion. For example:
>
> 1 Using the force constant for example 1 of Minimum Learning Goal (2) of Section 51.2 calculate the frequency of oscillation of the car about its equilibrium position. The car's mass is 0.20 kg. (Answer, page 51-24, Line 13)

For an oscillatory system having a linear restoring force, the mathematical description of the motion is especially simple. To obtain this description analytically, one can apply Newton's second law to the system and then obtain the solution of the resulting equation. We develop this solution here for the mass-spring system shown in Figure 51-3, which serves as a prototype for any system having a linear restoring force.

Our starting point is Newton's second law (section 42.3),

$$\Sigma\vec{F} = m\,\vec{a}.$$

Substituting the restoring force $F = -\mathrm{k}x$ and the acceleration $a = d^2x/dt^2$, the equation of motion for the mass-spring system becomes

$$-\mathrm{k}x = m\frac{d^2x}{dt^2}, \qquad \text{or} \qquad \frac{d^2x(t)}{dt^2} + \frac{\mathrm{k}}{m}x(t) = 0,$$

where the second equation shows explicitly the fact that the position of the mass is a function of time, $x(t)$. Writing the acceleration as d^2x/dt^2 expresses the fact that acceleration is the second derivative of the position, just as velocity is the first derivative of position (Appendix A.2). The equation we have obtained belongs to the class called *differential equations;* systematic methods for solving such equations are presented in some mathematics textbooks.

The equation for the mass-spring system can be solved readily using a physical approach, because we know that the motion of the system is an oscillation along the x axis centered at the equilibrium point. To describe this motion we need a function of time, $x(t)$, which alternates between positive and negative values periodically in time. Two such functions are the sinusoidal functions, $\sin \omega t$ and $\cos \omega t$, described in Appendix A.1. The quantity ωt corresponds to an angle which increases linearly with time at the rate ω; ω is called the *angular frequency*. The angle ωt is usually expressed in radians: 2π radians = $360°$. As ωt is dimensionless, ω has units of radians/second = rad/s = s^{-1}. Substitution of $\sin \omega t$ or $\cos \omega t$ into the differential equation shows that either function is a solution, provided the angular frequency ω is related to k and m by

★ $$\omega = \sqrt{k/m}.$$

Any linear combination of the functions $\sin \omega t$ and $\cos \omega t$ is also a solution. The requirement for ω shows that the frequency of oscillation increases as the stiffness of the spring or spring constant increases, but the frequency decreases as the mass increases.

The choice between $\sin \omega t$ and $\cos \omega t$ to describe the oscillation is related to the position of the system at the time $t = 0$. In this discussion, we choose to compress the spring from its relaxed configuration by an amount A, and release the system from this initial configuration at $t = 0$. The resulting motion of the block is then described by

★ $$x(t) = A \cos \omega t,$$

where A is the *amplitude* of the oscillation.

> The *amplitude* of any oscillation is the maximum displacement of the system from its equilibrium point.

The corresponding equations for the velocity $v(t)$ and acceleration $a(t)$ are obtained from $x(t)$ by differentiation, using the rules in Appendix A.2.

★ $$v(t) = dx(t)/dt = -\omega A \sin \omega t,$$

and

★ $$a(t) = dv(t)/dt = -\omega^2 A \cos \omega t.$$

It is readily seen that $a(t) = -\omega^2 x(t)$. It is not difficult to verify that our solutions for $x(t)$ and $a(t)$ satisfy the differential equation of motion for the mass-spring system, provided $\omega = \sqrt{k/m}$, thus showing that the solution we have intuited through physical reasoning is correct (see Exercise 51.3-2).

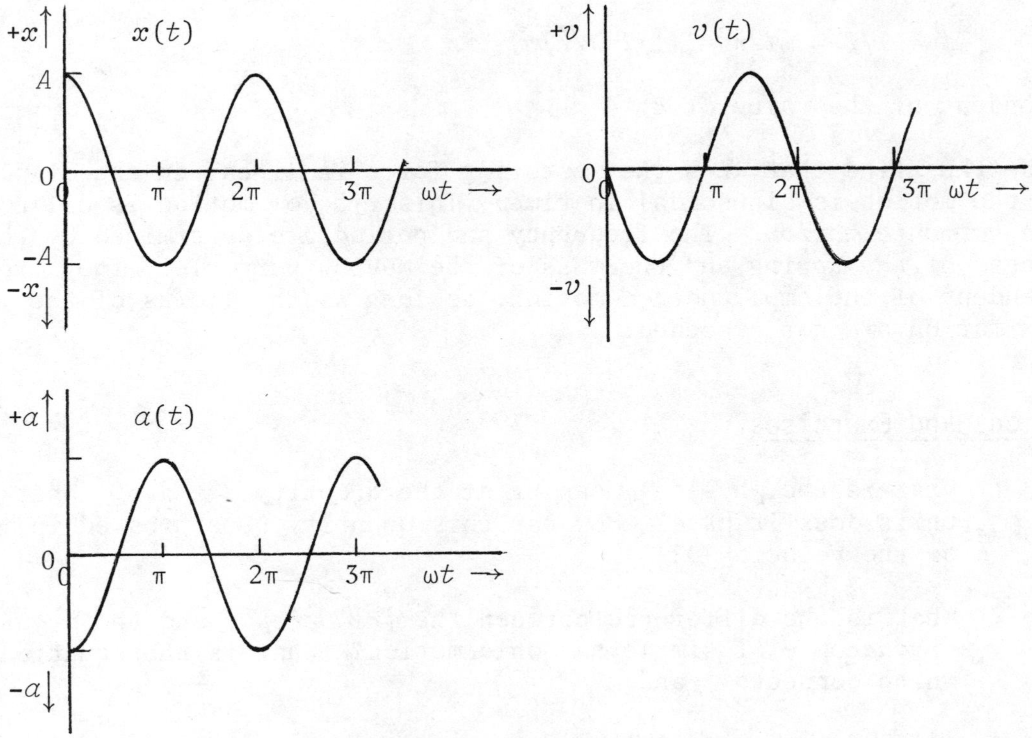

Figure 51-4. Graphs describing simple harmonic motion for which $x(t) = A \cos \omega t$. These graphs apply to any one-dimensional oscillatory system having a restoring force which is linear for all values of displacement from the equilibrium point $x = 0$.

Figure 51-4 contains graphs of our solutions for $x(t)$, $y(t)$ and $a(t)$ which show that the displacement, velocity and acceleration of the system reach their maximum positive values at different times during the motion. Each of these variables oscillates with the same angular frequency, ω, and the same frequency, f.

> The *frequency, f,* of oscillatory motion is the number of cycles of oscillation in one second.

Frequency and angular frequency are related by

★ $$f = \omega/2\pi \quad \text{and} \quad \omega = 2\pi f.$$

Frequency is measured in hertz, symbol Hz. One hertz equals one cycle per second.

The period, T, of oscillatory motion is the time for one complete cycle of oscillation.

From the definitions, we have

★ $$f = 1/T = \omega/2\pi = (1/2\pi)\sqrt{k/m},$$

independent of the value of A.

We thus find that when the restoring force is linear in the displacement, the motion is sinusoidal in time. This type of motion is called *simple harmonic motion.* The frequency and period are determined by the stiffness of the spring and the mass of the moving particle. They are independent of the amplitude of motion, as long as the limits of the linear approximation are not exceeded.

Questions and Exercises

Q 51.3-1 What is the physical meaning of the quantity d^2x/dt^2? What SI units does it have? How can this quantity be expressed in terms of the velocity v?

Q 51.3-2 What is the difference between the *frequency* f and the *angular frequency* ω of simple harmonic motion? What is the equation which connects f and ω?

Q 51.3-3 (a) Does a large value of the spring constant k correspond to a stiff spring or a limber spring?
(b) From your experience, do vibrating systems having stiff springs tend to have a lower, or higher, frequency of vibration compared with systems having limber springs?
(c) Correlate your answer to part (b) with the first equation for ω given in this section.

Q 51.3-4 For simple harmonic motion, explain in terms of forces why the velocity of the motion reaches its maximum only when the displacement is zero.

E 51.3-1 Using the SI units of spring constant k and mass m, show that angular frequency has units of radians/second or s^{-1}.

((E 51.3-2 Verify by substitution into the differential equation of motion for the mass-spring system that $x(t) = A\cos\omega t$ is a solution, provided $\omega = \sqrt{k/m}$.

E 51.3-3 Show that the equation $x(t) = A\sin\omega t + B\cos\omega t$ is a solution of the differential equation for an oscillating system given in this section, when $\omega = \sqrt{k/m}$, and A and B may have any constant values.))

E 51.3-4 The simple harmonic motion of a block moving at the end of a spring is described by the equation $x = A \cos \omega t$.
(a) Sketch a graph of x vs t for this motion.
(b) Sketch a graph of the acceleration a vs t for this motion.
(c) Sketch a graph of the restoring force $F = ma$ vs t for this motion.
(d) From your graphs, show that the restoring force is proportional to the negative of the displacement.

E 51.3-5 For the motion of the mass-spring system described in this section:
(a) What is the maximum speed of the block?
(b) At what value(s) of x does $v = v_{max}$ occur?
(c) What is the maximum acceleration of the block?
(d) At what value(s) of x does $a = a_{max}$ occur?
(e) What are the values of v and a at the end points of the block's motion?

E 51.3-6 For the mass-spring system described in this section, when $x = +\frac{1}{2} A$ at $\omega t = \pi/3$, what are the magnitude and direction of:
(a) the velocity?
(b) the acceleration?
(c) the restoring force?

E 51.3-7 For the mass-spring system described in this section, when $x = -\frac{1}{2} A$ at $\omega t = 4\pi/3$, what are the magnitude and direction of:
(a) the velocity?
(b) the acceleration?
(c) the restoring force?

((E 51.3-8 Consider the situation in which a constant force is superimposed on a linear restoring force. An example is the oscillatory motion of a mass suspended from a linear spring; the mass is acted on by its weight, in addition to the spring force.
(a) Use Newton's second law to obtain the differential equation for this system.
(b) Find a solution to this equation, assuming the displacement from equilibrium equals zero and the velocity is upward at $t = 0$.))

E 51.3-9 It is desired to construct a mass-spring system which oscillates in a vertical line and has a period of one second.
(a) Using a mass of ten grams, what spring constant is required?
(b) Using a mass of one kilogram, what spring constant is required?

E 51.3-10 The North American Air Defense Command (NORAD) is housed in a "city" located in large chambers blasted out of solid granite within Cheyenne Mountain, Colorado. The buildings of this city, with an aggregate weight in excess of 14×10^6 kg (14,000 tons), sit on an array of 941 huge coil springs wound of 3-inch wire. These giant springs carry this load in order to protect delicate equipment from the shock of an earthquake or a nuclear blast, and deflect nearly one foot from their unloaded height of 47 in. under the load.

(a) Assuming the springs are identical and each carries an equal share of the load, calculate the spring constant of one of these springs.

(b) Compute the natural frequency of vibration of NORAD city.

Check your understanding of the material above by testing yourself on the minimum learning goals at the beginning of this section.

51.4 Energy in Simple Harmonic Motion

Minimum Learning Goal. Given a description of a particle in simple harmonic motion, including values of its mass, the force constant, and the maximum amplitude of oscillation, be able to write equations for its displacement from equilibrium, potential energy, kinetic energy, and total energy, each as a function of time. For example:

1 A toy car having a mass of 0.20 kg is connected to two rubber bands (as in Section 51.1, MLG (2), Example 1) so it can oscillate on a frictionless table top. The effective force constant of the system is 220 N/m and the equilibrium position of the car is $x = +0.10$ m. Suppose this car is displaced to the position $x = 0.15$ m and released at $t = 0$. Write equations as functions of time for the car's (a) displacement from equilibrium, (b) velocity, (c) potential energy, (d) kinetic energy, and (e) total energy. (Answer, page 51-24, Line 3)

2 The two atoms in the diatomic molecule HBr are connected by an electronic bond so that the equilibrium distance between nuclei is 1.41×10^{-10} m, and the effective restoring force constant for small displacements from equilibrium is 200 N/m. Suppose the H atom (mass = 1.67×10^{-27} kg) is displaced 1.0×10^{-11} m from equilibrium and released at $t = 0$. Assuming the other atom remains fixed, write equations as functions of time for the H atom's (a) displacement from equilibrium, (b) velocity, (c) potential energy, (d) kinetic energy, and (e) total energy. (Answer, page 51-24, Line 11)

As we found in Section 51.2, in an oscillating system there is a continual trade-off between kinetic and potential energy. The energy of the system is all potential when the displacement from equilibrium is a maximum, and all kinetic when the displacement is zero. In a system having no dissipative forces, the sum of the kinetic and potential energies remains constant.

Using the results obtained in Section 51.3, we can develop equations for the kinetic and potential energies as functions of time for our prototype mass-spring system. The potential energy of the system for any displacement x is $\frac{1}{2}$ k x^2. Substituting the equation for $x(t)$, we have

$$PE(t) = \tfrac{1}{2}\,k\,A^2 \cos^2 \omega t.$$

The kinetic energy of the system for any speed of the oscillating mass is $\frac{1}{2}\,mv^2$. Substituting the equation for $v(t)$, and using $k = m\omega^2$, we have

$$KE(t) = \tfrac{1}{2}\,m\omega^2 A^2 \sin^2 \omega t = \tfrac{1}{2}\,kA^2 \sin^2 \omega t.$$

$\sin^2 \theta + \cos^2 \theta = 1$ for any angle θ, so

$$\sin^2 \omega t + \cos^2 \omega t = 1$$

for any value of t, and the total energy of the system is therefore

★ $$TE = PE(t) + KE(t) = \tfrac{1}{2}\,kA^2 = \tfrac{1}{2}\,m\omega^2 A^2,$$

a constant which does not depend on t. The total energy of the system *does* depend on the amplitude of oscillation, A.

Questions and Exercises

Q 51.4-1 Consider a harmonic oscillator consisting of a mass on the end of a mass-less spring.
- (a) Is the potential energy of this system associated with the mass, or the spring, or both?
- (b) Is the kinetic energy of this system associated with the mass, or the spring, or both?

E 51.4-1
- (a) Use the equations for $PE(t)$ and $KE(t)$ for the mass-spring system to sketch graphs of PE vs t and KE vs t for a complete cycle.
- (b) From your graphs, estimate the average values of PE and KE during one complete vibration. How do these averages compare with each other?

E 51.4-2 A glider having $m = 0.60$ kg is attached to the end of a horizontal air track by means of a spring (assumed massless), and is set into oscillation on the track. A force of 0.3 N is required to compress the spring by 10 cm. The amplitude of vibration is 0.2 m.
- (a) Compute the force constant of the spring.
- (b) Compute the maximum values of v, a, and F for the glider.
- (c) What is the frequency of this vibration?
- (d) What is the period of the vibration?
- (e) What is the maximum kinetic energy of the glider, and where does it occur?
- (f) What is the maximum potential energy of the system, and where does it occur?

Check your understanding of the material above by testing yourself on the minimum learning goals at the beginning of this section.

51.5 Damped Harmonic Motion

Minimum Learning Goal. Given a description of an oscillator, with a graph of its displacement as a function of time, be able to specify whether the motion is damped or undamped harmonic motion. For example:

1 Is the motion described in the following graph damped, or undamped? The graph shows the angular displacement as a function of time for a torsion pendulum used in the Cavendish experiment to measure the constant of universal gravitation. (Answer, page 51-24, Line 7)

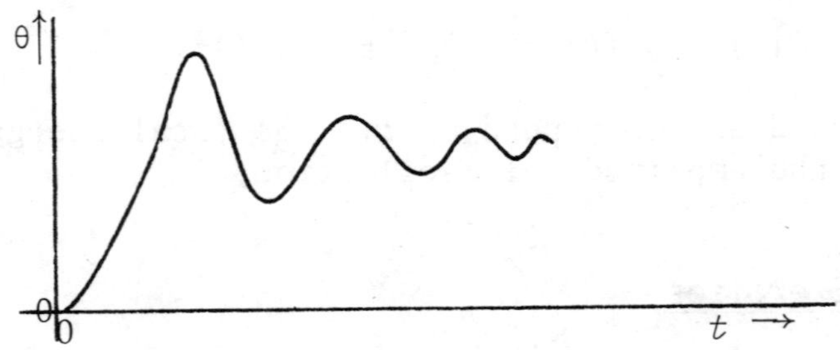

2 The graph below shows the current through the series ballast resistor of an automotive ignition system as a function of time. Considering only the section of the graph designated by *B* (where the time scale has been expanded by a factor of 100), is the oscillation damped, or undamped? (Answer, page 51-24, Line 17)

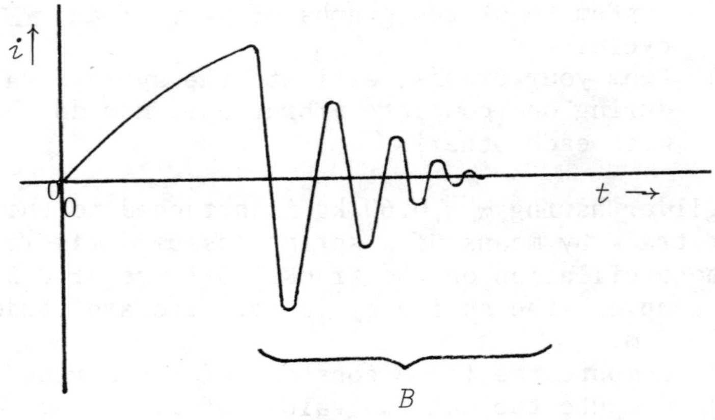

3 The graph below shows the angular displacement of the pendulum of a grandfather clock as a function of time. Is the motion damped, or undamped? (Answer, page 51-24, Line 14)

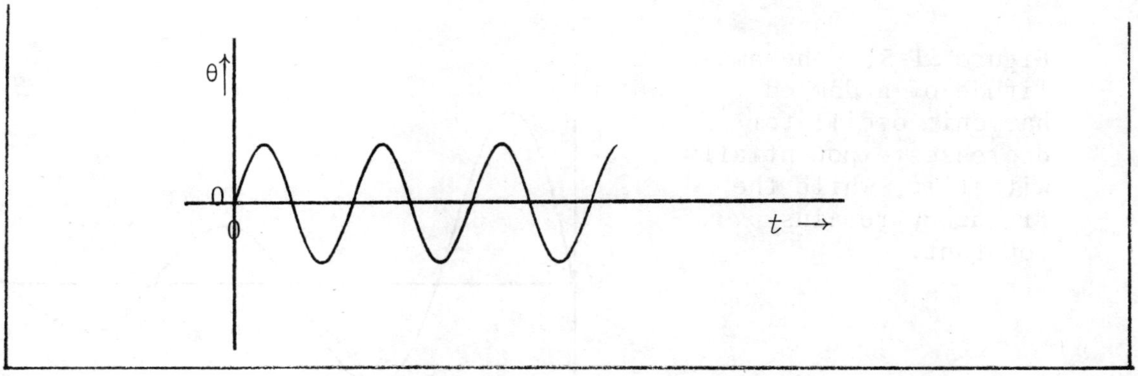

In the preceding discussion of oscillatory motion, we have assumed that the conservative restoring force is the only force acting in the system. For such an oscillator the total mechanical energy, frequency and amplitude remain constant. In most macroscopic systems, however, friction or some other dissipative force is present in addition to the restoring force. Even if the effect of the dissipative force is small, so that it does not appreciably change the motion during a single cycle, over a large number of cycles the cumulative effect is a noticeable dissipation of macroscopic energy to microscopic internal energy. The resulting motion, which may look like simple harmonic motion on the time scale of one or two cycles but which exhibits a progressive decrease in the amplitude of oscillation, is called *damped harmonic motion*. An oscillator subject to a dissipative force in addition to the conservative restoring force is called a *damped* oscillator.

The damping force has a direction opposite to that of the instantaneous velocity during all or part of each cycle of oscillation. For the case where the damping force is directly proportional to the instantaneous speed of the system, the equation of motion becomes

$$-kx - bv = ma \qquad \text{or} \qquad a(t) + \frac{b}{m}\,v(t) + \frac{k}{m}\,x(t) = 0.$$

Because $a = d^2x/dt^2$ and $v = dx/dt$, this is actually a differential equation. We leave the analytical solution to a more advanced textbook. The resulting solution for $x(t)$ shows that if the damping coefficient $\gamma = b/m$ is not too large, the amplitude of oscillation decreases exponentially with time, although the frequency remains constant. The energy and amplitude of a strongly damped oscillator (large b) quickly decrease to a small value, while the energy and amplitude of a weakly damped oscillator (small b) diminish at a slower rate. A graph of $x(t)$ for a damped oscillator is shown in Figure 51-5.

If we wish the harmonic motion to continue with constant amplitude, as in the case of a clock pendulum, we need a mechanism (the "escapement") that feeds a small amount of energy into the oscillator during each cycle so as to compensate for frictional losses.

Figure 51-5, The amplitude of a damped harmonic oscillator decreases exponentially with time, while the frequency remains constant.

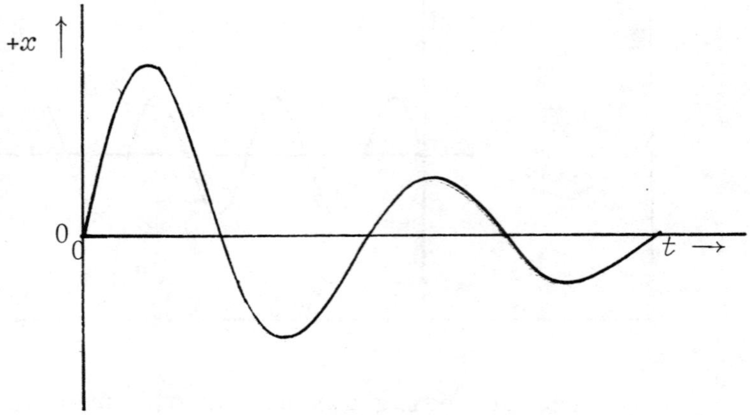

Questions

Q 51.5-1 What type of dissipative force acts in each of the following oscillating systems?
- (a) a simple pendulum in the laboratory;
- (b) an auto whose body bounces up and down when the wheels go over a severe bump;
- (c) a mass oscillating on the end of a spring.
- (d) What is the electrical analogue of the dissipative force in an *LRC* series circuit with an initially charged capacitor, in which the current oscillates after the switch is closed?

Q 51.5-2 From your experience, what is the closest macroscopic approximation to an oscillating system completely free from dissipative forces?

Q 51.5-3 Why does a dissipative force *always* act in such a way that it reduces the macroscopic energy of an oscillating system?

Q 51.5-4 What happens to the mechanical energy removed from an oscillating system by a dissipative force?

Q 51.5-5 What happens to the frequency of a damped harmonic oscillator while the energy of oscillation is decreasing?

Q 51.5-6
- (a) What are the SI units of b, in the dissipative force $-bv$?
- (b) Does a large value of b correspond to a rapid decrease, or a slow decrease, in the energy of oscillation?

Check your understanding of the material above by testing yourself on the minimum learning goals at the beginning of this section.

51.6 Forced Harmonic Oscillator. Resonance

Minimum Learning Goal. Given two graphs of the displacement amplitude (or an analogous quantity) as a function of the driving frequency for a forced harmonic oscillator, with the force amplitude constant and the damping weak, be able to infer from the graphs (a) the natural frequency of the system, and (b) which graph represents the case of least damping. For example:

1 A glider is positioned between two springs on a level air track. A mechanism is attached which can drive the system with varying values of applied force and frequency. The graphs below show the displacement amplitude A of the glider as a function of the driving frequency (at constant force amplitude), for two different conditions of damping. What is the natural frequency of the system, and for which graph is the damping least? (Answer, page 51-24, Line 6)

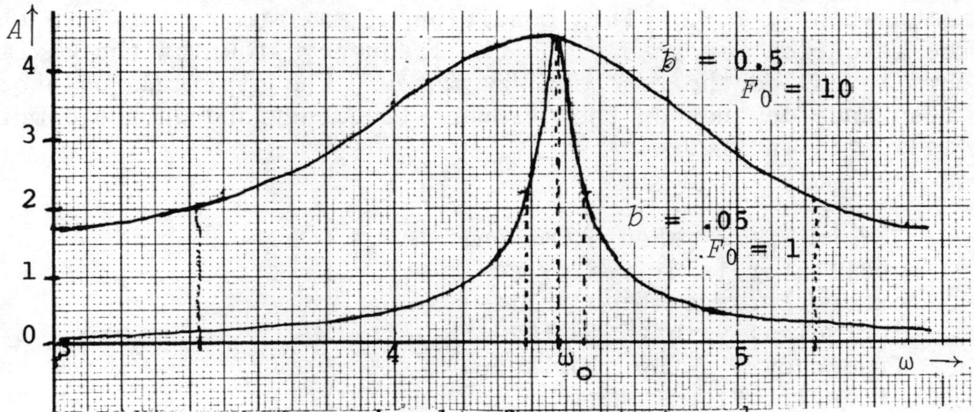

Angular frequency in s^{-1}

2 An inductor, resistor, capacitor, and source of a sinusoidally varying electromotance are included in a series electric circuit. The circuit is an electric analog to a forced harmonic oscillator, with electric current being analogous to velocity. The graphs below show the current amplitude i_0 in this circuit as a function of the driving frequency f (at constant voltage amplitude), for two different conditions of damping (that is, different resistances). What is the natural frequency of the circuit, and for which graph is the damping least? (Answer, page 51-24, Line 9)

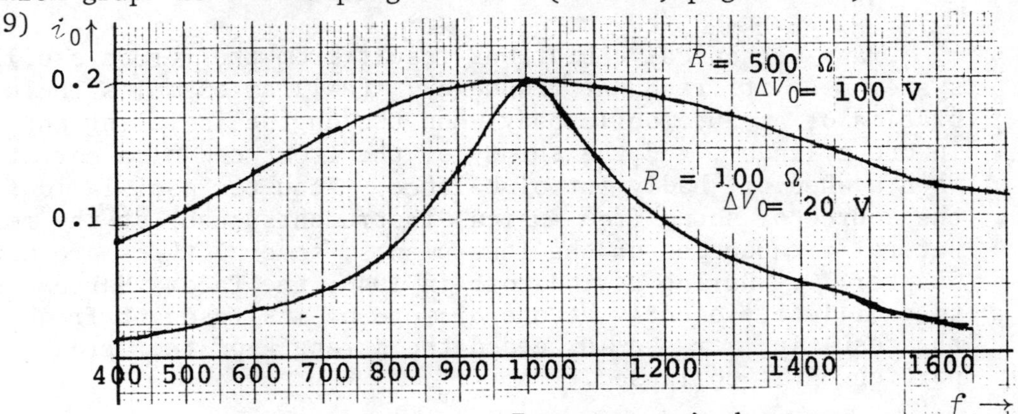

Frequency, in hertz

What happens to a harmonic oscillator, initially at rest, when it is subjected to a series of external "pushes" that repeat periodically in time? The system begins to oscillate at the frequency of the external force. A system of this type is called a *forced* harmonic oscillator. If the work put into the oscillator by the external force during each cycle is positive, this force continually supplies energy to the oscillator and the amplitude of oscillation steadily increases.

If no damping force were present in the oscillator, the amplitude would increase without limit. Physically, we know that this usually does not happen, which means that every physical oscillator contains some dissipative mechanism for unloading energy before it "blows up" or is destroyed. The amplitude of a forced oscillator increases until it reaches an equilibrium value at which the damping mechanism dissipates energy at the same average rate as the forcing mechanism supplies energy.

If the forcing frequency f_e is varied so that it approaches the natural frequency f_o of the oscillator, the equilibrium amplitude increases, reaching a maximum when $f_e = f_o$. This condition of maximum response to the driving force is called *resonance*, and the frequency at which the maximum amplitude occurs is called the *resonant frequency*. The amplitude of oscillation at resonance depends on the resonant frequency and on the damping coefficient, as indicated in Figure 51-6.

Figure 51-6. When a harmonic oscillator is driven by an external force which varies periodically in time at a frequency f_e, the amplitude of oscillation A varies with f_e. When the damping factor (b) is small, the amplitude is a maximum when f_e equals the resonant (natural) frequency of the system, f_o. At the resonant frequency, the amplitude increases as the damping factor decreases.

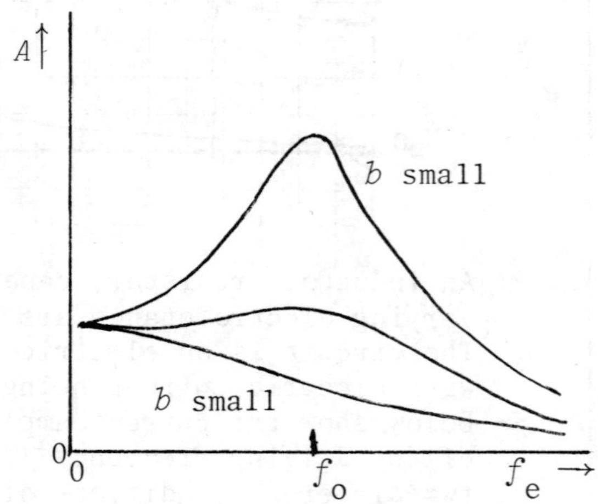

Many mechanical vibrations result from the forced oscillation of a system near its resonant frequency. Familiar examples include the vibration of an automobile body when the engine is idling and is "missing" in one cylinder, and the shimmy of the front end of a car at certain speeds when a wheel is badly out of balance. Acoustic oscillations which are the source of sound from musical instruments occur at the resonant or natural frequencies of the instruments; several types are discussed in Chapter 29. During a windstorm in 1940, the Tacoma Narrows Bridge began to oscillate violently at one or more of its resonant frequencies, until the resulting stresses exceeded the strength of the bridge and a large portion of it collapsed.

Questions

Q 51.6-1 What is the nature of the "push" or driving force which causes each of the following oscillatory motions?
- (a) a youngster "pumps himself up"in a garden swing;
- (b) the front end of a car shimmies at a speed of 50 miles/hour;
- (c) tree branches sway on a windy day;
- (d) the electric current in a desk lamp fluctuates with a frequency of 60 Hz when the lamp is turned on.
- (e) a certain window in an old church rattles when the organ plays the note of high G.

Q 51.6-2 From your experience, what happens to an initially motionless oscillator when it is subjected to an *impulsive force* (a large force applied suddenly for a short time)? Give an example.

Q 51.6-3 When a periodic driving force is out of step with the motion of the oscillator it is forcing, so that energy is removed from the oscillator, what happens to this energy?

Q 51.6-4 When a harmonic oscillator is subjected to a forcing frequency different from its natural frequency, what determines the frequency at which the system actually oscillates in the steady state?

Q 51.6-5 Give an example of a forced oscillator for which the amplitude increases to such a large value that the system itself is physically broken apart.

Q 51.6-6 How could one find the resonant frequency of a mass-spring system?

Q 51.6-7 When a forced, damped harmonic oscillator reaches its steady-state value of amplitude, what is the shortest interval of time for which it is correct to say that the energy loss due to damping is equal to the energy gain from the driving force?

Q 51.6-8 Which musical instruments include oscillating systems? What part does resonance play in their performance?

Q 51.6-9 How could such a massive structure as the Tacoma Narrows Bridge be destroyed by air molecules?

Q 51.6-10 For a system having a natural frequency of oscillation f_o, for what value of forcing frequency will the amplitude of oscillation be the largest (assuming constant force amplitude)?

Check your understanding of the material above by testing yourself on the minimum learning goals at the beginning of this section.

Answers to Minimum Learning Goals

Line 1

Line 2 Yes; about $z = 0.40$ m

Line 3 (a) $\Delta x(t) = 0.05 \text{ m} \cos(33.2\text{ s}^{-1})t$;
(b) $v(t) = -1.66 \text{ (m/s)} \sin(33.2\text{ s}^{-1})t$;
(c) $PE(t) = 0.275 \text{ J} \cos^2(33.2\text{ s}^{-1})t$;
(d) $KE(t) = 0.275 \text{ J} \sin^2(33.2\text{ s}^{-1})t$; (e) $TE = 0.275$ J.

Line 4

Line 5 $KE = 4.9 \times 10^{-2}$ J; $PE = 4.9 \times 10^{-2}$ J; $TE = 9.8 \times 10^{-2}$ J

Line 6 $\omega_0 = 4.5$ radian/sec, and the upper curve shows least damping

Line 7 damped

Line 8 Yes; about $x_0 = 0.10$ m

Line 9 $f_0 = 1000$ hertz, and the upper curve shows least damping

Line 10 220 N/m

Line 11 (a) $\Delta r(t) = 1.0 \times 10^{-11} \text{ m} \cos(3.46 \times 10^{14}\text{ s}^{-1})t$;
(b) $v(t) = -3.46 \times 10^{3} \text{ (m/s)} \sin(3.46 \times 10^{14}\text{ s}^{-1})t$;
(c) $PE(t) = 1.0 \times 10^{-20} \text{ J} \cos^2(3.46 \times 10^{14}\text{ s}^{-1})t$;
(d) $KE(t) = 1.0 \times 10^{-20} \text{ J} \sin^2(3.46 \times 10^{14}\text{ s}^{-1})t$;
(e) $TE = 1.0 \times 10^{-20}$ J.

Line 12

Line 13 $\omega = 33.2\text{ s}^{-1}$; $f = 5.28$ Hz

Line 14 undamped

Line 15 10,000 N/m

Line 16 $KE = 0.1125$ joule; $PE = 0.0375$ joule; $TE = 0.150$ joule

Line 17 damped

Line 18 (a) oscillatory; the equilibrium psoition is that occupied by the ball when the string hangs vertical;
(b) not oscillatory;
(c) oscillatory; the equilibrium position is that of the car at rest;
(d) not oscillatory.

ANSWERS to Questions and Exercises in Chapter

Section 51.1

Q 51.1-2 Any object supported from above by a thread or wire may be set into torsional oscillation by initially turning it from its equilibrium position about a vertical axis. Some 400-day clocks contain an oscillating system of this sort.

Q 51.1-4 The equilibrium position is the one taken by the system when it is not oscillating or subjected to external forces.

Q 51.1-6 The momentum of the system as it approaches the equilibrium point carries it beyond this point.

Section 51.2

Q 51.2-2 All except (c) and (e).

Q 51.2-4 (a) Graphs (a) and (h); (b) graph (h).

Q 51.2-6 (a) E_1 ; 0; E_1 ; (b) 0; E_1; 0; (c) $x_1 \leqq x \leqq x_3$; (d) x_2 ; (e) x_1 and x_3.

Q 51.2-8 (a) x_3, x_6, and x_9; (b) For x_3, $E_2 < E < E_3$; for x_6, $E > 0$; for x_9, $E_4 < E < E_5$; (c) $x_1 \leqq x \leqq x_7$; (d) $x_1 < x < x_3$ and $x_4 < x < x_6$.

Q 51.2-10 zero

Q 51.2-12 (2)

E 51.2-2

Section 51.3

Q 51.3-2 The frequency f is the number of cycles of oscillation per second, while the angular frequency ω is the rate at which the angle ωt increases, in radians per second. $\omega = 2\pi f$.

Q 51.3-4 As the motion proceeds from a point of maximum displacement, the restoring force continuously accelerates the system from rest (thus continuously increasing the velocity) until the system reaches the equilibrium point (where the displacement is zero, by definition). Beyond this point, the restoring force reverses its direction and slows the system.

E 51.3-4 (a) See Figure 51-4(a).
(b) See Figure 51-4(c).
(c) This graph has the same time-dependence as the one for $a(t)$.

E 51.3-6 (a) $-0.866\ \omega A$; (b) $-0.5\ \omega^2 A$; (c) $-0.5\ m\omega^2 A$.

E 51.3-8 (a) $md^2y/dt^2 + ky = -mg$.
(b) $y(t) = A \sin \omega t - mg/k$.

E 51.3-10 (a) 4.9×10^5 N/m; (b) 0.91 Hz

Section 51.4

E 51.4-2 (a) 3.0 N/m; (b) $v_{max} = 0.447$ m/s, $a_{max} = 1.00$ m/s², $F_{max} = 0.60$ N; (c) 0.36 Hz; (d) 2.80 s;
(e) 0.060 J, at the equilibrium point;
(f) 0.060 J, at the maxima of displacement.

Section 51.5

Q 51.5-2 The oscillator (torsional pendulum) in some "400 day" clocks has very small dissipative forces.

Q 51.5-4 It is transformed to microscopic internal energy.

Q 51.5-6 (a) kg/s; (b) a rapid decrease.

Section 51.6

Q 51.6-2 It begins to oscillate at its natural frequency. An example is a bell.

Q 51.6-4 It oscillates at the forcing frequency, in the steady state.

Q 51.6-6 Displace the system from equilibrium, then let go and observe the natural frequency of oscillation.

Q 51.6-8 All musical instruments include oscillating systems. Each instrument oscillates at one of its resonant frequencies.

Q 51.6-10 f_o

CHAPTER 53 REFLECTION AND REFRACTION OF LIGHT

Introduction

In other chapters we use the wave model for light to interpret experiments on interference and polarization, and the particle model to interpret experiments on atomic spectra and the photoelectric effect. Here we introduce a third model, in which we regard a beam of light as a bundle of *rays*. The ray model is applicable to optical phenomena in which the wavelength of light is small compared to the dimensions of objects which interact with the light.

The pattern of light reflected from a surface depends on whether the surface is optically smooth, rough, or retro-reflective. Light rays striking a smooth surface are partially reflected and partially refracted, at angles governed by the laws of reflection and refraction. The property of dispersion by a refracting medium is used to advantage in a prism, which can separate white light into its component colors.

When you have completed your study of the material in this chapter, you should be able to answer questions and work exercises like the examples in the minimum learning goals assigned by your instructor.

53.1 Models for Light. Light Rays

Minimum Learning Goal. Given the wavelength of a wave, and the significant dimensions of the physical system with which the wave interacts, be able to state whether or not the propagation of the wave fronts can be adequately described using rays. For example:

1 A 512-Hz tuning fork produces sound with a wavelength in air of 0.68 m. This sound wave passes through a window of a house. If the window is 0.80 m side and 1.2 m high, can the propagation of this sound wave be adequately described using rays? (Answer, page 53- 22, Line 5)

2 An ultrasonic tone has a wavelength of 1 mm. This sound passes through the window of example 1 above. Can the propagation of this sound wave be adequately described using rays? (Answer, page 53- 22, Line 2)

3 Green light having a wavelength of 5.5×10^{-7} m interacts with a pencil having a width of 0.5 cm. Can the propagation of this light wave be adequately described using rays?
(Answer, page 53-22, Line 3)

In discussing the total range of physical phenomena in which light takes part, it is advantageous to use three different models for light. Young's double-slit experiment and other experiments which indicate that light consists of waves are described or referred to in Chapters 28 and 55. The results of these experiments which show interference and polarization can be interpreted satisfactorily using a *wave model* for light. Experiments on atomic spectra and the photoelectric effect, however, indicate that light consists of particles, called photons; some of these experiments are described in Sections 30.2, 30.6 and 30.7. Experiments on the emission and absorption of light can be interpreted most satisfactorily using a *particle model* for light.

In this section we introduce a third model for light, called the *ray model*, in which we regard a beam of light as consisting of a bundle of rays. Each ray of light travels in a straight line as long as it remains within a medium for which the index of refraction is constant; such a medium is called a *homogeneous medium*. When a light ray arrives at the surface of the medium, or passes from one medium to another having a different refractive index, the direction of the ray can change. The following sections of this chapter discuss several examples of the behavior of light rays at surfaces.

The model of light which is most appropriate depends on the physical phenomenon one is considering, and in particular it depends on the ratio of the wavelength to the physical dimensions of the material system interacting with the light. To illustrate the conditions under which the ray model serves better than the wave model, we summarize briefly some of the results obtained in earlier chapters.

Using wave fronts we find that when an infinite plane wave traveling in one medium impinges on an infinite surface separating the first medium from a second in which the wave can also propagate, there will be a wave reflected back into the first medium, and also a wave refracted forward into the second medium. We also find the laws that determine the angles at which the reflected and refracted waves emerge. The law of reflection, for example, states that the angles of incidence and reflection are equal; see Section 53.3. Of course, in nature there are no infinite surfaces or infinite plane waves. So we must ask under what circumstances we can use the infinite-wave picture as a reasonable approximation of the real situation.

If the extent of the wave front, or of the surface, is not infinite, the reflected (or refracted) wave comes off not only at the one angle given by the law, but in a range of angles around that one. This range of angles, measured in radians, is approximately equal to the ratio of the wavelength to the size of the surface. In other words, if the extent of the surface and the wave front is many wavelengths, then the spread in angles of reflection and refraction is small.

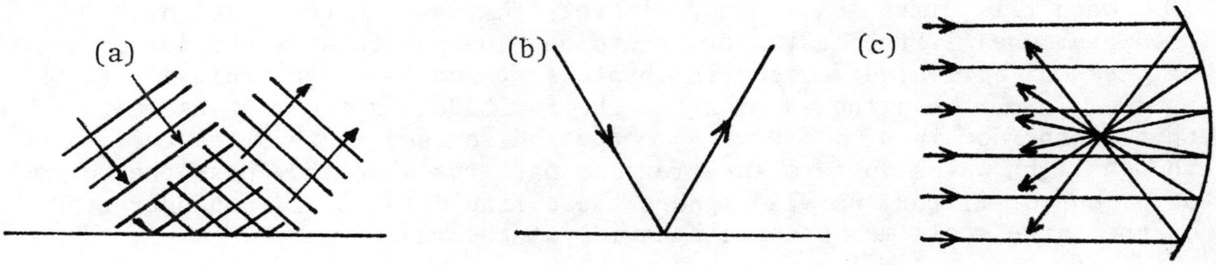

Figure 53-1. (a) A series of straight wave fronts strikes a plane mirror and is reflected as a series of straight wave fronts. (b) A ray is used to represent the incoming wave fronts shown in (a); the reflected ray represents the reflected wave fronts. Each ray is labeled with an arrowhead which indicates its direction of travel. (c) Parallel light rays striking a curved mirror are reflected in many directions; each reflected ray has the same direction as if it were reflected from a small plane mirror tangent to the curved surface at the ray's point of impact.

Figure 53-1(a) shows a "beam" having fairly narrow wave fronts incident on and reflected from a surface which separates two media. If the width of the beam is much more than a wavelength, it behaves pretty much as if it had infinite wave fronts. In such a case the behavior is more conveniently represented as in Figure 53-1(b), where the propagating beam is represented by a line perpendicular to its wave fronts. This line is called a *ray*.

Figure 53-1(c) shows a broad beam of light, represented by several parallel rays, hitting a curved mirror. Each ray is reflected according to the law of reflection applied at the point where that particular ray strikes the mirror. This representation is a good description of reality provided the mirror is not small and not too strongly curved. The size and curvature of the mirror must be such that we can subdivide the mirror into pieces so small that each is practically flat, and yet so large that each has a width equal to a large number of wavelengths.

The method of describing waves by using rays, and applying the laws of reflection and refraction to each ray, is called *geometrical optics*. Geometrical optics is a good description of what waves actually do if their wavelength is small enough compared to the geometrical dimensions of the physical objects with which the waves interact.

When a wave arrives at and interacts with a physical system, the way in which the wave interacts is determined by the *significant physical dimension* of the system. This dimension may be the width of a slit or a lens or a mirror, the spacing of lines in a grating, the spacing of atoms in a solid, or the diameter of a single atom. When this dimension is *much smaller* than the wavelength (one-tenth of the wavelength or smaller), the interaction of light with the system is best understood using the photon model; this is the domain of *quantum optics*, discussed in Chapter 30. When this dimension is approximately *the same* as the wavelength, the wave model is the most appropriate one to use; this is the domain of *physical optics*, discussed in Chapters 28 and 55. When this dimension is *much larger* than the wavelength, the ray model is the most useful; this is the domain of *geometrical optics*, discussed in this chapter. Although light waves do form interference patterns when they pass the edges of a lens or mirror, we will ignore these fine details and concentrate on the large-scale macroscopic behavior of the rays.

Questions and Exercises

Q 53.1-1 Suppose an "infinite" plane wave, such as a wave of visible light reaching the earth from the sun, is reflected from a window pane of flat glass. Most of the wave front after reflection from the surface is a plane wave, but at the edges each wave front is curved. Is this curvature a consequence of some property of the incident wave, or some property of the window, or is it a result of the interaction between the wave and the window?

Q 53.1-2 Three models used for light are the wave model, ray model, and particle model. Which model is most useful in explaining (a) the ability of x rays to produce a "picture" of a broken bone? (b) the pattern which results when x rays strike a crystal? (c) the emission of x rays in an x-ray tube? (d) Young's double-split experiment? (e) the inability to communicate by radio with an Apollo spaceship on the far side of the Moon?

Q 53.1-3 Name a region of the electromagnetic spectrum in which the waves would produce a sharply-defined shadow of (a) a sewing needle; (b) an automobile; (c) a planet. (Reference: Section 28.6)

Q 53.1-4 What are the advantages of using rays (instead of wavefronts) to represent a beam of light as it interacts with reflecting and refracting surfaces?

Q 53.1-5 The reflecting surface of a good-quality plane mirror is optically smooth; you can look into it and see your reflected image without distortion. A surface such as a sheet of aluminum or the surface of a body of water is optically rough: reflected images are distorted. From the discussion in this section, what can you infer as to the maximum size (in relation to the wavelength) of irregularities in a surface which serves as a good-quality mirror?

NOTE: Exercises 53.1-1, -2, and -3 require some knowledge of diffraction; see Chapter 55.

E 53.1-1 A plane radio wave having a wavelength of 3 meters is reflected from the flat side of a metal building 4 meters high and 20 meters long. What is the approximate range of angles for the direction of the reflected wave? Give your answer in radians, and in degrees.

E 53.1-2 A plane light wave having a wavelength of 600 nm is reflected from the side of the building described in E 53.1-1. What is the approximate range of angles for the direction of the reflected wave?

E 53.1-3 A plane radar wave having a wavelength of 3 cm is reflected from the side of the building described in E 53.1-1. What is the approximate range of angles for the direction of the reflected wave?

Check your understanding of the material above by testing yourself on the minimum learning goal at the beginning of this section.

53.2 Types of Reflecting Surfaces

An automobile driver on the highway at night must be able to see objects on and near the roadway in order to drive safely. Some objects illuminated by the headlights are just barely visible, while others seem to glow and are easily seen. What accounts for the differences in visibility of the things we see?

First, we need to distinguish between objects which are self-luminous and those which are not. An object is called *self-luminous* when it is a source of light, emitting light rays which originate in the object. Examples of self-luminous objects are the sun and stars, electric lamps, and flames. Self-luminous objects are seen when light rays which originate in the object travel out and enter the eye of an observer. Objects which are not self-luminous, on the other hand, can only be seen by reflected light or, in the case of a translucent object, by transmitted light. For an object to be seen by reflection, light rays which originate in some light source must be reflected from the object into the eye of an observer. One of the factors which determines the visibility of a non-self-luminous object is the nature of its reflecting surfaces.

There are three basic types of reflecting surfaces, each giving rise to a different kind of reflection. A surface is called *optically smooth* when the irregularities in the surface are much smaller than the wavelength of the light. Light rays striking an optically smooth surface are reflected from it with mirror-like or *specular reflection,* obeying the law of reflection (Section 53.3). Examples of optically smooth surfaces include the surface of an undisturbed liquid at rest, a cleavage plane of a crystal, and polished optical surfaces such as those on high-quality mirrors. Specular reflection is illustrated in Figure 53-1.

A surface may be called optically rough when the irregularities in the surface are comparable with or larger than the wavelength of light. Light rays striking such a surface are reflected in many directions, as shown in Figure 53.2. This type of reflection is called *diffuse reflection*. A diffuse reflector may appear to be dull or bright, depending on the percentage of the incident light which is reflected. Examples of diffuse reflectors include flat paint, concrete, earth, snow, paper, and most fabrics. Most non-luminous objects are seen by diffuse reflection.

Figure 53-2. Diffuse reflection of light occurs when incident parallel light rays which strike an optically rough surface are reflected through a wide range of directions.

Retro-reflection (also called *reflex-reflection)* occurs when light rays strike a surface so constructed as to reflect most of the incident rays directly back toward their source in a narrow range of angles. Retro-reflecting surfaces are of two basic kinds, as illustrated in Figure 53-3. Some reflectorized surfaces contain thousands of tiny glass beads per square inch, while others contain larger glass beads or reflecting cube corners. They are widely used to improve the nighttime visibility of signs, billboards, roadside markers and highway structures.

Questions and Exercises

Q 53.2-1 The aluminum skin of an airplane is useless as a mirror for the stewardess, but is an excellent reflector for radar. Explain.

Q 53.2-2 What kind of reflection occurs at an enameled surface?

Q 53.2-3 Discuss the value of reflectorized auto license plates as a safety measure.

E 53.2-1 Find several examples of reflectorized surfaces in your locality. Why are they reflectorized?

E 53.2-2 Find some places in your locality where the use of reflectorized material would improve visibility and highway safety at night. What could you do to stimulate effective action to reduce these hazards?

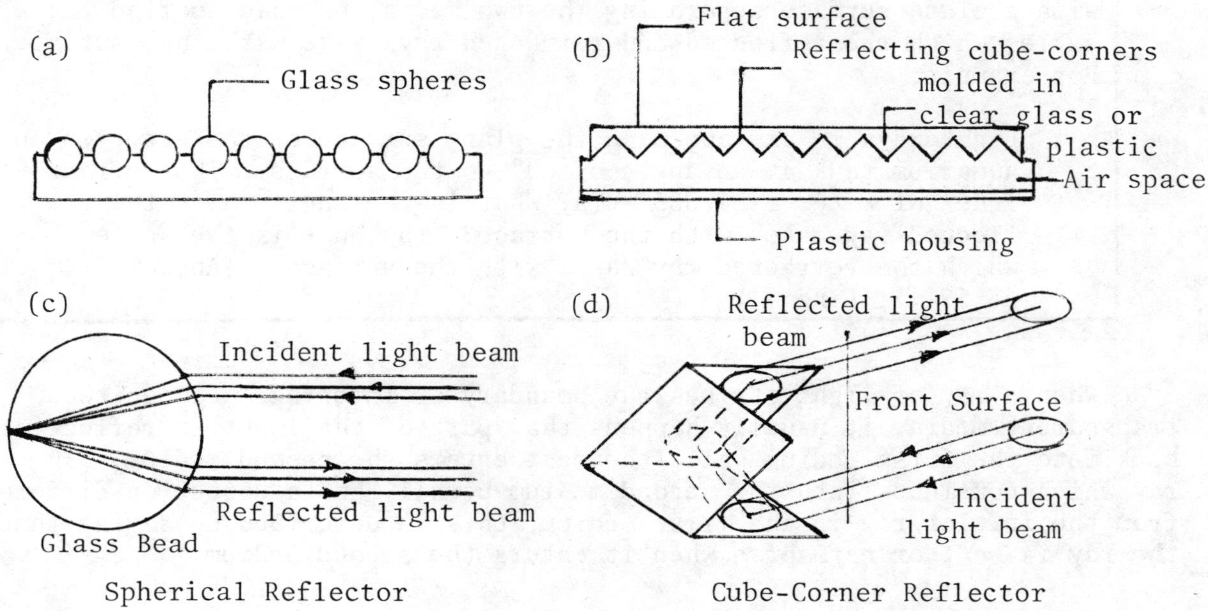

Figure 53-3. Optical systems used in retro-reflecting surfaces. (a) Surfaces having an exposed optical system often contain glass spheres or beads which are partially embedded in a substrate of plastic or paint. The optical elements which focus the incident light rays and reflect them back toward the source are exposed to the air. Rain drops on the surface alter the optical focus and tend to greatly reduce the desired retro-reflection. (b) Surfaces having a non-exposed optical system are constructed so that the optical elements which perform the focusing function are not exposed in use. The retro-reflecting optical elements may consist of reflecting cube corners (as shown here) or spheres, enclosed in a transparent housing. Because the surface exposed to the incident light rays is flat, there is a minimum loss of reflectance during rainfall. (c) In a spherical glass bead, incident light rays are refracted at the front surface, reflected from the back surface and then refracted again at the front surface so the reflected beam returns toward the source of the incident beam. (d) In a cube-corner reflector, incident light rays are reflected three times, once from each face of the cube corner. After the third reflection the rays return toward the source of the incident light beam. (Courtesy of 3M.)

53.3 Laws of Reflection and Refraction

Minimum Learning Goal. Given values for the refractive index in each of two media, and the angle which a ray in one of the media makes with a plane surface separating the two media, be able to find the angles which the reflected and refracted rays make with that surface. For example:

1 Light rays in air approach the plane surface of the water in an aquarium tank at an angle of 60° to the surface. The refractive index of water is 1.333. (a) What is the angle which the reflected ray makes with the surface? (b) What is the angle which the refracted ray makes with the surface? (Answer, p. 53-22, Line 7.)

When a ray of light strikes the boundary or interface between two transparent media, it usually happens that part of the light is reflected back into the first medium while the rest enters the second medium. The ray which continues into the second medium usually has a direction different from the initial ray in the first medium; this is described by saying that the ray is bent or *refracted* when it enters the second medium.

Figure 53-4. When an incident ray of light strikes the interface between two transparent media, part of the ray is reflected back into the first medium while part is refracted into the second medium. The angle of incidence is labeled θ_1, the angle of reflection is θ_1', and the angle of refraction is θ_2.

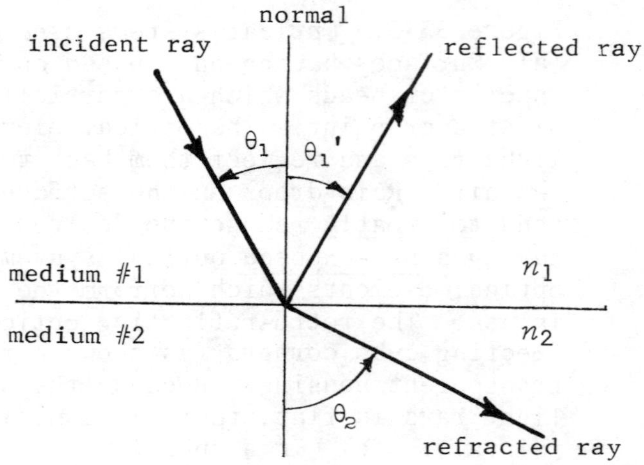

In order to state the laws which govern the angles of reflection and refraction, we assume that the interface between the two media is smooth and that it is planar, at least over a small area which includes the point at which the incoming or *incident* ray strikes the interface. The incident ray approaches the interface through the first medium, having index of refraction n_1 (Section 28.8), while the refracted ray is in the second medium, having refractive index n_2. As shown in Figure 53-4, the angles of incidence and refraction are defined with reference to the line (usually called the *normal*) which is perpendicular or normal to the interface at the point where the incident ray strikes it. The quantities needed to state the laws of reflection and refraction are defined as follows:

The *plane of incidence* is the plane containing the incident ray and the normal to the interface at the point of incidence.

The *angle of incidence* is the angle between the incident ray and the normal.

The *angle of reflection* is the angle between the reflected ray and the normal.

The *angle of refraction* is the angle between the refracted ray and the normal.

In Figure 53-4, the angle of incidence is denoted by θ_1, the angle of reflection by θ_1', and the angle of refraction by θ_2. Experimental observations show that the angles of incidence and reflection are related by the following law:

★ *Law of Reflection.* The reflected ray lies in the plane of incidence, and the angle of reflection equals the angle of incidence.

Using the notation of Figure 53-4, this law is expressed as

$$\theta_1 = \theta_1'.$$

The law of refraction was discovered about 1621 by Willebrord Snell of the University of Leyden in Holland, and also by René Descartes in France; it is usually referred to as Snell's law or Descartes' law.

★ *Law of Refraction (Snell's Law):* The refracted ray lies in the plane of incidence, and the angle of refraction (θ_2) is related to the angle of incidence (θ_1) by $n_1 \sin \theta_1 = n_2 \sin \theta_2$.

In this equation, n_1 is the refractive index of medium #1 and n_2 is the index of refraction of medium #2. As the path of a light ray is reversible, these laws may be applied to a ray which approaches a smooth interface from either the optically denser side (i.e., the side having a higher refractive index) or the optically rarer side (the side having a lower refractive index).

The laws of reflection and refraction are the foundation for all of geometrical optics. Although these laws may be established directly by experiments, they can also be derived by considering the motion of wave fronts at the interface (Section 55-1). On a more fundamental but more advanced level, they may also be derived using Maxwell's equations, which express the fundamental physical laws of electricity and magnetism. The laws of reflection and refraction do not depend on the electromagnetic properties of light, however, but only on the wave properties of light. As a result, the laws of reflection and refraction also apply to sound and other types of mechanical waves reaching the interface between two media in which the speeds of the waves are different.

Although the law of refraction is valid for glass and other materials most often used in optical systems, it is not valid for crystals which are anisotropic, that is, which have an internal structure such that the speed of light within the crystal depends on the direction in which the light is traveling. The optical properties of anisotropic media are discussed in more advanced textbooks on optics.

Worked Example 53-1

A ray of light enters a glass prism with an angle of incidence equal to 45°. The glass has a refractive index of 1.523.

(a) What is the angle of reflection?
(b) What is the angle of refraction?
(c) What is the angle through which the incident ray is deviated at the prism face?

Solution

The incident, reflected, and refracted rays may be visualized using the diagram shown here. The notation is the same as that in Figure 53-4.

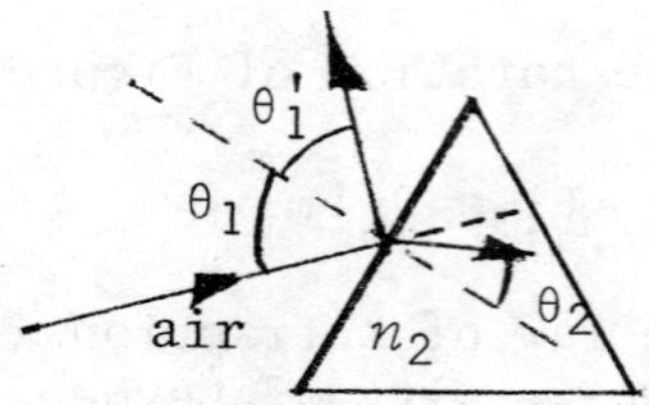

(a) Because the angle of reflection equals the angle of incidence, we can immediately state that $\theta_1' = 45°$.

(b) The angle of refraction must be computed using Snell's law, which requires that we know the refractive indexes of the two media, air and glass. The index of refraction of air is about 1.00028 for visible light, so for all but highly precise calculations we may use the approximate value $n_{air} = 1.000$. Applying Snell's law, we take $n_1 = 1$ and write

$$\sin \theta_1 = n_2 \sin \theta_2 \qquad \text{or} \qquad \sin \theta_2 = \frac{\sin 45°}{1.523} = 0.464,$$

from which the angle of refraction $\theta_2 = 27.6°$.

(c) If the incident ray were not refracted when it enters the prism, it would continue undeviated as indicated by the dashed line in the drawing. The angle between this dashed line and the normal is θ_1, and therefore the angle through which the incident ray is deviated is the difference $\theta_1 - \theta_2 = 45° - 27.6° = 17.4°$.

Questions and Exercises

Q 53.3-1 List several examples of the specular reflection of visible light which you observe in your daily experience.

Q 53.3-2 List several examples of the refraction of visible light which you observe in your daily experience.

Q 53.3-3 How can the laws of reflection and refraction be applied when the boundary between two media is not planar?

Q 53.3-4 A meter stick is partly submerged in the water in an aquarium tank.

(a) Describe the appearance of the stick when it is viewed by looking downward into the water at an angle.

(b) Describe the appearance of the stick when it is viewed through the side of the tank, looking perpendicular to the side.

Q 53.3-5 Consider a ray of light in air which is refracted into a transparent solid material. If the angle of incidence is greater than zero, which of the following statements are true?

(a) The ray is always refracted toward the normal when it enters the solid.

(b) The ray is always refracted away from the normal when it enters the solid.

(c) The angle of incidence always equals the angle of reflection.

(d) The angle of incidence always equals the angle of refraction.

E 53.3-1 Show that if a plane mirror is rotated about an axis in the plane of the mirror through an angle ϕ, a ray of light reflected by the mirror rotates through the angle 2ϕ.

E 53.3-2 Show that the index of refraction of a second medium relative to a first medium, $n = n_2/n_1$, is equal to the ratio of the speeds of light in the two media, v_1/v_2.

E 53.3-3 Using straight lines to represent light rays, make a diagram which shows the angle of incidence, the angle of reflection, and the angle of refraction for a ray incident at an angle of 40° to the normal of a medium having a relative index of refraction of 1.50.

E 53.3-4 Light rays pass from air into water at an angle of incidence equal to 30°. Find the angle of reflection and the angle of refraction.

E 53.3-5 A ray traveling initially in a medium having a refractive index of 1.64 enters a second medium having an index of 1.00. What is the angle of incidence for this wave which leads to an angle of refraction equal to 90°?

E 53.3-6 When light is refracted from air into water, for what value(s) of angle of incidence does the angle of refraction equal the angle of incidence?

E 53.3-7 Light rays strike the surface of a diamond (n = 2.418) in such a way that the angle of reflection is 60°. Compute the angle of incidence and the angle of refraction for this isotropic crystal.

E 53.3-8 Light rays strike the surface of a crystal of rutile (TiO_2, n = 2.90) at an angle of incidence of 57°. Find the angles of reflection and refraction for this isotropic crystal.

E 53.3-9 One way to measure the angle between two polished faces of a prism is to shine a beam of parallel rays at the vertex between these two faces in such a way that some of the rays are reflected from one face and some are reflected from the other face, as shown here. Show that the angle θ between the two reflected beams is twice the prism angle A.

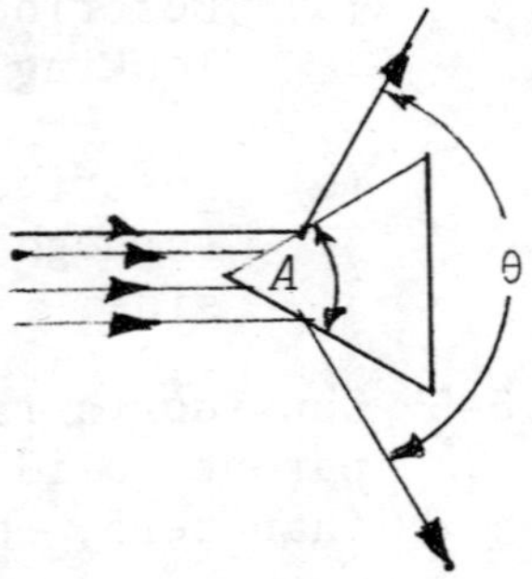

E 53.3-10 A beam of parallel light rays in air strikes the polished surface of a plate of fused quartz (n = 1.46) at an angle of 40° to the surface.

(a) What is the angle between the refracted beam and the surface?

(b) What angle of incidence is required so that the angle of refraction will be half the angle of incidence?

E 53.3-11 Light rays in air strike a polished block of translucent nylon at an angle of incidence of 56.8°. It is observed that the reflected and refracted rays make an angle of 90° with each other. What is the refractive index of nylon?

E 53.3-12 A sheet of transparent polystyrene having parallel sides and a refractive index of 1.60 is placed in contact with a sheet of glass having a refractive index of 1.51. A ray of light in the polystyrene strikes the interface between the two sheets at an angle of incidence of 42°. What is the angle of refraction in the glass?

E 53.3-13 A horizontal ray of light passes through the glass side wall (n = 1.52) of an aquarium tank and then into the water in the tank (n = 1.33). The ray in air has an angle of incidence of 60° at the outer surface of the glass.

(a) What is the angle of refraction of this ray in the water?

(b) How does the angle of refraction in the water depend on the refractive index of the glass?

Check your understanding of the material above by testing yourself on the minimum learning goal at the beginning of this section.

53.4 Total Internal Reflection. Critical Angle

Minimum Learning Goal. When you are given the index of refraction of one medium relative to another, be able to calculate the critical angle. For example:

1 The index of refraction of oil relative to water for sound waves is 1.10. For what range of angles of incidence is there total internal reflection of sound, and from which medium must the sound be traveling toward the boundary for this to occur? (Answer, p. 53-22, Line 1)

2 The index of refraction of water relative to air for light waves is 1.33. For what range of angles of incidence is there total internal reflection of light, and in which medium must the light approach the interface for this to occur? (Answer, p. 53-22, Line 8)

When a light ray approaches the interface between two optical media from the side having the lower refractive index (i.e., from the optically rarer side), the refracted ray is always bent toward the normal. As a result, the angle of refraction is always smaller than the angle of incidence, and refraction into the optically denser medium can occur for any angle of incidence up to 90°. When the ray approaches the interface from the optically denser side, however, the refracted ray is always bent away from the normal, so that the angle of refraction is always larger than the angle of incidence. The angle of refraction cannot exceed 90°, however, so there is an upper limit to the angle of incidence for which refraction can occur. This upper limit is called the *critical angle*, θ_c.

> The *critical angle* θ_c is the largest angle of incidence for which refraction can occur when light travels from an optically denser to an optically rarer medium.

For an angle of refraction equal to 90°, $\sin\theta_2 = 1$. Using Snell's law, we have

★ $$n_1 \sin\theta_c = n_2 \qquad \text{or} \qquad \sin\theta_c = \frac{n_2}{n_1}.$$

The incident, reflected, and refracted rays at the critical angle are shown in Figure 53-5. When the angle of incidence is larger than the critical angle, refraction cannot occur and the incident ray has *total internal reflection*. In this case, of course, the angle of reflection equals the angle of incidence.

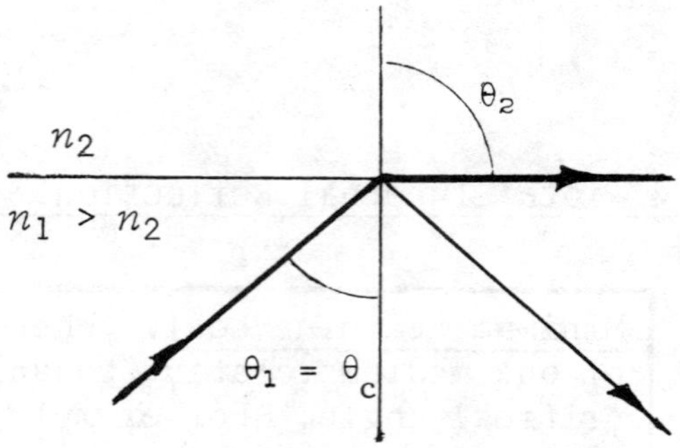

Figure 53-5. When a ray is traveling in the denser of two media and approaches a plane interface between the two media at an angle of incidence equal to the critical angle θ_c, the refracted ray travels along the interface. For angles of incidence larger than the critical angle, the ray is totally reflected back into the denser medium; there is no refracted ray.

Although this discussion is given in terms of light, it may be noted that the phenomenon of total internal reflection occurs for all types of waves when the wave speed in the initial medium is smaller than the wave speed in the second medium.

The phenomenon of total internal reflection is utilized in the design of some optical instruments, such as refractometers used to measure the index of refraction of substances. Total internal reflection is also used in optical fibers, which have been developed rapidly in recent years and now find use in a wide variety of applications. An optical fiber is a very thin rod or fiber, which in principle may be made of any transparent material, and which can conduct light entering at one end through the length of the fiber until it emerges at the other end. This phenomenon of the "light pipe" was demonstrated as early as 1870 by John Tyndall using a stream of water.

Light pipes in use today are usually made as a bundle of very thin, flexible glass fibers which can be used to conduct light around corners and into places which would be inaccessible by other means. In many cases each fiber is made as a composite, consisting of an inner fiber of glass having a high index of refraction and a uniform diameter typically in the range from 1 to 25 micrometers, surrounded by a thin coating of glass having a lower refractive index. The coating serves to protect the polished surface of the inner, conducting fiber and to separate adjacent fibers in the bundle. Light enters the flat end of each fiber over a range of angles. All of the entering light rays which strike the surface of the conducting inner fiber at an angle greater than the critical angle are totally reflected, and continue to bounce from the sides of the fiber by total internal reflection as they travel along the fiber in zigzag paths. For conducting fibers having smooth surfaces, the loss of light from the sides of the fiber is extremely small even though the light may have thousands of internal reflections per foot of fiber length. Although the optical properties of modern light pipes seem remarkable, it is of interest to note that the rods and cones of the human retina have always functioned as light pipes, because their refractive index is higher than that of the surrounding material.

Optical fibers are used in many applications, ranging from illumination of automobile instrument panels to the "fiberscope" which enables a physician to examine the interior of a patient's stomach. Because each optical fiber collects light over a range of angles, fibers are used to increase the brightness of some photographic subjects such as the display on an oscilloscope screen. Several applications in scientific work and in communications are discussed in the references at the end of this section.

Questions and Exercises

Q 53.4-1 Can total internal reflection occur for a light ray which strikes the surface of an optically denser medium? Explain.

Q 53.4-2 Can total internal reflection occur for a light ray which strikes the surface of an optically rarer medium? Explain.

Q 53.4-3 If a light ray in a prism is totally reflected at one surface, how can this ray ever escape from the prism?

Q 53.4-4 Is it proper to speak of the critical angle of a specific material? Explain.

E 53.4-1 The index of refraction of a second medium relative to a first medium is 1.10. The interface between the two media is a plane.
(a) From which medium must light approach the interface in order to have total internal reflection?
(b) For what range of angles of incidence will total internal reflection occur?

E 53.4-2 (a) What must be the relative refractive index of two media in order to have a critical angle of 50°?
(b) For these two media, for what range of angles does total internal reflection occur?

E 53.4-3 The index of refraction of glass relative to air for visible light waves is about 1.5. What is the critical angle for the air-glass interface? Sketch an experimental setup which could be used to demonstrate total internal reflection in glass.

E 53.4-4 A boat on a lake has a red spot painted on its side, just above the surface of the water. A fish is located 10 feet below the surface of the lake and 12 feet from a vertical line through the red spot. (a) Can the fish see the red spot? (b) Explain how, or why not.

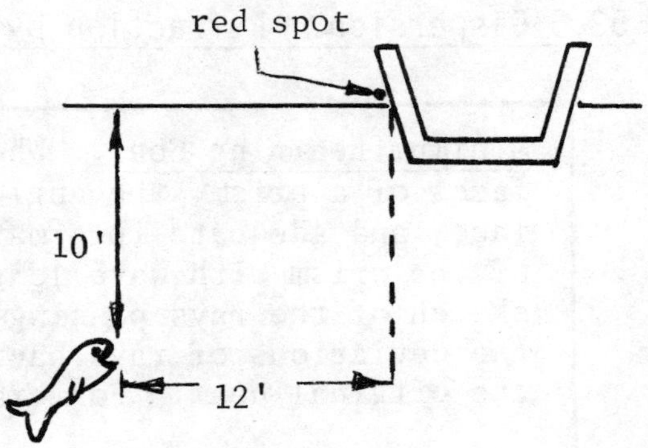

E 53.4-5 A 45°-45°-90° prism in air is used to deviate a parallel beam of light through an angle of 90° by internal reflection, as shown. The incident light beam strikes the prism perpendicularly. (a) What is the minimum value of n (index of refraction) which the prism may have if total internal reflection is to occur? (b) If the hypotenuse face of the prism is covered with a thin plane-parallel film of water, what must be the minimum n of the prism to ensure total internal reflection within the prism? (c) If the hypotenuse face of the prism is covered with a thin plane-parallel film of water, and the index of the prism is 1.5, make a neat diagram showing the paths of the light through the system.

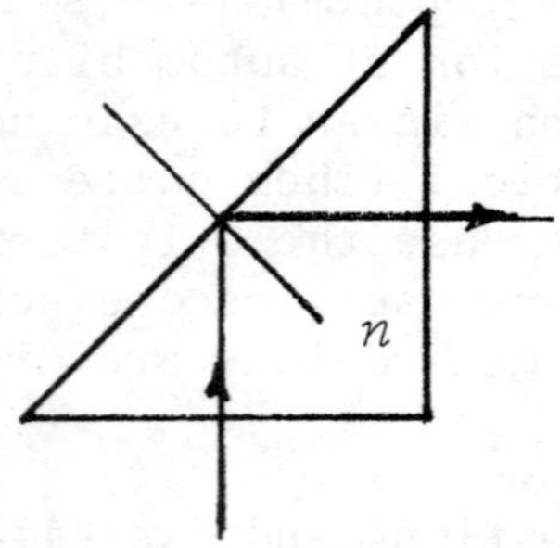

E 53.4-6 A glass fiber has a uniform diameter of 2.00 micrometers and a reflective index of 1.62. If a light ray enters one end of the fiber at an angle of 40° to the axis and is totally reflected at the fiber surface, how many reflections of this ray occur per cm of length of the fiber?

Check your understanding of the material above by testing yourself on the minimum learning goal at the beginning of this section.

For Further Reading

Cook, J. S. "Communication by Optical Fiber," *Scientific American* 229, 28-35 (November 1973)

Kapany, Narinder S. "Fiber Optics," *Scientific American* 203, 72-81 (November, 1960)

53.5 Dispersion. Refraction by a Prism

Minimum Learning Goal. When you are shown the angle between two plane faces of a prism, the angle at which a beam of light strikes one face, and adequate information on the variation of refractive index of the prism with wave length, be able to make a qualitatively correct sketch of the rays passing through and leaving the prism, showing the deviations of rays having different wave lengths contained in the original beam. For example:

1 Light from an incandescent lamp is incident on a glass prism as

shown below. The graph shows the variation of the index of refraction of the glass as a function of wavelength. Sketch the paths followed by the blue and red components of light as they pass through the prism. (Answer, page 53-22, Line 6)

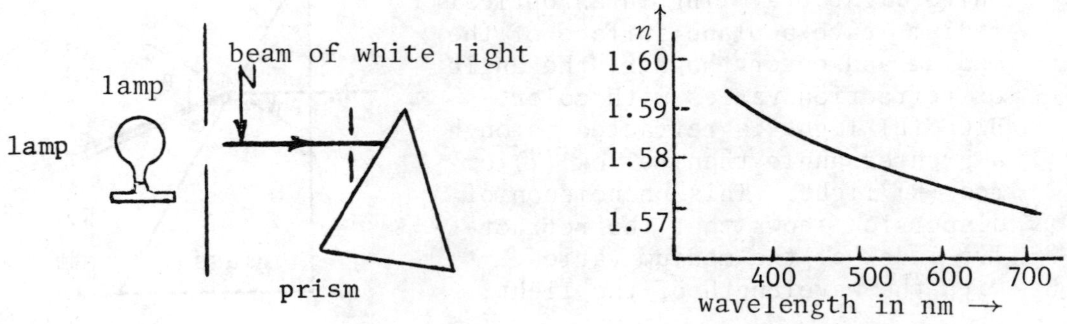

2 A neon discharge tube produces light of the following wavelengths: 540 nm, 585 nm, and 640 nm. For dense flint glass, the indices of refraction for these wavelengths are respectively: 1.657, 1.650, and 1.645. Sketch the paths followed by these components of neon light as they pass through the dense flint glass prism illustrated below. (Answer, page 53-22, Line 4)

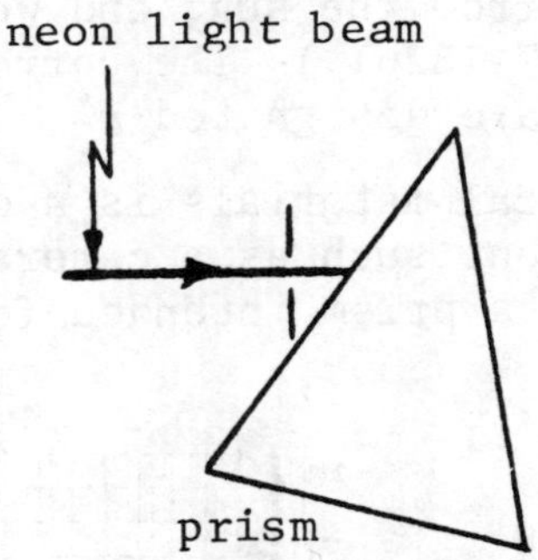

According to Snell's law, parallel light rays having the same angle of incidence as they arrive at a plane interface should all have the same angle of refraction in the second medium, provided the refractive index of the second medium relative to the first, n_2/n_1, is the same for all the rays. Observations using white light show that the angle of refraction when light leaves (or enters) a transparent medium depends on the color or wavelength of the light, as illustrated in Figure 53-6. From such observations we can conclude that the refractive index of a transparent medium has a different value for each wavelength of light. This phenomenon is called dispersion.

> *Dispersion* is the variation of refractive index of a medium with wavelength.

A medium showing high dispersion is one in which the refractive index changes rapidly with change in wavelength.

Figure 53-6. When parallel rays of white light traveling in an optical medium leave a plane surface of the medium and enter the air, the angle of refraction varies with color. Blue (B) light is refracted through a greater angle than yellow (Y) or red (R) light. This phenomenon of dispersion shows that the refractive index of the medium varies with the wavelength of the light.

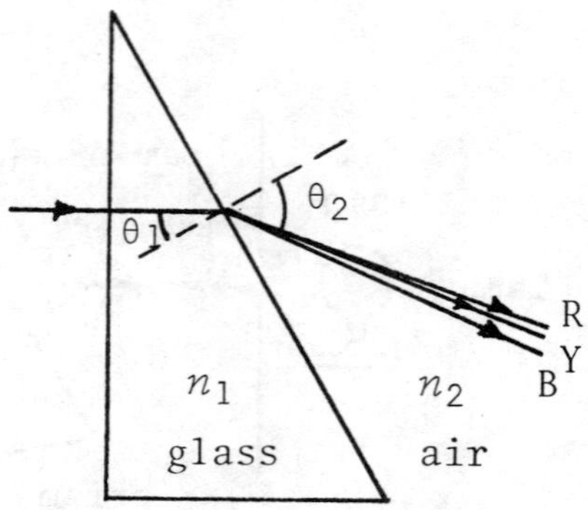

Two dispersion curves typical of a colorless transparent substance are shown in Figure 53-7, which includes the visible and adjacent regions of the spectrum. Designers of optical systems to be used for visible light usually choose three representative wavelengths as the basis for their calculations of the paths of light rays. Wavelengths commonly used for this purpose are those of three of the "Fraunhofer lines": the red C line at 653.3 nm, the yellow D line at 589.0 nm, and the blue-green F line at 486.1 nm. (The Fraunhofer lines were first observed as dark lines in the spectrum of light from the sun, and were studied by the Bavarian Joseph Fraunhofer (1787-1826).) The corresponding refractive indexes of an optical material are designated n_C, n_D, and n_F, respectively.

Although the dispersion of optical materials is a disadvantage when one is designing an optical instrument such as a camera or a microscope, dispersion is used to advantage in a prism intended for spectroscopy.

Figure 53-7. The variation of refractive index for two kinds of glass. These are typical dispersion curves for colorless transparent substances. Three wavelengths used in optical design are labeled C, D, and F.

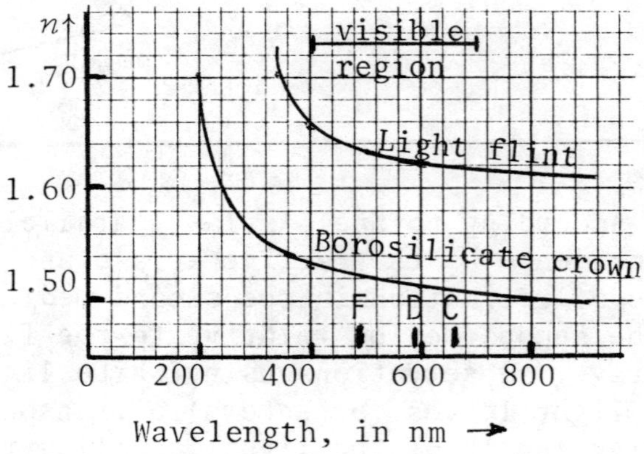

A prism is a piece of homogeneous transparent material having two plane surfaces that are not parallel to each other. Prisms used in spectroscopy are usually made to have a triangular cross-section (Figure 53-8), with flat parallel upper and lower surfaces. The two prism faces used for refraction, polished so they are very nearly planar, meet at an angle called the *prism angle*, α. The third face is often ground but not polished so that light rays striking this surface suffer diffuse reflection and do not form undesired optical images. When a ray enters a prism through one of its polished faces and leaves through the other, its final direction differs from the original one. The angle between the two rays is called the *angle of deviation*, δ; it depends on the prism angle, the angle of incidence, and the refractive indices of the prism and the surrounding material. In the following discussion we assume the prism is in air, which for practical purposes has a refractive index of unity.

Figure 53-8. Path of a monochromatic light ray through a prism having prism angle α . The normals to the prism faces where the ray intersects these faces are indicated by solid lines. At the first prism face the angle of incidence is θ_1 and the angle of refraction is θ_3. At the second prism face the angle of incidence is θ_4 and the angle of refraction is θ_2.

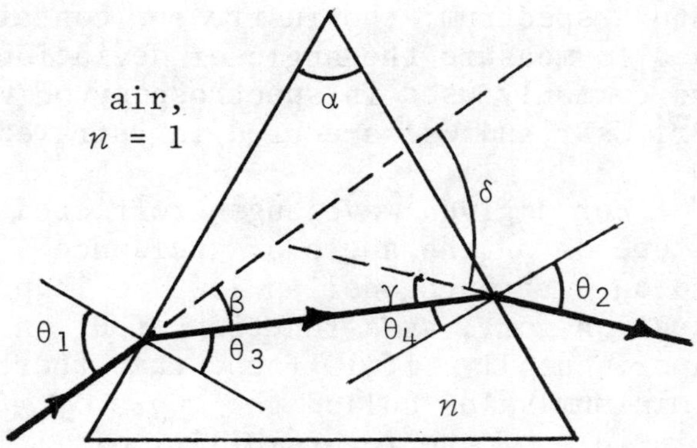

Consider the path of a monochromatic light ray through a prism, as shown in Figure 53-8. The term *monochromatic* means "single color," and refers to a light ray having a single wavelength. At the first prism face, the change in direction or deviation of the ray is

$$\beta = \theta_1 - \theta_3.$$

At the second prism face, the deviation of the ray is

$$\gamma = \theta_2 - \theta_4.$$

The total deviation of the ray is therefore

$$\delta = \beta + \gamma = \theta_1 + \theta_2 - (\theta_3 + \theta_4).$$

In the triangle formed by the two faces of the prism and the ray within the prism, the sum of the interior angle is 180°, so

$$\alpha + (90° - \theta_3) + (90° - \theta_4) = 180°,$$

or

$$\alpha = \theta_3 + \theta_4.$$

Combining these results, we find that the total angle of deviation is given by

$$\delta = \theta_1 + \theta_2 - \alpha.$$

Because the angles of refraction inside the prism depend on the refractive index n, we see that δ also depends on n, and therefore on the wavelength of the light ray. The individual colors present in an incoming parallel beam of white light are therefore deviated through different angles by a prism, forming a spectrum in the light leaving the prism. This ability to form a spectrum is the basis for the use of a prism in spectroscopy. A *spectrometer* is an optical instrument in which a prism or diffraction grating (Section 28.9) is used to disperse incoming light into a spectrum; the instrument contains a numerical scale which can be used to measure the angle of deviation for each wavelength. Glass prisms are commonly used in spectroscopy of visible light, while prisms of NaCl, KBr, CsBr and CsI are used in infrared spectroscopy.

For a given wavelength refracted by any particular prism, the deviation δ depends on the angle of incidence θ_1. The angle θ_2 also depends on θ_1 and on n through Snell's law, so in principle one could derive a theoretical equation for δ as a function of θ_1, n and α. This is not a simple equation. Experimentally, it is found that there is one value of θ_1 for which δ reaches a minimum value called the *angle of minimum deviation*, δ_m. Using the principle of the reversibility of light rays, it can be shown that δ has its minimum value when the refracted ray inside the prism is perpendicular to the bisector of α. Then $\theta_1 = \theta_2$, and

$$\delta_m = 2\theta_1 - \alpha.$$

Using Snell's law, for a prism in air where $n = 1$ it can be shown that the index of refraction is related to δ_m by

$$n = \frac{\sin \frac{\delta_m + \alpha}{2}}{\sin \frac{\alpha}{2}}.$$

Using monochromatic light, one can measure δ_m and α for a prism, and then compute the refractive index of the prism material for the wavelength of this light. This is a very accurate method for determining n, one of the basic optical properties of any material.

Questions and Exercises

Q 53.5-1 The Fraunhofer D line in the sun's spectrum results from sodium atoms in the sun's outer atmosphere. Which element accounts for the Fraunhofer C and F lines? (Hint: see Section 32.3.)

Q 53.5-2 Suppose you have a piece of clear glass, of which all the surfaces except two are rough; two of the surfaces are smooth planes which make an angle of 60° with each other, but these surfaces do not meet at a common edge. Can this piece of glass function as a prism? If it can, make a sketch showing the path of a light ray through the prism.

Q 53.5-3 Two plane surfaces of a glass prism make an angle of 45° with each other. Is it possible for a ray of visible light to travel through both of these surfaces with zero net angular deviation? Explain.

Q 53.5-4 Is it possible for a light ray which passes through a prism to be refracted only once? Explain.

Q 53.5-5 If a light ray in a prism is totally reflected at one surface, how can this ray ever escape from the prism?

Q 53.5-6 (a) List the major colors present in white light in the order of their position in a prism spectrum, beginning with the color having the smallest deviation. (b) Compare this with the spectrum formed by a diffraction grating. (See Section 28.9.)

Q 53.5-7 A prism spectrometer is to be used to measure the wavelengths of spectrum lines from a helium spectrum lamp. How one could calibrate the spectrometer, that is, determine the wavelength which corresponds to each angle of deviation?

E 53.5-1 A beam of yellow sodium light (λ = 589.3 nm) strikes an equilateral flint glass prism (n = 1.65) at an angle of incidence of 60°. Compute the deviation of this beam.

E 53.5-2 An equilateral hollow prism having thin glass sides (n = 1.50) is used to measure the index of refraction of a liquid, methylene iodine, which fills the hollow prism. The angle of minimum deviation for the sodium D line (wavelength 589.0 nm) is measured as 60°, using a spectrometer.
(a) Calculate the index of refraction of CH_2I_2 for the D line.
(b) Assuming CH_2I_2 has a typical dispersion curve, will δ_m for blue light be greater or less than 60°, for CH_2I_2?

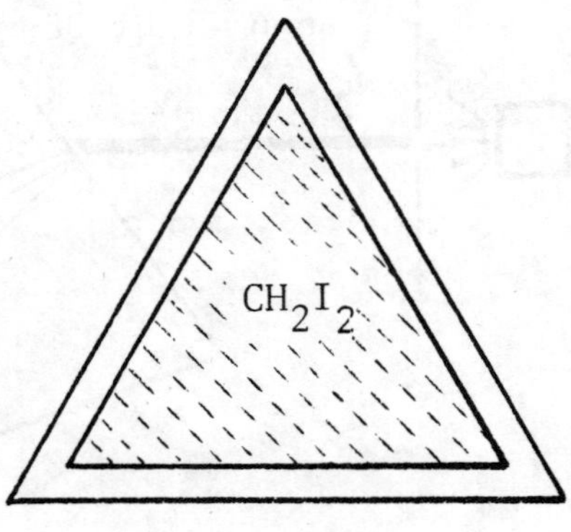

E 53.5-3 Beginning with the equation for minimum deviation, $\delta_m = 2\theta_1 - \alpha$, derive this equation for refractive index:

$$n = \sin\left[(\delta_m + \alpha)/2\right]/\sin(\alpha/2).$$

E 53.5-4 For a glass prism having α = 60°, δ_m was measured as 55° for sodium yellow light.

(a) Compute *n* at this wavelength for this glass.

(b) Compute the angle of incidence on the first prism face in this experiment.

(c) Verify by calculation that the ray inside the prism travels perpendicular to the bisector of α.

E 53.5-5 Two parallel beams of light travel in a borosilicate crown glass prism and strike the same plane prism surface. (See Figure 53-7.) One is red light from a He-Ne laser and the other is blue light from an Argon laser. Only the red beam emerges from the prism surface.

(a) Explain.

(b) Compute the approximate value of the angle of incidence of the beams on the prism face.

Check your understanding of the material above by testing yourself on the minimum learning goal at the beginning of this section.

Answers to Minimum Learning Goals

Line 1 angles ⩾ 65.4°; from the oil toward the water.

Line 2 Yes, because the dimensions of the window are much larger than a wavelength.

Line 3 Yes, because the dimensions of the pencil are much larger than a wavelength.

Line 4

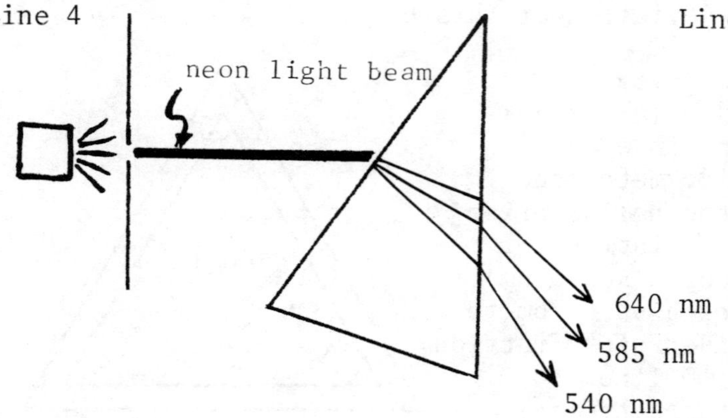

Line 6

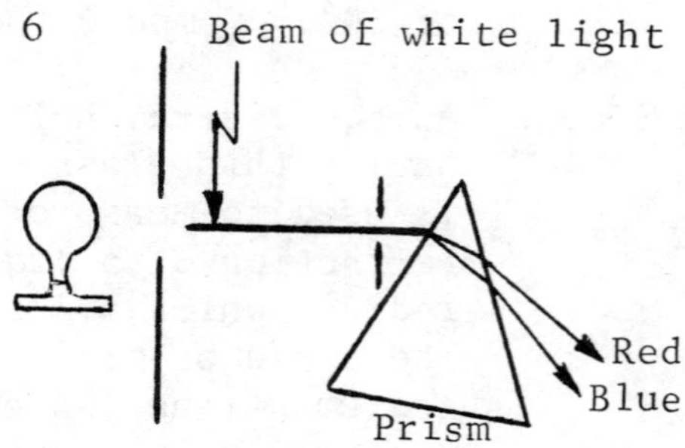

Line 5 No, because the dimensions of the window are not much larger than a wavelength.

Line 7 (a) 60°; (b) 68.0°

Line 8 angles ⩾ 48.8°; in the water.

Answers to Questions and Exercises in Chapter 53

Section 53.1

Q 53.1-2 (a) ray model; (b) wave model; (c) particle model; (d) wave model; (e) ray model.

Q 53.1-4 It is easier to draw ray diagrams than diagrams showing wavefronts, and it is easy to calculate angles of reflection and refraction.

E 53.1-2 Approximately 1.7×10^{-5} degree or 3.0×10^{-7} radian.

Section 53.2

Q 53.2-2 A combination of specular and diffuse reflection.

E 53.2-2 Effective action must be taken by agencies concerned with or responsible for highway safety, such as the police department traffic division, the state and county highway departments, the traffic safety council, automobile clubs, etc.

Section 53.3

Q 53.3-2 Examples include refraction of light reflected from objects immersed in water (in a wash bowl, swimming pool, river or lake), or in other clear liquids (salad oil, cleaning fluids), or objects viewed by looking through a container of such a liquid, when this light leaves the liquid and enters the air as it travels to the eye.

Q 53.3-4 (a) The stick will appear to be bent at the surface of the water.
(b) The stick will appear to be straight.

E 53.3-4 Angle of reflection = 30°; angle of refraction = 22.0°.

E 53.3-6 Only for 0°.

E 53.3-8 Angle of reflection = 57°; angle of refraction = 16.8°.

E 53.3-10 (a) 58.4°; (b) 86.2°.

E 53.3-12 45.2°.

Section 53.4

Q 53.4-2 Yes, it occurs when the angle of incidence exceeds the critical angle.

Q 53.4-4 No, because the critical angle is always a function of the *relative* index of refraction of two materials in contact.

E 53.4-2 (a) 1.305; (b) for angles between 50 and 90 degrees.

E 53.4-4 (a) yes; (b) The critical angle for the air-water interface is 48.6°. If the fish looks toward the boat at an angle of 48.5° to the vertical, light rays which begin at the red spot can enter its eye.

E 53.4-6 4,195.

Section 53.5

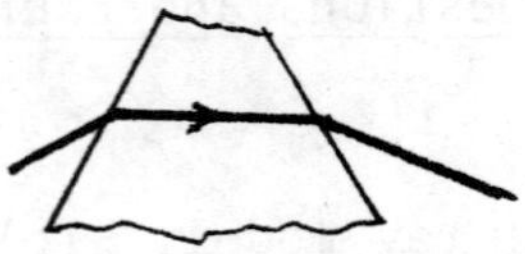

Q 53.5-2 Yes. See sketch at right.

Q 53.5-4 Yes, if it strikes one of the prism faces at an angle of incidence equal to 90°.

Q 53.5-6 (a) red, orange, yellow, green, blue, violet.
(b) The order of colors formed by a prism is just reversed from the order formed by a grating.

E 53.5-2 (a) 1.732; (b) greater than 60°.

E 53.5-4 (a) 1.69; (b) 57.5°

CHAPTER 54 MIRRORS AND LENSES

Introduction

The laws of reflection and refraction of light at a smooth surface were obtained in Chapter 53 using the ray model for light. In this chapter, ray diagrams and the mirror equation are used to study the way in which the reflecting properties of convex and concave mirrors produce optical images. The refracting properties of lenses also produce optical images, studied using ray diagrams and the lens equation. Lenses and mirrors are combined in optical systems of instruments such as the telescope and the microscope.

When you have completed your study of the material in this chapter, you should be able to answer questions and work exercises like the examples in the Minimum Learning Goals assigned by your instructor.

54.1 Objects and Images

Minimum Learning Goal. Given a diagram of light rays passing through an optical system, be able to state where objects and images are located, and for each object or image be able to state whether it is real or virtual, and whether it is erect or inverted. For example:

1 The diagram below shows some light rays which are reflected from an object (far to the left of lens 1) and then pass through two lenses. On the diagram, draw the images produced by the lenses in their correct locations, and state whether each one is real, or virtual. Also tell whether each image is erect, or inverted, relative to the object. (Answer, page 54-27, Line 4)

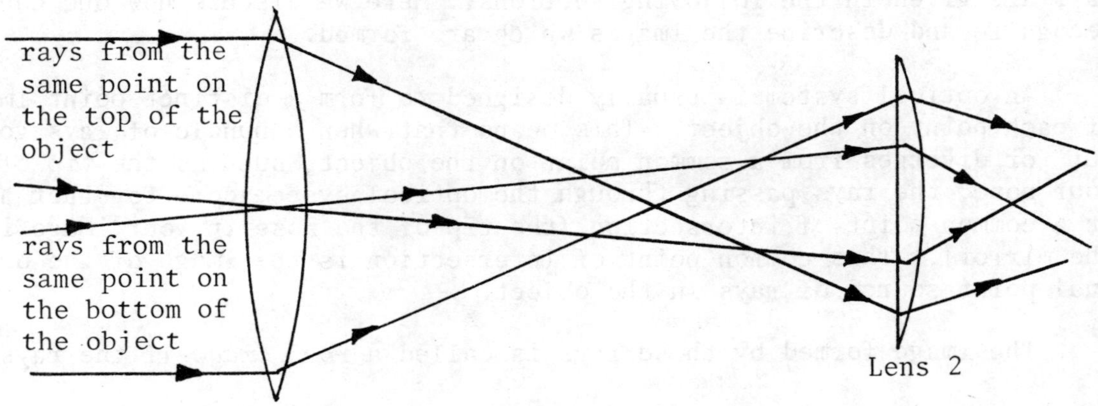

2 The diagram below shows some light rays which come from an object and are then reflected from a plane mirror and refracted by a lens. Draw the images produced by the mirror and lens in their correct locations, and state whether each one is real, or virtual. (Answer, page 54-26, Line 1)

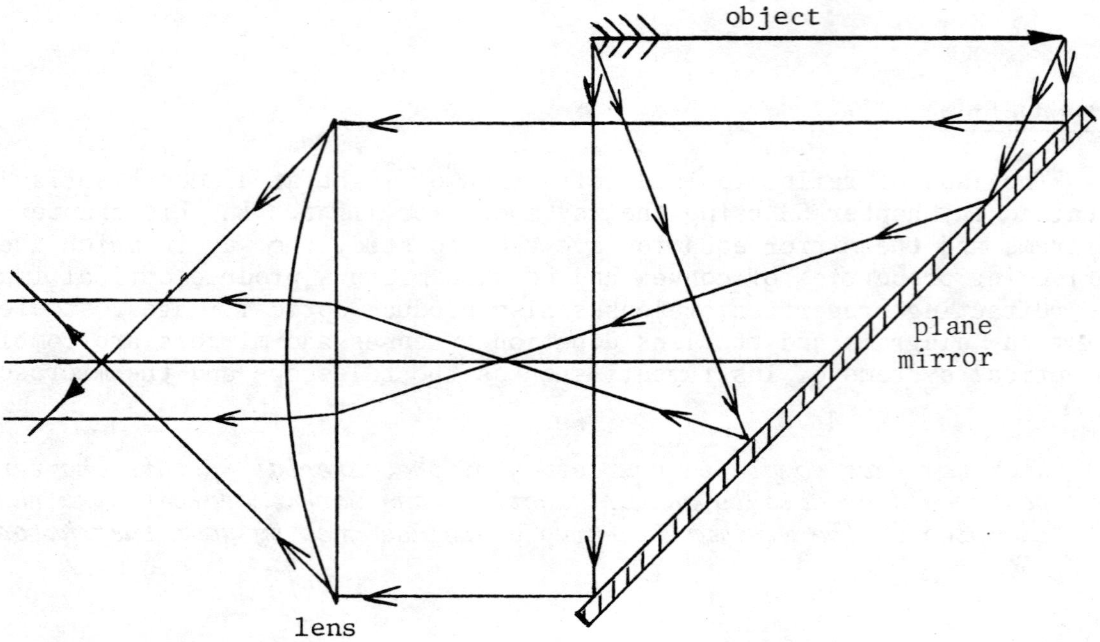

In the coming sections we discuss the paths of light rays as they are reflected by mirrors and refracted by lenses while passing through an optical system. The light rays of interest begin at points on a physical *object* and proceed through the system to form one or more *images*. In the terminology of geometrical optics, when you look at yourself in a mirror *you* are the object, and the representation of yourself that you see in the mirror is called the image. For any object having a specified location with respect to an optical system, the position(s) of the optical image(s) formed by the system are determined by the properties of the lenses and mirrors in the system, and by their locations in the system. One method of finding these positions is to trace the paths of light rays through the system. Procedures for making drawings which show these rays are given in the following sections. Here we discuss how one can recognize and describe the images which are formed.

An optical system is usually designed to form a distinct point image of each point on the object. This means that when a bundle of rays goes out or diverges from a common point on the object, such as the tip of your nose, the rays passing through the optical system come together again at a common point of intersection (the tip of the nose in your image in the mirror). This common point of intersection is the image of the original point source of rays in the object.

The image formed by these rays is called a *real image* if the rays come together as a *converging* beam, so that they actually intersect at the image. A real image can be seen by an observer if a *screen,* such as a projection screen used to view slides or simply a flat sheet of white paper, is placed at the location of the image and viewed from the same side as that from which the light rays are coming. Sometimes the rays which form an image do not actually come together as a converging beam, but rather they come out of the optical system as a *diverging* beam. Viewed as a geometrical lines, the rays still have a common point of intersection, but that point is found by extending the rays backward into a region where physically they do not exist. In such a case, the common point of intersection is called a *virtual image*. The rays do not actually go through the virtual image, but appear to be coming from it. A virtual image can be seen by an observer who looks into the optical system so that these diverging rays enter his eye, focusing his vision on the point where the rays seem to originate. This is what you do when you look at yourself in a mirror. A virtual image cannot be projected on a screen.

When light rays go into an optical system from an extended object (not simply a point), the system forms a point image of each point on the object, so that the image of the object is an extended image. When the image has the same orientation in space as the object, it is called an *erect image* When the image is upside down or has the opposite orientation in space, it is called an *inverted image* When making ray diagrams to portray objects and images, it is convenient to represent an object by an arrow having a head and a tail. Then it is easy to see whether an image of the arrow is erect or inverted with respect to the object.

Worked Example 54-1

An optical system consisting of a converging lens L and a plane mirror M forms two successive images of an object O (represented by an arrow pointing upward). The actual paths of some light rays proceeding through the system from the head and tail of the arrow are shown in the diagram below.

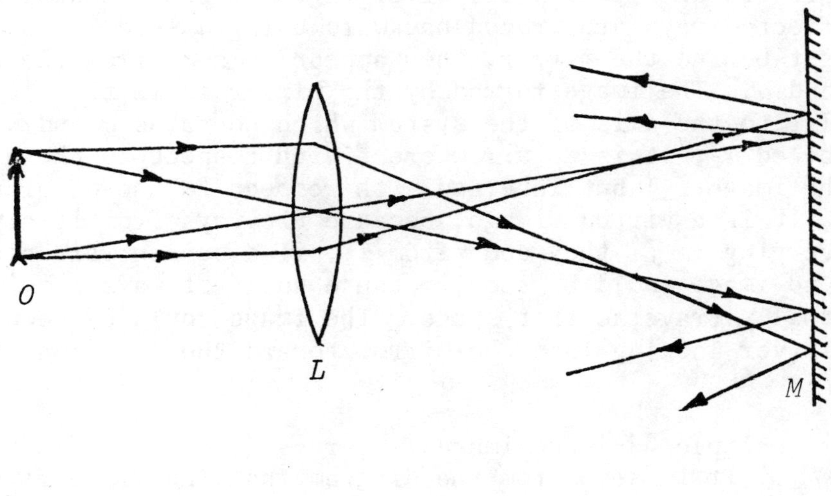

(a) What is the location of the image formed by the lens? Is this image real, or virtual? Is it erect, or inverted?

(b) The image formed by the lens serves as a *real object* for the mirror, which then forms an image of this object. What is the location of this image? Is it real, or virtual? Is it erect, or inverted with respect to this object?

Solution

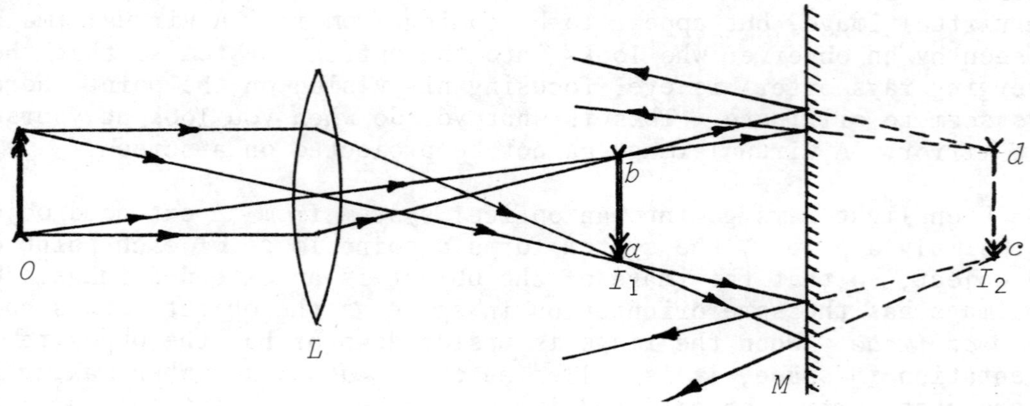

(a) The two rays proceeding from the head of the arrow object converge at a point to the right of the lens, labeled a. The two rays proceeding from the tail of the arrow object converge at a point labeled b. The image formed by the lens is in the plane perpendicular to the axis of the lens which contains a and b,and is labeled I_1; it is an arrow inverted with respect to the object. It is a real image, because the rays actually come together to form this image; if a screen were placed in this image plane, the image would be seen clearly projected on the screen.

(b) The rays from I_1 diverge to the right and are reflected by the mirror to the left, still diverging. If the paths of these reflected rays are traced backward using dashed lines, to the right behind the mirror, they appear to come from the points c and d. The image formed by the mirror is in the plane perpendicular to the axis of the system which contains c and d,and is labeled I_2; it is an arrow erect with respect to the object (the image I_1) but inverted with respect to the original object O. It is a virtual image, because the rays from it are diverging *as if* they come from I_2; if a screen were placed at I_2 no image would be seen, because no light rays from the object actually traverse that space. The image could be seen by an observer looking into the mirror toward the location of I_2.

In Worked Example 54-1 the image I_1 serves as a real object for the plane mirror M_1. It is seen from the diagram that the image of this object is located behind the mirror surface at a distance equal to the distance of the object from the front of the mirror. This result holds for any image formed by a plane mirror: the image distance is equal to the object distance. It is also evident that the object is erect with respect to the image, and of the same size. These results for a plane mirror are a special case of image formation by a curved mirror (discussed in the following sections), in the limit where the radius of curvature reaches infinity.

Questions and Exercise

Q 54.1-1 Give three examples of a household or laboratory device which contains an optical system. In which of your examples is the system used to form an image of an object?

Q 54.1-2 If a slide projector is not properly focused, so that each point on the object (the slide) does not form a point image on the screen, how would you describe the appearance of the image or picture on the screen?

Q 54.1-3 Which can be viewed on a projection screen: a real image, or a virtual image? How can a person see the kind of image which cannot be projected on a screen?

Q 54.1-4 When you look at your face in a mirror, do you see a real image or a virtual image? Do you see a point image, or an extended image?

Q 54.1-5 Give an example of an optical instrument that produces (a) a real image; (b) a virtual image.

Q 54.1-6 Suppose you form an image of a candle flame in a plane mirror. How can this image be made to serve as an object which will produce a second image?

E 54.1-1 Show that if a plane mirror is rotated about an axis in the plane of the mirror through an angle ϕ, a ray of light reflected by the mirror rotates through 2ϕ.

Check your understanding of the material above by testing yourself on the minimum learning goal at the beginning of this section.

54.2 Image Formation by Convex Mirrors

Minimum Learning Goal. Given the radius of curvature or focal length of a convex mirror, and the position of a real or virtual object relative to the mirror, (1) be able to make a qualitatively correct diagram showing the positions of the object, mirror and image and the paths of at least three rays as they proceed through the system, and (2) be able to calculate the position of the image formed by the mirror and state whether this image is real or virtual, and erect or inverted. For example:

1 A convex spherical mirror has a radius of curvature of 1.00 m. An arrow 12 mm in height is placed with its tail on the axis of the mirror 20 cm from the vertex, its head 12 mm from the axis, and is illuminated. Construct a ray diagram using at least three rays to locate the image of this arrow. Calculate the position of the image, and state whether it is real or virtual, and erect or inverted with respect to the actual arrow. (Answer, page 54-27, Line 3)

Mirrors having curved surfaces are widely used in scientific work, and some are also used in everyday life. A mirror used for shaving has its center dished or caved inward, to give an enlarge image of the shaver's face; mirrors shaped in this way, called *concave* mirrors, are discussed in Section 54.3. Some rearview mirrors for cars and trucks have a circular portion in which the center bulges outward toward the observer, to give a wide-angle field of view; a mirror shaped in this way is called a *convex* mirror. Curved mirrors used in scientific instruments may be made with spherical, paraboloidal, or ellipsoidal surfaces. Our discussions in this chapter are limited to spherical mirrors, assuming the mirrors are ideal and neglecting their aberrations. We begin our study with convex mirrors, because their image-forming properties are most similar to those of plane mirrors.

There are three methods for locating an image formed by a mirror or lens: (a) experiment, (b) geometrical construction of a ray diagram, and (c) calculation using an appropriate equation. In constructing ray diagrams, we place the object at the left and trace the paths taken by rays which travel to the right and enter the optical system under study.

This section gives methods for finding images formed by a convex mirror. The rays from any object point diverge as they approach the mirror, and after reflection they diverge even more because of the convex surface of the mirror. This behavior is illustrated in Figure 54-2 by four rays reflected from the mirror. Because the reflected rays always diverge when the object is at the left of the mirror, the image in such a case is always virtual, located behind (at the right of) the mirror, and smaller than the object. If the object moves closer to the mirror, the image also moves closer to the mirror.

In a ray diagram such as Figure 54-1 or 54-2, a spherical mirror is represented by a circular arc, a cross-section of the mirror through its center. An imaginary line perpendicular to the surface of the mirror at its center is called the *axis* of the system containing the mirror. The axis intersects the mirror at a point called the *vertex, V*. Object and image distances are measured from *V* along the axis.

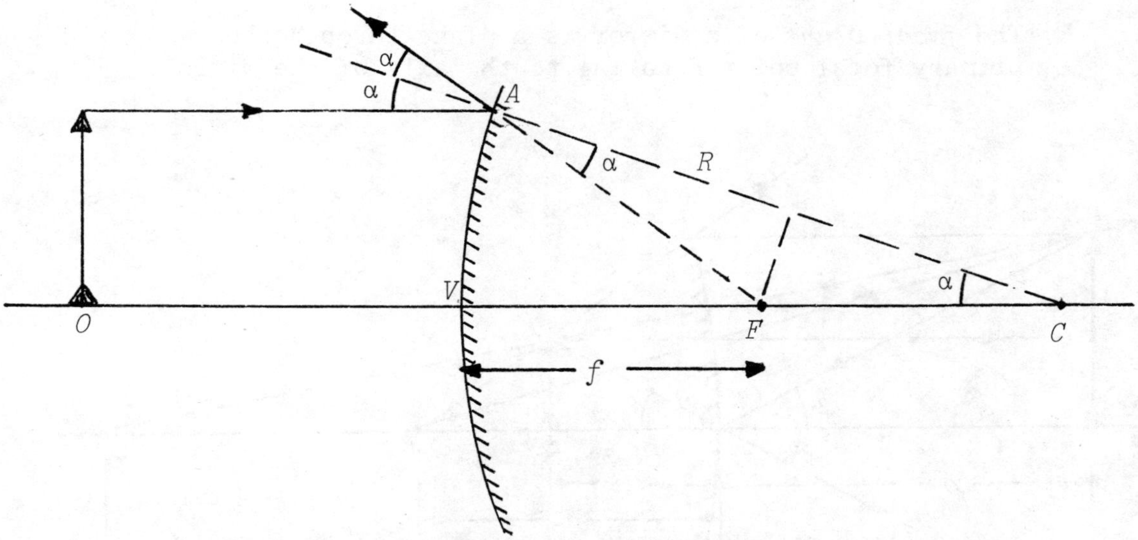

Figure 54-1. A ray which travels parallel to the axis as it approaches a convex mirror is reflected from the mirror in the same direction as if it originated at the primary focal point, F. The focal length f is half the radius of curvature R.

Consider a ray from the head of an object at O traveling parallel to the axis as it strikes the surface of a spherical convex mirror at point A (Figure 54-1). Point C is the *center of curvature* of the mirror, so the line CA is perpendicular to the surface at A and has a length equal to the radius of curvature, R. When the incoming parallel ray is reflected at A it obeys the law of reflection (Section 53.3), so the angle of reflection α is equal to the angle of incidence. When the reflected ray is projected backward using a dashed line, it intersects the axis at the point labeled F.

In the traingle AFC, the angle at A is equal to α because it is a vertical angle of the angle of reflection, and the angle at C is equal to α because its two sides are parallel to the sides of the angle of incidence. Therefore, triangle AFC is isosceles, and a line drawn from F normal to line AC intersects AC at its midpoint. When α is small, the length of $FC \cong R/2$. Because the length $VC = R$, by subtraction the length $VF \cong R/2$. The distance from V to F is called the *focal length* of the mirror, designated by f; we see that

$$f = R/2 \quad \text{when } \alpha \text{ is small.}$$

This discussion applies to any ray approaching the mirror parallel to the axis, provided its distance from the axis is small compared to R (so that α is small). We conclude that for all these rays which approach the mirror parallel to its axis, the reflected rays diverge as if they came from F, as shown in Figure 54-4(b) on page 54-14. F is called the *primary focal point* of the mirror, and the plane through F normal to the axis is called the *focal plane*.

> The *focal plane* of a mirror is a plane through the primary focal point F normal to the axis of the mirror.

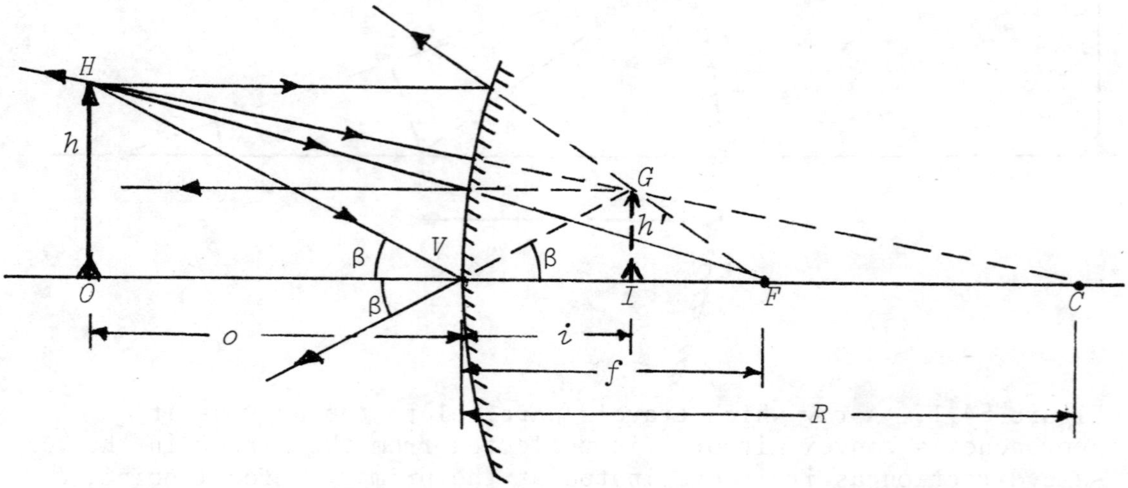

Figure 54-2 . A ray diagram used to locate the image of a real object formed by a convex mirror. From the point H on the object, four rays are drawn to the mirror and then reflected. All these reflected rays appear to come from the image point G.

Figure 54-2 depicts the same object and mirror as Figure 54-1, and shows four rays from the head H of the arrow which strike the mirror and are reflected. It is seen that all four reflected rays diverge as if they came from the same point, the head G of the image at I. The use of a ray diagram to find the image of an object is sometimes called the *graphical method*. In making such a diagram for a mirror, four rays may be drawn from any point on the object located off the axis of the system:

★ (1) A ray parallel to the axis, which travels directly away from (through) F after reflection.

★ (2) A ray directly toward (through) F, which travels parallel to the axis after reflection.

★ (3) A ray directly toward (through) C, which is reflected back on itself.

★ (4) A ray to the vertex of the mirror, which is reflected at an equal angle on the other side of the axis.

These statements apply to a *convex* mirror, if the parenthetical words are ignored. The alternative words in parentheses apply to a *concave* mirror (Section 54.3). While any two reflected rays which intersect (after being projected backward) are enough to determine the location of a point on the image, it is recommended that the student use at least three rays in order to improve accuracy and avoid errors.

Using Figure 54-2, we now derive an equation relating the focal length f, *object distance* o from the object to V, and *image distance* i from the image to V. The height of the object (measured normal to the axis) is h, and the height of the image is h'. In triangle HOV, $\tan \beta = h/o$. In triangle GIV, $\tan \beta = h'/i$. Equating these values for $\tan \beta$ gives

$$\frac{h}{o} = \frac{h'}{i}, \qquad \text{from which} \qquad h' = hi/o.$$

Next consider the ray from H normal to the mirror; the backward projection of the reflected ray passes through G and intersects the axis at C. Triangles HOC and GIC are similar, so the ratios of corresponding sides are equal:

$$\frac{HO}{OC} = \frac{GI}{IC} \qquad \text{or} \qquad \frac{h}{o + R} = \frac{h'}{R - i}.$$

Substituting $h' = hi/o$ and $R = 2f$ into the last equation above, cancelling h from both sides and cross-multiplying to clear the fractions, leads to

$$i(o + 2f) = o(2f - i).$$

Eliminating parentheses, combining similar terms and solving for f then leads to

$$\frac{1}{f} = \frac{1}{i} - \frac{1}{o}.$$

It is advantageous to modify this equation to a form convenient for use with both convex and concave mirrors. To do this, we adopt the following *sign conventions:*

★ (1) Rays from the object O initially travel from left to right.

★ (2) o is positive when O is at the left of the mirror;
o is negative when O is at the right of the mirror.

★ (3) i is positive when the image I is at the left of the mirror;
i is negative when I is at the right of the mirror.

★ (4) f is positive for a concave mirror;
f is negative for a convex mirror.

Using these conventions, an object distance is negative when the object is virtual, and an image distance is negative when the image is virtual. In a similar vein, the focal length is negative for a convex mirror because the primary focal point is a *virtual focus* for the rays; the rays do not actually meet there, but only appear to do so.

For the example in Figure 54-2, o is positive but both i and f are negative quantities. The equation derived above will correspond to this if f is changed to $-f$ and i is changed to $-i$. The equation then becomes

★ $$\frac{1}{f} = \frac{1}{i} + \frac{1}{o}.$$ Mirror Equation

This equation is applicable to any calculation of an image formed by a convex or concave mirror, observing the sign conventions given above.

Worked Example 54-2

An object is placed 50 cm in front of a convex mirror having a focal length of 40 cm. What is the location of the image?

Solution

The image distance can be computed using the mirror equation, rearranged to read

$$\frac{1}{i} = \frac{1}{f} - \frac{1}{o}.$$

In this example, $f = -40$ cm and $o = +50$ cm. Substituting, we obtain

$$\frac{1}{i} = \frac{1}{-40} - \frac{1}{50} = \frac{-5}{200} - \frac{4}{200} = \frac{-9}{200},$$

from which $i = -22.2$ cm. This example corresponds to the distances in Figure 54-2, drawn to a scale of 1 mm = 1 cm.

From a ray diagram it is easy to see whether an image is erect or inverted with respect to its object, and also whether an image is larger or smaller than its object. Using an object height h and image height h' measured normal to the axis, the *magnification* M of an optical system is defined as

$$M = h'/h.$$

The magnification is positive when the image is erect, negative when the image is inverted. One can trace the ray from any object point (such as H in Figure 54-2) to the vertex of the mirror, and then to the corresponding image point, and obtain similar triangles. Then using our sign conventions it can be shown that

★ $$M = -i/o.$$

This equation gives the magnification for convex, plane, or concave mirrors for any object distance.

The mirror equation can be used to draw some general conclusions about the image-forming properties of convex mirrors; proofs of these statements are left to the Exercises. If the object distance is positive the object is real; a real object always has a virtual image, erect and smaller than the object. If the object distance is negative the object is virtual; this can occur if the object of the convex mirror is an image formed by another mirror or lens. A virtual object located between the mirror and its primary focal point always has a real image, erect and larger than the object. A virtual object located beyond (at the right of) the primary focal point always has an inverted virtual image.

Questions and Exercises

Q 54.2-1 (a) Is the outer surface of a sphere convex, or concave?
(b) Is the inner surface of a sphere convex, or concave?

Q 54.2-2 What is the difference between a concave mirror and a convex mirror?

Q 54.2-3 Which of the rays striking a mirror obey the law of reflection?

Q 54.2-4 What is required to form a real image using a convex mirror?

Q 54.2-5 (a) For a convex mirror, what is the range of object distances for which the image is real?
(b) For a convex mirror, what is the range of object distances for which the image is virtual?

Q 54.2-6 Is it possible for the same convex mirror to form both real and virtual images at the same time? Explain.

E 54.2-1 Beginning with the equations $h' = hi/o$ and $h/(o + R) = h'/(R - i)$, fill in the missing steps in the derivation of the mirror equation given in this section.

E 54.2-2 Using the mirror equation, show that for a plane mirror (having $R = \infty$) the image distance equals the negative of the object distance.

E 54.2-3 Beginning with the equation $M = h'/h$, show that $M = -i/o$ for a convex mirror.

E 54.2-4 Use a ray diagram to show that a real object in front of a convex mirror always has a virtual image, erect and smaller than the object.

E 54.2-5 (a) Use the mirror equation to show that for a virtual object located between a convex mirror and its primary focal point, the image is real.
(b) Use a ray diagram to show that for this case the image is erect and larger than the object.

E 54.2-6 (a) Use the mirror equation to show that for a virtual object located farther behind a convex mirror than its primary focal point, the image is virtual.

(b) Use a ray diagram to show that for this case the image is inverted.

E 54.2-7 Using a convex mirror, where must an object be located so that its image distance equals the object distance, with both image and object on the same side of the mirror?

E 54.2-8 A convex mirror having a radius of curvature of 1.20 m forms a virtual image 25 cm behind the mirror. Where is the object?

E 54.2-9 The manager of a supermarket wants to install mirrors so that from his office he can observe people in the aisles between the shelves. He wants the images of the most distant people to be not more than 50 cm behind the mirror. What type of mirror should be used, and what radius of curvature should it have?

E 54.2-10 A convex rear-view mirror on a truck has a radius of curvature of 10 cm. The driver looks into the mirror and sees a police car 15 meters behind him.

(a) Where does this mirror form an image of the police car?
(b) Is the image real, or virtual?
(c) Where does the truck driver have to focus his vision to see the car clearly?
(d) Is the image erect, or inverted?

Check your understanding of the material above by testing yourself on the minimum learning goal at the beginning of this section.

54.3 Image Formation by Concave Mirrors

Minimum Learning Goal. Given the radius of curvature or focal length of a concave mirror, and the position of a real or virtual object relative to the mirror, (1) be able to make a qualitatively correct diagram showing the positions of the object, mirror and image and the paths of at least three rays as they proceed through the system, and (2) be able to compute the position of the image formed by the mirror and state whether this image is real or virtual, and erect or inverted. For example:

1 A concave spherical mirror has a radius of curvature of 1.00 m. An arrow 12 mm in height is placed with its tail on the axis of the mirror 120 cm from the vertex, its head 12 mm from the axis, and is illuminated. Construct a ray diagram using at least three rays to locate the image of this arrow. Compute the position of the image, and state whether it is real or virtual, and erect or inverted with respect to the actual arrow. (Answer, page 54-27, Line 5)

Images formed by concave mirrors may be found by experiment, by construction of ray diagrams, or by calculation using the mirror equation. The rays from any object point diverge as they approach the mirror, but after reflection they diverge less because of the concave surface of the mirror. This behavior is illustrated in Figure 54-5 by four rays reflected from the mirror. If the convergence caused by the mirror is greater than the divergence of the rays from the object, the reflected rays actually converge in space and form a real image in front of the mirror. If the convergence caused by the mirror is less than the divergence of the rays from the object, however, the reflected rays continue to diverge and do not form a real image. In this case, the reflected rays appear to come from a virtual image located behind the mirror. As shown below, the nature of the image is determined by the position of the object with respect to the primary focal point of the mirror.

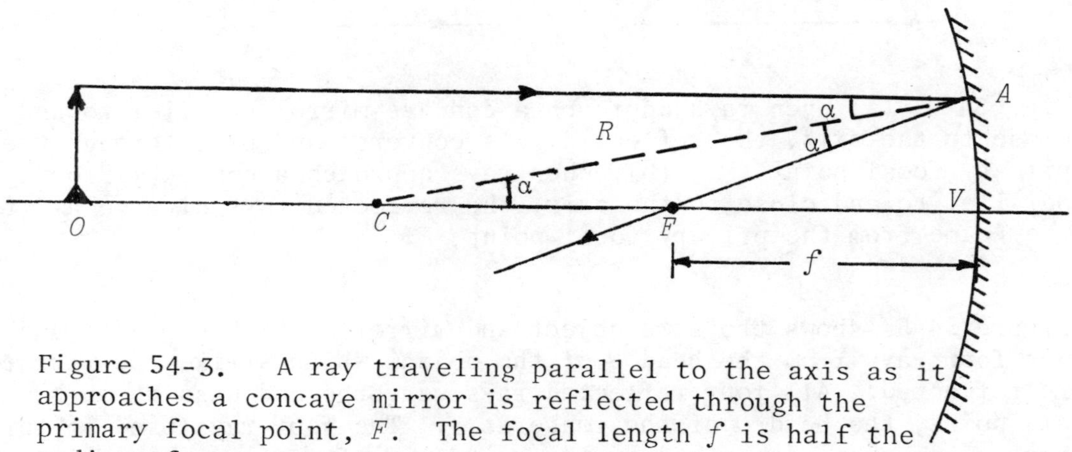

Figure 54-3. A ray traveling parallel to the axis as it approaches a concave mirror is reflected through the primary focal point, *F*. The focal length *f* is half the radius of curvature *R*.

Consider a ray from the head of an object at *O* traveling parallel to the axis as it strikes the surface of a spherical concave mirror at point *A* (Figure 54-3). Point *C* is the *center of curvature* of the mirror, so the line *CA* is normal to the surface at *A* and has a length equal to the radius of curvature, *R*. When the ray is reflected at *A*, the angle of reflection α is equal to the angle of incidence. The reflected ray intersects the axis at the point *F*. In the triangle *AFC*, the angle at *C* is equal to α because it is an alternate interior angle with the angle of incidence. Therefore, triangle *AFC* is isosceles. The distance from *V* to *F* is the *focal length* of the mirror, designated by *f*. Using reasoning similar to that applied to Figure 54-1 in Section 54.2 , it is not hard to show that

$$f = R/2 \quad \text{when } \alpha \text{ is small.}$$

This discussion applies to any ray approaching the mirror parallel to the axis, provided its distance from the axis is small compared to *R*. We conclude that all the rays which approach the mirror close to and parallel to the axis are reflected through *F*, the *primary focal point* of the mirror, as shown in Figure 54-4 (a). The plane through F normal to the axis is called the *focal plane*. For a paraboloidal mirror, all rays which approach the mirror parallel to the axis are reflected through the primary focal point, regardless of whether they are close to the axis or farther from it.

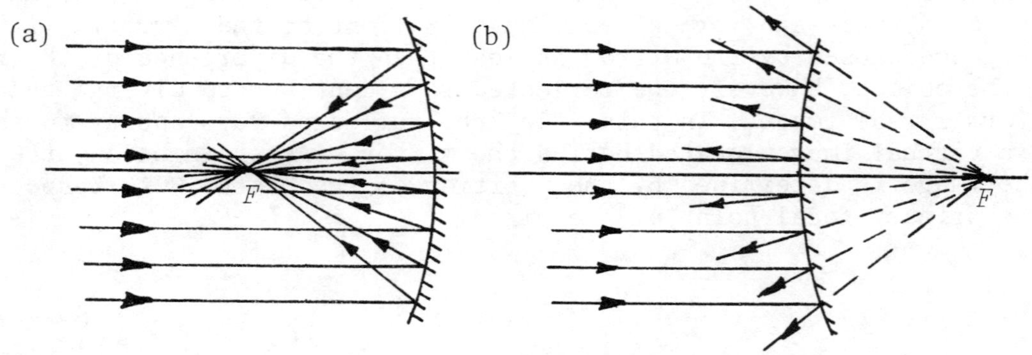

Fig. 54-4. (a) When rays approach a concave mirror parallel to and close to the axis, the reflected rays converge and pass through the primary focal point, F. (b) When rays approach a convex mirror parallel to and close to the axis, the reflected rays diverge as if they came from the primary focal point, F.

Figure 54-5 shows the same object and mirror as Figure 54-3, and includes four rays from the head H of the object which strike the mirror and are reflected. All four reflected rays converge and pass through the same point, the head G of the image at I. The four rays used in this ray diagram are those described in the guidelines for ray diagrams listed on page 54-8, where the alternative words in parentheses apply to a concave mirror.

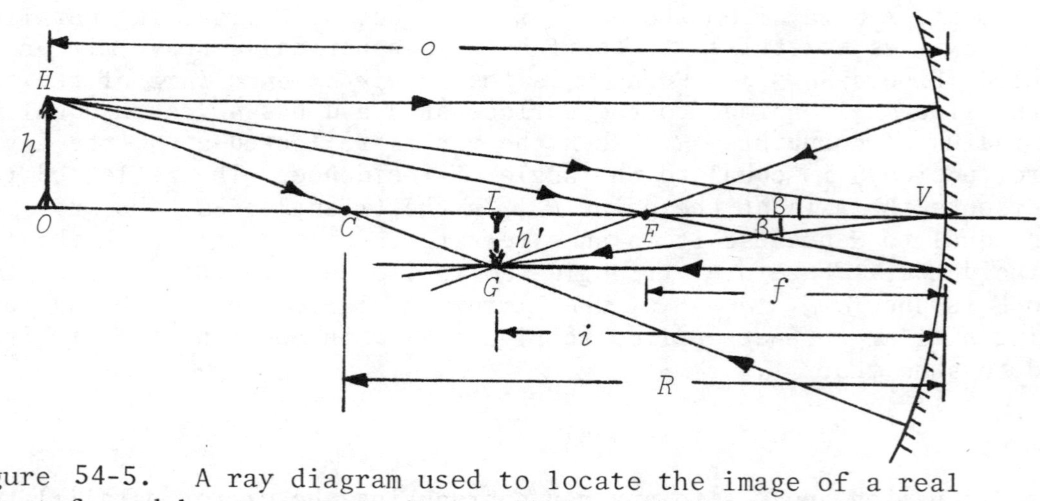

Figure 54-5. A ray diagram used to locate the image of a real object formed by a concave mirror. From point H on the object, four rays are drawn to the mirror and then reflected. All these reflected rays pass through the image point G.

Using Figure 54-5 , we now show that the mirror equation derived in Section 54.2 also applies to concave mirrors. As in Figure 54-2 , we denote the height of the object by h, and the height of the image by h'. In triangle *HOV*, $\tan \beta = h/o$. In triangle *GIV*, $\tan \beta = h'/i$. Equating these values for $\tan \beta$ readily leads to

$$h' = hi/o.$$

Next consider the ray from H normal to the mirror, which passes through C and then through G. Triangles *HOC* and *GIC* are similar, so the ratios of corresponding sides are equal:

$$\frac{HO}{OC} = \frac{GI}{IC} \qquad \text{or} \qquad \frac{h}{o - R} = \frac{h'}{R - i} \,.$$

Substituting $h' = hi/o$ and $R = 2f$ into the last equation above, cancelling h from both sides, and cross-multiplying to clear the fractions leads to

$$o(2f - i) = i(o - 2f).$$

Eliminating parentheses, combining similar terms, and solving for f then leads to

★ $$\frac{1}{f} = \frac{1}{i} + \frac{1}{o} \,. \qquad \text{Mirror equation}$$

This is the same equation we derived in Section 54.2 for convex mirrors, and the sign conventions stated on page 54-9 for convex mirrors also apply to its use for concave mirrors.

Worked Example 54-3

An object is placed 120 cm in front of a concave mirror having a focal length of 40 cm. What is the location of the image?

Solution

Using the mirror equation, we compute

$$\frac{1}{i} = \frac{1}{f} - \frac{1}{o} = \frac{1}{40\ \text{cm}} - \frac{1}{120\ \text{cm}} = 0.0167\ \text{cm}^{-1} = \frac{1}{60\ \text{cm}} \,,$$

from which $i = +60$ cm. This example corresponds to the distances in Figure 54-5 , drawn to a scale of 1 mm = 1 cm.

It can be shown that all the rays in a narrow parallel beam which approaches a spherical concave mirror at an angle to the axis are reflected to a common focus in the focal plane. For an object at a distance much greater than the focal length, the rays from any object point are practically parallel when they strike the mirror, so the image of such an object is a real image formed in the focal plane. The image can easily be projected on a screen held in the focal plane, so this offers a convenient method of measuring the focal length of a concave mirror.

The mirror equation can be used to draw some general conclusions about the image-forming properties of concave mirrors; proofs of these statements are left to the Exercises. When the object distance is positive and greater than the focal length, the image is real and inverted. As a real object moves from an infinite distance toward the mirror, the real image moves away from the mirror. When the object distance is positive and smaller than the focal length, the image is virtual, erect, and enlarged; this is the way a shaving mirror is used. When the object distance is negative the object is virtual. A virtual object always has an erect real image, reduced in size.

Questions and Exercises

Q 54.3-1 (a) For a concave mirror, what is the range of object distances for which the image is real?
(b) For a concave mirror, what is the range of object distances for which the image is virtual?

Q 54.3-2 Describe three methods for finding the focal length of a concave mirror.

E 54.3-1 Beginning with the equations $h' = hi/o$ and $h/(o - R) = h'/(R - i)$, fill in the missing steps in the derivation of the mirror equation.

E 54.3-2 Beginning with the equation for magnification $M = h'/h$, show that $M = -i/o$ for a concave mirror.

E 54.3-3 (a) Use a ray diagram to show that a real object in front of a concave mirror, and farther from it than the primary focal point, always has a real and inverted image.
(b) For what range of object distances is the image smaller than the object?
(c) For what range of object distances is the image larger than the object?

E 54.3-4 (a) Use the mirror equation to show that for a real object located between a concave mirror and its primary focal point, the image is virtual.
(b) Use a ray diagram to show that for this case the image is erect and larger than the object.

E 54.3-5 Use a ray diagram to show that when a virtual object is located behind a concave mirror, the image is real, erect, and reduced in size.

E 54.3-6 Using a concave mirror, where must an object be located so that its image distance equals its object distance, with both image and object on the same side of the mirror?

E 54.3-7 A lamp is 200 cm in front of a concave mirror having a focal length of 60 cm. Where should a white screen be placed so that the image of the lamp is focused on the screen?

E 54.3-8 The designer of an optical system desires to place a concave mirror 2.5 meters from a light source, so that the real image of the source formed by the mirror is also located 2.5 meters from the mirror. What focal length is required?

E 54.3-9 A man uses a shaving mirror having a focal length of 75 cm, and stands with his chin 30 cm from the mirror.
(a) Where is the image of his chin located?
(b) What is the magnification of the image?

E 54.3-10 For the mirror described in E 54.3-9, where must a (virtual) object be formed so that its image is located 40 cm in front of the mirror?

E 54.3-11 An infrared spectrometer contains a concave mirror having a focal length of 12 cm, which is used to form an image of the radiation source.
(a) If the source is 30 cm from the mirror, where is the image? Locate the image by the graphical method.
(b) Calculate the image position using the mirror equation.
(c) Is the image in part (a) real, or virtual?
(d) What is the ratio of image height/source height?
(e) Where must the source be located to produce an image having the same size as the source?
(f) How can the optical system for part (e) be arranged so the image does not coincide spatially with the source of its mounting?

Check your understanding of the material above by testing yourself on the minimum learning goal at the beginning of this section.

54.4 Image Formation by Thin Lenses

Minimum Learning Goals. (1) Given the focal length of a converging or diverging lens, and the position of a real or virtual object relative to the lens, (a) be able to compute the position of the image formed by the lens and to state whether the image is real or virtual, erect or inverted; and (b) be able to make a qualitatively correct ray diagram showing the positions of the object, lens, and image and showing the paths of several rays as they proceed through the system. For example:

1 A thin converging lens having a focal length of 5 cm is used as a simple magnifier. An arrow 1 cm in height standing normal to the axis of the lens, 3 cm from its center, represents an object being viewed through the magnifier. Construct a ray diagram to locate the image and state whether it is real or virtual, and erect or inverted. Calculate the image position. (Answer, page 54-28, Line 6)

2 An arrow serves as an object located 100 cm from a thin converging lens having a focal length of 50 cm. Construct a ray diagram to locate the image, and state whether it is real or virtual, and erect or inverted. Calculate the image position. (Answer, page 54-26, Line 2)

Lenses are used in eyeglasses, cameras, magnifiers, and a variety of scientific instruments to control the paths of light rays and form images. Some of the shapes in which lenses are made are illustrated in Figure 54-6 . The optical properties of a lens depend on whether it is a converging lens or a diverging lens.

A lens thicker at the center than at the edges is called a *converging* or *positive* lens.

A lens thinner at the center than at the edges is called a *diverging* or *negative* lens.

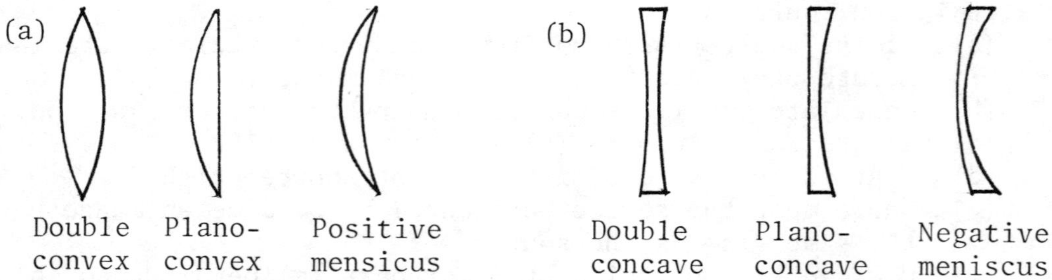

Figure 54-6. Cross-sections of some standard forms of lenses. (a) Converging or positive lenses. (b) Diverging or negative lenses.

In this discussion, we consider only idealized lenses of the type called a *thin lens*. For most practical purposes a lens may be regarded as thin when its ratios of $\frac{\text{thickness}}{\text{diameter}} \ll 1$ and $\frac{\text{thickness}}{\text{object distance}} \ll 1$. For actual lenses which approach this ideal, calculations using the simple lens equation (developed below) are accurate enough for many purposes. We consider only lenses for which both surfaces are spherical. We neglect the aberrations which result from the lens thickness, the dispersion of the lens material, and any non-spherical curvature of the lens surfaces. In the precise design of optical instruments a more exact approach is required, in which all sources of aberration are fully considered.

An imaginary line perpendicular to the surface of a lens at its center is called the *axis* of the system containing the lens. The following statements about rays converging at a focal point, and the lens equation derived below, are strictly correct only for paraxial rays of monochromatic light. *Paraxial rays* are rays which make very small angles with the axis and lie close to the axis throughout their travel from object to image. In spite of this limitation, the methods and equations developed in this section are commonly used in optical work when close approximations are acceptable.

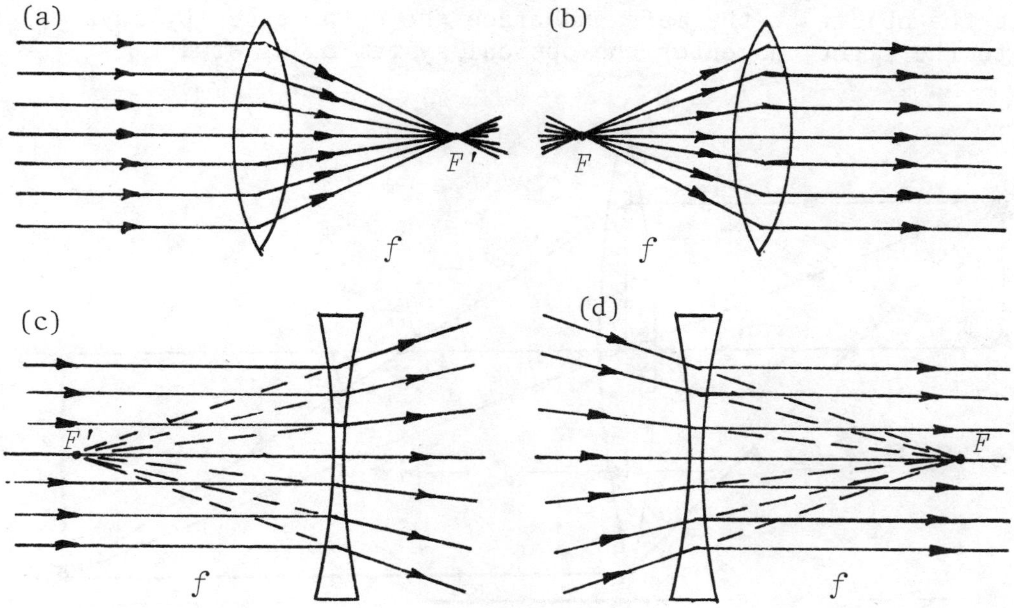

Figure 54-7. Refraction of incoming parallel rays, and formation of outgoing parallel rays, by converging and diverging lenses.

It is found that rays of light which approach a converging thin lens parallel to its axis are refracted and converge at a point called the secondary focal point, F' (Figure 54-7 a). Rays which pass through the *primary focal point* F of a converging thin lens as they approach the lens are refracted by the lens into a parallel beam (Figure 54-7 b). The distance from the center of any lens to F or F' is called the *focal length* of the lens, designated by f. Parallel rays which approach a diverging thin lens parallel to its axis are refracted by the lens and diverge as if they had come from the secondary focal point F' (Figure 54-7c). Rays which are headed directly toward the primary focal point F of a diverging thin lens as they approach the lens are refracted by the lens into a parallel beam (Figure 54-7d). It is seen from Figure 54-7 that for any lens, F and F' are located on the axis on opposite sides of the lens. For convenience in using the lens equation to be obtained below, it is customary to regard the focal length of a converging lens as a positive number, but the focal length of a diverging lens as a negative number.

The *focal plane* of a lens is a plane through either focal point, normal to the axis of the lens.

Idealized thin lenses having the same focal length and the same diameter have the same optical properties, regardless of the curvature of the lens surfaces.

There are three methods for locating an image formed by a lens or mirror: (a) experiment, (b) geometrical construction of a ray diagram, and (c) calculation using the lens equation. In constructing ray diagrams, we place the object at the left and trace the paths taken by rays which travel to the right and enter the optical system under study.

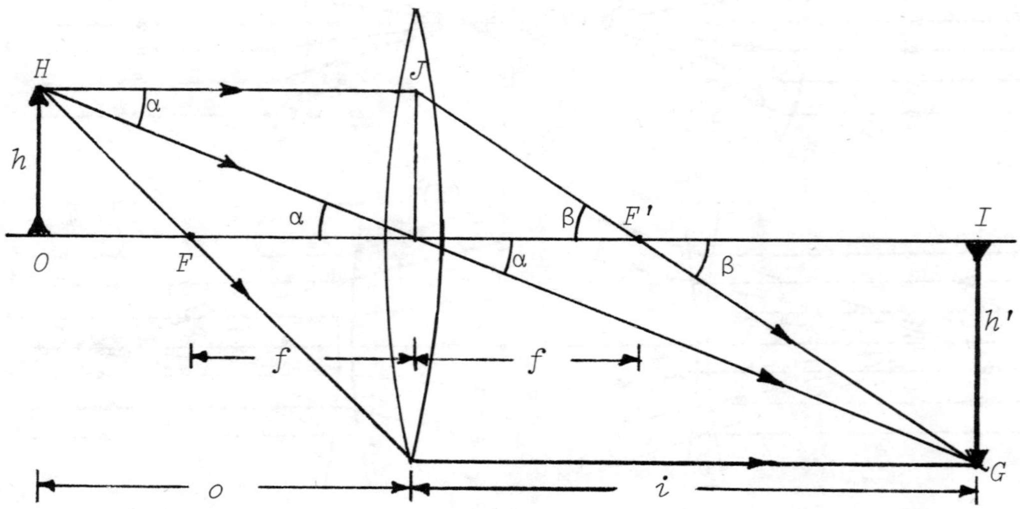

Figure 54-8. A ray diagram used to locate the image of a real object formed by a converging lens. From the point *H* on the object, three rays are drawn to the lens and then refracted. All these refracted rays pass through the image point *G*.

Figure 54-8 shows how a ray diagram may be used to locate the image of an object formed by a converging lens. The object is represented by an arrow normal to the axis, having its tail on the axis. The distance from the object to the center of the lens is the *object distance o*, and the distance from the center of the lens to the image is the *image distance i*, both distances being measured along the axis. In this method, sometimes called the graphical method, three rays may be drawn from any point on the object located off the axis:

★ (1) A ray parallel to the axis, which travels through (directly away from) *F'* after refraction.

★ (2) A ray through (directly toward) *F*, which travels parallel to the axis after refraction.

★ (3) A ray through the center of the lens which continues undeviated.

These statements apply to a *converging lens*, if the parenthetical words are ignored. The alternative words in parentheses apply to a *diverging lens*. While any two refracted rays which intersect (after being projected backward, if necessary) are enough to determine the location of a point on the image, it is recommended that the student use all three rays in order to improve accuracy and avoid errors.

From Figure 54-8 it is seen that for a converging lens, when $o > f$ then $i > f$, and the image is real but inverted. The ratio of image height to object height depends on the location of the object.

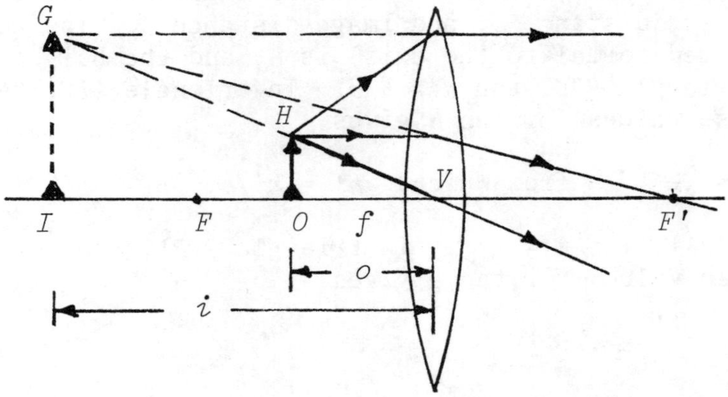

Figure 54-9. A ray diagram used to locate the image of a virtual object formed by a converging lens. From the point *H* on the object, three rays are drawn to the lens and then refracted. All these refracted rays are projected backward to pass through the image point at *G*.

Figure 54-9 shows a ray diagram for a converging lens, for a real object located between the primary focal point and the lens. In this case the refracted rays diverge, so the image must be located by projecting the refracted rays backward to their intersection at a point. The actual light rays do not meet at an image point, so the image is virtual. The image is also erect and enlarged. A single converging lens used in this way functions as a magnifier; examples include the common magnifying glass and reading glass. A virtual image can be seen only when light rays from it enter the eye, so the observer must look through the lens at the image in order to see the enlarged image.

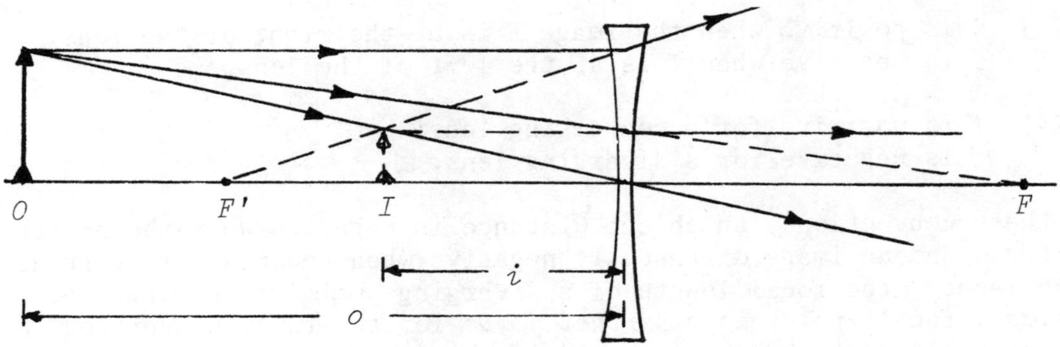

Figure 54-10. A ray diagram used to locate the image of a real object formed by a diverging lens. The image is virtual, erect, smaller than the object, and located between the object and the lens.

Figure 54-10 shows a ray diagram for a real object at the left of a diverging lens. The rays from any object point diverge as they approach the lens, and diverge even more after refraction. The image point is located by projecting these rays backward to their point of intersection. For any object position at the left of the lens, the image is virtual, erect, smaller than the object, and located between the object and the lens.

Using Figure 54-8,, we now derive an equation relating the focal length f, object distance o, and image distance i. The height of the object (measured normal to the axis) is h, and the height of the image is h'. In triangle HOV, $\tan \alpha = h/o$. In triangle GIV, $\tan \alpha = h'/i$. Equating these values for $\tan \alpha$ gives

$$\frac{h}{o} = \frac{h'}{i}, \qquad \text{from which} \qquad h' = hi/o.$$

In triangle JVF', $\tan \beta = h/f$. In triangle GIF', $\tan \beta = h'/(i - f)$. Equating these values for $\tan \beta$ gives

$$\frac{h}{f} = \frac{h'}{i - f}.$$

Substituting $h' = hi/o$ into the last equation above, cancelling h from both sides, and cross-multiplying to clear of fractions, leads to

$$o(i - f) = if.$$

Dividing by the product oif and simplifying then leads to

★ $$\frac{1}{f} = \frac{1}{o} + \frac{1}{i}. \qquad \text{Lens Equation}$$

This equation is applicable to any calculation of an image formed by a converging or diverging thin lens, provided the following *sign conventions* are observed:

★ (1) All actual light rays travel from left to right.

★ (2) o is positive when the object O is at the left of the lens; o is negative when O is at the right of the lens.

★ (3) i is positive when the image I is at the right of the lens; i is negative when I is at the left of the lens.

★ (4) f is positive for a converging lens; f is negative for a diverging lens.

Using these conventions, an object distance is negative when the object is virtual, and an image distance is negative when the image is virtual. In like manner, the focal length of a diverging lens is negative because the primary focal point is a *virtual focus* for the rays; converging rays do not actually meet there, but would if the lens did not refract them.

Worked Example 54-4

An object is placed 50 cm from a converging lens having a focal length of 30 cm. Where is the image located?

Solution

The lens equation can be rearranged to solve for image distance:

$$\frac{1}{i} = \frac{1}{f} - \frac{1}{o} \quad .$$

In this example, f = +30 cm and o = +50 cm. Substituting, we obtain

$$\frac{1}{i} = \frac{1}{30 \text{ cm}} - \frac{1}{50 \text{ cm}} = \frac{5}{150 \text{ cm}} - \frac{3}{150 \text{ cm}} = \frac{2}{150 \text{ cm}} ,$$

from which i = +75 cm. This example corresponds to the distances in Figure 54-8, drawn to a scale of 1 mm = 1 cm.

When an object is a long way from a converging lens, $o \cong \infty$ and $i \cong f$. The real image may be seen in sharp focus on a white card or screen in the secondary focal plane. This offers a convenient method for measuring the focal length of a converging lens.

From a ray diagram it is easy to see whether an image is erect or inverted with respect to its object, and also whether it is larger or smaller than its object. Using an object height h and image height h' measured normal to the axis, the *magnification* M of an optical system is defined as

$$M = h'/h.$$

The magnification is positive when the image is erect, negative when the image is inverted. One can trace the ray from any object point (such as H in Figure 54-8 or 54-9) through the center of the lens and then to the corresponding image point, and obtain similar triangles. then using our sign conventions it can be shown that

★ $$M = -i/o.$$

This equation gives the magnification for converging or diverging lenses, for any object distance.

Worked Example 54-5

An object is placed 60 cm to the left of a diverging lens having a focal length of 40 cm.

(a) Where is the image located?

(b) What is the magnification?

Solution

(a) Solving the lens equation for image distance, we have

$$\frac{1}{i} = \frac{1}{f} - \frac{1}{o} = \frac{1}{-40\ \text{cm}} - \frac{1}{+60\ \text{cm}} \quad ,$$

from which $i = -24$ cm. This example corresponds to the distances in Figure 54-10, drawn to a scale of 4 mm = 3 cm.

(b) The magnification $M = -\frac{i}{o} = -\frac{-24\ \text{cm}}{+60\ \text{cm}} = +0.4$.

Questions and Exercises

Q 54.4-1 Give an example of a lens which occurs in nature.

Q 54.4-2 (a) Is a converging lens thicker at the middle, or at the edges?
(b) Is a diverging lens thicker at the middle, or at the edges?

Q 54.4-3 Under what conditions is it justifiable to regard all of the rays coming to a lens or mirror from a "distant" object as parallel rays?

Q 54.4-4 How can a real image be formed with a diverging lens?

Q 54.4-5 When a bundle of parallel rays passes through a diverging lens, where is the image formed? Is this image real, or virtual? Erect, or inverted?

Q 54.4-6 (a) In what respects is a lens similar to a prism?
(b) In what ways are they different?

Q 54.4-7 Explain how the object distance for a lens can be negative.

Q 54.4-8 Consider a bundle of rays approaching a lens from an object.
(a) For a converging lens, what is the range of object distances for which the image is real?
(b) For a converging lens, what is the range of object distances for which the image is virtual?
(c) For a diverging lens, what is the range of object distances for which the image is real?
(d) For a diverging lens, what is the range of object distances for which the image is virtual?

Q 54.4-9 Does a camera lens form a real image or a virtual image of the object being photographed?

Q 54.4-10 Can a camera take a picture of a virtual image? Explain.

Q 54.4-11 What kind of lens should be used in a magnifying glass? Where should it be placed in relation to the object under examination?

Q 54.4-12 Describe two methods for determining the focal length of a converging lens.

Q 54.4-13 Why does a lens have two focal points, while a mirror has only one?

Q 54.4-14 When using the graphical method to locate the image of a point, why can four rays be drawn for a mirror but only three rays for a lens?

E 54.4-1 Beginning with the equations $h' = h/io$ and $h/f = h'/(i - f)$, fill in the missing steps in the derivation of the lens equation.

E 54.4-2 Beginning with the equation $M = h'/h$, show that $M = -i/o$ for a thin lens.

E 54.4-3 Use a ray diagram to show that when an object is farther than $2f$ from a converging lens, the image is less than $2f$ from the lens, and the image height is smaller than the object height.

E 54.4-4 Use a ray diagram to show that when $f < o < 2f$ for a converging lens, the image is farther than $2f$ from the lens, and the image height is greater than the object height.

E 54.4-5 How far from a converging lens must a real object be, so that $M = -1$?

E 54.4-6
(a) Use the lens equation to show that when an object has any position at the left of a diverging lens, the image is virtual and located between the object and the lens.
(b) Use a ray diagram to show that for this case the image is erect and smaller than the object.

E 54.4-7 An ordinary slide projector is used to project a real image of a slide on a screen.
(a) Does the projector contain a converging, or diverging, lens?
(b) If the projector lens has $|f| = 8$ cm, where should the slide be positioned in relation to the lens in order to form a clear image on a screen 10 m from the lens?
(c) Show the formation of the image in part (b) using the graphical method (not necessarily drawn to scale).
(d) How can the same projector be used when the screen is only 3 m from the lens?
(e) What extreme separation of the lens from the slide would be necessary in order to use this projector at the rear of a very long room?
(f) Is this extreme a minimum, or a maximum? Explain.

E 54.4-8 A magnifier forms a real image of a lamp 18.0 cm from the lens when the lamp is 200 cm from the lens. (a) What is the focal length? (b) Is the image erect, or inverted? (c) If the magnifier is held 8.0 cm from some fine print, where is the image of the print formed? (d) What magnification is achieved under the conditions of part (c)?

E 54.4-9 A real object is placed at the secondary focal point of a diverging lens, 20 cm from the lens.
(a) Where is the image located?
(b) Is the image erect, or inverted?
(c) Is the image real, or virtual?

E 54.4-10 A lens fixed in position forms a real image 16 cm from the lens, but it is necessary to modify the optical system to produce an image located closer to the fixed lens. This is done by inserting a second lens between the fixed lens and its image, 4.0 cm from the fixed lens. Then the initial image of the fixed lens serves as a virtual object for the second lens. The second lens has a focal length of 12.0 cm.
(a) Should the second lens be a converging, or diverging, lens?
(b) Where is the image of the second lens located?

Check your understanding of the material above by testing yourself on the minimum learning goal at the beginning of this section.

Answers to Minimum Learning Goals

Line 1

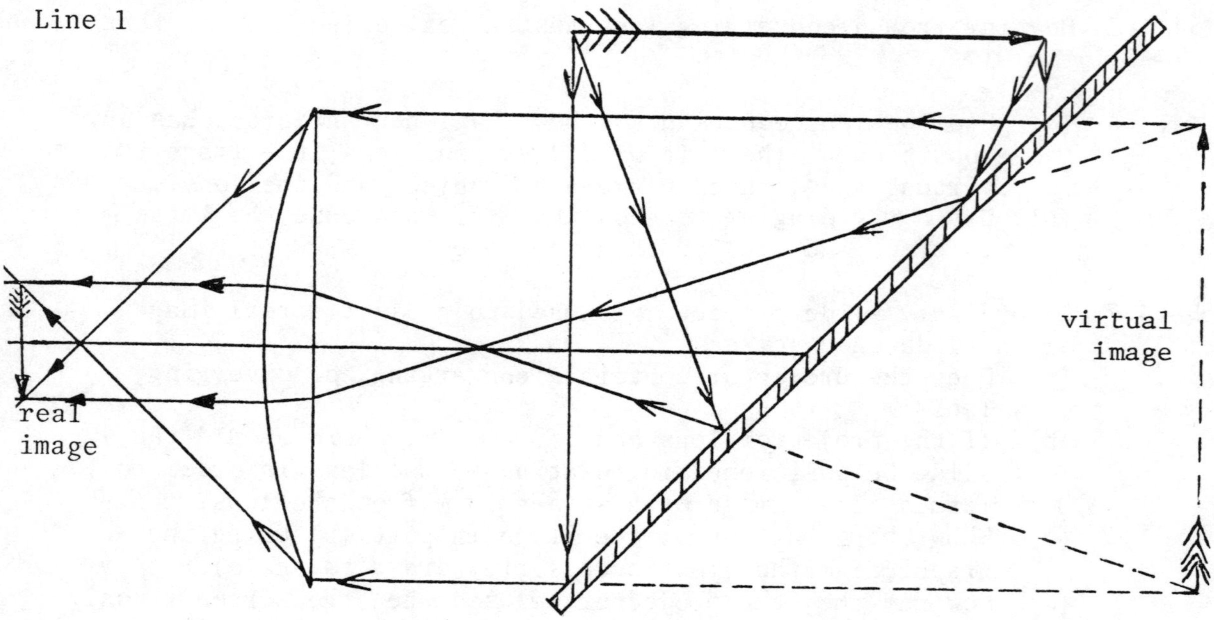

Line 2 Yes, because the dimensions of the window are much larger than a wavelength.

Line 3 i = -14.3 cm. In the drawing, the vertical scale is greater than the horizontal scale.

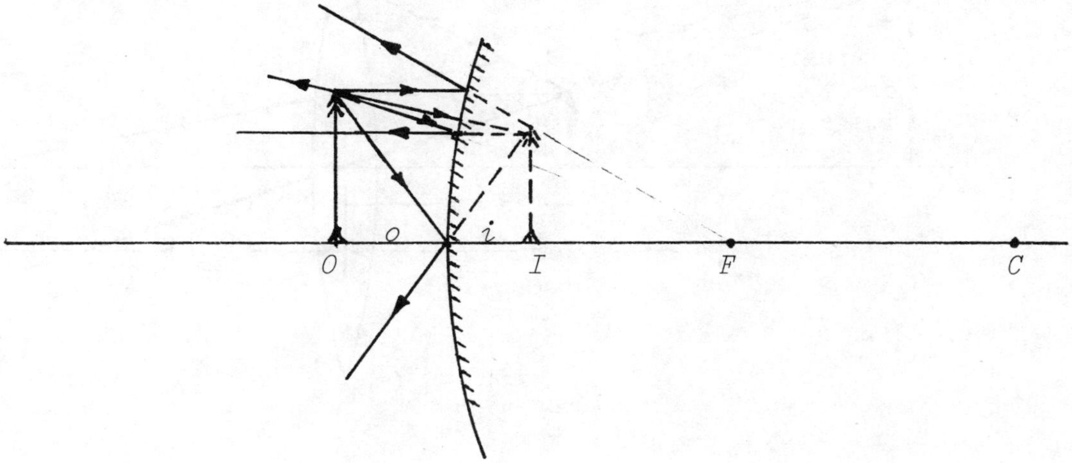

Line 4

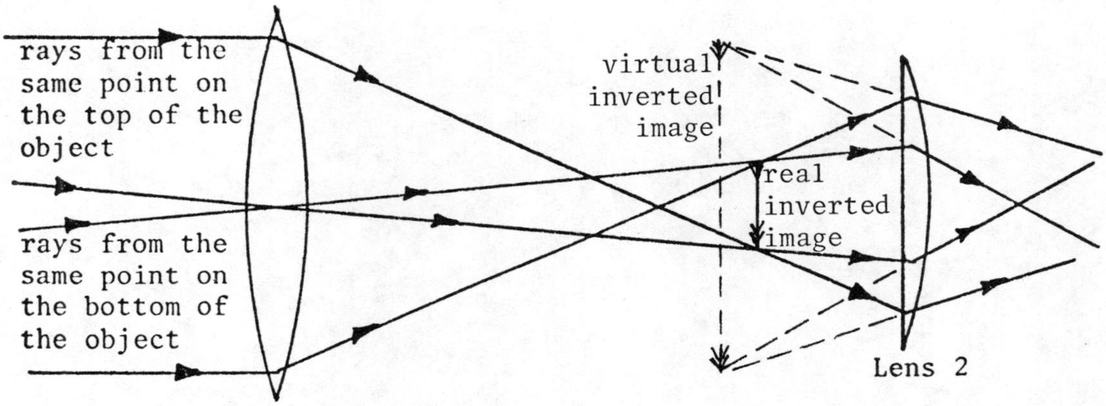

Line 5 i = 85.7 cm. In the drawing, the vertical scale is larger than the horizontal scale.

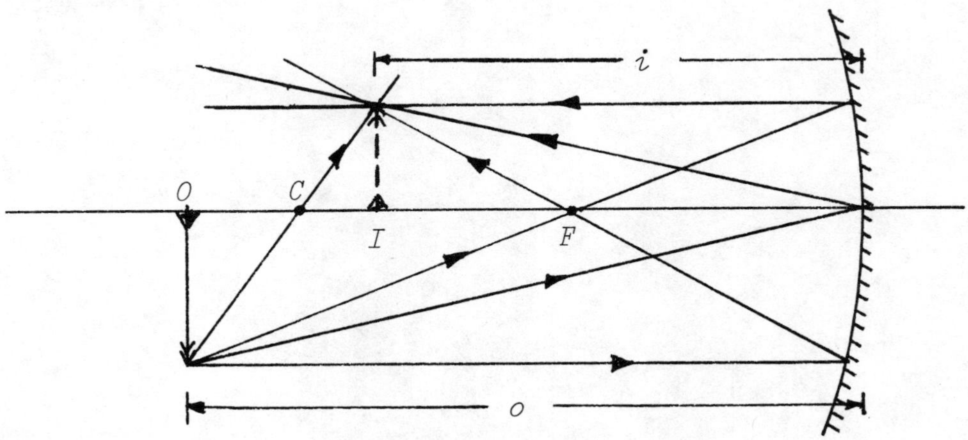

Line 6

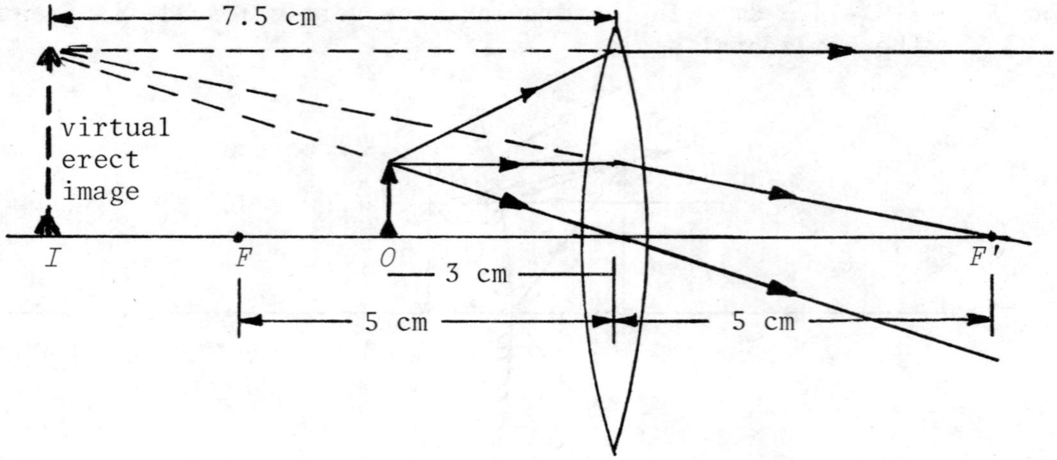

ANSWERS to Questions and Exercises in Chapter 54

Section 54.1

Q 54.1-2 The image is blurred or fuzzy.

Q 54.1-4 A virtual, extended image

Q 54.1-6 Look into a second mirror at the image produced by the first mirror.

Section 54.2

Q 54.2-2 A mirror having its center dished or caved inward is concave. A mirror for which the center bulges outward toward the observer is convex.

Q 54.2-4 Rays approaching the mirror must be converging rapidly so that in the absence of the mirror they would form an image between the location of the mirror and its primary focal point.

Q 54.2-6 Yes. For a convex mirror, all real objects reflected in the mirror form virtual images, while virtual objects located between the vertex and the primary focal point form real images. Existence of these virtual objects would require images formed by a lens or concave mirror placed in front of the convex mirror.

E 54.2-8 42.9 cm in front of the mirror.

E 54.2-10 (a) 4.98 cm behind the vertex; (b) virtual;
(c) On the image, 4.98 cm behind the mirror surface;
(d) erect.

Section 54.3

Q 54.3-2 (a) experiment: when the mirror is used to form a sharp image of a far-off light source such as the sun on a screen, the image is in the focal plane.
(b) use a ray diagram;
(c) use the mirror equation to compute f.

E 54.3-6 At the center of curvature.

E 54.3-8 1.25 m

E 54.3-10 85.7 cm behind the mirror

Section 54.4

Q 54.4-2 (a) at the middle; (b) at the edges.

Q 54.4-4 Use another converging lens or mirror to form a real image, then place the diverging lens at the left of this image so the image is between the lens and its primary focal point.

Q 54.4-6 (a) Both are made of a transparent material such as glass, and both show the effects of the dispersion of this material. Both are used to refract light rays.
(b) A lens forms images of objects, but a prism does not.

Q 54.4-8 (a) $o > f$, and $o < 0$; (b) $0 < o < f$; (c) $0 < o < |f|$;
(d) $o > 0$ or $o < f$.

Q 54.4-10 Yes, if the camera is focused on the image.

Q 54.4-12 (1) Use the lens to focus the image of a distant object on a screen; the distance from lens to screen is the focal length.
(2) Use the lens to focus the image of any object on a screen; then using the lens equation, compute the focal length from measured values of object and image distances.

Q 54.4-14 A mirror has a unique center of curvature, so a fourth ray may be drawn through that point. A lens does not have a unique center of curvature.

E 54.4-8 (a) 16.5 cm; (b) inverted; (c) i = - 15.5 cm; (d) 1.94.

E 54.4-10 (a) converging lens; (b) 6.0 cm from the second lens.

CHAPTER 55 PHYSICAL OPTICS

Introduction

Physical optics deals with optical phenomena which can be explained best using the wave model for light. These phenomena include interference, diffraction, polarization, refraction, dispersion and scattering. Some of these topics are discussed in this chapter.

The mathematical description of wave motion is here applied to phenomena observed using electromagnetic waves. The displacements of an electromagnetic wave are the transverse electric and magnetic fields which constitute the wave.

In the classical view of radiation an electric charge oscillating with simple harmonic motion generates a sinusoidal electromagnetic wave. We assume in this chapter that atoms and other microscopic sources of light emit sinusoidal waves, with each wave long enough to include a large number of wavelengths. This assumption is justified by its success in accounting for many phenomena of physical optics. The individual waves from the myriads of atoms in any macroscopic light source combine according to the principle of superposition.

Interference phenomena occur when two or more light waves combine after following different paths. The amplitude and intensity of the resulting light waves depend on the phase difference of the combining waves, and therefore on the difference in optical paths followed by the waves. Mathematical addition of combining waves accounts for the variation of intensity in the double-slit interference pattern, and in the diffraction pattern of a single slit. These concepts are applied to the practical problems of the resolving power obtainable by a microscope and by a diffraction grating.

Patterns of color observed in the light reflected from thin films such as soap bubbles are accounted for by interference between the light waves reflected from the front and back surfaces of the film. A description of plane-polarized light leads to an explanation of how such light can be produced by reflection, refraction, and transmission using suitable materials.

When you have completed your study of the material in this chapter, you should be able to answer questions and work exercises like the examples in the minimum learning goals assigned by your instructor.

55.1 Optical Path

When two waves combine with each other the resulting interference may be constructive or destructive, depending on the relative phases of the two waves (Section 28.3). This phase difference depends on how far the two waves travel from their common source, and on the refractive indexes of the media in which the waves are traveling. When light passes from one optical medium into another having a different refractive index n (Section 28.8), the frequency of the waves remains constant while the speed and wavelength change. As the refractive index increases, often expressed by stating that the *optical density* increases, the wavelength decreases so that a greater number of wavelengths fit into a given geometrical path length d.

The use of light *rays* to indicate the path taken by light waves is introduced in Section 53.1. To keep track of the numbers of wavelengths for light rays which follow two different paths, it is convenient to use the concept of the *optical path* traversed by a light ray between two points. When a ray travels a distance d in a medium having a refractive index n, its optical path in this medium is defined as the product nd. If the ray travels a distance d_1 in a medium having index n_1 and a distance d_2 in a medium having index n_2, its total optical path is the sum $n_1d_1 + n_2d_2$.

★ The *optical path* traveled by a light ray which goes through a succession of media having refractive indexes $n_1, n_2, n_3, \ldots$ is $\Sigma\, n_i d_i = n_1d_1 + n_2d_2 + n_3d_3 + \ldots$, where d_1, d_2, etc. are the lengths of the geometrical paths in the various media.

The significance of optical path is that for light having a given frequency, equal optical paths contain the same number of wavelengths. To show this, we consider plane light waves which travel from a medium having refractive index n across a plane boundary into a second medium having index n', as shown in Figure 55-1. The angle between the wave fronts and the boundary is ϕ in the first medium, and ϕ' in the second. Each wave front advances slower in the second medium than in the first, so the direction of the advancing wave fronts changes as they cross the boundary, and ϕ differs from ϕ'. This phenomenon is called *refraction;* it is discussed further in Section 53.3.

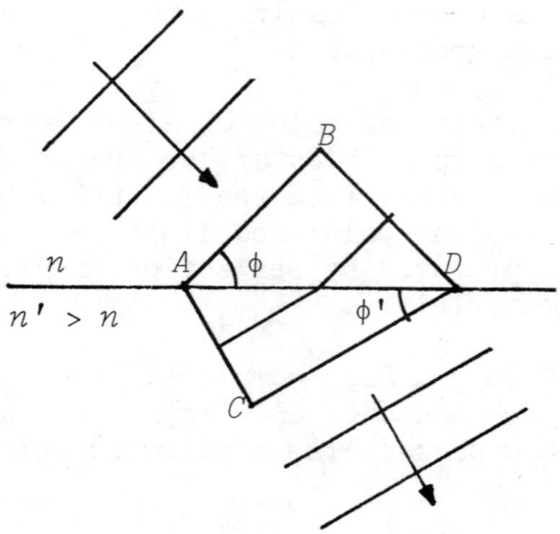

Figure 55-1. Plane wave fronts advancing in a medium having refractive index n are bent or refracted when they cross the boundary into a second medium having refractive index n'.

In Figure 55-1, $\overline{AB}$ is a portion of one wave front and $\overline{CD}$ is another. Each wave front represents one continuous crest of the wave, so $\overline{AB}$ and $\overline{CD}$ are separated by two wavelengths. That is,

$$\overline{BD} = 2\lambda \quad \text{and} \quad \overline{AC} = 2\lambda' ,$$

where the wavelength is λ in the first medium and λ' in the second. The ratio

$$\frac{\overline{AC}}{\overline{BD}} = \frac{2\lambda'}{2\lambda} = \frac{\lambda'}{\lambda} .$$

Multiplying both λ' and λ by the frequency ν, we have

$$\frac{\overline{AC}}{\overline{BD}} = \frac{\nu\lambda'}{\nu\lambda} = \frac{v'}{v} ,$$

where the wave speed is v in the first medium and v' in the second. The index of refraction of a medium is the ratio of the speed of light in free space to the speed of light in the medium. Therefore, $n = c/v$ and $n' = c/v'$, and the ratio

$$\frac{n}{n'} = \frac{c/v}{c/v'} = \frac{v'}{v} .$$

Combining these results, we obtain

$$n'\overline{AC} = n\,\overline{BD}.$$

But $n\overline{BD}$ is the optical path from B to D, and $n'\overline{AC}$ is the optical path from A to C. We have thus found that these optical paths are equal because they contain the same number of wavelengths.

Using ℓ to denote the distance from A to D in Figure 55-1, we have

$$\overline{AC} = \ell \sin\phi' \quad \text{and} \quad \overline{BD} = \ell \sin\phi .$$

Then the ratio

$$\frac{\overline{AC}}{\overline{BD}} = \frac{\sin\phi'}{\sin\phi} .$$

From our work above, this ratio is also equal to n/n'. Combining these results, we have

★ $$n \sin\phi = n' \sin\phi' . \qquad \text{Snell's law}$$

Snell's law is extensively used in geometrical optics, as shown in Chapters 53 and 54.

In physical optics we consider a number of phenomena where light waves follow different paths and then arrive at the same point. The difference between the optical paths taken by the two waves is the *optical path difference,* denoted by the symbol Δ (capital Greek delta). Note that Δ was used with this meaning in the discussion of Young's double-slit experiment (Section 28.4). Associated with the optical path difference between two waves is a *phase difference*, δ (lower-case Greek delta). To develop a relation between δ and Δ, we consider the waves traveling from the two openings in a double slit to a common screen point P, shown in Figure 55-2. The difference between the geometrical paths taken by these two waves is $\ell_2 - \ell_1$. For light waves in air, $n \cong 1$ so the optical path difference between the two waves when they reach P is

$$\Delta = \ell_2 - \ell_1.$$

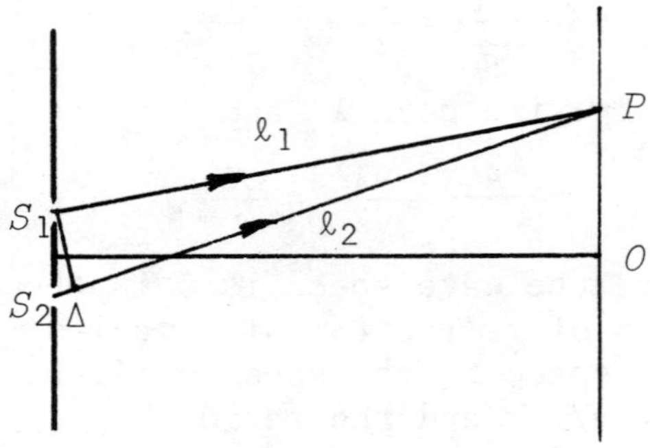

Figure 55-2. A portion of a schematic diagram of a double-slit experiment. (Figure 28.7 on page 28-11 is a more complete diagram.) The distance from slit S_1 to screen point P is ℓ_1, while the distance from slit S_2 to P is $\ell_2 = \ell_1 + \Delta$.

To find the phase difference between these waves when they arrive at P, we note that the two waves have a phase difference of zero as they emerge from S_1 and S_2. Then we may use the equation for a sinusoidal wave to describe the waves reaching P. The wave from S_1 is

$$E_1 = E_{10} \cos (k\ell_1 - \omega t),$$

while the wave from S_2 is

$$E_2 = E_{20} \cos (k\ell_2 - \omega t),$$

where $E_{10} = E_{20}$ is the amplitude of either wave at P. Comparing these two equations, we see that the phase difference between the two waves at P is

$$\delta = k(\ell_2 - \ell_1).$$

Combining this with the equation above for Δ leads to the relation

★ $$\delta = k\Delta = \frac{2\pi}{\lambda}\Delta,$$

where λ is the wavelength in free space. This gives the phase difference (in radians) at any point P, between two waves from a single source which follow paths having an optical path difference Δ and then recombine at point P.

Questions and Exercises

Q 55.1-1 Why is the optical path between two points always greater than or equal to the geometrical distance between the points?

Q 55.1-2 When computing the phase difference between two waves which arrive at the same point by different paths, why can't we simply use the geometrical path difference instead of the optical path difference? Give an example of a real situation in which a calculation of the phase difference using the geometrical path difference would lead to a substantial error.

Q 55.1-3 In this section it is assumed that the two waves reaching a given screen point in a double-slit experiment have the same amplitude. For which screen points is this a reasonable assumption?

E 55.1-1 In Figure 55-1, what is the optical path difference between the optical path from A to C, and the optical path from B to D?

E 55.1-2 A plate of glass 6.35 mm thick and having refractive index 1.65 is placed between a sodium lamp and a detector located 5.00 cm apart.

(a) What is the optical path length from lamp to detector?

(b) For sodium light having a wavelength of 589 nm, how many wavelengths are there between the source and the detector?

E 55.1-3 Figure 55-2 represents a double-slit experiment in which the center-to-center slit separation is 0.010 mm and the screen is 1.00 meter from the slits. The distance from O to P is 5.0 cm.

(a) What is the optical path difference between the rays to P from the two slits, assuming these rays travel through air?

(b) For a wavelength of 500 nm, what is the phase difference at P?

E 55.1-4 Two parallel slits 4.55×10^{-2} cm apart are illuminated with monochromatic light, and interference fringes are observed on a screen at a distance of 80.5 cm from the plane of the slits.

(a) For a screen point 2.30 mm from the center of the pattern, what is the optical path difference between the rays traveling in air to this point from the two slits?

(b) If the phase difference at the screen point is 4π radians for these two waves, what is the wavelength?

E 55.1-5 In a double-slit experiment using visible white light a thin film of transparent material is placed over *one* of the slits. This causes the fringe pattern to be displaced by three fringes from its usual position. The refractive index of the film is 1.5, and the effective wavelength of the light is 550 nm in air.

(a) What is the value of the *increase* in optical path for rays which travel through the film?

(b) What is the thickness of the film?

(c) What phase difference results from displacing some air by placing the film in position?

55.2 Intensity in a Double-Slit Interference Pattern

Young's double-slit experiment was described in some detail in Section 28.4. Using monochromatic light, the interference pattern observed on a screen consists of uniformly spaced alternating light and dark fringes, reproduced in Figure 28-6 on page 28-10. The brightness of the fringe pattern at any point on the screen is proportional to the intensity of the light reaching that point. When the intensity variation of the fringes near the center of the pattern is measured using a photometer, a graph like that in Figure 55-3 is obtained. In this graph, x is the position of any point on the screen measured from $x = 0$ at the center of the pattern, as shown in Figures 28-7 and 28-8. The measured intensity variation is well represented by the equation

$$I = I_0 \cos^2 (\gamma x),$$

where I_0 is the maximum intensity of the central bright fringe. How can we account for this variation in intensity?

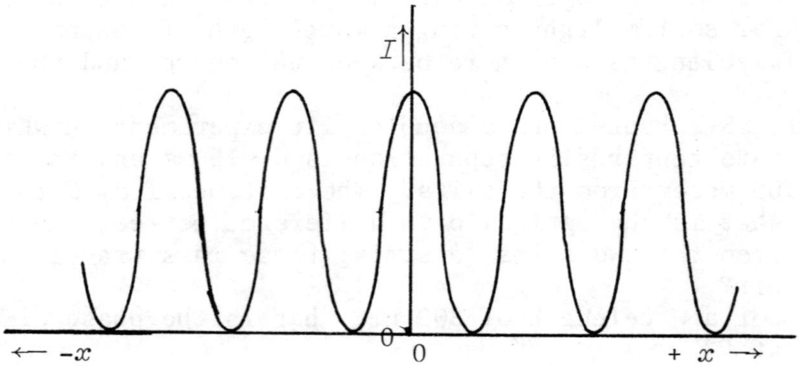

Figure 55-3. Variation of intensity I with position x near the center of a double-slit interference pattern. The maxima of intensity correspond to bright fringes, while the minima correspond to dark fringes. A photograph of this pattern is shown in Figure 28-6 on page 28-10.

The intensity of a wave refers to the energy transported by the wave per second per unit of area:

> The *intensity* I of a traveling wave is the average time rate at which the wave transports energy across a unit of area normal to the velocity of the wave.

The SI Unit of wave intensity is $J/(m^2 \cdot s) = W/m^2$. In Section 29.6 it was shown that the intensity of a wave is proportional to the square of the amplitude, for mechanical waves. Using more advanced electromagnetic theory, this result can also be derived for electromagnetic waves:

$$I \propto E_0{}^2.$$

To compute the variation of intensity with position in the double-slit pattern, we need to consider the phase difference between two waves arriving at any point *P*.

Figure 55-4. Two electromagnetic waves having equal amplitude E_{10} and a phase difference δ combine to form a sum wave having amplitude E_0.

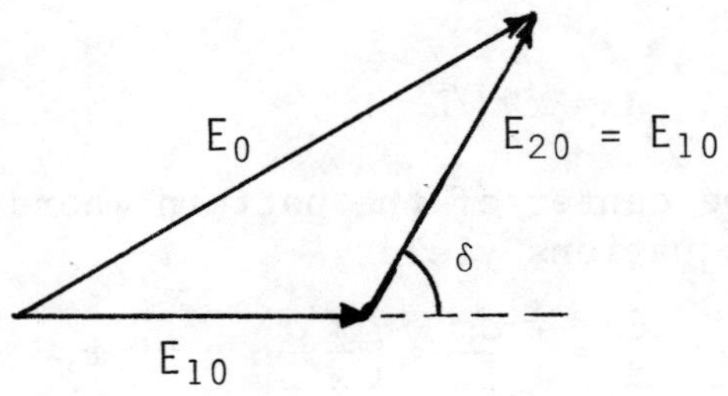

The displacement of the sum wave reaching any point *P* in the pattern may be written as the sum of two sinusoidal waves arriving at *P* from S_1 and S_2. As shown on page 55-4, these waves may be assumed to have the same amplitude E_{10}, and their phase difference is

$$\delta = k\Delta.$$

Each wave arriving at *P* has an instantaneous displacement which varies sinusoidally with time, so it may be represented by a rotating vector of phasor.
The amplitude of the sum wave may be obtained from the diagram in Figure 55-4, where the angle between the individual amplitudes is the phase angle δ. Using the law of cosines, we have

$$E_0{}^2 = E_{10}{}^2 + E_{10}{}^2 + 2E_{10}{}^2 \cos\delta$$

$$= 2E_{10}{}^2 (1 + \cos\delta).$$

Using the identity $\cos 2\theta = 2\cos^2\theta - 1$ (Appendix A.1), we let $\theta = \delta/2$ and obtain

$$1 + \cos\delta = 2\cos^2(\delta/2).$$

Combining our results leads to

$$E_0{}^2 = 4E_{10}{}^2 \cos^2(\delta/2).$$

The intensity of the sum wave at *P* is therefore

$$I \propto 4E_{10}{}^2 \cos^2(\delta/2)$$

or

★ $$I = I_0 \cos^2(\delta/2),$$

where I_0 is the maximum intensity of the sum wave.

To find the relation of δ to the position x on the screen, we recall that

$$\delta = \frac{2\pi}{\lambda} \Delta.$$

Also,

$$\Delta = x \, d/L$$

near the center of the pattern where the angle θ is small. Combining these equations yields

$$\frac{\delta}{2} = \frac{\pi}{\lambda} \frac{dx}{L} = \left(\frac{\pi}{\lambda} \frac{d}{L}\right) x = \gamma x,$$

where γ is a convenient symbol for the product $\pi d/\lambda L$. The derived equation for the intensity is then

$$I = I_0 \cos^2 (\gamma x),$$

in agreement with the result of experimental measurements.

When the intensity variation is measured farther from the center of the double-slit pattern, it is found to depart from the simple cosine-squared equation derived above. This occurs because the simple double-slit interference pattern is modified by the diffraction pattern of each individual slit (Section 55.4). The analysis is left to more advanced books on optics, where it is shown that the observed pattern can be fully accounted for.

Questions and Exercises

E 55.2-1 (a) Sketch the variation of intensity with position in the double-slit interference pattern.
(b) Compare your result with the equation for constructive interference derived in Section 28.4, and label each maximum of intensity with the proper value of m.

E 55.2-2 Compare the intensity variation near the center of the double-slit pattern with the intensity which would result from using only one slit of the double slit, and show that the maximum intensity in the double-slit pattern is four times the intensity produced by one slit.

E 55.2-3 (a) Referring to Figure 55-3, show that the average intensity near the center of a double-slit pattern is half the maximum intensity.
(b) Considering the result of E 55.2-2, how does the average intensity in the double-slit pattern compare with the sum of the intensities of the two individual slits used separately?

E 55.2-4 The equation for the intensity of an electromagnetic wave in free space is $I = (\varepsilon_0/\mu_0)^{\frac{1}{2}} E_0{}^2/2$, where E_0 is the amplitude of the wave. Show that this equation has the correct physical units of intensity. For ε_0 and μ_0 see Appendix F.

E 55.2-5 Combine the equation for phase difference given in this section with the equation for constructive interference derived in Section 28.4, to show that the phase difference at the centers of the bright fringes on one side of the center is given by $\delta = 2\pi m$, where $m = 0, 1, 2, \ldots$.
(b) What is the phase difference at the center of the first intensity maximum away from the center of the pattern?

55.3 Coherent Sources

The double-slit interference pattern described in Section 55.2 results when a single source of light provides the illumination for *both* slits of the double slit, as shown in Figure 28-7. If the single source and slit S are removed, and two separate sources are placed one in front of slit S_1 and the other in front of slit S_2, the interference pattern of alternating light and dark fringes is not observed. This is true even if the two separate sources are identical in the sense of being made in the same way, such as two lamps coming off the same production line one after the other. Why is it necessary to illuminate the two slits using light from the same original source, in order to observe an interference pattern?

The interference pattern resulting from the superposition of two waves depends on the phase relationship between the waves. An interference pattern is stable only if the interfering waves maintain a phase difference which is constant over an extended period of time. The light wave radiated from any one source originates in the multitude of radiating atoms in the source. The composite wave sent out by the source is the superposition of a very large number of individual waves from these individual atoms. Each atom radiates for only a very brief interval (on the order of 10^{-8} second or less) at a time. In the composite wave, the resultant phase changes in a random manner at irregular intervals on the order of 10^{-8} second. When two separate sources are compared, the *phase difference* between them also changes in this random manner. When the phase difference changes the interference pattern fluctuates as well, so no stable interference pattern can be observed.

When light from a single source falls on two slits simultaneously as shown in Figure 28-7, the waves entering the two slits remain exactly in phase with each other. Each radiating atom in the single source contributes equally to the waves entering the two slits. When the phase changes in the composite wave from the single source, this phase change arrives at both slits simultaneously. Under these conditions, the interference pattern is stable; we say that a stable pattern results from the superposition of waves from coherent sources.

> Two light sources are *coherent* when corresponding points in both sources always radiate in phase, or with the same phase difference.

With visible light, coherent sources may be derived from a single physical source by either of two methods: (1) division of the wave front, as in a double-slit apparatus, or (2) division of the amplitude, as in partial reflection from a thin film (Section 55.7). Two continuous-wave lasers having the same wavelength may also be used as coherent sources.

Questions

Q 55.3-1 What pattern would be observed on the screen if two sodium lamps were used to illuminate a double slit, one lamp next to each slit?

Q 55.3-2 Can two fluorescent lamps serve as coherent sources when they are both plugged in to the same 60-Hz power outlet?

Q 55.3-3 Consult reference materials or your local radio station to learn how an array of transmitting antennas is used to establish an interference pattern in the broadcast beam. Are these antennas coherent sources?

55.4 Fraunhofer Diffraction by a Single Slit

The diffraction of waves passing through a single slit was described briefly in Section 28.2, where it was noted that the wave fronts passing through the slit fan out to form a pattern wider than the slit. When light waves pass through a double slit, a pattern of alternating bright and dark fringes is formed which can be accounted for in terms of interference of the wave passing through one slit with the wave passing through the second slit. When the pattern of the wave passing through a *single slit* is examined more closely, it is found that it too has maxima and minima (Figure 55-5), suggesting that there is interference between different parts of the wave passing through the single slit. It is seen that the central bright fringe is twice as wide as any of the other bright fringes. How can we account for this single-slit interference pattern?

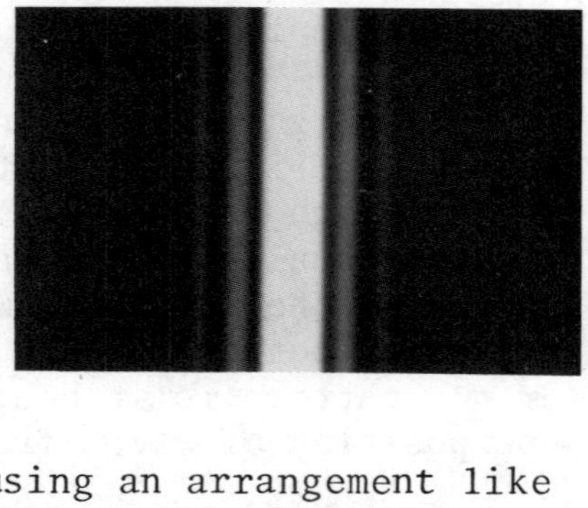

Figure 55-5. Fraunhofer diffraction pattern of a single slit. The width of any bright fringe is measured between the centers of the dark fringes on either side of it. The width of the central bright fringe is twice the width of each other bright fringe. Photographic positive courtesy of LaPine Scientific Company and Leybold-Heraeus GMBH & Co.

A single-slit diffraction pattern is formed using an arrangement like that shown in Figure 55-6. Plane wave fronts arriving at the barrier containing a single slit are represented by rays. The wave passing through

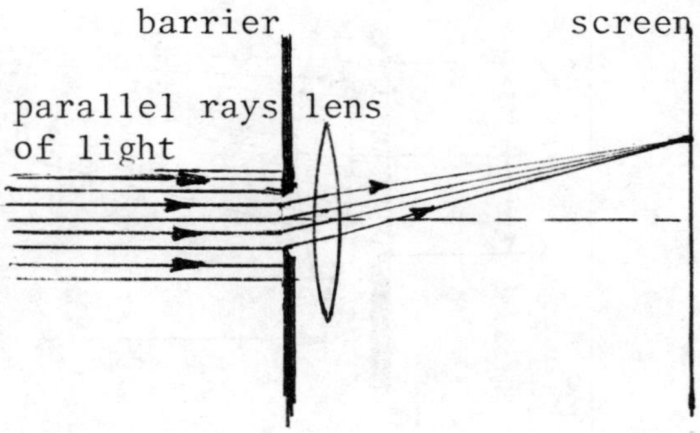

Figure 55-6. Experimental arrangement for observing the Fraunhofer diffraction pattern of a single slit. Parallel rays (or plane wavefronts) of light approach the slit, where they are diffracted and then focused by the lens so that the pattern is observed on a screen placed in the second focal plane of the lens. The long dimension of the slit is perpendicular to the plane of the figure. Care is required so that only the light which passes through the slit is allowed to reach the screen.

the slit is diffracted; if the lens were not present, the diffracted rays would continue beyond the slit and form a pattern at an infinite distance from the slit. The converging lens serves to focus these rays so that the pattern is formed in the focal plane of the lens, where it may be observed on a screen or photographed directly by placing a photographic plate in this focal plane. When a lens is used in this way, the resulting pattern is called the *Fraunhofer diffraction pattern* of the slit. The distance between the lens and its focal plane is called the *focal length* f of the lens, discussed further in Section 54.4 . (Actually, a second lens is required on the left of the slit so that rays from a light source located a finite distance to the left will be made parallel as they reach the slit.)

To account for the central bright fringe and the locations of the minima or dark fringes on either side of it, we imagine that the slit opening is divided into narrow strips whose length is parallel to the long dimension of the slit, as shown in Figure 55-7. The rays which are focused at the center of the pattern, point O, are those which arrived at the slit traveling parallel to the axis or centerline of the apparatus. Each of these rays represents a wavelet, or portion of the wavefront, which passes through one of the narrow strips comprising the slit opening. Because all of these wavelets are initially part of the same plane wave front, they arrive at the slit in phase with each other, and all of them also arrive at point O in phase, so that they interfere constructively and combine to give a maximum of intensity (a bright fringe) at O. This is the central bright fringe in the single-slit diffraction pattern.

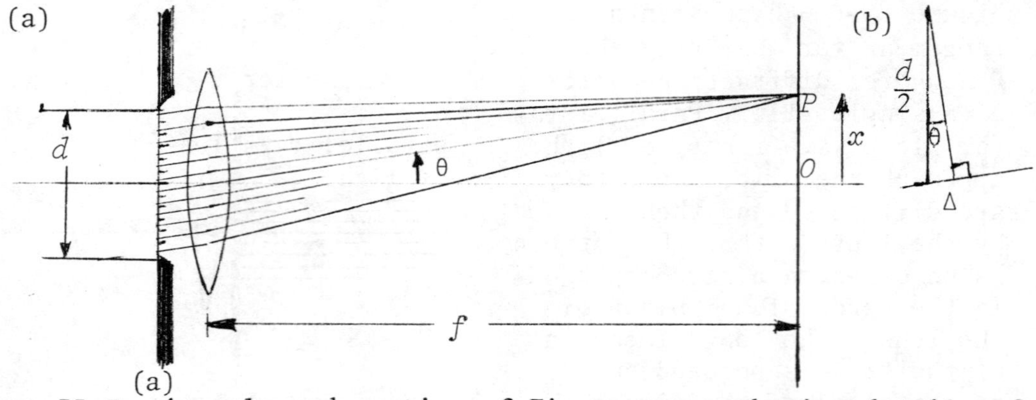

Figure 55.7. (a) An enlarged portion of Figure 55-6, showing details of the apparatus for observing the Fraunhofer diffraction pattern of a single slit. The slit opening is divided into a large number of narrow strips having equal width, so that the amplitudes are the same for all wavelets passing through the strips at an angle θ. These wavelets are focused by the lens at the screen point *P*, located a distance *x* from the screen point *O* at the center of the pattern. (b) A line perpendicular to the central ray makes an angle θ with the slit.

Next consider the parallel rays which emerge from the slit at an angle θ with the axis, as shown in Figure 55-7, and are then focused at the screen point *P* which corresponds to the first minimum from the center of the pattern. We divide these rays into two sets, with one set consisting of the rays which pass through the upper half of the slit, and the second set consisting of the rays which pass through the lower half of the slit. The ray passing through the first narrow strip below the centerline travels one half wavelength (of optical path) farther than the ray passing through the uppermost narrow strip of the slit opening. These two rays represent wavelets which arrive at *P* one half wavelength out of phase, and they therefore interfere distructively and cancel each other. In a similar way, the wavelet passing through the second narrow strip below the centerline travels one half wavelength farther than the wavelet passing through the second strip from the top of the slit, and so these two wavelets also cancel eachother when they arrive at *P*. In like manner, the other wavelets passing through strips in the upper and lower halves of the slit combine in pairs which cancel each other. The result is that the set of wavelets passing through the upper half of the slit just cancels the set passing through the lower half of the slit, so the resulting intensity of the light reaching *P* is zero. The first minimum in the pattern is therefore located at *P*, a distance *x* from the center point at *O*.

To find the value of x in terms of other quantities, we note that $x/f = \tan\theta$. The rays passing through the uppermost strip and the first strip below the centerline are separated by $d/2$ as they pass through the slit, and the optical path difference between these two rays is $\Delta = \lambda/2$. Figure 55-7(b) is an enlarged view of the upper half of the slit; the lower side of this triangle is part of the ray passing through the centerline of the slit. From the triangle it is seen that $\sin\theta = \Delta/(d/2)$. For small values of θ, the sine and tangent are very nearly equal. Combining the equations above then leads to the results

$$x = \lambda f/d \quad \text{and} \quad \sin\theta = \lambda/d$$

for the first minimum.

To account for the second minimum from the center of the diffraction pattern, we divide the narrow strips into four sets, each set comprising one fourth of the slit opening. Consider the parallel rays which pass through these strips at an angle such that they are focused at the second minimum. This requires that the ray passing through the uppermost strip of the second set travel one half wavelength farther than the ray passing through the uppermost strip of the first set. The two wavelets corresponding to these rays will cancel each other at the screen. Using reasoning similar to that in the preceding paragraphs, it is found that the set of wavelets passing through the upper quarter of the slit just cancels the set passing through the second quarter, and the set passing through the third quarter just cancels the set passing through the lowest quarter. The result is that the intensity reaching the corresponding screen point is zero. For this condition, the small triangle which is analogous to Figure 55-7 (b) has a hypotenuse equal to $d/4$, and a short side $\Delta = \lambda/2$. Algebra then leads to the results that

$$x = 2\lambda f/d \quad \text{and} \quad \sin\theta = 2\lambda/d \quad \text{for the second minimum.}$$

This analysis can be continued by dividing the strips into six sets to account for the third minimum in the pattern, eight sets for the fourth minimum, etc. The general result of the derivation may then be written

★ $$x = m\lambda f/d \quad \text{and} \quad \sin\theta = m\lambda/d \quad \text{for the } m\text{th minimum.}$$

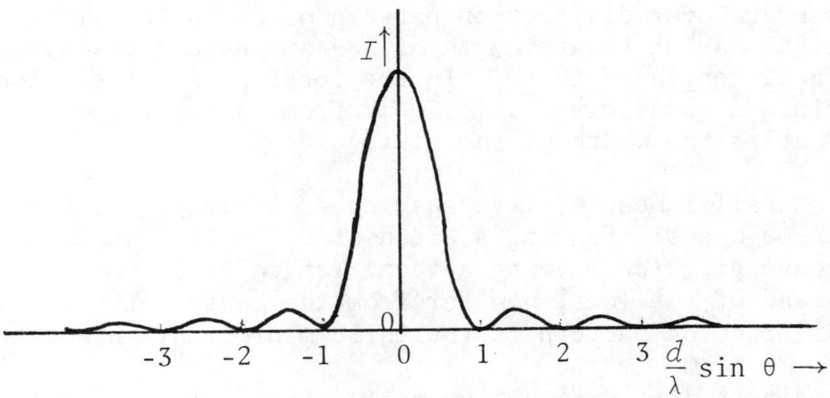

Figure 55-8. Variation of intensity I with angular position in a single-slit diffraction pattern. The maxima of intensity correspond to the bright fringes in Figure 55-5, while the minima correspond to dark fringes.

Figure 55-8 shows the intensity variation in a single-slit diffraction pattern computed using a more advanced analysis. The abscissa in the graph is the angular position measured in units of $(d/\lambda)\sin\theta$. This is related to the linear position x as shown in Figure 55-7 by $x = f\tan\theta = m\lambda f/d$. The numerical values of $(d/\lambda)\sin\theta$ listed on the graph are the values of m at the minima. It is seen that the width of the central maximum, measured between the two minima adjacent to it, is twice the width of each other maximum.

Questions and Exercises

Q 55.4-1 Sketch the fringe pattern observed in Fraunhofer diffraction by a single slit.

Q 55.4-2 Consider the optical path difference between a ray which travels through the uppermost narrow strip of the slit opening, and a ray which travels through the strip just below the midpoint of the slit, both rays arriving at the same point on the screen. For what values of this optical path difference is there a minimum in the pattern on the screen?

Q 55.4-3 In order to observe fringes in the Fraunhofer diffraction pattern of a single slit, is it necessary to have a slit width d which is large, or small, in comparison with the wavelength?

Q 55.4-4 Suppose the narrow strips comprising the slit are divided into three equal sets, with each set making up one third of the slit opening. If the ray passing through the uppermost strip of the second set travels one half wavelength farther than the ray passing through the uppermost strip of the first set, will the complete set of rays passing through the slit at this angle combine at the screen to form a minimum, or a maximum, of intensity? Explain.

E 55.4-1 Sketch an experimental arrangement which may be used to observe the Fraunhofer diffraction pattern of a single slit on a screen, using monochromatic visible light.

E 55.4-2 A Fraunhofer diffraction pattern of a single slit is formed using light having a wavelength of 589 nm and a focusing lens having a focal length of 80 cm. In the focal plane of the lens, the first minimum is observed at 4.70 mm from the center of the pattern. What is the width of the slit?

E 55.4-3 A parallel beam of light having a wavelength of 633 nm passes through a slit having a width of 5.0×10^{-5} m and then through a converging lens having a focal length of 50 cm. In the focal plane of the lens, how far from the center of the resulting diffraction pattern is the third minimum of intensity?

E 55.4-4 A single-slit diffraction pattern is formed using a wavelength of 546 nm, a slit width of 0.1 mm and a lens having a focal length of 40 cm. In the plane 40 cm from the lens, what is the width of the central maximum; i.e., how far apart are the two minima adjacent to the central maximum?

55.5 Resolving Power of Optical Instruments

When an optical instrument such as a microscope is used to study the details in an object of interest, it is found that there is a limit to the fineness of the detail which can be seen clearly. When one attempts to increase the magnification in order to see clearly parts of the object which are very close together, the image becomes larger but also fuzzier or less distinct. The ability of an optical instrument to produce distinct images of objects or parts of an object very close together is called the *resolving power* of the instrument. High resolving power requires that the image of a point in the object, or a point source of light, be as small as possible. The upper limit to the resolving power results from the finite size of the diffraction pattern of the source formed by the instrument.

Consider first the diffraction pattern of a rectangular aperture such as a long narrow slit having width d, illustrated in Figure 55-8. As we showed in Section 55.4, the central maximum or central bright fringe in this pattern has a width (measured between the two adjacent minima) of $2\lambda f/d$. If one attempts to make this fringe narrower by making the slit narrower, the effect is just the opposite: as d is decreased, the fringe width is increased!

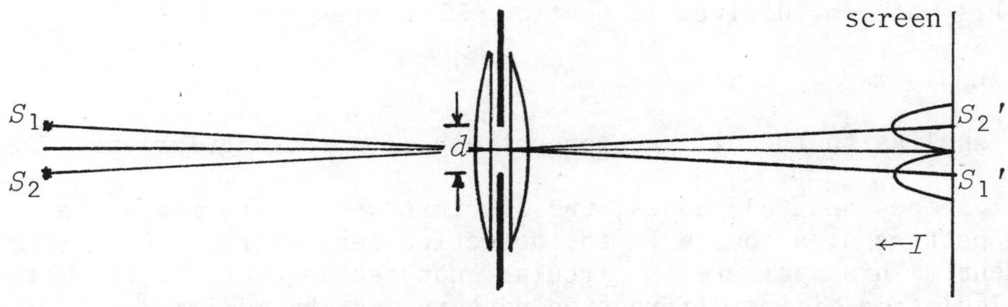

Figure 55-9. Experimental arrangement for studying the single-slit diffraction images of two narrow line sources, S_1 and S_2, located close together. The lenses are required to send parallel rays through the slit and to focus the images on the screen as sharply as possible. Each image is a diffraction pattern formed by the slit. The images S_1' and S_2' have one of their first minima superposed, so these images are clearly resolved.

What is required to produce distinct images of two sources located very close together in some object under study? Figure 55-9 shows the slit and the images it produces of two very narrow line sources, S_1 and S_2. The image of source S_1 is the diffraction pattern S_1' formed by the slit, for which only the central maximum is shown. The image of source S_2 is the central maximum of its diffraction pattern, S_2'. If the separation of the sources is such that the first minima of the two diffraction patterns occupy the same position on the screen, the images will be clearly resolved, as shown in Figure 55-10(a). In this case, the resultant intensity is zero between the two central maxima. If the separation of the sources is such that a first minimum of one image coincides with the maximum of the other, the images are just barely resolved, as shown in Figure 55-10(b). This condition, where a first minimum coincides with a central maximum, was chosen by Lord Rayleigh as the criterion for the resolution of two diffraction images, and is known as "Rayleigh's criterion."

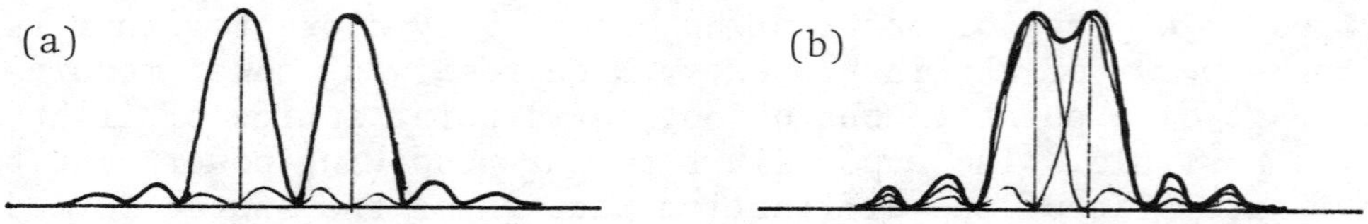

Figure 55-10. Diffraction images of two narrow line sources, obtained using the arrangement shown in Figure 55-9. (a) When the first minima coincide, the images are well resolved. (b) When a first minimum of one image coincides with the maximum of the other, the intensity of the resultant pattern has a slight dip at the center, so the images are just resolved.

To find the angular separation of two sources, or two images, which are just resolved, we refer to Figure 55-7. The angle θ denotes angular position in the single-slit diffraction pattern. The positions of the minima in this pattern, derived in Section 55.4, are given by

$$\sin\theta = m\lambda/d, \qquad m = 1, 2, 3, \ldots$$

This result applies to the diffraction pattern of a rectangular aperture.

In microscopes and telescopes, the aperture which determines the diffraction pattern of a source is the objective lens where light enters the instrument. This aperture is circular, not rectangular, so the mathematical calculation of the diffraction pattern must be modified. It turns out that the equation above is still applicable, but for a circular aperture the values of m are not integers. At the first minimum,

★ $$\sin\theta_1 = 1.220\ \lambda/d \qquad \text{or} \qquad \theta_1 \cong 1.220\ \lambda/d.$$

Here d is the diameter of the aperture, i.e. of the objective lens. This diameter is much larger than the wavelength of light, so $\sin\theta_1$ is very small and can be replaced by θ_1.

For an astronomical telescope, θ_1 is the minimum angular separation of distant stars or other objects which can just be resolved; θ_1 is called the *minimum angle of resolution*. For a telescope used with visible light, θ_1 may be calculated using λ = 560 nm for the effective wavelength of visible light.

For an optical microscope, one is more interested in the linear separation of objects which can just be resolved in the specimen under study. In this case, one needs to consider the difference in optical paths taken by rays passing through opposite edges of the objective lens. Analysis then shows that the linear separation s of objects just resolved is given by

$$s \cong 0.3\ \lambda.$$

Using λ = 560 nm for visible light, the minimum value of s is about 0.2 μm. For the human eye to see these objects distinctly, a magnification (page 54-10) in the range from 500 to 1000 diameters is required. Some optical microscopes provide a higher magnification, but this does not increase the resolving power which is limited by the finite size of the diffraction pattern.

To distinguish finer details in specimens studied using a microscope, it is necessary to use smaller wavelengths. The wave properties of electrons are used in electron microscopes (Section 31.5). The scanning electron microscope can be used with a magnification up to 20,000 diameters or more, and can resolve details of the specimen down to 10 nanometers in size. The transmission electron microscope can achieve a magnification of some 3,000,000 diameters and resolve features smaller than 0.4 nm.

Exercise

E 55.5-1 The unaided human eye can distinguish objects about 0.2 mm apart at the distance of most distinct vision, taken as 25 cm.
(a) Using these data, what is the minimum angle of resolution of the unaided eye?
(b) Taking the eye pupil diameter as 3.0 mm and using an effective wavelength for visible light, what is the theoretical minimum angle of resolution of the eye?

55.7 Interference in Thin Films

Most people have observed the patterns of color which result when light is reflected from a thin film such as a soap bubble or an oil film on the surface of water. How can we account for these colors? Because these colors are observed most readily for light reflected from *thin* films, rather than thick films, we are led to consider the light reflected from *both* the front and back surfaces of the film. As we shall see, it is interference between these two sets of reflected waves that gives rise to the observed patterns of color. We will discuss this interference using both the ray model (Section 53.1) and the wave model for light, using rays in several of the illustrations for simplicity.

In our discussion of waves on a string (Section 29.2), we saw that when a wave traveling along a string reaches a fixed end of the string, the wave is reflected back along the string with a reversed phase or displacement. That is, an incoming crest is reflected as a trough, and vice versa. The theory of electromagnetic waves predicts the same result for light waves reflected from a dielectric medium which is *optically denser* than the medium in which the light is traveling initially. This result is confirmed by experiment. Although neither the absolute phase nor the change of phase of a light wave can be observed directly, this *change of phase at reflection* is demonstrated by interference phenomena in thin films.

This change of phase at reflection is illustrated by light waves in air reflected from the surface of a sheet of glass. When a light wave is incident on the surface of a dielectric material such as glass, in general part of the wave is reflected and part is refracted into the dielectric medium. The wave reflected from the glass surface back into the air has a phase change of half a wavelength (sometimes expressed as a phase change of π radians), so that an incoming crest is reflected as a trough, and vice versa. The wave refracted into the glass travels to the second surface of the glass, where again part of the wave is reflected and part is refracted. In this case, it is found that the wave reflected from the *optically rarer* medium (the air) back into the glass does *not* have a change in phase.

Figure 55-11. A light ray in a medium having refractive index n' is incident on a thin layer of a medium having refractive index $n > n'$. A series of reflections and refractions leads to reflected rays *1* and *3*, and transmitted ray *2*. The interference pattern observed in the reflected light depends on the optical path difference between rays *1* and *3*.

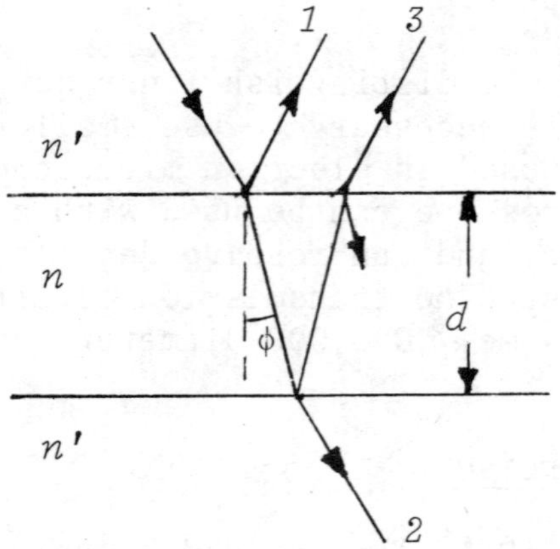

A light ray incident on a transparent sheet or film is successively reflected and refracted many times at the surfaces of the film. Figure 55-11 pictures the first few rays to emerge from the film, when a light ray is incident on a plane-parallel film at an angle to the normal. The intensity in the reflected light is the result of interference between rays *1*, *3*, and the other rays which leave the upper surface. The intensity in the transmitted light is the result of ray *2* and the other rays emerging from the second surface of the film. To account for the qualitative features of the interference patterns in the reflected and transmitted light, it is sufficient to consider only rays *1* and *3* and the optical path difference between them. Including the other reflected and transmitted rays would permit a quantitative calculation of the intensities in these patterns, but would not change the general conclusions obtained below. Using a little geometry and considering the reflected wave fronts, it can be shown that the optical path difference between rays *1* and *3* is given by

★ $$\Delta = 2nd \cos \phi.$$

When light is incident on the film at normal incidence, the angle of refraction ϕ is equal to zero, and the optical path difference is simply $\Delta = 2nd$.

The reflected light may appear bright or dark, depending on the thickness of the film. Consider a film of glass or soapy water in air, having a refractive index greater than that of the media on either side of it. Where the film is so thin that Δ is practically zero, rays *1* and *3* are out of phase by half a wavelength because there was a phase change for ray *1* but there was none for ray *3*. As a result, rays *1* and *3* interfere destructively and the reflected intensity is a minimum. Actually, the reflected intensity is zero, as shown by a more advanced analysis, and the transmitted intensity is a maximum.

Where the optical path difference Δ is a whole number of wavelengths for some wavelength,

$$\Delta = m\lambda, \qquad m = 0, 1, 2, 3, \ldots,$$

again rays *1* and *3* are out of phase by half a wavelength. So, again the reflected intensity is zero and the transmitted intensity is a maximum. Where the optical path difference is a half-integer number of wavelengths,

$$\Delta = (m + \tfrac{1}{2})\lambda\ , \qquad m = 0, 1, 2, 3, \ldots,$$

the optical path difference alone would put ray *3* out of phase with ray *1* by half a wavelength. But, the additional change of phase of ray *1* at reflection gives the result that *1* and *3* are *in phase*, so the reflected intensity is a maximum and the transmitted intensity is a minimum.

Questions and Exercises

Q 55.7-1 From your experience, give three examples of thin films which exhibit colored interference patterns in white light.

Q 55.7-2 Can you give an example of a thin transparent film which does not show a visible interference pattern? Suggest an explanation for the absence of a pattern.

Q 55.7-3 If visible light strikes a plate of glass and only part of the light energy is reflected, what happens to the rest of the energy? Is some of it always transmitted through the plate?

Q 55.7-4 (a) When observing interference in the light reflected from the two sides of a thin film, what is the condition for constructive interference?
(b) What is the condition for destructive interference?

E 55.7-1 Derive the equation $\Delta = 2nd \cos\phi$ for the optical path difference between rays *1* and *3* in Figure 55-11.

E 55.7-2 (a) Using glass having a refractive index of 1.56 and light having a wavelength of 589 nm, what thickness of glass would be required so that the optical path difference between light rays reflected normally from the two surfaces of the glass is 5 wavelengths of light?
(b) What would be the reflected intensity under these conditions?

E 55.7-3 A plastic film has a refractive index of 1.60 at a wavelength of 486 nm. The film thickness is 76 nm.
(a) What is the optical path difference between rays of light having this wavelength, reflected normally from the two surfaces of the film?

(b) How many wavelengths of this light are contained along the path of the light within the film?
(c) What is the phase difference between rays reflected normally from the two surfaces, as they emerge into the air?
(d) Would the film appear dark, or bright, in reflected light?

E 55.7-4 The theory of electromagnetic waves leads to the result that when a light wave in air is incident normally on a dielectric surface having refractive index n, the fraction of the intensity reflected (called the *reflectance*) at the surface is $[(n - 1)/(n + 1)]^2$.
(a) Compute the reflectance at normal incidence for glass, $n = 1.50$.
(b) Compute the reflectance at normal incidence for diamond, $n = 2.42$.

55.8 Fringes Observed in Thin Films

We now apply the theory developed in the preceding section to several examples of interference in thin films. The simplest example is a film having plane-parallel surfaces illuminated with monochromatic light (Figure 55-12). Such a film may consist of a thin flat sheet of transparent plastic or cellophane, or it may be formed using two flat sheets of glass separated by thin spacers so there is an air film between the glass plates.

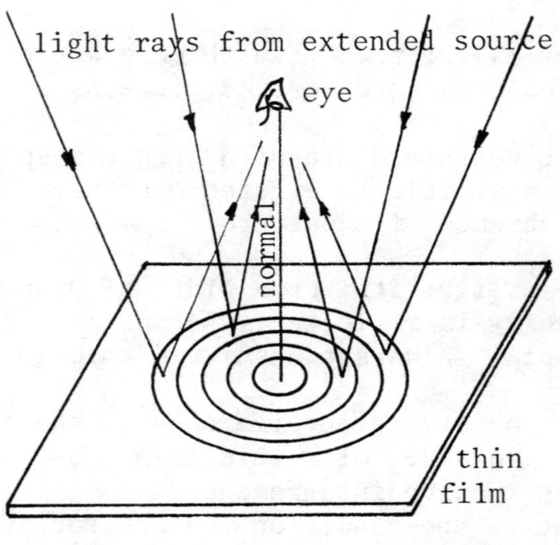

Figure 55-12. Schematic drawing of thin plane-parallel film viewed by reflected monochromatic light. The observer sees an interference pattern consisting of concentric circles, alternately bright and dark, centered at the foot of the line from the observer's eye normal to the film. The observer's eye must be focused on "infinity".

When such a film is viewed by allowing the *reflected* light to enter the eye, a pattern of circular rings or "fringes" is observed. The center of the circles is at the foot of the perpendicular from the eye to the film. Each bright fringe corresponds to one value of m in the equation $\Delta = (m + 1/2)\lambda$. The corresponding Δ is a constant at all points in the field of observation where $\cos \phi$ is a constant, that is, all points where the ray refracted within the film has the same angle of refraction ϕ. These points lie on a circle centered at the foot of the line from the eye normal to the film. At angles midway between those which give bright fringes, the reflected pattern is dark, corresponding to values of m in the equation $\Delta = m\lambda$.

To observe these fringes, it is necessary to use a light source which is extended or spread over some region of two-dimensional space (rather than a point source), so that light rays can be reflected from the film into the eye over a range of angles. The rays which interfere to form each part of a dark or bright fringe leave the film parallel to each other (as shown in Figure 55-11), so the eye must be focused on "infinity." In addition, it may be necessary to use a dark mask such as a sheet of black paper to prevent light from extraneous sources from entering the film from the side opposite the eye. Unless the film is very thin, the observer must look nearly perpendicular to the film so that individual fringes can be seen distinctly.

When a thin plane-parallel film is observed in *transmitted* monochromatic light, the film appears bright for each value of ϕ corresponding to $\Delta = m\lambda$, and less bright (but not dark) for intermediate values of ϕ.

When a thin plane-parallel film is observed in *reflected white light*, an interference pattern of alternating bright and dark colored fringes is formed *for each wavelength* in the light. For a given value of m, the angle of refraction (ϕ) corresponding to a bright fringe depends on the wavelength, so the bright fringes for various colors have different angular positions. If the film is thin enough a pattern of colors can be seen, with a spacing which depends on the thickness of the film. These are the colored bands seen in the light reflected from oil films and soap bubbles.

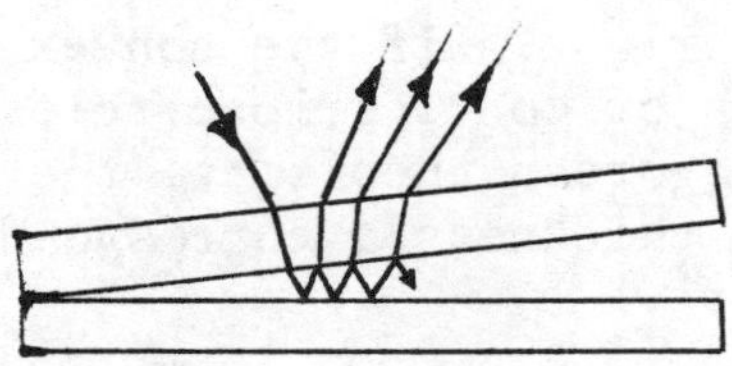

Figure 55-13. A wedge-shaped film of air between two glass plates. The rays which emerge from the upper plate, after being reflected within the film, diverge *as if* they had originated at a location near the film.

Interference patterns are also observed for thin films having two surfaces which are not parallel planes. The simplest example is a wedge-shaped film, such as is formed by the air between two glass plates which touch each other along one edge and are separated by a thin spacer along the opposite edge, as shown in Figure 55-13. Because of the angle between the plates, the rays which emerge from the upper plate after successive reflections within the film are not parallel to each other. Instead, they diverge as if they had originated at some location near the film, rather than at "infinity" as with the plane-parallel film. As before, the optical path difference between these successive rays depends on the film thickness, remaining constant along lines of equal film thickness. In reflected monochromatic light, therefore, the fringes follow these "contours" of equal film thickness. The quantity $\Delta/2 = nd \cos \phi$ changes by half a wavelength from the center of one bright fringe to the center of the next bright fringe. If one of the glass plates forming the wedge is known to be "optically flat," the other plate may be tested for flatness by observing the fringe pattern. Straight fringes indicate that the second plate is also flat, while curved fringes reveal the "high" and "low" spots on the plate.

Figure 55-14. A spherical convex lens L resting on a flat plate gives rise to an air film between the lens and the plate. When the film is illuminated from above using monochromatic light, the interference pattern consists of circles called Newton's rings, concentric with the point where the lens touches the plate. The radius r of each ring is related to the film thickness d and the radius of the spherical lens surface, R.

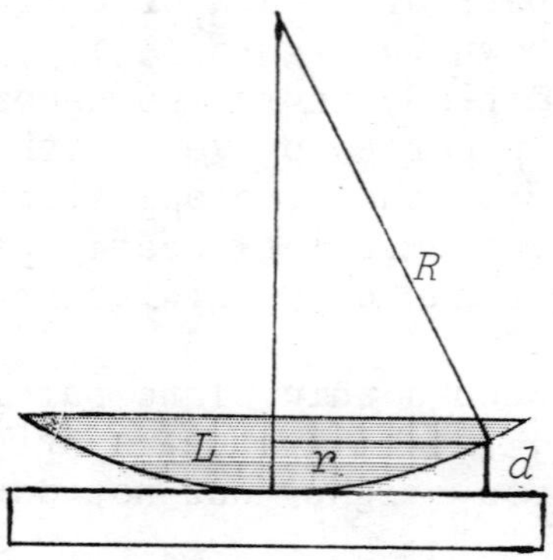

A wedge-shaped air film of special interest is formed when a convex lens rests on an optically flat plate, as shown in Figure 55-14. When monochromatic light enters the lens from above, an interference pattern is formed by light reflected from the lower lens surface and the upper surface of the flat plate. If the light approaches the film normal to the flat plate, to a very good approximation the optical path difference within the film is given by $\Delta = 2d$. Where $2d = m\lambda$, $m = 0, 1, 2, \ldots$, the fringes observed by reflected light are dark. Where $2d = (m + \frac{1}{2})\lambda$, $m = 0, 1, 2, \ldots$ the reflected fringes are bright. If the convex lens surface is spherical, the fringe pattern consists of concentric circles called *Newton's rings*. By applying the Pythagorean theorem to the triangle in Figure 55-14, it can be shown that the film thickness d corresponding to a circular fringe having radius r is given by

$$d = r^2/2R;$$

See Exercise 55.8-6.

Questions and Exercises

Q 55.8-1 How could one visually observe the maxima of intensity in reflected light corresponding to $\Delta = (m + \frac{1}{2})\lambda$, $m = 0, 1, 2, \ldots$ for a thin plane-parallel film? What would one see, i.e., what is the visual appearance of the interference pattern?

Q 55.8-2 Why are interference fringes observed by light reflected from a thin plane-parallel film *circular?* Where is the center of the circle in relation to the observer and the light source?

Q 55.8-3 Why are fringes not observed when light from a *point source* is reflected from a thin transparent film?

Q 55.8-4 Why is it often difficult or impossible to observe interference fringes in the light reflected from a *thick* plane-parallel transparent film, even though an extended monochromatic source is used?

Q 55.8-5 Given an extended monochromatic light source and a thin plane-parallel transparent film, why do you not see fringes by reflected light when your eye is focused on the film itself?

Q 55.8-6 For a plane-parallel film illuminated by monochromatic light the reflected intensity corresponding to $\Delta = m\lambda$ is zero, and the dark fringes are black. The transmitted intensity corresponding to $\Delta = (m + \frac{1}{2})\lambda$ is not zero, however, and the "dark fringes" in the transmitted light may be fairly bright. Explain.

Q 55.8-7 When a thin plane-parallel film is illuminated with white light, why can fringes be observed in reflected light only for small values of m?

Q 55.8-8 (a) What is meant by an "optically flat" surface? (b) How could an optical flat be used to determine whether a piece of plate glass is also flat?

Q 55.8-9 When Newton's rings are observed by reflected light, will the center of the pattern be dark or bright if the glass surfaces are in physical contact (i.e., not separated by dust particles)? Explain.

E 55.8-1 Monochromatic light ($\lambda = 600$ nm) from an extended source is reflected from both sides of a plane-parallel film, .001 mm thick, having a refractive index $n = 1.50$.
- (a) Will the center of the resulting circular interference pattern (seen when these reflected rays enter the observer's eye) be dark, or bright? Why?
- (b) Does the order of interference m increase, or decrease, as one moves outward from the center of the pattern? Explain.

E 55.8-2 Two optical flats (transparent glass plates) 30 cm in length are placed one on the other. A thin sheet of aluminum foil placed between the plates at one edge creates a wedgeshaped air film. When this assembly is illuminated with sodium light (λ = 589 nm), ten parallel bright fringes are observed in a total space of 90.0 mm. What is the thickness of the foil?

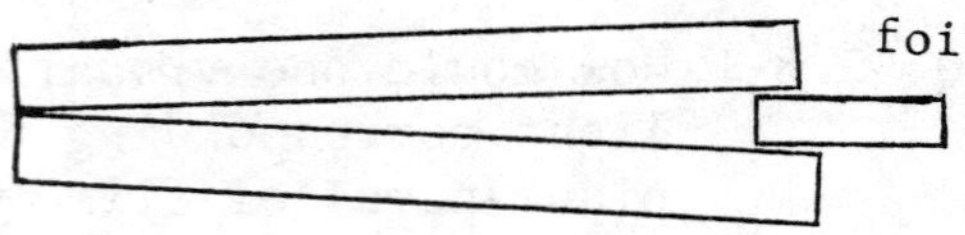

E 55.8-3 A glass plate having a refractive index of 1.50, a thickness of 0.8 micrometers and plane-parallel surfaces is illuminated by white light directed normal to the surfaces. What wavelengths within the visible range will have the greatest intensity in the reflected light?

E 55.8-4 Two optically flat glass plates are parallel to each other and are illuminated normal to their surfaces with light having a wavelength of 536 nm. Using a fixed separation of the plates equal to the wavelength of the light, several substances having differing refractive indexes are inserted into the space between the plates, one at a time.

(a) Using a substance having an index of 1.5 for the "film," will the pattern appear dark, or bright, by reflected light?

(b) Repeat part (a) for an index of 1.25.

(c) Repeat part (a) for an index of 1.00.

E 55.8-5 Two flat glass plates having a length of 10 cm are in contact along one edge, while they are separated at the opposite edge by a single sheet of paper having a thickness of 0.01 cm. If the plates are viewed in monochromatic light having a wavelength of 536 nm, incident normal to the surfaces of the upper plate, what is the spacing between adjacent bright fringes in the reflected interference pattern formed by the air film between the plates?

E 55.8-6 (a) Apply the Pythagorean theorem to the triangle shown in Figure 55-14 and show that for values of d much smaller than R, the film thickness for a given ring radius r is given by $d = r^2/2R$.

(b) Assuming the arrangement shown in Figure 55-14 is illuminated from directly above using monochromatic light, show that in the reflected interference pattern the radius r of the mth bright ring from the center is given $r^2 = (m - \frac{1}{2})\lambda R$, when the center of the lens makes good contact with the flat plate.

E 55.8-7 For the arrangement shown in Figure 55-14 the radius of the fifth bright ring in reflected light having a wavelength of 633 nm is measured to be 1.9 mm. Assuming the glass surfaces are clean and free from dust, compute the radius of curvature of the lens surface.

55.9 Plane-Polarized Light

An electromagnetic wave consists of an electric field vector $\vec{E}$ and a magnetic field vector $\vec{B}$ oscillating in phase with each other and traveling at the speed of light. Figure 55-15 pictures an idealized sinusoidal wave in which the $\vec{E}$ vector remains in one plane, while the $\vec{B}$ vector remains in the plane normal to it and normal to the velocity. For every electromagnetic wave the $\vec{B}$ vector is always perpendicular to the $\vec{E}$ vector at every point, so in this discussion of polarized light it will be sufficient to represent the complete light wave by only one of its field vectors.

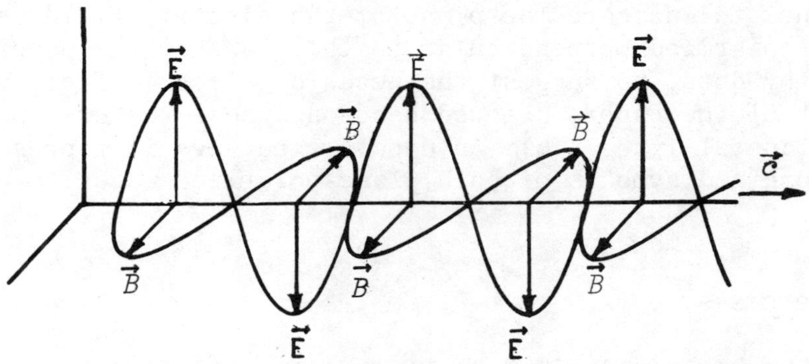

Figure 55-15. Schematic representation of a plane-polarized monochromatic electromagnetic wave. The wave is traveling toward the right with velocity $\vec{c}$. It consists of an electric wave in which the displacement is the electric field strength, $\vec{E}$, shown in the vertical plane, and a magnetic wave in which the displacement is the magnetic induction (field strength), $\vec{B}$, shown in the horizontal plane. These two parts of the electromagnetic wave travel together, always at right angles to each other and to the direction of travel.

A light wave like the one pictured in Figure 55-15, having its displacement (field) vector always in the same plane, is said to be plane-polarized.

> A *plane-polarized* or linearly-polarized wave has its displacement vector in the same plane at every point.

A plane-polarized light wave can be represented simply in a diagram by drawing a series of arrows or straight lines to indicate the plane of the $\vec{E}$ vector, as shown in Figure 55-16.

Except for light emitted by a continuous-wave laser, light from natural and artificial sources does not have an idealized sinusoidal waveform but has an amplitude vector which changes rapidly and irregularly with time. These rapid fluctuations occur because each radiating atom in the source emits its wave for only a very brief interval. The resultant unpolarized wave is the superposition of myriads of short pieces of waves from atoms in the source. We can account for the polarization phenomena of light, however, by regarding unpolarized light as the superposition of two plane-polarized waves vibrating in two mutually perpendicular planes. Plane-polarized light has its $\vec{E}$ vector entirely within one plane.

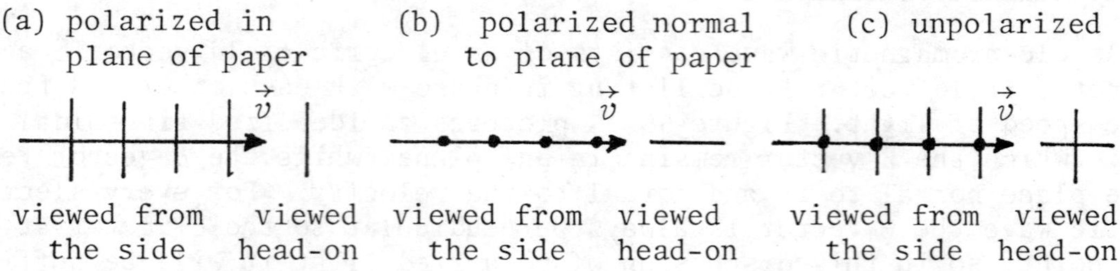

Figure 55-16. Simplified representations of light waves. (a) A wave polarized in the plane of the paper is represented by short straight lines to indicate the plane of the electric field vector. (b) A wave polarized perpendicular to the plane of the paper is represented by dots, to suggest the heads of electric field vectors directed out of the paper. Viewed head-on, these vectors appear as a short horizontal line. (c) An unpolarized wave is represented using the combined symbols of both plane-polarized waves.

Questions and Exercises

Q 55.9-1 (a) Is a plane-polarized light wave necessarily sinusoidal?
(b) Is a plane-polarized light wave necessarily monochromatic?

Q 55.9-2 Can a longitudinal wave be plane-polarized? Explain.

Q 55.9-3 What is the justification for pretending that unpolarized light is a superposition of two plane-polarized sinusoidal waves vibrating in two mutually perpendicular planes?

E 55.9-1 An electric charge oscillates sinusoidally along the z axis about the origin as an equilibrium position. Consider the variation with time of the resulting electric field vector at a field point some distance away on the x axis. Subtracting the more-or-less steady value of the $\vec{E}$ vector in the x direction, show qualitatively that the time-varying component of $\vec{E}$ corresponds to the passage of a sinusoidal transverse wave.

E 55.9-2 For the oscillating electric charge described in E 55.9-1, show qualitatively that the time-varying component of the resulting $\vec{B}$ vector at a field point some distance away on the x axis corresponds to the passage of a sinusoidal transverse $\vec{B}$ wave.

55.10 Polarization by Reflection or Refraction

When a parallel beam of unpolarized light is incident on a polished dielectric surface, it is found by experiment that both the reflected and refracted beams are partially plane-polarized. This result holds for any angle of incidence θ_1, and is shown in Figure 55-17. The amount of polarization of the two beams depends on the angle of incidence and on the refractive index of the dielectric.

Figure 55-17. When a parallel beam of unpolarized light is incident on a plane surface of a dielectric, in general both the reflected and refracted beams are partially plane-polarized.

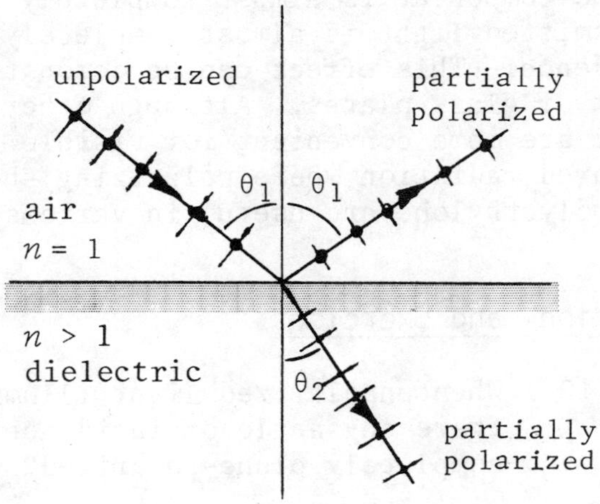

There is one angle of incidence for which the *reflected* beam is completely plane-polarized in the plane normal to the plane of incidence. This angle of incidence is called the polarizing angle or Brewster's angle, and is designated by $\bar{\theta}$ (theta-bar). The corresponding angle of refraction is $\bar{\theta}_2$. Experiments also show that at the polarizing angle, $\bar{\theta} + \bar{\theta}_2 = 90°$ for any dielectric. Applying Snell's law (Section 53.3) then leads to

★ $\tan \bar{\theta} = n,$ Brewster's law.

For glass, $\bar{\theta}$ is about 57°

Light reflected at an angle from the surface of a paved road is partially plane-polarized. When a driver encounters excessive glare in the light reflected from the road and from the hood of the car, the majority of the reflected light may be removed from the driver's vision by using sunglasses which absorb the component of the reflected light polarized in the horizontal plane (Section 55.11).

The phenomenon of polarization by reflection and Brewster's law can be derived from the theory of electromagnetic waves. They are consequences of the fact that the field vectors in the light wave must satisfy certain physical conditions at the surface of the dielectric.

We may think of the unpolarized light arriving at the surface of the dielectric as comprising two separate components. One component is plane-polarized in the plane of incidence, while the other is plane-polarized normal to the plane of incidence. When unpolarized light is incident on a dielectric plate at the polarizing angle, part of the second component is reflected at each surface, and so is removed from the beam transmitted by the plate. Suppose several flat dielectric plates are arranged parallel to each other in a stack, with an air space between each pair of plates. With light incident at the polarizing angle, part of the second component is reflected by each plate, while all of the component polarized in the plane of incidence is transmitted by each plate (assuming no absorption by the plate). If enough plates are used, the second component is almost completely removed from the beam, so that the transmitted light is almost completely plane-polarized in the plane of incidence. This effect can be demonstrated for visible light using a stack of glass plates. Although other methods of producing polarized light are more convenient for visible light, this method is used with infrared radiation where polarizing sheets of selenium, silver chloride, and polyethylene are useful in various regions of the spectrum.

Question and Exercises

Q 55.10-1 When unpolarized light illuminates a dielectric surface, is there any angle of incidence for which the refracted beam is completely plane-polarized?

E 55.10-1 Using the fact that $\bar{\theta} + \bar{\theta}_2 = 90°$ at the polarizing angle, and Snell's law, derive Brewster's law.

E 55.10-2 Sunlight reflected from a smooth water surface at an angle of reflection of 53.1° is observed to be completely plane-polarized. Find the refractive index of water.

E 55.10-3 Using polaroid sunglasses or some other analyzer of polarized light, observe the degree of polarization of the sunlight reflected from or transmitted by the windows or your residence or your college buildings at various times of the day. How could you use your observations to determine an approximate value for the refractive index of the window glass?

E 55.10-4 Consider the use of a stack of glass plates to produce plane-polarized light from incident unpolarized light by transmission of p light, at Brewster's angle. "p light" is the component of incident light having its electric vector in or parallel to the plane of incidence, while "s light" is the component of incident light having its electric vector perpendicular (senkrecht) to the plane of incidence. Neglect absorption by the glass. The reflectance (fraction of incident intensity reflected) at each surface for s light is $r_s = 0.15$.

(a) What is the reflectance r_p for p light?

(b) Compute the ratio of the intensity I_p of p light to the intensity I_s of s light in the transmitted beam, for 10 plates.

55.11 Dichroism. Polarization by Transmission

One of the most convenient polarizing devices for visible light is a sheet of the dichroic material known as Polaroid®, invented in 1932 by Edwin H. Land. A *dichroic substance* is one which resolves incident unpolarized light into two components polarized in perpendicular planes, and absorbs one component more strongly than the other. As a result, the light transmitted by a sheet of dichroic material is partially or completely plane-polarized. A dichroic substance acts as a plane-polarizer only at selected wavelengths, determined by the atomic or molecular structure of the material.

Figure 55-18. The effect of a sheet of Polaroid® on incident unpolarized light traveling from left to right. The transmission plane of the sheet is parallel to the plane of the paper.

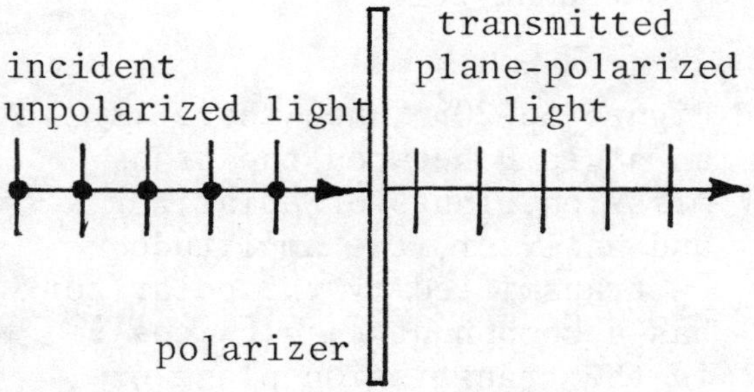

Polaroid® sheet contains a film of polyvinyl alcohol which is aligned by stretching and is impregnated with iodine crystals. A thin film of this material absorbs one component of the incoming light almost completely, and transmits most of the intensity in the other component (Figure 55-18). This effect may be compared to that of a picket fence having ropes stretched between the pickets and normal to the plane of the fence. If rope waves are generated at the ends of the ropes by shaking them, the waves vibrating perpendicular to the pickets are blocked by the fence, but the waves vibrating parallel to the pickets pass through the openings with little loss in intensity. The *transmission plane* (TP) of a polarizer (or a picket fence) is the plane containing the transmitted ray (or the rope) and the amplitude vector of the transmitted wave. A thin film of type HN 32 Polaroid® transmits 64% of the intensity of the component polarized in the transmission plane, but absorbs practically all of the other component of incident visible light, so the total transmitted intensity is 32% of the initial unpolarized light.

Another (non-dichroic) polarizing device used in the laboratory is a Nicol prism, invented in 1828 by Scotchman William Nicol and made from a crystal of calcite.

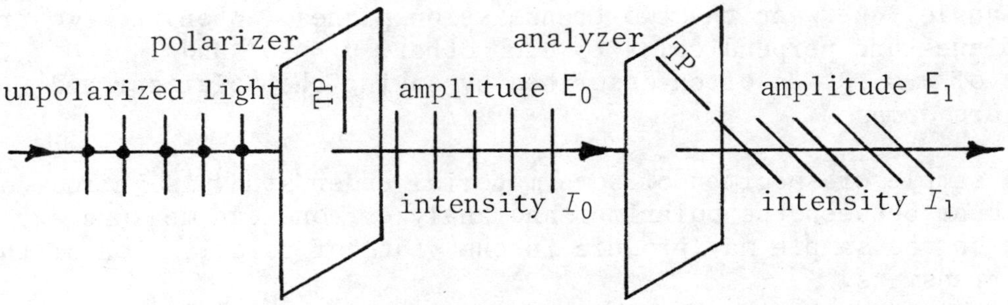

Figure 55-19. Phenomena of polarized light may be studied using two plane-polarizing devices in series. The first device serves as a polarizer while the second functions as an analyzer.

Plane-polarizing devices are often used in pairs, one serving as a *polarizer* and the other as an *analyzer* (Figure 55-19). The light transmitted by the polarizer has its amplitude vector in the TP of the polarizer, while the light transmitted by the analyzer has its amplitude vector in the TP of the analyzer.

Figure 55-20. When there is an angle θ between the transmission planes of polarizer and analyzer, the amplitude E_0 transmitted by the polarizer has a component $E_1 = E_0 \cos\theta$ in the transmission plane of the analyzer.

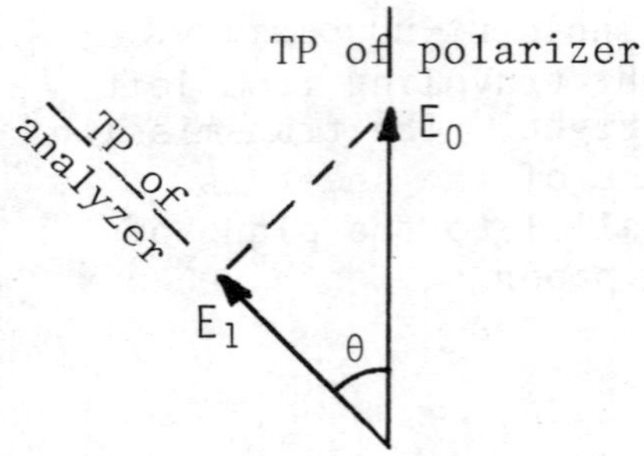

Suppose a polarizer and analyzer are arranged as in Figure 55-19, with an angle θ between the TP of the analyzer and the TP of the polarizer. Then if one looks opposite to the direction of travel of the oncoming light beam, the amplitudes of the wave have the relationship shown in Figure 55-20. The amplitude vector E_0 transmitted by the polarizer is in the TP of the polarizer. The analyzer then resolves E_0 into two perpendicular plane-polarized components, and transmits only the component E_1 in its own TP. If the analyzer absorbs none of the component in its own TP, from Figure 55-20 it is seen that $E_1 = E_0 \cos\theta$. The intensity of the light is proportional to the square of the amplitude, so

$$I_1 = I_0 \cos^2 \theta. \qquad \text{Law of Malus}$$

The law of Malus as stated here is rigorously correct only if there are no losses in intensity as a result of reflection from the surfaces of the analyzer, and no losses through absorption of part of the transmitted component by the analyzer. Even when such losses do occur, the $\cos^2\theta$ factor in the law correctly gives the reduction in intensity resulting from the angle θ between the two transmission planes. When the two transmission planes are perpendicular to each other, $\theta = 90°$ and $I_1 = 0$. The condition of $\theta = 90°$ is often described by saying the polarizer and analyzer are *crossed*.

If a sample or specimen of some material under study is introduced into the beam between the polarizer and analyzer, one can measure any change which the sample may produce in the state of polarization of the light it transmits.

Questions and Exercises

Q 55.11-1 (a) Why are Polaroid® sunglasses especially effective in reducing glare from the highway surface?

(b) Would it be desirable to include a sheet of Polaroid® in the entire windshield of an automobile?

Q 55.11-2 Suppose a salesman offers to sell you a new product "Polium," which he claims to be a perfect polarizing device throughout the spectral range from 300 nm to 1000 nm. I.e., he claims that a sheet of Polium absorbs none of the transmitted component of polarized light in this range, but absorbs all of the other component. Describe the laboratory equipment you would need to confirm (or disprove) the salesman's claim, and the procedure you would follow in doing so.

E 55.11-1 (a) If two sheets of type HN32 Polaroid® are crossed, what fraction of the incident light is transmitted by the second sheet?

(b) If the two sheets have parallel transmission planes, what fraction of the light intensity incident on the first sheet is transmitted by the second?

(c) If the two sheets are crossed, and a third sheet of HN32 Polaroid® is inserted between them at the angle which maximizes the intensity of the light transmitted by the combination, find the intensity of the transmitted light relative to the total intensity of the light incident on the first sheet.

ANSWERS to Questions and Exercises in Chapter 55

Section 55.1

Q 55.1-2 The geometrical path difference cannot be used when the refractive index is significantly greater than unity for part of the path(s), because interference between two waves depends on the relative positions of their crests and troughs. This, in turn, depends on how many *wavelengths* (not simply how many cm) are contained in the paths of the two waves. One example is a thin film of water or glass.

E 55.1-2 (a) 5.41 cm; (b) 9.19×10^4

E 55.1-4 (a) 1300 nm; (b) 650 nm.

Section 55.3

Q 55.3-2 No.

Section 55.4

Q 55.4-2 Minima occur for $\Delta = m\lambda/2$, $m = 1, 2, 3, \ldots$

Q 55.4-4 A maximum, because although the first and second sets cancel each other, the third set is not cancelled and so these wavelets combine to form the first bright fringe aside from the central fringe.

E 55.4-2 10 micrometers

E 55.4-4 4.37 mm

Section 55.7

Q 55.7-2 Examples include thin sheets of plastic such as Saran, cellophane, and polyethylene, and transparent "magic" tape. These films are too thick to display distinct interference fringes in visible light; that is, the optical path in the film corresponds to constructive interference for some visible wavelengths and destructive interference for others, with the result that no distinct pattern is observed.

Q 55.7-4 (a) $\Delta = (m + \frac{1}{2})\lambda$, $m = 0, 1, 2, \ldots$

(b) $\Delta = m\lambda$, $m = 0, 1, 2, \ldots$.

E 55.7-2 (a) 0.944 micrometers; (b) zero.

E 55.7-4 (a) 0.040; (b) 0.172

Section 55.8

Q 55.8-2 Because $\Delta = 2nd \cos \phi$, each bright fringe which corresponds to a fixed value of Δ has a fixed value of ϕ and a fixed angle of refraction into the air for rays which go to the observer's eye. Therefore the rays which enter the eye from this fringe form a cone with its vertex at the eye and its axis normal to the film. The circular fringe is the intersection of this cone with the film. Assuming an extended light source behind the observer which supplies rays for a complete cone, the center of the circle is at the foot of the normal from the observer's eye to the film.

Q 55.8-4 A thick film contains such a large optical path that the angles ϕ which correspond to successive values of m lie very close together, so that the angular interval between adjacent fringes is very small and it may not be possible to see the alternation of light and dark fringes.

Q 55.8-6 Only a small fraction of the intensity in the incident light is reflected to form the bright fringes, which leaves nearly all of the intensity in the incident light to be transmitted into the "minima" of the transmitted interference pattern. As a result, the maxima and the minima in the transmitted pattern have nearly equal intensities, and the resulting fringes have very little visual contrast which makes them difficult to observe.

Q 55.8-8 (a) An optically flat surface is one which is geometrically flat to within a small fraction of a wavelength of light.
(b) The plate glass to be tested may be placed on top of the optical flat, forming a thin film of air between them. Interference fringes formed by this film using monochromatic light will reveal variations in the thickness of the film. If the plate glass is also flat and is not separated from the optical flat by dust particles, no fringes will be seen.

E 55.8-2 9.82 micrometers.

E 55.8-4 (a) dark; (b) bright; (c) dark

Section 55.9

Q 55.9-2 No, because it does not have an amplitude vector directed normal to the direction of travel of the wave.

Section 55.10

E 55.10-2 1.33

E 55.10-4 (a) zero; (b) 25.8

Section 55.11

Q 55.11-2 One would need a device to produce plane-polarized light, and a device to measure the intensity of light. Procedure: Place a sheet of Polium normal to a beam of plane-polarized light, rotate the Polium so its TP is (a) parallel, and then (b) perpendicular, to the plane of polarization, and measure the intensities transmitted in each case. Also, measure the intensity of light with no Polium in the beam. Repeat at several wavelengths in the range from 300 to 1000 nm. With these data, it would be easy to compute the transmittance of Polium at each wavelength for both components of polarized light.

CHAPTER 56 LAWS OF THE ELECTROSTATIC FIELD

Introduction

In Chapters 24 and 25 the concepts of electric field strength and potential were developed using the methods of algebra. In order to express these concepts using more powerful equations, we use more advanced methods of mathematics: unit vectors, the scalar product of two vectors, and definite integrals. Using this notation, this chapter develops two highly useful principles which apply to electrostatic fields: Gauss's law and the circulation condition. These principles do not represent new physical laws, but they do allow us to deal with complex problems which cannot be handled using Coulomb's law alone.

The electric field lines passing through a surface are described using the concept of electric flux, closely related to the electric field strength at the surface. Gauss's law relates the electric flux through a surface to the charge enclosed by the surface, and leads to a quantitative statement of the number of electric field lines which emanate from a charge. Gauss's law is a powerful aid in finding the electric field resulting from a given charge distribution in cases where there is a high degree of symmetry, such as a charged sphere and a charged flat surface. The results are applied to draw conclusions about electric fields in and near a conductor.

The relation between electric field strength and potential difference is developed for the general case of a non-uniform field. This leads to a discussion of the use and evaluation of line integrals for simple cases. The circulation condition for electric field strength expresses the fact that an electric potential must exist in an electrostatic field.

When you have completed your study of the material in this chapter, you should be able to answer questions and work exercises like the examples in the Minimum Learning Goals assigned by your instructor.

56.1 The Field Concept - An Analogy

Minimum Learning Goal: Given an analogy between the electric field due to a charge and a mechanical model, be able to use the model to illustrate properties of the electric field. For example:

1 A large thin sheet of rubber is stretched in a horizontal plane. A large steel ball is placed on the sheet, resulting in a depression. Then a very small steel ball is placed on the sheet somewhere in the "region of influence" of the larger ball.
   (a) Describe what happens to the two balls.
   (b) To what configuration of electric charges, in terms of relative sign and magnitude, is this analogous? (Answer, page 56-38, Line 3)

2 Suppose the small steel ball of Example 1 is the analog of a small positive test charge, and the heavy steel ball is the analog of a large negative electric charge which establishes an electric field. Keeping the small steel ball as the analog of a positive test charge, how could one produce on the rubber sheet an analog of the field of a large positive electric charge? (Answer, p. 56-38, Line 9)

This chapter develops the theory of electrostatic fields and applies it to a number of problems. Most of the calculations involve vector quantities, so we will make extensive use of unit vectors (Section 43.1) and the dot product of two vectors (Section 47.1). The reader of this chapter needs a working knowledge of these vector methods.

The laws of electrostatics were introduced in Chapter 24 with Coulomb's law for the electric force between two point charges q_1 and q_2 separated by a distance d:

★ $$F_{elec} = k\frac{q_1 q_2}{d^2}, \qquad \text{Coulomb's law.}$$

The SI unit for electric charge is the *coulomb*, symbol C. In SI units, the proportionality constant in Coulomb's law is

$$k = 9.0 \times 10^9 \ \text{N}\cdot\text{m}^2/\text{C}^2.$$

The direction of the electric force in any instance must be found using the rule that like charges repel each other, while unlike charges attract each other. *Do not* depend on the positive or negative sign of the product to indicate the direction of the force.

The *electric field* is defined as a certain condition of space in the region surrounding electric charge. The electric force on a charge q is then the force exerted by the electric field at the position where q is located. The field itself is produced by all the charges other than q. The electric field strength at a point is the force per unit charge that would be exerted on a small positive test charge at that point if the test charge were there.

> *Electric field strength.* If a (small) test charge q is placed at a point where it is acted on by an electric force $\vec{F}_{elec}$,
> ★ the electric field strength at that point is $\vec{E} \equiv \vec{F}_{elec}/q$.

We think of the field as existing even if there is no charge at that point to "feel" it. The electric field concept serves as a convenient *way of thinking* about electrostatics.

As an analogy, consider a large thin sheet of rubber stretched in a horizontal plane (Figure 56-1). If we place two steel balls on such a sheet, they will roll toward each other. We could say that there is a force of attraction between the balls, but that is not the usual way to think about it. Rather, we say that the weight of one ball causes the rubber sheet to form a "crater" around it, and the second ball rolls downhill into that crater. The "force of attraction" thus turns out to be only an indirect effect. One ball modifies the condition of space around it; the second ball responds to that condition of space. If there were only one ball, there would still be a crater, even though there is nothing to roll down into it. *This is the field point of view.*

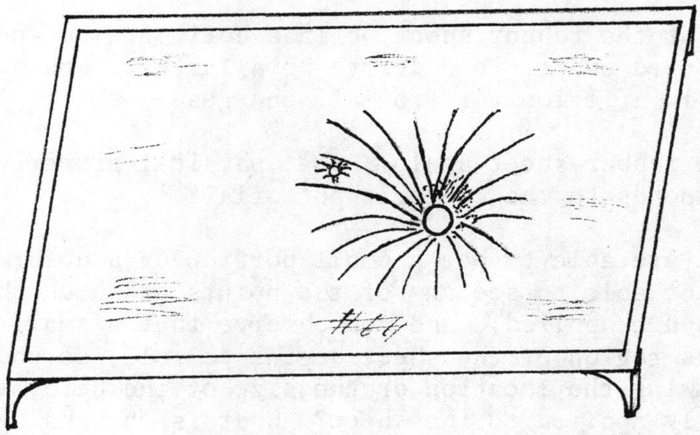

Figure 56-1. A large sheet of thin rubber, stretched uniformly and attached to a horizontal frame. A large steel ball placed on the sheet forms a crater. A small ball placed in the "field" created by the large ball rolls downhill into the crater. The deformed sheet is analogous to an electric field.

Even if the electric force law is stated in terms of the electric field due to a point charge, rather than in terms of the force exerted by one point charge on another, we are still omitting an important conceptual aspect of the field -- how a charge "there" causes a field "here." In the rubber sheet analogy, if we place a single steel ball on the sheet it gives rise to a crater that extends a large distance out from the ball. We do not think of the depression of the sheet at some distant location as being due directly to the ball. Rather, the sheet is dragged down *by its interaction with neighboring parts of the same sheet*. The ball only depresses the small part of sheet in contact with it; the other points on the sheet acquire their lowered positions because of the elastic forces that connect each piece of the sheet with its neighboring pieces.

In this chapter we treat the electrostatic field in a similar fashion. That is, we formulate the laws of the field in such a way that a charge has a *direct* effect only in its immediate vicinity, and its effect is communicated to other places by the laws that govern the field.

Questions

Q 56.1-1 In discussions of the electric field, a "small positive test charge" is often introduced as a device for exploring the properties of the field. In the rubber-sheet analog to the electric field, what would be the equivalent of a small positive test charge? How could you ensure that this equivalent corresponds to a "small" charge?

Q 56.1-2 In the rubber-sheet analog to the electric field, what is the electric equivalent of the slope of the sheet at any point?

Q 56.1-3 We say that an electric field exists at a point even if there is no charge at that point to "feel" the field. What is the analogous statement for the rubber-sheet analog to the field?

Q 56.1-4 Why must the rubber sheet be in a horizontal plane when no balls are placed on it, in order to be a faithful analog of the electric field in a region far from all charges?

Q 56.1-5 In the rubber-sheet analog, what physical property or condition corresponds to the electric potential?

Q 56.1-6 If you are able to see a small portion of a deformed rubber sheet (but not able to see any of the points at which the deforming forces are applied), and you observe that a small beebee placed in this region of the sheet begins to roll, do you have any way of knowing the location or the size of the deformations which are directly applied to the sheet? What is the analogy for an electric field?

Q 56.1-7 How could one set up a rubber-sheet analog to the physical system consisting of two parallel conducting plates bearing opposite charges separated by a distance d?

Check your understanding of the material above by testing yourself on the minimum learning goal at the beginning of this section.

56.2 Flux of the Electric Field

In preparation for Gauss's law (Section 56.3), we introduce the concept of the flux of the electric field. This quantity is defined in terms of the electric field strength at various points on a surface. In some cases the surface we use will be the actual surface of a physical object, while in others it will be merely an imaginary or mathematical surface in space.

(a) (b)

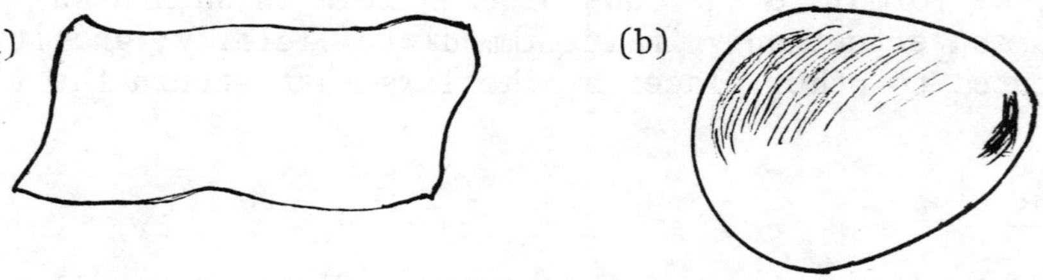

Figure 56-2. Open and closed surfaces. (a) An open surface is bounded by a single continuous edge, and does not enclose any region of space. (b) A closed surface is not bounded by an edge, but it does completely enclose a three-dimensional region of space. Both kinds of surface occur in an infinite variety of shapes.

In general, there are two kinds of surfaces: open surfaces and closed surfaces. An *open surface* is one that is bounded by a single continuous edge; it does not surround or completely enclose any volume of space. One side of a sheet of paper is an open surface, regardless of whether the sheet lies in a plane or is curved or wrinkled. However, if the sheet of paper is folded and glued to itself in such a way that it completely surrounds a volume of space, it forms a closed surface. A *closed surface* is one that is not bounded by an edge but is rather continuous; it encloses or contains a volume of space. For a closed surface, the region *inside* the surface is separated by the surface from the region *outside* the surface. A cube, for example, is bounded by a closed surface; although the three-dimensional cube is said to have twelve "edges," these are not boundary edges of the surface because one can trace a continuous linear path from one face of the cube across any of these "edges" and continue on to the next face, remaining all the while on the same surface.

We use the symbol S to refer to a surface of either kind, open or closed. A very small (infinitesimal) element of area on the surface is designated by the symbol dS. Figure 56-3 shows an element of surface area on a surface S having an arbitrary shape.

Figure 56-3. On a surface S, a very small element of area is designated by dS. The unit vector which is directed perpendicular to dS is labeled $\hat{n}$. The electric field strength at dS is labeled $\vec{E}$.

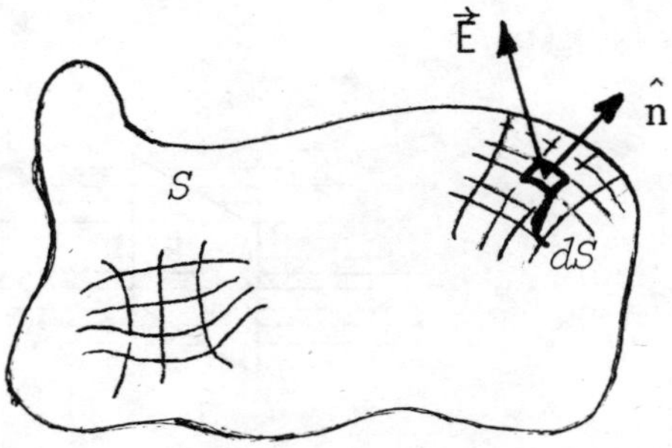

To indicate the spatial orientation of the area element dS, it is most convenient to use the direction perpendicular or *normal* to dS. In general every part of a surface may be curved, but we imagine that each area element is so small it approaches a plane, so that the direction normal to this plane is uniquely defined for each element of area. This direction, chosen either "up" or "down" from the surface, is then indicated by a *unit normal vector*, $\hat{n}$; the normal unit vector is pictured for the area element in Figure 56-3. For a closed surface, it is customary to choose the unit vector in the *outward* direction. If one were to draw a closed surface and include the outward normal unit vector for every element of surface area, the result would resemble a pincushion stuck full of pins, in which the pinheads correspond to the tips of all the unit vectors.

On a surface located in an electric field there is some value of electric field strength at every point. In Figure 56-3, the electric field vector is drawn for the point located in the center of dS. At other points on the surface, we may expect the electric field strength to have differing magnitudes and directions. For some surfaces, $\vec{E}$ is directed outward at some points and inward at others. We are often interested in the component of $\vec{E}$ normal to the surface in the directing of $\hat{n}$. For every area element, this component is given by the dot product

$\vec{E} \cdot \hat{n}$; this results from the definition of the dot product and the fact that the unit vector has a magnitude of unity.

We are now ready to define the flux of the electric field, or electric flux, through the area element dS.

★ The *electric flux* through dS in the sense of $\hat{n}$ is $d\Phi_E \equiv \vec{E} \cdot \hat{n}\, dS$. Electric flux is a scalar quantity; its SI Unit is the volt·meter or newton·meter^2/coulomb.

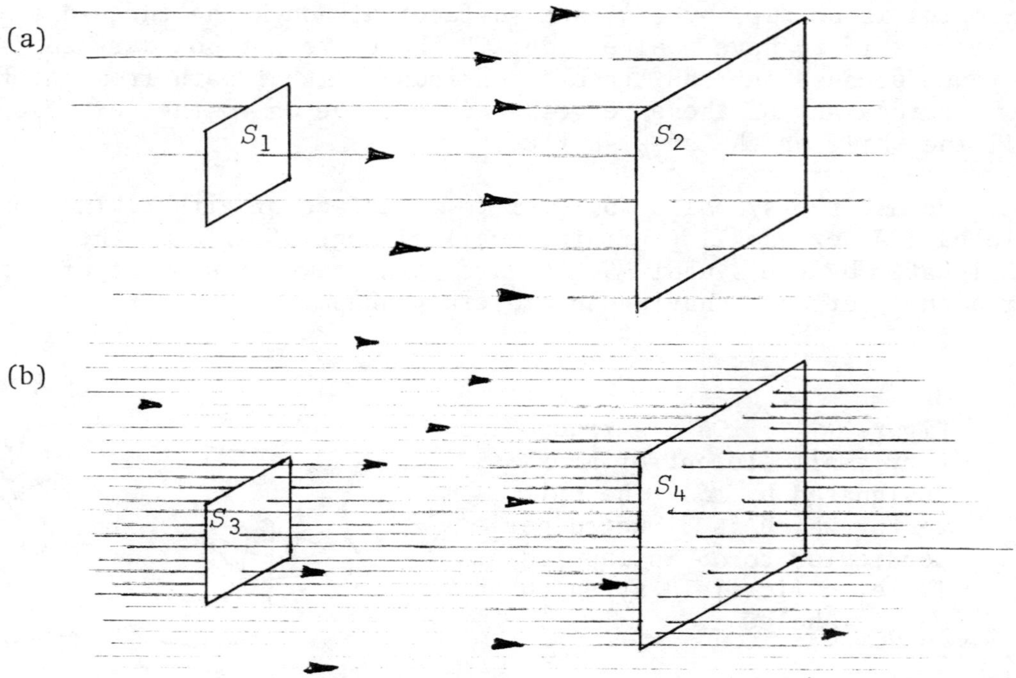

Figure 56-4 The total electric flux through a surface may be visualized as the number of electric field lines which pierce the surface. In both (a) and (b) the electric field is uniform, but in (b) the electric field strength is four times as large as in (a), and the density of field lines in (b) is also four times as large as in (a). Each of the four surfaces is a plane normal to the field lines. (a) The area of surface S_2 is four times as large as the area of surface S_1, so the flux through S_2 is four times the flux through S_1. The *flux density* or flux per unit area is the same for S_1 and S_2. (b) The area of surface S_4 is four times the area of surface S_3, so the flux through S_4 is four times the flux through S_3. The flux density is the same for S_3 and S_4. The *total flux* is the same for S_2 and S_3.

To obtain the *total* electric flux through the entire surface S, it is necessary to add the contributions to this flux which pass through all the area elements of the surface. Because electric flux is a scalar, the total flux is obtained by simply summing the small elements of flux through all the area elements. Rather than using the conventional summation symbol Σ for this purpose, it is advantageous to borrow the "summation symbol" used in calculus: the *integral sign*, written ∫ or $\int$. Simple calculations using this symbol are discussed in this and the following sections.

★ The *total electric flux* through a surface S in the sense of $\hat{n}$ is

$$\Phi_E \equiv \int_S d\,\Phi_E \equiv \int_S \vec{E} \cdot \hat{n}\, dS.$$

The symbol S under the integral sign indicates that the sum or integral is to be taken over the entire surface S. In cases where the surface is a closed surface, it is customary to indicate this fact by writing the integral sign with a small circle superposed: $\oint_S$. Although the word flux is derived from the Latin word for *to flow*, the existence of electric flux does not imply any actual flow of a material substance or of electric charge or of the electric field.

As an example, we compute the electric flux for a spherical surface concentric with a point charge. In preparation for this and other calculations, we introduce a change in the constant used to express Coulomb's law. From here on, the Coulomb constant k will be replaced by the quantity $1/(4\pi\varepsilon_0)$. Although initially this is more cumbersome, in the long run it leads to a simplification of equations used in the study of electricity. The constant ε_0 (epsilon-zero) is called the *permittivity of free space* (or the permittivity of a vacuum), and in order to be consistent with our earlier value for k it must have the numerical value

$$\varepsilon_0 = 8.85 \times 10^{-12}\ \mathrm{C^2/(N \cdot m^2)}.$$

Using this constant, Coulomb's law is written

★ $$F_{\text{elec}} = \frac{1}{4\pi\varepsilon_0}\frac{q_1 q_2}{d^2} .$$

Worked Example 56-1

Compute the total electric flux through a spherical surface concentric with a positive point charge q.

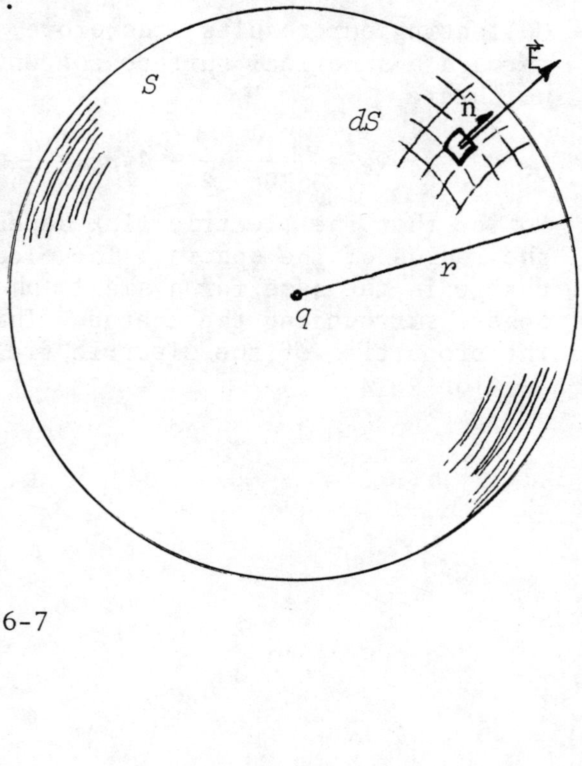

Solution

At the right is pictured a sphere having radius r and concentric with the positive point charge q. Because of the spherical symmetry, at each area element dS both the unit vector $\hat{n}$ and the electric field strength $\vec{E}$ point radially outward (Section 25.3).

Because $\vec{E}$ and $\hat{n}$ are exactly parallel to each other at every point on the surface S of the sphere, the dot product $\vec{E} \cdot \hat{n}$ is simply the magnitude of the electric field strength, E. The electric field strength at any point is the electric force per unit test charge at that point. Referring to the equation for Coulomb's law above, if q_1 refers to the source charge q, and q_2 refers to a unit test charge, for any point on the sphere the distance d is given by r, and so the electric field strength at any point on this sphere is given by

$$E = \frac{F_{elec}}{q_2} = \frac{1}{4\pi\varepsilon_0}\frac{q}{r^2} .$$

The total electric flux outward through S, the surface of the sphere, is

$$\Phi_E = \oint_S \vec{E} \cdot \hat{n}\, dS.$$

The summation symbol or integral sign $\oint_S$ in this equation means that the quantity $\vec{E} \cdot \hat{n}\, dS$ is to be evaluated for each area element dS, and then all of these quantities are to be added. We have already shown, however, that the factor $\vec{E} \cdot \hat{n}$ is a constant for all the area elements of the sphere, given by the equation above for E. In this summation (or *integration,* as it is usually called) therefore, the factor E is a constant while the factor dS is the "variable." In the equation for Φ_E, therefore, we may write the constant factor on the left of the integral sign, and leave the variable dS on the right of the integral sign showing that the integration applies only to the area elements dS:

$$\Phi_E = E \oint_S dS.$$

The meaning of the integral $\oint_S dS$ is simply the sum of all the elements of surface area dS over the entire surface of the sphere. The surface area of a sphere having radius r is

$$\oint_S dS = 4\pi r^2.$$

Collecting our results, therefore, for the total electric flux through a spherical surface concentric with a positive point charge we obtain

★ $$\Phi_E = \frac{1}{4\pi\varepsilon_0}\frac{q}{r^2} \cdot 4\pi r^2 = \frac{q}{\varepsilon_0} .$$

Notice that the electric flux depends on the value of q, but not on the radius of the sphere. The electric flux emanating from a point charge is the same for a small sphere surrounding the charge as for a large sphere surrounding the charge. This result will be correlated with the properties of the electric field lines proceeding from q, in Section 56.4.

Questions and Exercises

Q 56.2-1 (a) Is the peel of a whole, unpeeled potato an open surface or a closed surface?
(b) If the potato is peeled so that the peeling is all in one piece, is this peeling an open surface, or a closed surface?

Q 56.2-2 What is the SI unit of $\vec{E} \cdot \hat{n}$?

E 56.2-1 Show that 1 volt·meter = 1 newton·meter^2/coulomb.

E 56.2-2 Two small pith balls are held 0.2 meter apart. Ball *A* has a charge of $+ 5 \times 10^{-8}$ C, and ball *B* has a charge of $- 10 \times 10^{-8}$ C. What is the force of ball *A* on ball *B*?

E 56.2-3 An electron and a proton are 0.5×10^{-10} meter apart. Compute the electric force of the proton on the electron.

E 56.2-4 (a) Compute the total electric flux through a sphere concentric with a point charge of +3 nC.
(b) How does this flux depend on the radius of the sphere?

E 56.2-5 In Figure 56-4, the electric field strength in (a) is 5×10^3 V/m and in (b) it is 2×10^4 V/m. Surfaces S_1 and S_3 each have an area of 0.15 m^2.
(a) Compute the total electric flux through S_1.
(b) Compute the total electric flux through S_2.
(c) Compute the total electric flux through S_3.
(d) Compute the total electric flux through S_4.

56.3 Gauss's Law

In Section 56.2 it was shown that when a positive point charge is surrounded by a concentric spherical surface having any radius, the total electric flux outward through this surface is given by

$$\Phi_E = \oint_S \vec{E} \cdot \hat{n}\, dS = q/\varepsilon_o .$$

We now extend this result to a surface of any shape which surrounds a point charge, culminating in the statement of a fundamental law of electric fields known as Gauss's law. Our derivation of Gauss's law proceeds in two stages. We compute the electric flux first through a surface formed from segments of concentric spheres, and then through a surface having an arbitrary shape.

The first surface may be visualized by imagining an orange which has two layers of peel, one outside the other. If two thirds of the outer peel were removed, and then one third of the inner peel were removed, the surface of the orange would look something like that shown in Figure 56-5. The closed surface shown in Figure 56-5 consists of one segment from each

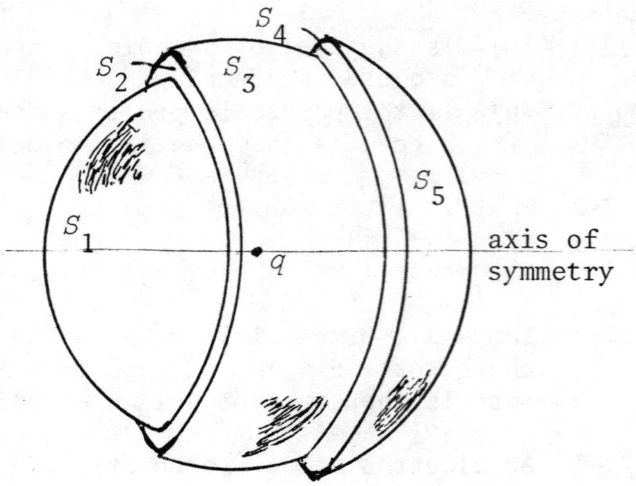

Figure 56-5. A closed surface consisting of five connected parts. S_1 is one-third of a sphere. S_3 is one-third of a sphere concentric with S_1. S_5 is one-third of a sphere concentric with S_1 and S_3. S_2 is a circular ring connecting S_1 and S_3; it is everywhere tangent to radii from the center of the spheres. S_4 is a similar ring connecting S_3 and S_5, tangent to radii from the center.

of three concentric spheres (labeled S_1, S_3 and S_5), connected by two narrow circular bands or rings (labeled S_2 and S_4). Each of the three spherical segments has an area equal to one-third of a sphere. Each of the two ring surfaces lies on a cone whose vertex is at the center of the spheres. The complete closed surface S thus consists of five connected portions:

$$S = S_1 + S_2 + S_3 + S_4 + S_5 .$$

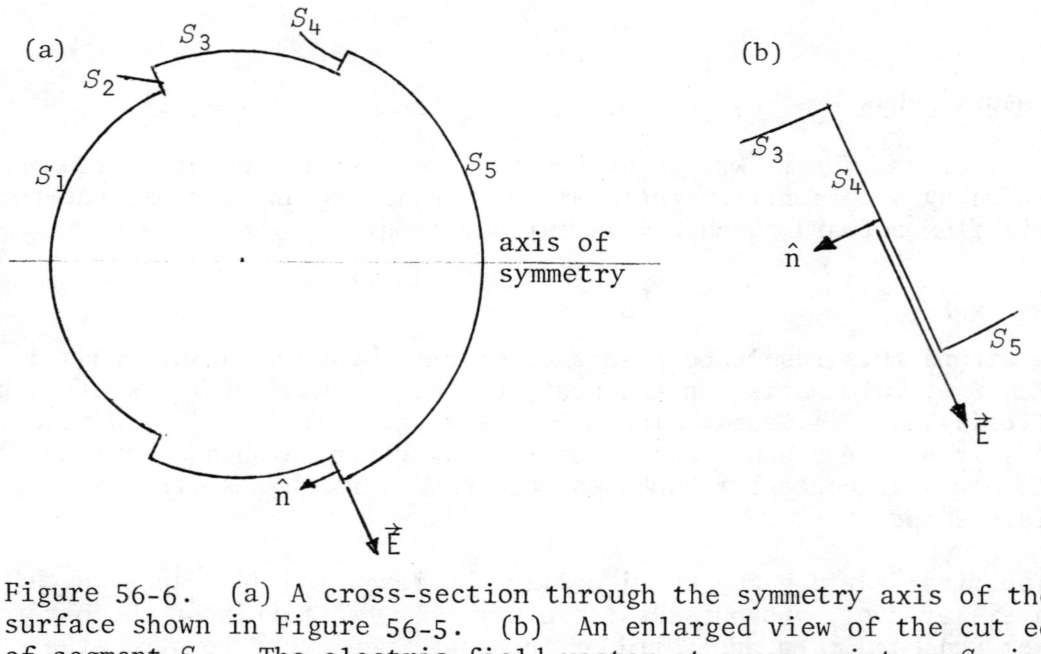

Figure 56-6. (a) A cross-section through the symmetry axis of the surface shown in Figure 56-5. (b) An enlarged view of the cut edge of segment S_4. The electric field vector at every point on S_4 is directed radially outward from the center, while the unit vector $\hat{n}$ normal to S_4 is also normal to $\vec{E}$.

The surface has an axis of symmetry indicated by the horizontal line in Figure 56-5. If the surface were sliced into two equal halves by a plane through the symmetry axis, the cut edge of each half would look like Figure 56-6(a).

We now compute the total electric flux outward through this surface from a positive point charge q at the center. The electric field vector $\vec{E}$ is directed radially outward at every point on the surface. For all points on the surfaces S_1 and S_3 and S_5, $\vec{E}$ is parallel to $\hat{n}$ so that $\vec{E} \cdot \hat{n} = E$, just as we found for the single sphere considered in Section 56.2. For all points on the surfaces S_2 and S_4, however, the outward unit normal $\hat{n}$ is directed normal to the surface while the $\vec{E}$ vector is parallel to the surface and thus perpendicular to $\hat{n}$; on S_2 and S_4, therefore, $\vec{E} \cdot \hat{n} = 0$ at every point.

Because the total surface consists of five parts, we may compute the total electric flux as the sum of five terms:

$$\begin{aligned}\Phi_E &= \oint_S \vec{E} \cdot \hat{n}\, dS \\ &= \int_{S_1} \vec{E} \cdot \hat{n}\, dS + \int_{S_2} \vec{E} \cdot \hat{n}\, dS + \int_{S_3} \vec{E} \cdot \hat{n}\, dS \\ &\quad + \int_{S_4} \vec{E} \cdot \hat{n}\, dS + \int_{S_5} \vec{E} \cdot \hat{n}\, dS.\end{aligned}$$

The second and fourth integrals in this equation are each equal to zero, because $\vec{E} \cdot \hat{n} = 0$ at every point on the surfaces S_2 and S_4. The first integral is just equal to one-third of the flux (computed in Section 56.2) for a complete spherical surface, because $\int_{S_1} dS$ is one third the surface area of a sphere. We thus obtain

$$\int_{S_1} \vec{E} \cdot \hat{n}\, dS = q/(3\varepsilon_0).$$

By similar calculations,

$$\int_{S_3} \vec{E} \cdot \hat{n}\, dS = q/(3\varepsilon_0) \quad \text{and} \quad \int_{S_5} \vec{E} \cdot \hat{n}\, dS = q/(3\varepsilon_0).$$

Adding our results for the five integrals, we have

$$\oint_S \vec{E} \cdot \hat{n}\, dS = \frac{q}{3\varepsilon_0} + 0 + \frac{q}{3\varepsilon_0} + 0 + \frac{q}{3\varepsilon_0} = \frac{q}{\varepsilon_0}.$$

This result shows that the total electric flux outward through this five-part surface is the same as for a single spherical surface concentric with q.

We now consider the electric flux outward from a positive point charge through a closed surface *of any shape* which surrounds the charge. The actual surface may be approximated by a "stepped" surface consisting of segments of concentric spheres, connected by rings parallel to radial lines from the center and thus perpendicular to the spherical segments. Figure 56-7 shows a cross-section of a smooth closed surface, and a stepped surface which lies close to the smooth surface, in a plane through a point charge taken as the center of the spherical segments. On each of the ring

Figure 56-7. A closed surface (smooth curve) may be approximated by a stepped surface consisting of concentric spherical segments joined by bands having surfaces parallel to radial lines from the center. The drawing shows a cross-section of both surfaces, in a plane through a point charge chosen as the center of the spherical segments.

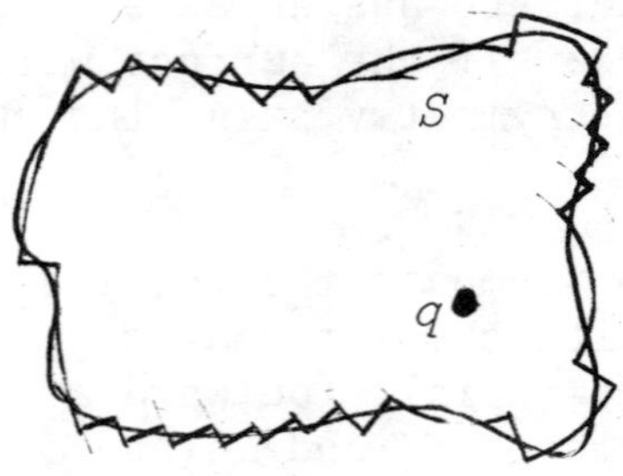

surfaces, the electric field vector is perpendicular to the unit normal vector and so $\vec{E} \cdot \hat{n} = 0$ at every point. On each of the spherical portions, $\vec{E}$ is parallel to $\hat{n}$ and so $\vec{E} \cdot \hat{n} = E$. Using the same reasoning applied to the stepped surface shown in Figure 56-5 the electric flux outward through each spherical segment is equal to q/ε_0 times the fraction of a complete sphere contained in the segment. The total electric flux through the stepped surface is the sum of the flux through all the spherical segments, which is q/ε_0. By using smaller and smaller steps, i.e. narrower and narrower rings, the stepped surface can be made to approach the smooth surface as closely as we please. We thus conclude that when a positive point charge is surrounded by a closed surface having any shape, the total electric flux outward through this surface is given by

★ $$\oint_S \vec{E} \cdot \hat{n}\, dS = q/\varepsilon_0 .$$ Gauss's law

This result is called Gauss's law, in recognition of the work of Karl Friedrich Gauss (1777-1855), a German Scientist and mathematician who made numerous contributions to physics and mathematics.

If the closed surface contains more than one point charge, each of these charges will contribute to the electric field at the surface and therefore to the electric flux through the surface. In Gauss's law, therefore, the charge q refers to the *total charge* contained within the surface S. Gauss's law enables us to compute the total electric flux through any closed surface, provided that we know the total electric charge within the surface. It is not necessary to know the shape of the surface, nor the locations of the charges within it.

It may seem that Gauss's law implies that the electric field strength at various points on a surface does not depend on the charges located *outside* of the surface. Such a conclusion is not correct, however. If the charges within a surface are kept fixed, and then some charge is brought up and placed close to the surface but outside of it, the electric field strength on the surface is altered by the presence of this outside charge. The result is that the electric flux leaving the surface is greater in some regions, but smaller in others, than it was before. Gauss's law tells us that in such a case the *net* or *total* electric flux leaving the surface does not change; the *total flux* through the surface depends only on the net charge *inside* the surface. This property of the electric flux will be derived from another point of view in Section 56.4.

Gauss's law does not give us the electric flux through any particular small area of surface, but only the total flux. So, one may ask how the law can have any practical value in computing the electric field which results from a known distribution of charge. The answer is that in cases where a high degree of symmetry prevails, it is often possible to choose a surface in such a way that Gauss's law *can* be applied to give the electric field strength in a localized region of interest. In these cases, Gauss's law becomes a simple but powerful tool for computing $\vec{E}$. Some examples are given in Sections 56.5 and 56.6

Gauss's law is one of the four famous Maxwell equations which form the basis of all advanced discussions of electric and magnetic fields. It does not, however, represent a new law of physics derived from theory alone. The physical basis of Gauss's law can be traced back to Coulomb's law for the force between two charges. When dealing with relatively simple problems of the interaction between two or a few point charges, it is more convenient to use the Coulomb form of this law, but when dealing with more complex charge distributions it is usually advantageous to use Gauss's form of the law:

★ $$\oint_S \vec{E} \cdot \hat{n}\, dS = \frac{1}{\varepsilon_0} \text{(total charge inside } S\text{)}$$

Questions and Exercises

Q 56.3-1 A closed surface contains a positive charge $+q$ near the center, and a negative point charge $-q$ close to the surface. Near the negative charge the electric field is strong and directed inward. On the surface farthest from the negative charge, the field is directed outward but it is much weaker, because the positive charge is farther from this part of the surface. What is the total electric flux through the surface?

Q 56.3-2 Suppose the surface and enclosed charges described in Q 56.3-1 are surrounded by a cloud of negative point charges. Then what is the total electric flux through the surface?

E 56.3-1 A surface encloses $q_1 = +5$ mC and $q_2 = -3.6$ mC. What is the total electric flux through the surface?

E 56.3-2 A steel bottle encloses 1.74×10^{21} hydrogen molecules each containing two electrons and two protons. What is the total electric flux through the surface of the bottle?

56.4 Electric Field Lines and Gauss's Law

Minimum Learning Goal: Be able to use Gauss's law to point out the locations, signs, and values of electric charges, when you are given a two-dimensional field map. For example:

1 The field map below shows part of the electric field around two parallel wires which run perpendicular to the drawing and are represented by circles. Assume that the diagram shows all of the field lines in a region 1 cm in depth (the direction normal to the drawing). Using Gauss's law, find the magnitude and the sign of charge per unit length on each wire. (Answer, page 56-38, Line 1)

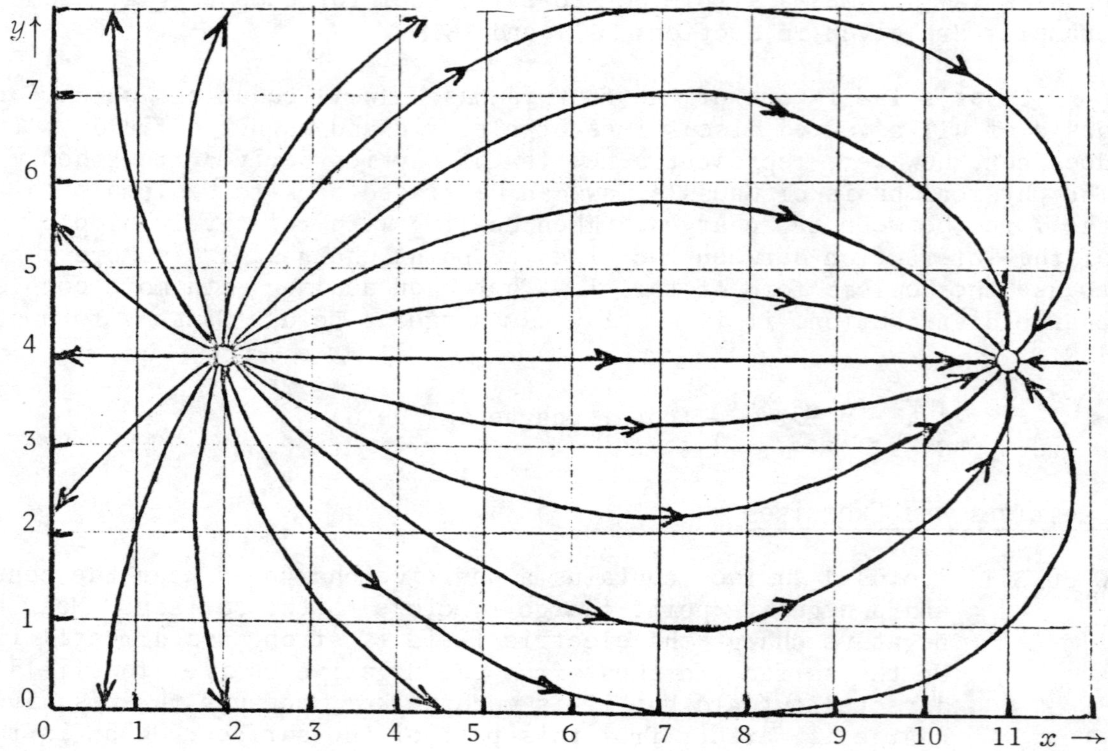

2

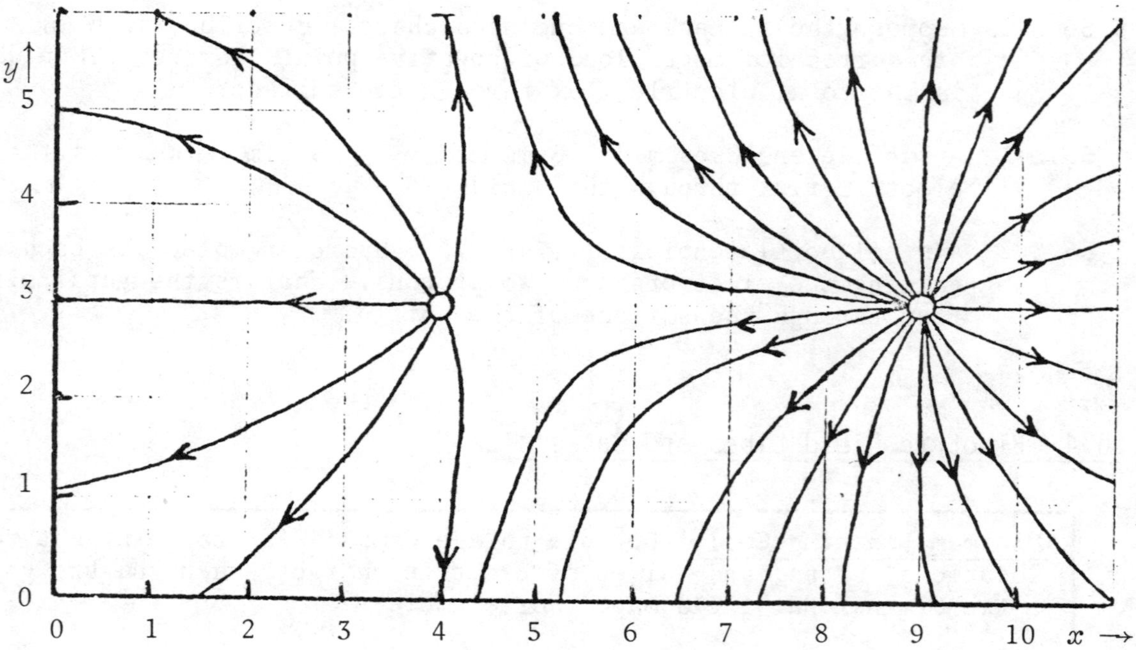

2 The field map above shows electric field lines around two charged parallel wires which run perpendicular to the drawing and are represented by circles. Assume the drawing shows all field lines within a depth of 1 cm (the direction normal to the drawing). Using Gauss's law, find the charges per unit length which produce the field, and specify their signs and magnitudes. (Answer, page 56-38, Line 4).

In Section 25.3 we introduced the concept of electric field lines as a pictorial aid in visualizing the electric field. The direction of a field line at any point is the direction of the $\vec{E}$ vector at that point. We found that field lines are crowded where the electric field strength is large, but are spaced farther apart where the field strength is small. We now develop a quantitiative connection between field lines and electric flux.

In this quantitative interpretation of electric field lines, the *density* of field lines (the number of lines crossing a unit of area normal to the lines) is chosen to equal the magnitude of the electric field strength.

> The number of electric field lines per unit area crossing a small element of area oriented perpendicular to the electric field strength is *equal* to the magnitude of the field strength at that element of area.

Figure 56-8 illustrates some field lines intersecting small surfaces in two regions of an electric field.

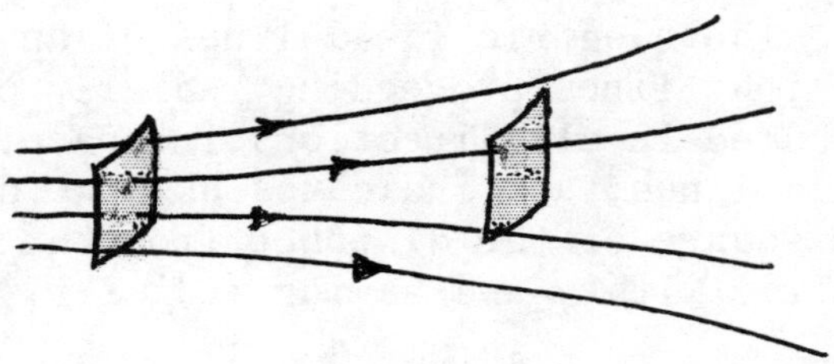

Figure 56-8. The number of electric field lines crossing a small element of area oriented normal to the electric field strength is equal to the product of the field strength times the area. The field strength is higher in a region where field lines are crowded than in a region where they are spaced farther apart.

This quantitative definition of electric field lines is expressed in the equation

★ $N/A = E$,

where N is the number of field lines passing through the area A. Using $N = EA$, it is evident that

$$1 \text{ field line} = (1 \text{ N/C}) \times 1 \text{ m}^2 = 1 \text{ N}\cdot\text{m}^2/\text{C}.$$

A field line has the same units as electric flux (Section 56.2), and in fact *the total flux through a surface S is just equal to the number of field lines through that surface.* This is true regardless of the angle which the unit vector normal to the surface makes with the field lines. The use of field lines as defined in this section gives us a quantitative graphical representation of electric flux. It also gives us an alternative approach to Gauss's law, which we develop next.

We already know that the electric field strength at any distance from a point charge is proportional to the magnitude of the charge. The quantitative interpretation of field lines then implies that the number of lines emanating from any charge has a definite numerical relationship to that charge. To compute the number of field lines which begin or end on a unit charge, we consider the field of an isolated point charge.

Figure 56-9. Electric field lines radiating outward from a point charge. The density of lines decreases inversely with the square of the distance from the charge.

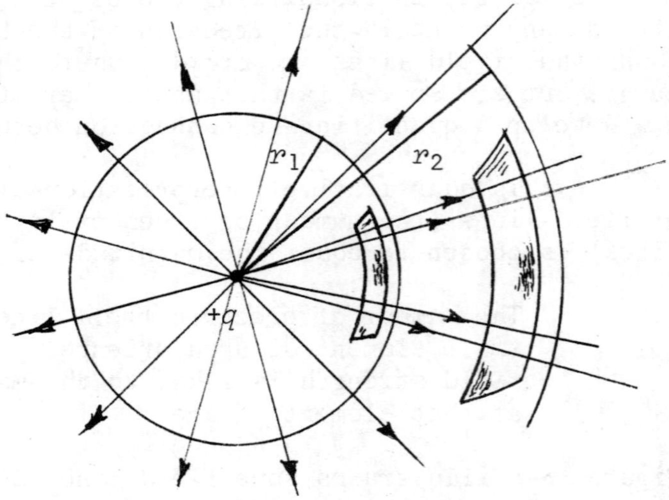

The electric field lines of an isolated positive point charge are straight lines proceeding radially outward from the charge, uniformly distributed in all directions in space. Figure 56-9 portrays some of these field lines, and indicates how the density of lines varies inversely with the square of the distance from the source charge. On the sphere concentric with q and having radius r_1, the electric field strength is

$$E = \frac{1}{4\pi\varepsilon_0}\frac{q}{r_1^2}$$

(see Worked Example 56-1 on page 56-7). The area of this sphere is $A_1 = 4\pi r_1^2$. The total number of field lines passing through the sphere is the product of the field strength times the area, or

$$N = EA = \frac{1}{4\pi\varepsilon_0}\frac{q}{r_1^2} \cdot 4\pi r_1^2 = \frac{q}{\varepsilon_0} .$$

If we consider the larger sphere concentric with q and having radius $r_2 = 2r_1$, the electric field strength is reduced to 1/4 its value on the smaller sphere but the total area of the sphere is quadrupled, so the number of field lines passing through the larger sphere is also q/ε_0. In fact, the number of field lines passing through any sphere (or any closed surface) which contains a total charge q is a constant, consistent with the observation made in Section 25.3 that each field line begins on a positive charge and ends on a negative charge. Since q is the only charge in view here, every line which begins on q proceeds outward until it finds a negative charge on which to end.

The calculation above shows that the number of field lines which begin on a charge q is $q/\varepsilon_0 = 1.13 \times 10^{11}\, q$ lines. An equal number of lines must end on a charge $-q$. The proportionality factor $1/\varepsilon_0$ is a consequence of our having chosen the density of field lines to equal the magnitude of the field strength. Given this choice, Gauss's law may be expressed as follows:

> *Gauss's law:* An electric field can be represented by field lines which begin only on positive charges and terminate only on negative charges. The number of field lines that begin or end on a charge q is equal to q/ε_0. The total number of field lines passing through any closed surface which contains a net charge q is equal to q/ε_0.

(This proportionality factor $1/\varepsilon_0$ applies only in the SI system of units.) The magnitude of a charge from which *one* field line emanates is numerically equal to ε_0, or 8.85×10^{-12} coulomb. The charge of a proton is much smaller than this, so only a small fraction of a field line proceeds from each proton.

In a field map or drawing which depicts an electric field, the number of lines drawn is not usually set exactly equal to q/ε_0. (An exception is found in the Learning Goals and in some of the Exercises in this section.) When the field is strong, it is customary to include only a representative sample of the total number of field lines. If the field is weak, the number of field lines computed from q/ε_0 may be too small to give an adequate representation of the field. In such a case, one may subdivide the field lines by drawing, say, a total of one hundred or one thousand lines to represent a field line, in order to convey adequately the pattern of the field. Further, in drawing and interpreting field maps the placement of particular field lines has no meaning; it is only their directions and density that matters.

If we are given the electric field in a region, Gauss's law allows us to determine the distribution of charges in the region. We do that by representing the field by field lines, and seeing where they begin and end. For example, if the electric field points always away from a particular central point and varies inversely with the square of the distance from that point, it is an easy matter to determine that there is a point charge at the central point, and no charge anywhere else.

Questions and Exercises

Q 56.4-1 Can electric field lines begin or end on an object which is electrically neutral? Explain.

Q 56.4-2 Is it possible for two electric field lines to merge together and become one line? Explain.

Q 56.4-3 In an electric field, does the field strength increase, or decrease, along a path entering a region where the field lines crowd closer together?

Q 56.4-4 How can one calculate the electric field at a point through which no field lines happen to pass?

Q 56.4-5 A student is asked to draw a field map for an isolated point charge which is the source of twelve field lines. He conscientiously draws a map in which twelve straight lines radiate from the charge, spaced uniformly at ten-degree intervals. Is this a correct field map? Explain.

Q 56.4-6 If you know the electric field at a point due to a certain charge configuration, and also the electric field at the same point due to a different charge configuration, how can you find the electric field (still at the same point) when both charge configurations are simultaneously present?

E 56.4-1 How many field lines originate on an isolated positive point charge of 1.77×10^{-11} coulomb? How should the lines be placed to give an adequate representation of this field?

E 56.4-2 (a) In the SI system of units, how many field lines terminate on a negative charge of one microcoulomb? (b) If this charge is spread uniformly on the surface of a sphere having a radius of 15 cm, how many field lines end on each square cm of the sphere?

E 56.4-3 In the region between two oppositely charged plane parallel plates the electric field strength is 10^5 volt/meter.
(a) What is the density of field lines in this region?
(b) What is the charge density on the positive plate in coulomb/meter^2?

E 56.4-4 The field lines in the region of two long parallel charged cylindrical conductors are shown in this figure. Assume that the diagram shows all field lines in a region 5 mm in depth (the direction normal to the drawing). What are the magnitude and sign of the net charge per unit length on each of these conductors?

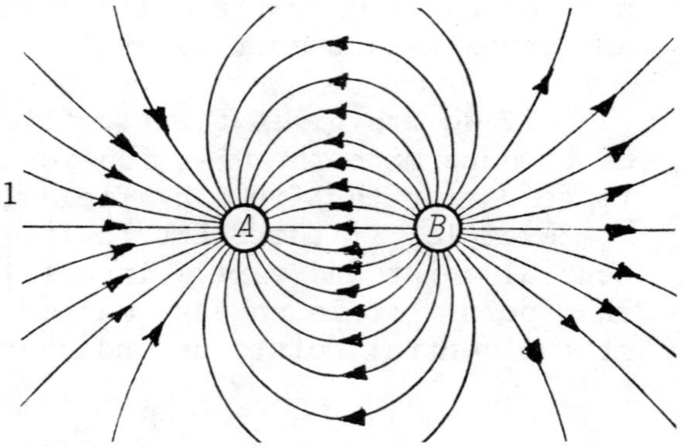

E 56.4-5 The field lines in the region of two long parallel charged cylindrical conductors are shown in this figure. Assume that the diagram shows all field lines in a region 1 cm in depth (the direction normal to the drawing). What are the magnitude and sign of the net charge per unit length on each of these conductors?

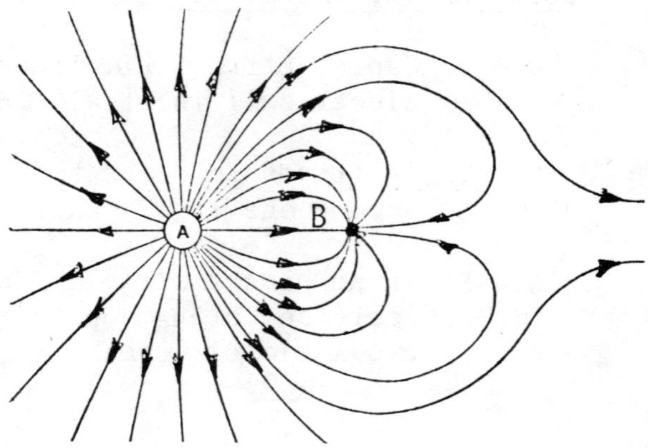

E 56.4-6 This full-scale drawing represents part of the electric field around a wire cage which is used to provide electrostatic shielding for a scientific experiment. The wires of the cage run perpendicular to the drawing, and are shown by circles in the figure. Assume that the diagram shows all of the field lines in a region 3 mm in depth (the direction normal to the drawing). (a) What are the magnitude and sign of the net charge per cm of length on each wire? (b) What is the electric field strength inside of the cage (at the top of the figure)? (c) What is the electric field strength outside the cage (at the bottom of the figure)?

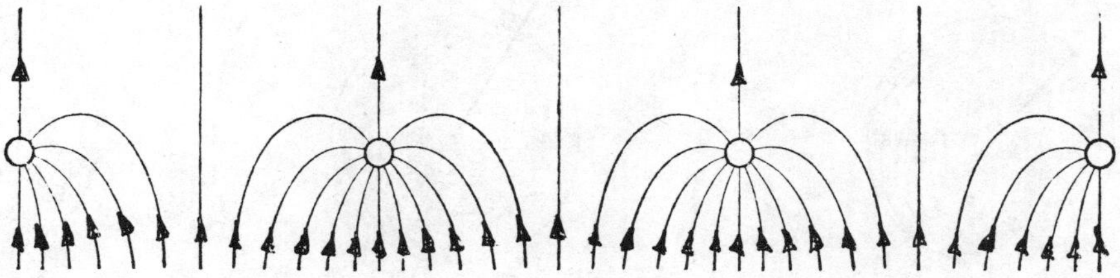

Check your understanding of the material above by testing yourself on the minimum learning goal at the beginning of this section.

56.5 Electric Field of a Uniformly Charged Sphere

Gauss's law enables us to determine the locations and magnitudes of the charges when we are given the electric field throughout a region of space. This may be done by constructing the field lines for the field. The places where field lines begin are locations of positive charges, and the places where they end are locations of negative charges. The numbers of lines that begin or end tell us the magnitudes of these charges.

Suppose, however, that we are given the charge distribution and are to find the electric field. In most cases Gauss's law alone is not sufficient to solve this problem, because it tells us where the field lines begin and end but not how they arrange themselves in the intervening space. In a few important cases where there is a high degree of symmetry, however, we can use Gauss's law to find the electric field. Examples are given in this section and the next.

Suppose the charge distribution consists of a total positive charge Q spread uniformly over a sphere having radius a, and we wish to find the field both inside and outside the surface. Consider some particular point in space P, as shown in Figure 56-10a. If we rotate the charge distribution through any angle around the line joining P to the center O, its appearance does not change; it always remains the same charge distribution. Therefore, the electric field at P must not change as a result of such a rotation. In other words, the electric field must be directed along the line OP. Also, if we consider any other point P' which is the same distance from O as P is, the electric field at P' must have the same magnitude as at P. The reason is that we can rotate the charge distribution around an axis perpendicular to the paper so as to bring P to P' -- and yet the charge distribution is unchanged by such a rotation. By these arguments we have established that the field lines must all have a radial direction and be uniformly distributed in angle. That is, the electric field of a uniformly charged sphere is spherically symmetric.

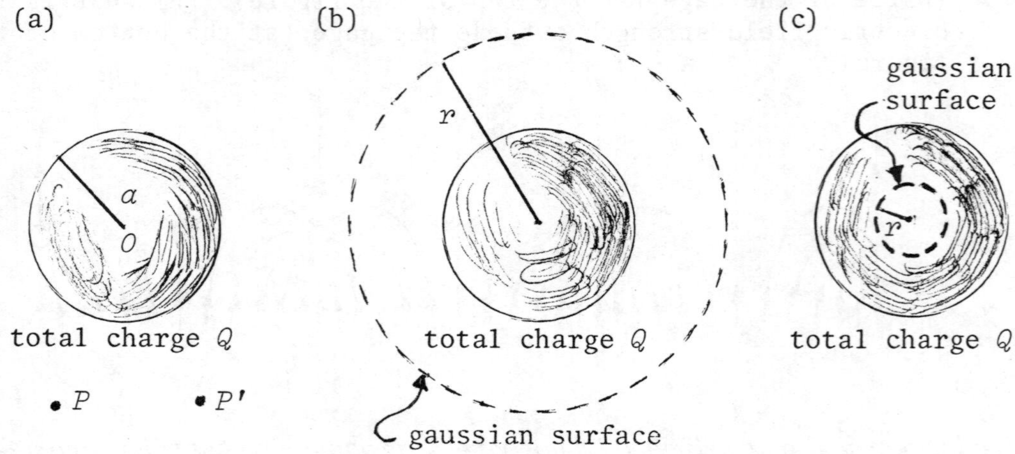

Figure 56-10. A charged sphere contains a total charge Q spread uniformly over its surface. (a) As shown in the text, the electric field which results from this charge distribution is spherically symmetric. (b) Using a spherical gaussian surface concentric with and surrounding the charged sphere, one can find the electric field strength in the region outside the sphere. (c) Using a spherical gaussian surface concentric with but located inside the charged sphere, one can find the electric field strength in the region inside the sphere.

Once this is known, we can find the electric field strength using Gauss's law alone. To find the field strength at a point outside the positively charged sphere, we construct (in our imagination) a "gaussian surface" consisting of a sphere having radius r and concentric with the charged sphere (Figure 56-10(b)). We then apply Gauss's law,

$$\oint \vec{E} \cdot \hat{n}\, dS = (\text{total charge inside } S)/\varepsilon_0,$$

to this surface. The unit normal vector $\hat{n}$ is directed radially outward and is therefore everywhere parallel to $\vec{E}$, so that $\vec{E} \cdot \hat{n} = E$ at every point on the surface. Furthermore, because the field has spherical symmetry the magnitude of E is constant on the gaussian surface. Therefore the total flux through this surface is simply E times the area of the sphere, $4\pi r^2$, from which

$$E \cdot 4\pi r^2 = Q/\varepsilon_0 \qquad \text{or} \qquad E = Q/4\pi\varepsilon_0 r^2.$$

For values of $r \geq a$, the field is the same as for a point charge Q (page 56-8).

To find the field strength at a point *inside* the charged sphere, we construct a spherical gaussian surface having a radius r smaller than a, as shown in Figure 56-10(c). Once again the magnitude of E is the same everywhere on this surface, but in this case the total charge *inside* of the surface is zero, and therefore E = 0 everywhere inside the spherical shell of charge. The total number of field lines passing through this inner gaussian surface is zero, so there is no field at all inside the sphere of charge.

At the surface of the charged sphere, at a distance a from the center, the result obtained above tells us that the field strength is

$$E = Q/4\pi\varepsilon_0 a^2.$$

In a case where electric charge is spread over the surface of an object, it is often convenient to speak of the charge density on the surface.

★ Where a charge dq is located on an element of surface area dS, the *surface charge density* $\sigma \equiv dq/dS$.

The symbol for surface charge density is σ (Greek, sigma), and the SI unit is coulomb/meter^2 = C/m^2. The surface charge density on the sphere described above is

$$\sigma = Q/4\pi a^2.$$

Combining this result with the value for E at the surface, we find that

★ $$E = \sigma/\varepsilon_0 .$$

just outside the surface of the charged sphere. This result holds because all the field lines proceed outward from the charged sphere.

The symmetry argument works only because we know that the solution to the problem is *unique*. Otherwise, we could only conclude that rotation about the axis *OP*, which leaves the charge distribution unchanged, changes a possible electric field into *another* possible electric field. In that case, we could not say that the field has to be in the radial direction. But because we know that the solution must be unique, we can be certain that the rotation changes the electric field into the *same* electric field.

Questions and Exercises

Q 56.5-1 Physical objects and systems sometimes have one or more *elements of symmetry*. One such element is a *plane of symmetry*. For an object which has a plane of symmetry, every point on the object is matched by an equivalent point on the opposite side of the plane of symmetry. That is, if every point on the object is reflected in the plane of symmetry, the reflected object is identical to the original object. An example of an object having a plane of symmetry is an ordinary paper clip. Which of the following objects have at least one plane of symmetry? (a) a coffee cup having a handle but no decorations; (b) a dime; (c) a new baseball; (d) a closed book; (e) a folding chair; (f) a picket on a picket fence; (g) the human body; (h) an automobile; (i) a ping pong ball; (j) a bicycle; (k) an ordinary lead pencil; (l) an ordinary drinking glass; (m) a checkerboard; (n) a billiard cue: (o) a solid cube of aluminum; (p) the planet earth; (q) an electric dipole, consisting of a point $+q$ and a point charge $-q$ a short distance apart.

Q 56.5-2 Another element of symmetry is an *axis of symmetry*. An object which has an n-fold axis of symmetry can be rotated about this axis through an angle of $360/n$ degrees into a configuration identical to its original configuration. An example of an object having a five-fold axis of symmetry is a five-pointed star. Which of the objects listed in Q 56.5-1 have an axis of symmetry? In each case, state the value of n for this axis.

Q 56.5-3 If an arrangement of charges has an axis of symmetry, will the electric field which results from these charges have the same axis of symmetry? Given an example to justify your answer.

Q 56.5-4 If an arrangement of charges has a plane of symmetry, will the electric field which results from these charges have the same plane of symmetry? Give an example to justify your answer.

Q 56.5-5 If a sphere has a charge which is not distributed uniformly over the surface, will the external electric field be the same as if all the charge were located at the center of the sphere?

Q 56.5-6 What is the SI unit of surface charge density?

E 56.5-1 A sphere has a radius of 4.0 cm and a total charge of -3.6 nC, uniformly spread over the surface.
(a) What is the electric field strength just outside the surface?
(b) What is the electric field strength 8.0 cm from the center?
(c) What is the electric field strength 2.0 cm from the center?
(d) What is the surface charge density?

E 56.5-2 A sphere has a radius of 3.2 cm and a uniform surface charge density of $+4.4\ \text{nC/cm}^2$.
(a) What is the total charge on the sphere?
(b) What is the electric field strength 2.0 cm from the center?
(c) What is the electric field strength just outside the surface?
(d) What is the electric field strength 10.0 cm from the center?

56.6 Electric Field Near a Charged Conducting Surface

Minimum Learning Goal: Be able to use symmetry, uniqueness, and Gauss's law to deduce an expression for the electric field due to a simple configuration of charges. For example:

1 What is the electric field *everywhere* above an isolated infinite flat sheet of uniform electric charge density (where σ is the charge per unit area)? (Answer, page 56-38, Line 5)

2 What is the electric field at *any* point a distance r from the axis of an isolated infinite straight line of uniform electric charge density (where λ, a constant, is the charger per unit length)? (Answer, page 56-38, Line 12)

The use of Gauss's law requires the evaluation of the integral of the total electric flux through a gaussian surface. Many problems of practical interest require the analytical methods of integral calculus. By a suitable choice of the gaussian surface, however, it is often possible to simplify this calculation considerably. This approach is illustrated by the use of Gauss's law to find the electric field near a charged conducting surface.

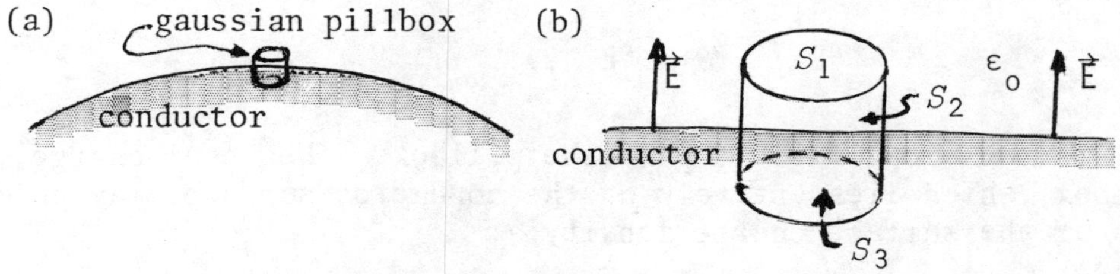

Figure 56-11. (a) On the curved surface of a conductor, a very small region is approximately a plane. A gaussian pillbox may be constructed there having sides normal to the conductor surface, and top and bottom parallel to the conductor surface. (b) An enlarged view of the pillbox, whose lower half is located within the material of the conductor. The top surface of the pillbox is labeled S_1, the side surface is labeled S_2, and the bottom surface S_3.

Although in general the surface of a conductor is curved, in a magnified view a small portion of the surface appears to be approximately a plane (Figure 56-11). On this localized plane we construct a gaussian surface, S, in the shape of a "pillbox," a right circular cylinder having sides perpendicular to the conductor surface, and top and bottom parallel to the conductor surface. The total gaussian surface S comprises three parts: S_1 (the top of the pillbox), S_2 (the sides), and S_3 (the bottom). The total flux outward through S is the sum of three separate terms:

$$\oint_S \vec{E} \cdot \hat{n}\, dS = \int_{S_1} \vec{E} \cdot \hat{n}\, dS + \int_{S_2} \vec{E} \cdot \hat{n}\, dS + \int_{S_3} \vec{E} \cdot \hat{n}\, dS = q/\varepsilon_0,$$

where q is the total charge inside the pillbox.

To evaluate these integrals, we need to find the value of $\vec{E} \cdot \hat{n}$ on each of the three surfaces. Assuming no electric current exists in the conductor, the electric potential (Section 25.5), is constant everywhere within the material of the conductor and on its surface; if the potential were not constant, charge carriers in the conductor would move in response to the potential difference until a uniform potential were established within the conductor. The conductor surface is therefore an equipotential surface. Because an electric field line is perpendicular to every equipotential surface it intersects (Section 25.4), the field lines just outside the conductor surface are normal to this surface. On S_2 therefore, $\vec{E} \perp \hat{n}$ and $\vec{E} \cdot \hat{n} = 0$ everywhere, so that the integral over S_2 is equal to zero.

It is shown in Section 56.7 that in the absence of current the electric field strength is zero everywhere within a conductor. Using this result, the integral over S_3 must be equal to zero. The only flux emerging from S, therefore, is that which passes through the top of the pillbox. On S_1, the outward normal unit vector points upward parallel to $\vec{E}$, so that $\vec{E} \cdot \hat{n} = E$, which may be assumed constant over the very small area of S_1.

Collecting the results given above, we have

$$\oint_S \vec{E} \cdot \hat{n}\, dS = \int_{S_1} \vec{E} \cdot \hat{n}\, dS + 0 + 0$$

$$= E \int_{S_1} dS = E A_1,$$

where A_1 is the area of the top of the pillbox. The total charge q inside the pillbox, which lies entirely on the conductor surface, may be expressed in terms of the surface charge density σ:

$$q = \sigma A_1.$$

Gauss's law then gives us

$$EA_1 = q/\varepsilon_0 = \sigma A_1/\varepsilon_0, \qquad \text{or} \quad \bigstar\ E = \sigma/\varepsilon_0$$

just outside the surface of the conductor.

The field strength seems to be independent of other charges outside the conductor. Actually, the field strength near any part of a conductor surface is *not* independent of these external charges. If an external negative charge (for example) were located near some part of the surface, charge carriers within the conductor would be attracted or repelled from that region in accord with Coulomb's law, so that there would be a local concentration of positive charge on that part of the conductor. As a result, the field just off the surface would be stronger there. In an electrostatic situation (no current), the surface density of charge may vary from point to point on the conductor, but at each point the field strength resulting from the local charge density is given by $E = \sigma/\varepsilon_0$.

Exercises

E 56.6-1 A flat sheet of dielectric, 0.60 m by 0.80 m in size, contains 24 nC of positive charge spread uniformly over its surface. (a) What is the surface charge density? (continued on p.56-25)

(b) Do the electric field lines emanate from this charge on one side, or on both sides, of the sheet?
(c) Compute the electric field strength 6 cm from the sheet near its center.

E 56.6-2 A very long straight wire has a diameter of 5.0 mm and a uniform linear charge density of -18 pC/m.
(a) Compute the electric field strength at the surface of the wire.
(b) Compute the electric field strength at a point 1.0 meter from the wire.

E 56.6-3 The electric field strength in the atmosphere near the earth's surface is about 100 V/m toward the earth. What does this imply for the value of the total net charge on the earth's surface? The earth's radius is 6.4×10^6 m.

Check your understanding of the material above by testing yourself on the minimum learning goal at the beginning of this section.

56.7 Conductors in Electrostatic Fields

We consider here conductors carrying no current and not connected to any source of electromotance. Using Gauss's law, several important conclusions can be drawn about the electric field in such a conductor.

A conductor is a substance or material containing charge carriers (electrons or other particles) which can travel readily within the conductor in response to any applied electric field or potential difference. We assume there is no limit to the number of charge carriers available to distribute themselves in response to any applied electric field. Because the charge carriers are free to move swiftly within the conductor whenever a macroscopic electric field exists there, when such a field is applied the charge carriers promptly move to new positions so that the applied field is cancelled out within the conductor.

For example, if a charged rubber rod is used to place a quantity of negative charge at one end of a metal bar, electrons within the bar are repelled by this extra negative charge and move toward the other end of the bar. At every point within the bar where the macroscopic field is not zero, the electrons move in response to the electric field in such a direction as to reduce the field, and this motion continues until the field diminishes to zero. The time required for this transient rearrangement of charge is very small, typically on the order of 10^{-8} second. Because of this short response time, the conductor quickly arrives at a state of static equilibrium in which the macroscopic field at every point within the conductor is zero. (Within each atom, of course, the *microscopic* electric field of the nucleus and the bound electrons continues indefinitely, because the electric charges in each atom and molecule obey the laws of quantum physics. It is only the mobile charge carriers in the conductor, which are free to run on the macroscopic scale, that move to produce the *macroscopic* field-free condition we are describing in this section.

As we saw in Section 25.5, in any region where a macroscopic electric field $\vec{E}$ exists between two points separated by a displacement $d\vec{s}$ there will be a difference of potential between the two points given by $dV = -\vec{E} \cdot d\vec{s}$. In a conductor which carries no current the electric field strength is everywhere zero, so the difference of potential between any two points is also zero. In other words, the electric potential V is the same at every point within the conductor, including points on its surface. The surface of any conductor is therefore an equipotential surface when there is no current. The value of the potential depends on the net charge on the conductor's surface; see, for example, Worked Example 25-2 on page 25-10.

Figure 56-12. The closed gaussian surface S lies entirely within the material of a conductor. Because the electric field within the conductor is zero, there are no field lines within the conductor and so the net charge within S is zero.

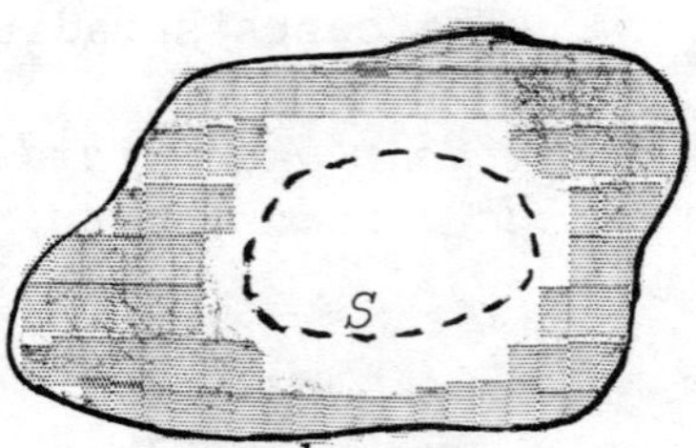

Consider a conductor which occupies some volume of space; it may be a solid piece of metal or a conducting solution in a beaker or some other kind of conductor. Imagine a closed gaussian surface S which lies entirely within the volume of conducting material (Figure 56-12). Because $\vec{E} = 0$ at every point in the conductor, there are no field lines there. As a result, the electric flux through every portion of S is zero and so the total flux through S is also zero. By Gauss's law, the net charge within S must also be zero. Every closed surface which lies entirely within the material of a conductor contains zero net charge.

Figure 56-13. A negatively charged pith ball suspended from a thread inside a conducting container. When a charge is placed inside such an enclosure, an equal charge of the opposite sign appears on the inner surface of the container. In addition, a charge is induced on the outer surface of the container equal to the charge suspended inside.

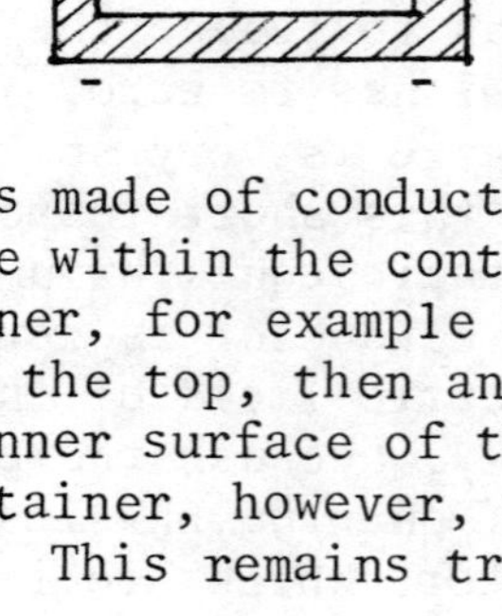

If a container such as a closed metal can is made of conducting material, the foregoing result tells us that the net charge within the container is zero. If a charge is introduced into the container, for example by hanging a charged pith ball on an insulating thread from the top, then an equal charge of the opposite sign must appear on the inner surface of the container. If no charge is put into the interior of the container, however, there will be no electric flux and no electric field there. This remains true even when there is a strong electric field outside the container. The container thus acts as an *electrostatic shield;* external electric fields cannot penetrate into the region shielded by the closed conductor. (If the external field varies with time at a very high frequency, however, the situation is more complex and the conducting container no longer serves as a perfect shield.)

We found in Section 25.4 that electric field lines always intersect equipotential surfaces at right angles. Because the conductor has an equipotential surface (in the absence of current), the field lines which begin or end at the surface all intersect the surface at right angles. At all points adjacent to the surface of the conductor, therefore, the electric field vector $\vec{E}$ is directed normal to the surface. The direction of $\vec{E}$ at any point depends on the sign of the charge at that point on the surface.

56.8 Potential Difference in Electrostatic Fields

In addition to the $\vec{E}$ vector, the electric field is characterized by another quantity complementary to $\vec{E}$, the scalar electric potential V (Section 25.4). We now derive an equation for the difference of potential between two points in an electric field.

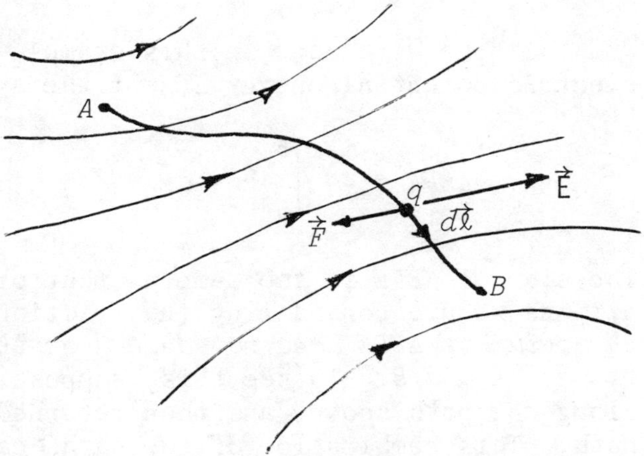

Figure 56-14. The line from point A to point B represents a path through an electric field. When a small positive test charge q is moved from A to B along this path at constant speed, a force $\vec{F}$ must be exerted on the particle which is equal in magnitude but opposite in direction to the force exerted on q by the electric field.

Consider a region of space permeated by a non-uniform electric field, with some field lines as shown in Figure 56-14. The line from A to B represents a *path* between these two points in the field. Suppose a small positive test charge q is moved from A to B along this path at a constant low speed. To prevent the particle from accelerating, the electric force $\vec{E}q$ which the field exerts on q must be opposed at every point by an equal but oppositely directed force $\vec{F} = -\vec{E}q$. (We assume the particle's speed is so small that the tiny force required to accelerate it around the curves in its path may be neglected.) During the motion of q along the path, work is done on the charge by the force $\vec{F}$, and as a result the energy of the charge changes. Assuming the gravitational force on the test charge is negligible and that there is no friction, all the work done goes into an increase in electric potential energy.

While the charge q is moved through the small displacement $d\vec{\ell}$, the work done on the charge by the force $\vec{F}$ is equal to

$$dW = \vec{F} \cdot d\vec{\ell} = -\vec{E}q \cdot d\vec{\ell}.$$

((To compute the *total* work done while q is moved from A to B, it is necessary to add the small increments of work done along all of the small displacements or elements of path from A to B. For some portions of the path $\vec{E}$ has a component parallel to $d\vec{\ell}$ and the increment of work is negative, while for other portions the component of $\vec{E}$ is directed opposite to $d\vec{\ell}$ so the increment of work is positive. To represent the sum of all these elements of work, we use an integral sign. Then the total work done by $\vec{E}$ while q is moved from A to B is equal to

$$W_A^B = \int_A^B dW = \int_A^B - q\vec{E} \cdot d\vec{\ell} = -q \int_A^B \vec{E} \cdot d\vec{\ell}.$$

In this notation, the subscript A (the beginning of the path) and the superscript B (the end of the path) next to the integral sign are called the *limits* of the integral because they mark the beginning and end points of the path, respectively. An integral of this type, in which the integrand ($q\vec{E} \cdot d\vec{\ell}$ in this case) is evaluated and summed along a path, is called a *line integral*. Some suggestions for calculating or evaluating line integrals are given in Section 56.9.))

All the work done in this example goes into an increase in the electric potential energy *EPE* of the system, so we have

$$\Delta EPE = -q \int_A^B \vec{E} \cdot d\vec{\ell}.$$

The sign of ΔEPE is the same as that of the total work. Under the conditions we are considering (no friction and negligible gravitational forces), *the value of* ΔEPE *does not depend on the path which the charge follows between* A *and* B. To see this, suppose the charge is moved from A to B along the path shown, and then returned from B to A along a different path. This combination of two paths can be viewed as a single *closed path* from A to B and back to A. The line integral of $-q\vec{E}$ around this closed path equals the total work done in moving q around the path. The law of conservation of energy requires that this total work equal zero. If the work were not equal to zero, we could move q around the path and return the system to the same total electric potential energy, with a net work which is negative. This would mean that the system did work on the outside world with no loss of energy. By repeating this procedure indefinitely, the system would act as an inexhaustible supply of energy, which is impossible. It is only because the work done around a closed path equals zero that we can define a potential energy for this system. A field of force like the electrostatic force, which has a potential energy associated with it, is called a *conservative* force field. Conservative forces are discussed further in Chapter 49.

The electric potential V at a point where a small test charge q is located is defined (Section 25.1) as the ratio of the mutual electric potential energy EPE of the system to the charge:

★ $$V \equiv EPE/q.$$

The potential difference between points A and B is obtained by dividing ΔEPE by q:

$$\Delta V = V_B - V_A = -\int_A^B \vec{E} \cdot d\vec{\ell}.$$

This is the "integral form" of the equation for potential difference which we used earlier (Section 25.5) in "differential form":

★ $$dV = -\vec{E} \cdot d\vec{s} \quad \text{or} \quad dV = -\vec{E} \cdot d\vec{\ell}.$$

The equations above give the fundamental connection between the two electric field variables of potential and electric field strength. Because the change in electric potential energy from A to B is independent of the path, the potential difference between points A and B is also independent of the path. If the value of $\vec{E}$ at every point in an electric field is known, the equations above allow us to compute the potential V at every point, provided the potential has been specified at any one point in the field. In electrostatics problems, it is customary to choose the zero of potential "at infinity." In electric circuit problems, it is customary to connect some point in the circuit to the ground and choose this as the zero of potential.

Questions and Exercises

Q 56.8-1 Suppose you are given a description of a vector field, $\vec{E}$, which is alleged to be an electric field. You then choose three closed paths and compute the line integral of $\vec{E} \cdot d\vec{\ell}$ around each of the three paths separately. For two of the paths the line integral is zero, but for the third path it is not zero. Can this field be an electric field?

Q 56.8-2 Is the electric potential a vector, or a scalar?

Q 56.8-3 Does the potential at a point P depend on the positions and values of the charges which establish the electric field at P, or on the physical properties of the test particle, or both?

Q 56.8-4 (a) If in a certain region the electric field strength points north, and we move a slight distance to the north, does the potential increase, decrease, or remain the same?
(b) What if we move a slight distance south? (e) east?
(d) west? (e) up? (f) down?

Q 56.8-5 Is there any electrostatic field for which a potential does not exist? If there is, give an example of such a field.

E 56.8-1 (a) Write the equation for the force $\vec{F}$ on a charge q in an electric field $\vec{E}$ (b) Show that if the line integral of $\vec{F} \cdot d\vec{\ell}$ around a closed path (which is equivalent to the work done by the force on this charge during this motion) is zero, then the line integral of $\vec{E} \cdot d\vec{\ell}$ around this path is also zero.

E 56.8-2 Given two point charges, q_1 and q_2, located a distance d apart:
- (a) What is the mutual electric potential energy of these two charges?
- (b) Find the equation for the electric potential at a point which is a distance d from each of the two charges.
- (c) Suppose we take the potential energy calculated in part (a) and divide it by q_2. What electric potential does this represent?

((E 56.8-3 Consider a fixed charge $q_1 = 10^{-7}$ C, and a small exploring charge q = 1 pC which is brought to a point P at a distance of 2 cm from q .
- (a) Compute the work required to bring q from "infinity" to P.
- (b) Compute the mutual potential energy of q_1 and q, when q is at P.
- (c) Compute the potential due to q_1 at P .
- (d) Which of your previous answers would remain the same if the value of q were doubled?))

E 56.8-4 Show that if a test charge is moved through an electric field in a direction perpendicular to the field lines, no work is done on the charge.

E 56.8-5 Two large parallel conducting plates are 0.30 cm apart, and are charged to a potential difference of 240 volts.
- (a) Find the value of E between the plates.
- (b) How much work is required to move a charge q = +4 nC from point A on the negative plate to point B one cm from A on the negative plate?
- (c) Point C is on the positive plate directly opposite A, and point D is on the postive plate directly opposite B. How much work is required to move q from A to C?
- (d) Compute the work required to move q from A to C and then to D.
- (e) Compute the work required to move q from A to B and then to D.
- (f) How does the work required to move q from A to D depend on the path followed by q?
- (g) Compute the work required to move q from D to B and then to A. How does this work compare with your result for part (e)?
- (h) Suppose the charge is moved from A to D and then back to A again by different path. What is the net change in potential energy of the charge?

56.9 Evaluation of Line Integrals

Minimum Learning Goal. Given an expression for the electric field strength in a region of space, be able to calculate the line integral of $\vec{E}$ between two designated points. For example:

1 Between two parallel plates, one charged uniformly with $+\sigma$ coulombs per unit area and the other with $-\sigma$ coulombs per unit area, there is an electric field given by

$$\vec{E} = (\sigma/\varepsilon_0)\hat{y}$$

where $\hat{y}$ is a unit vector normal to the plates.

(a) What is the line integral $\vec{E} \cdot d\vec{\ell}$ from (x_0, y_0) to (x_1, y_1) along path 2 shown below?

(b) Given the result of part (a), what is the value of this line integral along path 1? (Answer, p. 56-38, Line 10)

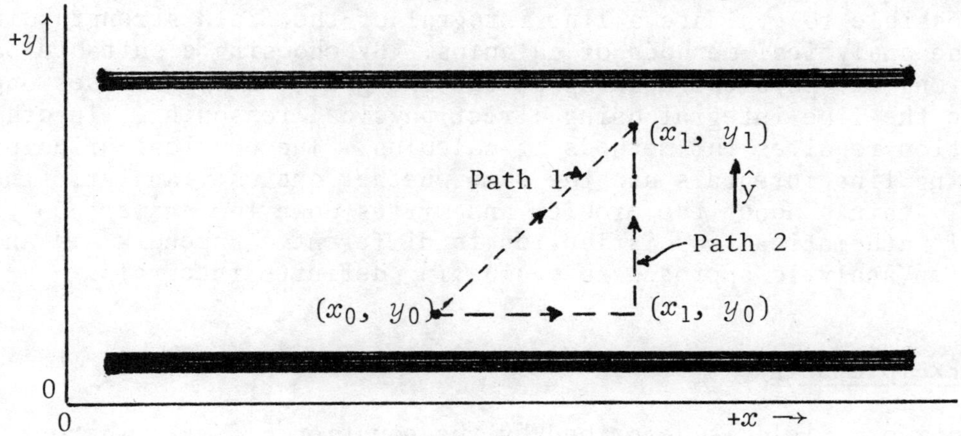

((2 Between two infinite coaxial conductors, one charged uniformly with $+\sigma$ coulombs per unit area and the other with $-\sigma$ coulombs per unit area, there is an electric field given by

$$\vec{E} = (\sigma/\varepsilon_0)(a/r)\hat{r},$$

where a is the radius of the inner conductor and $\hat{r}$ is a unit vector in the direction of increasing r. What is the line integral of $\vec{E} \cdot d\vec{\ell}$ from point A to point B along each of the paths shown below? (Answer, p. 56-38, Line 8)))

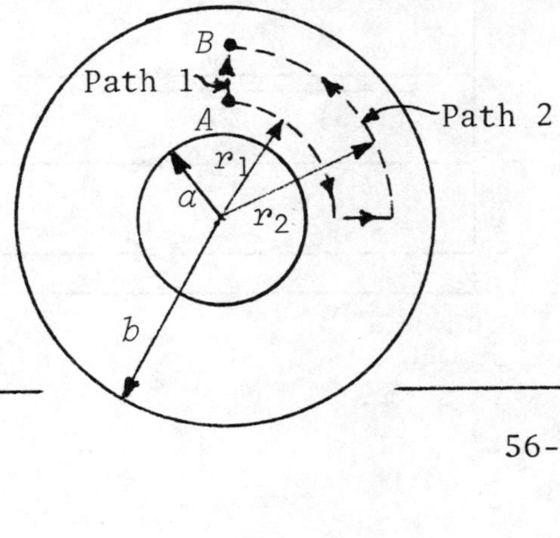

To compute the difference of potential between two points, it is often necessary to evaluate the line integral $\int \vec{E} \cdot d\vec{\ell}$ between the points. Because the value of this integral does not depend on the path between the points, when we are free to choose the path it is advantageous to select one which simplifies the calculation.

The vector $d\vec{\ell}$ in this integral always lies along the path, but the $\vec{E}$ vector may make any angle with $d\vec{\ell}$. Where a portion of the path lies along an equipotential, $\vec{E}$ is perpendicular to the path so the dot product $\vec{E} \cdot d\vec{\ell}$ is zero everywhere along this portion and the integral $\int \vec{E} \cdot d\vec{\ell}$ is zero for this portion. Where the path lies along a field line (or another line parallel to $\vec{E}$ at every point), $\vec{E}$ is parallel to $d\vec{\ell}$ so the dot product $\vec{E} \cdot d\vec{\ell}$ becomes the simple scalar quantity $E d\ell$. When the expression for E for the particular field is substituted into the integral, it may be evaluated using the standard methods of integral calculus, and in some cases by simply thinking about the meaning of the integral and writing out the answer.

As is evident from the worked examples below, for some electric fields it is possible to evaluate a line integral of the field strength without using the analytical methods of calculus. By choosing a suitable path and considering the physical meaning of the integral, in these cases one can evaluate the line integral using direct physical reasoning. In other cases, integration requires the methods of calculus. The physical principles underlying line integrals are the same whether one uses analytic integration or simply thinks about the problem and writes down the answer; only the level of mathematical sophistication is different. Appendix A.4 and A.5 outline an analytic approach to evaluating definite integrals.

Worked Example 56-2

A uniform field is described by the equation $\vec{E} = \hat{x}E_0$, where $E_0 = 3 \times 10^3$ volt/meter. In this field, point *A* is located at (x = 1 cm, y = 8 cm) and point *B* is located at (x = 6cm, y = 3cm). Compute the potential difference between these points, $V_B - V_A$.

Solution

A sketch of this situation is shown at the right. The potential difference is given by

$$V_B - V_A = -\int_A^B \vec{E} \cdot d\vec{\ell}.$$

This line integral must be computed over some specific path between *A* and *B*, but the Example does not specify the path so we are free to choose one which will simplify the calculation.

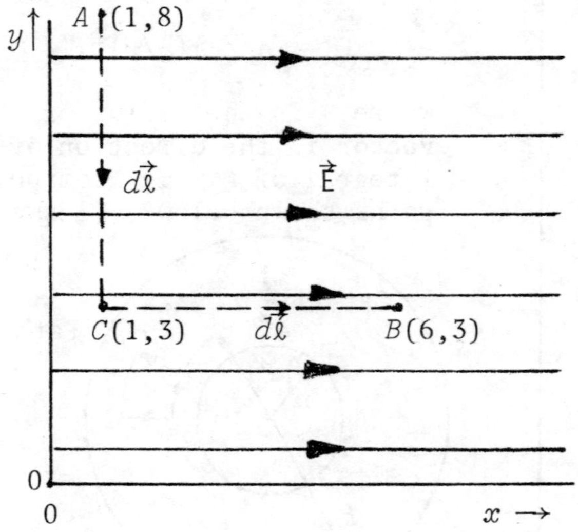

ΔV is independent of the path, so the result obtained does not depend on the path chosen. We use C to label the point at (x = 1 cm, y = 3 cm) and then choose the path ACB. We can then write

$$V_B - V_A = -\int_A^B \vec{E} \cdot d\vec{\ell} = -\int_A^C \vec{E} \cdot d\vec{\ell} - \int_C^B \vec{E} \cdot d\vec{\ell}.$$

Along the path from A to C, $\vec{E} \perp d\vec{\ell}$ so that $\vec{E} \cdot d\vec{\ell} = 0$ for every $d\vec{\ell}$ and therefore

$$\int_A^C \vec{E} \cdot d\vec{\ell} = 0.$$

Along the path from C to B, $\vec{E} \parallel d\vec{\ell}$ so that $\vec{E} \cdot d\vec{\ell} = E_0\, d\ell$

$$\text{and} \int_C^B \vec{E} \cdot d\vec{\ell} = \int_C^B E_0\, d\ell = E_0 \int_C^B d\ell.$$

In the last step above, the quantity E_0 may be written at the left of the integral sign because it is a constant at every point on the path. The integral $\int_C^B d\ell$ is simply the path length from C to B, which equals 6 cm - 1 cm = 5 cm = 0.05 m. Collecting the results above, we have

$$V_B - V_A = 0 - \int_C^B \vec{E} \cdot d\vec{\ell} = -E_0 \times 0.05 \text{ m}$$

$$= -3 \times 10^3 \text{ volt/m} \times 0.05 \text{ m} = -150 \text{ volts}.$$

That is, the potential at B is 150 volts below the potential at A.

Worked Example 56-3

Compute the line integral of the electric field strength to find the potential difference between two points at distances a and b from a point charge q.

Solution

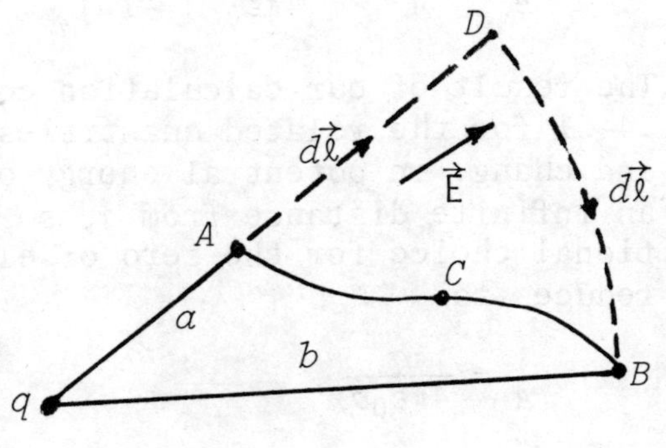

Our first step is to sketch the physical situation, as shown at the right. The point A is a distance a from the source charge q, and the point B is a distance b from this charge. The charge is taken as the origin of a coordinate system. We know the electric field strength from a positive point charge is directed radially outward, having the value

$$E = \frac{q}{4\pi\varepsilon_0\, r^2} .$$

We need to compute

$$V_B - V_A = -\int_A^B \vec{E} \cdot d\vec{\ell}$$

along some path between A and B.

((We might choose a path like ACB, but this would have the disadvantage that $\vec{E}$ is neither parallel nor perpendicular to the path at any point, with the result that the integral could be very difficult to evaluate. Our calculation is simplified if we choose the path ADB, on which $\vec{E}$ is parallel to $d\vec{\ell}$ from A to D and $\vec{E}$ is perpendicular to $d\vec{\ell}$ from D to B. We then compute

$$V_B - V_A = -\int_A^D \vec{E} \cdot d\vec{\ell} - \int_D^B \vec{E} \cdot d\vec{\ell}$$

$$= -\int_A^D E\, d\ell - 0 .$$

The line integral from D to B is zero because $\vec{E} \cdot d\vec{\ell} = 0$ everywhere on this portion. In the integral from A to D, the direction of $d\vec{\ell}$ is the direction of increasing r at every point, where r is the distance from the source charge. We therefore substitute dr for $d\ell$ in this integral, and change the limits of the integral to correspond to the new variable of integration; the value of r at point A is a, and the value of r at point B is b. We then write

$$V_B - V_A = -\int_a^b \frac{q}{4\pi\varepsilon_0} \cdot \frac{1}{r^2}\, dr = -\frac{q}{4\pi\varepsilon_0} \int_a^b r^{-2}\, dr .$$

The variable of integration is r, and because the quantity $q/4\pi\varepsilon_0$ does not contain this variable it is written in front of the integral sign. Following the procedure given in Appendix A.4, we first evaluate the indefinite integral of $r^n\, dr = r^{-2}\, dr$, and then substitute the upper and lower limits to obtain

$$V_B - V_A = -\frac{q}{4\pi\varepsilon_0}\left[\frac{r^{-1}}{-1}\right]_a^b = \frac{-q}{4\pi\varepsilon_0}\left[\frac{-1}{b} - \frac{-1}{a}\right] = \frac{q}{4\pi\varepsilon_0}\left[\frac{1}{b} - \frac{1}{a}\right] .$$))

The result of our calculation corresponds to the equations on page 24-31 for the related quantities of work required to move the charge, and change in potential energy, of the system. If point B is taken at an infinite distance from q, so that $V_B = 0$ in accord with the conventional choice for the zero of electric potential energy, the result reduces to

$$V_A = \frac{q}{4\pi\varepsilon_0 a} ,$$

which may be generalized to the familiar equation

$$V = q/4\pi\varepsilon_0 r$$

for the potential in the field of a point charge.

Exercises

E 56.9-1 Using the cartesian unit vectors $\hat{x}$, $\hat{y}$, and $\hat{z}$, write equations for $\vec{E}$ and $d\vec{\ell}$ in terms of their cartesian components, then multiply them using the rules for the dot product to show that

$$\vec{E} \cdot d\vec{\ell} = E_x\, dx + E_y\, dy + E_z\, dz.$$

E 56.9-2 Consider a uniform field $\vec{E} = \hat{y}E_0$. For the rectangular path *abcd* in the *xy*-plane shown here, compute

$$\oint_C \vec{E} \cdot d\vec{\ell}\ .$$

The rectangle has sides of length g and h. Hint:

$$\oint_C = \int_a^b + \int_b^c + \int_c^d + \int_d^a .$$

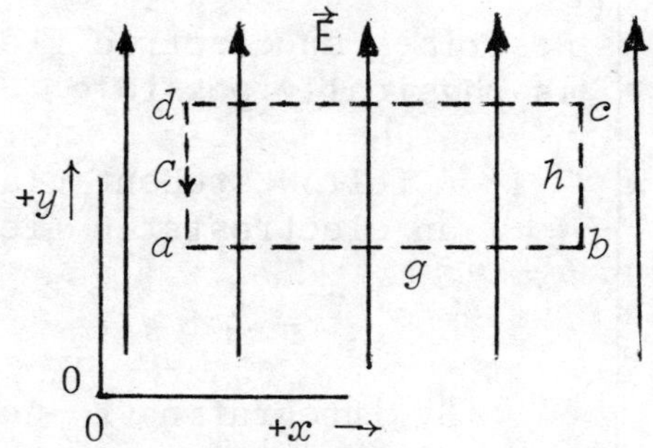

Check your understanding of the material above by testing yourself on the minimum learning goal at the beginning of this section.

56.10 The Circulation Condition for Electrostatic Fields

Minimum Learning Goals: (1) Given a diagram of a plane region of space containing vectors whose length and direction represent the magnitude and direction of a field at the points in space where the vectors originate, be able to state whether the line integral of the field variable around a designated closed path is zero or nonzero, and whether the field could, or could not, be electrostatic. For example:

1 Suppose the arrows shown at the right represent values of a field, $\vec{A}$ (at the origin of each arrow).
(a) Is the line integral of $\vec{A} \cdot d\vec{\ell}$ around the designated path zero?
(b) Could this represent an electro-static field?
(Answer, p. 56-38, Line 11)

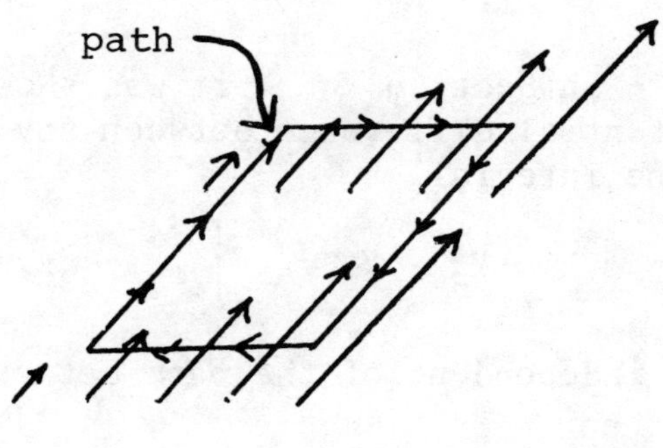

2 Suppose the arrows shown below represent values of a field, $\vec{A}$ (at the origin of each arrow).

(a) Is the line integral of $\vec{A} \cdot d\vec{\ell}$ around the designated path zero?

(b) Could this represent an electrostatic field? (Answer, p. 56-38, Line 7)

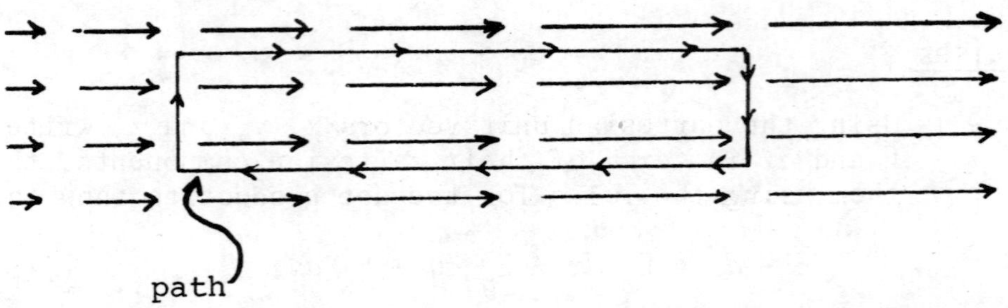

(((2) Given an expression which supposedly describes an electric field, be able to determine whether the field is physically possible. For example:

1 A fellow student claims that he can arrange charges to produce an electrostatic field given by the expression

$$\vec{E} = kx^2 y\hat{x}.$$

By integrating $\vec{E} \cdot d\vec{\ell}$ around the path shown below, determine whether this equation describes an electrostatic field. (Answer, p. 56-38, Line 6)

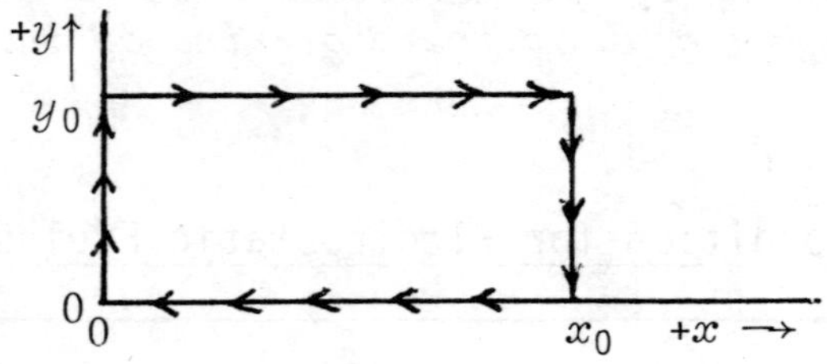

2 An electrical engineer states that he has a device which produces an electric field given by the expression

$$\vec{E} = k\hat{x} + k\hat{y} + k\hat{z}.$$

Can this equation describe an electrostatic field? Why, or why not? (Answer, p. 56-38 , Line 2)))

In Section 56.9 it was shown that in an electrostatic field the potential difference between any two points A and B, computed using the line integral

$$V_B - V_A = -\int_C^B \vec{E} \cdot d\vec{\ell},$$

is independent of the path between A and B. If we now consider any *two* paths between *A* and *B*, these two paths can be viewed as a single *closed path* where we go from point *A* to point *B* on one path and return on the other. Adding the line integrals for the two separate paths gives us the potential difference between the beginning point (*A*) and the ending point (*A*) of the closed loop, which must be zero:

$$V_A - V_A = -\underbrace{\int_A^B \vec{E} \cdot \vec{d\ell}}_{\text{path } 1} + -\underbrace{\int_B^A \vec{E} \cdot \vec{d\ell}}_{\text{path } 2},$$

or

$$0 = -\oint \vec{E} \cdot \vec{d\ell}.$$

The symbol $\oint$ is used to indicate that the line integral is computed around a complete closed path.

Points *A* and *B* may be any points in the field, and the closed path may consist of any two paths between these points, so it is not necessary to specify any limits or end points in writing the equation

★ $$\oint \vec{E} \cdot \vec{d\ell} = 0. \quad \text{Circulation Condition for Electrostatic Fields}$$

This equation, called the *circulation condition for electrostatic fields*, is a fundamental physical law of electrostatics. (It must be extended with an additional term when the field is changing with time, or when a changing magnetic field is present along with an electrostatic field; see Section 59.1). The fact that the integral is written using $\vec{d\ell}$ instead of $\vec{dS}$ indicates that it is a line integral rather than a surface integral as in Gauss's law. The circulation condition is equivalent to the requirement that an electrostatic potential *must* exist.

Combined with Gauss's law, the circulation condition allows one to compute any electric field configuration when the locations and magnitudes of the charges which establish the field are given. Gauss's law alone tells us where the field lines begin and end, but not how they arrange themselves in the intervening space. The circulation condition is needed to find the complete field.

Using this circulation condition, together with Gauss's law, to find the electrostatic field due to a given charge distribution is often mathematically very complicated. Large parts of advanced courses in electromagnetism are concerned with this problem. For our purposes, it is sufficient to know the basic condition plus one fact which can be proved: Gauss's law and the circulation condition *together* determine a *unique* electrostatic field configuration once the charge distribution is given. In other words, there is only one field configuration which can result from a given charge distribution. This *uniqueness theorem* is a powerful aid in the solution of electrostatics problems.

Answers to Minimum Learning Goals

Line 1 A net positive charge of 1.59×10^{-10} C/cm on the wire at (2,4); a net negative charge of 8.85×10^{-11} C/cm on the wire at (11,4).

Line 2 Yes, because the line integral around any closed path is zero.

Line 3 (a) The large ball remains fixed; the smaller ball rolls toward the larger one. (b) If the large steel ball is analogous to a large negative electric charge than the small steel ball is analogous to a postive test charge.

Line 4 A positive charge of 6.2×10^{-11} C/cm at (4,3); another positive charge of 1.95×10^{-10} C/cm at (9,3).

Line 5 By symmetry the field is everywhere perpendicular to the sheet. Considering a part of the sheet having area A and charge σA, we find by Gauss's law $E = \frac{Q/\varepsilon_0}{2A}$. But $Q = \sigma A$, so $E = \sigma/2\varepsilon_0$.

Line 6 It does not describe an electrostatic field because the line integral around this closed path is $(1/3)\ kx_0\ y_0$, not equal to zero.

Line 7 The line integral is zero, so this could represent an electrostatic field.

Line 8 Path 1 or Path 2: $(\sigma/\varepsilon_0)\ a\ \ln(r_2/r_1)$.

Line 9 The analog of a large positive electric charge could be produced by pushing upward on the sheet to produce a hill.

Line 10 Path 1 or Path 2: $(\sigma/\varepsilon_0)(y_1 - y_0)$.

Line 11 The line integral is non-zero, so this could not represent an electrostatic field.

Line 12 By symmetry, the field is everywhere radial and the same in all directions. Considering a cylinder of radius r and height h surrounding the line charge, we find

$$E = \frac{\text{number of lines crossing cylinder}}{\text{lateral area of cylinder}} = \frac{\lambda h/\varepsilon_0}{2\pi r h} = \lambda/2\pi\varepsilon_0 r.$$

ANSWERS to Questions and Exercises in Chapter 56

Section 56.1

Q 56.1-2 The electric field strength; its magnitude is proportional to the slope of the sheet, and its direction is down-hill.

Q 56.1-4 Far from all charges the electric field strength is negligible; the analogy is that the sheet must have negligible slope far from any ball which deforms it.

Q 56.1-6 No. The analogy is that even if one observes the existence of an electric field by its effect on a test charge, the location and magnitude of the sources of this field may be unknown.

Section 56.2

Q 56.2-2 volt/meter or newton/coulomb

E 56.2-2 1.13×10^{-3} newton, toward ball *A*.

E 56.2-4 (a) 3.39×10^{2} N·m²/C; (b) it is independent of the radius.

Section 56.3

Q 56.3-2 Zero

E 56.3-2 Zero.

Section 56.4

Q 56.4-2 No. If this could happen, at the point where they merged the electric field strength would have two different directions, which is physically impossible.

Q 56.4-4 Choose an area normal to the field lines in the region around the point in question, through which several field lines pass, and compute the average field strength over that area.

Q 56.4-6 The field from both charge configurations is the sum of the two separate fields.

E 56.4-2 (a) 1.13×10^{5} lines. (b) 40.0 lines.

E 56.4-4 On *A*, -4.60×10^{-8} C/m; on *B*, $+4.60 \times 10^{-8}$ C/m

E 56.4-6 (a) -2.95×10^{-10} C/cm; (b) 1.8×10^{4} N/C upward; (c) 10.8×10^{4} N/C, upward.

Section 56.5

Q 56.5-2 (c) a 2-fold axis, (f) a 2-fold axis, (i) an ∞-fold axis, (k) a 6-fold axis, (l) an ∞-fold axis, (m) a 2-fold axis, (n) An ∞-fold axis, (o) 2-fold, 3-fold and 4-fold axes, (q) an ∞-fold axis.

Q 56.5-4 Yes. Examples are a charged sphere and a charged disc.

Q 56.5-6 Coulomb/meter^2.

E 56.5-2 (a) $+5.7 \times 10^{-7}$ C; (b) zero; (c) 4.0×10^6 V/m, radial outward; (d) 5.1×10^5 V/m, radial outward.

Section 56.6

E 56.6-2 (a) 1.29×10^2 V/m radial inward; (b) 0.32 V/m radial inward.

Section 56.8

Q 56.8-2 a scalar

Q 56.8-4 (a) decreases; (b) increases; (c - f) for motions east, west, up and down it remains the same.

E 56.8-2 (a) $EPE = (1/4\pi\varepsilon_0)\, q_1q_2/d$.

(b) $V = (1/4\pi\varepsilon_0)(q_1 + q_2)/d$

(c) The potential at the location of q_2, due to q_1.

Section 56.9

E 56.9-2 zero.

CHAPTER 57 CAPACITANCE

Introduction

The topic of capacitance provides a link between the theoretical ideas of electrostatics and the more concrete study of electric circuits. Capacitance is a property of every pair of conductors, and also of an isolated conductor. Practical devices which utilize capacitance are called capacitors. The idealized parallel-plate capacitor is considered in some detail, leading to a discussion of dielectric materials. An important function of capacitors in electric circuits is the storage of electric energy. The chapter concludes with a mathematical description of the flow of charge and energy in a circuit containing a capacitor which is being charged or discharged.

When you have completed your study of the material in this chapter, you should be able to answer questions and work exercises like the examples in the minimum learning goals assigned by your instructor.

57.1 The Meaning of Capacitance

Minimum Learning Goals. (1) When you are given information about two conductors, be able to specify whether or not the information is sufficient to determine the capacitance between the conductors, and when it is not, be able to state the additional information which is required. For example:

1 Which of the following descriptions are complete enough to determine the capacitance between the given conductors? For those descriptions which are incomplete, state the additional information which must be known for the capacitance to be completely determined. (Answer, page 57-25, Line 17)
(a) Two parallel plates, each having an area of 100 cm^2, are separated by 5 mm.
(b) Two parallel conducting plates are placed in nitrobenzene.

2 Repeat the instructions of example 1 for the following descriptions. (Answer, page 57-25, Line 18)
(a) Two metal spheres having radii of 10 cm are in air, with their centers 30 cm apart.
(b) Two conductors are placed in air.

(2) Be able to calculate the capacitance between two conductors having given charges and a given potential difference between them. For example:

1 A girl walking across a carpet acquires an electric charge of 6×10^{-9} C. If the resulting potential difference between her body and the ground is 1000 V, what is the capacitance between her and the ground? (Answer, page 57-24, Line 10.)

2 A 12-volt battery is connected across two parallel plates 2 cm apart. If the resulting charge on the upper plate is 6×10^{-11} C, and the charge on the lower plate is -6×10^{-11} C, what is the capactiance between the plates? (Answer, page 57-25, Line 16.)

If there are two conductors located in empty space, having a charge $+q$ on one and a charge $-q$ on the other, there will be a potential difference ΔV between the two conductors. If the charge on each conductor is doubled the electric field strength at every point in space will also be doubled, so that the potential difference between the conductors will be twice as great as before. If the initial charges on the two conductors are both multiplied by any other factor, the potential difference will also be multiplied by that factor. This leads to the conclusion that the potential difference ΔV between the conductors is directly proportional to the magnitude of the (equal and opposite) charges on the conductors:

$$\Delta V \propto q \qquad \text{or} \qquad q \propto \Delta V.$$

The constant ratio $q/\Delta V$ is denoted by the symbol C and is called the capacitance between the two conductors.

★ The *capacitance* between a pair of isolated conductors having charges $+q$ and $-q$, and having a potential difference ΔV, is $C \equiv q/\Delta V$.

The SI unit of capacitance is given a special name, the *farad,* in honor of Michael Faraday (1791-1867). One farad (symbol F) is equal to one coulomb/volt; 1 F = 1 C/V.

For an isolated pair of conductors in empty space, it is found that the capacitance between them is determined entirely by the geometrical arrangement, that is, by the sizes and shapes of the two conductors and by the spacing between them. This is illustrated for the case of two flat parallel plates in Section 57.3. If the space between two conductors is filled with some material or medium other than empty space, the capacitance between the conductors depends on the geometry and on a property of the medium called the dielectric constant (Section 57.4).

The concept of capacitance as defined above applies to two isolated conductors which interact with each other, but which are so far apart from all other physical objects that any effect which the others may have is negligible. A pair of conductors like this, in which all the field lines which begin on the positive conductor end on the negative one, is called an *ideal capacitor*.

Any pair of conductors insulated from each other may be regarded as a capacitor, even though these conductors interact electrically with other objects as well as with each other. Every capacitor has some value of capacitance, but when only *some* of the field lines beginning on one of the conductors end on the other, the simple definition of capacitance given above cannot be applied to this more complicated case. In general, the voltage between any pair of conductors will depend on the charges on *all* the conductors. In this text we shall not work with non-ideal capacitors, however, because most of the capacitors we study will be those connected in electric circuits. These are close enough to being ideal capacitors so that their capacitance is well defined.

Occasionally, as in Section 57.2, we refer to the capacitance of a *single* conductor. By this is meant the ratio of its charge to the potential difference between it and a point infinitely far away.

Questions and Exercises

Q 57.1-1 Suppose you have two parallel conducting plates separated by a small distance, and bearing charges $+q$ and $-q$. You then double the charge on each plate. Why doesn't the increase in the positive charge just cancel the effect of the increase in the negative charge, so that the electric field and the potential difference between the plates remains the same?

Q 57.1-2 Suppose you have two conductors which are near each other but are relatively far from all other objects. What factors will affect the potential difference between the conductors?

Q 57.1-3 How does the capacitance between two conductors depend on the charges on the conductors?

Q 57.1-4 What is the SI unit of capacitance? How is it defined?

Q 57.1-5 If the charge on an isolated conductor "leaks off" and decreases to one-third of its initial value, what happens to the potential of the conductor? to its capacitance?

E 57.1-1 The two conductors of a nearby pair have charges of ±3 μC, and a potential difference of 100 volts. What is the capacitance of this pair?

E 57.1-2 Two conductors have a capacitance of 100 pF and a potential difference of 1000 volts. What is the charge on the positive conductor?

E 57.1-3 Two conductors have charges of ±12 μC and a capacitance of 6 nF. What is the voltage between these conductors?

Check your understanding of the material above by testing yourself on the minimum learning goals at the beginning of this section.

57.2 Capacitance of an Isolated Sphere

Minimum Learning Goal. When you are given descriptions of objects, be able to select the one object which is most nearly an isolated conducting sphere, and calculate its capacitance assuming it is an isolated sphere. For example:

1 Calculate the capacitance of the object in the following list which is most nearly a charged, conducting, isolated sphere. (Answer, page 57-25, Line 13.)
(a) An aluminum sphere of radius 10 cm whose center is located 30 cm above the center of a flat sheet of aluminum 30 meters square. The sphere has a charge of -3×10^{-6} C and the sheet a charge of $+3 \times 10^{-6}$ C.
(b) The aluminum sphere of (a), located 10 m above a flat sheet of aluminum 100 m square, both of which are charged as described in (a).
(c) A copper sphere of radius 1 m located 2 m above the ground, and having a charge of 6×10^{-6} C.
(d) A polyethylene sphere of radius 5 cm located 100 m above ground and having a charge of 2×10^{-9} C.
(e) A silver sphere of radius 5 cm located 100 m above ground and having a charge of 2×10^{-9} C.

2 Calculate the capacitance of the object in the following list which is most nearly a charged, conducting, isolated sphere. (Answer, page 57-24, Line 7.)
(a) A silver sphere of radius 10^{-1} cm whose center is located 10 m above the center of a flat sheet of aluminum 100 meters square. The sphere has a charge of $+6 \times 10^{-14}$ C and the sheet has a charge of -6×10^{-14} C.
(b) A copper sphere of radius 1 m with the same location and charge as in (a).
(c) A rubber sphere of the same radius, location, and charge as in (a).
(d) A gold sphere of the same radius, and having the same charge as in (a), but located 2 cm above the flat sheet.

If we imagine a single conductor which is at a very great distance from all other objects, this isolated conductor has a capacitance even though there is no one other conductor which combines with it to form a capacitor. When the isolated conductor is charged, all of the field lines which begin (or end) on it may be said to end (or begin) "at infinity." In order to compute the capacitance of such an isolated charged conductor, it is most convenient to consider a conductor having the shape of a sphere.

As shown in Section 56.5, the electric field outside an isolated charged conducting sphere bearing a total charge q is the same as if all of the charge were concentrated at the center. If we make the customary choice that the zero of potential is at infinity, the potential at any distance d from such a point charge is given by $V = (1/4\pi\varepsilon_0)q/d$ (Sec. 25.2), where $\varepsilon_0 = 8.85 \times 10^{-12}\ C^2/(N{\cdot}m^2)$. To find the potential at the surface of a charged sphere having a radius a, we substitute a for d and write

$$V = \frac{1}{4\pi\varepsilon_0}\frac{q}{a}.$$

It is evident that the *ratio* of q to V is a constant which depends only on the radius of the sphere. This ratio is the capacitance of the sphere; it is

$$C = 4\pi\varepsilon_0 a.$$

For an isolated conductor of *any* shape, it is found that q/V is a constant whose value depends only on the geometry, i.e., on the size and shape of the conductor.

★ The *capacitance of an isolated conductor* having a total charge q and a potential V is $C \equiv q/V$.

If the shape of the conductor has some resemblance to a sphere, one could use the "average radius" of the conductor and compute an approximate value for its capacitance using the equation for a sphere. For other shapes, a precise equation for capacitance would be more complicated.

Because the field lines from an isolated conductor go to infinity, one may suppose that the conductor is one part of a capacitor whose other conductor is a sphere or large conducting surface located at infinity. In judging whether a spherical conductor is sufficiently isolated to justify using the equation for the capacitance of an isolated sphere, the important factor is the ratio of the radius of the sphere to the distance from it to the nearest conducting object. The smaller this ratio is, the more clearly the sphere approaches the idealized condition of being "isolated."

Questions and Exercises

Q 57.2-1 What could you do to set up in your lab the closest possible approximation to an isolated charged conducting sphere in free space?

Q 57.2-2 Why does the charge on an isolated conducting sphere distribute itself uniformly over the surface? Why doesn't it cluster predominantly on the north side, or on the shady side?

Q 57.2-3 The potential of a charged sphere relative to infinity is $q/4\pi\varepsilon_0 a$, because the potential at infinity is chosen to be zero. If the potential at infinity is assigned a different value, how will this affect the capacitance of a charged sphere?

E57.2-1 The 12-inch-diameter spherical dome of a small van de Graaff generator is charged to a potential of 400 kV.

(a) What is the charge density on the dome?
(b) What is the value of $|\vec{E}|$ next to the dome?
(c) What is the capacitance of the dome assuming it is an isolated sphere?

E 57.2-2 Compute the capacitance of the earth, using an average radius of 6.38×10^6 meters.

Check your understanding of the material above by testing yourself on the minimum learning goal at the beginning of this section.

57.3 Parallel-plate Capacitor

Minimum Learning Goal. When you are given the capacitance of a parallel plate capacitor, and changes in its dimensions, be able to specify the new capacitance. For example:

1 A parallel-plate capacitor made of two rectangular sheets of copper having width a, length b, and a separation d, has a capacitance of 6×10^{-12} F. What is the capacitance of a parallel-plate capacitor having plates of width $3a$, lenght $4b$, and plate separation $d/3$? (Answer, page 57-24, Line 4.)

2 Using data from example (1), what is the capacitance of a parallel plate capacitor having plates of width $a/2$, length $b/3$, and separation $8d$? (Answer, page 57-24, Line 1.)

A parallel-plate capacitor consists of two parallel conducting plates separated by a distance small compared to the transverse dimensions of the plates. If such a pair of plates are charged with equal and opposite charges $\pm q$, their electric field will be almost completely confined to the region between them, where it will be uniform and directed perpendicular to the plates. The field lines will "bulge" outward slightly at the edges of the capacitor, but this bulging or fringing effect may be neglected in our derivation.

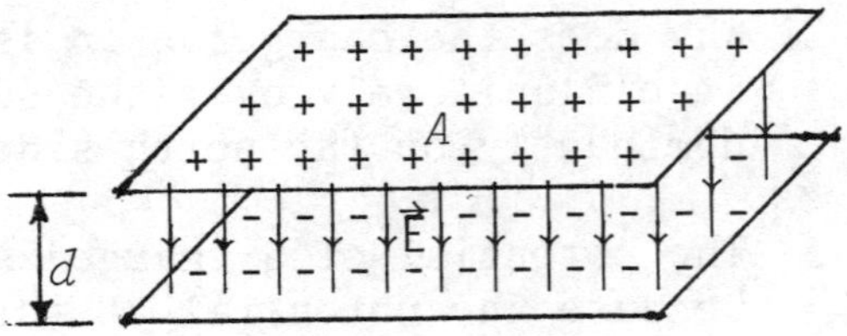

Figure 57-1. A parallel-plate capacitor consists of two oppositely charged parallel conducting plates, separated by a distance d and each having area A. On each plate, the charge is located on the side closest to the opposite plate. The electric field between the plates is directed perpendicular to the plates.

To compute the capacitance of an air-filled parallel-plate capacitor, we note that the permittivity of air at a pressure of one atmosphere is nearly identical to the permittivity ε_0 of a vacuum. Each plate has an area A (on each side), and because the total charge q is spread uniformly over this area the charge density on each plate is $\sigma = q/A$. The field strength between the plates is given by $\mathrm{E} = \sigma/\varepsilon_0$ (Section 56.6). The potential difference ΔV between the plates may be computed using a line integral of the electric field strength (Section 56.8), giving the result $\Delta V = \mathrm{E}d$. Substituting these equations into the definition for capacitance, we compute

$$C = \frac{q}{\Delta V} = \frac{\sigma A}{\mathrm{E}d} = \frac{\sigma A}{(\sigma/\varepsilon_0)d} = \frac{\varepsilon_0 A}{d} .$$

In order to obtain a large capacitance, it is desirable to have a large plate area and a small separation between plates. For a given size and spacing of the plates, the capacitance can be increased using a material between the plates which has a permittivity higher than air (Section 57.4). Practical capacitors are available in many configurations, usually made as compact as possible. They are used in electric circuits for many different purposes: charge and energy storage, "smoothing" a fluctuating voltage, etc. Most capacitors have values of capacitance which are conveniently expressed in μF (microfarads) or pF (picofarads). The usual symbol for a capacitor in a circuit diagram is 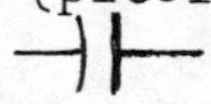 or 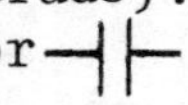 .

Questions and Exercises

Q 57.3-1 In an electronics store one can buy a variety of capacitors; most of them consist of a compact unit with two terminals which can be connected electrically to something else. Is it possible to have a capacitor which is not a unit having two terminals? If it is, give an example.

Q 57.3-2 What is the capacitance of the "condenser" in the ignition circuit of your car?

Q 57.3-3 How could a parallel-plate capacitor be used to produce controlled deflection of a beam of charged particles?

E 57.3-1 Two flat metal plates, each 2.0 cm in width and 5.0 cm in length, are 1.0 mm apart. They are connected to a 6.0-volt battery.
(a) What is the capacitance of this combination?
(b) What is the total charge on the positive plate?
(c) What is the charge density on the positive plate?

E 57.3-2 An air-filled capacitor consists of parallel plates having an area of 18.0 cm^2. Its capacitance is 25 pF.
(a) What is the separation of the plates?
(b) What voltage between the plates is required to give a surface charge density of 6.25×10^{-7} $\mathrm{C/m^2}$?
(c) If the plate separation is increased, will the capacitance increase, or decrease? What will be the limiting value of capacitance when the separation is very large?

E 57.3-3 Dan desires to make a parallel-plate capacitor having a plate separation of 0.5 mm and a capacitance of 40 pF.

(a) What area is required for each of the plates?
(b) What total charge on either plate is required for a capacitor voltage of 90 volts?

E 57.3-4 For the capacitor described in E 57.3-1, if the length, width, and separation of the plates are all doubled, what is the new value of the capacitance?

E 57.3-5 For the capacitor described in E 57.3-3, if the length of the plates is doubled, their width is tripled, and their separation is quadrupled, what is the new value of capacitance?

E 57.3-6 Consider a parallel-plate capacitor in which the plates are initially charged with $\pm q$. If the capacitor is now disconnected from the battery and the separation of the plates is doubled, what will be the effect on:
(a) the charge on either plate?
(b) the voltage between the plates?
(c) the capacitance?

E 57.3-7 Suggest an experiment by which you could determine the extent to which the electric field lines bulge out from the space directly between the plates of a parallel-plate capacitor.

E 57.3-8 Consider a parallel-plate capacitor which is connected to a battery so that the plates acquire charges of $\pm q$. If the capacitor now remains connected to the battery while the separation of the plates is doubled, what will be the effect on
(a) the voltage between the plates? (b) the capacitance?
(c) the charge on either plate?

Check your understanding of the material above by testing yourself on the minimum learning goal at the beginning of this section.

57.4 Dielectrics

Minimum Learning Goal. When you are given the potential difference between the plates of a capacitor and the charge on the plates for each of two different dielectric substances placed between the plates, be able to determine the ratio of the dielectric constants of the two substances. For example:

1 A parallel-plate capacitor filled with air has a charge of $+q$ on one plate and $-q$ on the other plate when the potential difference between the plates is 30 volts. When the air is replaced by rubber without changing the charge on the plates, the potential difference changes to 10 volts. What is the ratio of the dielectric constant of rubber to that of air? (Answer, page 57-24, Line 9.)

2 A parallel-plate capacitor filled with air has a charge of $+q$ on one plate and $-q$ on the other plate when the potential difference between the plates is 42 volts. When the air is replaced by oil, there is a charge of $+2.1q$ on one plate and $-2.1q$ on the other plate, with the same potential difference between them. What is the ratio of the dielectric constant of oil to that of air? (Answer, page 57-25, Line 14.)

In our study of electrostatics thus far, the only materials we have considered are conductors, and air which has electric properties very close to those of a vacuum. We are now ready to consider some of the electric properties of insulating materials other than a vacuum. Such a material is called a *dielectric*. Examples of familiar dielectrics include rubber, dry paper, mica, porcelain, glass, plastics, and oil.

Suppose we have a parallel-plate capacitor filled with air and connected to a battery having a terminal voltage ΔV. The charge on the plates separated by air is denoted by $\pm q_0$, and the capacitance is $C_0 = q_0/\Delta V$. Now while the capacitor remains connected to the battery, imagine that a slab of some dielectric material is inserted so that it just fills the region between the plates. Then it is found by experiment that while the dielectric is being inserted, additional charge flows from the battery to the plates until the total charge on each plate has increased to the value $q = \kappa q_0$. The proportionality factor κ (Greek kappa) is called the *dielectric constant* of the dielectric material; it is a dimensionless number greater than unity. Because the charge on the capacitor has increased while the potential difference remained the same, inserting the dielectric has increased the capacitance; the new value is

$$C = \frac{q}{\Delta V} = \kappa C_0 = \frac{\kappa\varepsilon_0 A}{d}\,.$$

Table 57-1. Properties of Some Dielectrics

Material	Dielectric Constant
Vacuum	1.00000
Air	1.00054
Pure water	78
Paper	3.5
Mica	5.4
Porcelain	6.5
Pyrex glass	4.5
Polyethylene	2.3
Teflon	2.1
Neoprene	6.9

To write an equation for the capacitance of the dielectric-filled capacitor similar in form to the equation derived in Section 57.3 for a vacuum-filled capacitor, we introduce a quantity called the *permittivity of the dielectric*, denoted by the symbol ε (Greek epsilon) and defined by the equation

★ $$\varepsilon \equiv \kappa\,\varepsilon_0 .$$

The SI unit of permittivity is $C^2/(N{\cdot}m^2)$. The capacitance of the parallel-plate capacitor filled with a dielectric is then

★ $$C = \frac{\varepsilon A}{d} .$$

In the experiment described above, inserting the dielectric slab drew added charge from the battery to the plates even though the voltage across the capacitor remained constant. The increase in charge means that the surface charge density has increased by the factor κ, and also that the number of field lines which begin at the positive plate has increased by κ. The electric field strength $\vec{E}$ *within the dielectric,* however, must be the same as it was in the vacuum-filled capacitor, because the line integral of $\vec{E}$ from the negative to the positive plate is still equal to the battery voltage, ΔV. How can the number of field lines increase while $\vec{E}$ remains the same?

A clue comes from asking what draws additional charge to the capacitor plates when the dielectric is inserted. With no dielectric, the plates are already "saturated" with charge in the sense that the potential difference between them has risen to the full voltage which can be supplied by the battery. When additional charge flows to the plates, it must be because the dielectric slab is attracting this charge, and this in turn implies that the dielectric slab (which was initially uncharged) acquires charge at its surface while it is being inserted. This newly acquired surface charge does *not* rub off from the plates, however. In fact, the surface charge adjacent to the positive plate of the capacitor must be *negative,* in order to draw more positive charge from the battery, and the surface charge adjacent to the negative plate must be *positive.* After the dielectric is inserted the new physical state of the capacitor persists as long as the slab remains between the plates, which shows that the surface charge does not flow on to the capacitor plates. Rather, the surface charge remains *bound* to the dielectric.

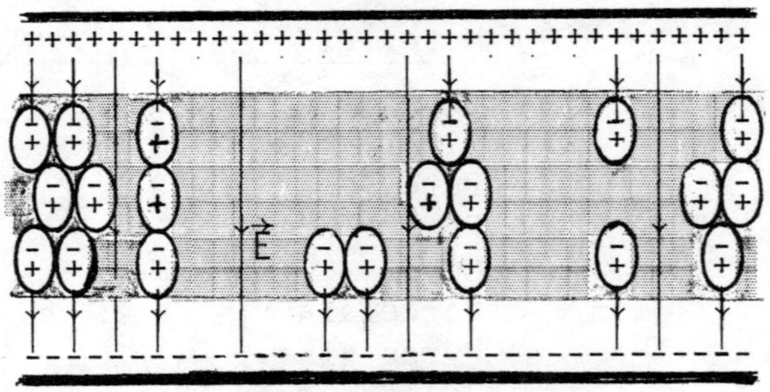

Figure 57-2. In a dielectric slab placed between oppositely charged plates, the individual molecules become polarized by the electric field of the plates. In this figure, the metal plates are shown slightly separated from the dielectric, for clarity. Only a few of the polarized molecules and a few of the field lines are shown. The polarized molecules supply a layer of negative bound surface charge next to the positive plate, and a layer of positive bound surface charge next to the negative plate.

How can we account for the appearance of this bound surface charge on the dielectric? We must consider what happens to the individual molecules of the dielectric when the slab is placed between the capacitor plates.

Figure 57-2 shows a charged parallel-plate capacitor having its positive plate at the top and its negative plate at the bottom. The shaded area between the plates represents the dielectric. Shown within the dielectric (drawn to a greatly enlarged scale) are a few molecules which have been *polarized* by the electric field of the plates. This means that the charged particles composing the molecules have their average positions displaced by the electric field. From their positions in the absence of a field, the electrons are shifted slightly toward the positive plate, while the protons are shifted slightly toward the positive plate. Of course, the electrons and protons cannot move very far, since they are bound to the atoms by the forces which hold the atoms together. (If the electrons could move long distances, the material would be a conductor rather than an insulator.)

Because the electric field pushes the protons one way and the electrons the opposite way, a charge called *polarization charge* appears at the surfaces of the dielectric. This polarization charge is the bound surface charge which we inferred above through the argument concerning the field lines. Most of the field lines which begin on the positive charge *on the conductor* end on the negative polarization charge, and most of the field lines which end on the charge on the negative conductor begin on the positive bound surface charge adjacent to it. Only a fraction ($1/\kappa$) of the field lines which begin on the positive plate end on the negative plate. As a result, the electric field strength in the dielectric is smaller than would be calculated from the charge on the plates alone.

Suppose we have a parallel-plate capacitor initially filled with air and given a charge $\pm q$, completely disconnected from any circuit. If a neutral dielectric slab is then inserted between the plates, polarization charge appears on the slab and the electric field strength within the dielectric becomes less than its prior value in the air-filled capacitor. The distance between plates remains constant, however, and so the potential difference between the plates decreases. If the dielectric slab fills the capacitor, it can be shown that the field strength in the dielectric is given by $E = \sigma/\varepsilon$, where σ is the density of free charge on either plate of the capacitor.

A *linear dielectric* is one for which the polarization is proportional to the applied field. Naturally, no dielectric is linear at very high fields, because there is only so much charge in it which can move only so far. But for typical laboratory fields, most dielectrics can be considered linear.

Questions and Exercises

Q 57.4-1 Give some examples of dielectric materials which are used in capacitors.

Q 57.4-2 When a dielectric becomes polarized, is the average displacement of the charged particles in the dielectric more nearly equal to the thickness of the dielectric, or the diameter of an atom? Justify your answer.

Q 57.4-3 If a charged capacitor (containing a polarized dielectric) is disconnected from a battery and the capacitor terminals are connected together, what happens to: (a) the charge on the plates? (b) the polarization charge on the dielectric? (c) the electric field in the capacitor?

Q 57.4-4 When a capacitor is filled with a polarized dielectric, why doesn't the negative polarization charge adjacent to the positively charged capacitor plate cancel some of this positive charge and thus reduce the net charge on the positive plate?

Q 57.4-5 If a slab of dielectric material, such as rubber, is placed on the table top, is it polarized? If not, what must be done to the material to polarize it?

Q 57.4-6 Would it be possible to have a dielectric in a parallel-plate capacitor such that *all* the field lines which begin on the positive plate end on the negative polarization charge? What would be the value of $\vec{E}$ within the dielectric for such a situation?

E 57.4-1 Consider two parallel-plate capacitors, one air-filled and the other filled with a solid dielectric having dielectric constant κ.
(a) If the two capacitors have the same A, d, and q, what is the ratio of their potential differences?
(b) What is the ratio of their capacitances?

E 57.4-2 A parallel-plate capacitor having C = 30 nF is connected to a 12-volt battery. Later a sheet of pyrex glass is inserted to fill the region between the plates, while the capacitor remains connected to the battery.
(a) What is the charge on the positive capacitor plate before the glass is inserted?
(b) What is the capacitance after the glass is inserted?
(c) What is the voltage between the plates after the glass is inserted?
(d) What is the charge on the positive capacitor plate after the glass is inserted?
(e) What is the total bound charge on the surface of the glass sheet opposite the positive plate?

E 57.4-3 A parallel-plate capacitor having C = 30 nF is connected to a 12-volt battery. After receiving a charge, the capacitor is disconnected from the battery, and then a sheet of pyrex glass is inserted to fill the region between the plates.
(a) What is the charge on the positive capacitor plate before the glass is inserted?
(b) What is the capacitance after the glass is inserted?
(c) What is the voltage between the plates after the glass is inserted?
(d) What is the charge on the positive capacitor plate after the glass is inserted?
(e) What is the total bound charge on the surface of the glass sheet opposite the positive capacitor plate?

E 57.4-4 In a parallel-plate capacitor filled with dielectric the charge density on either plate is 1.0 μC/m², and the polarization charge density is 0.8 μC/m².
(a) What is the electric field strength between the plates?
(b) If the plates are 0.6 cm apart, what is the potential difference between the capacitor plates?
(c) If the dielectric material is removed from this capacitor without any change in the capacitor geometry, while the capacitor is not connected to a battery, what will be the new value of electric field strength?

E 57.4-5 Show that for a parallel-plate capacitor filled with a linear dielectric the ratio of the polarization charge on the dielectric to the charge on the adjacent plate is $1/\kappa - 1$. This ratio is always negative for any dielectric having $\kappa > 1$.

E 57.4-6 Suppose we have a parallel-plate capacitor in air which remains connected to a battery and has an initial charge of $\pm q_1$ on its plates, an initial electric field strength E_1 and an initial capacitance C_1. We then insert a slab of dielectric material (dielectric constant κ) which fills the region between the plates. In relation to the earlier values, what are the new values of:
(a) voltage between the plates?
(b) charge on the plates?
(c) electric field strength between the plates?
(d) capacitance?

Check your understanding of the material above by testing yourself on the minimum learning goal at the beginning of this section.

57.5 Energy Stored in a Charged Capacitor

Minimum Learning Goal. Be able to calculate the energy stored in a charged capacitor, when you are given its capacitance and potential difference. Be able to estimate the work which this charged capacitor could do. For example:

1 A 250-μF capacitor is connected to a 100-volt battery until the capacitor is fully charged. If the battery were then disconnected and the capacitor correctly wired to an electric motor, the most work the motor could do would be to:
(a) lift 1 kg a distance of 1 m
(b) lower 1 kg a distance of 0.5 m
(c) lift 0.1 kg a distance of 1 m
(d) lower 0.1 kg a distance of 0.5 m
(e) lift 10 kg a distance of 2 m
(Answer, page 57-24, Line 8).

2 A 450-μF capacitor is charged to 50 volts. If the capacitor were then connected to an electric motor of a toy car having a total mass of 0.1 kg, the highest hill which the car could climb would be:

(a) 20 cm high
(b) 2 cm high
(c) 50 cm high
(d) 30 cm high
(e) 10 cm high
(Answer, page 57-25, Line 12.)

Consider a process by which a capacitor becomes charged to a potential difference ΔV, beginning from a state in which it is completely discharged. How much work is done in this charging process?

The capacitor may be charged by connecting it to a battery, as shown in Figure 57-3(a). In order to compute the work done, we assume that the negative plate is kept at the ground potential, taken as zero. Then the charge on the negative plate has an electric potential energy equal to zero.

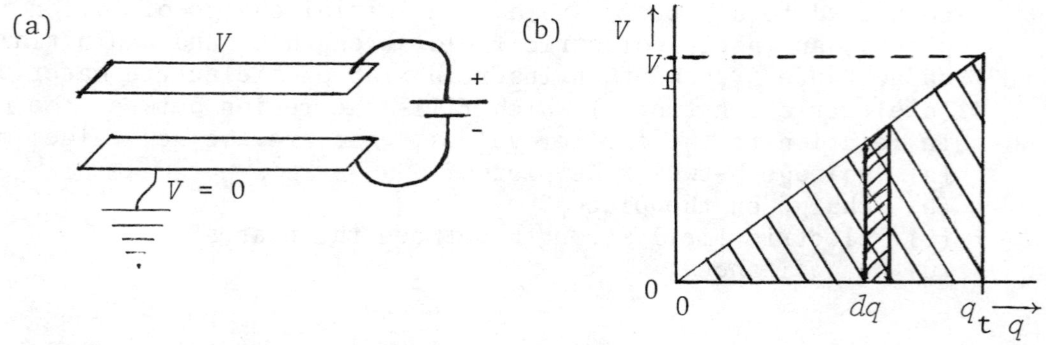

Figure 57-3. (a) A capacitor (pictured as two parallel plates) may be charged by connecting it to a battery. If the negative plate is connected to the ground, its potential will remain equal to zero while the potential of the positive plate increases to the voltage supplied by the battery. (b) During the charging process, the potential V of the positive plate increases in direct proportion to the total charge q on this plate. The work done in charging the capacitor is represented by the area under the graph of V vs. q.

Work is done by the battery in transferring charge from the negative plate to the positive plate. All of this work goes into an increase in electric potential energy *EPE* of the system (Section 24.6). During the charging process, the charge grows from an initial value of zero to a final value q_t. We use q to refer to the charge on the plate at any instant during the charging process; the corresponding instantaneous potential of the positive plate is $V = q/C$, where C is the capacitance.

To compute the work, we apply the definition of electric potential (Section 25.1) and note that when a charge dq is brought from a place where the potential is zero (the negative plate) to a location where the potential is V (the positive plate), it acquires an electric potential energy given by

$$d\,EPE = V\,dq.$$

When the charge has reached some value q, therefore, the work required to add a small additional charge dq is equal to

$$dW = dEPE = V\,dq = q\,dq/C.$$

This work is represented by the narrow vertical strip in Figure 57-3 (b). The total work done during the entire charging process is the sum of all the increments of work as the charge increases from zero to q_t. This work may be represented as the total area under a curve of V vs. q, as shown in Figure 57-3 (b). This is the area of the shaded triangle, and is equal to

$$W = \frac{1}{2} q_t\, V_f\,.$$

((This total work may also be expressed as the summation or integral of all the increments of work dW; the integral is evaluated using the procedure given in Appendix A.4:

$$W = \int dW = \int_0^{q_t} V\,dq = \int_0^{q_t} \frac{q}{C}\,dq$$

$$= \frac{1}{C}\int_0^{q_t} q\,dq = \frac{1}{C}\left[\frac{q^2}{2}\right]_0^{q_t} = \frac{1}{C}\frac{q_t^2}{2}\,. \qquad))$$

This result equals that obtained using the area of the triangle, because the final potential reached by the positive plate is $V_f = q_t/C$.

For future reference and calculations, it is convenient to drop the subscripts and speak of the total electric potential energy stored in a capacitor having a charge q and a potential difference ΔV between its terminals. This is given by several equivalent expressions:

★ $$EPE = \frac{1}{2} q\,\Delta V = \frac{1}{2} q^2/C = \frac{1}{2} C\,\Delta V^2.$$

We have computed this energy as the work done to bring charge to the plates. In the future it will often be useful to regard this energy as being stored in the electric field of the capacitor.

Questions and Exercises

Q 57.5-1 When you charge a capacitor in the laboratory, you don't literally carry charge from one plate to the other. What does cause charge to move from one plate to the other? Is it correct to say that work is done by this agent in charging the capacitor?

Q 57.5-2 When a capacitor is partially charged, the charge already on the plates tends to repel the next bit of charge which is approaching the plates. Why does additional charge flow on to the plates against this repulsive force? What determines the state at which the capacitor is "fully" charged?

Q 57.5-3 How much work is required to charge a capacitor to a total charge q_t, if the *positive* plate is kept at ground potential during the charging process?

Q 57.5-4 Is it more appropriate to refer to the energy in a capacitor as potential energy, or as kinetic energy? Give a reason for your answer.

E 57.5-1 Consider a capacitor, initially uncharged, which is then charged to a total charge Q_1, a final voltage ΔV_1, and a total stored energy E_1. Choose the horizontal axis to be the charge axis, and sketch qualitatively correct graphs which describe the charging process as follows:
(a) capacitor voltage vs. charge;
(b) capacitance vs. charge; and
(c) stored energy vs. charge.

E 57.5-2 For the process described in E 57.3-6 what will be the effect on the energy of the capacitor? If the energy changes, what happens to the difference in energy?

E 57.5-3 For the process described in E 57.3-8, what will be the effect on the energy stored in the capacitor? If the energy changes, how can you account for the difference in energy?

E 57.5-4 For the process described in E 57.4-2, what is the change in energy of the capacitor when the glass is inserted? Account for this change in energy.

E 57.5-5 For the process described in E 57.4-3, what is the change in energy of the capacitor? Account for this change in energy.

E 57.5-6 Don wants to make a capacitor using two flat aluminum plates which are separated by an air layer 2 mm thick.
(a) What plate area is required to obtain a capacitance of 1.0 pF?
(b) What charge will the plates acquire at a voltage of 4000 volts?
(c) How much energy is stored in this capacitor when the voltage is 4000 V?
(d) If the plates are charged as in (b) and then disconnected from the battery, and a 2-mm layer of glass having $\kappa = 4.5$ is then inserted between the plates, what is the new value of C?
(e) What is the new value of charge?
(f) What is the new capacitor voltage?
(g) What is the new value of stored energy?

Check your understanding of the material above by testing yourself on the minimum learning goal at the beginning of this section.

57.6 Capacitors Connected in Series or Parallel

Minimum Learning Goals. (1) When you are given the values of several capacitances and a schematic of their locations in a circuit in which they are connected in series and/or parallel, be able to calculate the equivalent capacitance between any two points in the circuit. For example:

1 What is the equivalent capacitance between points *A* and *B* in the following circuit? (Answer, page 57-25, Line 15.)

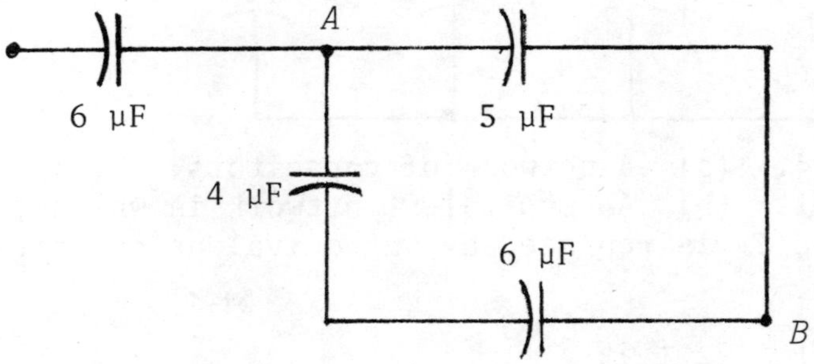

2 What is the equivalent capacitance between points *A* and *B* in the following circuit? (Answer, page 57-24, Line 2.)

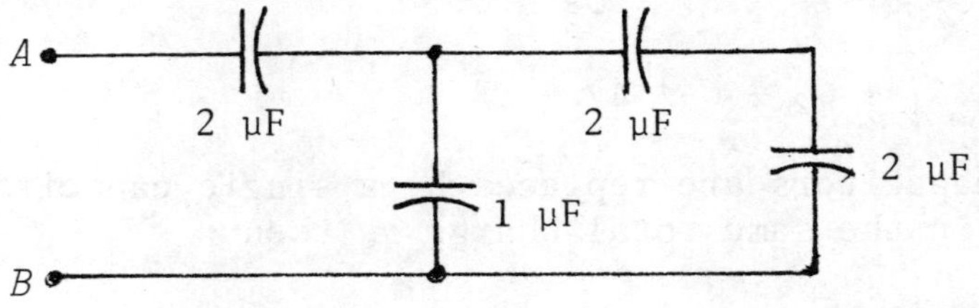

(2) When you are given values of capacitance for three to five capacitors, be able to draw a schematic diagram which shows how they could be connected to produce a specified equivalent capacitance. For example:

1 How could a 2-μF, a 5-μF, and a 7-μF capacitor be connected to produce an equivalent capacitance of 3.5 μF? (Answer, page 57-24, Line 5.)

2 How could one 5-μF, two 6-μF and two 4-μF capacitors be connected to produce an equivalent capacitance of 2.50 μF? (Answer, page 57-25, Line 11.)

Some electric circuits contain capacitors connected in series, in parallel, or both. In order to compute currents and voltages in such a circuit, we need a method for finding the equivalent capacitance of any part of the circuit and of the network as a whole. We begin by considering separately a set of capacitors in series, and then a set of capacitors in parallel.

When capacitors are connected in parallel, as in Figure 57-4, the same voltage ΔV exists across each capacitor. Then the charge on C_1 is $q_1 = C_1\,\Delta V$, the charge on C_2 is $q_2 = C_2\,\Delta V$ and the charge on C_3 is $q_3 = C_3\,\Delta V$. The sum of these charges must equal the total charge supplied by the source:

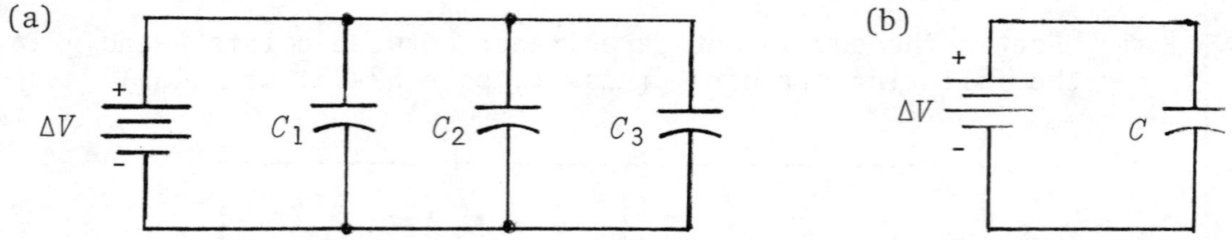

Figure 57-4. (a) A network of capacitors, C_1, C_2, and C_3 connected in parallel. (b) An equivalent network in which the combination of C_1, C_2, and C_3 is replaced by an equivalent capacitance, C.

$$q = q_1 + q_2 + q_3.$$

Substituting in this equation, we obtain

$$q = C_1\ \Delta V + C_2\ \Delta V + C_3\ \Delta V$$

$$= (C_1 + C_2 + C_3)\ \Delta V.$$

If the three capacitors are replaced by a single capacitor C, having a value which results in the same total charge q, then

$$q = C\ \Delta V$$

where the equivalent capacitance

★ $$C = C_1 + C_2 + C_3 \quad \text{for capacitors in parallel.}$$

This capacitance is *equivalent* to the parallel combination of C_1, C_2, and C_3 because it results in the same total charge on the single capacitor. In a circuit (or branch of a circuit) containing capacitors in parallel, the voltage is the same across each capacitor but the charges on the capacitors add to equal the total charge supplied to these capacitors by the source.

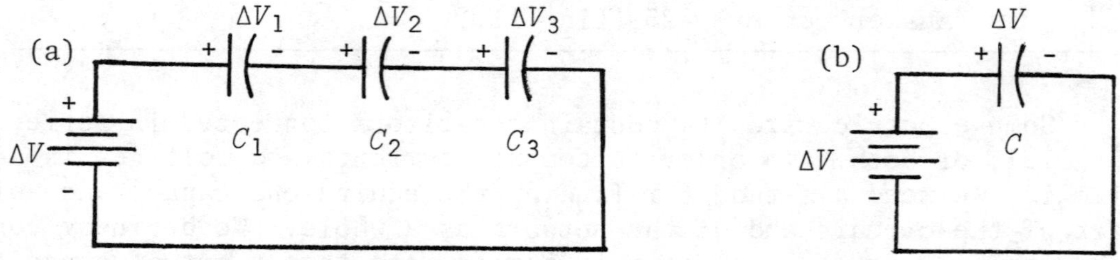

Figure 57-5. (a) A network of capacitors C_1, C_2, and C_3 connected in series. (b) An equivalent network in which the combination of C_1, C_2, and C_3 is replaced by an equivalent capacitance, C.

When capacitors are connected in series, as in Figure 57-5, it is the charge which is the same for each capacitor (assuming the capacitors were uncharged before being connected to the source), because the capacitors become charged simultaneously. Then the voltage across C_1 is $\Delta V_1 = q/C_1$, the voltage across C_2 is $\Delta V_2 = q/C_2$, and the voltage across C_3 is $\Delta V_3 = q/C_3$. The sum of these voltages must equal the potential difference supplied by the source:

$$\Delta V = \Delta V_1 + \Delta V_2 + \Delta V_3.$$

Substituting in this equation, we obtain

$$\Delta V = \frac{q}{C_1} + \frac{q}{C_2} + \frac{q}{C_3}.$$

If the three capacitors are replaced by a single capacitor, C, having a value which results in the same total charge q, then

$$\Delta V = \frac{q}{C} = q\left(\frac{1}{C_1} + \frac{1}{C_2} + \frac{1}{C_3}\right),$$

where the equivalent capacitance is given by

★ $$\frac{1}{C} = \frac{1}{C_1} + \frac{1}{C_2} + \frac{1}{C_3} \quad \text{for capacitors in series.}$$

In a circuit containing capacitors in series, the charge on each capacitor is the same but the voltages across the capacitors add to equal the total voltage supplied by the source.

The equations above for computing the capacitance of a single capacitor equivalent to three capacitors in series, or in parallel, may easily be extended to cases where there are more than three capacitors. To find the equivalent capacitance of a more complicated network, in which some capacitors are connected in parallel with each other while others are connected in series with each other, it is necessary to proceed one step at a time. At each step, a group of series capacitors is reduced to a single equivalent capacitor, *or* a group of parallel capacitors is reduced to a single equivalent capacitor. One should not try to include a capacitor as part of a parallel group and a series group in the same step of the solution. In solving a complicated network, it is advisable to begin with the part which is farthest from the terminals of the network or from the source of electromotance.

Question and Exercises

Q 57.6-1 This field map shows field lines for 3 conductors.

(a) Which conductor is at the highest potential?

(b) Which conductor is at the lowest potential?

(c) Draw an electric circuit in which the capacitances among these conductors are represented by a network of capacitors. Label your circuit to show clearly which portions correspond respectively to the conductors *A*, *B*, and *C*.

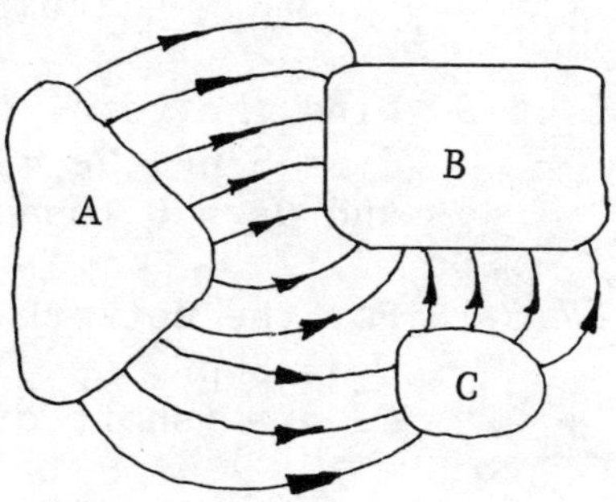

E 57.6-1 Three capacitors connected in series have capacitances of 2 nF, 4 nF, and 6 nF, respectively. What is the equivalent capacitance?

E 57.6-2 Three capacitors connected in series have capacitances of 3 nF, 4 nF, and 5 nF, respectively. What is the equivalent capacitance?

E 57.6-3 Three capacitors connected in parallel have capacitances of 2 nF, 4 nF, and 6 nF, respectively. What is the equivalent capacitance?

E 57.6-4 Three capacitors connected in parallel have capacitances of 3 nF, 4 nF, and 5 nF, respectively. What is the equivalent capacitance?

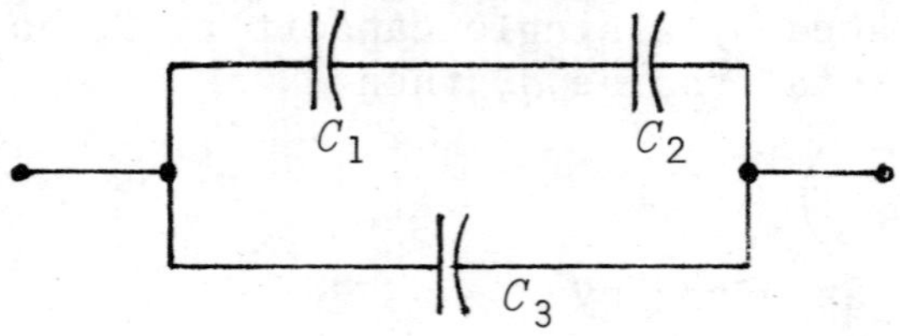

Circuit diagram for E 57.6-5 and E 57.6-6

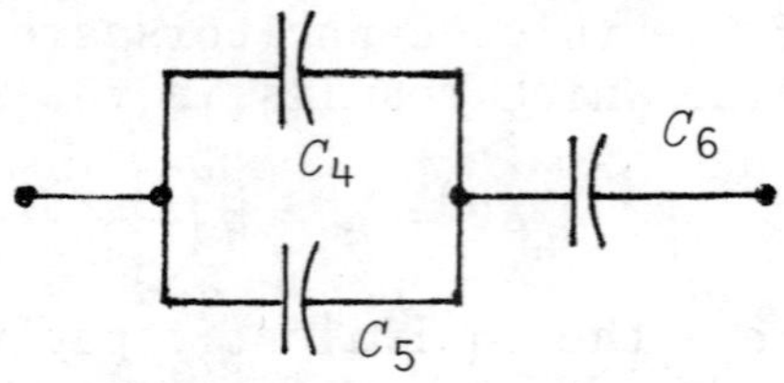

Circuit diagram for E 57.6-7 and 57.6-8

E 57.6-5 In the diagram above, C_1 = 2 nF, C_2 = 4 nF, and C_3 = 6 nF. What is the equivalent capacitance of the network?

E 57.6-6 In the diagram above, C_1 = 3 nF, C_2 = 4 nF, and C_3 = 5 nF. What is the equivalent capacitance of the network?

E 57.6-7 In the diagram above, C_4 = 2 nF, C_5 = 4 nF, and C_6 = 6 nF. What is the equivalent capacitance of the network?

E 57.6-8 In the diagram above, C_4 = 3 nF, C_5 = 4 nF, and C_6 = 5 nF. What is the equivalent capacitance of the network?

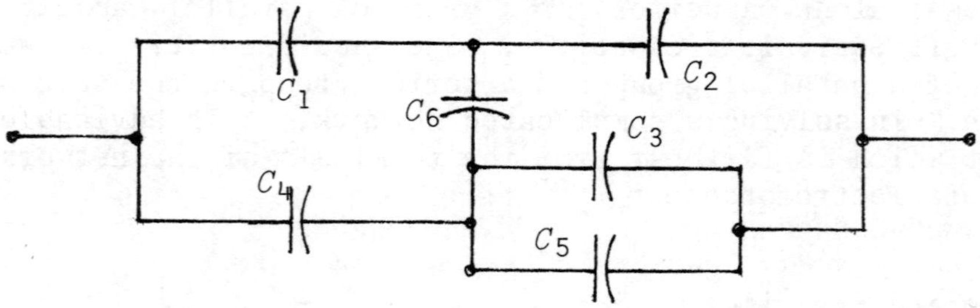

Circuit diagram for E 57.6-9 and 57.6-10

E 57.6-9 Find the total capacitance of the network shown above, if C_1 = 5 pF, C_2 = 20 pF, C_3 = 2 pF, C_4 = 6 pF, C_5 = 4 pF, and C_6 = 0 (open circuit).

E 57.6-10 For the network shown above, find the total capacitance if C_1 = 5 pF, C_2 = 20 pF, C_3 = 2 pF, C_4 = 6 pF, C_5 = 4 pF, and $C_6 = \infty$ (short circuit).

E 57.6-11 Many radios include a tuning capacitor made of interleaved parallel plates which are connected as shown here. The capacitance is varied by rotating one set of plates (connected to the rotor *R*) into the space between the other set of fixed plates (connected to the stator *S*). Consider a tuning capacitor in which the effective area (when *R* is fully rotated into *S*) of each side of each stator plate is 8 cm^2, and the distance between adjacent stator and rotor plates is 1.0 mm.

(a) If the stator *S* has two plates and the rotor *R* has one plate, positioned midway between the two stator plates, what is the total capacitance *C* of the capacitor?
(b) If *S* has 6 plates and *R* has 5 plates, what is the total *C*?
(c) If the capacitor of part (b) has *R* rotated out of *S* to the position where the total *C* is only one tenth of the maximum value, and the capacitor is then connected to a 90-volt battery, what is the total charge on *C*?
(d) If the capacitor remains connected to the battery while *R* is rotated into the position of maximum capacitance, how much charge flows in the circuit, and where does it go?
(e) Discuss the energy changes which occur during the rotation of part (d).

Check your understanding of the material above by testing yourself on the minimum learning goals at the beginning of this section.

57.7 Transient Behavior of a Capacitor. Time Constant

Minimum Learning Goal. Be able to calculate the time constant of a given circuit containing resistance and capacitance, and state the physical significance of this time constant. For example:

1 A circuit has a 12-volt battery and a capacitor with a value of 2×10^{-7} F. The resistance of the circuit is 4 ohms. When a switch connected across the capacitor is closed the capacitor is short-circuited, and when it is open the capacitor begins to charge. What is the time constant of the circuit after the switch is opened, and what is the significance of this constant? (Answer, page 57-24, Line 6.)

2 In electronics, a combination of resistance and capacitance (as shown below) is called a differentiating circuit if the duration of the pulse to be differentiated is much greater (say 100 times greater) than the time constant. When periodic pulses having a given shape are fed into this circuit, the output

pulses have the time-dependence of the derivative of the input pulses. For input pulses of 10-millisecond duration, what maximum value of resistance could be used in the circuit shown here to produce a differentiating circuit? (Answer, page 57-24, Line 3.)

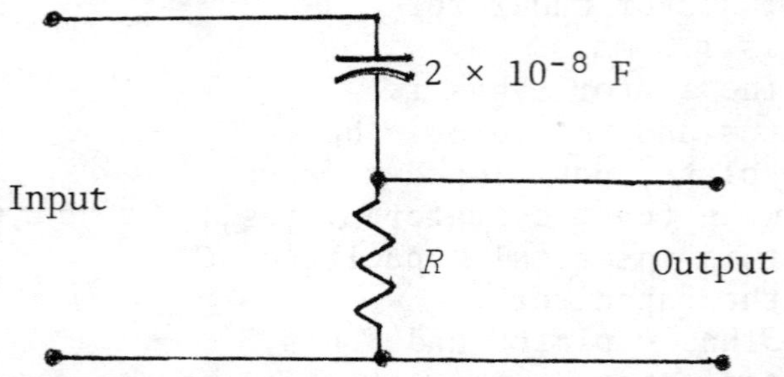

In considering charge and voltage in circuits having capacitors (Section 57.6), we have been speaking of the steady state in which charge and voltage remain constant with time. While a capacitor is being charged or discharged, however, these quantities vary with time in a definite way. How do they vary, and how can we account for this time-dependence?

When an uncharged capacitor is connected to a battery, measurements taken while the capacitor is being charged show that the charging current and the charge acquired vary with time as shown in Figure 57.6. In the circuit diagram, R represents the combined resistance of the battery, the connecting wires, and any resistors which may be in the series circuit. We use $q(t)$ to represent the charge (as a function of time) on the capacitor, and $i(t)$ to represent the charging current (as a function of time). While the switch S remains open, $i = 0$ and $q = 0$. If S is closed at $t = 0$,

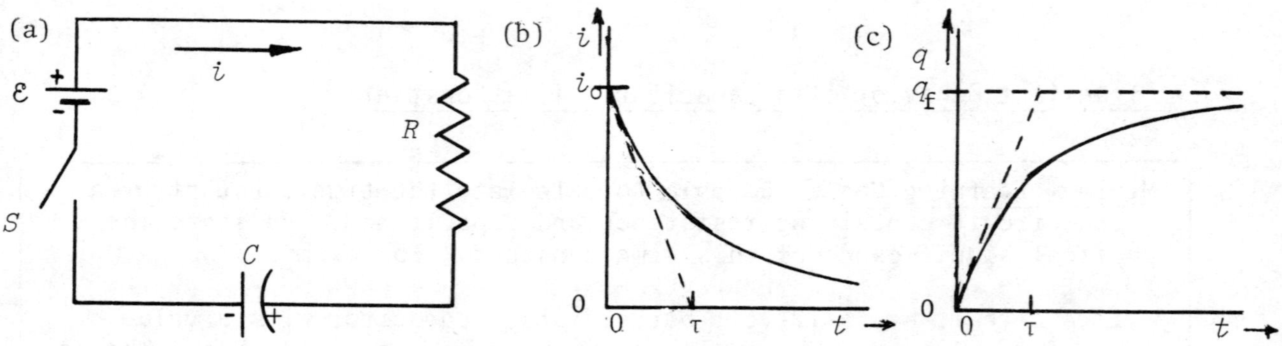

Figure 57.6. (a) A circuit used to charge a capacitor which initially is uncharged. When the switch S is closed (at $t = 0$), current from the battery delivers charge to the capacitor until it is fully charged. (b) Beginning with an initial value i_0, the current decreases with time and approaches zero as the charge approaches its maximum value. (c) The charge grows rapidly at first, but the rate decreases as the charge approaches its maximum value. Both the current and charge have an exponential variation with time.

measurements yield graphs of $i(t)$ and $q(t)$ like those in Figure 57-6 (b) and (c).

Just after the switch is closed there is no charge on the capacitor and therefore no voltage across the capacitor, so that the total voltage supplied by the battery is across the resistance, leading to an initial current $i_o = \xi/R$. This current begins to charge the capacitor, which thus acquires a voltage $\Delta V_c = q/C$. As q increases, ΔV_c also increases so that the potential difference $\xi - \Delta V_c$ available to drive current through the resistance decreases, and therefore the current decreases. The current, by charging up the capacitor, "bites off the voltage that feeds it." As the current diminishes, the rate at which charge builds up also diminishes, thus decreasing the rate at which ΔV_c grows and therefore decreasing the *rate* at which the current drops off. In this way we can account qualitatively for the variation of $i(t)$ and $q(t)$ shown in Figure 57-6. Eventually, the charge approaches a "final" value q_f while the current approaches zero; in both cases these values are approached asymptotically.

In order to acquire a voltage equal to the full electromotance ξ, the capacitor must obtain an amount of charge $q_f = C\xi$. *If* the charge flow continued at its initial rate $i_o = \xi/R$, the time necessary to charge the capacitor would be given by

$$t = q_f/i_o$$

$$= \frac{C\xi}{\xi/R} = RC,$$

independent of the final voltage $\Delta V_c = \xi$ of the capacitor. The quantity $RC \equiv \tau$ (Greek, tau) thus emerges as a natural *time constant* of an *RC* series circuit. The time constant τ is indicated on the graphs in Figure 57-6. The SI unit of the product *RC* is

$$\text{ohm}\cdot\text{farad} = \frac{\cancel{\text{volt}}}{\text{ampere}} \cdot \frac{\text{coulomb}}{\cancel{\text{volt}}} = \frac{\cancel{\text{coulomb}}}{\cancel{\text{coulomb}}/\text{second}} = \text{second}.$$

This built-in time constant of an *RC* series circuit is analogous to the time constant $\tau = 1/\lambda$ in radioactive decay (Section 34.7). As shown in Section 60.4, the curves in Figure 57-6 (b) and (c) have the mathematical form of exponential equations. The quantity τ specifies the time required for the current to die down from any particular value to one which is smaller by a factor of e ($e = 2.71828...$ is the base of natural logarithms), that is to a value of 0.368 times the initial value.

Our study of capacitance began with electrostatic concepts of potential difference and electric field lines, and has led to electric circuits containing capacitors. Further study of the role of capacitance in electric circuits is taken up in Chapter 60.

Questions and Exercises

Q 57.7-1 Can a capacitor be charged instantaneously if no resistors are in the circuit? Explain.

Q 57.7-2 If a capacitor is initially charged and is not connected to anything else, will the capacitor retain its charge forever? Explain.

Q 57.7-3 For an *RC* circuit, make a reasonable estimate of the shortest practicable time constant which can be achieved.

Q 57.7-4 Describe a method of measuring the charge on a capacitor as a function of time.

E 57.7-1 An initially-uncharged capacitor (C = 2 μF) is charged through a resistor $R = 10^6$ ohms, using a 12-volt battery.
(a) Find the time constant of this circuit.
(b) What is the initial value of charging current?
(c) What is the final value of the charge acquired?

Check your understanding of the material above by testing yourself on the minimum learning goal at the beginning of this section.

<u>Answers to Minimum Learning Goals</u>

Line 1 1.25×10^{-13} F

Line 2 1 μF

Line 3 $R \leq 5{,}000$ ohms

Line 4 2.16×10^{-10} F

Line 5

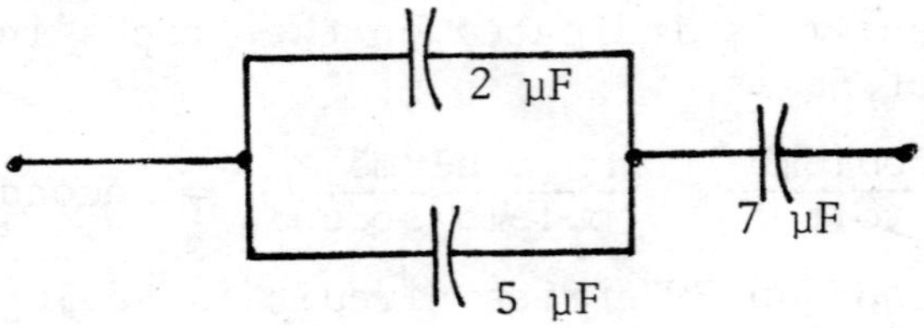

Line 6 8×10^{-7} s. This is the time required to charge the capacitor in the circuit to 63% of its final charge.

Line 7 (a) 1.11×10^{-13} F

Line 8 (c) $mg\ \Delta h \leq \frac{1}{2}\ C\ \Delta V^2$

Line 9 3

Line 10 6×10^{-12} F

Line 11

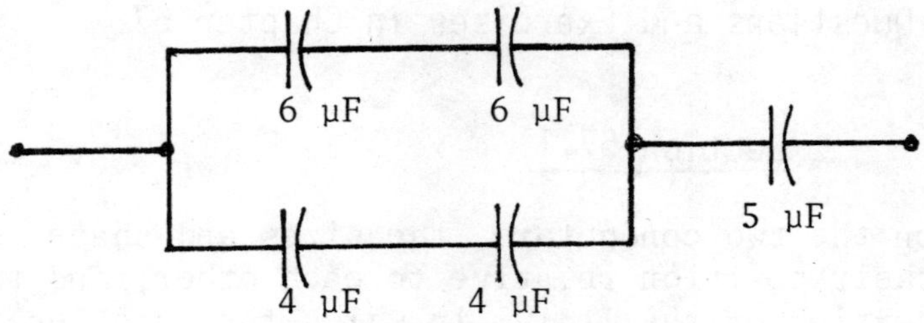

Line 12 (c) $mg\ \Delta h \leq \frac{1}{2}\ C\ \Delta V^2$

Line 13 (e) 5.6×10^{-12} F

Line 14 2.1

Line 15 7.4 μF

Line 16 5×10^{-12} F

Line 17 Description (a) needs information on the medium between the plates. Description (b) needs the separation of the plates, and their surface area.

Line 18 Description (a) is complete. Description (b) needs information on the geometry of the conductors and on their relative orientations in space.

ANSWERS to Questions and Exercises in Chapter 57

Section 57.1

Q 57.1-2 The charges on the two conductors, the sizes and shapes of the conductors, their position relative to each other, and the electric properties of the medium in which they are located.

Q 57.1-4 The farad; one farad equals one coulomb per volt.

E 57.1-2 10^{-7} C

Section 57.2

Q 57.2-2 The electric force law causes the individual charge carriers to move as far from each other as they can, while remaining on the sphere. This results in a uniform charge distribution on the surface.

E 57.2-2 7.2×10^{-4} F

Section 57.3

Q 57.3-2

E 57.3-2 (a) 0.64 mm; (b) 45 V; (c) decrease; zero

E 57.3-4 17.7 pF

E 57.3-6 (a) Charge remains the same; (b) voltage doubles; (c) capacitance decreases by one half.

E 57.3-8 (a) voltage remains the same; (b) capacitance decreases by one half; (c) charge decreases by one half.

Section 57.4

Q 57.4-2 It is less than the diameter of an atom, because each displaced proton or electron remains bound to its atom.

Q 57.4-4 The polarization charge remains bound to the dielectric, and is not free to flow to the plate.

Q 57.4-6 No. If that could happen, there would be no field lines and therefore no electric field strength in the dielectric.

E 57.4-2 (a) 3.6×10^{-7} C; (b) 135 nF ; (c) 12 V ; (d) 1.62×10^{-6} C; (e) -1.26×10^{-6} C.

E 57.4-4 (a) 2.26×10^{4} V/m; (b) 1.36×10^{2} V ; (c) 1.13×10^{5} V/m.

E 57.4-6 (a) the same; (b) $q_2 = \kappa q_1$; (c) $E_2 = E_1$; (d) $C_2 = \kappa C_1$.

Section 57.5

Q 57.5-2 Additional charge flows to the plates because the electric force supplied by the battery exceeds the repulsive force of the charge on the plates. The capacitor is fully charged when its potential difference equals that of the battery.

Q 57.5-4 It is electric potential energy, because it is not energy of motion but rather is stored energy.

E 57.5-2 Energy doubles. The additional energy is supplied by mechanical work done while moving the capacitor plates.

E 57.5-4 $+7.56 \times 10^{-6}$ J. Charge and energy flow from the battery to the capacitor while the glass is being inserted.

E 57.5-6 (a) 2.3×10^{-4} m^2; (b) 4 nC; (c) 8×10^{-6} J; (d) 4.5 pF; (e) 4 nC; (f) 889 V; (g) 1.8×10^{-6} J

Section 57.6

E 57.6-2 1.28 nF

E 57.6-4 12 nF

E 57.6-6 6.71 nF

E 57.6-8 2.92 nF

E 57.6-10 7.8 pF

Section 57.7

Q 57.7-2 No, because air conducts electric current, so the capacitor will discharge through the air at a rate dependent on the air's conductivity.

Q 57.7-4 Measure the voltage as a function of time using a cathode-ray oscilloscope; compute $q(t)$ using $q = C\,\Delta V$.

[illegible]

CHAPTER 58 MAGNETIC FIELDS

Introduction

Magnetic fields are described using a field vector called the magnetic induction, defined in terms of the force it exerts on a moving charged particle. A magnetic field map consists of magnetic field lines which indicate the direction of the induction at every point. The sources of magnetic fields are electric currents, which may be macroscopic currents in wires or microscopic currents within atoms. If the current which produces a magnetic field is known, Ampere's law enables us to compute the field easily in situations having a high spatial symmetry, such as the field near a long straight wire and the field in a long solenoid.

When you have completed your study of the material in this chapter, you should be able to answer questions and work exercises like the examples in the minimum learning goals assigned by your instructor.

58.1 Electricity and Magnetism

Minimum Learning Goal. Given a description of an experiment involving electric and magnetic phenomena, be able to state whether the electric and magnetic fields are coupled, or uncoupled. For example:

1 A small charged sphere on the end of an insulating rod is held motionless in a constant magnetic field, and no force is exerted on the sphere by the field. Is the electric field of the charged sphere coupled to the magnetic field? (Answer, page 58-36, Line 2)

2 A length of conducting wire is formed into a closed circular loop, with its ends soldered together. While a magnet is being brought up to the loop in a certain way, a current exists in the wire. Is the magnetic field coupled to the electric field inside the wire? (Answer, page 58-36, Line 8)

Some electric and magnetic effects have been known for thousands of years. Until 1820, one would have said that electricity and magnetism have many analogous properties, but that there is no *dynamical* connection between them. In other words, it was thought that one could do any experiment on magnetism without involving electric phenomena, and *vice versa*.

Today we know that is not true. Rather, electric and magnetic fields are coupled. This means that *time-varying* magnetic fields are always accompanied by electric fields, and *time-varying* electric fields are always associated with magnetic fields. The reason this was not known earlier was that investigation had been more or less limited to static phenomena (not changing in time). Under static conditions, electric and magnetic fields can be viewed as separate entities. But when conditions are changing with time, it makes more sense to regard the two as manifestations of the same dynamical entity, the *electromagnetic field.*

In Chapter 56 we studied the electrostatic field. In this chapter we shall first describe some properties of a *magnetostatic field* -- that is, a magnetic field due to a configuration of currents that is constant in time -- and then study the modifications that take place when time-variable phenomena are introduced.

Questions

Q 58.1-1 The geomagnetic field varies from one place to another on the surface of the earth. Is an electric field associated with this changing magnetic field? Explain.

Q 58.1-2 In the region around a charged pith ball, the electric field strength varies with distance from the center of the ball. Is a magnetic field associated with this changing field? Explain.

Q 58.1-3 If small ceramic permanent magnets are taped to the skirts of an air glider and the glider is then allowed to move down an inclined air track, what kind of field (if any) would one expect to be produced as a result of this motion?

Q 58.1-4 If a charged pith ball is set into oscillation as a pendulum, what kind of field (if any) would you expect to be produced as a result of this motion?

Q 58.1-5 Is there any dynamical connection between electricity and gravitation? For example, is there any evidence that a changing gravitational field is accompanied by an electric field, or that a changing electric field is accompanied by a gravitational field?

Check your understanding of the material above by testing yourself on the minimum learning goal at the beginning of this section.

58.2 Magnets and Magnetic Fields

Minimum Learning Goal. Be able to draw correctly some lines of a magnetic field when you are given a drawing of a permanent magnet which shows the poles of the magnet. For example:

1 Draw lines for a magnetic field outside the magnet shown below. (Answer, page 58-36, Line 9)

Our introduction to magnetism differs from the development we used in studying electricity. There, we first studied electric forces and the electric force law, and then the alternative description using the concept of electric field. In the study of magnetism, it is more instructive to do the reverse: first learn some things about the magnetic field, and then obtain the equation for the magnetic force.

When small objects cling to an iron magnet they tend to cluster in two clumps, one at each end of the magnet. The points on the magnet where the clumps are centered are called the *poles* of the magnet, and it is customary to refer to them as the *north pole* and the *south pole*.

Out of doors away from the steel of cars and buildings, the needle of a magnetic compass indicates the direction of the magnetic north. If the compass is tipped or rotated for a short time the needle is turned away from the magnetic north, but if the compass is then held steady and level the needle quickly swings back to the north. At any place on earth except the magnetic poles, the compass needle turns to the direction determined by the earth's magnetism. The compass thus serves as a suitable probe for investigating the earth's magnetism; wherever the needle is placed, a magnetic force aligns it in the direction dictated by the earth. By analogy with electric fields, we say that the earth gives rise to a certain condition in the space around it, which we call the *magnetic field* of the earth. The magnetic forces on the compass needle are the forces exerted by the magnetic field at the location of the needle.

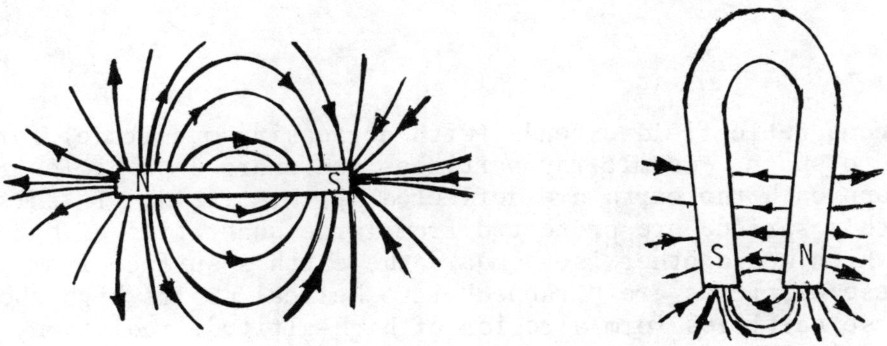

Figure 58-1 Patterns of magnetic field lines for two types of permanent magnets. The direction of the magnetic induction vector at any point is tangent to the field line through that point, and is directed from the north pole to the south pole of the magnet.

The symbol used by physicists for the strength of a magnetic field is $\vec{B}$, and the name given to this quantity is the *magnetic induction*. Every magnet produces its own magnetic field, and the pattern of this field can be indicated by drawing some magnetic field lines. These lines are usually labeled with arrows to indicate the direction of the field. The direction of the magnetic induction vector at any point in the field is tangent to the field line through that point, in the direction shown by the arrow. Portions of the field patterns produced by an iron bar magnet and by a horseshoe magnet are shown in Figure 58-1. Note that in the region outside the magnet the direction assigned to $\vec{B}$ is always from the north pole to the south pole of the magnet.

The direction of the $\vec{B}$ vector associated with the earth's magnetic field is in a northerly direction at most places on the earth. The distribution of the earth's magnetic field is approximately what we might expect if a huge bar magnet were embedded within the earth. Places where the magnetic field is horizontal lie on the *magnetic equator*, and the places where it is exactly vertical are the *magnetic poles* (see Figure 58-2). When examined closely, however, the earth's field is found to be rather irregular. The magnetic equator is south of the geographic equator in South America, and north of it in Africa, Asia, and the western Pacific.

Figure 58-2. A schematic drawing of the earth's magnetic field, showing the magnetic equator, the magnetic poles, and magnetic field lines in space. N and S indicate the geographic poles. The two figures which resemble crowns indicate the regions where the aurora is formed. Photo courtesy of National Aeronautics and Space Administration.

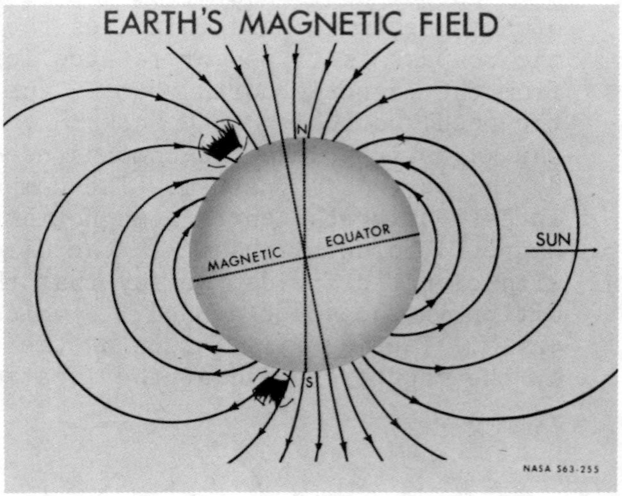

The geomagnetic field extends (with decreasing magnitude) for thousands of miles into space. Cosmic-ray particles and charged particles from the sun which approach the earth are deflected by this field. In this way, we on the earth's surface are protected from large quantities of hazardous radiation which would otherwise bombard the earth's surface from space. Some of these particles are "trapped" into helical orbits high above the earth. These particles form a region of high-altitude radiation, called the Van Allen radiation zone, which surrounds the earth except near the magnetic poles.

The magnetic storms on the sun which cause marked fluctuations in the earth's magnetic field are sometimes so intense that terrestrial communications by radio, telephone, and television are disrupted. The same disturbances, however, often trigger some of the most beautiful and awe-inspiring spectacles in nature, the *aurora borealis* or "northern lights" in northern latitudes, and the *aurora australis* in the southern hemisphere (Section 58.5).

One might suppose that the magnetic poles of an iron magnet would offer a suitable starting point for studying the fundamental nature of magnetism. When a bar magnet is broken or cut into two pieces, however, one finds that *each* piece of the magnet actually contains two poles: a north pole and a south pole. No one has found a way to separate a magnet and obtain an individual north or south pole. Every magnet is a *magnetic dipole*. In recent years there have been extensive and sophisticated searches for *magnetic monopoles*, but thus far no monopoles have been found. The study of magnetic poles is not a fruitful approach to understanding the fundamental nature of magnetism; it is much better to study the relation of magnetism to the motions of electric charges.

Questions and Exercises

Q 58.2-1 (a) List some examples of magnets which you meet in your daily experience.
(b) List some examples of magnetic forces which you meet in your daily experience.

Q 58.2-2 A horseshoe magnet has a north pole and a south pole. What happens to these poles when the magnet is broken into two pieces?

E 58.2-1 Draw some lines to show the magnetic field of a horseshoe magnet. Include arrows to show the directions of the lines.

E 58.2-2 Draw some lines to show the magnetic field of a bar magnet. Draw arrows to show the directions of the field lines.

Check your understanding of the material above by testing yourself on the minimum learning goal at the beginning of this section.

58.3 Magnetic Flux

Minimum Learning Goals. Given a diagram containing field lines which represent a field in a plane region of space, be able to state whether the field could be an electrostatic field, a magnetostatic field, or neither one. For example:

1 The lines in the following diagrams are field lines. Electric charges may be present in the region of each field pattern, but they are not shown. Which field diagram(s) could represent an electrostatic field? Which diagram(s) could represent a magnetostatic field? (Answer, page 58-36 Line 11)

(a)

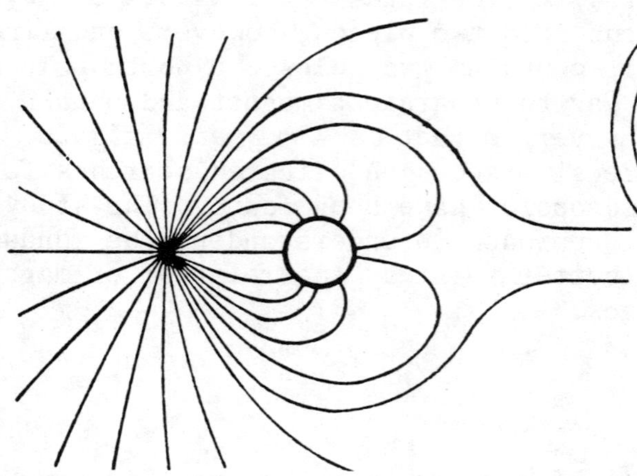

(b)

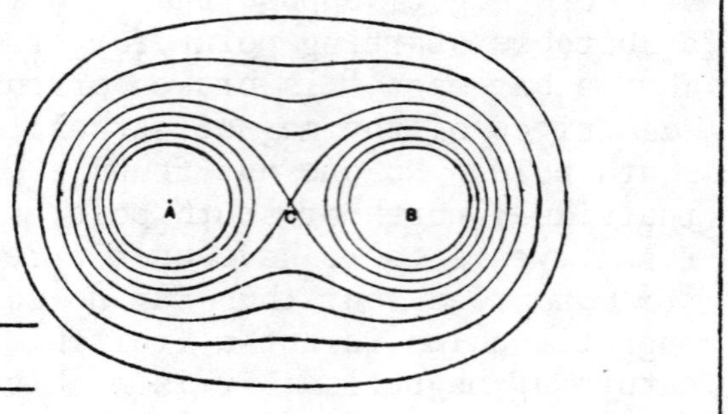

(c)

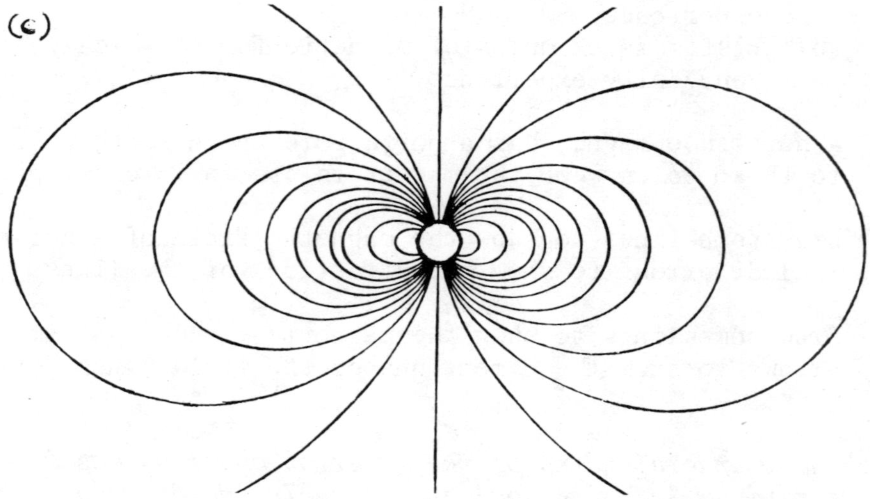

2 The arrows in the following two figures are vectors which represent fields having these magnitudes and directions at the points where the vectors originate. Electric charges may be present but are not shown. Which diagram could represent an electrostatic field, and which a magnetostatic field? (Answer, p. 58-36, Line 4.)

(a)

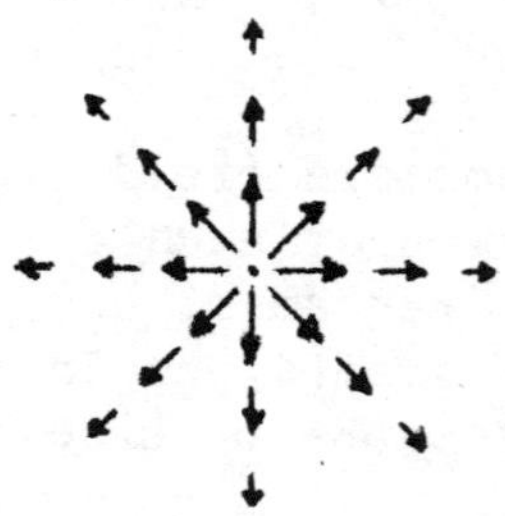

(b)

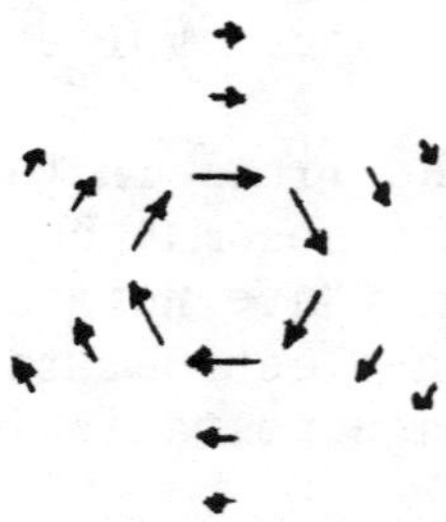

(2) Given an expression for the magnetic induction over a specified surface, be able to calculate the magnetic flux through that surface. For example:

((1 The circular pole faces of a laboratory electromagnet are parallel and are 3 cm apart, with radii of 5 cm. On a surface parallel to the pole faces and midway between them, the magnetic induction at a distance r from the axis has a direction normal to the surface and a magnitude in teslas given by the equation,

$$B = 0.25\ (1 - 100\ r^2) \quad (\text{valid for } r \leq 0.05\ \text{m}),$$

where r is the radius in meters measured from the axis. What is the magnetic flux through a circular surface of radius 5 cm, midway between the pole faces and centered on the axis? (Answer, page 58-36, Line 14).))

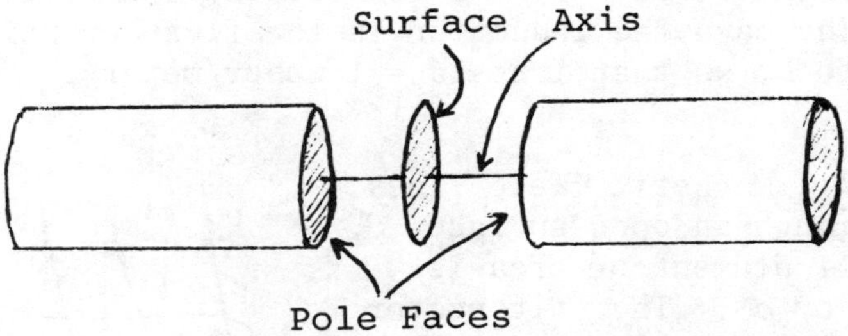

2 The magnetic induction of the earth (in teslas) at a certain geographical location is given by the expression

$$\vec{B} = 10^{-5}\ (0.16\hat{x} + 1.81\hat{y} - 4.98\hat{x}),$$

where the direction of $\hat{x}$ is east, $\hat{y}$ is north, and $\hat{z}$ is upward. What is the total magnetic flux passing downward through a level, rectangular surface measuring 1.5 meters by 1.3 meters? (Answer, page 58-36, Line 7)

Following the qualitative description given above, we now begin a quantitative discussion of the magnetic field. Mathematically, it can be shown that any vector field is uniquely determined when two rules are specified: (a) the rule that states where the field lines begin and end, and (b) the rule that gives the value of the line integral of the field around any closed path.

In the case of physical fields, these rules must be empirically determined for each field separately. For electrostatic fields, rule (a) is Gauss's law: Electric field lines originate on positive charges and end on negative charges (Section 56.3). For magnetic fields, on the other hand, the corresponding rule is that field lines *never* begin or end anywhere; that is, they always close on themselves. Each magnetic field line is a continuous closed loop. In Figures 58-1 and 58-2 the field lines appear to begin and end at the surfaces of the magnets and of the earth, but in actuality they continue within the magnets and the earth to form continuous closed loops or lines. Figure 58-6 on Page 58-19 shows the magnetic field lines resulting from current in a straight wire; each field line is a closed loop surrounding the wire.

To express this property of magnetic fields in equation form, we introduce the concept of *magnetic flux*. Figure 58-3 shows an open surface (Section 56.2) S in a magnetic field. We use dS to designate a very small (infinitesimal) element of surface area on S, and the unit normal vector $\hat{n}$ to indicate the direction perpendicular to dS. There is some value of magnetic induction $\vec{B}$ at every point on the surface. The component of $\vec{B}$ normal to the surface in the direction of $\hat{n}$ is given by the dot product $\vec{B} \cdot \hat{n}$. (Dot products are discussed in Section 47.1.)

★ The *magnetic flux* through dS in the sense of $\hat{n}$ is $d\Phi_m \equiv \vec{B} \cdot \hat{n}\, dS$.

Magnetic flux is a scalar quantity; its SI unit is the *weber*, symbol Wb. The SI Unit of magnetic induction is the *tesla*, symbol T, defined in Section 58.4. The magnetic induction is the flux per unit area through a surface normal to $\vec{B}$, so that 1 tesla = 1 weber/meter^2.

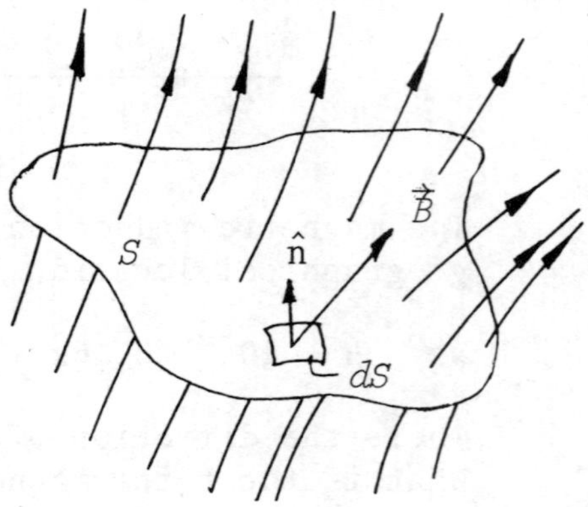

Figure 58-3. Magnetic field lines passing through an open surface, S. A very small element of area is designated by dS. The unit vector directed perpendicular to dS is labeled $\hat{n}$. The magnetic induction at dS is labeled $\vec{B}$.

To obtain the total magnetic flux through the surface S, it is necessary to add the contributions to this flux which pass through all the area elements of the surface. Magnetic flux is a scalar quantity, so the total flux is obtained simply by summing the small elements of flux through all the area elements. For a surface which is everywhere perpendicular to the magnetic field lines and has an area A, the total magnetic flux through the surface is given by

$$\Phi_m = B\,A,$$

where $\vec{B}$ is the *average* magnetic induction at the surface. Where the surface is a plane but the field lines are not perpendicular to the surface, the flux is computed using the component of $\vec{B}$ normal to the surface, $B_\perp = \vec{B} \cdot \hat{n}$. In this case, the total flux is given by

★ $$\Phi_m = \vec{B} \cdot \hat{n}\,A,$$

where once again $\vec{B}$ is the average induction at the surface. If the magnetic induction varies over the surface the total flux is computed using a double integral sign, $\iint$, as a summation symbol. The double integral indicates that the summation is performed over the full range of two variables, and corresponds to the fact that dS is a two-dimensional element of integration. That is, the summation or integration of the small elements of flux is carried out over the full length and full width of the surface.

The *total magnetic flux* through a surface S in the sense of $\hat{n}$ is

★ $$\Phi_m \equiv \iint_S d\,\Phi_m \equiv \iint_S \vec{B} \cdot \hat{n}\, dS.$$

The symbol S under the integral sign indicates that the sum or integral is to be taken over the entire surface S.

Each magnetic field line is a continuous closed loop, so for a closed surface (Section 56.2) the total number of magnetic field lines leaving the surface is exactly equal to the total number of lines entering the surface. That is, the total magnetic flux through a closed surface is zero. Expressed as an integral, for any closed surface S we have

★ $$\oint\!\!\oint_S \vec{B} \cdot \hat{n}\, dS = 0. \qquad \text{Gauss's law for magnetism}$$

Contrasted with Gauss's law for electrostatics (Section 56.3), we see that for magnetic fields there is no "magnetic charge" which serves as the source of magnetic fields. That is, as far as we know magnetic monopoles do not exist as the source of magnetic fields; Gauss's law for magnetism expresses this experimental fact in equation form.

The failure of all efforts to find magnetic monopoles in nature raises the question, what is the source of magnetic fields? All magnetic fields are believed to result from electric currents, which may be macroscopic currents in wires or microscopic currents within atoms or even within individual particles such as protons and neutrons. The relationship of a macroscopic magnetic field to the current which produces it is discussed in Section 58.6, where we give "rule (b)" for the line integral of the magnetic induction around a closed path.

The microscopic currents within atoms are of two kinds: (a) currents associated with the *orbital motions* of electrons within the atom and of nucleons within the nucleus, and (b) currents associated with the *spins* of fundamental particles which make up the atom. An atomic electron in orbital motion around its nucleus can be visualized as a microscopic analog of a current in a loop of wire, producing a magnetic dipole field and thus acting as a tiny magnet. The spin of a particle can be visualized as a rotational motion of the particle about its own axis, similar to the rotational motion of the earth about its polar axis. Protons, neutrons, and electrons contain internal charges which rotate with these particles as they spin, so that each of these fundamental particles has an internal current loop and behaves as a tiny individual magnet.

When protons and neutrons combine to form a nucleus, some of their internal spin magnets are aligned "up" while others are "down." As a result, these nuclear spin magnets partially cancel each other (as far as external effects are concerned), but most nuclei have a net non-zero magnetism. The exceptions are the "even-even" nuclei, those containing an even number of protons and an even number of neutrons (Section 33.7), which have no net nuclear magnetism.

In an atom containing more than one electron, some of the spin magnets and orbital magnets of the electrons are oriented "up" while others are "down," so in each atom the magnetism of some electrons may cancel that of other electrons, leaving only a small net magnetism for the atom as a whole. Furthermore, in most substances the individual atomic magnets are oriented randomly, so their effects cancel each other and a macroscopic piece of the material is not in itself a magnet. However, in *ferromagnetic materials* -- the elements iron, cobalt, nickel and many of their alloys -- neighboring atoms interact with each other in such a way that the electron spins of many atoms become aligned spontaneously. This local region of magnetically aligned atoms, called a *magnetic domain*, has a thickness on the order of hundreds to thousands of atomic diameters. A domain may contain a million atoms, each of which acts as a tiny permanent magnet. In an unmagnetized piece of iron the magnetism of the domains is oriented randomly so that the iron as a whole does not act as a magnet, but when the iron is placed within a strong magnetic field many of the domains acquire the same alignment, and retain some of this magnetism after the magnetizing field is removed.

Questions and Exercise

Q 58.3-1 Cite some experimental evidence that magnetic field lines are continuous in space, with no beginning or end.

Q 58.3-2 The magnetic field lines associated with a bar magnet appear to begin on or near one end of the magnet and end near the other end. How then can we maintain that magnetic field lines form closed loops?

Q 58.3-3 Define the magnetic flux through a closed loop in terms of the magnetic induction at points on a surface bounded by the loop.

Q 58.3-4 What, if anything, "flows" in a region of magnetic flux?

Q 58.3-5 Does the total magnetic flux through a surface necessarily equal zero if the surface is open? Explain.

Q 58.3-6 (a) How would you define the flux of $\vec{E}$, by analogy to the flux of $\vec{B}$?
(b) Is the flux of $\vec{E}$ equal to zero through a closed surface which contains a net positive charge?
(c) Is the flux of $\vec{B}$ equal to zero through a closed surface which contains a net positive charge?

E 58.3-1 In Flint, Michigan the earth's magnetic field has a magnitude of about 5.9×10^{-5} teslas, and a dip angle of about 75° (below the horizontal). (a) Find the horizontal and vertical components of of the earth's $\vec{B}$ at Flint. (b) Compute the total magnetic flux which enters one square kilometer of the earth's surface at Flint.

Check your understanding of the material above by testing yourself on the minimum learning goals at the beginning of this section.

58.4 Magnetic Force on a Moving Charged Particle

Minimum Learning Goal. (1) Be able to recall and use the equation for the magnetic force on a moving charge to find the charge, force, velocity, or magnetic induction, as required, when you are given sufficient information. For example:

1 An electron travels with a velocity of 2.0×10^7 m/s eastward in a magnetic field having an induction of 5×10^{-5} teslas northward. What is the force on this electron? (Answer, p. 58-36, Line 13)

2 A particle having a charge of $+6 \times 10^{-8}$ C is moving with a velocity $\vec{v} = (2 \times 10^5 \text{ m/s})\hat{z}$ perpendicular to a magnetic field, $\vec{B}$. Acting on this charge is a magnetic force $\vec{F} = (2 \times 10^{-5} \text{ N})\hat{x} + (5 \times 10^{-8} \text{ N})\hat{y}$. What are the magnitude and direction of the magnetic induction in the neighborhood of the charge? (Answer, p. 58-36, Line 3)

To investigate the law which governs the force of a magnetic field on a moving charged particle, it is necessary to observe the paths of charged particles under a variety of conditions. When a beam of electrons is sent through a magnetic field, under some conditions there is a clear deflection of the electron beam while under other conditions the beam is not deflected at all! To examine this unusual behavior, consider what happens in the specific case when the magnetic field is directed away from us. We need to use three mutually perpendicular axes, shown in Figure 58-4 and labeled x, y, and z. The field is directed along the z-axis in the $-z$ direction.

With $\vec{B}$ in the $-z$ direction, experiments yield the following results: Electrons traveling in the $+x$ direction are deflected downward. Electrons traveling in the $-x$ direction are deflected upward. Electrons traveling downward are deflected toward the left. Electrons traveling upward are deflected toward the right. Electrons traveling in the $+z$ or $-z$ direction are not deflected at all! In Figure 58-4 each of the curved arrows lies entirely in the plane of the paper, and indicates part of the path of an electron which was initially moving parallel to one of the axes.

It is obvious that the effect of a magnetic field on charged particles is more complex than the effect of an electric field. An electric field exerts a force on a positive charge in the *same* direction as the field,

Figure 58-4. Paths of electrons moving in a magnetic field, with reference to mutually perpendicular x, y, and z axes. An electron traveling perpendicular to the direction of $\vec{B}$ is deflected at right angles to its path. The curved arrows represent the paths of electrons moving in the xy-plane. An electron moving parallel or anti-parallel to $\vec{B}$ is not deflected.

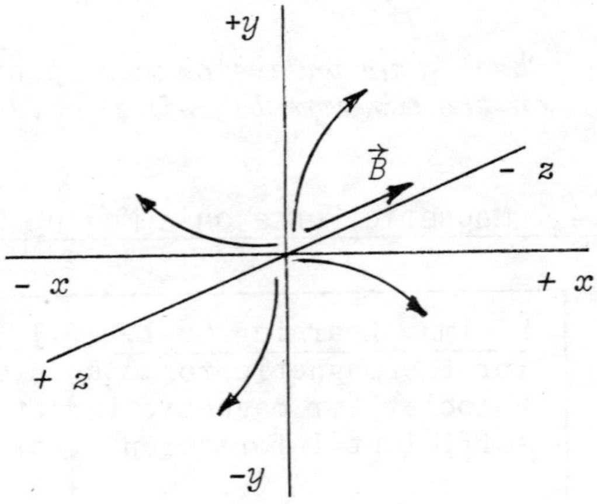

while the force which the field exerts on a negative charge is in the direction exactly *opposite* to the field. In one and the same *magnetic* field, however, the forces on charged particles act in many different directions, depending on the direction of the particle's motion. How can we find any order or pattern in the results described above?

From the paths of the electrons, we can infer the directions of the magnetic forces acting on them. With the magnetic induction directed into the paper, for electrons moving upward the magnetic force is to the right; for electrons moving to the right it is downward; for electrons moving downward the force is to the left; and for electrons moving to the left the force is upward. Clearly, the magnetic field cannot have the same direction (nor the opposite direction) as the magnetic force. Rather, we conclude that the magnetic force on a moving charged particle is always *perpendicular* to the velocity of the particle. Furthermore, we conclude that the magnetic force is perpendicular to the direction of $\vec{B}$.

$$\vec{F}_m \perp \vec{v} \quad \text{and} \quad \vec{F}_m \perp \vec{B} \, .$$

If one knows the direction of a magnetic field and the velocity of a charged particle moving in the field, the statements above are sufficient to find the line along which the magnetic force acts, but not to find the direction of the force along this line. Some rule of thumb is needed to specify the correct direction, and there are a number of rules in use which are equally correct. It is customary to state each rule in terms of the force on a moving *positive* charge; for a negative charge the force has the opposite direction. A convenient rule is the following *Right Hand Rule:*

★ Hold your right hand flat. For a moving positive charge, point your thumb in the direction of $\vec{v}$. Direct your fingers parallel to $\vec{B}$. Then $\vec{F}_m$ has the direction in which your right hand would naturally push.

For a moving *negative* charge, one may follow the same rule but use the left hand instead of the right hand.

Figure 58-5. A Right Hand Rule for finding the direction of the magnetic force on a moving *positive* charge. For a negative charge, the direction of the force is opposite to that given by the Right Hand Rule.

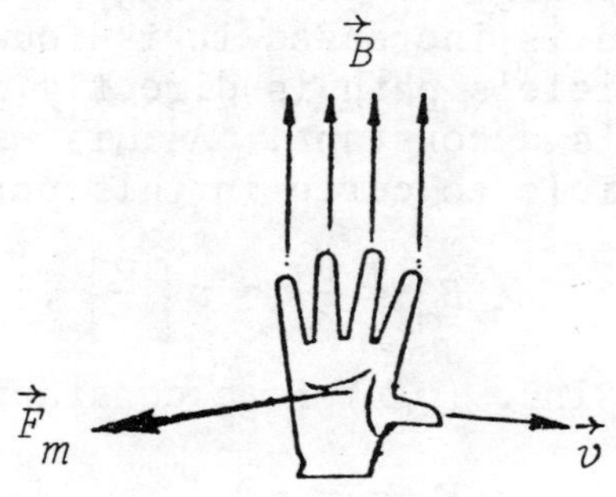

What physical quantities or variables would we expect in the magnetic force law? It seems reasonable that the magnetic force $\vec{F}_m$ on a moving charged particle would depend on the particle's velocity $\vec{v}$, and on its charge q, and also on the magnetic induction $\vec{B}$. In addition, it is clear from the experiments described above that the force depends on the angle between the directions of $\vec{v}$ and $\vec{B}$.

To obtain a quantitative relationship between $\vec{F}_m$ and $\vec{B}$, we note that when the particle's velocity is exactly perpendicular to the induction, both $\vec{F}_m$ and $\vec{v}$ are perpendicular to $\vec{B}$ and hence both are in the same plane. The magnetic force causes the particle to move in a curved path, and moreover the force is constant in magnitude and always perpendicular to the path, so this force meets all the requirements for a centripetal force (Section 43.4). If the magnetic force is the only significant force acting on the particle, the particle moves with uniform circular motion. The magnetic force thus serves as a centripetal force $\vec{F}_c$, and we may write

$$F_m = \frac{m\,v^2}{r}\,.$$

Therefore, when m and v are constant, the magnetic force is inversely proportional to the radius of the circle traced by the moving particle. Experiments show that the radius r is inversely proportional to the induction $\vec{B}$. Because both the force and the induction are inversely proportional to the radius, they must be directly proportional to each other:

$$F_m \propto \frac{1}{r} \qquad \text{and} \qquad r \propto \frac{1}{B}\ ; \ \text{therefore,}\ \ F_m \propto B\,.$$

To obtain a quantitative relationship between F_m and q, experiments may be done using particles having differing values of charge. Examples include the electron (charge $-e$), proton (charge $+e$), and alpha particle (charge $+2e$). The result of such measurements is that

$$F_m \propto q.$$

The magnetic force on a moving negative charge has a direction opposite to the force on a positive charge moving in the same direction.

To obtain a quantitative relationship between $\vec{F}_m$ and $\vec{v}$, experiments are needed using particles having different velocities. The speed acquired by an electron shot from an "electron gun" depends on the accelerating voltage applied to the gun (Chapter 25). If the magnetic induction and the charge on the moving particles are kept constant, when the particle speed is increased it is found by experiment that the radius r of the particle's path is directly proportional to the speed. That is, the ratio v/r is a constant. As discussed above, the centripetal force causing the particle to curve in this path is the magnetic force, and so we have

$$F_m \propto \frac{v^2}{r} \propto v\left[\frac{v}{r}\right],$$

but since (v/r) is a constant we obtain simply

$$F_m \propto v\ .$$

Combining the results obtained above, we find that the magnetic force F_m is directly proportional to the product Bqv. The proportionality factor depends on the units used, and on the angle between the directions of $\vec{B}$ and $\vec{v}$. We have already found that the magnetic force

$$\vec{F}_m = 0 \text{ when } \vec{v} \parallel \vec{B}.$$

A general equation for the magnetic force which embodies all the results obtained above is written as a vector cross product (Section 46.6):

★ The *magnetic force* $\vec{F}_m$ on a charge q moving with velocity $\vec{v}$ in a field of magnetic induction $\vec{B}$ is $\vec{F}_m = q\,\vec{v} \times \vec{B}$.

The last equation above has a proportionality constant of unity, which results from the definition of the SI unit of magnetic induction:

> One *tesla* (symbol T) is that value of magnetic induction which exerts a force of one newton on a charge of one coulomb moving perpendicular to the induction with a speed of one meter/second.

When the angle between the directions of $\vec{B}$ and $\vec{v}$ is some angle θ between 0° and 90°, the magnitude of the magnetic force is given by

$$F_m = Bqv \sin\theta.$$

The magnetic force may also be computed using

$$F_m = B_\perp qv,$$

where $B_\perp$ is the component of $\vec{B}$ perpendicular to $\vec{v}$.

Worked Example 58-1

An experiment is done using a uniform magnetic field having an induction of 0.03 tesla directed vertically upward. In this field, an alpha particle (charge $+2e$) is traveling eastward with a speed of 1.6×10^7 m/s. What are the magnitude and direction of the magnetic force on this alpha particle?

Solution

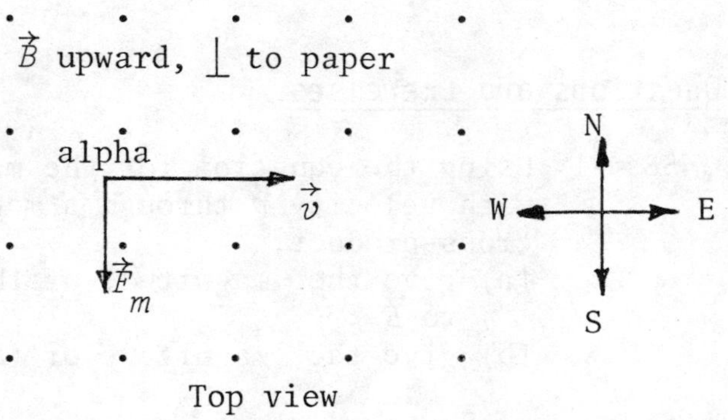

As always, a sketch is helpful in visualizing the physical situation. Because a charged particle moving in a magnetic field is always acted on by a force perpendicular to the field, this force and the path of the particle can be visualized most readily using a sketch in which the magnetic induction is perpendicular to the plane of the drawing. Magnetic field lines are drawn with arrows showing the direction of $\vec{B}$, as in Figure 58-1. It is customary in drawings of this sort to represent an arrow directed downward into the paper by an X, and an arrow directed upward out of the paper by a dot, • (suggesting an arrowhead). The uniform magnetic field directed upward is indicated by an array of dots in the sketch which depicts the alpha particle moving eastward as seen from above. Using the Right Hand Rule, with fingers pointing upward and thumb toward the east, it is found that the magnetic force on the particle is directed toward the south. The magnitude of this force is calculated using

$$F_m = B\,q\,v = 0.03\ \text{T} \times 2 \times 1.6 \times 10^{-19}\ \text{C} \times 1.6 \times 10^{7}\ \text{m/s}$$

$$= 1.5 \times 10^{-13}\ \text{N}.$$

The magnetic force and the electrostatic force are two manifestations of a single fundamental electromagnetic force. At the introductory level it is desirable to present these as two separate forces, but is it not hard to see that they are closely related, as we shall now show.

When two charged particles are located any distance r from each other, both a gravitational force and an electrostatic force exist between these particles. When the two particles are *moving* relative to each other, we find by experiment that an *additional* force exists between them which we call the *magnetic* force. The fact that the charge of the particles is involved shows that magnetic interactions and electrostatic interactions are related.

The relationship can be pointed out more specifically as follows: If two particles have the same charge, they experience the same electrostatic force in a given electric field. They need *not* experience the same force in a given gravitational field, however, since the latter force is determined by a particle's mass rather than its charge. The charge of a particle does not determine its mass, nor vice versa. That is why we say that gravitational forces and electrostatic forces are separate phenomena. But it is *not* possible to adjust independently the electrostatic force which a particle having a given velocity feels in an electric field and the magnetic force which it feels in a given magnetic field. Both forces are determined by the same quantity, electric charge.

Questions and Exercises

Q 58.4-1 Using the equation for the magnetic force $\vec{F}_m$ on a charge q moving with velocity $\vec{v}$ through a magnetic induction $\vec{B}$, expressed as a cross-product,
(a) give the magnitude of the force when $\vec{v}$ is perpendicular to $\vec{B}$.
(b) give the magnitude of the force when $\vec{v}$ is parallel to $\vec{B}$.

Q 58.4-2 Two charges are moving toward each other along the same straight line. Can you find an argument, using the geometrical symmetry of the situation, to show that there cannot be a magnetic force between them?

Q 58.4-3 State a rule which you can always use to determine the *direction* of the magnetic force on a moving charged particle given the direction of $\vec{B}$. Is your rule the same for negative and positive charges?

Q 58.4-4 Consider a particle having mass m and charge q moving through a magnetic field (induction $\vec{B}$) with a velocity $\vec{v}$. Which of these statements correctly describes the magnetic force $\vec{F}_m$ on the particle?

(a) $F_m \propto m$	(f) $F_m = mv^2/r$	(k) $\vec{F}_m \perp \vec{B}$
(b) $F_m \propto q$	(g) $F_m \propto B$	(ℓ) $\vec{F}_m = q\vec{E}$
(c) $F_m \propto q^2$	(h) $\vec{F}_m \perp \vec{v}$	(m) $\vec{F}_m = 0$ when $\vec{v} \parallel \vec{B}$
(d) $F_m \propto v$	(i) $\vec{F} \parallel \vec{v}$	(n) $\vec{F}_m = 0$ when $\vec{v} \perp \vec{B}$
(e) $F_m \propto 1/r^2$	(j) $\vec{F}_m \parallel \vec{B}$	

Q 58.4-5 It seems annoying that unlike the electric force, which is in the direction of the electric field vector $\vec{E}$, the magnetic force is *not* in the direction of the magnetic field vector $\vec{B}$. Since the magnetic field can only be observed through the force it exerts on charges, let us consider whether we could abandon our earlier definition of this field and discard the right-hand rule, start from the beginning to explore the magnetic force on a moving charge when it enters a magnetic field from various directions, and then redefine this field in such a way that it *does* point in the direction of the force. For example, consider the following set of experimental observations on the deflection of electrons in a particular magnetic field: (1) electrons moving toward the east are deflected downward; (2) electrons moving toward the west are deflected upward; and (3) electrons moving upward are deflected toward the east. From the first observation we might say that the magnetic force is downward and therefore this magnetic field is directed downward. Can this redefinition of the direction of the magnetic field be correlated satisfactorily with observations (2) and (3) above? Discuss clearly.

Q 58.4-6 Suppose a stream of electrons is projected through a chamber in the laboratory, and it is observed that there is no deflection of these electrons.

(a) Clearly describe three different possible combinations of electric and/or magnetic fields in the chamber which would be consistent with this observation.

(b) What further experiments could be done to decide which of these possibilities is actually the case?

E 58.4-1 A magnetic field of magnitude 0.5 tesla is directed toward the west. Find the magnetic force on an electron traveling in this field with $v = 10^7$ m/s: (a) due north; (b) due east.

E 58.4-2 An electron moving with a velocity of 10^5 m/s due west is traveling at right angles to a magnetic field. The magnetic force on the electron is 2×10^{-14} newton, straight up. Find the direction and magnitude of the magnetic induction.

E 58.4-3 An electron has a velocity $\vec{v} = 10^6\ \hat{x}$ m/s in a magnetic field having $\vec{B} = -0.8\hat{z}$ tesla. Compute the magnetic force on this electron.

E 58.4-4 A proton has a velocity $\vec{v} = (\hat{x} + 2\hat{y})\ 10^5$ m/s in a magnetic field having $\vec{B} = (\hat{x} - 3\hat{z})\ 0.5$ tesla. What is the magnetic force on this proton?

E 58.4-5 Compute the magnetic force on a proton traveling eastward at 10^6 m/s in the earth's field, at a point where $\vec{B} = 0.7$ T directed downward and to the north at an angle of 70° with the vertical.

E 58.4-6 A high-speed proton enters a region permeated by a uniform magnetic field $\vec{B}$ and a uniform electric field $\vec{E}$. $B = 0.5$ T, and $\vec{E} = 5 \times 10^5$ N/C. The directions of $\vec{E}$, $\vec{B}$, and $\vec{v}$ are such that the proton feels no resultant force. (Neglect gravitation.)

(a) Make a simple but clear sketch of this situation, showing the directions of $\vec{E}$, $\vec{B}$, $\vec{F}_m$, $\vec{F}_{elec}$, and $\vec{v}$. For convenience, in your diagram take $\vec{v}$ directed toward the right, and $\vec{E}$ directed toward the top of the paper.

(b) Compute the speed of the proton when it is undeflected.

E 58.4-7 Suppose you are given an apparatus which produces a uniform magnetic field in a spherical volume of space approximately 30 cm in diameter, but you receive no information which gives the spatial direction of this field. You also have an "*e*/*m* tube" within which a narrow beam of electrons is produced, making a visible trail, which reveals the path of these electrons. Describe a procedure by which you could use the *e*/*m* tube to determine the direction of the magnetic field.

Check your understanding of the material above by testing yourself on the minimum learning goals at the beginning of this section.

58.5 Motion of a Charged Particle in a Magnetic Field

Minimum Learning Goal. For a charged particle moving perpendicular to a uniform magnetic field, be able to recall or deduce a relationship among the magnetic induction, the charge-to-mass ratio, the speed, and the radius of the circular path, and use that relationship to find one of these quantities, given the others. For example:

1 In a laboratory e/m apparatus, electrons ($e/m = 1.76 \times 10^{11}$ C/kg) are launched into a magnetic field at 10^7 m/s. What value of B is required to deflect the electrons into a circle of 0.1 m diameter? (Answer, p. 58-36, Line 17)

2 An ion with charge $+e$ travels at a speed of 3×10^4 m/s in a semicircle with $r = 0.089$ m in a mass spectrometer having $B = 0.02$ T. Compute q/m for this ion. (Answer, page 58-36, Line 16)

As we have seen in Section 58.4, when a charged particle is moving perpendicular to the field lines in a uniform magnetic field, the magnetic force is perpendicular to the instantaneous velocity of the particle and this force meets all the requirements for a centripetal force. If the magnetic force is the only significant force acting on the particle, it moves with uniform circular motion. Setting the magnetic force equal to the centripetal force, we have

$$\vec{F}_m = \vec{F}_c , \qquad or \qquad B\,q\,v = \frac{mv^2}{r} .$$

Cancelling one v from each side and rearranging terms, we obtain

★ $$\frac{q}{m} = \frac{v}{Br} .$$

The quantity q/m is the *charge-to-mass ratio* of the particle, and it can be found from measurements made in a magnetic deflection experiment. Ordinarily the magnetic induction $\vec{B}$ is known by the experimenter, and the velocity of the particle can be found from other information. Then a measurement of the radius of the particle's path allows one to calculate the value of q/m.

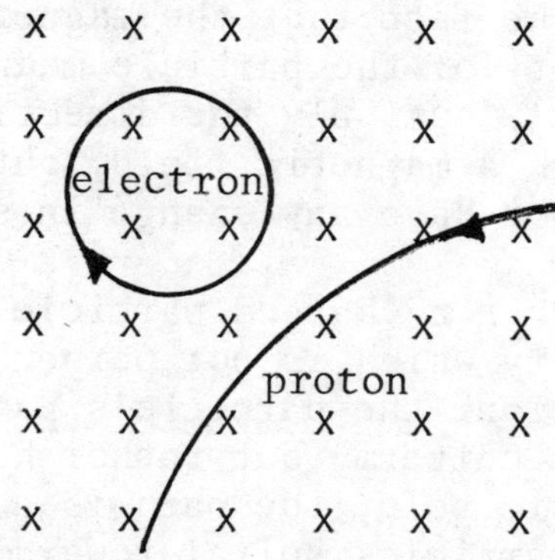

Figure 58-6. A charged particle moving perpendicular to the field lines in a uniform magnetic field travels in a circular path. The radius of the path depends on the magnetic field strength and on the particle's speed and charge-to-mass ratio. In this figure, the field lines are directed downward into the paper and are denoted by X's, so an electron circles clockwise while a proton circles in a counterclockwise direction.

Worked Example 58-2

An electron traveling in a uniform magnetic field having an induction of 4×10^{-3} tesla is observed to travel in a circular path having a diameter of 4.54 cm. What is the speed of the electron?

Solution

We can use the equation for q/m given above, because we know the values of all the quantities in that equation except the speed, v.

Solving the equation for v leads to

$$v = B\, r\, q/m.$$

In this equation, q/m is the charge-to-mass ratio of the electron, which is

$$\frac{q}{m} = \frac{1.60 \times 10^{-19} \text{ coulomb}}{9.11 \times 10^{-31} \text{ kilogram}} = 1.76 \times 10^{11} \text{ C/kg}.$$

Substituting into the equation for v, we obtain

$$v = 4 \times 10^{-3} \text{ T} \times 2.27 \times 10^{-2} \text{ m} \times 1.76 \times 10^{11} \text{ C/kg}$$

$$= 1.60 \times 10^{7} \text{ m/s}.$$

Observing the path of a charged particle in a magnetic field is an important method for measuring the q/m of all kinds of charged fundamental particles, and of ions. Chapter 34 showed how this method contributed to the identification of alpha, beta, and gamma rays from radioactive materials. Using magnetic deflection alone, however, it is not possible to find the value of either q or m separately. The equation for q/m given above is fundamental for the design of charged-particle accelerators such as cyclotrons, synchrotrons, and the larger accelerators in which particles travel in circular paths.

The fact that the magnetic force is always perpendicular to the velocity of the particle means that it can never change the speed of the particle but only the direction of its motion. A charged particle moving through a magnetic field, but not subject to any non-magnetic forces, does not have any change in speed.

When a charged particle moves in a uniform magnetic field with a velocity which is not perpendicular but at some angle to the magnetic induction, the particle's path is not a circle but a helix. If the field is not uniform, but rather has field lines which are converging toward a magnetic pole, the path is a helix with a gradually decreasing pitch, so that particles spiral toward the pole around the field lines but are then "reflected" and spiral back in the opposite direction along the field lines. Many of the charged particles from the sun and from outer space which enter the earth's magnetic field have this type of motion; the paths of such particles are illustrated in color on the front cover of this book. Some of these particles become trapped in the earth's magnetic field and form the Van Allen radiation belts. Other particles spiraling in toward the earth's magnetic poles have collisions with atoms in the earth's upper atmosphere. At times when these collisions are very numerous, they give rise to the beauthiful displays called the *northern lights* (aurora borealis) and *southern lights* (aurora australis). The spiraling of charged particles toward a magnetic pole may be demonstrated using a fine beam tube and an iron magnet held near the tube.

Questions and Exercises

Q 58.5-1 What kind of path would be described by a charged particle in a uniform magnetic field if the particle's velocity is originally *parallel* to the field?

Q 58.5-2 For a charged particle moving in a $\vec{B}$ field, why does F_m remain constant while $\vec{F}_m$ is changing?

Q 58.5-3 Consider a particle having mass m and charge q moving in a uniform magnetic field with $\vec{v} \perp \vec{B}$. Which of the following equations apply to this particle?

(a) $\vec{F}_c = \vec{F}_m$; (b) $B\,q\,v = mv^2/r$; (c) $r = mv/Bq$; (d) $q/m = v/Br$.

Q 58.5-4 (a) Can a uniform magnetic field be used to change the speed of charged particles?
(b) Can it be used to change the velocity of charged particles?

Q 58.5-5 Check all the true statements.
A proton having mass m, charge q, and speed v, moves through a region of uniform electric and/or magnetic fields. (Neglect gravitation.)
(a) In a magnetic field alone, it can move in a circular path of radius $r = mv^2/qB$.
(b) In a magnetic field alone, its speed is unaltered regardless of the direction of its motion.
(c) In an electric field alone, it experiences no force when the field is parallel to its direction of motion.

(d) If $\vec{E}$ and $\vec{B}$ are parallel, pointing in opposite directions, and $v = E/B$, the proton is undeflected in its motion.

E 58.5-1 A proton travels in a circular path in a uniform magnetic field, with a speed of 3.4×10^4 m/s. The path has a radius of 0.04 m.
(a) Compute the strength of the magnetic field.
(b) Make a diagram showing the proton's path and direction of the uniform magnetic field. Use arrows if the field is in the plane of the paper, X's if the field is into the paper, and •'s if the field is directed out of the paper.

E 58.5-2 An electron travels in a circular path in a uniform magnetic field, with a speed of 1.0×10^7 m/s. The path has a radius of 0.03 m.
(a) Compute the strength of the magnetic field.
(b) Make a diagram showing the electron's path and direction of motion, clearly indicating the direction of the magnetic induction using symbols as described in E 58.4-1.

E 58.5-3 An ion having charge $+e$ travels at a speed of 3×10^4 m/s in a semicircle with $r = 0.109$ m, in a mass spectrometer having $B = 0.02$ T.
(a) Compute q/m for this ion.
(b) What is the atomic mass number of this atom?
(c) To what element does the atom probably belong?

E 58.5-4 The following particles travel in circles in the same magnetic field: electron, neutron, proton. Assume each particle has the same speed, and list the particles in the order of increasing radius of curvature.

Check your understanding of the material above by testing yourself on the minimum learning goal at the beginning of this section.

58.6 Ampere's Law

Minimum Learning Goal. Given an equation for the magnetic induction in a region of space, be able to use Ampere's law to calculate the current enclosed by a specified path of integration. For example:

((1 Using Ampere's law, calculate the current which must thread the area enclosed by the path designated in the figure below. The magnetic induction in teslas is $\vec{B} = 10^{-12}\ (x^2y\ \hat{x} - 2xyz\ \hat{z})$. (Answer, Page 58-36, Line 10)

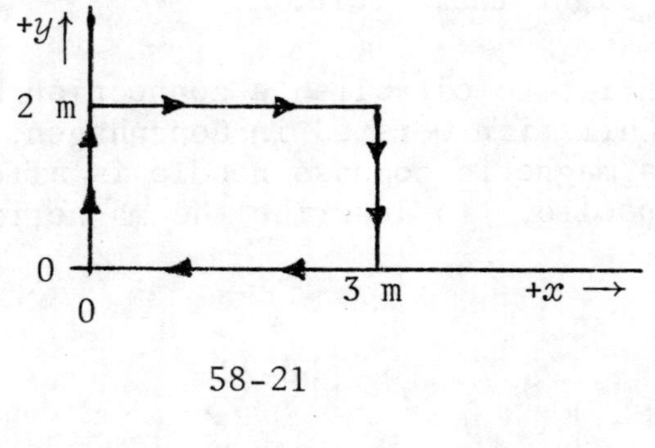

2 Using Ampere's law, calculate the current which must thread the area enclosed by the path shown in the figure below. The magnetic induction (in teslas) is $\vec{B} = 10^{-12}\,[r(\cos\theta)(-\hat{r}) + 2z\cos\theta\,\hat{z}]$, where θ is the angle measured in the xy plane between $\vec{r}$ and the positive x-axis, $\hat{r}$ is a unit vector in the direction of increasing r, and r is the distance from the origin (measured in meters). (Answer, page 58-36 Line 1)

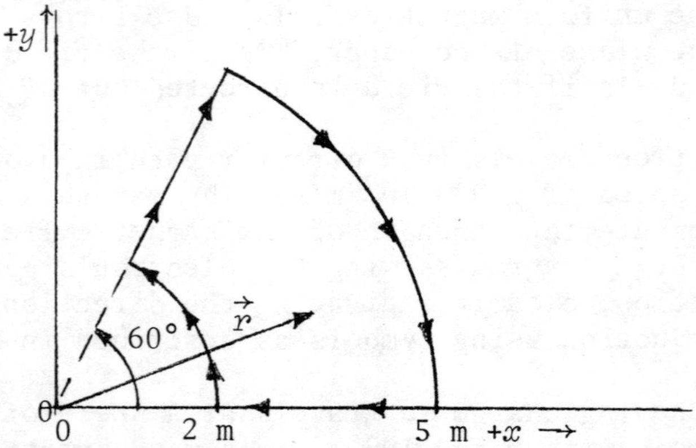

Experiments show that in a situation where currents, charges, and fields do not vary in time, the magnetic fields are uniquely determined by the currents, and there are no magnetic fields if there are no currents. In the case of permanent magnets such as a horseshoe magnet, the currents are microscopic currents due to motions of charges on an atomic scale. In magnetized material, these microscopic currents line up in an organized fashion so that they have macroscopic effects.

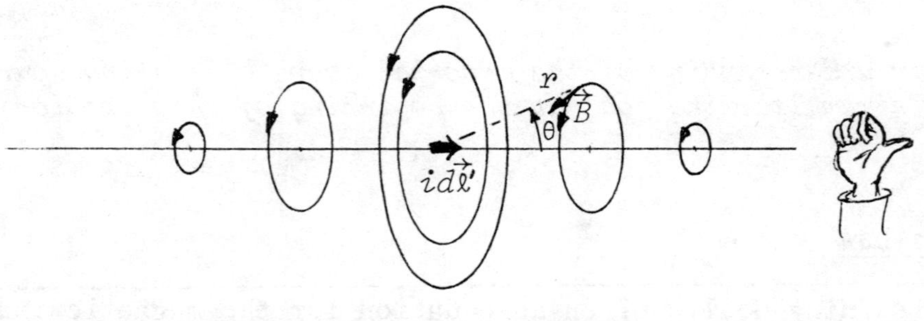

Figure 58-7. The magnetic field which results from a small current element $i\,d\vec{\ell}'$ has circular field lines which are concentric with a straight line along the current element. The magnetic induction $\vec{B}$ at any point is tangent to the circle through that point. In moving away from the current element in any direction, the magnitude of $\vec{B}$ decreases inversely with the square of the distance r from the current element. The field lines encircle the current in the direction indicated by a right thumb rule.

The first scientist to establish a connection between electricity and magnetism was Hans Christian Oersted in Copenhagen, in 1820. He observed that the orientation of a magnetic compass needle is affected by electric current in a wire near the needle. To describe the magnetic field which results from electric current, we consider first a steady current i along a straight line, as shown in Figure 58-7. The length of a short segment of the current's path is denoted by $d\vec{\ell}'$, a vector having the same direction as the current in this segment. The product of the current times this length is called a *current element*, $i\, d\vec{\ell}'$. The pattern of magnetic induction produced in the surrounding region by the current element is the following: at each point, the vector $\vec{B}$ is perpendicular both to the line connecting it to the position of the current element, and to the direction of the current element. This may be visualized by first drawing the straight line along (or tangent to) the current element, and then imagining a lot of circles in space which have that line as their common axis. The magnetic induction $\vec{B}$ at a point in space is then tangent to the circle that passes through that point.

Figure 58.8. Arrows showing the magnetic induction $\vec{B}$ at points on a sphere, resulting from a current element $id\vec{\ell}'$ at its center. The magnitude and direction of the induction depends on the direction of the field point from the current element. At a fixed distance r from the current element, the induction is largest in the plane through the current element and normal to it, but it is zero on the line collinear with the element.

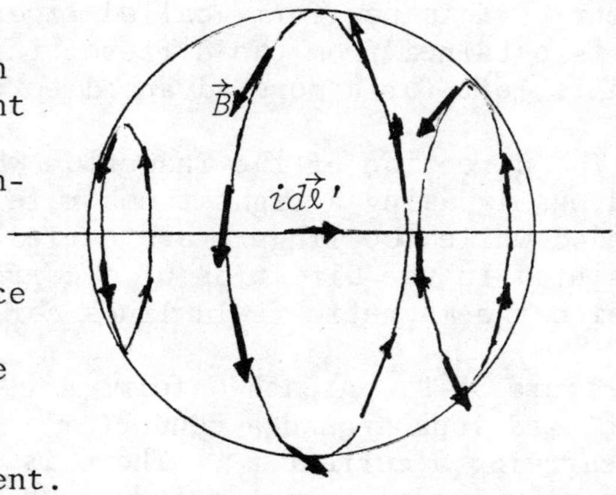

The magnitude of $\vec{B}$ decreases inversely with the square of the distance r from the current element, as one moves away from the current element in a fixed direction. Unlike the electrostatic field, however, the magnitude of $\vec{B}$ also has a directional dependence (Figure 58-8). If a sphere is drawn with the current element at its center, the magnetic induction is zero at the "poles" where the straight line intersects the sphere; it is a maximum on the "equator," and in other directions it varies as the sine of the angle θ between $id\vec{\ell}'$ and the displacement vector $\vec{r}$ from the current element to the point where $\vec{B}$ is measured. The magnetic induction is also proportional to the current element $id\vec{\ell}'$ which is causing it.

These experimental results are summarized in the following equation for the small contribution to the induction $d\vec{B}$ which results from the small current element $id\vec{\ell}'$:

> The element of magnetic induction $d\vec{B}$ at a point located at position $\vec{r}$ with respect to a current element $id\vec{\ell}'$ is
>
> $$d\vec{B} = \frac{\mu_0}{4\pi} \frac{id\ell' \times \hat{r}}{r^2} ,$$

where $\hat{r}$ is the unit vector in the direction of $\hat{r}$. It is customary to write the proportionality constant in this equation as $\mu_0/4\pi$. The factor μ_0 is called the *permeability of free space,* and it has the precise numerical value

$$\mu_0 = 4\pi \times 10^{-7} \text{ newton/ampere}^2 = 4\pi \times 10^{-7} \text{ weber/ampere·meter}$$

in the SI system of units. (This choice of the value for μ_0 is closely linked to the definition of the ampere, defined in terms of the magnetic force between currents.)

The equation above for $d\vec{B}$ contains the infinitesimal or differential quantities $d\vec{B}$ and $d\vec{\ell}'$; it is known as the *differential form of Ampere's law*. To use this equation to compute the total magnetic induction at some point which results from the current, it is necessary to integrate over the complete path taken by the current. This integration is difficult even for current paths having a simple geometric shape, so we shall not attempt any calculations of this sort. Rather, we go to another form of Ampere's law which is not difficult to use in situations having a high degree of symmetry. This new form, called *Ampere's circuital law* or simply Ampere's law, is obtained from the differential form by a long and complex derivation which is left for a more advanced textbook.

The direction of the induction which results from a current may be found easily using a right thumb rule: if the thumb of the right hand is extended while the fingers are curled (as in hitch-hiking), when the thumb is pointed in the direction of a current the fingers indicate the direction in which the magnetic field lines curl around the current.

Figure 58-9. A path C forms a closed loop around a conductor carrying a current i_C. There is some value of magnetic induction $\vec{B}$ at every point on the path. The line integral of $\vec{B}$ around the path is related to the current i_C linking this path by Ampere's law.

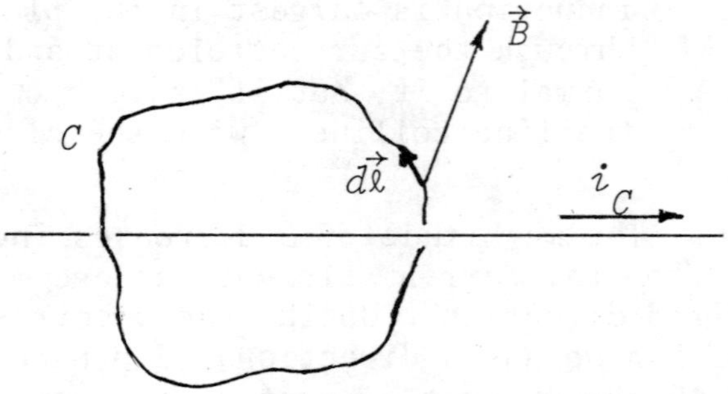

Consider a conductor carrying a constant current i_C, and a closed curve or path C which encircles the conductor (Figure 58-9). We speak of i_C as the current *linking* the path. This current establishes some value of magnetic induction $\vec{B}$ at every point on C. We use the symbol $d\vec{\ell}$ to describe a small displacement along (tangent to) the path C, the path of integration for computing the line integral of $\vec{B}$. The senses of i_C and $d\vec{\ell}$ are related by another *right thumb rule:* if the right hand grasps the conductor with the thumb extended in the direction of i_C, the fingers will encircle the conductor in the sense of $d\vec{\ell}$. The component of $\vec{B}$ parallel to the path at any point times $d\vec{\ell}$ is $\vec{B} \cdot d\vec{\ell}$. The summation of $\vec{B} \cdot d\vec{\ell}$ around the complete path C is $\sum_C \vec{B} \cdot d\vec{\ell}$, computed by adding the values of $\vec{B} \cdot d\vec{\ell}$ for all the segments making up the path. This sum is also expressed as $\oint_C \vec{B} \cdot d\vec{\ell}$, the line integral of $\vec{B}$ around the closed path C. Line integrals are discussed in Appendix A.5 and Section 56.9.

Because magnetic field lines which result from a current form closed loops encircling the current (Figure 58-7), a line integral around a field line does not equal zero. In fact, a line integral of $\vec{B}$ around *any* closed path which encircles a current does not equal zero. The fact that the field lines encircle the current may be expressed by saying there is a *circulation* of $\vec{B}$ around the current. A line integral of $\vec{B}$ around a path encircling the current forms part of the *circulation condition* for $\vec{B}$, which relates the line integral to the current producing the field. It is stated as follows:

$$\bigstar \qquad \sum_C \vec{B} \cdot d\vec{\ell} = \mu_0\, i_C \quad \text{or} \quad \oint_C \vec{B} \cdot d\vec{\ell} = \mu_0\, i_C , \quad \text{Ampere's law.}$$

Ampere's law states that *the line integral of $\vec{B}$ around any closed path is proportional to the total electric current linking that path.* The proportionality constant in SI units is μ_0 (page 58-24). Ampere's law in this form is valid for paths which do not pass through magnetic material, when the current i_C is constant in time.

Ampere's law may be used to compute the current linking a path if the induction is known everywhere on the path. This path does not need to coincide with a field line, but the integration is easier if the path is chosen as a series of segments which have the same symmetry as the field. Having chosen a path, the product $\vec{B} \cdot d\vec{\ell}$ is computed to give a scalar function along the path, and this function is then summed or integrated along each segment of the path to obtain the total of the line integral and from this to compute the current.

<u>Worked Example 58-3</u>

In the region shown in this figure, the magnetic induction is given by

$$\vec{B} = (10^{-6}\ \mathrm{T/m^3})\ (\hat{y}\ 3x^2y\ -\ \hat{z}\ 3x^2z).$$

Compute the total current which threads the area enclosed by the path shown.

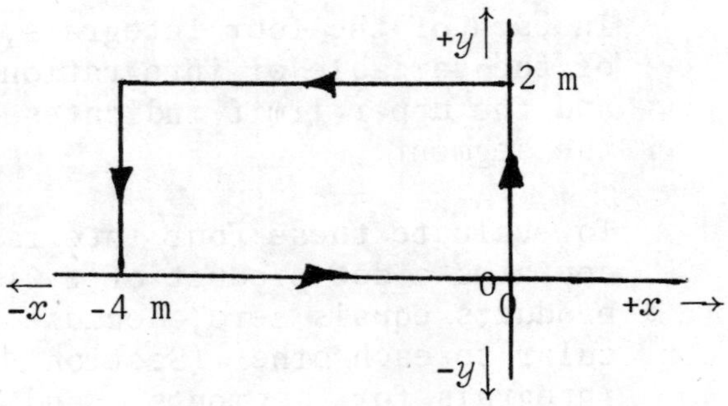

<u>Solution</u>

At first glance this may appear to be a complicated problem, but by working one step at a time we shall find that the solution is not too difficult. Because we are given the induction in the region of interest, we can find the current which threads any loop in this region by using Ampere's law. This law requires one to compute the line integral of $\vec{B}$ around a closed path. In this example, it is convenient to label the four segments of the path using the letters *a*, *b*, *c*, and *d* as shown in the figure below. Then the total line integral of $\vec{B}$ around the closed path may be written as the sum of four separate integrals, one for each segment:

$$\oint \vec{B} \cdot d\vec{\ell} = \int_a \vec{B} \cdot d\vec{\ell} + \int_b \vec{B} \cdot d\vec{\ell} + \int_c \vec{B} \cdot d\vec{\ell} + \int_d \vec{B} \cdot d\vec{\ell}.$$

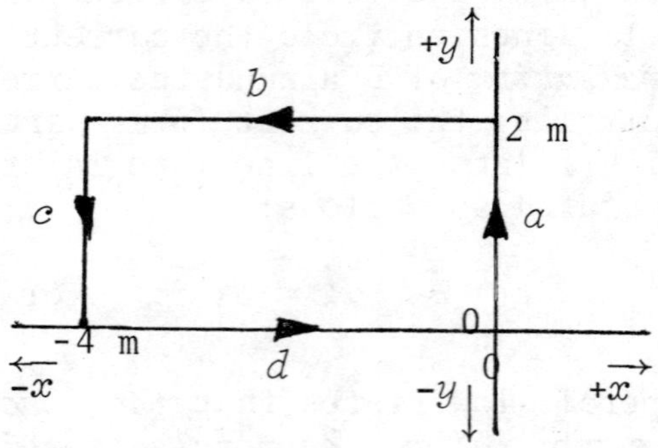

The symbol $\vec{d\ell}$ represents a small displacement along the path. In this Example, each segment of the path is parallel to either the x or the y axis. Therefore, on segment a, $\vec{d\ell}$ may be replaced by $\hat{y}\ dy$; on segment b, $\vec{d\ell}$ may be replaced by $-\hat{x}\ dx$; on segment c, $\vec{d\ell}$ may be replaced by $-\hat{y}\ dy$; on segment d, $\vec{d\ell}$ may be replaced by $\hat{x}\ dx$. Writing out the complete line integral, we then have

$$\oint \vec{B} \cdot \vec{d\ell} = \int_{0\ (a)}^{2\text{m}} (10^{-6}\ \text{T/m}^3)\ (\hat{y}\ 3x^2y - \hat{z}\ 3x^2z) \cdot \hat{y}\ dy$$

$$+ (10^{-6}\ \text{T/m}^3) \int_{0\ (b)}^{-4\text{m}} (\hat{y}\ 3x^2y - \hat{z}\ 3x^2z) \cdot (-\hat{x})\ dx$$

$$+ (10^{-6}\ \text{T/m}^3) \int_{2\text{m}\ (c)}^{0} (\hat{y}\ 3x^2y - \hat{z}\ 3x^2z) \cdot (-\hat{y})\ dy$$

$$+ (10^{-6}\ \text{T/m}^3) \int_{-4\text{m}\ (d)}^{0} (\hat{y}\ 3x^2y - \hat{z}\ 3x^2z) \cdot \hat{x}\ dx.$$

In each of the four integrals, the lower limit indicates the value of the variable of integration at the beginning of the path segment, and the upper limit indicates the value of this variable at the end of the segment.

To evaluate these four integrals, we first notice that each of them contains a dot product of $\hat{z}$ with either $\hat{x}$ or $\hat{y}$. Each of these dot products equals zero, because it contains two unit vectors perpendicular to each other (Section 47.1). The remaining term in the integrals for segments b and d contains the dot product $\hat{y} \cdot \hat{x}$, which equals zero for the same reason. We thus find that the integrals for segments b and d are both equal to zero, which results from the fact that $\vec{B}$ is perpendicular to $\vec{d\ell}$ along these segments.

This leaves only the following integrals to be evaluated for segments a and c:

$$\oint \vec{B} \cdot \vec{d\ell} = (10^{-6}\ \text{T/m}^3) \times 3x^2 \int_{0\ (a)}^{2\text{m}} \hat{y} \cdot \hat{y}\ y\ dy$$

$$+ (10^{-6}\ \text{T/m}^3) \times 3x^2 \int_{2\text{m}\ (c)}^{0} -\hat{y} \cdot \hat{y}\ \ y\ dy.$$

In these two remaining integrals, the product $\hat{y} \cdot \hat{y} = 1$. In both of these integrals, the factor x^2 is a constant along the path of integration, so it is written to the left of the integral sign; only the factors of y need to be written behind the integral sign, because the integration is over the variable y.

Along path segment a the value of $x = 0$, so the term for path a in the equation above is equal to zero. Along path segment c, the constant value of $x = -4$ meters. Carrying out this integration, we have

$$\oint \vec{B} \cdot d\vec{\ell} = (10^{-6}\ \text{T/m}^3) \times 3(-4\text{m})^2 \int_{2\text{m}}^{0} -y\ dy$$

$$= -4.8 \times 10^{-5}\ \text{T/m}\ [y^2/2]_{2\text{m}}^{0}$$

$$= -4.8 \times 10^{-5}\ \text{T/m}\ [0 - 4\text{m}^2/2]$$

$$= +9.6 \times 10^{-5}\ \text{T} \cdot \text{m} = 9.6 \times 10^{-5}\ \text{Wb/m}.$$

(Note that the units of tesla/meter^3, attached to the factor 10^{-6} from the beginning, are needed to make the units of the integral come out correctly.) The final value of the line integral for $\vec{B}$ is positive, which means that a component of $\vec{B}$ must be parallel to the path along segment c. Referring to the original expression for $\vec{B}$, we see that the y-component of $\vec{B}$ is indeed negative where $y = -4$ meters.

The current which threads the path *abcd* in this Example is computed using Ampere's Law:

$$i = \oint \vec{B} \cdot d\vec{\ell}/\mu_0$$

$$= \frac{9.6 \times 10^{-5}\ \text{Wb/m}}{4\pi \times 10^{-7}\ \text{Wb/A}\cdot\text{m}} = 7.64 \times 10^{1}\ \text{A}.$$

Applying the right-thumb rule for Ampere's law, when the fingers encircle the path in the counter-clockwise sense to match the arrows along the path, the thumb extends in the $+z$ direction, indicating that the current is in the $+z$ direction.

Ampere's law may also be used to compute the magnetic induction in some region if the current producing this magnetic field is known. It might seem that Ampere's law is not really sufficient to determine $\vec{B}$, because many different configurations of $\vec{B}$ can be imagined which would result in the same value of the line integral on a given path. The point is, however, that Ampere's law must hold on *every* path. It turns out that this *does* determine the field uniquely at each point. As a practical method of calculation, however, Ampere's law is useful only if the problem has sufficient symmetry to allow us to reason from some fairly simple paths.

If the locations of the current-conducting paths and the currents in these paths are known, one may use the right-thumb rule for Ampere's law to obtain a qualitative description of the field. Then if the field has an evident symmetry, one may use Ampere's law to compute the magnitude of the field; this is done for the field of a long straight wire in Section 58.7.

We can also use Ampere's law to compute the field when it is constant in some regions and zero or nearly zero in others; this is done for the field of a solenoid in Section 58.8.

TABLE 58-1 Fundamental Laws of Electrostatics and Magnetostatics

	Electrostatics	Magnetostatics
Gauss's law	$\oint\!\!\oint_S \vec{E} \cdot \hat{n}\, dS = q/\varepsilon_0$	$\oint\!\!\oint_S \vec{B} \cdot \hat{n}\, dS = 0$
Circulation Condition	$\oint_C \vec{E} \cdot d\vec{\ell} = 0$	$\oint_C \vec{B} \cdot d\vec{\ell} = \mu_o i_c$

Note how magnetostatics is just the reverse of electrostatics, as far as the fundamental laws are concerned (Table 58-1). In electrostatics, Gauss's law relates the field to its sources (charges), and says that the total electric flux outward through a closed surface is equal to the enclosed charge divided by ε_0. The circulation condition for an electrostatic field says that the line integral of $\vec{E}$ around a closed path is zero. In magnetostatics, on the other hand, the line integral of $\vec{B}$ is not zero, and it is the circulation condition which relates the magnetic field to its sources (currents). Gauss's law for magnetism says that the total magnetic flux outward through a closed surface is zero, corresponding to the experimental fact that no isolated magnetic poles have been found.

In the electrostatic case, the zero in the circulation condition is related to the conservation of energy. No such argument is applicable to the magnetic field, because the magnetic induction itself is not the force on a unit magnetic charge.

Questions and Exercises

Q 58.6-1 Consider the region between the poles of an electromagnet, where the field is uniform and normal to the pole faces.

(a) Give an example of a closed path in this region around which $\oint \vec{B} \cdot d\vec{\ell}$ equals zero.

(b) Is there any closed path in this region around which $\oint \vec{B} \cdot d\vec{\ell}$ does not equal zero? Explain.

Q 58.6-2 In what sense is it correct to say that the source of magnetic field lines is electric current?

E 58.6-1 Show that both sides of the equation for Ampere's law have the same physical units.

E 58.6-2 What is the net current that threads each of the paths C_1, C_2, C_3 and C_4 in this figure?

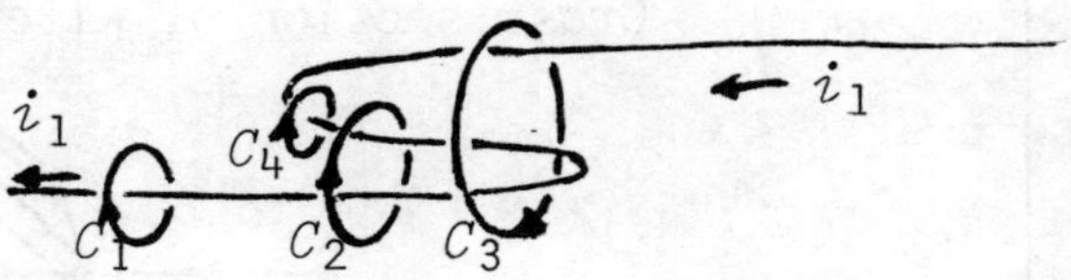

E 58.6-3 Consider a uniform field $\vec{B} = \hat{y}B_0$.

(a) For the rectangular path *abcd* in the *xy*-plane shown here, compute $\oint \vec{B} \cdot d\vec{\ell}$. The rectangle has sides of length *g* and *h*. Hint:

$$\oint = \int_a^b + \int_b^c + \int_c^d + \int_d^a .$$

(b) Compute the total magnetic flux passing through the surface bounded by this rectangle.

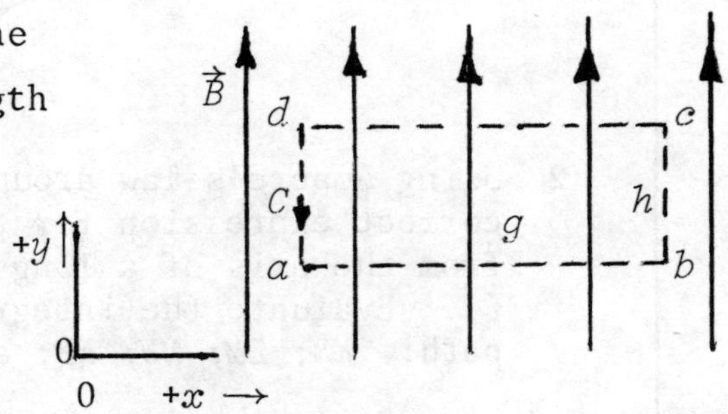

E 58.6-4 (a) For a circular loop (carrying current *i*) located in the *xy*-plane with its center at the origin, sketch a map of the $\vec{B}$ field in the *xz*-plane.

(b) Can Ampere's law be used to compute $\vec{B}$ at the center of this loop? Explain.

Check your understanding of the material above by testing yourself on the minimum learning goal at the beginning of this section.

58.7 Magnetic Field Near a Long Straight Wire

Minimum Learning Goal. Be able to deduce the correct expression for the magnetic induction around a long straight wire carrying an electric current, by using Ampere's law and summing or integrating over a specified path. For example:

1 Using Ampere's law around the closed path shown below, deduce the correct expression for the magnetic induction at a distance *r* from the axis of a long straight wire carrying an electric current *i*. Evaluate the integrals separately for each specified segment of the path: *KL; LM; MN;* and *NK*. (Answer, Page 58-36, Line 12).

Cross section of wire carrying i into the paper

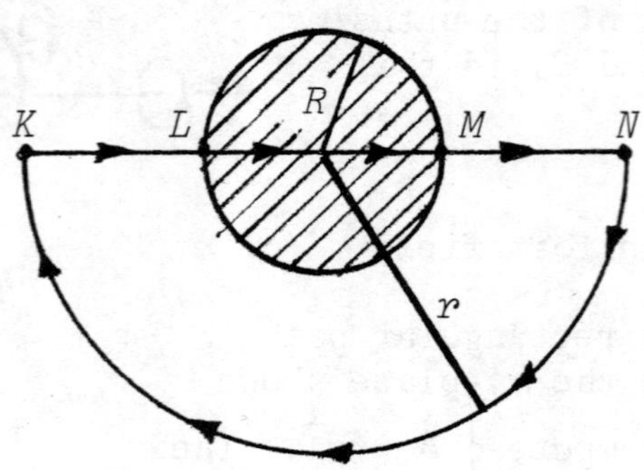

2 Using Ampere's law around the closed path shown below, deduce the correct expression for the magnetic induction at a distance r from the axis of a long straight wire carrying an electric current i. Evaluate the integrals for each specified segment of the path: *KL; LM; MN; NP;* and *PK*. (Answer, Page 58-36, Line 5.)

Cross section of wire carrying i into the paper

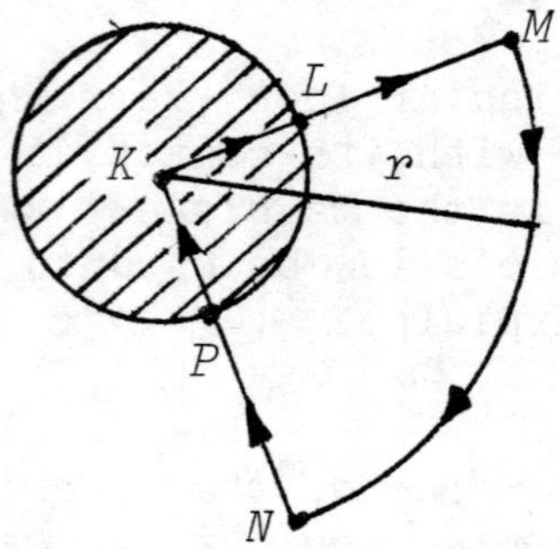

To illustrate the use of Ampere's law, we compute the magnetic induction which results from a steady current i in a long straight wire. By a "long" wire, we mean one having a length much greater than the radial distance r from the center of the wire to the field point at which we compute the induction. First, we use arguments from symmetry to show that the field lines are circles concentric with the wire.

The wire constitutes an *axis of symmetry*, which means that the magnetic induction is symmetric with respect to the wire. In general, we may resolve the $\vec{B}$ vector at any point into three mutually perpendicular components (Figure 58-10). At any specified distance r from the wire, the radial component of $\vec{B}$ must be constant in magnitude; if this component were not equal to zero, there would be a net magnetic flux originating or ending on the wire, which would contradict Gauss's law for magnetism. We therefore conclude that the radial component of $\vec{B}$ equals zero.

The axial component of $\vec{B}$, parallel to the wire, must also be constant in magnitude at any given distance r from the wire. It can be shown that this component is also zero, but a proof from symmetry considerations alone requires the distinction between polar and axial vectors, discussed in a more advanced text.

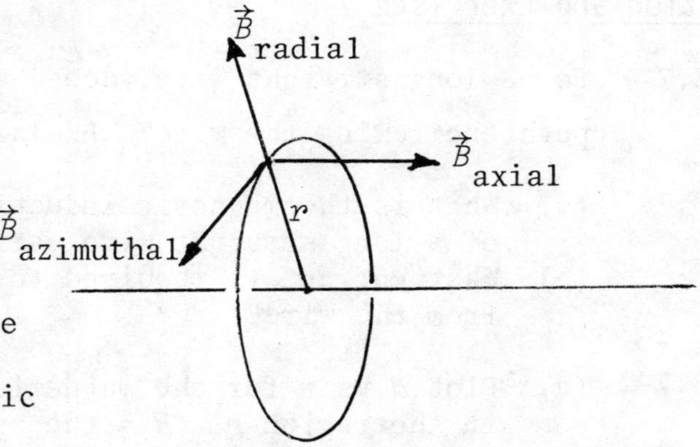

Figure 58-10. The magnetic induction at any point near a straight wire may be resolved into three mutually perpendicular components: (a) a radial component directed radially outward perpendicular to the wire; (b) an azimuthal or tangential component, in a plane perpendicular to the wire, tangent to a circle concentric with the wire; and (c) an axial component parallel to the wire. Using symmetry, it can be shown that the radial and axial components are both equal to zero.

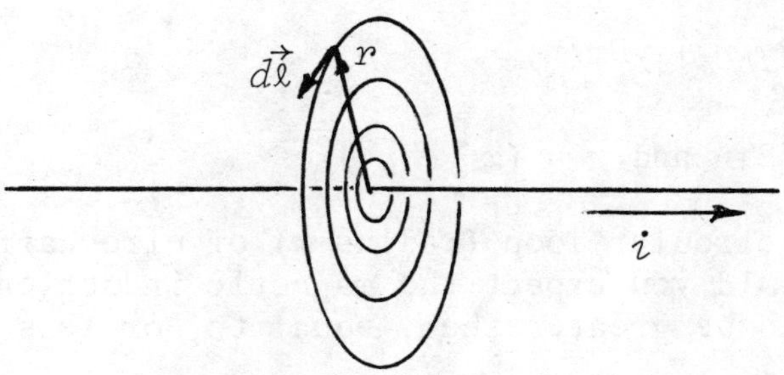

Figure 58-11. The magnetic field lines which result from a steady current in a long straight wire are circles concentric with the wire. The magnitude of the magnetic induction at a distance r from the center of the wire may be computed using Ampere's law.

To visualize the third component of $\vec{B}$, consider a circle in a plane normal to the axis of the wire, concentric with the wire and having radius r. The azimuthal component of $\vec{B}$ is tangent to this circle. Symmetry tells us that the azimuthal component of $\vec{B}$ must be constant in magnitude at a given distance r from the center of the wire. This means that each magnetic field line is a circle concentric with the wire.

We can find the azimuthal component, as a function of r, by applying Ampere's law to a circular path concentric with the wire and lying outside of the wire itself (Figure 58-11), a path which coincides with a circular field line having radius r. At every point on the path, $\vec{B}$ is parallel to $d\vec{\ell}$ so that $\vec{B} \cdot d\vec{\ell} = B\, d\ell$. Substituting in Ampere's law yields

$$\mu_0 i = \oint_C \vec{B} \cdot d\vec{\ell} = \oint_C B\, d\ell = B \oint_C d\ell = B\; 2\pi r \;,$$

from which

$$B = \mu_0 i/2\pi r, \qquad \text{for } r > \text{radius of the wire.}$$

Question and Exercises

Q 58.7-1 For a long straight wire, does $\oint \vec{B} \cdot d\vec{\ell} = \mu_0 i$ for a *non-circular* path encircling the wire? Explain.

E 58.7-1 (a) What is the magnetic induction at $r = 1$ cm from the center of a long straight wire carrying a current of one ampere?
(b) What current is required to produce $B = 10^{-4}$ T at $r = 1$ cm from the wire?

E 58.7-2 (a) Plot B vs r for the magnetic field of a long straight wire, in the region $r \geq R$ = the radius of the wire.
(b) How do you think B varies with r for $r < R$?

E 58.7-3 Consider a long straight wire along the z-axis carrying current i in the $+z$ direction. Show that at any point having coordinates (x,y,z) the total magnetic induction resulting from this current is given by

$$\vec{B} = -\hat{x}\, By/r + \hat{y}Bx/r,$$

where

$$B = \mu_0 i/2\pi r \text{ and } r = (x^2 + y^2)^{\frac{1}{2}}.$$

E 58.7-4 For a single circular loop (radius r) of wire carrying a steady current i, would you expect the magnetic induction B at the center of the loop to be greater than, equal to, or less than $\mu_0 i/2\pi r$? Explain.

Check your understanding of the material above by testing yourself on the minimum learning goal at the beginning of this section.

58.8 Magnetic Field in a Long Solenoid

Minimum Learning Goal. Given a description of a solenoid having a finite length, and an equation for the magnetic induction inside this solenoid on its axis, be able to calculate the relative difference between the magnetic induction in the middle of this "real" solenoid and the induction inside a "long" solenoid. (In other words, be able to compute how closely an actual solenoid approximates a "long" solenoid.) For example:

1 The magnetic induction inside and on the axis of any real solenoid is given by the equation

$$B = (\mu_0/2)(N/L)i\ (\cos\theta_1 + \cos\theta_2)$$

where N is the total number of turns along the solenoid length L, i is the current, and θ_1 and θ_2 are the angles shown below.

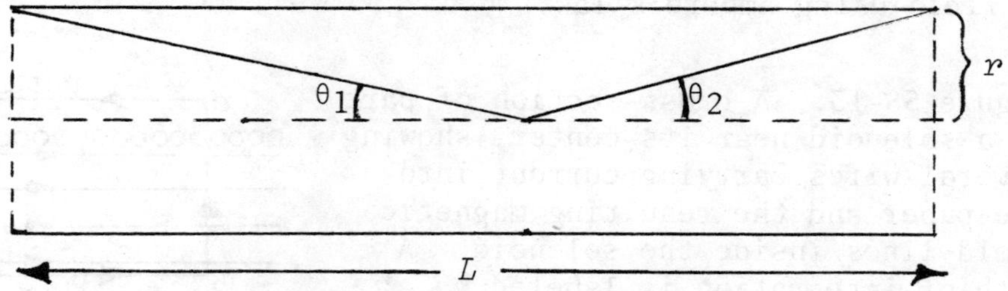

Consider a real solenoid having L = 10 cm and r = 2 cm. What is the relative difference, $(B_{long} - B_{real})/B_{long}$, between the magnetic induction on the axis, midway between the ends of this real solenoid, and that induction inside a "long" solenoid having the same number of turns per unit length and carrying the same current? (Answer, page 58-36, Line 15)

2 Consider a real solenoid having L = 100 cm and r = 1 cm. Using the equation of example 1, what is the relative difference between the magnetic induction on the axis, midway between the ends of this solenoid, and the induction inside a "long" solenoid having the same number of turns per unit length and carrying the same current? (Answer, page 58-36, Line 6)

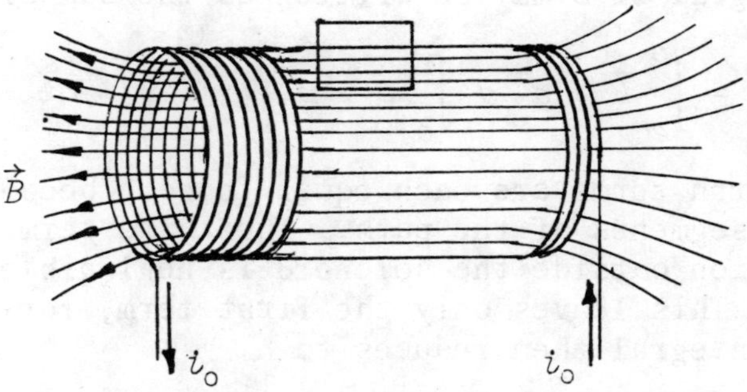

Figure 58-12. A solenoid contains a large number of closely spaced turns of wire, wound on a form such as a cylindrical tube. Current in the solenoid establishes a magnetic field having field lines which run through the solenoid approximately parallel to its axis and return outside the solenoid in loops spaced relatively far apart.

A *solenoid* is a cylindrical tube wound with many turns of wire carrying a current. Figure 58-12 is a cutaway view showing some of the turns of wire and some of the field lines in the region of a solenoid. Each turn of the solenoid produces field lines which encircle that turn. These fields lines are all approximately parallel to the axis inside the solenoid, but they diverge near the ends of the solenoid to form closed loops which spread relatively far apart in the region outside the solenoid. If the solenoid has a length much greater than its diameter, the field lines are much more dense inside than outside so that the magnetic induction outside is negligible compared to the induction inside. If we consider a region inside the solenoid far from the ends, the field is parallel to the axis. We can therefore readily calculate the magnitude of this field using Ampere's law.

Figure 58-13. A cross-section of part of a solenoid near its center, showing several wires carrying current into the paper and the resulting magnetic field lines inside the solenoid. A path of integration is labeled *bcde*; this path is also shown (but not labeled) in Figure 58-12.

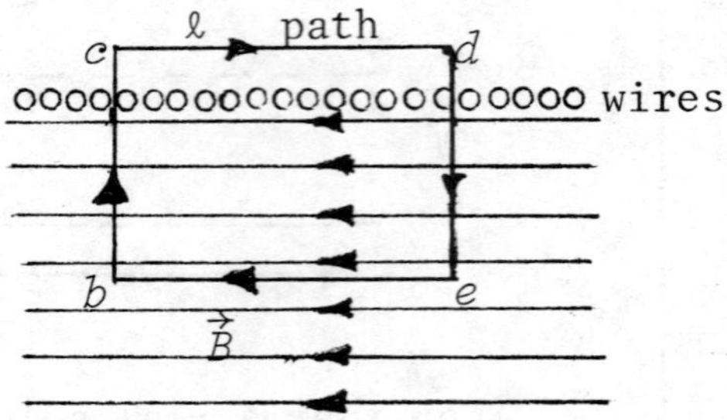

We use i to denote the current (which is the same in each turn), r to denote the radius of the solenoid, L for its length, and N for the total number of turns of wire. Then the number of turns per unit length is $n = N/L$. To apply Ampere's law, we choose a rectangular path C as shown in Figure 58-13, having corners labeled a, c, d and e and a length ℓ parallel to the axis of the solenoid. Ampere's law for this closed path is

$$\oint_C \vec{B} \cdot d\vec{\ell} = \mu_0 \, i_C,$$

where i_C is the total current linking C. This current is i times the number of turns in the length ℓ, which gives $i_C = in\ell$.

The line integral of $\vec{B}$ may be written as the sum of four separate terms:

$$\oint_C \vec{B} \cdot d\vec{\ell} = \int_e^b \vec{B} \cdot d\vec{\ell} + \int_b^c \vec{B} \cdot d\vec{\ell} + \int_c^d \vec{B} \cdot d\vec{\ell} + \int_d^e \vec{B} \cdot d\vec{\ell}.$$

The second and fourth terms are each equal to zero because $\vec{B}$ is perpendicular to $d\vec{\ell}$ along these segments of the path. The third term is equal to zero because the induction outside the solenoid is negligible compared to the induction inside. This leaves only the first term, for which $\vec{B}$ is parallel to $d\vec{\ell}$. The line integral then reduces to

$$\oint_C \vec{B} \cdot d\vec{\ell} = \int_e^b B \, d\ell = B \int_e^b d\ell = B\ell.$$

Combining our results, we obtain

$$B\ell = \mu_0 \, in\ell \,, \qquad \text{or} \qquad B = \mu_0 \, ni$$

for the magnetic induction near the center of a *long* solenoid. This result does not depend on the diameter of the solenoid nor on the total number of turns, but only on the current and the number of turns per unit length. For a solenoid whose length is not much greater than its diameter, the equation for the induction on the axis is more complicated; it is given in the Minimum Learning Goal for this section.

The reader should confirm that the direction of the current in the winding and the direction of the induction inside the solenoid are related by a Right Thumb Rule, where the fingers curl around the solenoid in the direction of the current and the thumb indicates the direction of the induction.

Questions and Exercises

Q 58.8-1 State a rule which gives the direction of the magnetic induction in a solenoid, in terms of the direction of the current in the winding.

Q 58.8-2 The magnetic field inside a solenoid is approximately uniform. Why is it often advantageous to do experiments in a uniform field rather than in a non-uniform field?

Q 58.8-3 Find some examples of solenoids in equipment used in or near your residence. What are the functions of these solenoids?

E 58.8-1 Sketch a map of the $\vec{B}$ field in a plane containing the axis of a solenoid.

E 58.8-2 Suppose you are asked to design a solenoid which will produce a $\vec{B}$ field as large as possible throughout a volume of $(0.1\ \text{m})^3$. What factors must be considered in working out your design?

E 58.8-3 Show that the total magnetic flux inside a long solenoid is $\Phi_m = \mu_0 niA$, where A is the total cross-sectional area of the solenoid.

Check your understanding of the material above by testing yourself on the minimum learning goal at the beginning of this section.

Answers to Minimum Learning Goals

Line 1 $i = 4.17$ microampere

Line 2 No.

Line 3 $\vec{B} = (4.17 \times 10^{-6}\ \hat{x} - 1.67 \times 10^{-6}\ \hat{y})$ tesla

Line 4 (a) The diagram could represent an electrostatic field.
(b) The diagram could represent a magnetostatic field.

Line 5 $$\int_P^K (B \cos 90°)d\ell + \int_K^L (B \cos 90°)d\ell + \int_L^M (B \cos 90°)d\ell$$

$$+ (B \cos 0°) \int_M^N d\ell + \int_N^P (B \cos 90°)d\ell = 0 + 0 + 0 + B(2\pi r/4) + 0 = \mu_0 i/4.$$

Then $B = \mu_0 i/(2\pi r)$.

Line 6 $\Delta B/B = 2 \times 10^{-4}$ (i.e. 0.02% error)

Line 7 9.7×10^{-5} Wb

Line 8 Yes.

Line 9 See diagram at right.

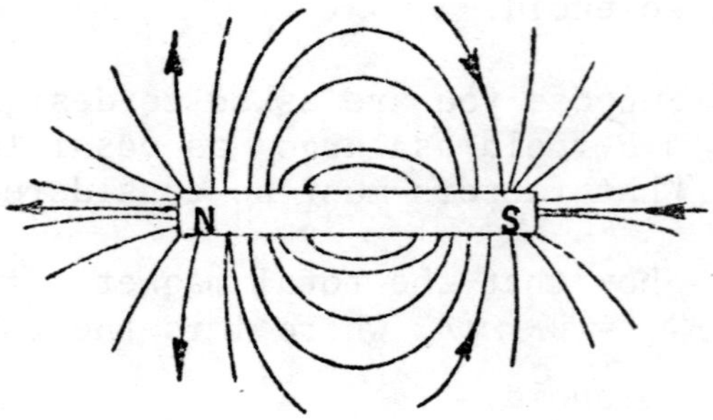

Line 10 14.3 microampere

Line 11 (a) The field could be electrostatic but not magnetostatic.
(b) The field could be a magnetostatic field.
(c) Could be either depending on what is inside the central region (not shown on the graph).

Line 12 $$\int_K^L (B \cos 90°)d\ell + \int_L^M (B \cos 90°)d\ell + \int_M^N (B \cos 90°)d\ell + (B \cos 0°) \int_N^K d\ell$$

$= 0 + 0 + 0 + B(2\pi r/2)$, all of which must equal $\mu_0 i/2$; therefore $B = \mu_0 i/2\pi r$.

Line 13 1.6×10^{-16} N, downward

Line 14 1.72×10^{-3} Wb (Hint: the area element $dS = 2\pi r dr$).

Line 15 $\Delta B/B = .07$ (i.e., 7% error)

Line 16 $q/m = 1.69 \times 10^{7}$ C/kg

Line 17 $B = 1.14 \times 10^{-3}$ T

ANSWERS to Questions and Exercises in Chapter 58

Section 58.1

Q 58.1-2 No, not if the pith ball is stationary, because then its electric field does not change with time.

Q 58.1-4 A magnetic field.

Section 58.2

Q 58.2-2 Each piece, whether two or more, will have its own north pole and south pole, with a "pole strength" smaller than that of the original magnet.

E 58.2-2 (See Figure 58-1).

Section 58.3

Q 58.3-2 Because there are no magnetic monopoles, field lines must be continuous loops, so in the case of a bar magnet these loops must pass through the interior of the magnet.

Q 58.3-4 Nothing flows that corresponds to the magnetic flux.

Q 58.3-6 (a) The electric flux through dS in the sense of $\hat{n}$ is

$$d\Phi_e \equiv \vec{E} \cdot \hat{n}\, dS$$

(b) No. (c) Yes.

Section 58.4

Q 58.4-2 The line along which the charges are moving is a line of symmetry for this situation. If charge *A* establishes a magnetic field at the location of charge *B*, this field cannot be perpendicular to the line of symmetry because such a field would not have the same symmetry as the line on which the charges are moving. Therefore, if charge *A* establishes a magnetic field at charge *B*, this field must lie along the line of symmetry. As shown in this section, a charged particle moving parallel to a magnetic field has no deflection, because there is no magnetic force on such a particle. Therefore, the two charges described in this question cannot exert magnetic forces on each other.

Q 58.4-4 (b), (d), (g), (h), (k), (m).

Q 58.4-6 (a) (1) A magnetic field alone, parallel (or anti-parallel) to the electrons' velocity.
(2) Neither a magnetic nor an electric field.
(3) Electric and magnetic fields at right angles to each other, and also at right angles to the electrons' velocity, having magnitudes such that the resulting electric force on the electrons just cancels the magnetic force on them.
(b) Direct the stream of electrons through the same field, but moving in different directions. If an electric field is present the electric force which it exerts will remain unchanged, whereas if a magnetic field is present the magnetic force will depend on the angle between the velocity and the field.

E 58.4-2 1.25 tesla, toward the north.

E 58.4-4 8.0×10^{-15} $[-6\hat{x} + 3\hat{y} - 2\hat{z}]$ newton

E 58.4-6 (a) $\vec{F}_e$ is parallel to $\vec{E}$, and $\vec{F}_m$ has the opposite direction.
$\vec{B}$ is directed upward, out of the paper.
(b) 10^6 m/s

Section 58.5

Q 58.5-2 F_m, the magnitude of the force, remains constant; the direction of $\vec{F}_m$, however, continually changes so as to remain perpendicular to $\vec{v}$.

Q 58.5-4 (a) no; (b) yes.

E 58.5-2 (a) 1.9×10^{-3} T;
(b)

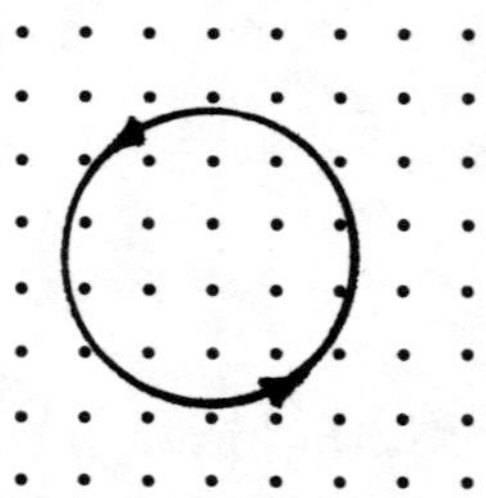

$\vec{B}$ up, out of paper

E 58.5-4 electron, proton, neutron.

Section 58.6

Q 58.6-2 Magnetic field lines exist only when current exists, and the field lines always encircle the current which produces them.

E 58.6-2 i_1 threads C_1; zero current threads C_2; i_1 threads C_3; $-i_1$ threads C_4.

E 58.6-4 (a)

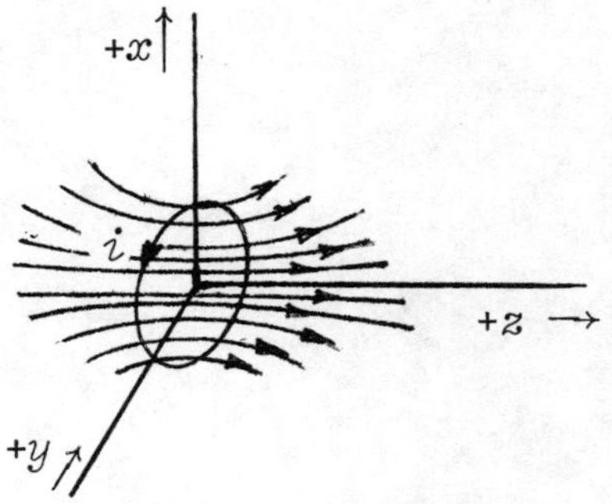

(b) No, because there is no closed path encircling the current on which $\vec{B}$ has enough symmetry to permit calculation of $\vec{B}$ at the center of the loop.

Section 58.7

E 58.7-2 (a)

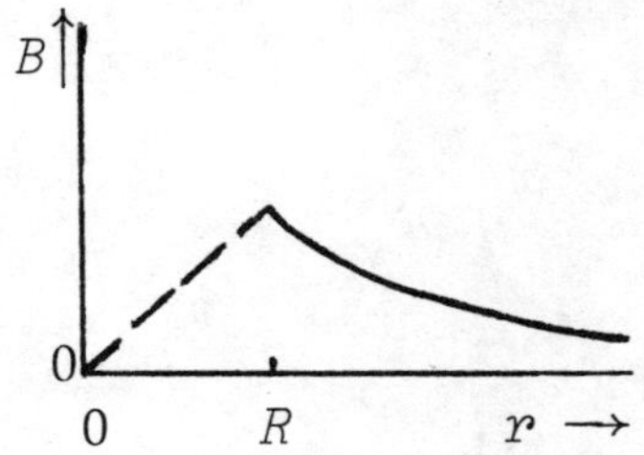

(b) B must be finite for all values of r. (In fact it varies linearly as shown by the dashed line in the graph.)

E 58.7-4 One might expect B to be greater than $\mu_0 i/2\pi r$, because all parts of the wire loop are at the same distance r from the center. (The actual value is $B = \mu_0 i/2$, which is greater than $\mu_0 i/2\pi r$ only for $r > 1/\pi$ meter.)

Section 58.8

Q 58.8-2 In a uniform magnetic field the magnetic force on a moving charged particle does not depend on the particle's location in the field, so it is much easier to compute the effects of the field in this case than when the field is non-uniform.

E 58.8-2 Factors include the desired shape of the working region of the magnetic field, the number of turns of wire needed, the wire diameter, the current, the heating effect of the current and provision for safe dissipation of the joule heat developed, the source of current, connecting the windings in series or parallel of both, the cost of materials and of electric power, etc.

CHAPTER 59 ELECTROMAGNETISM

Introduction

In Chapter 58 we considered magnetic fields which result from electric currents which are steady, not changing with time. Here we consider some interactions of electric and magnetic fields which result when electric currents vary with time. The concept of magnetic flux is used to state Faraday's law of electromagnetic induction: a changing magnetic field gives rise to an electric field. Lenz's law indicates that the direction of the induced electric field always opposes the change in magnetic field which induces the electric field. The magnetic flux linking an electric circuit element gives rise to a property of the circuit element called the inductance, which depends on the geometry of the circuit element but not on the current in it. In a circuit containing an inductor and resistor where the current is changing, the induced voltage leads to a variation of current with time characterized by a natural time constant.

When you have completed your study of the material in this chapter, you should be able to answer questions and work exercises like the examples in the minimum learning goals assigned by your instructor.

59.1 Electromagnetic Induction. Faraday's Law

Minimum Learning Goals. (1) Given the orientation of a changing magnetic field, be able to specify the direction of another field which results from the one which is changing. For example:

1 A betatron has a doughnut-shaped, evacuated cavity placed so that the axis is parallel to a magnetic field as shown below. Describe the field inside the doughnut cavity when the magnetic field through its hole is changing at a uniform rate per unit of time. (Answer, page 59-15, Line 1)

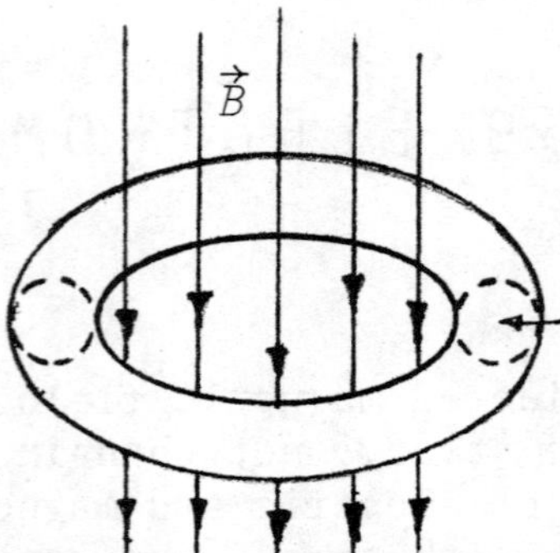

Magnetic field lines passing through a doughnut-shaped cavity

(2) Be able to calculate the tangential electric field strength on the perimeter of a specified surface, given the dimensions of the surface and the change in magnetic induction through the surface in a given period of time. Also, be able to specify the direction of the magnetic field produced by the current, if any, which results from this electric field. For example:

1 A long solenoid has a radius of 5 cm. Suppose a high-resistance wire forming a small closed circular loop having a radius of 2 cm is placed inside the solenoid, far from either end, and aligned with its axis coincident which the solenoid axis. The magnetic induction inside the solenoid is then changed uniformly from 0 to 0.050 tesla in 2 seconds.
(a) What are the magnitude and direction of the electric field strength in the wire loop during those 2 seconds?
(b) What is the direction of the magnetic field produced by the resulting current in the wire loop, relative to the direction of the solenoid magnetic field? (Answer, page 59-15, Line 3)

2 A circular wire loop of radius 4 cm is placed between the parallel pole faces of an electromagnet with its axis normal to the pole faces.
(a) If the magnetic field is increased uniformly from zero to 0.024 tesla in 5 seconds, what are the magnitude and direction of the electric field strength in the wire loop during those 5 seconds?
(b) What is the direction of the magnetic field produced by the resulting current in the loop, relative to the direction of the magnetic induction of the electromagnet?
(Answer, page 59-15, Line 2)

In electrostatics, the only sources of electric field are charges, a fact expressed in Gauss's law for electricity. In magnetostatics, the only sources of magnetic field are currents, a fact expressed in Ampere's law. Experiments show, however, that *the fields can also act as sources for each other*. Specifically, the *time-rate of change* (or time-derivative) of $\vec{B}$ acts as a source of $\vec{E}$. This cross-coupling of electric and magnetic fields occurs in the phenomenon of *electromagnetic induction,* discussed in this section and illustrated using several examples. It is also found that the *time-rate of change* of $\vec{E}$ acts as a source of $\vec{B}$, a different phenomenon to be discussed later. These new types of "sources" are included in the circulation conditions for the electric and magnetic fields.

Figure 59-1. While a conductor forming part of a circuit is moved through a magnetic field so as to change the total magnetic flux linking the circuit, current exists in the circuit. The changing magnetic field creates an electric field which causes this current.

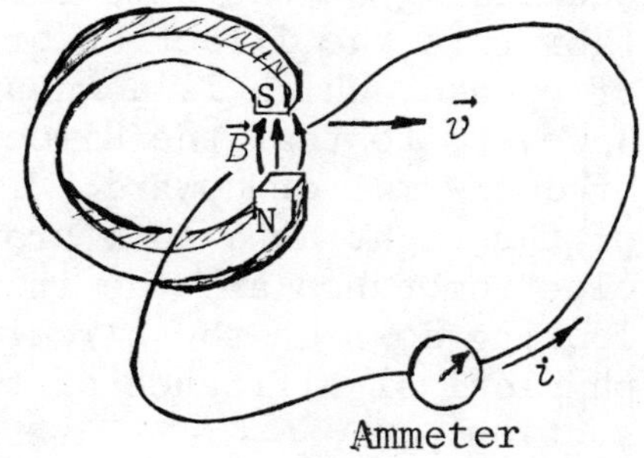

Suppose we have a length of wire connected to the terminals of an ammeter, such that the loop formed by the wire is linked with some of the magnetic flux of a permanent magnet (Figure 59-1). If the wire is moved so that it cuts across some magnetic field lines, there is a change in the total magnetic flux linking the circuit, and the ammeter indicates a current in the wire. The current exists only while the wire is moving, that is, only while the magnetic flux through the loop is changing. An electromotance $\mathcal{E}$ is required to produce a current (Section 26.3), so we conclude that while the flux through the loop is changing an electromotance is created in the wire. Because there is no battery or other conventional source of electromotance in this circuit, this is called an *induced electromotance.* The experiment just described is said to display the phenomenon of *electromagnetic induction.*

From measurements, the equation connecting the electromotance with the time-rate of change of flux is found to be very simple:

★ $$\mathcal{E} = -\frac{d\Phi_m}{dt}, \qquad \text{Faraday's law.}$$

Applied to the experiment described above, $\mathcal{E}$ represents the total electromotance induced throughout the length of the circuit, and Φ_m is the total magnetic flux linking this circuit at any time. The reason no proportionality constant is required in the equation is that this law of electromagnetic induction is used to define the unit of flux, the weber.

> The *weber* is the magnetic flux which, linking a circuit of one turn, produces in it an electromotance of 1 volt as it is reduced to zero at a uniform rate in 1 second.

The equation above is often called *Faraday's law,* after Michael Faraday (1791-1867) who discovered it in England in 1831. This law was also discovered independently in the United States by Joseph Henry in the same year.

The negative sign in Faraday's law indicates the direction of the induced electromotance (and of the resulting current) relative to the direction of the changing flux. These directions are related by another Right Thumb Rule, in which the fingers curl around the circuit in the direction of the induced electromotance and the thumb indicates the direction *opposite* to that in which the flux is *increasing*, or *the same* as that in which the flux is *decreasing*. To illustrate this rule, in the example described above the direction of the magnetic flux Φ_m (or the magnetic induction $\vec{B}$) threading the circuit is upward, but this flux is decreasing so the time-rate of change of flux $d\Phi_m/dt$ is negative and therefore the direction of the change in flux is the negative or opposite of upward, which is downward. Faraday's law contains a negative sign, however, to that the direction of the negative of $d\Phi_m/dt$ is the negative of downward, or upward. If the right thumb is pointed upward, the fingers of the right hand curl around the circuit in the direction of the induced electromotance and the induced current. As shown in Figure 59-1, when looking down on the circuit this is the counter-clockwise direction. The physical significance of this negative sign is discussed further in the next section.

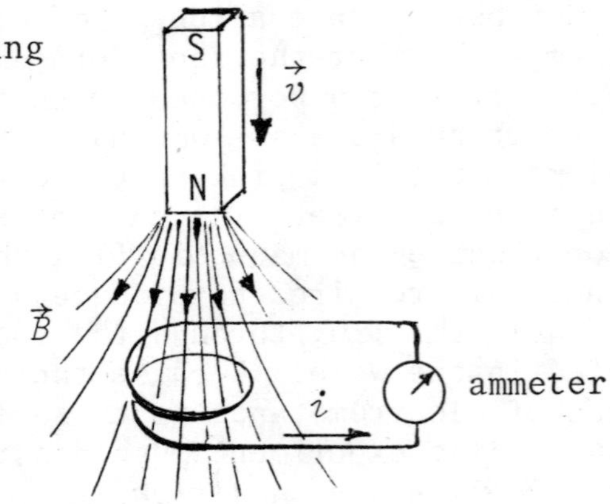

Figure 59-2. While a magnet is moving in such a way that the magnetic flux linking a circuit is changing, current exists in the circuit. The changing magnetic field creates an electric field in the circuit. Only a few of the field lines from the magnet are shown.

In the example described above, an electromotance is induced in a circuit while the wire is being moved with respect to a magnet. What happens if the circuit is kept stationary while the magnet is moved, as shown in Figure 59-2? In this figure, a magnet is positioned so that some of the magnetic flux from it passes through or "links" a circuit consisting of some loops of wire connected to an ammeter. When the magnet is moved downward toward the circuit, as shown, the total magnetic flux linking the circuit increases, and once again the ammeter indicates a current in the wire. As before, the current is there only while the magnetic flux through the circuit is changing. The direction of the induced current is related to the direction of the magnetic induction by the Right Thumb Rule stated in this section. The induced electromotance is equal to the time-rate of change of flux, by Faraday's law.

The first example above was described using the frame of reference in which the magnet is at rest and the circuit is moving. The second example was described using the frame of reference in which the circuit is at rest and the magnet is moving. The first example could also be described using a reference frame in which the circuit is at rest, and the second example could be described using a frame in which the magnet is at rest. In every case the ammeter indication would show that an electromotance is induced in the wire. That is, Faraday's law applies in a moving frame of reference as well as in a stationary frame. Faraday's law is *also* valid in situations where neither the circuit nor the magnet is moving, as shown in the next example.

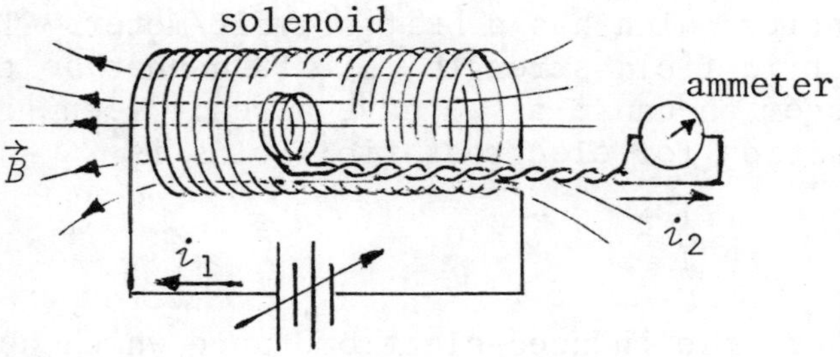

Figure 59-3. While the current i_1 in the solenoid is changing, another current i_2 exists in a separate circuit which is linked to the magnetic flux produced by the solenoid. The changing magnetic field of the solenoid creates an electric field in the secondary circuit containing the ammeter. The wires leading from the small secondary coil to the ammeter are twisted around eachother so that the net flux linking the lead wires will be negligible.

Consider a solenoid supplied with current from a variable-voltage source, as shown in Figure 59-3. While current exists in the solenoid, the flux density within the solenoid near its center is directly proportional to this current. If a small coil consisting of one or a few turns of wire is placed near the center of the solenoid so that field lines from the solenoid link the coil, and lead wires from the coil are connected to an ammeter, then it is found that while the magnetic flux within the solenoid *is changing* current exists in the secondary circuit containing the ammeter, even though this circuit has no direct electrical connection to the primary circuit containing the battery and the solenoid. The secondary current has one direction while the solenoid current is increasing, but the opposite direction while it is decreasing. Again it is found that the direction of the current induced in the coil is related to the direction of the changing induction by the Right Thumb Rule stated above, and the induced electromotance is correctly given by Faraday's law:

$$\xi = -d\Phi_m/dt \ .$$

The equation above expresses Faraday's law in terms of electromotance and magnetic flux. To see how this law is expressed in terms of electric and magnetic fields, we note that the total magnetic flux through a surface S in the sense of the normal unit vector $\hat{n}$ is

$$\Phi_m = \iint_S \vec{B} \cdot \hat{n} \, dS \ .$$

In each of the examples described above, we may imagine that the circuit containing the ammeter is the edge or boundary of an open surface S. The magnetic field lines which pierce this surface compose the magnetic flux which links the circuit. The time-rate of change of this flux corresponds to the time-rate of change of the magnetic induction $\vec{B}$ at points on the surface.

The electromotance which sends induced current through the ammeter in each example is not associated with a battery or other distinct source of electromotance. Rather, it is an electromotance produced or induced throughout the entire circuit for which the linking flux is changing. It is reasonable, therefore, to express this electromotance, which has the SI unit of volts, in terms of the line integral around the circuit of some quantity which has a unit of volt/meter. This quantity, of course, is an electric field strength, $\vec{E}$. It cannot be the electro*static* field strength, however, because as we have already seen (in Section 56.10) the circulation condition for electrostatic fields is

$$\oint_C \vec{E} \cdot d\vec{\ell} = 0.$$

Rather, the induced electromotance which appears in Faraday's law is associated with an *induced* electric field strength, $\vec{E}_{ind}$. The circulation condition for this induced field strength may then be written

$$\oint_C \vec{E}_{ind} \cdot d\vec{\ell} = \xi = -\frac{d}{dt}\iint_S \vec{B} \cdot \hat{n}\, dS,$$

where the path of integration, C, coincides (in our examples) with the circuit containing the ammeter.

In this way, the induced electromotance ξ may be expressed in terms of electric field strength, $\vec{E}_{ind}$, and the magnetic flux may be expressed in terms of the magnetic induction $\vec{B}$. We are thus led to the following statement of Faraday's law: *A changing magnetic field creates an electric field.* More specifically, the induced electric field strength is proportional to the time-rate of change of the magnetic field strength. Because one of the integrals above is a line integral while the other is a surface integral, it is not an easy task to transform these integrals to obtain an equation for $\vec{E}_{ind}$ in terms of $d\vec{B}/dt$ which does not involve integrals. This derivation and the resulting equation are left to a more advanced textbook.

In the general case, the total electric field strength at any point is the sum of the electrostatic field strength and the induced electric field strength. The time-derivative of $\vec{B}$ then appears in the circulation condition for the total electric field. In the presence of a changing magnetic field, therefore, the electric field has a radical "personality change." In the static case the line integral of the electric field strength around a closed path was always zero, but now it is not. In the presence of time-varying magnetic fields, *there is no electric potential*, and a test charge can have net work performed on it when it is carried around a closed path. The *induced* electric field lines run in closed loops; they do not begin and end on electric charges as do the electrostatic field lines.

The magnetic field, however, still consists of closed field lines. The fact that these lines do not begin or end is essential for defining the magnetic flux linking a circuit or a closed path in the field. If field lines begin or end, it doesn't make sense to ask whether a line "links" a loop.

We note that a changing magnetic field induces an electric field even if there is no electric circuit or wire which coincides with a closed path in the magnetic field. An electric field is induced at every point in space where the magnetic field is changing, and if a free charged particle is located at such a point it will be accelerated by this electric field. This principle is the basis for the design of a charged-particle accelerator called the *betatron*.

Questions and Exercises

Q 59.1-1 Suppose you have a wire, not connected to a battery or other voltage source, which is in a magnetic field. Is it necessary for the wire to move across lines of magnetic flux in order to produce an induced electric field in the wire? Explain.

Q 59.1-2 Given a loop of wire having a fixed size and shape, and a constant magnetic field, how could you induce an electromotance in the wire: (a) if the field is non-uniform? (b) if the field is uniform?

Q 59.1-3 Describe several different ways to demonstrate the production of an induced electromotance by Faraday's law.

Q 59.1-4 (a) Under what conditions does the electric field contain only field lines which have definite beginning and end points?
(b) Under what conditions can the electric field contain field lines which are closed loops?
(c) Under what conditions does the electric field contain *only* field lines which are closed loops?

Q 59.1-5 What is the complete equation for the line integral $\int \vec{E} \cdot d\vec{\ell}$ around a closed path?

Q 59.1-6 Consult reference material to learn how a betatron uses a changing magnetic field to accelerate charged particles.

E 59.1-1 (a) What are the physical units of $\int \vec{E} \cdot d\vec{\ell}$?
(b) Because the time-derivative of $\vec{B}$ acts as a source of $\vec{E}$, we may expect that some product $\beta d\vec{B}/dt$ has the same units found in part (a). Then what units must β have?

E 59.1-2 In an extended uniform field $\vec{B} = \hat{z}B_0$ there is a fixed U-shaped conducting wire having the configuration shown. A conducting rod parallel to the y-axis rolls along the U in contact with it, with velocity $\vec{v} = \hat{x}v_0$. $B_0 = 0.1$ T, and $v_0 = 0.2$ m/s.
(a) What electromotance is induced in the conducting loop?
(b) What is the direction of the resulting current?

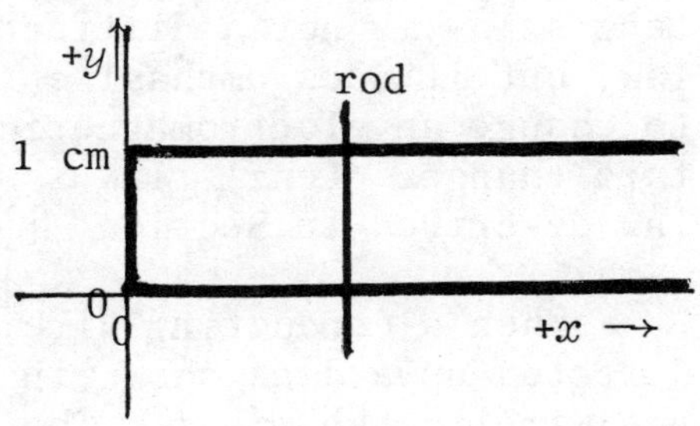

E 59.1-3 A localized uniform field $\vec{B} = -\hat{z}B_0$ penetrates an area 1 cm square centered at the origin. A loop of wire 1 cm square moves along the $+x$ axis with constant velocity $\vec{v}$.

(a) Plot Φ_m vs t for this motion, where Φ_m is the magnetic flux which pierces the loop.

(b) Plot $\mathcal{E}$ vs t for this motion, where $\mathcal{E}$ is the electromotance induced in the loop.

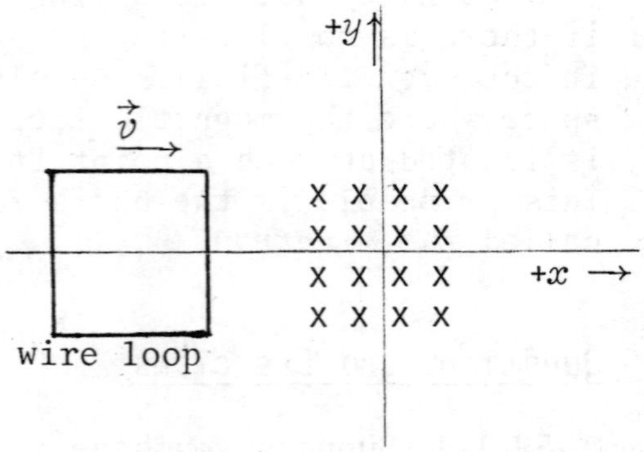

((E 59.1-4 In a uniform field $\vec{B} = \hat{z}B_0$ there is a rectangular wire loop having sides a and b. The loop is rotated about an axis through its center, parallel to the x axis and parallel to the long side a with uniform angular speed of ω radians/second.

(a) Compute the total magnetic flux through the loop, $\Phi_m(t)$.

(b) Compute the electromotance $\mathcal{E}(t)$ induced in the loop.))

Check your understanding of the material above by testing yourself on the minimum learning goals at the beginning of this section

59.2 Lenz's Law

As stated in the preceding section, the negative sign in Faraday's law indicates the direction of the induced electromotance and the resulting current, relative to the direction of the changing flux. These directions are related by a Right Thumb Rule. The significance of this negative sign goes far beyond its role in indicating the direction of the induced electric field, however. The more general import of the negative sign is summarized in a principle known as Lenz's law:

★ *Lenz's Law.* The effects of a current produced by electromagnetic induction are always in the sense which *opposes* the action causing the induction.

Lenz's law is not a distinct principle of physics separate from Faraday's law, but it does emphasize the physical fact that whenever some agent acts to change an electromagnetic field, nature responds in a way that opposes this change. Lenz's law is illustrated by the three examples of Faraday's law described in Section 59.1.

When a conducting wire is moved to the right in a magnetic field directed upward as shown in Figure 59-1, the "action" is the movement of the wire to the *right*. The induced current is counterclockwise in the loop of wire, with the result that the magnetic force on the induced current is toward the *left*, as may be verified by applying the Right Hand Rule described on page 58-13. The effect of this magnetic force to the left is to oppose the initial movement of the wire toward the right.

When a magnet is moved downward toward a stationary circuit or coil of wire as shown in Figure 59-2, the "action" is the downward movement of the magnet. The induced current is counterclockwise in the coil of wire, with the result that this current creates its own magnetic field. The field lines of this induced field are directed upward, as may be verified by applying the Right Thumb Rule for the solenoid (page 58-35) or referring to Figure 58-12 on page 58-33. The current in the coil thus acts like a magnet having its north pole upward and its south pole downward. The north pole of this induced magnet repels the north pole of the original magnet moving downward. The effect of this induced magnetic field is thus to oppose the initial motion of the permanent magnet toward the coil.

When the current in a solenoid is decreasing, the flux produced by the solenoid is also decreasing, and this *decreasing* flux may be considered as the "action" in Lenz's law. If the small coil within the solenoid in Figure 59-3 is viewed from the location of the ammeter, the sense of the electromotance induced in the coil is clockwise, and the resulting current induced in the coil is also clockwise. This induced current produces magnetic flux directed toward the left, in the same direction as the original flux in the solenoid. The induced flux, therefore, tends to offset the decreasing flux of the solenoid and to maintain the original flux density; it thus opposes the decrease in flux which caused the induction.

In these illustrations of Lenz's law, the "action" was identified as the movement of a wire in the first example, as the movement of a magnet in the second example, and as a change in flux in the third example. Alternatively, one can think of the action as a change in flux in all three examples. In each example, the induced current produces some induced flux which partially offsets the decrease in flux causing the induction. The reader should apply Lenz's law to the first two examples, considering the change in flux through the circuit as the "action."

Lenz's law is manifested whenever the magnetic flux linking a circuit or a closed path is changing.

It is instructive to speculate on the consequences which would result if the sign in Faraday's law were positive instead of negative. Then the opposite of Lenz's law would be true, and the effects of an induced current would always *reinforce* the action causing the induction. The induced current would cause a moving wire to move *faster* rather than slower in Figure 59-1. The induced current would create a magnetic field which attracts the moving magnet in Figure 59-2, instead of repelling it. The induced current would hasten the decrease in flux in Figure 59-3 if the solenoid current were decreasing, but would hasten the increase in flux if the solenoid current were increasing. If the opposite of Lenz's law were to prevail, nature would "run away with itself" whenever magnetic flux is changing. The fact that Lenz's law is operative is closely linked to the law of conservation of energy.

Questions

Q 59.2-1 State Lenz's law.

Q 59.2-2 For each example you found for Q 59.1-3, list the action producing the electromotance and the effects of the electromotance, in order to illustrate Lenz's law.

Q 59.2-3 What would happen if the sign of Lenz's law were the opposite of what it is?

59.3 Inductance and Inductors

Wherever electric current exists, it establishes a magnetic field (Section 58.6). The magnetic induction $\vec{B}$ at every point in the field is directly proportional to the current, and the magnetic flux through any surface is proportional to the induction (Section 58.3). This means that the flux Φ_m is directly proportional to the current i producing it, when no iron or other magnetic material is in the vicinity. The flux through magnetic material does *not* depend linearly on the current, so we limit all of our consideration of inductance to situations where no magnetic materials are present.

Figure 59-4. Electric current establishes magnetic flux which links the circuit where the current exists. In a circuit element containing closely spaced turns of wire, the flux produced by the current in each turn links all the turns of the coil or solenoid.

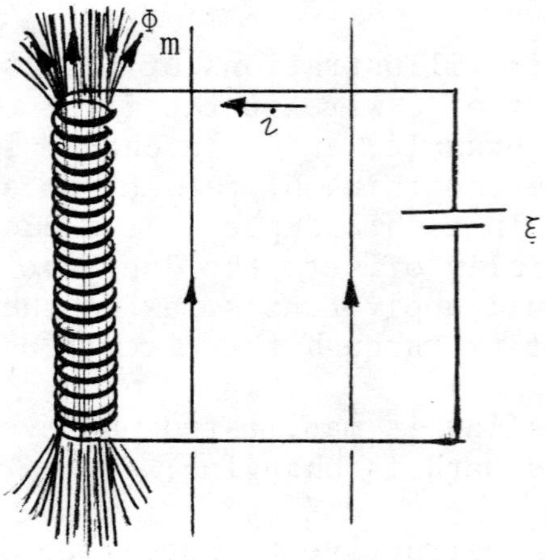

Magnetic flux is defined in terms of the field lines piercing an open surface S which has a definite boundary. When the flux is produced by current in a circuit, it is natural to choose a surface whose boundary coincides with the circuit, and to speak of the flux *linking* the circuit (Figure 59-4). When the circuit contains a solenoid or coil having closely spaced turns of wire, the flux in this circuit element has a relatively high density because the current in every turn of wire produces flux which links all of the turns. If the flux Φ_m in this circuit element links each of N turns of wire, the total flux linking the circuit element is $\Phi_\ell = N\Phi_m$. The self-inductance often called simply the *inductance*, of a circuit and of a circuit element are defined as follows:

★ The *self-inductance* of a circuit (element) is $L \equiv \Phi_\ell / i$, where Φ_ℓ is the *total* magnetic flux linking the circuit (element resulting from the current i in the circuit (element).

(While this definition is satisfactory for our purposes, it must be reconsidered when dealing with circuit elements containing iron or other magnetic materials.) The SI unit of inductance is called the *henry*, symbol H, in honor of the American physicist Joseph Henry (1797-1878). From the definition of inductance, it is seen that 1 henry = 1 weber/ampere = 1 volt·second/ampere.

Every circuit has some self-inductance, but in most cases we are concerned only with the self-inductance concentrated or lumped in circuit elements having closely spaced turns of wire, such as a solenoid. Such a circuit element is called an *inductor*. When one or more inductors are in a circuit, nearly all of the flux linkage is concentrated or lumped in the inductors, so that for practical purposes we need to consider only the inductance of the inductors while the inductance in other parts of the circuit may be neglected.

Consider an air-filled ideal solenoid (Section 58.8) having a total of $N = n\ell$ turns in its length ℓ, a cross-sectional area A and a curent i. The total flux inside the solenoid is given by $\Phi_m = \mu_0 \, niA$ (Exercise 58.8-3), and this flux links *each* turn of the solenoid. The total flux linking the solenoid is therefore $\Phi_\ell = N\Phi_m = \mu_0 \, ni^2 A\ell$. From the definition, we compute the inductance of the solenoid to be

$$L = \Phi_\ell / i = \mu_0 \, n^2 A \, \ell.$$

It is seen that the inductance depends only on μ_0 and on geometrical factors, but not on the current in the solenoid. The same is true for all inductors, in the absence of magnetic materials.

Practical inductors are often made in the form of a solenoid wound on a cylindrical form. Inductors are used in electric circuits to store energy in the magnetic field, and to reduce abrupt changes in current in circuits subjected to sudden changes in the applied electromotance or to high-frequency alternating voltages. Common inductors have values of inductance in the microhenry or millihenry range. The symbol for an inductor in a circuit is

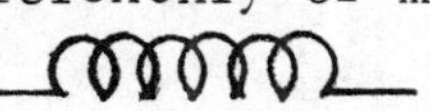 .

Questions and Exercise

Q 59.3-1 What is an inductor?

Q 59.3-2 Does an inductor have a self-inductance even though it carries no current? Explain.

Q 59.3-3 Consider a circuit loop which contains two inductors in series. In view of the definition of inductance given in this section, how could you determine the inductance of one inductor in distinction from the other?

Q 59.3-4 For a linear inductor, one in which the magnetic flux is proportional to the current, which of the following factors determine(s) the inductance? The current, the flux, the geometrical properties of the inductor.

Q 59.3-5 What is the function of the induction coil associated with the engine of a car?

E 59.3-1 A solenoid in the form of a circular cylinder has a length of 0.8 m and a diameter of 13 cm. It contains 460 m of wire and carries a steady current of 5.0 A.
(a) Compute the value of $\vec{B}$ in the center of the solenoid.
(b) Assuming an ideal solenoid, compute the total magnetic flux in the solenoid.
(c) Compute the self-inductance of the solenoid.

59.4 Transient Behavior of an Inductor. Time Constant

To study the current through an inductor and the voltage across it, the inductor can be connected to a battery through a switch as shown in Figure 59-5(a). The battery and the turns of wire or windings of the inductor have some resistance, shown separately in the circuit. Before the switch S is closed, the current through the inductor and the voltage across it are both zero. After the switch is closed (at $t = 0$) measurements of the current $i(t)$ in the circuit and of the voltage $\Delta V_L(t)$ across the inductor show that these quantities vary with time as shown in Figure 59-5 (b) and (c). The current does not instantaneously reach its final value, but increases gradually while the voltage across the inductor decreases exponentially with time. These results have some resemblance to the transient response of an RC circuit (Section 57.7). How can we account for them?

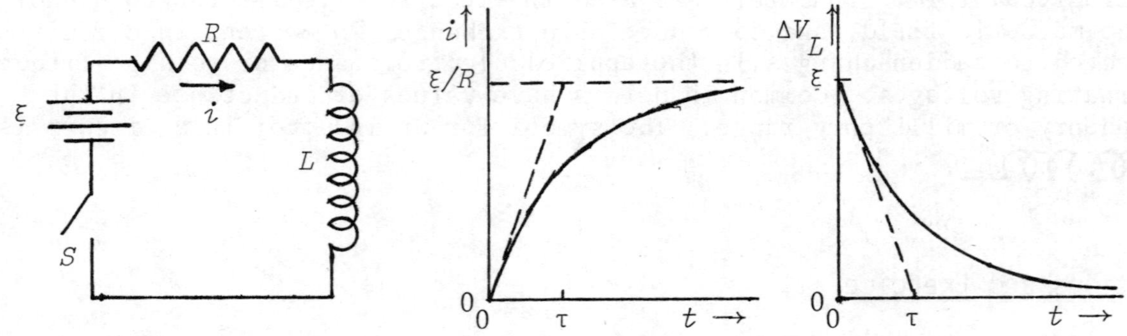

Figure 59-5. (a) A circuit used to send current through an inductor. (b) When the switch S is closed (at $t = 0$), the current rises rapidly at first, but the rate decreases with time as the current approaches its maximum value. (c) Beginning with its initial value ξ, the voltage across the (ideal) inductor decreases with time and approaches zero as the current approaches its maximum value. Both the current and the voltage across the inductor have an exponential variation with time.

For this discussion we assume the inductor is *ideal* in the sense that it has inductance L but no resistance, and we consider the total resistance R of the circuit to be lumped in a separate resistor, as shown in Figure 59-5(a). Just after the switch is closed, while the current is nearly zero the voltage iR across the resistor is also nearly zero, so that nearly the full electromotance of the battery is applied to the inductor. To see why the current in the inductor is initially very small, we need to consider the process of electromagnetic induction (Section 59.1) taking place there. When the inductor current is increasing at the rate di/dt, the linking magnetic flux is simultaneously increasing at the rate $d\Phi_\ell/dt$. Because the flux depends on the current, these two rates (or derivatives) are related by

$$\frac{d\Phi_\ell}{dt} = \frac{d\Phi_\ell}{di}\frac{di}{dt} = L\frac{di}{dt}.$$

In this equation, L (defined earlier as Φ_ℓ/i) may be substituted for $d\Phi_\ell/di$ because Φ_ℓ is directly proportional to i for inductors not containing magnetic material. From Faraday's law we know that $-d\Phi_\ell/dt$ is the electromotance induced in the inductor while its flux is changing, and from Lenz's law we know that the direction of this electromotance is such as to oppose the change in flux. Combining the results above, we have for the induced electromotance

★ $$\xi_{\text{ind}} = -L\,di/dt.$$

It is this induced voltage which opposes the portion $(\xi - iR)$ of the battery's electromotance applied to the ideal inductor, and which prevents sudden changes in the current in the RL series circuit.

As the current in this circuit increases, the voltage iR across the resistor also increases, leaving a smaller amount of the battery's electromotance applied to the inductor. The rate of increase of current (di/dt) through the inductor also diminishes, as do the rate of increase of flux $(d\Phi_\ell/dt)$ and the resulting voltage induced in the inductor. As the current rises further, the rate of increase of current di/dt decreases, and the rate of increase of flux $d\Phi_\ell/dt$ and the induced voltage ξ_{ind} become smaller and smaller. Ultimately the current approaches a limiting value and remains steady. Thereafter the magnetic flux is constant and the voltage ΔV_L is zero, so this steady current is determined simply by the ratio of the battery electromotance ξ to the total resistance R. In this way, we account qualitatively for the variations of $i(t)$ and $\Delta V_L(t)$ shown in Figure 59-5.

When the switch is first closed and the induced electromotance equals the negative of the battery's electromotance, it is seen from the equation above for ξ_{ind} that the initial value of $di/dt = \xi/L$. If the current increase continued at this rate, the time required to reach the final current ξ/R would be

$$t = \frac{i_{\text{f}}}{di/dt} = \frac{\xi/R}{\xi/L} = \frac{L}{R},$$

independent of the battery voltage and of the final value of current. The quantity $L/R = \tau$ thus emerges as a natural *time constant* of an LR series circuit, and is indicated on the graphs in Figure 59-5.

This built-in time constant of an *LR* series circuit is analogous to the time constant *RC* of an *RC* series circuit (Section 57.7) and to the time constant $\tau = 1/\lambda$ in radioactive decay (Section 34.7). As shown in Section 60.7, the curves in Figure 59-5 (b) and (c) have the mathematical form of exponential equations. The quantity τ specifies the time required for the induced voltage to change from any particular value to one which is smaller by a factor of *e*, that is to 0.368 times the initial value.

Our study of inductance began with magnetic flux linking turns of wire, and has led us to electric circuits containing inductors. Further study of inductance in electric circuits is taken up in Chapter 60.

Questions and Exercises

Q 59.4-1 Explain the behavior of $i(t)$ in an *LR* series circuit containing a switch and a battery, after the switch is closed.

Q 59.4-2 Evaluate this statement: "An inductor always opposes any change in the current in the inductor."

Q 59.4-3 How can $i(t)$ and $\Delta V_L(t)$ for an *LR* series circuit be measured without an appreciable effect on the "natural" behavior of the circuit?

Q 59.4-4 Consider a circuit loop containing an ideal inductor having an inductance *L*.
(a) If a voltmeter which draws no current from the circuit is connected across the terminals of the inductor at an instant when the rate of change of current is di/dt, what will be the voltage reading on the voltmeter?
(b) Is this the voltage across the inductor, or across the rest of the circuit loop?

Q 59.4-5 The time constant of an *RC* series circuit *increases* when *R* increases, but the time constant of an *LR* series circuit *decreases* when *R* increases. Explain, in terms of the physical effect of *R* on the current in each circuit.

E 59.4-1 Beginning with the SI units of *L* and *R*, show that the product L/R has units of seconds.

E 59.4-2 Consider the solenoid described in E 59.3-1.
(a) If the current in the solenoid is reduced to zero in 0.001 second by opening a switch, what is the average value of electromotance induced?
(b) How will this electromotance manifest itself?

E 59.4-3 An inductor having $L = 4.5$ H and $R = 300$ ohms is connected in series with a switch *S* and a 12-volt battery.
(a) What is the time constant of this circuit?
(b) What is the steady-state value of current when *S* is closed?
(c) What is the initial value of ΔV_L, just after *S* is closed?

Answers to Minimum Learning Goals

Line 1 The field would be an electric field directed in a closed loop inside the doughnut cavity.

Line 2 (a) -9.6×10^{-5} V/m around the wire; the minus sign indicates that the $\vec{E}$ field is opposite the direction of the integral.

(b) This current produces a magnetic field with a direction opposite to that of the electromagnet.

Line 3 (a) -2.5×10^{-4} V/m around the wire. The minus sign indicates that the $\vec{E}$ field is opposite the direction of the integral.

(b) This current produces a magnetic field in a direction opposite to that of the solenoid.

Section 59.1

Q 59.1-2 (a) The loop could be moved from one part of the field to another, or tilted to change its orientation in the field--and movement which changes the flux encircling the loop.
(b) The loop could be tilted or rotated to change the flux passing through the loop.

Q 59.1-4 (a) Under static conditions: no moving charges or magnets, and no currents.
(b) Any condition in which there is a changing magnetic field at some point or through some circuit.
(c) The conditions of (b), with the absence of any charged free particles or charged macroscopic bodies.

E 59.1-2 (a) -2.0×10^{-4} volt; (b) clockwise

E 59.1-4 (a) $\Phi_m(t) = B_0\, ab \cos(\omega t)$; (b) $\mathcal{E}(t) = \omega B_0 ab \sin(\omega t)$.

Section 59.2

Q 59.2-2 The examples described in the text of Section 59.1 are interpreted in terms of Lenz's law in Section 59.2.

Section 59.3

Q 59.3-2 Yes. Its inductance depends on the geometrical construction of the inductor, not on the current in it.

Q 59.3-4 The geometrical properties of the inductor.

Section 59.4

Q 59.4-2 The statement is correct.

Q 59.4-4 (a) The voltmeter reading is $-L\,di/dt$, assuming the response time of the meter is small enough so that it accurately indicates the instantaneous voltage across the inductor.
(b) It is both, because the terminals of the inductor are also the terminals of the rest of the circuit loop.

E 59.4-2 (a) 132 V; (b) through a small arc across the terminals of the open switch.

CHAPTER 60 VARIABLE - CURRENT ELECTRIC CIRCUITS

Introduction

Electric circuits are found in every item of industrial, commercial and household electrical equipment, as well as in vehicles and buildings of all kinds. In Chapter 27 we considered only circuits containing resistors and steady sources of electromotance such as batteries. This chapter begins by introducing the point of view which the physicist has in mind when discussing the theory or analysis of electric circuits. A variable-current circuit is one in which the current varies with time but does not change direction during the time interval being considered. The properties of capacitors and inductors as circuit elements are presented, with special attention to the variations of current and voltage with time which occur when one of these elements is connected to a battery.

When you have completed your study of the material in this chapter, you should be able to answer questions and work exercises like the examples in the minimum learning goals assigned by your instructor.

60.1 Concept of a Circuit

Minimum Learning Goal. Given several circuit diagrams, be able to select those which are equivalent, and be able to revise the remaining diagrams so that all are equivalent. For example:

1 Which three of the following four diagrams are equivalent? Revise the remaining diagram so that it is equivalent to the other three. (Answer, page 60-28, Line 3)

(a)

(b)

(c)

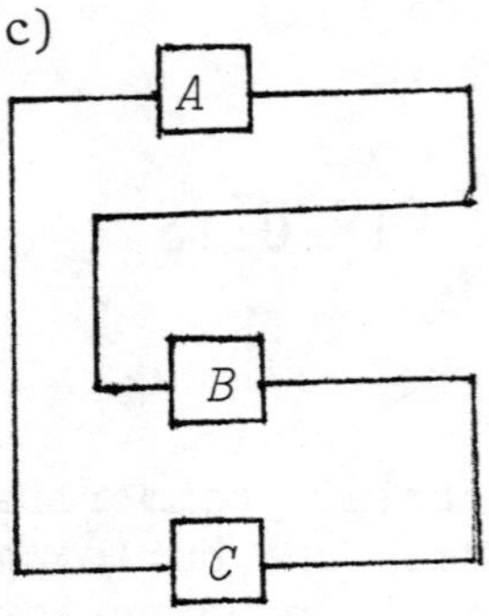

(d)

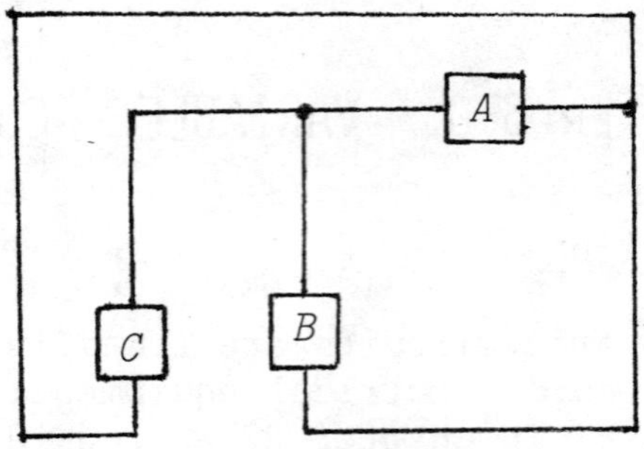

2 Which of the following four diagrams are equivalent? Revise the remaining diagram or diagrams so that all are equivalent. (Answer, Page 60-29, Line 14)

(a)

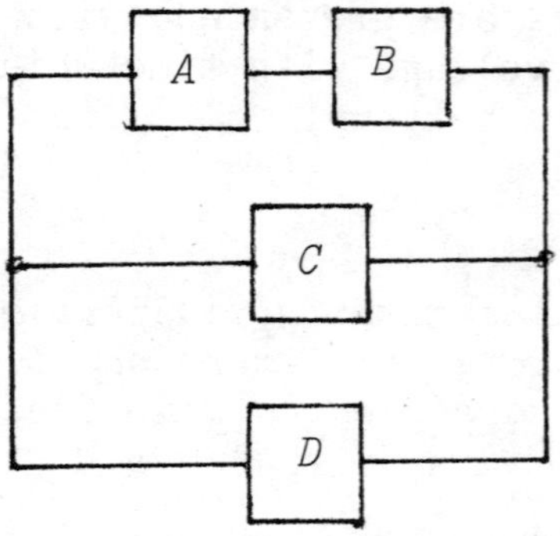

(b)

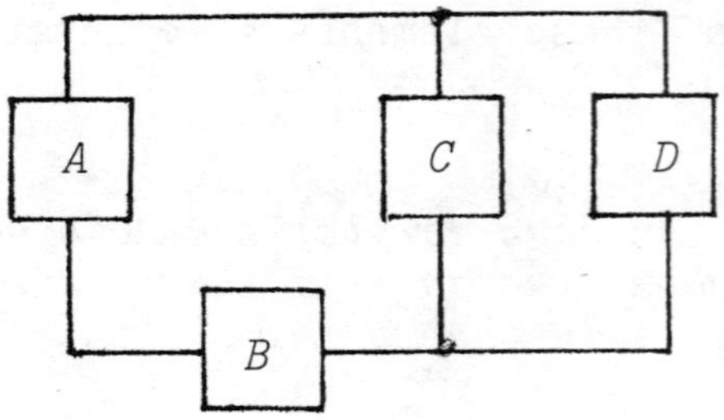

(c)

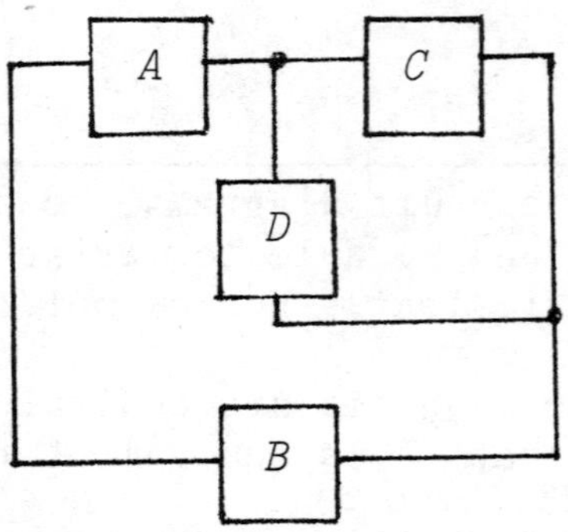

(d)

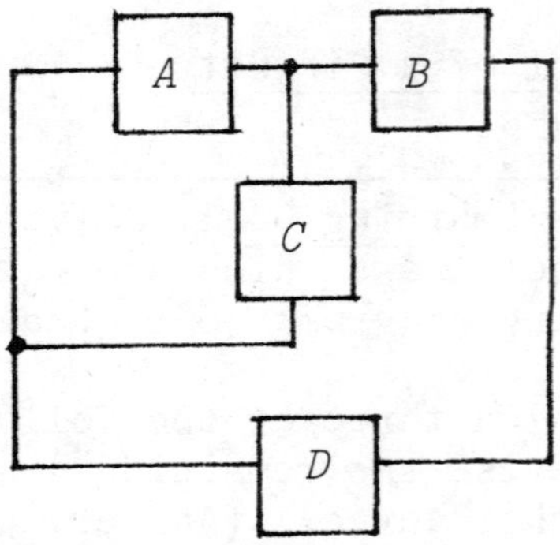

The word *circuit* denotes a certain point of view with respect to a physical system in which electric currents exist. Specifically, to talk about a circuit we have to imagine a number of *circuit elements* connected by wires or *leads* (rhymes with beads). Examples of circuit elements include resistors, inductors, and capacitors.

Circuit elements and leads are defined by the following properties. Each circuit element has two leads connected to it, such that current enters by one lead and leaves by the other. It is assumed that these two currents are equal, and that the current in a lead is the same at all points on that lead. It is also assumed that one can define a potential difference or voltage across each circuit element, but that there is no potential difference between points on the same lead. Each circuit element is characterized by some relation between its current and voltage, a relation assumed to be independent of the other circuit elements. The relation between current and voltage for each type of circuit element is discussed in the following sections.

Apart from the properties of the circuit elements, it is assumed that only the *topology* of the circuit, as shown in a *circuit diagram,* is significant. In other words, it matters what point is connected by a lead to what other point, but it does not matter how long the lead is, or into what shape it is bent.

In the following sections we discuss two questions: (a) How does one analyze circuits? (b) To what physical situations is the model of a circuit applicable?

The Questions and Exercises below review some of the concepts introduced in Chapter 27.

Questions and Exercises

Q 60.1-1 What is meant by an "open circuit?"

Q 60.1-2 What is meant by a "short circuit" between two points in a circuit?

E 60.1-1 (a) Which of the three circuit diagrams shown below are equivalent?
(b) On the diagram which is not equivalent to the other two, re-draw the lines which represent leads so the diagram is equivalent.

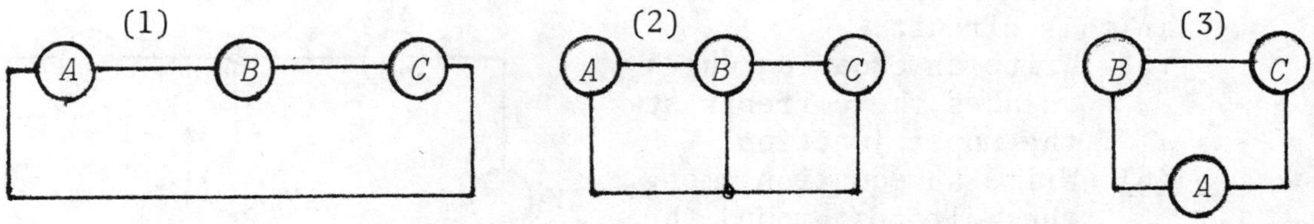

E 60.1-2 (a) Which of the circuit diagrams shown below are equivalent?
(b) On the diagram which is not equivalent to the others, redraw the lines which represent leads so the diagram become equivalent.

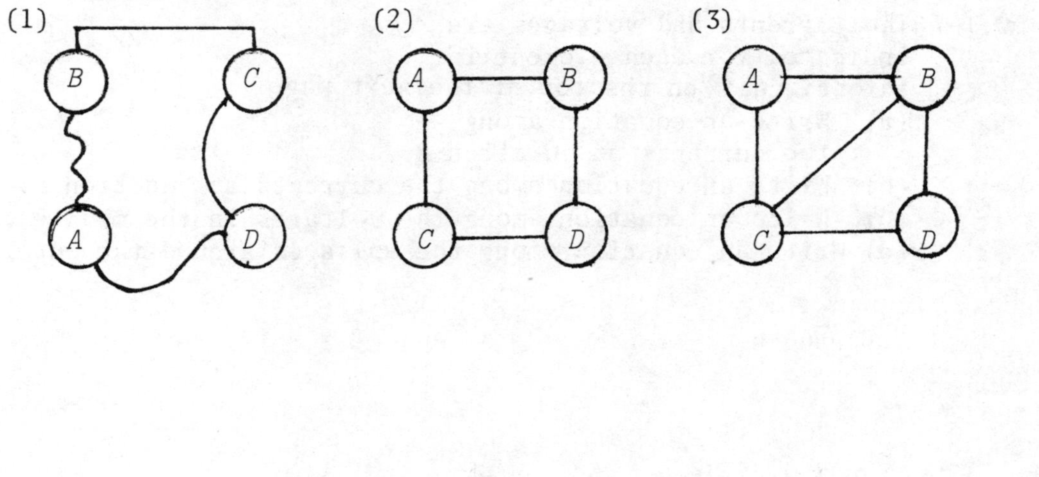

E 60.1-3 (a) Which of the circuit diagrams shown below are equivalent?

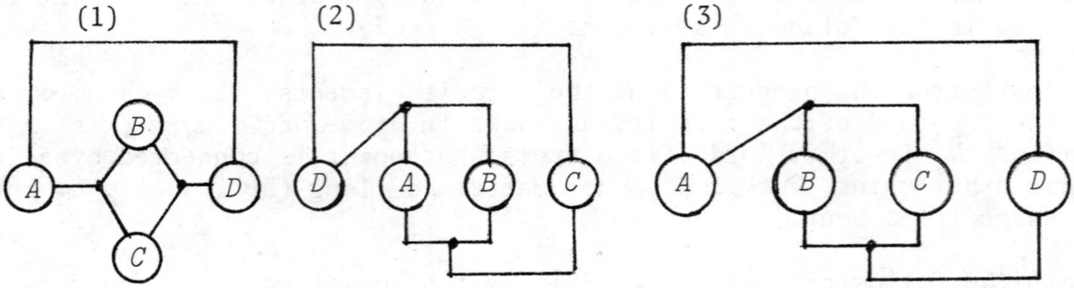

(b) On the diagram which is not equivalent to the other two, re-draw the lines which represent leads so the diagram becomes equivalent.

E 60.1-4 Draw a complete circuit diagram showing two circuit elements (*A* and *B*) connected in parallel with each other, with this combination connected in series with two more circuit elements *C* and *D* to form a complete circuit. Redraw this circuit in two different additional ways that look different on paper but which are the same electrically.

E 60.1-5 The voltage and current are indicated for each element in this circuit. Write an equation for the sum of the voltages around this circuit.

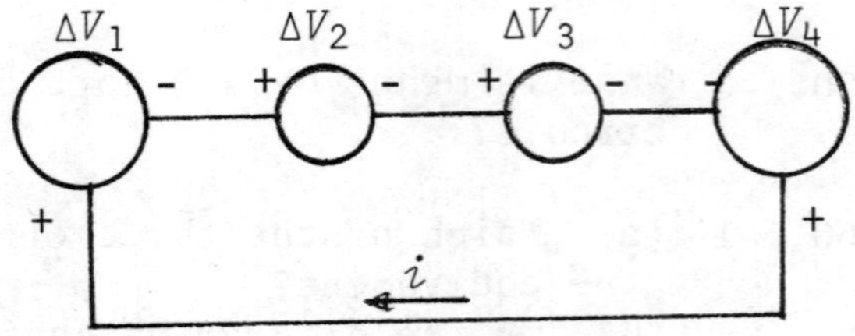

E 60.1-6 The current and voltage are indicated for each element in this circuit.

(a) Write an equation which relates the currents at the upper junction.

(b) Write an equation among the voltages around the right-hand loop.

(c) Write an equation among the voltages in the left-hand loop.

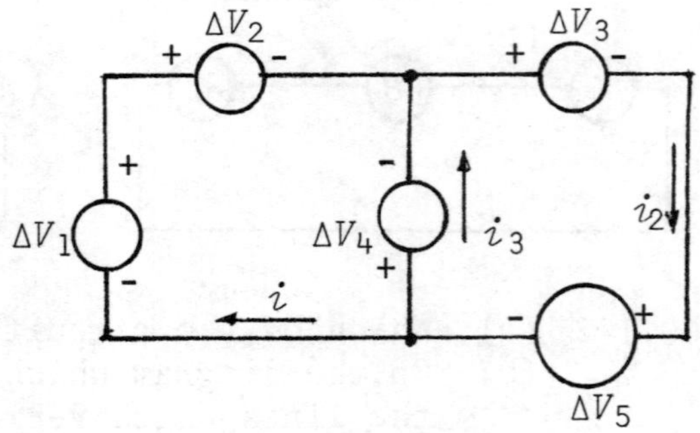

E 60.1-7 The currents and voltages are indicated for each element in the circuit on the top of the next page.

(a) Write an equation among the currents at junction *A*.

(b) Write an equation among the currents at junction *B*.

(c) Write an equation among the voltages in the middle loop.

(d) Write an equation among the voltages around the outer loop.

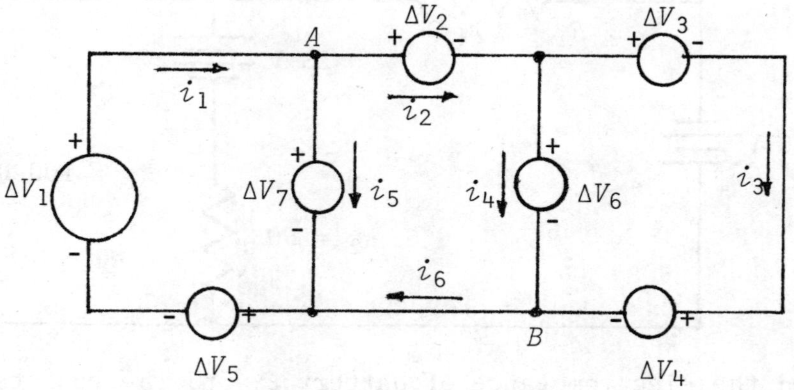

Check your understanding of the material above by testing yourself on the minimum learning goal at the beginning of this section.

60.2 Resistors, Batteries, and S-C Circuits

Minimum Learning Goals. (1) Given sufficient information about the terminal voltage of a battery as a function of current, be able to find its internal resistance. For example:

1 When the current in the primary circuit of an automobile ignition system is 4.0 A, the battery voltage is 12.2 volts. If the battery's electromotance (zero-current value of terminal voltage) is 12.6 volts, what is its internal resistance? (Answer, page 60-28, Line 2)

2 A current of 1 A is produced by a flashlight cell which is short-circuited with a heavy-gauge copper wire, so that the external current path between the terminals of the cell has zero resistance. If the cell electromotance (zero-current value of terminal voltage) is 1.6 volts, what is the internal resistance of the cell? (Answer, page 60-28, Line 8)

(2) Given a conventional diagram of an electric circuit (having up to six branches) containing resistors and batteries, and sufficient information on the values of circuit elements, be able to find the values of unknown voltages, resistance and currents. For example:

1 Find the value of the current in R_2 and the voltages across each 100-ohm resistor in the circuit below. (Answer, page 60-28, Line 9)

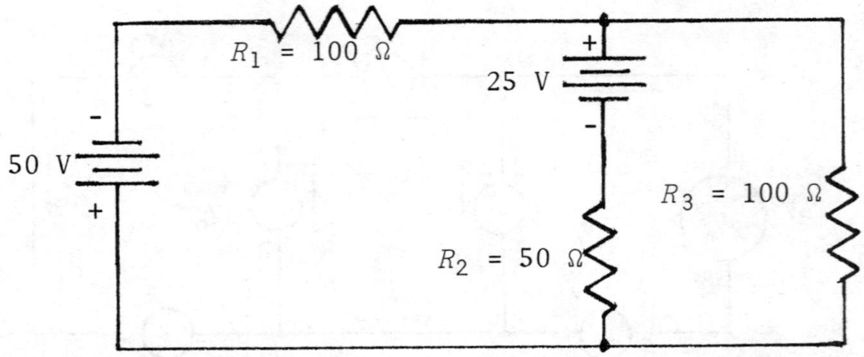

2 Find the electromotance of battery $\mathcal{E}_1$ and the resistance R_3 in the circuit below. (Answer, page 60-28, Line 4)

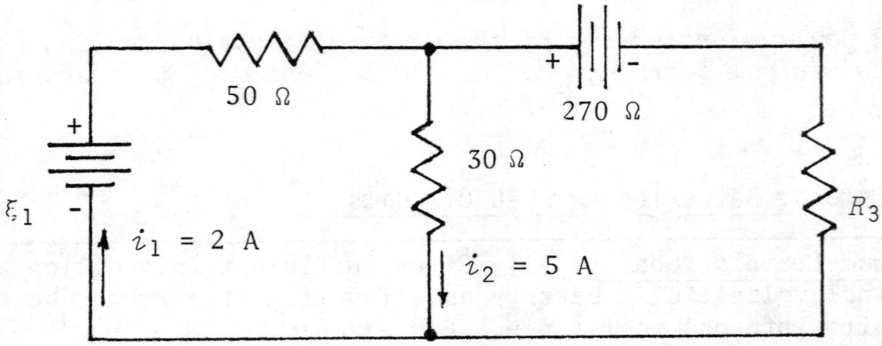

In developing methods for analyzing electric circuits it is convenient to classify circuits into several categories. A twofold classification which is often used includes direct-current (d-c) circuits and alternating-current (a-c) circuits. In this chapter we adopt a more useful classification, having *three* categories. A *steady-current* (s-c) circuit is one in which the current does not vary but is constant with time, both in magnitude and direction. S-c circuits are discussed in Chapter 27 and in this section. A *variable-current* (v-c) circuit is one in which the current varies with time but does not change direction during the time interval being considered. V-c circuits are considered in the rest of this chapter. An *alternating-current* (a-c) circuit is one in which the current changes direction and magnitude in a regular pattern, varying sinusoidally with time. A-c circuits are considered in Chapter 61.
The category to which a circuit belongs depends on the properties of the source of electromotance and of the other circuit elements.

Two principles which are highly useful for analyzing circuits, especially circuits having more than one source of electromotance and more than one branch, are known as *Kirchhoff's rules* (for Gustav Robert Kirchhoff, 1824-1887). These rules are stated with reference to junctions and loops of the network or circuit. A *junction* refers to any point in the circuit where three or more conducting paths (e.g. leads) meet at a common point. A *branch* refers to a single conducting path between a pair of junctions, and on which no other junction is located. A *loop* refers to any continuous conducting path from one point in the circuit through any number of circuit elements and back again to the starting point. Kirchhoff's rules are:

★ I. The algebraic sum of the currents arriving at any junction is zero.

★ II. The algebraic sum of the voltages around any closed loop of the circuit is zero.

These are sometimes referred to as the *junction rule* and the *loop rule*, respectively.

To apply Kirchhoff's first rule to a network, one should first indicate a direction (by guess, if necessary) for the current in each branch of the circuit. When the solution for the network is obtained, a negative value for any current simply means that the direction chosen was wrong for that current. If the network contains a number of junctions, one needs to apply Kirchhoff's first law to *all but one* of these junctions and write a separate equation for each, and this set of equations must contain each branch current at least once.

To apply Kirchhoff's second rule, one traces around a loop of the network either clockwise or counterclockwise, beginning at any point and returning to that point, and writes an expression for the voltage or potential difference associated with each element in the circuit. The term in the resulting equation associated with a source of electromotance ξ_n is $+\xi_n$ or $-\xi_n$. The term associated with a resistor R through which current i passes is $+iR$ or $-iR$. The correct sign to use for each voltage term depends on whether the path goes to a higher or lower value of potential in passing through the circuit element. A useful and consistent set of rules for determining the sign of each voltage term is the following:

★ If the path passes through a source of electromotance from its - to its + terminal, the term for this electromotance is positive.

★ If the path passes through a resistor in the direction opposite to the current, the iR term is positive.

If the path passes one of these circuit elements in the opposite direction, the corresponding term is negative. In applying Kirchhoff's second rule to the loops of the circuit, the loops must be chosen so that each branch of the circuit is included in at least one loop. The total number of equations needed from the use of Kirchhoff's two rules is equal to the number of branch currents. One can then solve for the unknown quantities in the network by systematically eliminating variables from these simultaneous equations.

An s-c circuit is one in which the currents, as well as the associated electric and magnetic fields, are constant in time. For such a circuit Kirchhoff's rules can be justified rather easily. If the algebraic sum of the currents arriving at any junction were not zero, then charge would be accumulating at that junction giving rise to an electric field growing in time, so that we would not be dealing with an s-c circuit. With regard to Kirchhoff's second rule, an electrostatic potential (i.e., a potential constant with time) must exist provided there are no time-varying magnetic fields (Section 59.1). If an electrostatic potential exists, then the algebraic sum of the voltages around any closed loop of the circuit must equal zero, since in tracing around a loop we begin and end at the same point and hence at the same value of potential.

Each circuit element is characterized by some functional relationship between the current through it and the voltage across it. We now discuss the two most common types of s-c circuit elements: resistors and batteries.

A linear *resistor* is a circuit element characterized by a voltage ΔV proportional to the current i; the proportionality constant is the *resistance* R of the resistor (Section 26.4).

★ $$\Delta V = iR.$$

A source of electromotance (Section 26.3) is a device or circuit element which can maintain a voltage between its terminals and supply current to an external circuit connected to them. When the source is supplying current, it is also supplying energy and as a result its internal store of energy is decreasing. For some sources of electromotance, such as an automotive storage battery, the internal supply of energy can be replenished by connecting the battery to a battery charger. Other sources, such as a dry cell or flashlight battery, cannot be recharged in this way. When the battery itself is the source of current through it, the battery terminal at higher potential is called the positive terminal, and the other terminal is the negative terminal. The current in the external circuit is then from the positive terminal to the negative one, while the current continues through the battery from the negative terminal to the positive. When current and energy are being supplied to the battery by a battery charger, the current in the battery has the opposite direction, from the positive terminal to the negative.

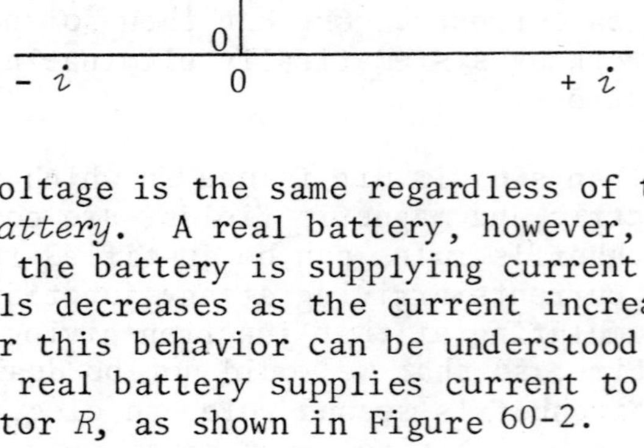

Figure 60-1. A graph of terminal voltage ΔV vs. current i for a rechargeable battery having a definite value of internal resistance. When there is no current, the terminal voltage equals the electromotance $\mathcal{E}$ of the battery. The internal resistance r is equal to the negative slope of the curve.

A circuit element whose terminal voltage is the same regardless of the current through it is called an *ideal battery*. A real battery, however, does not behave that way. Rather, when the battery is supplying current the voltage between the battery terminals decreases as the current increases, as shown in Figure 60-1. The reason for this behavior can be understood if we consider a simple circuit in which a real battery supplies current to an external circuit represented by a resistor R, as shown in Figure 60-2.

The real battery is represented as an ideal battery having electromotance ξ in series with a resistance r, called the *internal resistance* of the battery. Applying Kirchhoff's second rule to this circuit yields

$$+\xi - ir - iR = 0.$$

The terminal voltage, or potential difference between the battery terminals, is given by

$$\Delta V = \xi - ir.$$

As i increases, the terminal voltage decreases linearly as shown in Figure 60-1. Solving the equation above for r, we obtain

$$r = \frac{\xi - \Delta V}{i}$$

The value of r is equal to the negative of the slope of the graph in Figure 60-1. It is evident that if r is small enough, the slope approaches zero and the battery may be regarded as an ideal battery for practical purposes. To send current through the battery in the negative direction, opposite to the current which would be supplied by the battery's own electromotance, a voltage greater than ξ must be supplied to the battery terminals by an external device such as a battery charger.

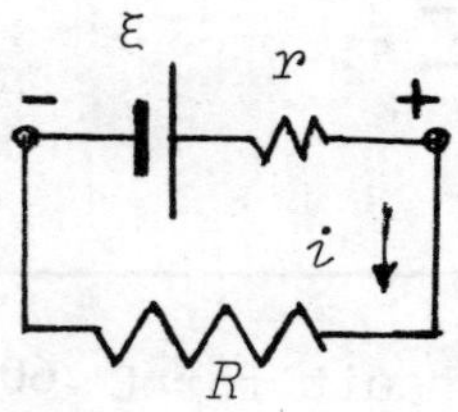

Figure 60-2. Diagram of a circuit in which a real battery having an electromotance ξ and an internal resistance r supplies current to a resistor R. The terminals of the battery are labeled + and - .

Questions and Exercises

Q 60.2-1 What is the conventional symbol for each of these circuit elements?
(a) resistor; (b) battery;
(c) switch.

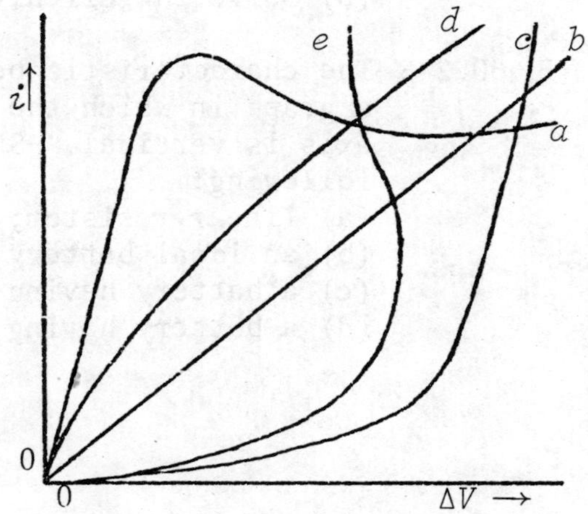

Q 60.2-2 (a) Which of the curves in this figure corresponds most closely to a linear resistor?
(b) A 100-watt tungsten-filament lamp becomes hot when operated at its rated voltage of 115 volts. Which of the curves in this figure is most likely to represent the behavior of this lamp over the voltage range from 0 to 115 volts?

Q 60.2-3 What numerical data do you need to find the resistance of a certain resistor at a voltage of 8 volts?

Q 60.2-4 In the description of a circuit given in Section 60.1, what is the resistance of the leads which connect the circuit elements?

Q 60.2-5 Is there any way to force current through a resistor from the terminal at lower potential to the one at higher potential? Explain.

E 60.2-1 Show that Kirchhoff's first rule is equivalent to the principle of conservation of charge.

E 60.2-2 Show that Kirchholff's second rule is equivalent to the principle of conservation of energy for the charge carriers in a circuit.

E 60.2-3 In the circuit below, $\mathcal{E}_1 = 1.5$ V, $R_1 = 1$ ohm, $\mathcal{E}_2 = 1.5$ V, and $R_2 = 29$ ohm.
(a) Use Kirchhoff's rule to write an equation for this circuit.
(b) Compute the current.

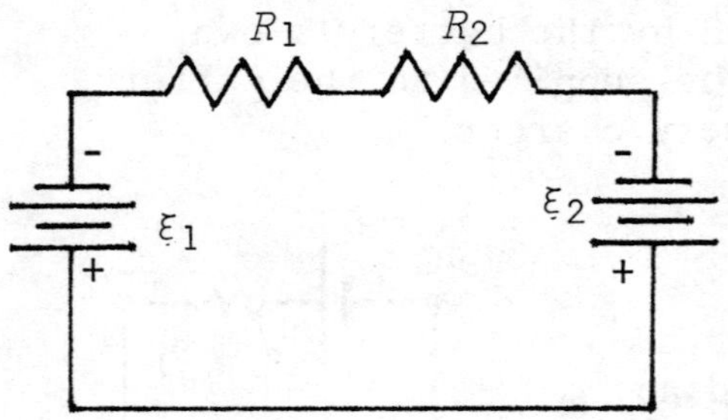

Circuit for E 60.2-3

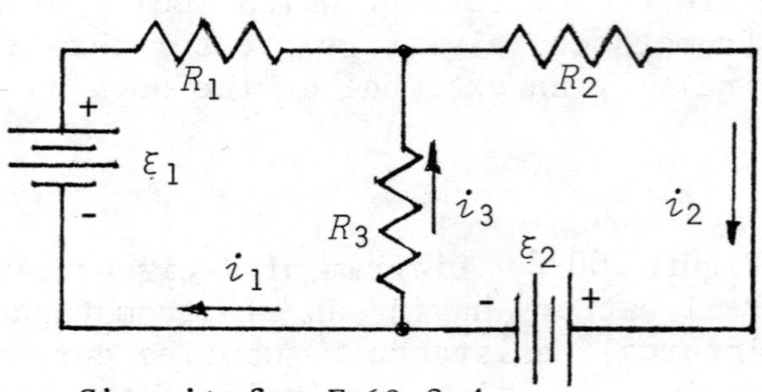

Circuit for E 60.2-4

E 60.2-4 In the circuit above, $\mathcal{E}_1 = 6$ V, $\mathcal{E}_2 = 3$V, $R_1 = 2$ ohm, $R_3 = 6$ ohm, and $i_2 = 0.27$A.
(a) Use Kirchhoff's rules to write three equations for this circuit.
(b) Solve the circuit equations for R_2, i_1, and i_3.

E 60.2-5 The characteristic behavior of a circuit element may be shown on a graph in which the voltage axis is horizontal and the current axis is vertical. Sketch a characteristic curve for each of the following:
(a) linear resistor;
(b) an ideal battery, having no internal resistance;
(c) a battery having a small internal resistance;
(d) a battery having a large internal resistance.

E 60.2-6 This graph shows the characteristic curve for a small battery.
(a) Find the internal resistance of the battery.
(b) For what value of current would the terminal voltage ΔV drop to zero, assuming the internal resistance is independent of the current drawn from the battery?

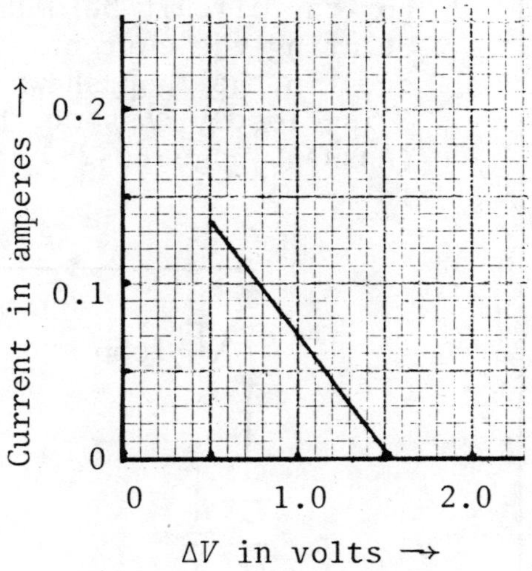

Check your understanding of the material above by testing yourself on the minimum learning goals at the beginning of this section.

60.3 Kirchhoff's Rules and Capacitance

Minimum Learning Goal. Given a diagram of an electric circuit containing batteries, resistors, capacitors (having given initial charges) and a switch, be able to recall Kirchhoff's rules and use them to write (a) equations relating the currents at a junction, and (b) equations relating the voltages across circuit elements, for times after the switch is either opened or closed. For example:

1 For the circuit below, use Kirchhoff's laws to write (a) an equation relating the currents at junction A, and (b) an equation relating the voltages in the loop shown as a function of time, t, after switch S is closed. Assume that C_1 and C_2 are initially uncharged. (Answer, page 60-28, Line 6)

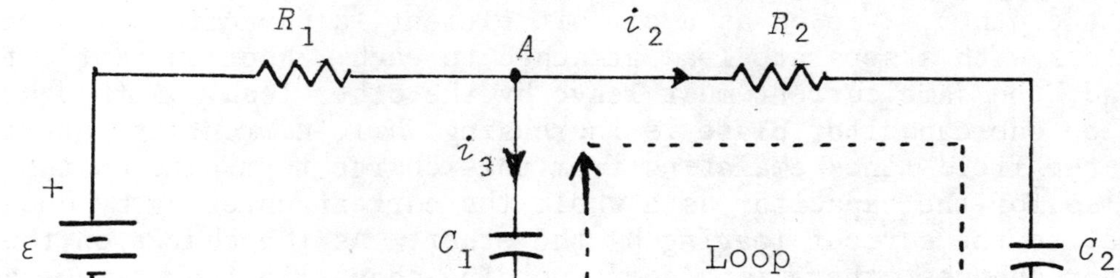

2 For the circuit below, use Kirchhoff's laws to write an equation relating the currents at junction B and an equation relating voltages in the loop shown as a function of time, t, after switch S is closed. Assume that C_1 and C_2 are initially uncharged. (Answer, page 60-28, Line 5).

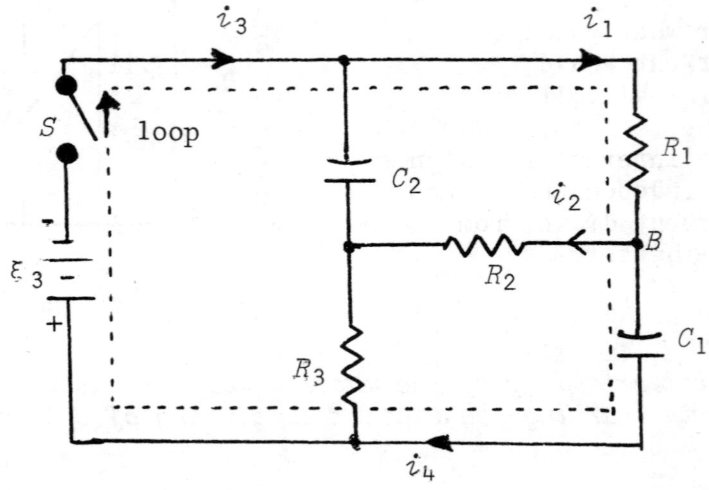

In Section 60.2 we applied Kirchhoff's rules to the analysis of steady current networks containing resistors and batteries. This approach to circuit analysis can be extended to the case where the currents vary in time, but this has to be done very carefully because Kirchhoff's rules cannot be justified in the same way as they are for the s-c case.

Kirchhoff's first rule says, in effect, that the current entering any part of the circuit must be equal to the current leaving that part of the circuit. In a circuit where the currents vary with time, there may be some parts of a circuit where there is a net influx of current with a resulting increase in the charge on those parts. This occurs, for example, when a capacitor receives charge from a battery, as discussed in Section 57.7.

A capacitor is a pair of conductors so arranged that when they have equal and opposite charges, all field lines which begin on one conductor end on the other. Viewed as a circuit element, a capacitor is a *pair* of conductors with a separate lead attached to each. When current enters by one lead, the same current must leave by the other lead. Although the net charge on one capacitor plate is increasing while current is entering it, all of the field lines emanating from this charge terminate on the other plate, so for the capacitor as a whole the current entering by one lead is equal to the current leaving by the other. As the charge on the capacitor changes, there is a corresponding change in the voltage across the capacitor.

Kirchhoff's first rule may be applied to circuits containing capacitors, provided it is applied to the currents arriving *at a junction*. It would not be correct to apply this rule to the currents arriving at either one of the capacitor plates. Kirchhoff's second rule, that the algebraic sum of the voltages around any closed loop of the circuit is zero, may also be applied to circuits containing capacitors. How can we write an expression for the voltage across a capacitor when the current is changing with time?

The voltage ΔV_C across a capacitor is proportional to the charge q on either electrode:

★ $$\Delta V_C = q/C.$$

The current i through a capacitor is equal to the time-rate change of this charge:

★ $$i = dq/dt.$$

Unlike a resistor, a capacitor has no simple direct relationship between the voltage across it and the current through it. At any moment while a capacitor is being charged, the potential across it increases in going from the negative to the positive terminal of the capacitor as one traces around the circuit in the direction *opposite* to the current through the capacitor. At any moment while a capacitor is being discharged, the potential across it increases as one traces around the circuit in the *same* direction as the current. In many cases one does not know whether the capacitor at some instant is gaining or losing charge, so it is useful to have a general guideline which can always be followed in applying Kirchhoff's loop rule to a capacitor.

When labeling a circuit diagram, indicate the current in a branch containing a capacitor in the same direction or sense as if this capacitor were a resistor. This current delivers positive charge to the plate where it enters the capacitor, so this plate is designated as the positive plate. That is, the terminal where current *enters* the capacitor is marked + while the other terminal is marked -. A correct equation for any loop containing a capacitor is then obtained if the following rule is used in conjunction with the rules on page 60-7:

★ If the path passes through a capacitor from its - to its + terminal, the q/c term for this capacitor is positive.

If the path passes a capacitor in the opposite direction, the q/c term is negative.

Figure 60-3. A simple *RC* series circuit, containing a resistor, capacitor, battery and switch. If the capacitor is initially uncharged, when the switch is closed current from the battery delivers charge to the capacitor with the polarity indicated.

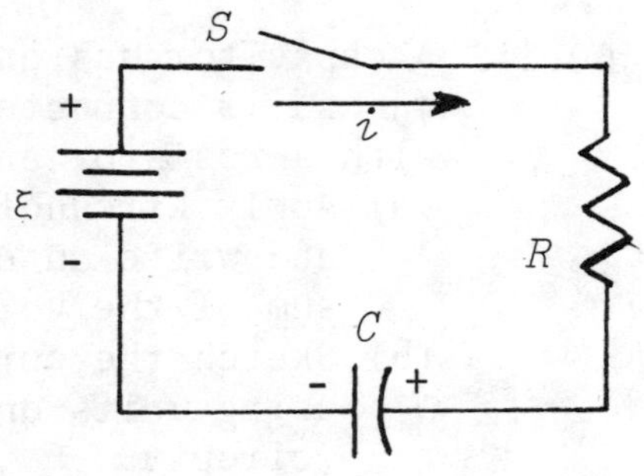

Figure 60-3 shows a simple *RC* series circuit. As we have seen in Section 57.7, both the current and the charge on the capacitor change with time in such a circuit, after the switch is closed. In preparation for computing these quantities, Kirchhoff's loop rule is applied to the circuit. Using the sign conventions we have adopted, at all times after the switch is closed the voltages around the loop (tracing clockwise) are

$$\xi - iR - q/C = 0.$$

Substituting $i = dq/dt$ and rearranging, this becomes

$$R\frac{dq}{dt} + \frac{q}{C} = \xi.$$

This equation is solved in Section 60.4.

Question and Exercises

Q 60.3-1 When a battery is connected in series with a resistor and a capacitor, the negative terminal of the battery sends out electrons which travel along the circuit as far as they can from this terminal.

(a) Where in the circuit does the net charge which flows from this terminal end up?

(b) What *force* causes an equal number of electrons to enter the battery at its positive terminal?

(c) Where do the electrons come from, which enter the positive terminal of the battery?

(d) In what sense is it proper to say that charge is flowing "through" the capacitor?

(e) Does the capacitor offer any hindrance or opposition to the flow of charge? Explain.

E 60.3-1 When a capacitor is being charged by a battery, the charging current varies with time as shown here. Assuming the capacitor was initially uncharged, sketch the total charge on the positive plate as a function of time.

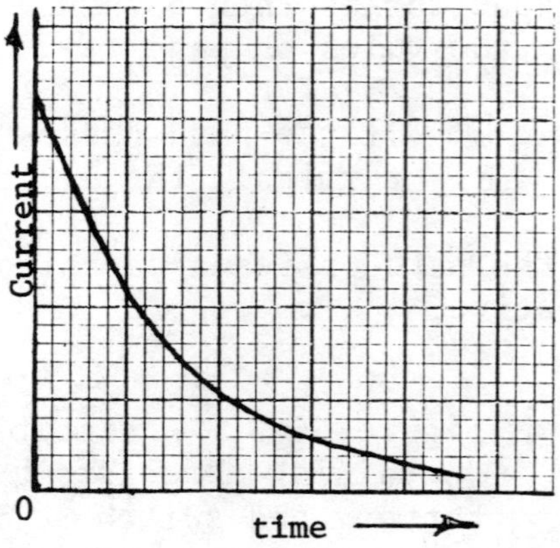

E 60.3-2 A capacitor has an initial charge q_0 and is connected in series with a resistor and a switch.

(a) Apply Kirchhoff's second law and write an equation for the sum of the voltages around the circuit after the switch is closed.

(b) Sketch the current in this circuit as a function of time, using units on the time-axis equal to the time constant of the circuit.

(c) Sketch the voltage across the capacitor as a function of time.

(d) Sketch the charge on the capacitor as a function of time.

Check your understanding of the material above by testing yourself on the minimum learning goal at the beginning of this section.

60.4 Charging a Capacitor

Minimum Learning Goal. (1) Given a diagram of a series circuit containing a battery, a switch, an initially uncharged capacitor, and a resistor, with specified values of electromotance, resistance, and capacitance, be able to sketch qualitatively correct graphs of voltage and current for each circuit element as functions of time after the switch is closed. On each graph, indicate the initial numerical value of the voltage or current, and the numerical value of the time constant of the circuit. For example:

1 Sketch graphs of voltage and current for each circuit element in the diagram at right, as functions of time after the switch is closed. Assume that the capacitor is initially uncharged. (Answer, page 60-29, Line 12)

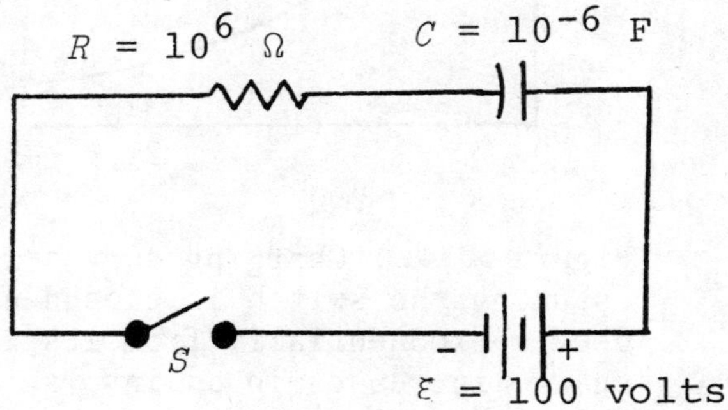

2 Sketch graphs of voltage and current for each circuit element in the diagram at right, as functions of time after the switch is closed. Assume that the capacitor is initially uncharged. (Answer, page 60-28, Line 11)

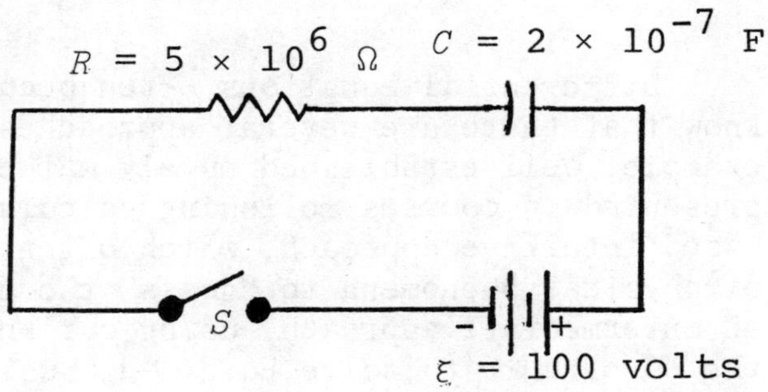

In Section 57.7 we considered the transient behavior of a capacitor being charged in a circuit like that of Figure 60-3. The experimental curves of current vs. time and charge vs. time were accounted for in a qualitative way by considering the effect of the rising capacitor voltage on the current. This section is devoted to obtaining and interpreting the mathematical equations for these curves, which are reproduced in Figure 60-4.

Using Kirchhoff's loop rule, in Section 60.3 we obtained the following equation for the circuit:

$$R\frac{dq}{dt} + \frac{q}{C} = \mathcal{E}. \qquad \text{Eq. (60-1)}$$

Our task now is to solve this equation to obtain solutions for $q(t)$ and $i(t)$. The quantity dq/dt is the time-rate of change of the variable q; in mathematics it is called the *time-derivative* or simply the *derivative* of q. Because Eq. (60-1) contains both the variable q and its derivative, it belongs to a family of equations called *differential equations*.

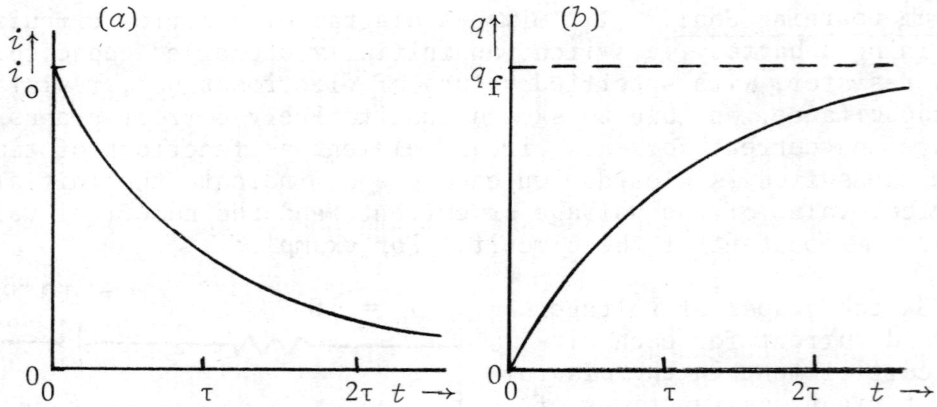

Figure 60-4. Charging curves for the circuit in Figure 60-3, assuming the switch is closed at time $t = 0$. (a) The current decays exponentially from its initial value $i_0 = \xi/R$. (b) The charge increases in an inverse exponential curve to a final value $q_f = C\xi$.

Differential equations often occur in physics, and it is useful to know that there are several approaches to solving them. There are, for example, well-established purely mathematical methods which are usually presented in courses following calculus. At the other extreme is a physics-based intuitive approach, which often enables a person having a good grasp of physical phenomena to "guess" the correct solution. We will employ here an intermediate approach, using our knowledge of the physical behavior of the *RC* circuit to solve Eq. (60-1) using a rather simple integration.

((The first step is to compute the time-rate-of-change of each term in the equation. (As a mathematical procedure this is called differentiation, discussed in Appendix A.2.) The quantities R, $1/C$, and ξ which appear in the equation are constants which do not change with time, so for each of them the time-rate of change is zero. The quantity dq/dt is simply the current i; the time-rate of change of i is written di/dt. Similarly, the time-rate of change of q in the second term is dq/dt. Using these substitutions, the time-rate of change (or the time-derivative) of Eq. (60-1) may be written

$$R\frac{di}{dt} + \frac{1}{C}\frac{dq}{dt} = 0 \qquad \text{or} \qquad R\frac{di}{dt} + \frac{i}{C} = 0 \qquad \text{Eq. (60-2)}$$

Note that by taking the derivative of Eq. (60-1) we have replaced a differential equation having three terms by one having only two terms. If we write Eq. (60-2) as $di/dt = -i/RC$, we see that the current decreases at a rate proportional to its value at that instant. Such a situation corresponds to an exponential decay, such as that found for radioactivity (Section 34.7).

The next step is to multiply both terms in Eq. (60-2) by the quantity dt, and then rearrange terms:

$$R\,\frac{di}{dt}\,dt + \frac{1}{C}\,i\,dt = 0 \qquad \text{or} \qquad R\,di = -\,\frac{1}{C}\,i\,dt.$$

Multiplying both sides by C/i leads to

$$RC\,\frac{di}{i} = -dt. \qquad \text{Eq. (60-3)}$$

We now have the variable i only on the left side of the equation, and the other variable t only on the right side; we have separated the variables.

We next integrate both sides of Eq. (60-3) using definite integrals (Section A.4). In doing this, we must use special care to choose limits for the two integrals which correspond physically to each other. Even before we do the integration, it is evident that this procedure will lead to an equation for i in terms of t. Since we want the result to express i at any value of t, the upper limit of integration on the integral for t will be the variable t itself; the corresponding upper limit on the integral for i must then be the variable i. We choose $t = 0$ (the time at which the switch is closed) for the lower limit on the integral for t; the lower limit on the integral for i must then be the value of current at $t = 0$, denoted by i_0. We integrate Eq. (60-3) by writing the integral signs with their limits:

$$RC\int_{i_0}^{i}\frac{di}{i} = -\int_0^t dt.$$

The product RC is written to the left of the integral sign because it is a constant, independent of the values of i and of t.

To evaluate these two integrals, we apply the recipes described in Appendix A.4 for the indefinite integrals of di/i and dt, and then substitute the upper and lower limits to obtain

$$RC\,\Big[\ln i\Big]_{i_0}^{i} = -\Big[t\Big]_0^t$$

$$RC\,[\ln i - \ln i_0] = -\,[t - 0] = -t. \qquad \text{Eq. (60-4)}$$

The symbol $\ln i$ is the abbreviation for "the natural logarithm of i"; it is the exponent which must be applied to the base of natural logarithms, $e = 2.71828$, to obtain the quantity i. Dividing both sides of Eq. (60-4) by RC yields

$$\ln i - \ln i_0 = -t/RC.$$

Then using each side of this equation as an exponent for the base e gives us

$$e^{\ln i - \ln i_0} = e^{-t/RC}.$$

Applying the rule for subtracting exponents, the left side becomes

$$e^{\ln i - \ln i_0} = \frac{e^{\ln i}}{e^{\ln i_0}} = \frac{i}{i_0}.$$

Substituting, we obtain finally

★ $$\frac{i}{i_0} = e^{-t/RC} \qquad \text{or} \qquad i = i_0 e^{-t/RC}. \qquad \text{Eq. (60-5)}$$

We have already found in Section 57.7 that an *RC* series circuit has a time constant given by $\tau = RC$, with units of seconds. The result obtained for i as a function of t is an exponential decay curve, in which the current decreases exponentially from an initial value i_0 at $t = 0$; this curve is shown in Figure 60-4(a). To evaluate i_0 in terms of the properties of the circuit, we refer back to Eq. (60-1), and note that the charge q on the capacitor is zero at $t = 0$; at $t = 0$ this equation then become $Ri_0 = \xi$ from which we obtain

$$i_0 = \xi/R.$$

((To obtain an equation for q as a function of time, we use the definition $i \equiv dq/dt$, multiply both sides by dt to obtain $dq = i\ dt$, and then integrate both sides of this equation beginning with the lower limits $q = 0$ and $t = 0$, which correspond to the initial uncharged state of the capacitor. For the upper limits, we use the variable value of q and the corresponding variable time t:

$$\int_0^q dq = \int_0^t i\ dt.$$

Evaluating these integrals using the recipes in Appendix A. 4, we have

$$\Big[q\Big]_0^q = \int_0^t i_0 e^{-t/RC}\ dt .$$

Comparing the last integral with the recipe

$$\int e^{at}\ dt = (1/a)\ e^{at} ,$$

We see that the general factor a in the recipe corresponds to the specific factor $-1/RC$ in our integral. We then write

$$q = i_0 \Big[(-RC)e^{-t/RC}\Big]_0^t .$$

We next substitute the limits 0 and t, and recalling that $e^0 = 1$ we obtain

$$q = i_0\ (-RC)\ [e^{-t/RC} - 1] .$$))

Finally, we substitute ξ/R for i_0 and obtain

★ $$q(t) = -\xi C\ [e^{-t/RC} - 1] = \xi C[1 - e^{-t/RC}] . \qquad \text{Eq. (60-6)}$$

To interpret this result, we observe that at $t = 0$ the exponential term

$e^{-t/RC}$ is equal to unity, so that the initial value of $q = 0$. This is one of the initial conditions stated at the beginning of the derivation, so obtaining this result at the end is evidence that our work has been correct. As time proceeds after $t = 0$ the exponential term steadily decreases from unity toward zero, so that the quantity in square brackets steadily increases from zero toward unity. After a time long compared with the time constant RC, the charge q approaches the limiting value

$$q_f = \xi C.$$

This inverse exponential growth corresponds to the curve in Figure 60-4(b). As the charge approaches its constant limiting value, the current approaches its limiting value of zero. When the current ceases, the capacitor is equivalent to an open circuit.

Exercises

E 60.4-1 Two resistors R_1 and R_2 are connected in parallel, and this combination is connected across a series combination of a battery, a capacitor and a switch. The switch is closed at $t = 0$.
- (a) Draw a diagram of this circuit.
- (b) Designate the current through the capacitor by i, the current through R_1 by i_1, and the current through R_1 by i_2. Apply Kirchhoff's laws and write three equations for this circuit: a voltage equation for each small loop, and a current equation for one junction.
- (c) Eliminate the variables i_1 and i_2, and obtain a differential equation for i as a function of time.
- (d) What is the time constant for the exponential decay of the current in this circuit?

E 60.4-2 An initially uncharged capacitor (C = 2 μF) is charged through a resistor having $R = 10^6$ ohms, using a 12-volt battery.
- (a) Find the time constant of this circuit.
- (b) What is the initial value of charging current?
- (c) What is the final value q_f of the charge acquired?
- (d) At t = 2 s, what is the current?
- (e) At t = 2 s, what is the charge on the capacitor?
- (f) At t = 2 s, what is the potential difference across R?
- (g) At t = 2 s, what is the potential difference across C?
- (h) At what value of t are those two potential differences equal?
- (i) How many time constants are required for the charge q to reach 0.9 q_f?
- (j) How many time constants are required for q to reach 0.99 q_f?
- (k) How many time constants are required for the capacitor to acquire half its final value of stored energy?

Check your understanding of the material above by testing yourself on the minimum learning goal at the beginning of this section.

60.5 Discharging a Capacitor

Suppose a capacitor is connected to a battery and charged to a voltage $\Delta V_0 = q_0/C$, after which the battery is disconnected leaving the capacitor with a charge q_0. If the capacitor is then connected in series with a resistor and allowed to discharge (Figure 60-5), measurements show that the current and charge both decrease exponentially with time as shown in Figure 60-6. Using circuit theory, how can we account for this behavior?

Figure 60-5. An *RC* series circuit containing a switch. If the capacitor has an initial charge q_o, when the switch is closed current flows until the capacitor is discharged.

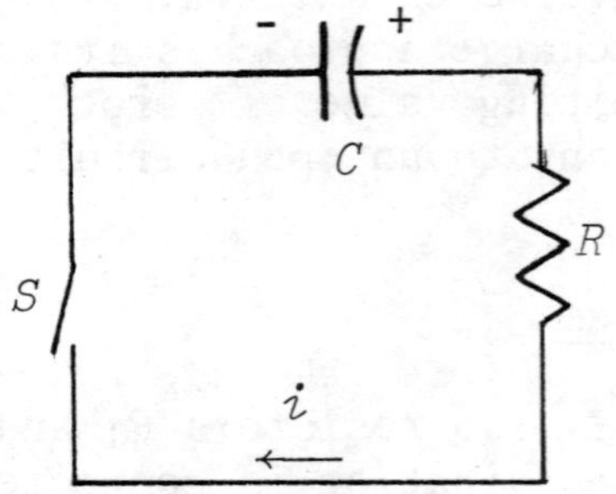

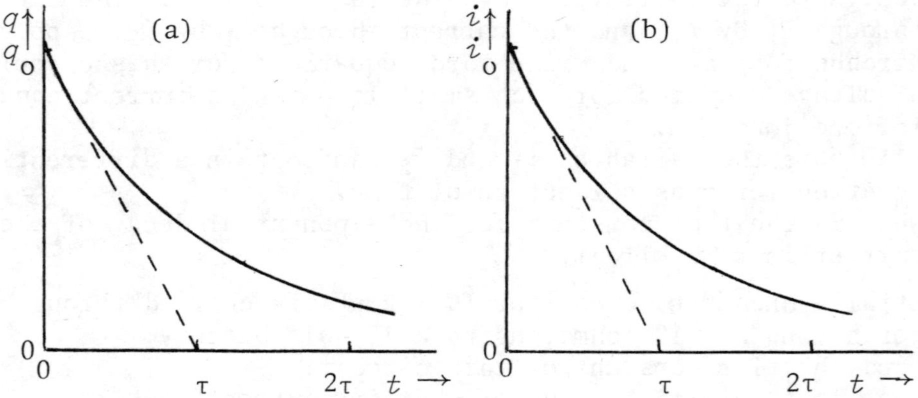

Figure 60-6. Discharging curves for the circuit in Figure 60-5, assuming the capacitor is initially charged to q_o and the switch is closed at time $t = 0$. (a) The charge decreases exponentially from its initial value q_o. (b) The current decreases exponentially from its initial value $i_o = q_o/RC$.

We choose $t = 0$ at the instant the switch is closed. Using Kirchhoff's loop rule, we trace clockwise around the circuit and obtain the following equation:

$$+\frac{q}{C} - iR = 0 \qquad \text{Eq. (60-7)}$$

In this case the current is *removing* positive charge from the positive terminal of the capacitor, so the connection between the current and the charge is given by

$$i = -dq/dt \; .$$

Substituting, we obtain an equation having only one unknown variable, q:

$$q/C + R\, dq/dt = 0.$$

Dividing by R and rearranging terms, we obtain

$$\frac{q}{RC} = -\frac{dq}{dt} \qquad \text{or} \qquad \frac{dq}{q} = -\frac{dt}{RC} \; . \qquad \text{Eq. (60-8)}$$

The time-rate of decrease of charge is proportional to the value of charge at any instant, which leads to an exponential decay of the charge. Eq. (60-8) is identical to Eq. (60-3) on page 60-17, except that here the variable is q instead of i. The integration of Eq. (60-8) for q follows the same procedure used to integrate Eq. (60-3), and leads to a similar result. Some key steps in the integration are:

((

$$RC \int_{q_0}^{q} \frac{dq}{q} = -\int_{0}^{t} dt \, ,$$

$$RC \Big[\ln q \Big]_{q_0}^{q} = -\Big[t \Big]_{0}^{t} = -t \, ,$$

$$\ln (q/q_0) = -\, t/RC,$$

$$e^{\ln (q/q_0)} = e^{-t/RC}.$$

))

and finally

★ $$q = q_0 \, e^{-t/RC}. \qquad \text{Eq. (60-9)}$$

This equation corresponds to the curve in Figure 60-6 (a), with a time constant $\tau = RC$ which is usual for a series RC circuit. A straight line tangent to the initial slope of the exponential decay curve intersects the time axis at $t = \tau$. The time constant is the time required for the charge to decay from its initial value q_0 to $1/e = 0.368$ of this value.

One can compute the equation for the current as a function of time by taking the derivative of Eq. (60-9), following the recipe given in Appendix A.2. For this circuit, however, an easier method of obtaining this result is simply to solve Eq. (60-7) for i:

$$i = q/RC$$

★ $$i = (q_0/RC)e^{-t/RC} \qquad \text{or} \qquad i = i_0 \, e^{-t/RC} \, , \qquad \text{Eq. (60-10)}$$

where

$$i_0 = q_0/RC = \Delta V_0/R \; .$$

Once again we have obtained an exponential decay curve for the current, with a time constant $\tau = RC$; this equation matches the curve in Figure 60-6(b).

Question and Exercises

Q 60.5-1 Why does $i(t)$ follow a decreasing curve both when a capacitor is charged and when it is discharged, while $q(t)$ follows a decreasing curve in one case and an increasing curve in the other?

E 60.5-1 For the $q(t)$ curve of a discharging capacitor, show that a straight line tangent to the curve at $t = 0$ intersects the t axis at $t = \tau$, where $\tau = RC$.

((E 60.5-2 Write out a complete derivation of the result $q = q_0 e^{-t/RC}$, beginning with the differential equation $dq/q = -dt/RC$.))

E 60.5-3 A capacitor has an initial charge q_0, and is connected in series with a resistor and a switch. After the switch is closed, how many time constants elapse before:
(a) the charge q on the capacitor decreases to 0.37 q_0?
(b) q decreases to 0.05 q_0?
(c) the current i decreases to 0.5 i_0?
(d) the energy stored in the capacitor decreases to half its initial value?
(e) the energy dissipated in the resistor reaches half its total value?
(f) the voltage ΔV_R across R equals the voltage ΔV_C across C?

60.6 Kirchhoff's Second Rule and Inductance

Minimum Learning Goals. Given a circuit diagram containing batteries, resistors, inductors, and a switch, be able to recall Kirchhoff's laws and use them to write (a) equations relating currents at a designated junction, and (b) equations relating voltages (expressed in terms of currents) around a specified loop after the switch is either opened or closed. For example:

1 For the circuit shown below: (a) write an equation relating the currents at junction A; (b) write an equation relating voltages (expressed in terms of current) across all circuit elements of the designated loop as a function of time. t, after switch S is closed. (Answer, page 60-28, Line 7)

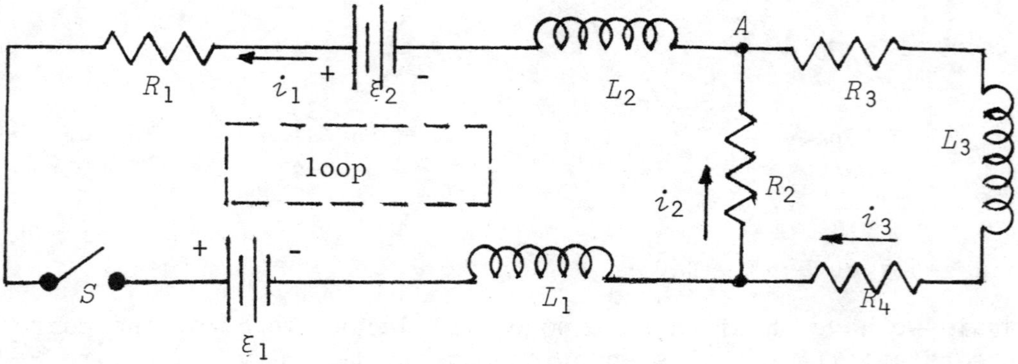

2 For the circuit shown at the right: (a) write an equation relating the currents at junction *A*; (b) write an equation relating voltages (expressed in terms of current) across circuit elements of the designated loop as a function of time, *t*, after switch *S* is closed. (Answer, page 60-28, Line 10)

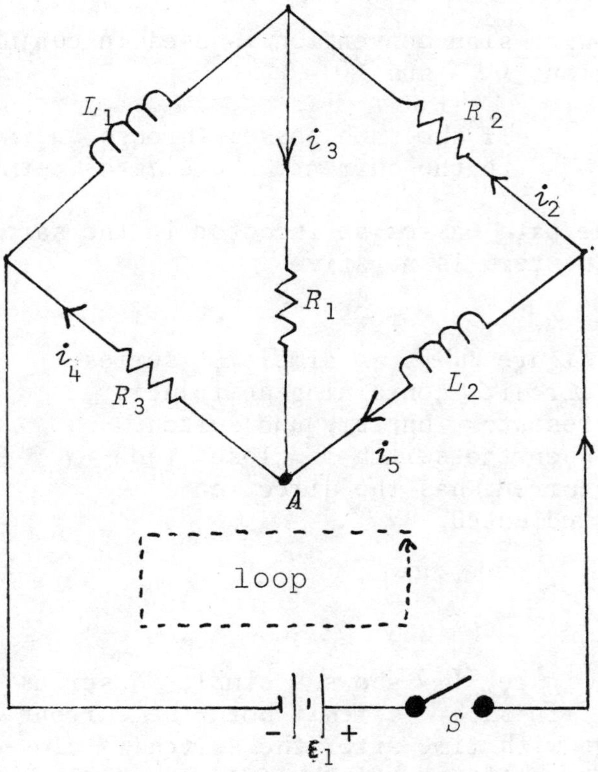

For s-c circuits the existence of an electrostatic potential was used to justify Kirchhoff's second rule (Section 60.2). When time-varying magnetic fields are present, however, there is no electrostatic potential (Section 59.1). The voltage across a circuit element can still be defined as the line integral of $\vec{E}$ through the element, but in going around a closed loop these voltages may not add up to zero. Rather, the sum of the voltages is equal to the time-rate of change of magnetic flux through the closed loop.

Kirchhoff's second rule may be applied to circuits containing inductors by taking account of the time-rate of change of magnetic flux in the inductors. This rate of change of flux is proportional to the rate of change of the current producing the flux. The changing flux induces an electromotance in the inductor, such that the voltage ΔV_L across an ideal inductor is proportional to the time-rate of change of current in the inductor (Section 59.4):

★ $$\Delta V_L = -L\, di/dt.$$

Therefore, the sum of the voltages around a closed loop is equal to the sum of the terms $- L\, di/dt$, one for each inductor in the loop.

When labeling a circuit diagram, it is convenient to indicate the current in a branch containing an inductor in the same direction or sense as if this branch were solely resistive. In using Kirchhoff's second rule, a correct equation for any loop containing an inductor is then obtained if the following sign convention is used in conjunction with those given in Sections 60.2 and 60.3:

★ If the path passes through an inductor in the direction opposite to the current, the $L\,di/dt$ term is positive.

If the path passes an inductor in the same direction as the current, the $L\,di/dt$ term is negative.

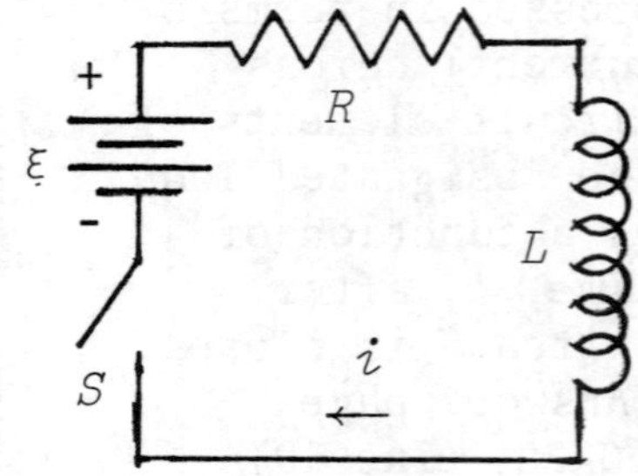

Figure 60-7. A simple *LR* series circuit, containing an inductor, resistor, battery and switch. When the switch is closed the current has the direction indicated.

Figure 60-7 shows a simple *LR* series circuit. As we found in Section 59.4, in such a circuit both the current and the voltage across the inductor change with time after the switch is closed. In preparation for computing these quantities, Kirchhoff's loop rule is applied to the circuit. Using the sign conventions we have adopted, at all times after the switch is closed the voltages around the loop (tracing clockwise) are

$$\xi - iR - L\,di/dt = 0.$$

This equation is solved in Section 60.7.

In some situations a current in one branch of a circuit gives rise to magnetic flux linking other parts of the circuit. The analysis then requires for such a branch an induced voltage term proportional to the rate of change of current in another branch of the circuit. The proportionality constant in such a term is called the *mutual inductance* between the two branches or loops of the circuit. Mutual inductance is the basis for the functioning of electrical transformers.

Questions

Q 60.6-1 When an inductor is included in a circuit loop and the magnetic field outside the inductor is negligible, the time-rate of change of magnetic flux in the inductor equals the sum of the voltages in the other elements of the circuit loop. Does the magnetic flux threading the inductor depend only on what goes on in the inductor, or on what goes on outside the inductor, or both? Explain.

Q 60.6-2 When applying Kirchhoff's second rule to a circuit loop, for each circuit element one must decide whether to write the term with a plus or minus sign. State a self-consistent set of sign conventions for the signs of these terms.

Check your understanding of the material above by testing yourself on the minimum learning goal at the beginning of this section.

60.7 LR Series V-C Circuits

Minimum Learning Goal. (1) Given a diagram of a series circuit containing a battery, switch, resistor, and inductor, with specified values of electromotance, resistance, and inductance, be able to sketch qualitatively correct graphs of voltage and current for each circuit element as functions of time after the switch is closed. On each graph, indicate the initial numerical value of the voltage or current, and the numerical value of the time constant of the circuit. For example:

1 Sketch graphs of voltage and current for each circuit element in the diagram at right as functions of time after the switch is closed. (Answer, page 60-29, Line 13)

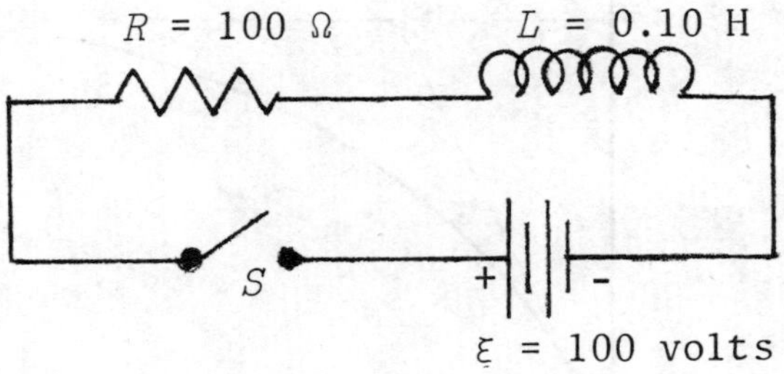

2 Sketch graphs of voltage and current for each circuit element in the diagram at right as functions of time after the switch is closed. (Answer, page 60-28, Line 1)

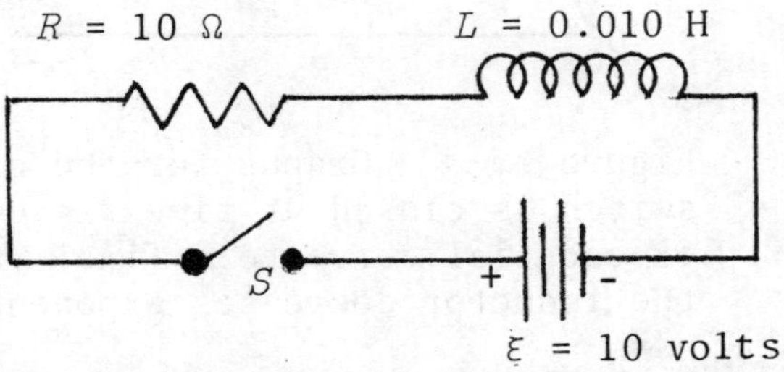

In Section 59.4 we considered the transient behavior of an inductor connected in a circuit like that of Figure 60-7 (page 60-24). In this variable-current (v-c) circuit the current varies with time but does not change direction during the time interval considered. The experimental curves of current vs. time and inductor voltage vs. time for this series *LR* circuit were accounted for in a qualitative way by considering the effect of self-inductance. In this section we obtain the mathematical equations for these curves, which are reproduced in Figure 60-8.

Using Kirchhoff's loop rule, in Section 60.6 we obtained an equation for the circuit which is easily transposed to the following:

$$L\, di/dt + iR = \xi . \qquad \text{Eq. (60-11)}$$

This equation has the same mathematical form as Eq. (60-1),

$$R\, dq/dt + q/C = \xi ,$$

for an RC series circuit when the capacitor is being charged (page 60-15). The LR circuit equation is obtained from the RC circuit equation if we replace q by i, R by L, and C by $1/R$. The solution of the RC circuit equation is

$$q(t) = \xi C[1 - e^{-t/RC}]. \quad \text{Eq. (60-6)}$$

By analogy, the solution of the LR circuit equation is

★ $$i(t) = (\xi/R)[1 - e^{-tR/L}]. \quad \text{Eq. (60-12)}$$

As RC is the time constant for a series RC circuit, L/R is the time constant for a series LR circuit. Eq. (60-12) corresponds to the curve in Figure 60-8(a).

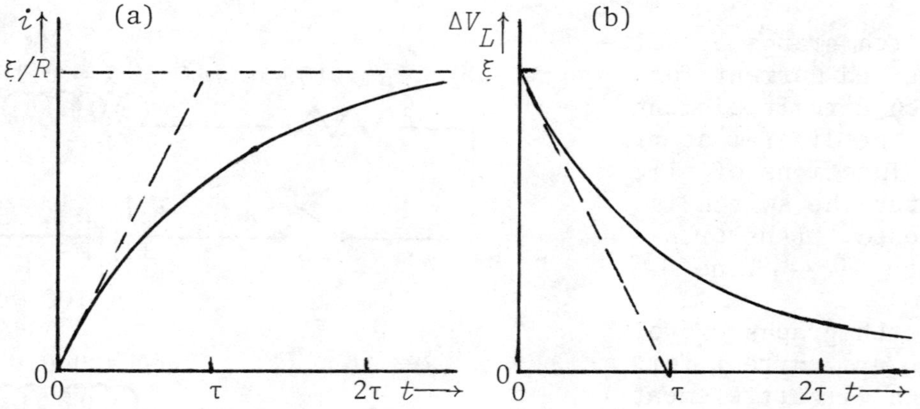

Figure 60-8. Graphs for the circuit in Figure 60-7, assuming the switch is closed at time $t = 0$. (a) The current rises in an inverse exponential curve to a final value $i_f = \xi/R$. (b) The voltage across the inductor decreases exponentially from its initial value ξ.

To interpret Eq. (60-12), we observe that at $t = 0$ the exponential term $e^{-tR/L}$ is equal to unity, so that the initial value of $i = 0$. This is one of the initial conditions of the solution, corresponding to having an open switch S in the circuit until $t = 0$. As time proceeds after $t = 0$, the quantity in square brackets steadily increases from zero toward unity. After a time long compared with the time constant L/R, the current approaches the limiting value $i_f = \xi/R$.

The voltage across the ideal inductor in this circuit is always given by $\Delta V_L = L\, di/dt$. Combining this with Eq. (60-11) and then substituting Eq. (60-12) for $i(t)$, we have

$$\Delta V_L = \xi - iR$$

$$= \xi - R(\xi/R)[1 - e^{-tR/L}] .$$

This is easily reduced to

★ $$\Delta V_L = \xi\, e^{-t/\tau}, \quad \text{Eq. (60-13)}$$

where $\tau = L/R$. This result matches the curve in Figure 60-8(b), where the inductor voltage decreases exponentially from its initial value $\mathcal{E}$ toward an eventual value of 0, reached after a time long compared to the time constant τ.

Questions and Exercises

Q 60.7-1 In the *LR* series circuit in Figure 60-7, why is the final value of the current (after closing the switch) independent of the value of the inductance?

Q 60.7-2 Consider a circuit containing a battery, resistor, inductor and switch in series. Why does the voltage across the inductor follow a decreasing curve just after the switch is *closed* and also just after the switch is *opened*, while $i(t)$ follows an increasing curve in one case and a decreasing curve in the other?

Q 60.7-3 Consider the circuit in Figure 60-7. Can the voltage across the inductor ever exceed the electromotance of the battery in this circuit? Explain.

((E 60.7-1 For Eq. (60-11), verify by substitution that the solution is $i(t) = (\mathcal{E}/R)(1 - e^{-t/\tau})$, where $\tau = L/R$.))

E 60.7-2 In the series *LR* circuit of Figure 60-7, after the switch is closed how many time constants elapse before:

(a) $i(t)$ reaches 0.9 of its final value?

(b) the voltage across the inductor reaches 0.9 of its initial value?

(c) iR reaches 0.9 of its final value?

E 60.7-3 An inductor having $L = 4.5$ H and $R = 300$ ohms is connected in series with a switch S and a 12-volt battery. The switch is closed at $t = 0$.

(a) Find the time constant of this circuit.

(b) At $t = 0.01$ s, what is i?

(c) At $t = 0.01$ s, what is the induced voltage across the inductor?

(d) At $t = 0.01$ s, what is iR?

(e) At $t = 0.01$ s, what is the total voltage across the inductor?

((E 60.7-4 Consider the circuit shown here, with the switch initially open and the current at its steady-state value. The switch is closed at $t = 0$.

(a) Apply Kirchhoff's rule to write a complete equation for the loop containing the inductor and the switch.

(b) Show that the solution of this equation is $i(t) = \left(\frac{\mathcal{E}}{R_1 + R_2}\right)e^{-t/\tau}$, where $\tau = L/R_2$.

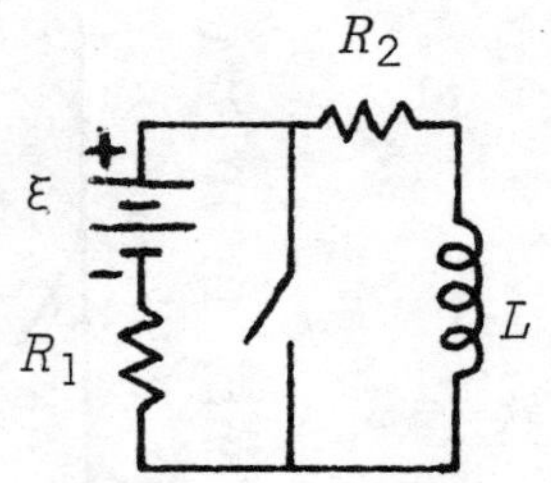

))

E 60.7-5 For the circuit in Figure 60-7, show that the initial rate of change of current is ξ/L.

Check your understanding of the material above by testing yourself on the minimum learning goal at the beginning of this section.

Answers to Minimum Learning Goals

Line 1

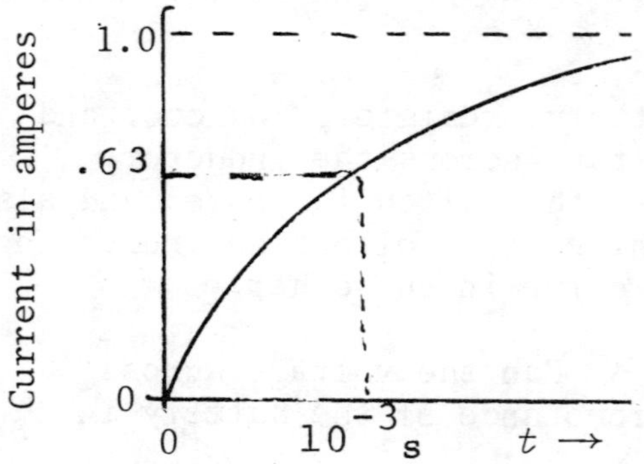

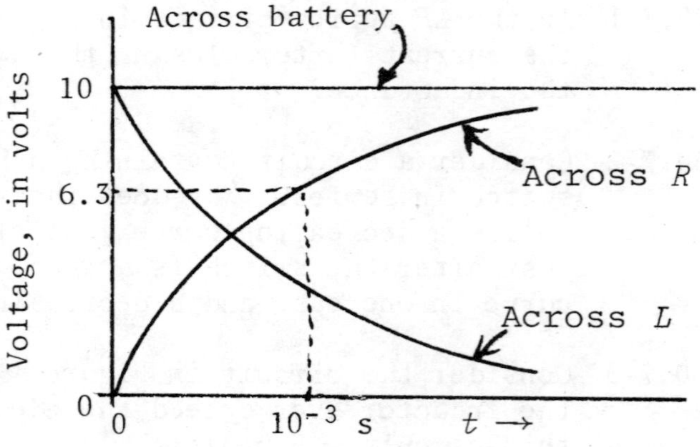

Line 2 0.1 ohm

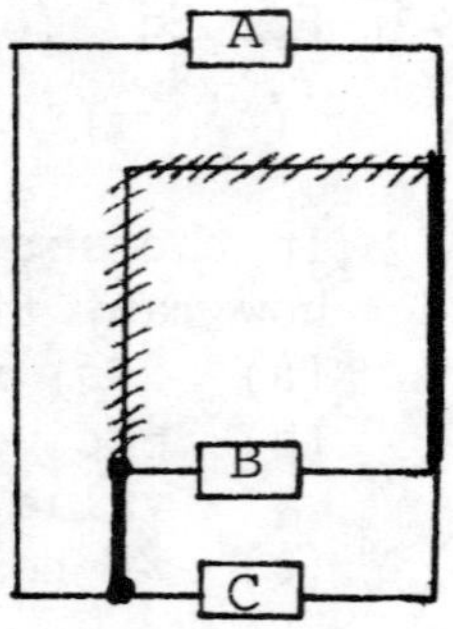

Line 3 (a), (b), and (d) equivalent.

Line 4 $\xi = 250$ volts; $R_3 = 40$ ohms

Line 5 $i_1 - i_2 - i_4 = 0;\quad -\xi_3 - i_1R_1 - \dfrac{q_4}{C_1} = 0$

Line 6 $i_1 - i_2 - i_3 = 0;$

$q_3/C_1 - i_2R_2 - \dfrac{q_2}{C_2} - i_2R_3 = 0$

Line 7 $i_1 - i_2 + i_3 = 0;\quad -L_2\dfrac{di_1}{dt} + \xi_2 - i_1R_1 - \xi_1 - L_1\dfrac{di_1}{dt} - i_2R_2 = 0$

Line 8 1.6 ohm

Line 9 0.5 A in R_2; 50 V across R_1; 0 volt across R_2.

Line 10 $i_5 + i_3 - i_4 = 0;\quad \xi_1 - L_2\dfrac{di_5}{dt} - i_4R_3 = 0$

Line 11

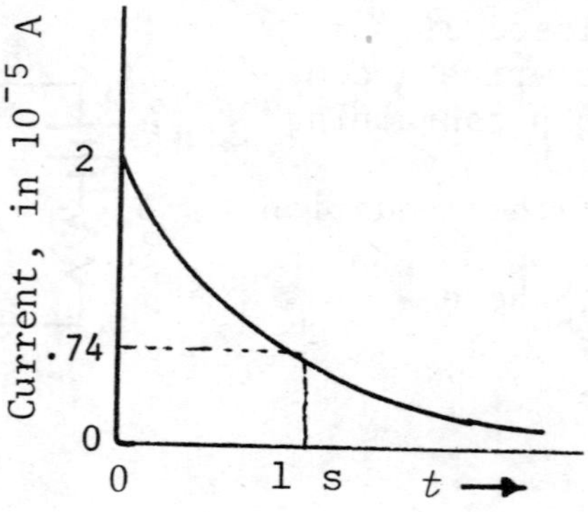

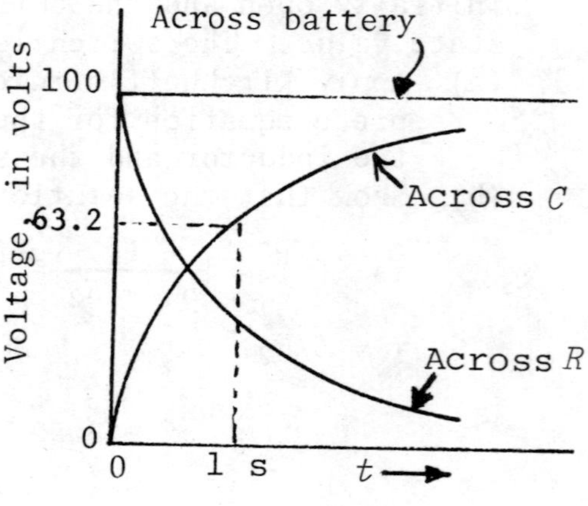

Line 12

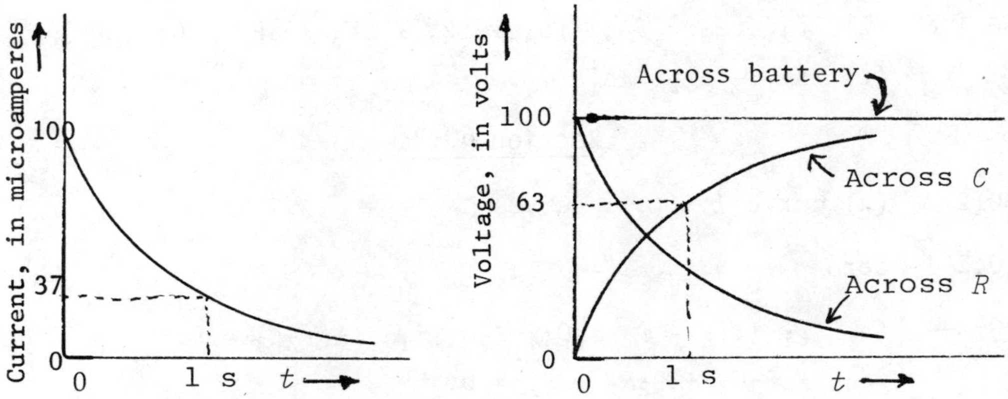

Line 13

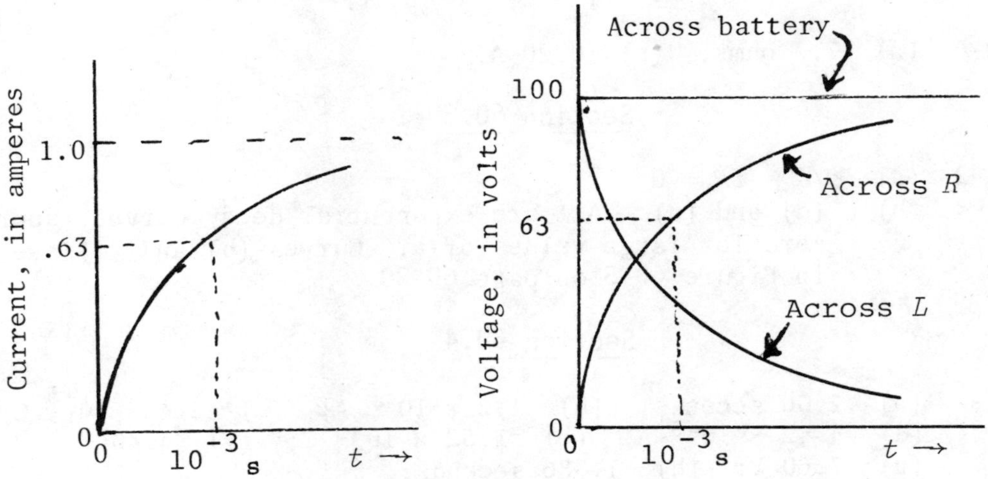

Line 14 (a), (b), and (c) are equivalent. (d) must be revised as shown at the right.

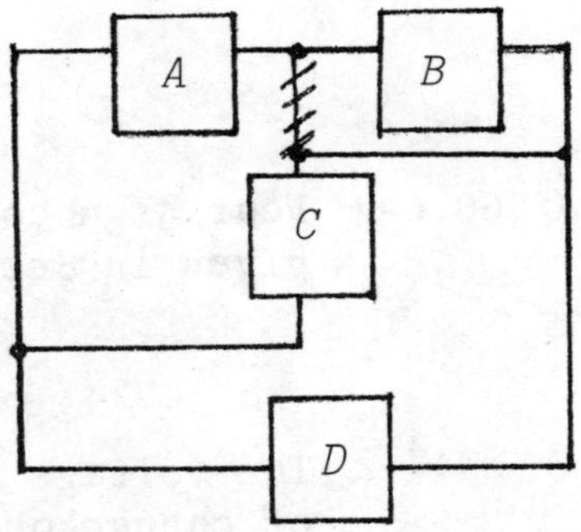

ANSWERS to Questions and Exercises in Chapter 60

Section 60.1

Q 60.1-2 When there is a short circuit between two points, a conducting path having negligible (effectively zero) resistance exists between the two points.

E 60.1-2 (a) (1) and (2). (b) On diagram (3), delete the direct lead from *B* to *C*.

E 60.1-4

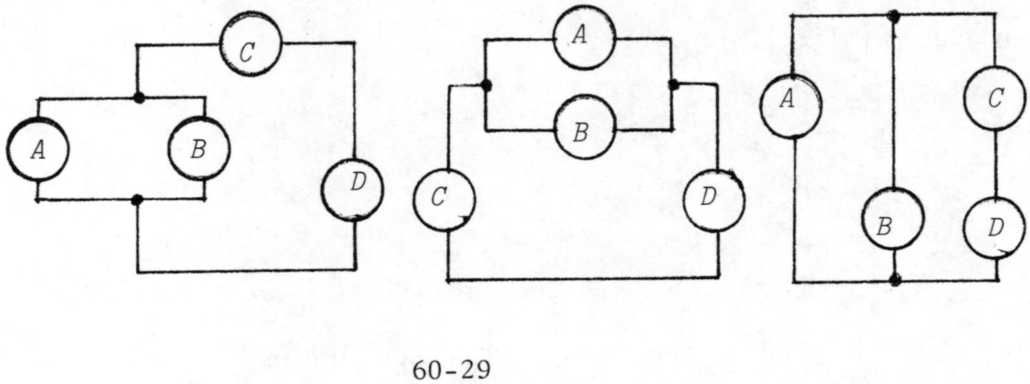

E 60.1-6 (a) $i_1 + i_3 = i_2$; (b) $+ \Delta V_4 + \Delta V_5 + \Delta V_3 = 0$;
(c) $\Delta V_1 - \Delta V_2 + \Delta V_4 = 0$.

Section 60.2

Q 60.2-2 (a) curve *b*; (b) curve *d*.

Q 60.2-4 zero

E 60.2-4 (a) $i_1 + i_3 - i_2 = 0$; $\xi_1 - i_1R_1 + i_3R_3 = 0$;
$+ i_3R_3 + i_2R_2 + \xi_2 = 0$.
(b) $R_2 = 4$ ohms; $i_1 = 0.95$ A, $i_3 = -0.68$ A

E 60.2-6 (a) 7.7 ohms; (b) 0.20 A.

Section 60.3

E 60.3-2 (a) $q/C - iR = 0$.
(b) (c) and (d). All are exponential decay curves, approaching zero for large values of t. Curves (b) and (d) are shown in Figure 60-5 on page 60-20.

Section 60.4

E 60.4-2 (a) 2.00 seconds; (b) 1.2×10^{-5} A; (c) 2.4×10^{-5} C;
(d) 4.42×10^{-6} A; (e) 1.52×10^{-5} C; (f) 4.40 V;
(g) 7.60 V; (h) 1.386 seconds;
(i) 2.30 time constants; (j) 4.60 τ; (k) 1.23 τ.

Section 60.6

Q 60.6-2 Four sign conventions are needed; a self-consistent set is given in Sections 60.2, 60.3, and 60.6.

Section 60.7

Q 60.7-2 The voltage across the inductor is proportional to the time-rate of change of current, which is always greatest just after a switch in an *LR* series circuit is either opened or closed. The current however, increases with time after the switch is closed but decreases with time after the switch is opened.

E 60.7-2 (a) 2.30 τ; (b) 0.105 τ; (c) 2.30 τ.

E 60.7-4 (a) $iR_2 + L\,di/dt = 0$

CHAPTER 61 ALTERNATING - CURRENT CIRCUITS

Introduction

An alternating-current (a-c) circuit is one in which the current changes direction and magnitude in a regular pattern, varying sinusoidally with time. A sinusoidal current results from a sinusoidal electromotance applied to the circuit. This chapter discusses the current, voltages, and impedance in series circuits containing inductance, resistance, and capacitance, driven by a sinusoidal voltage. Also considered are power transfer and resonance in series *LRC* circuits.

When you have completed your study of the material in this chapter, you should be able to answer questions and work exercises like the examples in the minimum learning goals assigned by your instuctor.

61.1 Voltages Sinusoidal in Time. Impedance

Electricity supplied by public utility companies for household, commercial and industrial use is generated by electric generators which produce an electromotance that varies sinusoidally with time. (A *sinusoidal* variation is one that follows a sine or cosine function of the independent variable.) When a sinusoidal electromotance is applied to a circuit containing resistors, capacitors, and inductors, the resulting steady-state current also has a sinusoidal variation with time. Graphs of sinusoidal electromotance and current are shown in Figure 61-1. It is seen that both electromotance and current alternate in a periodic manner between positive and negative values. For this reason the sinusoidal current is called *alternating current*, and circuits where it exists are called alternating-current circuits or simple *a-c circuits*.

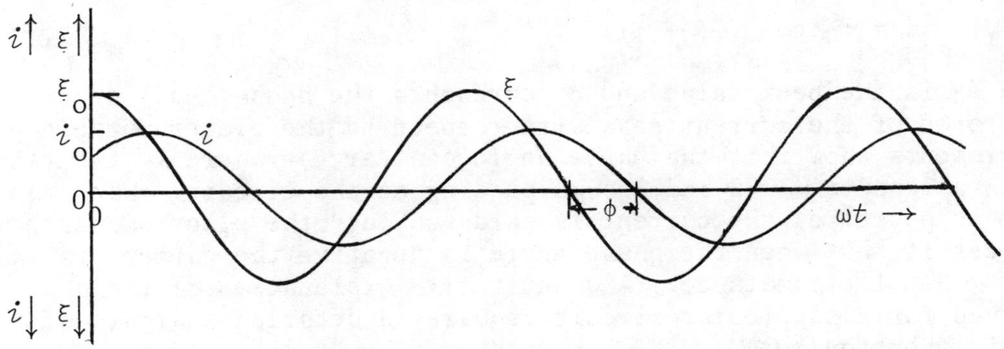

Figure 61-1. An electromotance ξ having a sinusoidal variation with time, represented by the function $\xi = \xi_0 \cos \omega t$. The resulting steady-state current in an a-c circuit is also sinusoidal, represented by the function $i = i_0 \cos(\omega t - \phi)$. In this figure, the current lags the electromotance by the phase angle ϕ.

An alternating electromotance $\xi(t)$ is represented by the function

★ $$\xi = \xi_0 \cos \omega t. \qquad \text{Eq. (61-1)}$$

ξ_0 is the amplitude or *peak value* of the electromotance, and ω is its *angular frequency*, expressed in radians per second. The angular frequency is related to the frequency of alternation, f, by

★ $$\omega = 2\pi f.$$

The SI unit of frequency is the hertz, symbol Hz (formerly expressed as a cycle per second). The SI unit of angular frequency is second^{-1}, symbol s^{-1}, because a radian is a pure number having no physical units. $\xi(t)$ is plotted in Figure 61-1 using as the abscissa ωt, which has units of radians. Our choice of the cosine function for $\xi(t)$ is arbitrary; we might equally well have chosen a sine function, in which case the graph of Figure 61-1 would have a similar appearance but both the electromotance and current curves would be shifted to the right or left by equal amounts. (Sinusoidal functions are discussed briefly in Appendix A.1.) In one complete *cycle* of the electromotance or the current, the instantaneous value of ξ or i varies from an initial value through its positive and negative peak values and again to its initial value, going through an angular variation of 2π radians. Each cycle occurs in a time equal to one period of the oscillation; the period is the reciprocal of the frequency, so that a total of f cycles occur each second. The frequency of commercial electric power in the United States and Canada is 60 hertz.

What happens when a sinusoidal electromotance is applied to a circuit where the circuit elements are resistors, inductors, and capacitors? Initially, just after the switch is closed, the current has a transient or short-term behavior similar to those discussed earlier for *RC* and *LR* circuits connected to a battery. Our concern in this and the following sections, however, is with the *steady-state* response of the circuit, in which each cycle is exactly like every other cycle. Observations using a cathode-ray oscilloscope show that the steady-state current is also sinusoidal, and that in most circuits it reaches its peak value earlier or later than the electromotance (Figure 61-1). The sinusoidal response comes as no surprise, because if one electrically "shakes" the circuit by applying an electromotance at a certain frequency we would expect the resulting flow of charge to oscillate at this same frequency. The equation for the current may be written

★ $$i = i_0 \cos(\omega t - \phi), \qquad \text{Eq. (61-2)}$$

where i_0 is the peak value and ϕ represents the *phase angle* or time difference of the current peak with respect to the electromotance peak. Measurements show that the phase angle can vary anywhere in the range from $+\pi/2$ radians to $-\pi/2$ radians, depending on the circuit. When the phase angle is positive, the current is said to "lag" the electromotance which produces it, but when the phase angle is negative the current is said to "lead" the electromotance. A quantitative explanation of the phase angle observed for a particular circuit requires a detailed analysis of the circuit (Section 61.2).

In s-c circuits containing only a battery and resistors, the ratio of the applied electromotance to the current is called the resistance of the circuit. In a-c circuits this ratio is called the impedance, symbol Z.

> The *impedance* Z of an a-c circuit is the ratio of the peak value of applied electromotance, ξ_0 to the peak value of the resulting current, i_0.

★ $$Z \equiv \frac{\xi_0}{i_0}. \qquad \text{Eq. (61-3)}$$

As with resistance, the SI unit of impedance is the ohm, symbol Ω. As we shall see in Section 61.2, the impedance of a circuit depends on the circuit elements and how they are connected, and also on the frequency of the applied electromotance. Impedance is defined only for circuits where the electromotance and current are sinusoidal.

Exercises

E 61.1-1 What is the angular frequency for commercial electric power in the U.S.A.?

E 61.1-2 In an a-c circuit, the peak value of current is 17.7 A when the peak value of applied electromotance is 165.5 V. What is the impedance of this circuit?

61.2 Series LRC Circuits

Minimum Learning Goal. (1) Given the values of R, L, and C in a series circuit containing an inductor, resistor, and capacitor, and the amplitude and angular frequency of the sinusoidal voltage applied to the circuit, be able to compute (a) the impedance of each circuit element and of the total circuit, and (b) the peak value of current in the circuit. For example:

1 An *LRC* series circuit has a resistance of 5.0 ohms, an inductance of 50 mH, and a capacitance of 2.0×10^{-4} F. A sinusoidal voltage $\mathcal{E} = 160 \sin (400\,t)$ is applied to this circuit. (a) What are the impedances of the circuit elements? (b) What is the impedance of the total circuit? (c) What is the peak value of current in the circuit? (Answer, page 61-19, Line 2).

2 A series circuit has $L = 1.0$ mH, $R = 10.0$ ohms, and $C = 2.0 \times 10^{-7}$ F. The applied voltage has a peak value of 20.0 volts and a frequency of 7.96×10^{4} Hz. (a) What are the impedances of the circuit elements? (b) What is the impedance of the total circuit? (c) What is the peak value of current in the circuit? (Answer, page 61-19, Line 4)

To analyze an a-c series circuit containing one or more resistors, inductors, and capacitors, the first step is to draw a circuit diagram. The symbol for a source of alternating electromotance is . In preparation for applying Kirchhoff's second rule to the circuit, the instantaneous polarity of the source at some instant is indicated on the diagram using + and - symbols. It is then customary to indicate the direction of the resulting instantaneous current as if the circuit contained only resistance. To designate the instantaneous polarity of a capacitor, the terminal to which positive charge is being delivered by the instantaneous current is labeled +, and the other terminal is labeled -. A circuit diagram labeled in this way is shown in Figure 61-2.

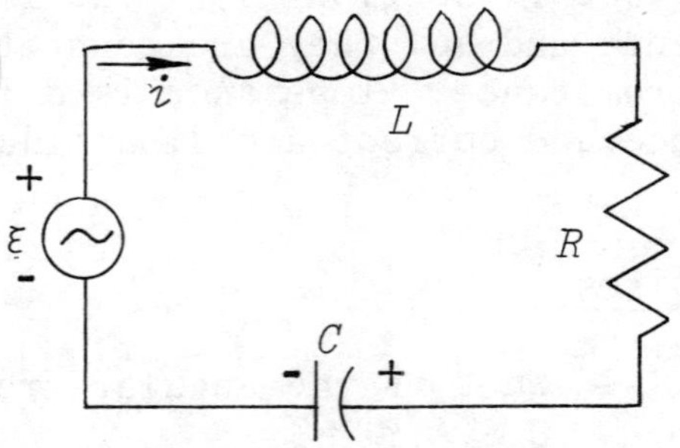

Figure 61-2. An a-c series circuit containing an inductor, a resistor, and a capacitor. For the purpose of circuit analysis, the direction of the instantaneous current and the polarity of the capacitor are determined by the instantaneous polarity chosen for the source of electromotance.

In a series circuit containing no parallel branches, Kirchhoff's second rule is the basis for obtaining the circuit equation:

★ II. The algebraic sum of the voltages around any closed loop of the circuit is zero.

To write the voltage terms with the proper signs, the following sign conventions are used:

★ If the path passes through a source of electromotance from its - to its + terminal, the term for this electromotance is positive.

★ If the path passes through an inductor in the direction opposite to the current, the $L\ di/dt$ term is positive.

★ If the path passes through a resistor in the direction opposite to the current, the iR term is positive.

★ If the path passes through a capacitor from its - to its + terminal, terminal, the q/C term for this capacitor is positive.

★ If the path passes through one of these circuit elements in the opposite direction, the corresponding term is negative.

These are the same sign conventions we have used in our earlier analyses of circuits.

Applying these guidelines to the circuit in Figure 61-2, we trace counterclockwise around the loop and obtain the following:

$$-\xi + q/C + iR + L\ di/dt = 0.$$

Rearranging terms and substituting Eq. (61-1) for ξ leads to

★ $$L\ di/dt + Ri + q/C = \xi_0 \cos \omega t. \qquad \text{Eq. (61-4)}$$

We next need to substitute expressions for di/dt, i, and q in Eq. (61-4). The current is given by

$$i = i_0 \cos(\omega t - \phi), \qquad \text{Eq. (61-2)}$$

where the phase angle ϕ (determined by the values of L, R, C, and ω in the circuit) is something we want to calculate in this analysis. The ((derivative di/dt is obtained by applying the appropriate recipe for differentiation (Appendix A.2) to Eq. (61-2), where i_0 is a constant:

$$di/dt = -i_0\omega \sin(\omega t - \phi).$$

The charge q delivered to the capacitor is related to the current by $i = dq/dt$; after multiplying by dt we obtain $dq = i\ dt$. We find q as the indefinite integral of $i\ dt$, using a recipe from Appendix A.4 and recognizing that i_0 and ω are constants:

$$q = \int i\ dt = \int i_0 \cos(\omega t - \phi)\ dt$$

$$= (i_0/\omega) \int \cos(\omega t - \phi)\omega\ dt$$

$$= (i_0/\omega) \sin(\omega t - \phi).$$))

Substituting the expressions obtained above for di/dt, i, and q in the basic circuit equation, Eq. (61-4), and combining the two terms in $\sin(\omega t - \phi)$, leads to

$$(i_0/\omega C - i_0\omega L) \sin(\omega t - \phi) + Ri_0 \cos(\omega t - \phi) = \mathcal{E}_0 \cos \omega t. \quad \text{Eq. (61-5)}$$

The next step is to expand the terms $\cos(\omega t - \phi)$ and $\sin(\omega t - \phi)$ using the identities

$$\cos(\omega t - \phi) = \cos \omega t \cos \phi + \sin \omega t \sin \phi,$$

$$\sin(\omega t - \phi) = \sin \omega t \cos \phi - \cos \omega t \sin \phi.$$

Substituting these expressions into Eq. (61-5) leads to

$$i_0[(\frac{1}{\omega C} - \omega L) \cos \phi + R \sin \phi] \sin \omega t$$

$$+ i_0[(\omega L - \frac{1}{C}) \sin \phi + R \cos \phi] \cos \omega t = \mathcal{E}_0 \cos \omega t. \quad \text{Eq. (61-6)}$$

Eq. (61-6), derived from the circuit equation (61-4) and representing the instantaneous sum of the voltages around the circuit, must be valid for all values of t. At instants when $\cos \omega t = 0$, this requires that the coefficient of the $\sin \omega t$ term must equal zero. Setting the expression inside the brackets equal to zero, and rearranging terms, leads to

★ $$\tan \phi = \frac{\omega L - 1/(\omega C)}{R} \quad \text{or} \quad \phi = \tan^{-1} \frac{\omega L - 1/(\omega C)}{R}. \quad \text{Eq. (61-7)}$$

At instants when $\sin \omega t = 0$ and $\cos \omega t = 1$, Eq. (61-6) becomes

$$i_0[(\omega L - \frac{1}{\omega C}) \sin \phi + R \cos \phi] = \mathcal{E}_0.$$

Removing the brackets and dividing by R leads to

$$i_0 \cos\phi + i_0 \frac{\left(\omega L - \frac{1}{\omega C}\right)}{R} \sin\phi = \frac{\xi_0}{R}.$$

Combining this result with Eq. (61-7), and using another common identity, then leads to

★ $$i_0 R = \xi_0 \cos\phi. \qquad \text{Eq. (61-8)}$$

Some details of these derivations are left as exercises for the reader.

Combining Eq. (61-8) with Eq. (61-3), $Z = \xi_0/i_0$, immediately leads to

★ $$\cos\phi = R/Z. \qquad \text{Eq. (61-9)}$$

Another equation for cos ϕ can be derived from Eq. (61-7), as follows: Squaring Eq. (61-7), and then adding unity to each side yields

$$\tan^2\phi + 1 = \left[\frac{\omega L - \frac{1}{\omega C}}{R}\right]^2 + 1,$$

which can be rearranged to give

$$\sec^2\phi = \frac{\left(\omega L - \frac{1}{\omega C}\right)^2 + R^2}{R^2}.$$

Inverting each side, we have

$$\frac{1}{\sec^2\phi} = \cos^2\phi = \frac{R^2}{R^2 + \left(\omega L - \frac{1}{\omega C}\right)^2}.$$

Taking the square root of each side gives

$$\cos\phi = \frac{R}{\sqrt{R^2 + \left(\omega L - \frac{1}{\omega C}\right)^2}}. \qquad \text{Eq. (61-10)}$$

Comparing Eq. (61-10) with Eq. (61-9), it is seen that the impedance of the series *LRC* circuit is given by

★ $$Z = \sqrt{R^2 + \left(\omega L - \frac{1}{\omega C}\right)^2}. \qquad \text{Eq. (61-11)}$$

How shall we interpret the results of these derivations? We recall that in any valid physical equation, the physical units of each term must be the same. Referring to Eq. (61-6), each term in brackets must have the same units. Sin ϕ and cos ϕ are dimensionless quantities, so the unit of each term in these brackets must be the unit of R, that is, ohms. From this it is reasonable to interpret the quantity $1/(\omega C)$ as the impedance of the capacitance C at the angular frequency ω; when C is expressed in farads, the unit of $1/(\omega C)$ is the ohm. Similarly, the quantity ωL is the impedance of the inductance L at the angular frequency ω; when L is expressed in henrys, the unit of ωL is also the ohm.

Eq. (61-11) shows that the impedance Z of the circuit is always a positive quantity, and since R is also positive, Eq. (61-9) shows that cos ϕ is always positive. As a result, ϕ always lies in the range $-\pi/2 \leq \phi \leq +\pi/2$. Eq. (61-9) does not indicate whether ϕ is positive or negative, however, because cos ϕ = cos $(-\phi)$; the sign (and the value) of ϕ can be found using Eq. (61-7) to compute tan ϕ. When ϕ is positive (as in Figure (61-1), the peak current i_0 in the circuit lags the peak electromotance ξ_0 by a phase angle between 0 and $\pi/2$, but when ϕ is negative the peak current leads ξ_0 by a phase angle between 0 and $\pi/2$.

Note that ϕ and Z do not depend on ξ_0, but only on the values of L, R, C and ω. It is not the resistance alone but the total impedance of the circuit, with contributions from the individual impedances of the inductance, resistance, and capacitance, which determines the amount of current that results from the applied electromotance:

$$i_0 = \xi_0/Z. \qquad \text{Eq. (61-3)}$$

Questions and Exercises

Q 61.2-1 When Kirchhoff's second rule is applied to a series LRC circuit, is one adding peak values, or instantaneous values, of voltage?

Q 61.2-2 When Kirchhoff's second rule is applied to a circuit loop, is the circuit equation obtained the same when the loop is traced clockwise as when it is traced counterclockwise? Explain.

Q 61.2-3 What is the impedance of a resistor, and how does it depend on the frequency of the current through the resistor?

Q 61.2-4 What is the impedance of a capacitor, and how does it vary with the frequency of the current through the capacitor?

Q 61.2-6 Write expressions for the combined impedance of the following:
(a) a resistor and a capacitor;
(b) a resistor and an inductor;
(c) a resistor, capacitor, and inductor.

Q 61.2-7 (a) How can one find the phase difference between the current and the applied electromotance in a series LRC circuit, by mathematical analysis?
(b) How can one find this phase difference by experiment?

Q 61.2-8 In a series LRC circuit with sinusoidal excitation:
(a) Which instantaneous voltage is proportional to the derivative of i with respect to t?
(b) Which instantaneous voltage is proportional to the integral of i with respect to t?

E 61.2-1 (a) Derive Eq. (61-6) from Eq. (61-5) using the identities for cos $(\omega t - \phi)$ and sin $(\omega t - \phi)$.
(b) Beginning with Eq. (61-6), fill in all the steps in the argument to show that tan $\phi = [\omega L - 1/(\omega C)]/R$.

E 61.2-2 Fill in the missing steps to complete the derivation of Eq. (61-8), beginning with Eq. (61-6).

E 61.2-3 In the circuit shown at the right, the electromotance may be written $\xi = \xi_0 \sin(\omega t + \phi)$, and the current $i = i_0 \sin \omega t$.

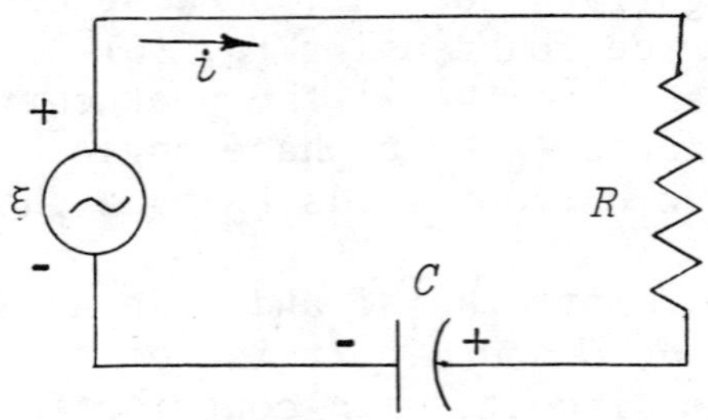

(a) Apply Kirchhoff's second rule and write an equation for the complete circuit.
(b) What is the peak value of the current, in terms of ξ_0 and the circuit parameters?
(c) What is the phase angle ϕ in terms of the circuit parameters?
(d) What impedance does this circuit present to the voltage source?

E 61.2-4 Apply Kirchhoff's second rule to write the circuit equation for the circuit below:

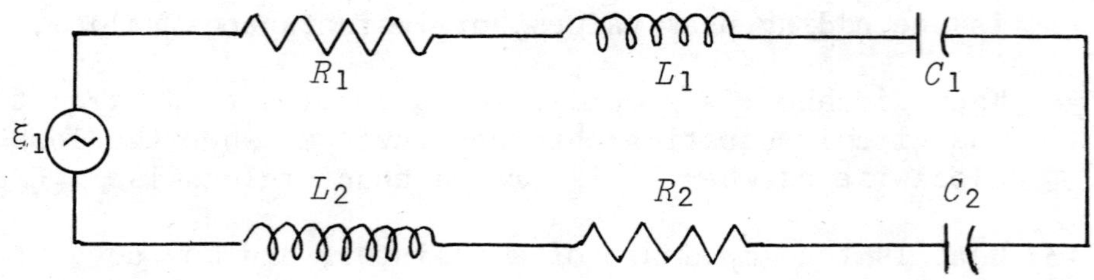

E 61.2-5 For the circuit in Figure 61-2, the electromotance may be written $\xi = \xi_0 \sin(\omega t + \phi)$, where ϕ is the phase angle between ξ and the resulting current.
(a) Apply Kirchhoff's rule and write an equation for the complete circuit.
(b) What impedance does this circuit present to the voltage source?
(c) What is the peak value of the current, in terms of ξ_0 and the circuit parameters?
(d) What is the phase angle ϕ in terms of the circuit parameters?

E 61.2-6 For the circuit of E 61.2-3, find $i(t)$ when $\xi_0 = 163 \cos \omega t$ volts, $= 60$ Hz, $R = 60$ ohms and $C = 100$ μF.

E 61.2-7 For the circuit in Figure 61-2: $\xi_0 = 180$ V, $\omega = 377\ \text{s}^{-1}$, $L = 0$, $R = 600$ ohms and $C = 1$ μF. Compute the impedance Z, the phase angle ϕ, and the peak value of the current i_0.

E 61.2-8 (a) What value of inductance must be added in series with the circuit in E 61.2-7, in order to reduce ϕ to zero?
(b) What value will i_0 then have?

E 61.2-9 Write an equation for the impedance of the circuit in E 61.2-4. (Hint: Inductors in series combine like resistors: $L = L_1 + L_2$. How do capacitors in series combine?)

E 61.2-10 For the circuit in Figure 61-2: $\xi = 9.0 \cos 377t$ volts, $R = 10$ ohms, $L = 1.0$ henry, and $C = 10\ \mu F$. Compute Z, ϕ, and $i(t)$.

E 61.2-11 An actual (non-ideal) inductor has $L = 4.5$ henry and $R = 300$ ohms.
(a) Compute Z and ϕ for this inductor at $f = 60$ Hz.
(b) Compute Z and ϕ for this inductor at $f = 1$ Hz.
(c) In what frequency range will the inductor act approximately like an ideal inductor?
(d) In what frequency range will this inductor act approximately like a resistor?

Check your understanding of the material above by testing yourself on the minimum learning goal at the beginning of this section.

61.3 Steady-State Voltages in a Series LRC Circuit

Minimum Learning Goal. (1) Given the values of R, L, and C in a series circuit containing an inductor, resistor, and capacitor, and the amplitude and angular frequency of the sinusoidal voltage applied to the circuit, be able to plot, as functions of time on the same axis, the instantaneous steady-state values of the current, the applied electromotance, and the voltages across the resistor, capacitor and inductor. For example:

1 An *LRC* series circuit has a resistance of 5.0 ohms, an inductance of 50 mH, and a capacitance of 2.0×10^{-4} F. A sinusoidal voltage $\xi = 160 \sin (400\ t)$ is applied to this circuit. (a) Plot a graph of current vs. time for this circuit. (b) Using the same time axis, plot curves of voltage vs. time for the applied electromotance, and for each circuit element. (Answer, page 61-19, Line 3)

2 A series circuit has $L = 1.0$ mH, $R = 10.0$ ohms, and $C = 2.0 \times 10^{-7}$F. The applied voltage has a peak value of 20.0 volts and a frequency of 7.96×10^4 hertz. (a) Plot a graph of current vs. time for this circuit. (b) Using the same time axis, plot curves of voltage vs. time for the applied electromotance, and for each circuit element. (Answer, page 61-18, Line 1)

Our analysis of steady-state conditions in series *LRC* circuits having a sinusoidal electromotance began with the use of Kirchhoff's second rule to write an equation for the instantaneous voltages of the circuit elements. It is instructive to plot a graph showing the phase relationships among the voltages and the current in such a circuit.

In Section 61.2 we began with $\xi = \xi_0 \cos \omega t$ and then found that $i = i_0 \cos(\omega t - \phi)$. Because the *current* is the same in all elements of a series circuit, is convenient here to shift the time origin so that the current is the reference waveform, and write

★ $$i = i_0 \sin \omega t \quad \text{and} \quad \xi = \xi_0 \sin(\omega t + \phi). \qquad \text{Eqs. (61-12)}$$

Using these functions for electromotance and current, it is still the case that when the phase angle ϕ (computed using Eq. (61-7) or (61-9)) is positive, $i(t)$ lags behind $\xi(t)$ by ϕ radians in phase and by ϕ/ω seconds in time. When ϕ is negative, $\xi(t)$ lags $i(t)$ by ϕ radians in phase and by ϕ/ω seconds in time.

A general equation for instantaneous voltage across any circuit element having impedance Z may be written as

★ $$\Delta V = \Delta V_0 \sin(\omega t + \phi), \qquad \text{Eq. (61-13)}$$

where ϕ is the phase angle of that particular circuit element computed using

★ $$\phi = \tan^{-1} \frac{\omega L - 1/(\omega C)}{R}. \qquad \text{Eq. (61-7)}$$

To apply Eq. (61-7) to individual ideal circuit elements, we may imagine that in the *LRC* series circuit of Fig. 61-2 (p. 61-4) two of the three circuit elements at a time are short-circuited, leaving only one functioning element in the circuit. When the capacitor is short-circuited it is unable to hold any charge, so in the equation $\Delta V_C = q/C$ the value of q is zero and therefore also ΔV_C is zero. From $\Delta V_C = q/C$, we may say that the value $\Delta V_C = 0$ corresponds to $C = \infty$, and this is the value to be used in Eq. (61-7) when computing the phase angle for an inductor or a resistor.

Applying Eq. (61-7), for a resistor alone we obtain $\tan \phi_R = 0/R = 0$, from which $\phi_R = 0$ for a resistor. For an ideal inductor alone, we obtain $\tan \phi_L = \omega L/0 = +\infty$, from which $\phi_L = +\pi/2$ for an ideal inductor. For a capacitor alone, we obtain $\tan \phi_C = -(1/\omega C)/0 = -\infty$, from which $\phi_C = -\pi/2$ for a capacitor.

Substituting these results in Eq. (61-13), we obtain the following equations for the instantaneous voltages across an ideal inductor, resistor, and capacitor respectively:

★ $$\Delta V_L = \Delta V_{L_0} \sin(\omega t + \pi/2),$$

★ $$\Delta V_R = \Delta V_{R_0} \sin \omega t,$$

★ $$\Delta V_C = \Delta V_{C_0} \sin(\omega t - \pi/2).$$

These instantaneous voltages may also be computed from the individual terms in the circuit equation (61-4), using Eq. (61-12) for $i(t)$. The calculations are:

$$\Delta V_L = L\frac{di}{dt} = L\frac{d}{dt}i_0 \sin \omega t$$

$$= \omega L i_0 \cos \omega t = \omega L\, i_0 \sin(\omega t + \pi/2);$$

$$\Delta V_R = Ri = Ri_0 \sin \omega t;$$

$$\Delta V_C = \frac{q}{C} = \frac{1}{C}\int i\, dt = \frac{1}{C}\int i_0 \sin \omega t\, dt$$

$$= -\frac{i_0}{\omega C}\cos \omega t = \frac{i_0}{\omega C}\sin(\omega t - \pi/2).$$

The equations above show that the instantaneous voltage ΔV_R across a resistor is in phase with the current, but the instantaneous voltage ΔV_L across an ideal inductor leads the current by $\pi/2$ radians in phase, and the instantaneous voltage ΔV_C across a capacitor lags the current by $\pi/2$ radians in phase. These phase relationships are shown in Figure 61-3. They may be studied experimentally using a cathode-ray oscilloscope.

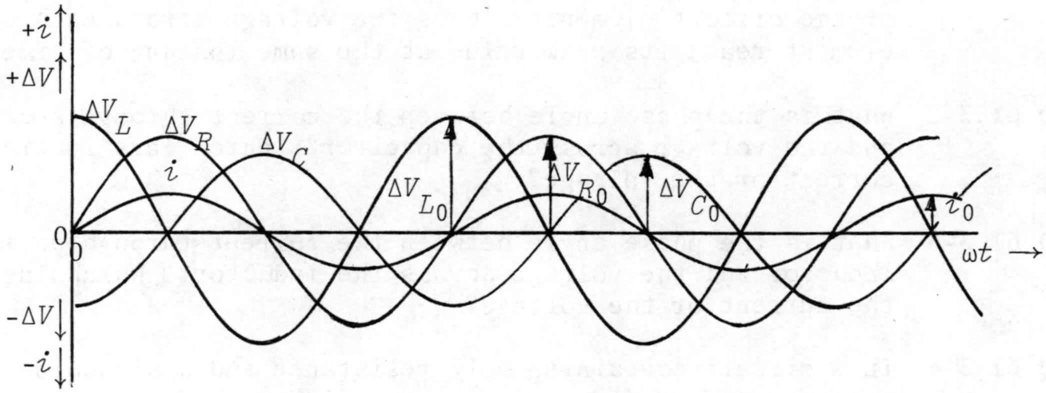

Figure 61-3. Phase relationships among the steady-state current and voltages in a series *LRC* circuit driven by a sinusoidal electromotance. The instantaneous voltage across a resistor is in phase with the current. The instantaneous voltage across an ideal inductor reaches its peak value $\pi/2$ radians earlier in time than the current peak, and thus leads the current by $\pi/2$ radians. The instantaneous voltage across a capacitor reaches its peak value $\pi/2$ radians later in time than the current peak and thus lags the current by $\pi/2$ radians.

Comparing the two sets of equations above for the instantaneous voltages, we see that the peak values of these voltages are:

★ $$\Delta V_{L_0} = \omega L i_0, \quad \Delta V_{R_0} = Ri_0, \quad \Delta V_{C_0} = i_0/(\omega C).$$

More generally, the steady-state peak voltage between any two points in a series *LRC* circuit is $\Delta V_Z = Zi_0$, where Z is the total impedance between those two circuit points. The instantaneous voltage between those two points is given by Eq. (61-13), where $\tan\phi$ is computed using Eq. (61-7), including only the values of L, R, and C located between the two points.

The *instantaneous* values of the individual voltages for the circuit elements between two points in a circuit always add to give the instantaneous voltage between the two points. The *peak* values of the individual voltages do *not* add to give the peak value of the voltage between the two points, however, because the individual voltages reach their peak values at different times. Indeed, it is possible for the voltage across the combination of two circuit elements to have a peak value smaller than the voltage on either one alone: If an ideal inductor and a capacitor have the same magnitude of impedance at a certain frequency, then their series combination has *zero* voltage across it.

Questions and Exercises

Q 61.3-1 When a sinusoidal current exists in a series *LRC* circuit, does the current reach its peak value at the same instant of time in each of the circuit elements? Does the voltage across each circuit element reach its peak value at the same instant of time? Explain.

Q 61.3-2 What is the phase angle between the current through a capacitor and the voltage across the capacitor? Which lags in time, the current or the voltage?

Q 61.3-3 What is the phase angle between the current through an ideal inductor and the voltage across the inductor? Which lags in time, the current or the voltage?

Q 61.3-4 In a circuit containing only resistance and a sinusoidal electromotance, $i_0 = \xi_0/R$.
- (a) If capacitance (but no inductance) is added to this circuit in series with the resistance, will i_0 increase, or decrease?
- (b) If inductance (but no capacitance) is added to this circuit in series with the resistance, will i_0 increase, or decrease?
- (c) Is there any way to increase i_0 above the value ξ_0/R by adding a combination of inductance and capacitance in series with the resistance?

Q 61.3-5 In a series *LRC* circuit in the steady-state, under what conditions can the peak value of the voltage across one circuit element exceed the peak value of the applied electromotance?

E 61.3-1 For a series *LRC* circuit with a sinusoidal current, sketch a graph showing at least two cycles of each of the following quantities as a function of time, in the correct phase relationships: current, ΔV_R, ΔV_C, ΔV_L.

Check your understanding of the material above by testing yourself on the minimum learning goal at the beginning of this section.

61.4 Power in A-C Circuits

When a source of electromotance does work on charge passing through it, this work goes into kinetic energy of the charge carriers, and this kinetic energy in turn is transformed into other forms of energy in the resistors, inductors, and capacitors in the circuit. Energy delivered to a capacitor is stored as electric potential enerty in the electric field of the capacitor. Energy delivered to an inductor is stored in the magnetic field of the inductor. In a resistor, the kinetic energy of the charge carriers is transformed to internal energy of the atoms by means of repeated collisions of the charge carriers with these atoms (Section 26.8). When the voltage ΔV_R sends a steady current i through a resistance R, the time-rate at which thermal energy is developed in the resistance is the power

★ $P = i\ \Delta V = i^2 R = (\Delta V)^2/R.$

In an a-c circuit having a sinusoidal electromotance, energy is supplied to the circuit by the generator or other source of electromotance. Because the current and the voltage across the circuit elements vary with time, and are often out of phase with each other, to compute the power supplied to various circuit elements it is necessary to consider the instantaneous values of voltage and current. We first develop some general equations for the power supplied by or to a designated portion of the circuit, which may consist of one or more circuit elements.

★ The *instantaneous power* P supplied by or to a portion of a circuit having an instantaneous voltage ΔV and carrying instantaneous current i is $P \equiv i\ \Delta V$.

Substituting Eqs. (61-12) and (61-13) we obtain for the instantaneous power

$$P = i_0\ \Delta V_0 \sin \omega t \sin (\omega t + \phi). \qquad \text{Eq. (61-14)}$$

The instantaneous power oscillates with time during each cycle.

Practical interest in power in a-c circuits centers in the *average power* $\bar{P}$, the instantaneous power averaged over the time required for one or more cycles. (A bar above the symbol for a quantity is used to denote the time-average value of that quantity). To compute the average power, we use the identity $\sin (\omega t + \phi) = \sin \omega t \cos \phi + \cos \omega t \sin \phi$ to expand Eq. (61-14) and obtain

$$P = i_0\ \Delta V_0 (\sin^2 \omega t \cos \phi + \sin \omega t \cos \omega t \sin \phi). \qquad \text{Eq. (61-15)}$$

To find $\bar{P}$, we must compute the time averages $\overline{\sin^2 \omega t}$ and $\overline{\sin \omega t \cos \omega t} = \frac{1}{2}\overline{\sin 2\omega t}$ during a cycle. Sin 2 ωt is alternately positive and negative with a time-average of zero. $\text{Sin}^2\ \omega t$ is always positive, however, oscillating between zero and unity with an average value of 1/2. Combining these results and substituting in Eq. (61-15), we obtain for the average power

$$\overline{P} = \frac{1}{2} i_0 \, \Delta V_0 \cos \phi. \qquad \text{Eq. (61-16)}$$

Combining Eq. (61-16) with equations obtained earlier for any portion of a circuit having an impedance Z and resistance R, $\Delta V_0 = Z i_0$ and $\cos \phi = R/Z$, we obtain equivalent equations for the average power in this portion of the circuit:

★ $$\overline{P} = \frac{1}{2} i_0^2 \, Z \cos \phi = \frac{1}{2} (\Delta V_0^2/Z) \cos \phi = \frac{1}{2} i_0^2 R. \qquad \text{Eq. (61-17)}$$

Eq. (61-17) can be applied to individual circuit elements. For a resistance alone, $\phi = 0$ and $\overline{P} = 1/2 \; i_0 \; R$. For an ideal inductor, $\phi = +\pi/2$ and $\overline{P} = 0$. For a capacitor, $\phi = -\pi/2$ and $\overline{P} = 0$. For a capacitor the instantaneous power oscillates between positive and negative values, so that during each cycle power flows alternately into and out of the electric field of the capacitor while the *average* power during each cycle is zero. For an ideal inductor, the instantaneous power oscillates between positive and negative values, so that during each cycle power flows alternately into and out of the magnetic field of the inductor while the *average* power during each cycle is zero. The only circuit elements for which the average power consumption during a cycle is not zero are those containing resistance.

The instantaneous power supplied to the circuit by the source of electromotance is

$$P = \xi_0 \sin (\omega t + \phi) \, i_0 \sin \omega t.$$

Expanding this equation and computing time-average quantities, we obtain for the average power supplied by the source

★ $$\overline{P} = \frac{1}{2} \xi_0 \, i_0 \cos \phi. \qquad \text{Eq. (61-18)}$$

Because $\xi_0 = i_0 Z$ for the circuit as a whole, Eq. (61-18) is equivalent to

$$\overline{P} = \frac{1}{2} i_0^2 \, Z \cos \phi.$$

Because energy is conserved, the average power supplied by the source equals the average power consumed by the circuit elements, Eq. (61-17).

In a steady-current circuit where the current is i, the electric power dissipated in a resistance R is $P = i^2 R$. In an a-c circuit where the peak current is i_0, the average power dissipated in a resistance R is $\overline{P} = 1/2 \; i_0^2 R$. For a-c circuits carrying sinusoidal currents, it is customary to speak of the *effective values* of current and voltage with reference to power consumption in a circuit. We use capital letters as symbols for these quantities, defined as follows:

★ $$\textit{effective current} = I = i_0/\sqrt{2}; \quad \textit{effective voltage} = \Delta V = \Delta V_0/\sqrt{2}.$$

Then $\Delta V = Z \, I$, and the average power consumption in any portion of the circuit is

★ $\overline{P} = I\,\Delta V \cos\phi = I^2 Z \cos\phi = (\Delta V^2/Z)\cos\phi = I^2 R.$

The current and voltage ratings of electrical appliances and laboratory equipment which operate on alternating current are expressed in terms of effective values of current and voltage. Ammeters and voltmeters designed for use with a-c circuits are calibrated to indicate effective values.

Questions and Exercises

Q 61.4-1 A capacitor carries a sinusoidal current $i = i_0 \sin \omega t$.
(a) During what portions of the cycle is energy flowing into the capacitor?
(b) During what portions of the cycle is energy flowing out of the capacitor?

Q 61.4-2 Suppose you have a sealed "black box" containing a series *LRC* network with unknown values of L, R, and C, and having no internal source of electromotance. When this black box is connected to a 115-V 60-Hz source, an a-c ammeter in series with the box indicates 750 mA. How could you experimentally find the phase angle between the voltage and current for the black box, using the fact that the instantaneous voltage across a resistor is in phase with $i(t)$?

((E 61.4-1 Show by calculation that the time-average $\overline{\sin^2 \omega t} = 1/2$.))

E 61.4-2 The nominal (effective) voltage in a household a-c circuit is 117 V. What is the peak voltage?

E 61.4-3 Show that the maximum value of instantaneous power supplied to an ideal inductor is $\frac{1}{2} i_0\,\Delta V_{L_0}$. Hint: plot $i(t)$, $\Delta V_L(t)$, and the product $i(t)\,\Delta V_L(t)$.

E 61.4-4 For the circuit in Figure 61-2: $\xi_0 = 9$ V, $\omega = 377\ \mathrm{s}^{-1}$, $R = 10$ ohms, $L = 1.0$ henry, and $C = 10\ \mu$F.
(a) Compute the average power dissipated in the resistor.
(b) Compute the average power supplied to the capacitor.
(c) Compute the average power supplied to the ideal inductor.
(d) Compute the average power supplied to the entire circuit.
(e) Compute the average power supplied by the source.
(f) Compute the effective current.
(g) Compute the effective electromotance of the source.

61.5 Resonance in a Series LRC Circuit

The current in a circuit is the "response" of the circuit to the applied electromotance. Under what conditions is this response a maximum for a series *LRC* a-c circuit? When a constant-voltage variable-frequency sinusoidal electromotance is applied to a series *LRC* circuit, measurements show that the effective current varies with frequency as shown in Figure 61-4(a). The frequency at which the current reaches its maximum value is called the *resonant frequency* f_r of the circuit. The phase angle between current and voltage also varies with frequency (Figure 61-4(b)). changing from $-\pi/2$ at frequencies near zero through $\phi = 0$ at the resonant frequency to $+\pi/2$ at very high frequencies. How can we account for this phenomenon of resonance and the variations in frequency and phase angle associated with it?

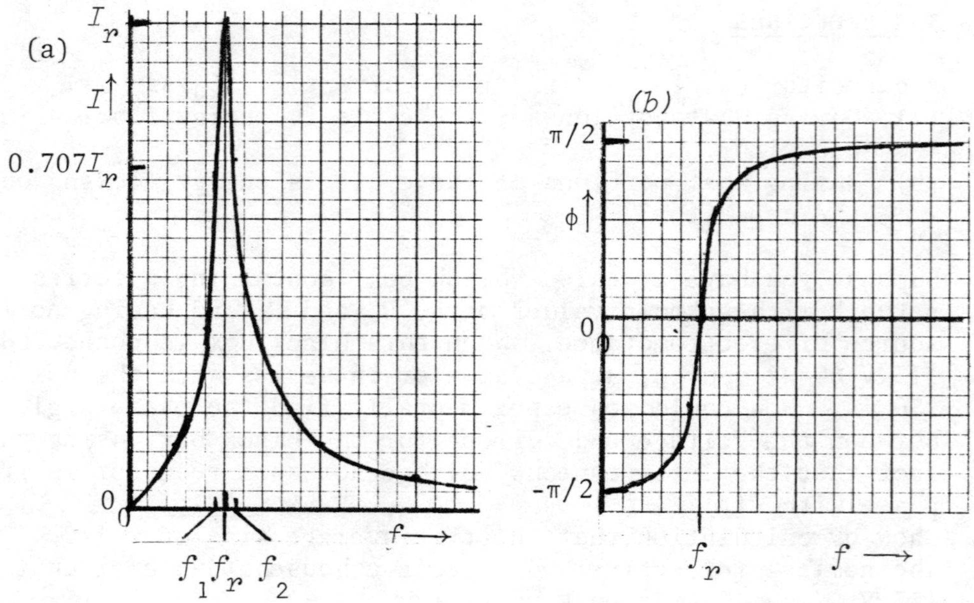

Figure 61-4. Resonance in a series *LRC* a-c circuit. (a) The current has the maximum value I_r at the resonant frequency f_r. At the half-power frequencies f_1 and f_2, $I = 0.707\ I_r$. (b) The phase angle between current and voltage equals zero at f_r. The phase angle approaches $-\pi/2$ at low frequencies and $+\pi/2$ at high frequencies. The scales in these graphs are linear.

When the effective current varies while the electromotance remains constant, this variation must be a result of variations in the impedance of the circuit. The variation of impedance with frequency is given by

$$Z = \sqrt{R^2 + \left(\omega L - \frac{1}{\omega C}\right)^2}. \qquad \text{Eq. (61-11)}$$

The impedance is a minimum when the quantity $[\omega L - 1/(\omega C)] = 0$, which occurs at the angular frequency

$$\omega_r = (LC)^{-1/2}$$

corresponding to the resonant frequency

★ $$f_r = 1/(2\pi\sqrt{LC}). \qquad \text{Eq. (61-19)}$$

At this frequency, the inductive impedance cancels the capacitive impedance so the resulting impedance of the entire *LRC* series circuit is simply R. The variation of phase angle with frequency is given by

$$\tan\phi = \frac{\omega L - 1/(\omega C)}{R}, \qquad \text{Eq. (61-7)}$$

from which it is seen that at the resonant frequency $\tan\phi = 0$ and $\phi = 0$. At resonance, the current is exactly in phase with the electromotance driving the circuit. For frequencies much smaller than f_r, $\omega L \ll 1/(\omega C)$ and the capacitive impedance dominates the circuit, limiting the current and giving a phase angle close to $-\pi/2$ between the electromotance and the current. For frequencies much larger than f_r, $\omega L \gg 1/(\omega C)$ and the inductive impedance dominates the circuit, again limiting the current but giving a phase angle close to $+\pi/2$.

The effective voltages across an ideal inductance and a capacitance are given by

$$\Delta V_L = I\,\omega L \quad \text{and} \quad \Delta V_C = I/(\omega C),$$

respectively. If $\omega_r L = 1/(\omega_r C)$ is larger than R at resonance, then the effective voltages across the inductor and the capacitor are both larger than the applied electromotance!

What determines the sharpness of the peak in the graph of current vs. frequency, Figure 61-4(a)? The frequencies f_1 and f_2 on either side of the resonance peak are defined as those for which $I = I_r/\sqrt{2}$. At f_1 and f_2, therefore, $Z = \sqrt{2}\,R$. At f_1,

$$-R = \omega_1 L - 1/(\omega_1 C),$$

and at f_2

$$R = \omega_2 L - 1/(\omega_2 C).$$

Using a little algebra, it can be shown that

$$\omega_2 - \omega_1 = R/L. \qquad \text{Eq. (61-20)}$$

A quantity called the *quality* of the circuit is defined as

$$Q \equiv \frac{f_r}{f_2 - f_1} = \frac{\omega_r}{\omega_2 - \omega_1}. \qquad \text{Eq. (61-21)}$$

Combining Eqs. (61-20) and (61-21), we obtain

$$Q = \omega_r L/R.$$

When Q is large, the frequencies f_1 and f_2 are relatively close together and the resonance displays a sharp current peak. To have a large Q, the circuit must have a value of R small compared with the inductive impedance at the resonant frequency.

Questions and Exercises

Q 61.5-1 What is the meaning of "resonance" in a series *LRC* a-c circuit?

Q 61.5-2 Does every series *LRC* circuit have a resonant frequency for sinusoidal oscillation? What determines the value of the resonant frequency?

Q 61.5-3 Describe two ways in which the condition of resonance in a series *LRC* circuit could be detected using a cathode-ray oscilloscope.

E 61.5-1 Complete the derivation of Eq. (61-20).

E 61.5-2 (a) Write a general expression for the impedance of a series *LRC* circuit.
(b) Sketch the variation of impedance with frequency.
(c) If the values of L, R, and C in the circuit are fixed, under what condition does Z reach its minimum value?
(d) If the values of L, R, C and the peak value of applied electromotance are kept constant, under what condition is the effective current a maximum?

E 61.5-3 A series *LRC* circuit has $\xi_0 = 9.0$ V, $R = 10$ ohms, $L = 1.0$ H, and $C = 10$ μF.
(a) What is the resonant frequency?
(b) If the sinusoidal electromotance is tuned to the resonant frequency, what is the peak voltage across the resistance?
(c) At resonance, what is the peak voltage across the (ideal) inductor?
(d) At resonance, what is the peak voltage across the capacitor?

E 61.5-4 An actual (non-ideal) inductor has $L = 4.5$ H and $R = 300$ ohms.
(a) What value of capacitance must be connected in series so that the circuit will resonate at 60 Hz?
(b) What is the quality of this circuit?

Answers to Minimum Learning Goals

Line 1

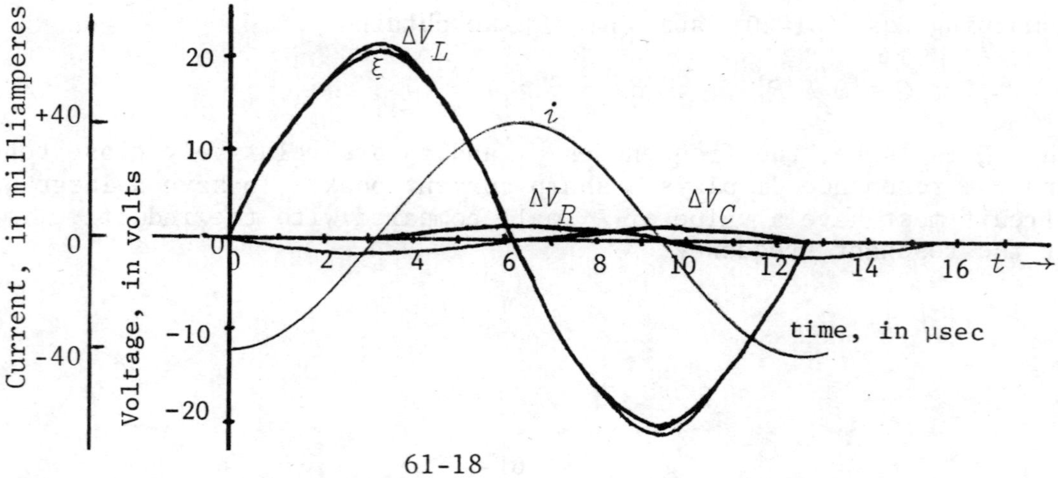

Line 2 (a) Z_L = 20 ohms, Z_C = 12.5 ohms, R = 5.0 ohms;
(b) Z = 9.01 ohms; (c) 17.8 A.

Line 3

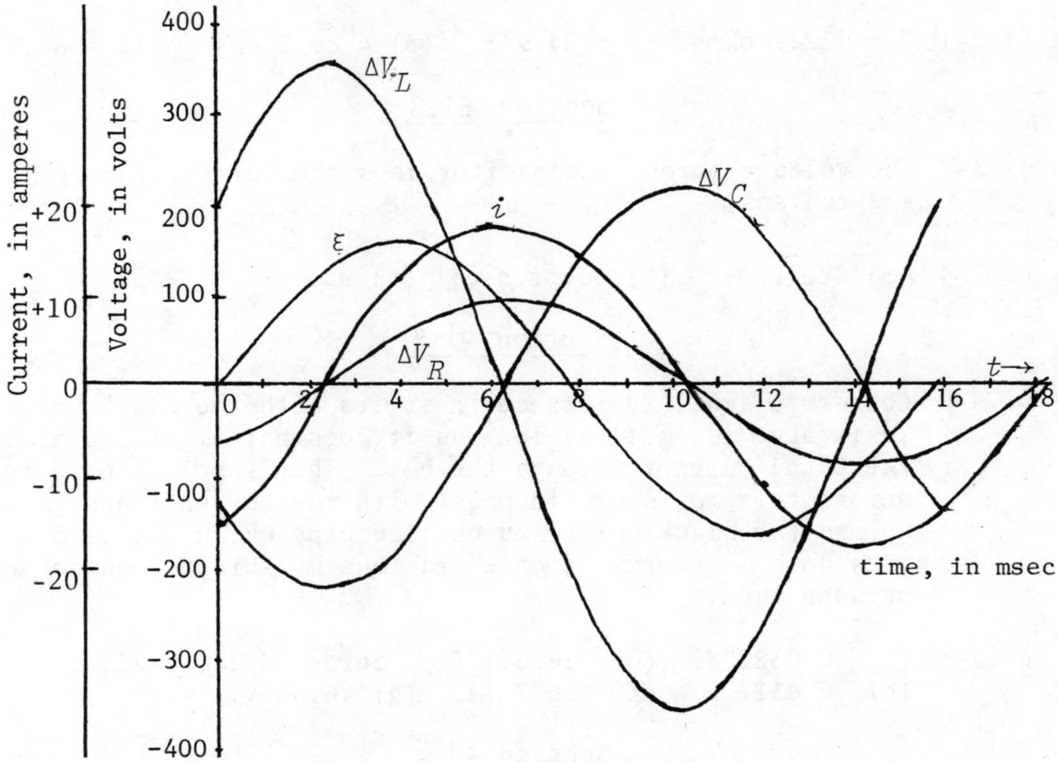

Line 4 (a) Z_C = 10.0 ohms, Z_L = 500 ohms, R = 10 ohms;
(b) Z = 490.1 ohms; (c) 0.041 A

ANSWERS to Questions and Exercises in Chapter 61

Section 61.1

E 61.1-2 9.35 ohms

Section 61.2

Q 61.2-2 Yes, except that if the direction of tracing is reversed the signs of all the terms are reversed, which is a trivial distinction.

Q 61.2-4 $(\omega C)^{-1}$, inversely proportional to the frequency

Q 61.2-6 (a) $Z = [R^2 + 1/(\omega C)^2]^{1/2}$; (b) $Z = [R^2 + (\omega L)^2]^{1/2}$;

(c) $Z = [R^2 + (\omega L - 1/(\omega C))^2]^{1/2}$

Q 61.2-8 (a) the voltage across the inductor; (b) the voltage across the capacitor.

E 61.2-4
$$\xi_1 - iR_1 - L_1\, di/dt - (1/C_1)\int i\,dt - (1/C_2)\int i\,dt - i\,R_2 - L_2\, di/dt = 0.$$

E 61.2-6 $i = 2.48 \cos(377\,t + 23.8°)$ amperes

E 61.2-8 $L = 7.04$ H ; $i_0 = 300$ mA.

E 61.2-10 $Z = 112.2$ ohms; $\phi = 84.9°$; $i(t) = 80.2 \cos(377t - 84.9°)$ mA.

Section 61.3

Q 61.3-2 The voltage across a capacitor lags the current through it by $\pi/2$ radians

Q 61.3-4 (a) decrease; (b) decrease; (c) no.

Section 61.4

Q 61.4-2 Connect a small resistance in series with the black box; this resistance is small as long as it does not significantly alter the total current through the box. Then supply the voltage across this resistor (in phase with the current) and the voltage across the black box to an oscilloscope which allows one to view both waveforms at once and thus measure the phase angle between them.

E 61.4-4 (a) 0.0322 W; (b) zero; (c) zero; (d) 0.0322 W; (e) 0.0322 W; (f) 56.7 mA; (g) 6.36 V.

Section 61.5

Q 61.5-2 Yes, at the frequency where $\omega L = 1/(\omega C)$.

E 61.5-2 (a) $Z = \sqrt{R^2 + [\omega L - 1/(\omega C)]^2}$

(b) From a minimum at the resonant frequency, the impedance rises monotonically toward both lower and higher frequencies.
(c) and (d) at the frequency $\omega L = 1/(\omega C)$.

E 61.5-4 (a) 1.56 microfarad; (b) 5.66

Appendix A. DIFFERENTIATION AND INTEGRATION

A.1 Sinusoidal Functions

A mathematical description of any oscillation requires a function $f(t)$ which alternates between positive and negative values periodically in time. In many cases, the most appropriate functions are $f(t) = \sin \omega t$ and $f(t) = \cos \omega t$, which oscillate between +1 and -1. The quantity ωt corresponds to an angle which increases linearly with time at the rate ω; ω is called the *angular frequency*. The angle ωt is usually expressed in radians: 2π radians = 360°. As ωt is dimensionless, ω has units of radians/second = rad/s = s^{-1}.

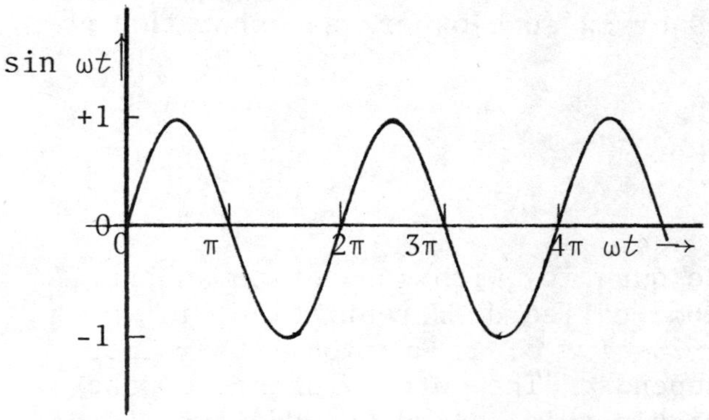

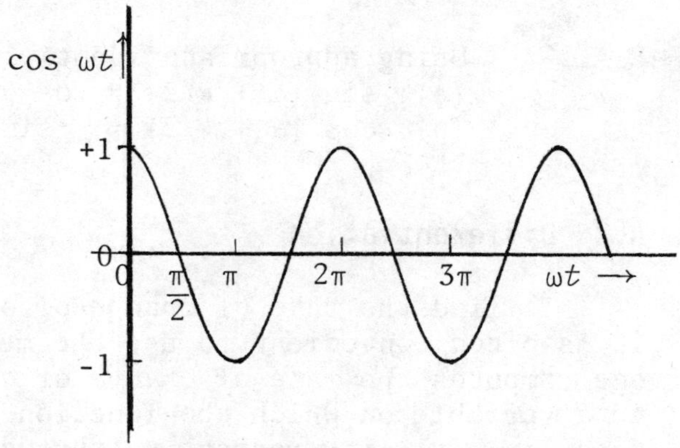

A table of numerical values of sines and cosines for angles from 0 to $\pi/2$ radians is given in Appendix B. Some useful identities relating sinusoidal functions include the following:

$$\sin^2 \theta + \cos^2 \theta = 1 \qquad \sin^2 \theta = (1 - \cos 2\theta)/2$$

$$\sin (\alpha + \beta) = \sin \alpha \cos \beta + \cos \alpha \sin \beta$$

$$\sin (\alpha - \beta) = \sin \alpha \cos \beta - \cos \alpha \sin \beta$$

$$\cos (\alpha + \beta) = \cos \alpha \cos \beta - \sin \alpha \sin \beta$$

$$\cos (\alpha - \beta) = \cos \alpha \cos \beta + \sin \alpha \sin \beta$$

$$\sin 2\theta = 2 \sin \theta \cos \theta$$

$$\cos 2\theta = \cos^2 \theta - \sin^2 \theta = 2 \cos^2 \theta - 1 = 1 - 2 \sin^2 \theta$$

$$\tan \theta = \sin \theta/(\cos \theta)$$

The rules for differentiating sine and cosine functions are given in Section A.2. The rules for integrating sine and cosine functions are given in Section A.4.

Questions and Exercises

Q A.1-1 For what positive values of ωt does $\sin \omega t$ equal:
(a) zero?
(b) +1?
(c) -1?

Q A.1-2 For what positive values of ωt does $\cos \omega t$ equal:
(a) zero?
(b) +1?
(c) -1?

E A.1-1 For $\omega = 377\ s^{-1}$, find the values of $\sin \omega t$ and $\cos \omega t$ for the following values of t:
(a) $t = 0$ (b) $t = 0.00139$ s (c) $t = 0.00417$ s
(d) $t = 0.00833$ s (e) $t = 0.010$ s (f) $t = 1/60$ s

E A.1-2 Using appropriate identities or in some other way, show that:
(a) $\sin(\alpha + \pi/2) = \cos \alpha$
(b) $\cos(\alpha + \pi/2) = -\sin \alpha$

A.2 Differentiation

To find the rate of change of one quantity with respect to another, it is often convenient to use the method called differentiation, in which one computes the rate of change or *derivative* of a function with respect to a variable on which the function depends. The rate of change of the function $x(t)$ with respect to the variable t is called the "derivative of x with respect to t," and is denoted by dx/dt. Using methods described below, an expression can be obtained for the derivative dx/dt if we have an expression for x as a function of t, $x(t)$. Examples of such functions are $x = 5t^2$, $x = 8t^3 - 4t$, $x = c_1 t + c_2 t^2$, etc. c_1 and c_2 (or simply c) are constants which have numerical values in each specific physical example.

The correct procedures for obtaining derivatives of many functions are developed in textbooks on calculus. The list below gives the procedures, or "recipes," for writing the derivatives of some common functions. In this list, x is used to represent any dependent variable, t to represent the independent variable, and c is a constant.

1. $dc/dt = 0$
2. $d(ct^n)/dt = c\,d(t^n)/dt = cnt^{n-1}$
3. If $x(t) = x_1(t) + x_2(t)$, then $dx/dt = dx_1/dt + dx_2/dt$.
4. $d(e^{ct})/dt = ce^{ct}$.
5. $d(\ln at)/dt = 1/t$.
6. $d(\sin \omega t)/dt = \omega \cos \omega t$.
7. $d(\cos \omega t)/dt = -\omega \sin \omega t$.

8. $\frac{d}{dt} f(x) = \frac{d}{dx} [f(x)] \cdot \frac{dx}{dt}$.

The meaning of a derivative may be visualized using a graphical representation. On a graph of $x(t)$ vs t, dx/dt is the slope of the curve at any point. As shown in the drawing at the right, dx/dt is given approximately by $\Delta x/\Delta t$, where Δx and Δt are small increments in x and t. In the limit as Δx and Δt approach zero, the ratio $\Delta x/\Delta t$ equals the derivative dx/dt. At values of t for which dx/dt is small, x is changing slowly as t increases. At values of t for which dx/dt is large, the rate of change of x with t is rapid.

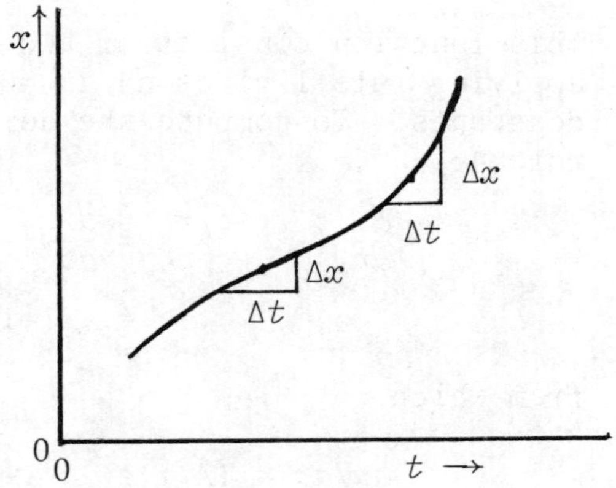

The derivative of the derivative of a function is called the *second derivative* of the function.

The rules for differentiating these functions are illustrated here by two examples. In the first example, we compute the first and second derivatives with respect to t of the function

$$x(t) = 6t^2 - 4t + 8.$$

The first derivative must be computed before the second derivative. To obtain dx/dt, we apply the rules for differentiation which are appropriate for the functional form of $x(t)$. $x(t)$ consists of three terms, so rule 3 indicates that the complete derivative dx/dt is the sum of the derivatives of the three separate terms. The first and second terms are powers of t, so these terms must be differentiated by applying rule 2, giving

$$\frac{d}{dt}(6t^2) = 6 \cdot 2t = 12t, \qquad \frac{d}{dt}(-4t) = -4 \cdot t^0 = -4.$$

The third term in $x(t)$ is a constant, so according to rule 1 its derivative is equal to zero. Combining these results,

$$\frac{dx}{dt} = 12t - 4.$$

To find the second derivative of $x(t)$,

$$\frac{d^2x}{dt^2} = \frac{d}{dt}\left(\frac{dx}{dt}\right) = \frac{d}{dt}(12t - 4),$$

we apply the appropriate rules for differentiating the two terms $12t$ and -4:

$$\frac{d}{dt}(12t) = 12 \cdot t^0 = 12, \quad \text{and} \quad \frac{d}{dt}(-4) = 0.$$

Combining these results,

$$d^2x/dt^2 = 12.$$

As a second example, we compute the first derivative of the function

$$y(t) = A[1 - e^{-t/\tau}] .$$

This function consists of two terms, A and $-Ae^{-t/\tau}$. A is a constant, so applying rule 1 gives $dA/dt = 0$. In the exponential terms, A and τ are constants. To compute the derivative of the exponential term, we apply rule 4:

$$\frac{d}{dt}(-Ae^{-t/\tau}) = \frac{d}{dt}[-Ae^{-(1/\tau)t}]$$
$$= -A(-1/\tau)e^{-(1/\tau)t},$$

from which

$$dy/dt = (A/\tau)e^{-t/\tau}.$$

Exercises

E A.2-1 The equations for $x(t)$ in this exercise describe particles in motion along the x axis. For each, compute $v(t) \equiv dx/dt$ and $a(t) \equiv dv/dt$.
(a) $x = 6t^3$.
(b) $x = (1/4)t^4 - (1/2)t^2 + 1$;
(c) $x = 10t^2$;
(d) $x = x_0 + v_0 t + at^2/2$.

E A.2-2 For each $x(t)$ in Exercise A.2-1, compute the value of x at $t = 0$ and the value of v at $t = 0$.

E A.2-3 Compute dV/ds for $V = as^3 + bs^2$.

E A.2-4 $V = 1000/s + 500$. What is the rate of change of V with s at $s = 20$?

E A.2-5 The charge q received by an electrode increases with time according to the equation $q = 10t$ coulombs. Compute $i(t) \equiv dq/dt$, in amperes.

E A.2-6 One equation which describes radioactive decay is $N = N_0 e^{-\lambda t}$.
(a) Show that $dN/dt = (dN/dt)_0\, e^{-\lambda t}$, where $(dN/dt)_0$ is the value of (dN/dt) at $t = 0$.
(b) Show that $(dN/dt)_0 = -\lambda N_0$.

E A.2-7 Recall that the derivative $df(t)/dt$ may be interpreted as the slope of the graph of $f(t)$ vs t. By comparing graphs of $\sin \omega t$ vs t and $\cos \omega t$ vs t: (a) show that $d(\sin \omega t)/dt$ corresponds to $+\omega \cos \omega t$. (b) show that $d(\cos \omega t)/dt$ corresponds to $-\omega \sin \omega t$.

A.3 Partial Derivatives

Some physical quantities are functions of two or more independent variables. An example is the potential energy PE of a system of two particles, which depends on the three spatial coordinates x, y, and z of the second particle relative to the first, and is written $PE(x, y, z)$. PE is the dependent variable, and x, y, and z are the independent variables. In such a case, we can still calculate the derivative or rate of change of the function with respect to any one of the independent variables separately, by regarding the other independent variables as constants during this calculation. Such a derivative is called a *partial derivative,* and it is written using a special notation. For example, $\frac{\partial PE}{\partial x}$ is the partial derivative of PE with respect to x; it is calculated by pretending that $PE(x,y,z)$ is a function of x only, regarding y and z as constants during the differentiation. Similarly, the symbol $\partial/\partial y$ is understood to describe a process of differentiation in which the values of x and z are held constant.

To compute a partial derivative of a function, one must have an expression for the function in terms of its independent variables, and must know which of these variables is regarded as varying and which variables are being held constant during the differentiation.

Exercises

E A.3-1 In each of the following examples, state which of the independent variables is regarded as varying, and which are being held constant:

(a) $\partial PE(x,y,z)/\partial z$;

(b) $\partial PE(r,\theta,\phi)/\partial r$;

(c) $\partial B(x,y,z,t)/\partial t$.

E A.3-2 The gravitational potential energy of a satellite in the field of the earth may be written $PE(x,y,z) = -GMm/(x^2 + y^2 + z^2)^{\frac{1}{2}}$.

(a) Compute $\partial PE/\partial x$.

(b) Compute $\partial PE/\partial y$.

(c) Compute $\partial PE/\partial z$.

A.4 Integration. Definite Integrals

The mathematical procedure called integration may be regarded as the inverse of differentiation. Whereas differentiation is used to compute the rate of change of one variable with respect to another, integration is used to compute the product of one variable multiplied by another.

We introduce the method of integration using the physical example of the work done by a variable force $F(x)$. Writing the force as $F(x)$ indicates that the value of F depends on the position x of the body acted on by this force.

Specifically, we compute the work W_{12} done by $F(x)$ while it acts parallel to the displacement of an object moving from x_1 to x_2. This work is represented by *the area under the curve* of $F(x)$ vs. x, between the points x_1 and x_2 (Section 47.2), as shown in the figure below.

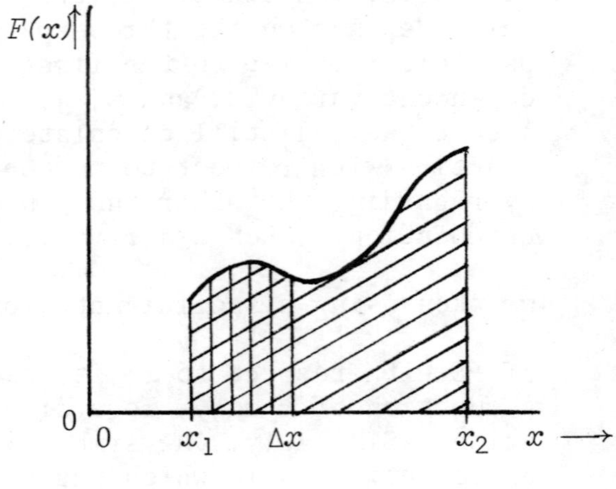

To compute W_{12}, the area is divided into a large number of narrow vertical strips, each having width Δx. Each strip is approximately a rectangle, so its area ΔW is approximately the local "height" $F(x)$ times the "width" Δx; that is, $\Delta W \simeq F(x)\Delta x$. The total area is the sum of the areas of the strips between x_1 and x_2, and may be written as the sum

$$W_{12} \simeq \sum_{x_1}^{x_2} F(x)\Delta x.$$

As we increase the number of strips, making all of them very narrow, this sum approaches the area more closely, and in the limit as $\Delta x \to 0$ we have

$$W_{12} = \lim_{\Delta x \to 0} \sum_{x_1}^{x_2} F(x)\Delta x \equiv \int_{x_1}^{x_2} F(x)dx.$$

The right-hand side of the equation above is called "the integral of $F(x)dx$ between x_1 and x_2," and it is defined in terms of the limiting process described above. For our purposes the integral sign $\int$ may be regarded as an abbreviation for "$\lim_{\Delta x \to 0} \Sigma$." The physical bounds or limits of the variable x in our example are x_1 and x_2; they are shown by writing the mathematical "limits" x_1 and x_2 next to the integral sign. An integral with limits such as x_1 and x_2 is called a *definite integral*, and the function $F(x)$ is called the *integrand*.

By a procedure similar to that described above, one can obtain a value for any definite integral by approximating the area under the curve of the integrand in the required interval of the variable x. When the function $F(x)$ is given by some known analytic expression, however, it is possible to evaluate the integral without drawing a graph or dividing the area under the curve into strips. To do this, one used the idea that the process of integration is analogous to *anti-differentiation*. This means that the integral function $f(x)$ defined by

$$f(x) \equiv \int_a^x F(t)dt,$$

where x is now understood to be a variable upper limit, has a derivative with respect to x given by

$$\frac{df(x)}{dx} = F(x).$$

In other words, the derivative of a definite integral with respect to the upper limit of integration is equal to the value of the integrand at this upper limit:

$$\frac{d}{dx}\int_a^x F(t)dt = f(x).$$

Using this theorem, it is shown in textbooks on calculus that for a continuous function $F(x)$,

$$\int_a^b F(x)dx = f(b) - f(a),$$

where

$$\frac{df(x)}{dx} = F(x).$$

That is, $f(x)$ is any function whose derivative is the given integrand $F(x)$. The function $f(x)$ is called the antiderivative or the *indefinite integral* of $F(x)$, and is written

$$f(x) = \int F(x)dx.$$

One of the first steps in evaluating a definite integral is to obtain the expression for the corresponding indefinite integral. Most of the indefinite integrals encountered in introductory physics can be evaluated readily using a table of integrals such as that found in a calculus textbook. In the short table of integrals given below, t is used for the independent variable while c and a are constants.

1. $\int c\, dt = ct$
2. $\int c\, t^n\, dt = \dfrac{ct^{n+1}}{n+1}$ for $n \neq -1$
3. $\int \dfrac{dt}{t} = \int t^{-1}\, dt = \ln t$
4. $\int e^{at}\, dt = (1/a)e^{at}$
5. $\int \sin \omega t\, dt = -(1/\omega) \cos \omega t$
6. $\int \cos \omega t\, dt = (1/\omega) \sin \omega t$
7. If $f(t) = f_1(t) + f_2(t)$, then $\int f(t)dt = \int f_1(t)dt + \int f_2(t)dt$.

Notice that in each case the integrand is the derivative of the function appearing on the right. An arbitrary constant may be added to any indefinite integral, because the derivative of any constant is zero.

The final basic step in evaluating a definite integral is to substitute into the expression for the indefinite integral the value of the *upper limit*, and then *subtract* from this result the quantity obtained when the value of the *lower limit* is substituted into the expression for the indefinite integral. This procedure is illustrated in the specific examples below.

The work W_{12} done by $F(x)$ as it acts through the displacement from x_1 to x_2 is equal to the definite integral of $F(x)dx$, between the limits x_1 and x_2. This integral is evaluated using the general procedure outlined above. As a specific example, suppose $F(x) = bx^2$ with $b = 0.1$ newton/meter^2, and we wish to evaluate the work done by $F(x)$ on a particle as it moves from $x_1 = 2$ meters to $x_2 = 5$ meters.

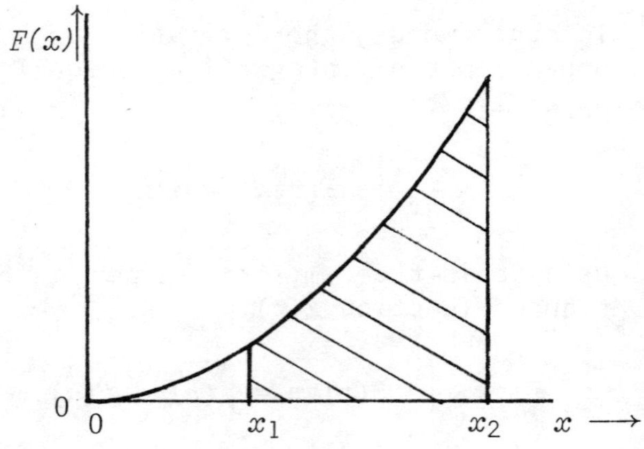

$$W_{12} = \int_{x_1}^{x_2} F(x)dx = \int_{x_1}^{x_2} bx^2\,dx = \left[\frac{bx^3}{3}\right]_{2\text{ m}}^{5\text{ m}}$$

$$= \left(\frac{125b}{3} - \frac{8\,b}{3}\right)\text{m}^3 = 39b\ \text{m}^3 = 39\ \text{m}^3 \times 0.1\ \text{N/m}^2 = 3.9\ \text{joules}.$$

The calculation of the work done by a variable force not parallel to the displacement is taken up in Section A.5.

As a second example we consider one-dimensional motion in which the velocity $v(t) = dx/dt$ and the position $x(t)$ are both functions of t. Taking the indefinite integral with respect to t of both sides of the equation $dx/dt = v(t)$, we obtain

$$x(t) = \int v(t)dt\ .$$

We can visualize this example with the aid of a graph of $v(t)$ vs. t. The value of the product $v(t)dt$ for any interval dt is represented by the area of the vertical strip under the curve between t and $t + dt$, shown shaded in the figure. The product $v(t)dt$ has physical units of distance, and equals the distance traveled during the time interval dt.

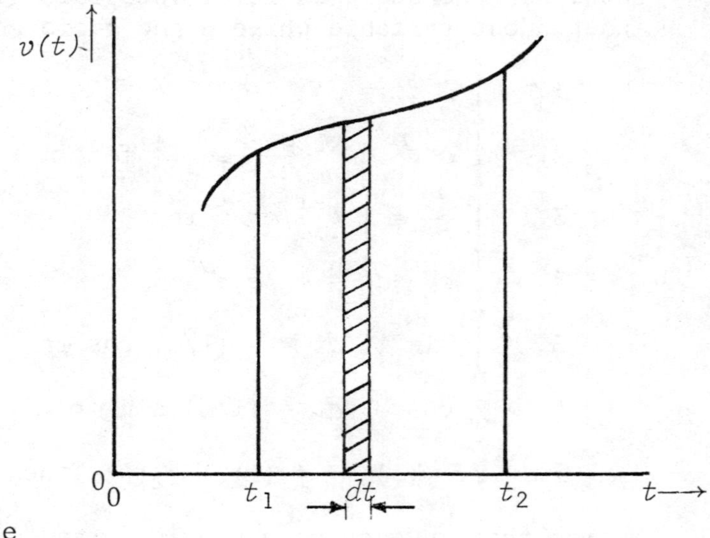

Suppose $v(t) = at$, where $a = 3.0$ m/s^2, and we need to compute the total distance x traveled between $t = 2$ seconds and $t = 4$ seconds. This distance is equal to the definite integral of $v(t)dt$, between the limits $t_1 = 2$ s and $t_2 = 4$ s. To compute this distance, we first evaluate the indefinite integral and then substitute the limits:

$$x = \int_2^4 v(t)dt = \int_2^4 at\ dt = \left[a\ \frac{t^2}{2}\right]_2^4 = \left(\frac{4^2\ a}{2} - \frac{2^2\ a}{2}\right) \text{s}^2$$

$$= (8a - 2a)\ \text{s}^2 = 6a\ \text{s}^2 = 6\ \text{s}^2 \times 3.0\ \text{m/s}^2 = 18.0\ \text{meters.}$$

Exercises

E A.4-1 Write (but do not evaluate) the integrals which represent:

(a) the work W_{34} done by a variable force $F(x)$ as it acts on a car moving along the x axis from x_3 to x_4;

(b) the work W_{ab} done by a force $F(y)$ as it acts on a ball moving along the y axis from y_a to y_b;

(c) the distance traveled by a vehicle having speed $v(t)$ during the time interval t_0 to t_f.

E A.4-2 Evaluate these indefinite integrals:

(a) $\int dt =$;

(b) $\int 3dx =$;

(c) $\int t^2\ dt =$;

(d) $\int 4t^3\ dt =$;

(e) $\int q\ dq =$;

(f) $\int kx\ dx =$;

(g) $\int (ax^2 + bx + c)dx =$.

E A.4-3 Evaluate these definite integrals:

(a) $\int_2^4 dt =$;

(b) $\int_0^2 3x\ dx =$;

(c) $\int_0^4 t^2\ dt =$;

(d) $\int_{-4}^4 t^2\ dt =$;

(e) $\int_{-4}^4 4t^3\ dt =$;

(f) $\int_0^Q q\ dq =$;

(g) $\int_0^{x_1} kx\ dx =$.

E A.4-4 Compute the work done by $F(x) = 3x^2$ newtons on a particle as the particle moves along the x axis from $x_3 = 1$ meter to $x_4 = 3$ meters.

E A.4-5 Compute the distance traveled by a car having speed $v(t) = 2.0\, t$ m/s during the time interval $t = 0$ to $t = 4$ sec.

E A.4-6 (a) Sketch a graph of $\cos^2 \omega t$ vs t.
(b) From your graph, estimate the average value of $\cos^2 \omega t$ during one cycle.
(c) Show that the definite integral $\int_0^{2\pi/\omega} \cos^2 \omega t\, dt = \frac{\pi}{\omega}$, using the idenity $\cos^2 \theta = \frac{1}{2}(1 + \cos 2\theta)$.
(d) What is the interpretation of your result for part (c), in terms of the graph? Compare with your answer to part (b).

A.5 Line Integrals

In this section we extend the concept of integration to certain kinds of integrals called line integrals. A *line integral* indicates an integration to be carried out along a *line* or path in space.

Suppose a particle moves along a path through a succession of small displacements $d\vec{\ell}$. We use C to label the path or curve in three-dimensional space which is traced out by the successive displacements. To find the geometrical *length* of the path C from point A to point B, we simply add all the *scalar* $d\ell$'s. In the limit as $d\ell \to 0$, the path length of C from A to B is

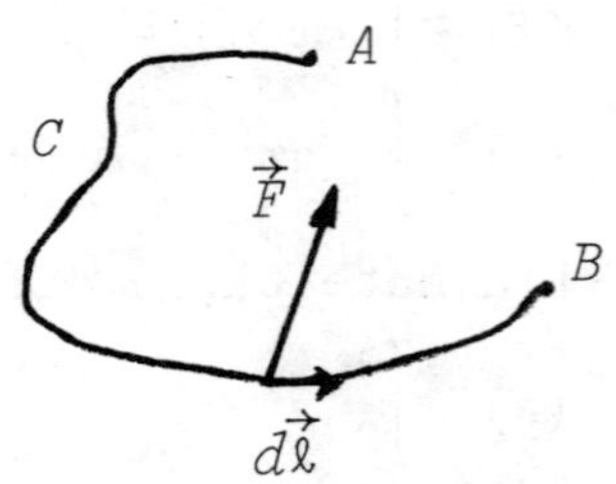

$$\sum_A^B d\ell = \int_A^B d\ell = \int_C d\ell.$$

Each *vector* $d\vec{\ell}$ is tangent to the path C. The sum

$$\sum_A^B d\vec{\ell} = \int_A^B d\vec{\ell} = \int_C d\vec{\ell}$$

equals the vector displacement from point A to point B. If point A coincides with point B, C is a *closed path*, i.e. a path which returns to its starting point. A closed path is indicated by writing the integral sign $\oint_C$. For and closed path C,

$$\oint_C d\ell = 0.$$

Now consider the case in which a force $\vec{F}$ acts on the particle as it moves along its path. In general, the force $\vec{F}(x, y, z)$ depends on all three spatial coordinates, and the angle between $\vec{F}$ and $d\vec{\ell}$ varies along the curved path. To compute the total work done by the force, we first note that the work dW done by $\vec{F}$ as it acts through the small displacement $d\vec{\ell}$ is given by the dot product

$$dW \equiv \vec{F} \cdot d\vec{\ell}$$

(Section 47.1). The total work W_{AB} done by $\vec{F}$ as it acts on the particle moving along path C from an initial point A to a final point B is the sum of all the products $\vec{F} \cdot d\vec{\ell}$ along this path. This sum is the line integral of $\vec{F} \cdot d\vec{\ell}$ along path C, written

$$W_{AB} = \int_A^B \vec{F} \cdot d\vec{\ell} = \int_C \vec{F} \cdot d\vec{\ell}.$$

Except in the special case of a conservative force (Section 49.1), the integration must be carried out along the actual line or path followed by the particle.

Most of the line integrals encountered in this textbook can be evaluated directly by thinking about the physical or geometrical relation between the integrand ($\vec{F}$ in this example) and $d\vec{\ell}$ along the path of integration. In any case, when the integral contains a dot product or cross product of two vectors one must convert this product to a *scalar* expression at each point along the path before the integral can be evaluated.

Sometimes one is given the initial point A and the final point B, but is free to select the path C of a line integral between points A and B. In these cases, evaluation of the integral can often be simplified by choosing a path consisting of a series of segments, such that on each segment either $\vec{F} \parallel d\vec{\ell}$, or $\vec{F} \perp d\vec{\ell}$, or $\vec{F} = 0$.

When $\vec{F} \cdot d\vec{\ell} = 0$ everywhere on a segment of the path, then of course the integral $\int \vec{F} \cdot d\vec{\ell} = 0$ over that segment. When $\vec{F} \parallel d\vec{\ell}$ everywhere on a segment, then $\vec{F} \cdot d\vec{\ell} = F d\ell$ for that portion. When $\vec{F} \parallel d\vec{\ell}$ and $\vec{F}$ is constant in magnitude over a segment, say from point D to point E, then this constant F may be placed in front of the integral sign so that

$$\int_D^E \vec{F} \cdot d\vec{\ell} = \int_D^E F\, d\ell = F \int_D^E d\ell$$

$$= F \times (\text{actual path length along } C \text{ from } D \text{ to } E).$$

Exercises

E A.5-1 Write (but do not evaluate) line integrals for the following:

(a) the work W_{34} done by a force $\vec{F}(x,y,z)$ as it acts on a bird moving from $\vec{r}_3$ to $\vec{r}_4$ along a path C;

(b) the work done by a force $\vec{F}(x,y,z)$ as it acts on a golf ball moving from $\vec{\ell}_0$ to $\vec{\ell}_f$ along a path C_1;

(c) the line integral of $\vec{B}$ along a path C_2;

(d) the line integral of $\vec{E}$ around the closed path C_3.

E A.5-2 Evaluate these line integrals:

(a) $\int_{r_1}^{r_2} \vec{F} \cdot d\vec{r}$, where $\vec{F} = \hat{x}\, F_0$ newtons, $\vec{r}_1 = \hat{y}3$ meters and $\vec{r}_2 = \hat{x}5 + \hat{y}\, 3$ meters, and you may choose the path of integration.

(b) $\int_{r_1}^{\infty} (k\, q_1 q/r^3)\vec{r} \cdot d\vec{r}$, where you may choose the path of integration.

E A.5-3 Simplify the integral $\int_C \vec{E} \cdot d\vec{\ell}$ for the case when C coincides with an electric line field and E is constant along the path.

E A.5-4 Show that $\oint_C d\vec{\ell} = 0$ for any closed path C.

ANSWERS to Questions and Exercises in Appendix A

Section A.1

Q A.1-1 (a) $n\pi$, where $n = 0, 1, 2, 3, \ldots$
(b) $2n\pi + \pi/2$, where $n = 0, 1, 2, 3, \ldots$
(c) $2n\pi - \pi/2$, where $n = 1, 2, 3, \ldots.$

Q A.1-2 (a) $n\pi + \pi/2$, where $n = 0, 1, 2, 3, \ldots$
(b) $2n\pi$, where $n = 0, 1, 2, 3, \ldots.$
(c) $2n\pi + \pi$, where $n = 0, 1, 2, 3, \ldots$

E A.1-1 (a) $\sin \omega t = 0$, $\cos \omega t = +1$; (b) $\sin \omega t = 0.500$, $\cos \omega t = 0.866$;
(c) $\sin \omega t = 1.000$, $\cos \omega t = 0$; (d) $\sin \omega t = 0$; $\cos \omega t = -1.000$;
(e) $\sin \omega t = -0.588$, $\cos \omega t = -0.809$;
(f) $\sin \omega t = 0$, $\cos \omega t = +1.000$.

Section A.2

E A.2-1 (a) $v = 18t^2$, $a = 36t$;
(b) $v = t^3 - t$, $a = 3t^2 - 1$;
(c) $v = 20t$; $a = 20$;
(d) $v = v_0 + at$, $a = a =$ constant.

E A.2-2 (a) $x(0) = 0$, $v(0) = 0$;
(b) $x(0) = 0$, $v(0) = 0$;
(c) $x(0) = 0$, $v(0) = 0$;
(d) $x(0) = \mathrm{x}_0$, $v(0) = v_0$.

E A.2-3 $3as^2 + 2bs$.

E A. 2-4 -2.5

E A.2-5 10 A

Section A.3

E A.3-1 (a) z varies, x and y are constant;
(b) r varies, θ and ϕ are constant;
(c) t varies, x, y and z are constant.

E A.3-2 (a) $G\, Mmx/(x^2 + y^2 + z^2)^{3/2}$;
(b) $GMmy/(x^2 + y^2 + z^2)^{3/2}$;
(c) $GMmz/(x^2 + y^2 + z^2)^{3/2}$.

Section A.4

E A.4-1 (a) $W_{34} = \int_{x_3}^{x_4} F(x)dx$; (b) $W_{ab} = \int_{y_a}^{y_b} F(y)dy$; (c) $s = \int_{t_0}^{t_f} v(t)dt$.

E A.4-2 (a) t; (b) $3x$; (c) $t^3/3$; (d) t^4; (e) $q^2/2$;

(f) $kx^2/2$; (g) $ax^3/3 + bx^2/2 + cx$.

E A.4-3 (a) 2; (b) 6; (c) 64/3; (d) 128/3; (e) 0; (f) $Q^2/2$; (g) $kx_1^2/2$.

E A.4-4 26 joules

E A.4-5 16 meters

E A.4-6 (a)

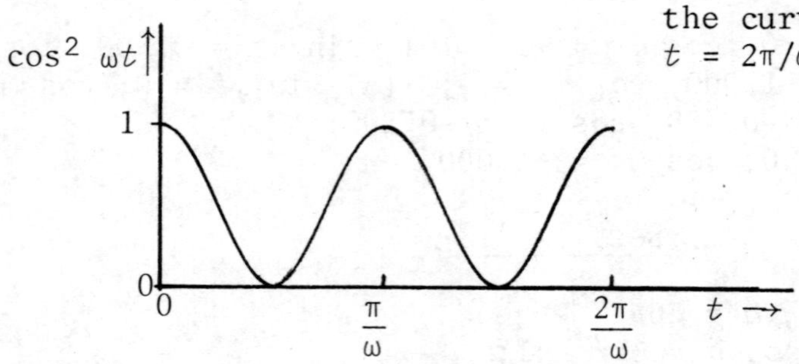

(b) 1/2; (d) The area under the curve, from $t = 0$ to $t = 2\pi/\omega$, is $\pi/\omega = (1/2)2\pi\omega$.

<u>Section A.5</u>

E A.5-1 (a) $W_{34} = \int_{C\ \vec{r}_3}^{\vec{r}_4} \vec{F}(x,y,z) \cdot d\vec{r}$; (b) $W = \int_{C_1\ \vec{\ell}_o}^{\vec{\ell}_f} \vec{F}(x,y,z) \cdot d\vec{\ell}$;

(c) $\int_{C_2} \vec{B} \cdot d\vec{\ell}$; (d) $\oint_{C_3} \vec{E} \cdot d\vec{\ell}$.

E A.5-2 (a) $5F_0$ joules; (b) $k\, q_1 q/r_1$.

E A.5-3 $E \int_C d\ell$

Appendix B

Values of Trigonometric Functions

Angle Degree	Angle Radian	Sine	Cosine	Tangent	Angle Degree	Angle Radian	Sine	Cosine	Tangent
0°	.000	.000	1.000	0.000					
1°	.017	.018	1.000	.018	46°	0.803	0.719	0.695	1.036
2°	.035	.035	0.999	.035	47°	.820	.731	.682	1.072
3°	.052	.052	.999	.052	48°	.838	.743	.669	1.111
4°	.070	.070	.998	.070	49°	.855	.755	.656	1.150
5°	.087	.087	.996	.088	50°	.873	.766	.643	1.192
6°	.105	.105	.995	.105	51°	.890	.777	.629	1.235
7°	.122	.122	.993	.123	52°	.908	.788	.616	1.280
8°	.140	.139	.990	.141	53°	.925	.799	.602	1.327
9°	.157	.156	.988	.158	54°	.942	.809	.588	1.376
10°	.175	.174	.985	.176	55°	.960	.819	.574	1.428
11°	.192	.191	.982	.194	56°	.977	.829	.559	1.483
12°	.209	.208	.978	.213	57°	.995	.839	.545	1.540
13°	.227	.225	.974	.231	58°	1.012	.848	.530	1.600
14°	.244	.242	.970	.249	59°	1.030	.857	.515	1.664
15°	.262	.259	.966	.268	60°	1.047	.866	.500	1.732
16°	.279	.276	.961	.287	61°	1.065	.875	.458	1.804
17°	.297	.292	.956	.306	62°	1.082	.883	.470	1.881
18°	.314	.309	.951	.325	63°	1.100	.891	.454	1.963
19°	.332	.326	.946	.344	64°	1.117	.899	.438	2.050
20°	.349	.342	.940	.364	65°	1.134	.906	.423	2.145
21°	.367	.358	.934	.384	66°	1.152	.914	.407	2.246
22°	.384	.375	.927	.404	67°	1.169	.921	.391	2.356
23°	.401	.391	.921	.425	68°	1.187	.927	.375	2.475
24°	.419	.407	.914	.445	69°	1.204	.934	.358	2.605
25°	.436	.423	.906	.466	70°	1.222	.940	.342	2.747
26°	.454	.438	.899	.488	71°	1.239	.946	.326	2.904
27°	.471	.454	.891	.510	72°	1.257	.951	.309	3.078
28°	.489	.470	.883	.532	73°	1.274	.956	.292	3.271
29°	.506	.485	.875	.554	74°	1.292	.961	.276	3.487
30°	.524	.500	.866	.577	75°	1.309	.966	.259	3.732
31°	.541	.515	.857	.601	76°	1.326	.970	.242	4.011
32°	.559	.530	.848	.625	77°	1.344	.974	.225	4.331
33°	.576	.545	.839	.649	78°	1.361	.978	.208	4.705
34°	.593	.559	.829	.675	79°	1.379	.982	.191	5.145
35°	.611	.574	.819	.700	80°	1.396	.985	.174	5.671
36°	.628	.588	.809	.727	81°	1.414	.988	.156	6.314
37°	.646	.602	.799	.754	82°	1.431	.990	.139	7.115
38°	.663	.616	.788	.781	83°	1.449	.993	.122	8.144
39°	.681	.629	.777	.810	84°	1.466	.995	.105	9.514
40°	.698	.643	.766	.839	85°	1.484	.996	.087	11.43
41°	.716	.658	.755	.869	86°	1.501	.998	.070	14.30
42°	.733	.669	.743	.900	87°	1.518	.999	.052	19.08
43°	.751	.682	.731	.933	88°	1.536	.999	.035	28.64
44°	.768	.695	.719	.966	89°	1.553	1.000	.018	57.29
45°	.785	.707	.707	1.000	90°	1.571	1.000	.000	∞

Appendix C

VALUES OF EXPONENTIAL FUNCTIONS

x	e^x	e^{-x}	x	e^x	e^{-x}	x	e^x	e^{-x}	x	e^x	e^{-x}
0.00	1.00	1.000	1.00	2.72	.368	2.00	7.39	.1353	4.5	90.0	.01111
0.02	1.02	.980	1.02	2.77	.361	2.05	7.77	.1287	4.6	99.5	.01005
0.04	1.04	.961	1.04	2.83	.353	2.10	8.17	.1225	4.7	110.0	.00910
0.06	1.06	.942	1.06	2.89	.346	2.15	8.58	.1165	4.8	121.5	.00823
0.08	1.08	.923	1.08	2.94	.340	2.20	9.03	.1108	4.9	134.3	.00745
0.10	1.11	.905	1.10	3.00	.333	2.25	9.49	.1054	5.0	148.4	.00674
0.12	1.13	.887	1.12	3.06	.326	2.30	9.97	.1003	5.1	164.0	.00610
0.14	1.15	.869	1.14	3.13	.320	2.35	10.49	.0954	5.2	181.3	.00552
0.16	1.17	.852	1.16	3.19	.313	2.40	11.02	.0907	5.3	200.3	.00499
0.18	1.20	.835	1.18	3.25	.307	2.45	11.59	.0863	5.4	221.4	.00452
0.20	1.22	.819	1.20	3.32	.301	2.50	12.18	.0821	5.5	244.7	.00409
0.22	1.25	.803	1.22	3.39	.295	2.55	12.81	.0781	5.6	270.4	.00370
0.24	1.27	.787	1.24	3.46	.289	2.60	13.46	.0743	5.7	298.9	.00335
0.26	1.30	.771	1.26	3.53	.284	2.65	14.15	.0707	5.8	330.3	.00303
0.28	1.32	.756	1.28	3.60	.278	2.70	14.88	.0672	5.9	365.0	.00274
0.30	1.35	.741	1.30	3.67	.273	2.75	15.64	.0639	6.0	403.4	.00248
0.32	1.38	.726	1.32	3.74	.267	2.80	16.45	.0608	6.1	445.9	.00224
0.34	1.40	.712	1.34	3.82	.262	2.85	17.29	.0578	6.2	492.8	.00203
0.36	1.43	.698	1.36	3.90	.257	2.90	18.17	.0550	6.3	544.6	.00184
0.38	1.46	.684	1.38	3.97	.252	2.95	19.11	.0523	6.4	601.9	.00166
0.40	1.49	.670	1.40	4.06	.247	3.00	20.09	.0498	6.5	665.1	.00150
0.42	1.52	.657	1.42	4.14	.242	3.05	21.12	.0474	6.6	735.1	.00136
0.44	1.55	.644	1.44	4.22	.237	3.10	22.20	.0450	6.7	812.4	.00123
0.46	1.58	.631	1.46	4.31	.232	3.15	23.34	.0429	6.8	897.9	.00111
0.48	1.62	.619	1.48	4.39	.228	3.20	24.53	.0408	6.9	992.3	.00101
0.50	1.65	.607	1.50	4.48	.223	3.25	25.79	.0388	7.0	1096.6	.000912
0.52	1.68	.595	1.52	4.57	.219	3.30	27.11	.0369	7.1	1212.0	.000825
0.54	1.72	.583	1.54	4.66	.214	3.35	28.50	.0351	7.2	1339.4	.000747
0.56	1.75	.571	1.56	4.76	.210	3.40	29.96	.0334	7.3	1480.3	.000676
0.58	1.79	.560	1.58	4.86	.206	3.45	31.50	.0317	7.4	1636.0	.000611
0.60	1.82	.549	1.60	4.95	.202	3.50	33.12	.0302	7.5	1808.0	.000553
0.62	1.86	.538	1.62	5.05	.198	3.55	34.81	.0287	7.6	1998.2	.000501
0.64	1.90	.527	1.64	5.16	.194	3.60	36.60	.0273	7.7	2208.3	.000453
0.66	1.93	.517	1.66	5.26	.190	3.65	38.48	.0260	7.8	2440.6	.000410
0.68	1.97	.507	1.68	5.37	.186	3.70	40.45	.0247	7.9	2697.3	.000371
0.70	2.01	.497	1.70	5.47	.183	3.75	42.52	.0235	8.0	2981.0	.000336
0.72	2.05	.487	1.72	5.58	.179	3.80	44.70	.0224	8.1	3294.5	.000304
0.74	2.10	.477	1.74	5.70	.176	3.85	46.99	.0213	8.2	3641.0	.000275
0.76	2.14	.468	1.76	5.81	.172	3.90	49.40	.0202	8.3	4023.9	.000249
0.78	2.18	.458	1.78	5.93	.169	3.95	51.94	.0193	8.4	4447.1	.000225
0.80	2.23	.449	1.80	6.05	.165	4.00	54.60	.0183	8.5	4914.8	.000204
0.82	2.27	.440	1.82	6.17	.162	4.05	57.40	.0174	8.6	5431.7	.000184
0.84	2.32	.432	1.84	6.30	.159	4.10	60.34	.0166	8.7	6002.9	.000167
0.86	2.36	.423	1.86	6.42	.156	4.15	63.43	.0158	8.8	6634.2	.000151
0.88	2.41	.415	1.88	6.55	.153	4.20	66.69	.0150	8.9	7332.0	.000136
0.90	2.46	.407	1.90	6.69	.150	4.25	70.11	.0143	9.0	8103.1	.000123
0.92	2.51	.399	1.92	6.82	.147	4.30	73.70	.0136	9.2	9897.1	.000101
0.94	2.56	.391	1.94	6.96	.144	4.35	77.48	.0129	9.4	12088	.0000827
0.96	2.61	.383	1.96	7.10	.141	4.40	81.45	.0123	9.6	14765	.0000677
0.98	2.66	.375	1.98	7.24	.138	4.45	85.63	.0117	9.8	18034	.0000555
1.00	2.72	.368	2.00	7.39	.135	4.50	90.02	.0111	10.0	22026	.0000454

Appendix G. Physical Data on Some Bodies in the Solar System*

$G = 6.67 \times 10^{-11}$ newton·meter^2/kilogram^2

Object	Mass, in kg	Radius of the Object, in meters	Mean Radius of Planet's Orbit around the Sun, in meters	Mean Radius of Satellite's Orbit around the Planet, in meters	Acceleration of Gravity at Object's Surface, in m/s^2
Sun	1.99×10^{30}	6.96×10^{8}			274
Mercury	3.18×10^{23}	2.43×10^{6}	5.80×10^{10}		3.6
Venus	4.88×10^{24}	6.05×10^{6}	1.08×10^{11}		8.9
Earth	5.98×10^{24}	6.38×10^{6}	1.50×10^{11}		9.8
Moon	7.35×10^{22}	1.74×10^{6}		3.84×10^{8}	1.62
Mars	6.42×10^{23}	3.39×10^{6}	2.28×10^{11}		3.7
Phobos		$\sim 2.2 \times 10^{4}$		9.4×10^{6}	
Deimos		$\sim 1.3 \times 10^{4}$		2.35×10^{7}	
Jupiter	1.90×10^{27}	7.14×10^{7}	7.78×10^{11}		24.9
Io	7.87×10^{22}	1.72×10^{6}		4.22×10^{8}	1.8
Europa	4.78×10^{22}	1.49×10^{6}		6.71×10^{8}	1.4
Ganymede	1.54×10^{23}	2.53×10^{6}		1.07×10^{9}	1.6
Callisto	7.35×10^{22}	2.42×10^{6}		1.88×10^{9}	.84
Saturn	5.69×10^{26}	6.04×10^{7}	1.43×10^{12}		10.4
Titan	1.19×10^{23}	2.9×10^{6}		1.22×10^{9}	1.4
Uranus	8.68×10^{25}	2.40×10^{7}	2.87×10^{12}		10.1
Titania	2.1×10^{21}	5.0×10^{5}		4.39×10^{8}	.56
Neptune	1.03×10^{26}	2.27×10^{7}	4.50×10^{12}		13.3
Triton	1.46×10^{23}	2.01×10^{6}		3.53×10^{7}	2.4
Pluto	1.08×10^{24}	5.7×10^{6}	5.91×10^{12}		2.2

*Most of these data are from *Handbook of Chemistry and Physics*, Chemical Rubber Co., 56th Edition, 1976.

Appendix H. CONVERSION FACTORS

Length

1 Å = 1 ångstrom = 0.1 nm = 10^{-10} m
1 foot = 12 inches ≐ 0.305 m = 30.48 cm
1 inch = 2.54 cm = 0.0254 m
1 light-year = 9.46×10^{12} km = 5.88×10^{12} mi
1 micron = 10^{-6} m
1 mile = 5280 ft = 1.61 km

Area

1 acre = 43,600 ft^2 = 0.405 hectare = 4.05×10^3 m^2
1 barn = 10^{-28} m^2
1 $foot^2$ = 0.0929 m^2
1 hectare = 1 ha = 10^4 m^2 = 2.47 acre
1 $inch^2$ = 6.452 cm^2 = 6.452×10^{-4} m^2

Volume

1 $foot^3$ = 2.832×10^{-2} m^3
1 $inch^3$ = 16.39 cm^3
1 liter = 1000 cm^3 = 0.264 gallon
1 U.S. fluid gallon = 231 in^3 = 3.79×10^3 cm^3 = 3.79 liter
1 U.S. fluid quart = 946.5 cm^3 = 9.465×10^{-4} m^3

Time

1 day = 86,400 s
1 year = 365 da = 3.15×10^7 s

Speed

1 ft/s = 0.6818 mi/hr = 0.3048 m/s
1 km/hr = 0.2778 m/s
1 mi/hr = 1.467 ft/s = 0.447 m/s

Mass

1 amu = 1.66×10^{-27} kg
1 metric ton = 1 tonne = 10^3 kg

Mass Density

1 g/cm^3 = 10^3 kg/m^3

Force

1 dyne = 10^{-5} N
1 long ton = 2240 pounds
1 newton = 0.225 pound
1 pound = 4.45 N

Pressure

1 atmosphere = 1.01×10^2 kPa = 14.7 lb/in² = 760 torr
1 lb/in² = 6.89×10^3 Pa = 6.80×10^{-2} atm = 51.7 torr
1 pascal = 9.90×10^{-6} atm = 1.46×10^{-4} lb/in² = 7.52×10^{-3} torr
1 torr = 1.333×10^2 Pa = 1.93×10^{-2} lb/in² = 1.32×10^{-3} atm

Energy, Work, Heat

1 amu = 931 MeV
1 British thermal unit = 1 Btu = 1055 J
1 calorie = 4.186 J
1 electron-volt = 1 eV = 1.60×10^{-19} J
1 foot-pound = 1.356 J
1 kilocalorie = 4.186×10^3 J
1 kilowatt-hour = 3.6×10^6 J

Power

1 horsepower = 550 ft·lb/s = 745.7 W

Radioactive Decay Rate

1 curie = 37 GBq = 3.70×10^{10} disintegration/sec.

Appendix J GREEK ALPHABET

Alpha	Α	α	Iota	Ι	ι	Rho	Ρ	ρ
Beta	Β	β	Kappa	Κ	κ	Sigma	Σ	σ
Gamma	Γ	γ	Lambda	Λ	λ	Tau	Τ	τ
Delta	Δ	δ	Mu	Μ	μ	Upsilon	Υ	υ
Epsilon	Ε	ε	Nu	Ν	ν	Phi	Φ	φ
Zeta	Ζ	ζ	Xi	Ξ	ξ	Chi	Χ	χ
Eta	Η	η	Omicron	Ο	ο	Psi	Ψ	ψ
Theta	Θ	θ	Pi	Π	π	Omega	Ω	ω

INDEX

Italic letters following page numbers are used with the following meanings: *d*, definition; *f*, figure; *m*, Minimum Learning Goal; *t*, table; and *w*, Worked Example. SI units of physical quantities are given in the text following their definitions, and inside the front cover. Values of physical constants are given in the text following their definitions, and inside the back cover. Parentheses following a page number enclosed the number of a question or exercise; such an item is indexed only if it contains factual information not given elsewhere in the book. Authors cited only in references "For Further Reading" are not indexed.